Nano- and Micro-Electromechanical Systems

Fundamentals of Nano- and Microengineering

Nano- and Microscience, Engineering, Technology, and Medicine Series

Series Editor
Sergey Edward Lyshevski

Titles in the Series

MEMS and NEMS:
Systems, Devices, and Structures
Sergey Edward Lyshevski

Microelectrofluidic Systems: Modeling and Simulation
Tianhao Zhang, Krishnendu Chakrabarty,
and Richard B. Fair

Nano- and Micro-Electromechanical Systems: Fundamentals
of Nano- and Microengineering, Second Edition
Sergey Edward Lyshevski

Nanoelectromechanics in Engineering and Biology
Michael Pycraft Hughes

Microdrop Generation
Eric R. Lee

Micromechatronics: Modeling, Analysis, and Design
with MATLAB®
Victor Giurgiutiu and Sergey Edward Lyshevski

SECOND EDITION

Nano- and Micro- Electromechanical Systems

Fundamentals of Nano- and Microengineering

Sergey Edward Lyshevski

CRC PRESS

Boca Raton   London   New York   Washington, D.C.

Library of Congress Cataloging-in-Publication Data

Lyshevski, Sergey Edward.
    Nano- and micro-electromechanical systems : fundamentals of nano- and microengineering
    / Sergey Edward Lyshevski.—2nd ed.
        p. cm. — (Nano- and microscience, engineering, technology, and medicine series ; 8)
    Includes bibliographical references and index.
    ISBN 0-8493-2838-1 (alk. paper)
    1. Microelectromechanical systems. 2. Nanotechnology. I. Title. II. Series.
TK7875.L96 2004
621.381—dc22   2004057931

Visit the CRC Press Web site at www.crcpress.com

© 2005 by CRC Press

No claim to original U.S. Government works
International Standard Book Number 0-8493-2838-1
Library of Congress Card Number 2004057931
Printed in the United States of America 1 2 3 4 5 6 7 8 9 0
Printed on acid-free paper

Dedication

I dedicate this book to the memory of my beloved father Edward Lyshevski

(1929–2003) and the blessed memory of his family: Peter, Adel, Jadwiga, and Helena.

Preface

This book introduces nano- and microsystems, devices, and structures to a wide audience. The author coherently documents basic fundamentals of nano- and microengineering as well as nano- and microtechnology. Nanotechnology can be viewed as a revolutionary technology that is based on nanoscience that will redefine engineering, medicine, security, etc. For example, evolutionary and revolutionary changes in electronics, electromechanics, optics, informatics, computers, and communication have significant immediate positive impacts on aerospace, biotechnology, electronics, health, informatics, medicine, power, etc. Nanotechnology is viewed as the most significant technological frontier to be developed and utilized in this century. Those developments are positioned on the cornerstone of science and engineering.

Nano-everything has become the buzzword for many fantasists and romantic dreamers who have offered many futuristic and highly speculative viewpoints. Their shocking speculative declarations will unlikely be supported by science, engineering, and technology simply due to fundamental and technological limits. These pseudo nanotechnology futurists/analysts have been optimistically painting rosy pictures and have presented stratospheric imaginary representations that cannot be supported and will probably never materialize. It will be no wonder if soon they shift their attention to *femtotechnology* (1×10^{-15}) or maybe *yoctotechnology* (1×10^{-24}), or armed with string theory and the 10^{-35} domain, maybe even the *negative-thirty-five-power-technology* (there is no prefix for 1×10^{-35}). Fortunately, society has approached nano- and microtechnology from the science and engineering mindset, and promising results have been achieved. However, further focused fundamental, applied, and experimental research is needed to support far-reaching engineering and technological developments without shocking speculations.

The system-, device-, and structure-level basics, applied (hardware- and software-oriented), and experimental analyses are introduced to the reader with an attempt to lead him or her to the nano- and microscience and engineering inroads. With a complete awareness of the different readers' backgrounds and interests, the author understands the challenges. Correspondingly, the multidisciplinary fundamentals are introduced and coherently covered. This book, written for a two-semester senior undergraduate or graduate course in microelectromechanical systems (MEMS) and nanoengineering (including courses in nanotechnology and nanoelectronics), intends to overcome the challenges. A typical reader's background should include calculus and physics.

The purpose of this book is to bring together the various concepts, methods, techniques, and technologies needed to attack and solve a wide array of problems including synthesis, modeling, simulation, analysis, design, and optimization of high-performance nano- and electromechanical systems (NEMS), MEMS, devices, and structures. These NEMS and MEMS are the subclasses of nano- and microsystems, and, in general, the book focuses on nano- and microsystems. Microfabrication aspects and some nanoscale fabrication technologies are covered to assist the readers. The availability of advanced fabrication technologies has been the considerable motivation for further developments. The emphasis of this book is on the fundamental multidisciplinary principles of NEMS and MEMS and practical applications of the basic theory in engineering practice and technology development.

It is evident that due to a wide spectrum of problems and issues, one can have some reservations and suggestions that will be very valuable. Please, do not hesitate to provide me with your feedback, and I will try to integrate the suggested topics and examples in the future. At the same time, it appears that it is impossible to cover all topics, areas, and technologies due to a wide spectrum of themes, the variety of unsolved problems, and some uncertainties regarding the applicability and affordability of emerging technologies particularly to fabricate nanosystems. This book is written in a textbook style, with the goal to reach the widest possible range of readers who have an interest in the subject. Specifically, the objective is to satisfy the existing growing demands of undergraduate and graduate students, engineers, professionals, researchers, and instructors in the fields of nano- and microengineering, science, and technologies. With these goals, the structure of the book was developed and significantly modified compared with the first edition.

Efforts were made to bring together fundamental (basic theory) and technology (fabrication) aspects in different areas that are important to study, understand, and research advanced nano- and microsystems in a unified and consistent manner. The author believes that the coherent coverage has been achieved. At the end of each chapter, the reader will find homework problems that will allow him or her to practice, apply, and assess the material.

Recent accelerating interest in nano- and microengineering and technologies is due to the 21st century nanotechnology revolution. This eventually will lead to fundamental breakthroughs in the way materials, devices, and systems are understood, utilized, designed, manufactured, and used. Nano- and microengineering will change or refine the nature of the majority of human-made structures, devices, and systems, revolutionizing or enhancing their performance and functionality. Current needs and trends include leading-edge fundamental, applied, and experimental research as well as technology developments. In particular, utilizing application-specific requirements, one needs to synthesize (discover), model, simulate, analyze, design, optimize, fabricate, and characterize nano- and microscale systems, devices, and structures. Recent developments have focused on analysis and design of molecular structures and devices that will lead to revolutionary breakthroughs in informatics, data processing, computing, data storage, imaging, intelligent automata, etc. Specifically, molecular computers, logic gates, switches, resonators, actuators, sensors, and circuits have been devised and studied. Nanoengineering and science lead to fundamental breakthroughs in the way devices and systems are devised, designed, and optimized. High-performance nano- and microscale structures and devices will be widely used in nanocomputers, medicine (nanosurgery, nanotherapy, nonrejectable artificial organs, drug delivery, and diagnosis), etc.

New nanomechanics phenomena, quantum physics and chemistry, novel nanofabrication technologies, control of complex molecular structures, and design of large-scale architectures and optimization, among other problems, must be addressed and examined. The major objective of this book is the development, coverage, and delivery of basic theory (through multidisciplinary fundamental and applied research and coherent studies) to achieve the highest degree of understanding regarding complex phenomena and effects, as well as development of novel paradigms and methods in optimization, analysis, and control of nano- and microsystem properties and behavior. This will lead to new advances and will allow the designer to comprehensively solve a number of long-standing problems in synthesis, analysis, control, modeling, simulation, virtual prototyping, fabrication, implementation, and commercialization of novel nano- and microsystems. In addition to technological developments, the ability to synthesize and optimize systems depends on analytical and numerical methods. Novel paradigms and concepts should be devised and applied to analyze and study complex phenomena and effects. Advanced interdisciplinary research must be carried out, and the objectives are to expand the frontiers of the nano- and microscale-based research through pioneering fundamental and applied multidisciplinary studies and developments.

This book develops and delivers the basic theoretical foundations in order to synthesize, design, analyze, and examine high-performance nano- and microsystems. In addition, the focus is centered on the development of fundamental theory for nano- and microsystems, as well as their components (subsystems and devices) and structures, using advanced multidisciplinary basic and applied developments. In particular, coherent synthesis and design are illustrated with analysis of the phenomena and effects at nano- and microscales, development of system architectures, physical representations, optimization, etc. It is the author's goal to substantially contribute to these basic issues, efficiently deliver the rigorous theory to the reader, and integrate the challenging problems in the context of well-defined applications addressing specific issues. The primary emphasis will be on the development of basic theory to attain fundamental understanding of nano- and microsystems, processes in nano- and microscale structures and devices, and devising novel devices, as well as the application of the developed theory.

It should be acknowledged that no matter how many times the material is reviewed and how many efforts are spent to guarantee the highest quality, the author cannot guarantee that the manuscript is free from minor errors and shortcomings. If you find something that you feel needs correction, adjustment, clarification, and/or modification, please notify me at Sergey.Lyshevski@rit.edu. Your help and assistance are greatly appreciated and sincerely acknowledged.

Author

Sergey Edward Lyshevski was born in Kiev, Ukraine. He received his M.S. (1980) and Ph.D. (1987) degrees from Kiev Polytechnic Institute, both in electrical engineering. From 1980 to 1993, Dr. Lyshevski held faculty positions at the Department of Electrical Engineering at Kiev Polytechnic Institute and the Academy of Sciences of Ukraine. From 1989 to 1993, he was the Microelectronic and Electromechanical Systems Division Head at the Academy of Sciences of Ukraine. From 1993 to 2002, he was with Purdue School of Engineering as an associate professor of electrical and computer engineering. In 2002, Dr. Lyshevski joined Rochester Institute of Technology as a professor of electrical engineering, professor of microsystems engineering, and Gleason Chair. Dr. Lyshevski serves as the Senior Faculty Fellow at the U.S. Surface and Undersea Naval Warfare Centers and Air Force Research Laboratories. He is the author of 11 books (including *Nano- and Microelectromechanical Systems: Fundamentals of Micro- and Nanoengineering*, CRC Press, 2000; and *MEMS and NEMS: Systems, Devices, and Structures*, CRC Press, 2002) and is the author or coauthor of more than 250 journal articles, handbook chapters, and regular conference papers. His current teaching and research activities include the areas of MEMS and NEMS (CAD, design, high-fidelity modeling, data-intensive analysis, heterogeneous simulation, fabrication), micro- and nanoengineering, intelligent large-scale microsystems, learning configurations, novel architectures, self-organization, micro- and nanoscale devices (e.g., actuators, sensors, logics, switches, memories), nanocomputers and their components, reconfigurable (adaptive) defect-tolerant computer architectures, and systems informatics. Dr. Lyshevski has made significant contributions in the design, application, verification, and implementation of advanced aerospace, automotive, electromechanical, and naval systems. He has made 29 invited presentations nationally and internationally and serves as editor of the CRC Press series *Nano- and Microscience, Engineering, Technology, and Medicine*. Dr. Lyshevski has taught undergraduate and graduate courses in NEMS, MEMS, microsystems, computer architecture, microelectromechanical motion devices, integrated circuits, and signals and systems.

Acknowledgments

Many people contributed to this book. First, thanks go to my beloved family. I would like to express my sincere acknowledgments and gratitude to many colleagues and peers. Special thanks to undergraduate and graduate students for whom I have taught courses in nanoscience and nanoengineering, nanotechnology, microsystems, MEMS, and microelectromechanical motion devices. It gives me great pleasure to acknowledge the help I received from many people in the preparation of this book. The outstanding CRC Press team, especially Nora Konopka (Acquisitions Editor, Electrical Engineering) and Regina Gregory (Project Editor), tremendously helped and assisted me by providing valuable and deeply treasured feedback. Many thanks to MathWorks, Inc., for supplying the MATLAB® environment (MathWorks, Inc., 24 Prime Park Way, Natick, MA 01760-1500, http://www.mathworks.com). Partial support of the NSF (NSF awards 0407281 and 0311588) is sincerely acknowledged. (Disclaimer. "Any opinions, findings, and conclusions or recommendations expressed in this material are those of the author(s) and do not necessarily reflect the views of the National Science Foundation.") The opportunities to perform the funded research for different agencies, laboratories, and companies under numerous grants and contracts have had a positive impact. Finally, I am sincerely acknowledging the Gleason professorship as well as assistance and encouragements I have received from the administration at the RIT Kate Gleason College of Engineering. Many thanks to all of you.

Sergey Edward Lyshevski
Department of Electrical Engineering
Rochester Institute of Technology
Rochester, New York 14623-5603
E-mail: Sergey.Lyshevski@rit.edu
Web site: www.rit/~seleee

Table of Contents

6 Modeling of Micro- and Nanoscale Electromechanical Systems and Devices

1

Nano- and Microscience, Engineering, and Technology

1.1 Introduction and Overview: From Micro- to Nanoscale and Beyond to Stringoscale

The author would like to start with some examples that may inform the reader of the sizing features of systems, devices, and structures under consideration in this book. Consider ants, which are familiar to everybody. These ants, 1 to 5 mm in length, have thousands of nano- and microbioscale subsystems and devices that have evolved and advanced over millions of years. These ants' systems are extremely small, efficient, and robust. The well-known *Escherichia coli* (*E. coli*) bacterium is 1000 times smaller than an ant, and the length of the bacteria is 2 μm. Though this is a single-cell bacteria, *E. coli* has nanoscale subsystems and devices. For example, nanobiomotor, which is 45 nm in diameter, is built using different proteins. In *E. coli*, many proteins form nanobiocircuits and nanosensors. *E. coli* and other bacteria (e.g., *Salmonella typhimurium* and *Helicobacter pylori*) have many almost identical nanobiosubsystems, devices, and structures (e.g., nanobiomotors, flagella, nanobiocircuitry). One can continue to scale down the examination of different biological systems and devices. In particular, these systems and devices can be examined at the protein, molecular, and atomic levels. The sizing features for nanosystems, devices, structures (proteins), molecules, and atoms are significantly different, and different units to describe the size are used. One is unlikely to use miles, kilometers, or meters to describe nano- and microsystems. Let us recall the basic units of length, that is, 1 μm = 1 × 10^{-6} m, 1 nm = 1 × 10^{-9} m, and 1 Å = 1 × 10^{-10} m.

The following prefixes are used:

- stringo (1 × 10^{-35}; this prefix has not been used)
- yocto (1 × 10^{-24})
- zepto (1 × 10^{-21})
- atto (1 × 10^{-18})
- femto (1 × 10^{-15})
- pico (1 × 10^{-12})
- nano (1 × 10^{-9})
- micro (1 × 10^{-6})
- milli (1 × 10^{-3})
- centi (1 × 10^{-2})

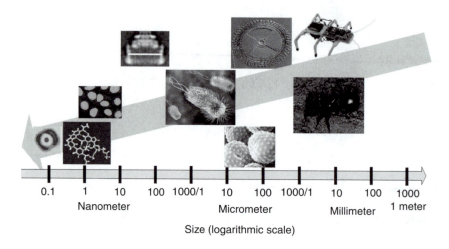

FIGURE 1.1 (See color insert following page 332.)
Scaling common items: hydrogen atom (atomic radius is 0.0529 nm), molecular electronic nanoassembly (1 nm), quantum dots (2 nm), *E. coli* nanobiomotor (45 nm), *E. coli* bacteria (2 μm), ragweed pollen (20 μm), stepper micromotor (50 μm), and ant (5 mm).

Because atoms are the units of chemistry, cells can be considered as the building blocks and fundamental units of life. However, recent studies have progressed beyond atomic and cell levels. The size of the hydrogen atom is known, and the calculated atomic radius is 0.0529 nm (e.g., 0.0529×10^{-9} m). All bioorganisms comprise μm-size cells that contain thousands of different molecules. Cells use these molecules to transform matter and energy, sense the environment, compute (calculations, decision making, processing, etc.), reproduce themselves, and so on. Figure 1.1 provides the reader with the appreciable features of organic and inorganic nano-, micro-, and minisystems. Everybody has seen ants, but can anybody see bacteria without a microscope? Single-cell *Helicobacter pylori* (*H. pylori* is the most prevalent bacterial pathogen) and *E. coli* bacteria, though only up to 5 μm in length, have billions of molecules. These bacteria exhibit a remarkable level of sophisti- cation and functionality. In particular, these bacteria integrate sophisticated three-dimensional nano-biocircuitry, nanobiomotors, nanobiosensors, and a high-performance propulsion system. For example, nanobiocircuitry performs computing, processing, and networking nanobio-electronics that perform decision making and control. Remarkable complexity and sophis-tication of the simplest bacteria is a subject of fascination, and Mother Nature is the best nanoassembler that is unlikely to be explicitly understood and coherently prototyped.

What is the difference between ant and atom? The hydrogen atom is approximately 10^{-10} m in size. For the 1-mm ant and the hydrogen atom, the difference in size is 10,000,000 times.

Now let us estimate the difference in weight.

Let the weight of ant be 1.7×10^{-5} kg (0.017 g). The weight of the hydrogen atom is 1.00794 amu or $1.66053873 \times 10^{-27}$ kg. Thus, the difference in weight is 10,000,000,000,000,000,000,000 times.

Atoms comprise subatomic particles (e.g., proton, neutron, and electron). Protons and neutrons form the nucleus with the diameter in the range of 10^{-15} m.

The electron (classical electron radius is $2.817940285 \times 10^{-15}$ m) has a mass of $9.10938188 \times 10^{-31}$ kg.

The electron charge is $1.602176462 \times 10^{-19}$ C.

The mass of the positively charged proton is $1.67262158 \times 10^{-27}$ kg, while the electrically neutral neutron has a mass of $1.67492716 \times 10^{-27}$ kg.

The atomic mass unit (amu) is commonly used. This (unified) atomic mass unit is 1/12 the mass of the carbon atom, which has six protons and six neutrons in its nucleus, i.e.,

$$1 \text{ amu} = \frac{1}{12} m(^{12}C)$$

and 1 amu = $1.66053873 \times 10^{-27}$ kg. On this scale, the mass of the proton is 1.00727646688 amu, the neutron mass is 1.00866491578 amu, and the mass of the electron is 0.000548579911 amu.

The proton is a positively charged particle, the neutron is neutral, and the electron has a negative charge. In an uncharged atom, there are an equal number of protons and electrons. Atoms may gain or lose electrons and in the process become ions. Atoms that lose electrons become cations (positively charged) or become anions when they gain electrons (negatively charged). Having mentioned protons, neutrons, and electrons, it should be emphasized that there exist smaller particles within the protons and neutrons, e.g., quarks and antiquarks. Novel physics will result in further revolutionary advances in nano-, pico-, femto-, and even stringosystems.

Please try to comprehend the size and mass we are discussing. One should realize that the difference between the atom and string is striking. The atomic diameter of the hydrogen atom is 1.058×10^{-10} m, while the size of the string is estimated to be 1.6161×10^{-35} m. Thus, there is a 6.6×10^{24} difference in size, i.e., the string is 6,600,000,000,000,000,000,000,000 times smaller than the hydrogen atom. In the future, science and engineering will not stop at the nanoscale level but will progress to pico- (1×10^{-12}), femto- (1×10^{-15}), atto- (1×10^{-18}), and stringoscale (1×10^{-35}) levels. Furthermore, the stringosystems may progress to even smaller dimensions, and there is no end to the advancements that science will experience. We experience abrupt revolutionary (not evolutionary) changes not only at the systems level, but at the fundamental level. It is the author's hope that we will wisely and responsibly use new discoveries to benefit humankind, not to destroy the world and civilization.

We have scaled things down to the atomic nanolevel (1×10^{-10} m) and mentioned *stringotechnology* (which is on the scale of 1×10^{-35} m), some interesting historical facts are reported below. Around 440 B.C., Leucippus of Miletus (Greece) envisioned the atom. Leucippus and his student Democritus (460–371 B.C.) of Abdera refined and extended this far-reaching concept. The paradigm of Leucippus and Democritus was further expanded by Epicurus (341–270 B.C.) of Samos. Aristotle (384–322 B.C.) argued and strongly opposed this concept. These Greek philosophers are shown in Figure 1.2. Though the original writings of Leucippus and Democritus are lost, their concept is known from a poem entitled *De Rerum Natura* (*On the Nature of Things*), written by Lucretius (95–55 B.C.).

FIGURE 1.2
Democritus (460–371 B.C.), Epicurus (341–270 B.C.), and Aristotle (384–322 B.C.).

The major points of their concept are as follows:

- All matter comprises atoms, which are units of matter too small to be seen. These atoms cannot be further split into smaller portions. (Democritus quotes Leucippus that "the atoms hold that splitting stops when it reaches indivisible particles and does not go on infinitely." In Greek, the prefix *a-* means *not* and the word *tomos* means *cut*. The word *atom* therefore comes from *atomos*, a Greek word meaning *uncut*. Democritus reasoned that if matter could be infinitely divided, it was also subject to complete disintegration from which it can never be put back together.)
- There is a void, which is empty space between atoms.
- Atoms are solid. Atoms are homogeneous, with no internal structure. (In 1897 Thomson discovered an electron departing from this hypothesis, and it is well known that atoms are made of neutrons, protons, and electrons.)
- Atoms are different in their sizes, shapes, and weight. (According to Aristotle, "Democritus and Leucippus say that there are indivisible bodies, infinite both in number and in the varieties of their shapes," and "Democritus recognized only two basic properties of the atom: size and shape. But Epicurus added weight as a third. For, according to him, the bodies move by necessity through the force of weight.")

Leucippus, Democritus, and Epicurus showed a genius in their predictions that is truly amazing. They can be regarded as the inventors of modern nanoscience. *Nanotechnology* refers to the science, engineering, and technologies involved in the synthesis, design, and fabrication of structures, devices, and systems from atoms and molecules with atomic specifications and precision. Everything that surrounds us is made of atoms, and the atoms' composition and arrangement determine the properties of the material, e.g., water, stone, paper, or plastic. Nanotechnology focuses on the process of building complex molecular systems, molecular machines, and molecular electronic devices from atoms or molecules. Nanotechnology requires one to apply fundamental physical laws, benchmarking engineering perspectives, and novel fabrication technologies. In fact, one must devise and design the system, device, or structure prior to building it. For example, it will be demonstrated that novel discoveries of molecular biology and genetics significantly contribute to nanotechnology and microtechnology. In particular, molecular biology provides a direct demonstration of principles that can be utilized in molecular machines and molecular electronic devices.

1.2   Introductory Definitions

This book examines nano- and microscale systems, devices, and structures. The multidisciplinary areas of nanoscience, engineering, and technology have experienced phenomenal growth. Recent fundamental, applied, and experimental developments have notably contributed to progress and have motivated further research, engineering, and educational developments and enhancements. These prominent trends signify the importance of nanoscience and engineering.[1–11] There is a need to coherently integrate and cover nano- and microscale engineering and science, possibly incorporating even liberal arts components within nano- and microengineering, science, and technology.

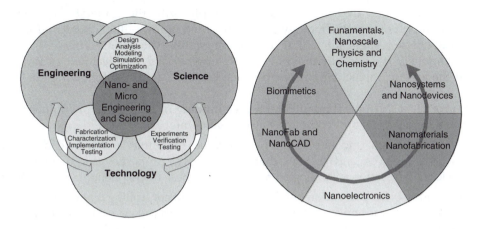

FIGURE 1.3 (See color insert.)
Representation of nano- and microengineering and science and core areas of nanotechnology.

There are many core areas that can be emphasized and examined. Some of those areas of importance, as illustrated in Figure 1.3, are:

- Fundamentals of nanoscience and engineering: new physical and chemical phenomena at nanoscale (these unique phenomena should be exclusively examined and utilized)
- Nanosystems and nanodevices (organic, inorganic, and hybrid)
- Nanomaterials and nanofabrication
- Nanoelectronics (biomolecular-, molecular-, carbon-, and silicon-based nanoelectronics)
- Nanofabrication (nanotechnologies, processes, techniques, and materials) and nano-CAD
- Biomimetics and bioinformatics

Other areas topics and tasks also can be emphasized and added. One can define two distinct paradigms: (1) nanoscience, engineering, and technology, and (2) microscience, engineering, and technology. It is important to emphasize that definitions for nanotechnology and research areas can be application-specific. For example, emphasizing nanotechnology-based nanoelectronics as a core nanotechnology area, one can state that: nanoelectronics is a fundamental/applied/experimental research that envisions and strives to devise (discover) new enhanced-functionality high-performance atomic- and molecular-scale structures, devices, and systems (discovering, understanding, and utilizing novel phenomena and effects) and fabricate them to develop novel nanotechnologies and processes for supercomplex high-yield, high-performance nanoscale integrated circuits (*ICs*) that will revolutionize electronics and information technologies.

These areas of nano- and microscience, engineering, and technology can be examined from different perspectives. However, the most common terminology to date uses the words *nanotechnology* and *microtechnology*. Although it seems that *nanoscience/engineering/technology* (nano-SET) is a more suitable term compared with *nanotechnology*, let us use the conventional word *nanotechnology* to avoid needless discussions and unnecessary arguments.

Although there have been attempts to define nanotechnology, many definitions are ambiguous, and some of those definitions are likely only partially accurate. A proposed definition is presented in the following paragraph, with the understanding that the author's definition of nanotechnology can be disputed (and I will be thankful for discussions and appreciate suggestions that can help crystallize the definition).

Nanotechnology is a combination of research, engineering, and technological developments at the atomic, molecular, or macromolecular levels with the following goals:

- To provide a coherent fundamental understanding of interactive phenomena and effects in materials and complexes at the nanoscale
- To devise, design, and implement new organic, inorganic, and hybrid nanoscale structures, devices, and systems that exhibit novel properties due to their basic physics or fabrication technologies, and to utilize these novel phenomena and effects
- To provide a coherent fundamental understanding of controlling matter at nanoscale and develop paradigms to coherently control and manipulate at the molecular level with atomic precision to fabricate advanced and novel complex devices and systems

The definition of nanobiotechnology can be formulated as a coproduct of the aforementioned nanotechnology definition by adding biological structures, devices, and systems. In particular, Nanobiotechnology is a combination of research, engineering, and technological developments at the atomic, molecular, or macromolecular levels with the following goals:

- To provide a coherent fundamental understanding of interactive phenomena and effects in biomaterials and complexes at the nanoscale
- To devise, design, and implement new organic bionanoscale structures, devices, and systems that exhibit novel properties due to their basic physics or fabrication technologies, and to utilize these novel phenomena and effects
- To provide a coherent fundamental understanding of controlling matter at nanoscale and develop paradigms to coherently control and manipulate at the molecular level with atomic precision to fabricate advanced and novel complex devices and systems

Nanoscale devices and systems will allow one to utilize and access the atomic scale phenomena, where quantum effects are predominant and can be uniquely utilized. Depending on the size and architecture, microsystems can be designed using a scaling paradigm. However, for 1-μm dimensions, even in microsystems, the secondary effects cannot be neglected. Hence, advanced synthesis and analysis concepts are needed for various nanosystems as well as microsystems that can integrate nanoscale devices and structures. Nanosystems usually are not scalable, which leads to numerous challenges. To support both nano- and microengineering, basic, applied, and experimental research as well as engineering and technological developments must be performed.

1.3 Current Developments and the Need for Coherent Revolutionary Developments

This chapter welcomes and introduces the reader to micro- and nanodomains of systems, devices, and structures. Imagine systems so small that they are impossible to see without a powerful microscope (see Figure 1.1). These micro- and nanoscale biosystems and man-made devices exist in enormous variety and sophistication. For example, without a microscope, one cannot see cells, bacteria, and microorganisms of nano- and micrometer size, which during millions of years evolved into very complex systems that display robust and adaptive decision making, self-repairing, adaptation, intelligence, and many other very complex vital tasks and functions. Though much speculation has been associated with nanorobots and nanoterminators, it is highly unlikely that real man-made nanosystems (100- to 1000-nm integrated functional systems that exhibit even primitive intelligence) will be fabricated in this century unless one utilizes nanobiotechnology, which is a quite dangerous technology. Though nano-ICs, semiconductors, and quantum nanodevices (nanotransistors, nanocapacitors, nanoinductors, nanoresonators, nanoswitches, etc.) have become a reality,[4] intelligent nanosystems will be a fanciful dream and visionary (but not realistic) fantasy of futurists for many years due to formidable challenges and unsolved problems. Design of organic and inorganic nanosystems is a delusional fantasy of romantic dreamers who manifested nanotechnology's power to accomplish their goals. The real danger has come from nanobiotechnology.

It seems that some speculative authors strive to prepare society for the anticipated advances and changes associated with nanobiotechnology, and many of their predictions are nonsense. For example, one must distinguish among

- Semiconductor, molecular, and quantum nanodevices
- Motion nanodevices
- Intelligent functional nanosystems

Motion microdevices can be considered a scaled-down version of minimachines that have been around for years—computer drivers, minifans, minipumps, minirelays, miniconveyers, etc. To control microdevices, ICs that integrate millions of nanodevices are used. Microdevices and ICs compose simple microsystems. It is envisioned that microsystems will be controlled by three-dimensional nano-ICs that will integrate molecular-scale nanodevices. Having mentioned microdevices, let us progress to the nanoscale, with an emphasis on nanosystems.

As will be discussed and illustrated later in this book, axial and radial nanomotors can be made and prototyped from DNA and RNA protein complexes. These nanometer-size machines can be powered by ATP. Imagine that four, six, or eight strands of RNA (stator) surround a central strand of DNA (rotor). Due to ATP, the RNA strands exert the rotational force on the DNA, leading to the rotational torque. For example, it was reported that the gamma-subunit of F_1 rotates with respect to the a_3b_3-hexamer. Due to their nanoscale size, these nanobiomotors could be foreseen to deliver genes or therapeutic molecules into live cells. However, these nanobiomachines, which could be envisioned to be fabricated through biochemical synthesis, have a number of serious limitations mainly due to packaging problems, synthesis limitations, restricted operating environment (temperature, pressure, pH, etc.), formidable controllability constraints, etc. One should understand that to deliver drugs or to actuate nanorobots, these nanomachines must be powered and controlled. Controlling nano-ICs are unavailable and have not even been examined in the literature. The stand-alone nano-ICs are at the forefront of nanoscience, engineering, and technology for the 21st century. It is the author's hope that the reader distinguishes illusionary speculative fantasy from reality.

To control nanomachines, nano-ICs (computing, signal processing, information propagation, amplification, signal conditioning, and other functions) and other nanosubsystems (power, sensors, etc.) should be designed and implemented. The integrated multifunctional nanobio-ICs that exist in nature in enormous variety and sophistication can be considered as built-in nanobiocomputers. The simplest bacteria have extremely sophisticated high-performance nanobio-ICs that have not been comprehended to date and likely

will be an unsolved problem for many decades. Nanocomputers, nanorobots, nanomachines, and other topics have been emphasized by Richard P. Feynman in his 1959 talk
"There's Plenty of Room at the Bottom: An Invitation to Enter a New Field of Physics" at
the annual meeting of the American Physical Society. I will cite just the beginning of Dr.
Feynman's presentation:

> I would like to describe a field, in which little has been done, but in which an enormous
> amount can be done in principle. This field is not quite the same as the others in that
> it will not tell us much of fundamental physics (in the sense of, "What are the strange
> particles?") but it is more like solid-state physics in the sense that it might tell us much
> of great interest about the strange phenomena that occur in complex situations. Fur-
> thermore, a point that is most important is that it would have an enormous number of
> technical applications.
>
> What I want to talk about is the problem of manipulating and controlling things on a
> small scale.
>
> As soon as I mention this, people tell me about miniaturization, and how far it has
> progressed today. They tell me about electric motors that are the size of the nail on
> your small finger. And there is a device on the market, they tell me, by which you can
> write the Lord's Prayer on the head of a pin. But that's nothing; that's the most
> primitive, halting step in the direction I intend to discuss. It is a staggeringly small
> world that is below. In the year 2000, when they look back at this age, they will wonder
> why it was not until the year 1960 that anybody began seriously to move in this
> direction.[3]

Among the concentration areas, Dr. Feynman emphasized the following fields: "information on a small scale," "better electron microscope," "marvelous biological systems," "min-
iaturizing the computer," "miniaturization by evapo-ration" (fabrication processes and
technologies), "problems of lubrication," "hundred tiny hands" (scale-down problem), "rear-
ranging the atoms," and "atoms in a small world."
This talk was published by many journals and can be easily found on their Web sites.
The reader is encouraged to read Dr. Feynman's presentation. Though quantum physics
and chemistry were largely developed, Dr. Feynman formulated the open problems in
nanotechnology emphasizing synthesis, design, analysis, and fabrication of micro- and
nanosystems with different applications. Drexler[1] provides further visionary possible directions for nanosystems and nanotechnology with emphasis rather on mechanical paradigms.
This book covers micro- and nanoscale systems, devices, and structures. Microsystems
and microfabrication (evolutionary progress that has been already achieved, widely
acknowledged, and recognized) significantly contribute to nanosystems, complementing
the emerging and revolutionary fields of nanoscience, nanoengineering, and nanotechnology. Microsystems need nano-ICs and other nanoscale subsystems. To date, some progress
has been made. For example, microscale machines, actuators, and sensors (in the μm range)
have been fabricated, tested, and implemented.[5] However, many formidable challenges
have not been resolved. These challenges comprise numerous serious fundamental, analytical, experimental, and technological problems. For example, from the fundamental
viewpoint, one cannot perform synergetic synthesis, high-fidelity modeling, data-intensive analysis, heterogeneous simulations, and coherent optimization in nanoscale due to
formidable analytical and numerical difficulties. In addition, high-yield, high-throughput,
scalable, robust, and affordable fabrication processes and technologies to make nanodevices and nanosystems must be developed. These fabrication technologies have not
matured and have barely been addressed.

More than two decades ago, Dr. Drexler introduced the intriguing molecular assembler concept (assemblers that will be able to guide chemical reactions by mechanically positioning reactive molecules with atomic precision).[2] This concept is supposed to guide the chemical synthesis of complex structures by mechanically positioning reactive molecules[2] but not by manipulating individual atoms, which is likely impossible. Atomic manipulating is an illusionary dream reported in many fantasy books. Heated arguments have emerged questioning the science, feasibility, and technological limits of these positioning assemblers[11,12] due to enormously complex synthesis. Let's say one needs to fabricate a 10-μm drug delivery propulsor (5 times bigger than *E. coli* bacteria) that weighs, say, 1×10^{-11} g. This may require at least 1×10^{12} bonds. Assuming a fantastic assembly rate of 1×10^{6} bonds per seconds (imagine that one million mechano bonds can be made per second), this "assembly" will take 1×10^{6} seconds, or 2 years. This may not be too bad for a microdevice, but if one makes the calculations for a 1-g microrobot, the number of years needed to "assemble" it will be 4×10^{11}. Correspondingly, the molecular assemblers, if successful, can be suited only for nanosystems. Trying to avoid any speculations, it seems that the answer regarding the feasibility of the molecular assembler will be open until a functional prototype is fabricated, characterized, and demonstrated. Promising technologies have been developed to fabricate nanostructures and electronic nanodevices, and some progress has been achieved in nanosystems and NEMS fabrication. Though society has been advanced and novel concepts have been developed, further developments are needed to support promising innovations and encouraging horizons.

Let us again recall the molecular assembler, nanofactories, and mechanosynthesis concept envisioned by Dr. Drexler.[2] The basic idea is that this mechanical molecular assembler will position reactive molecules with atomic precision, carrying them together to the specific location at the desired time computed by the super-performance computer. These molecules will bond to neighbors, implementing the bottom-up strategy utilizing computers for digitally precise control of machine-phase chemistry and atomic positioning. For example, Dr. Drexler has proposed the deposition of carbon through the mechanosynthetic reaction. In particular, a device moves a vinylidenecarbene along a barrier-free path to insert into the strained alkene, twists 90° to break a π bond, and pulls to cleave the remaining σ bond.[12] By holding and positioning molecules, this mechanical assembler and conveyer would control how the molecules react, building complex three-dimensional structures with atomically precise control. In particular, the process of mechanically guiding chemical reactions that typically add a few atoms at a time is proposed. There is no doubt that very complex biosystems exist and, programmed by genetic data (with a level of complexity that unlikely will be comprehended soon), these "molecular biomachines" are built through enzymatic solution-phase chemistry. As previously emphasized, while the basics of molecular manufacturing through mechanical molecular assemblers and mechanosynthesis have been reported in the literature for over 20 years, controversy still continues about the feasibility of this paradigm. For example, nanomachinery (nanoscale grips, robots, and conveyers) and positioning nanosensors must exist (but they do not exist) to accurately position reactive molecules with atomic precision, super-performance computers (probably nano-ICs) must be available (but their not available) to perform control, super-accurate sensors and actuators must be available, etc. The mechanosynthesis is an intriguing paradigm and has been implemented by microelectronics industry fabricate ICs for many years. It is a hope that the mechanosynthesis can be applied to nanosystems and nanodevices. Other concepts that radically depart from the mechanosynthesis-based robotic-manufacturing-like processes are available and under developments. For example, enzymatic- and solution-like chemistry may ensure feasibility and affordability with a high level of atomic precision and adaptability. This paradigm focuses on nanobiotechnology.

1.4   Societal Challenges and Implications

The scientific, engineering, technological, ethical, environmental, human resource, commercialization, and legal implications, as well as other societal challenges, issues, and problems, must be addressed. There are even demands to ban nanotechnology. Bans on nano-everything (science, engineering, and technology) will set the downhill trajectory to overall human progress, delaying future progress and scientific developments. It is also unclear how the nano-everything ban activists envision banning nanoscience and engineering. This book introduces micro- and nanoengineering and science with the belief that nanoengineering and nanoscience (the fundamentals of which have been largely developed within the last 100 years) will benefit and prosper each and every person and living system in the world. Multidisciplinary nanoscience, nanoengineering, and nanotechnology demand increasing levels of interaction and association. There is a critical need to deliver novel discoveries and technologies to consumers, customers, lawyers, legislators, policy makers, and media. There is an increasing need to forecast and evaluate the social contexts, ethical implications, and environmental consequences of nanoscience, nanoengineering, and nanotechnology. It is undeniable that nanotechnology spurs societal and ethical concerns that have some ground. Some nanotechnology ban activists mystify false and real nanotechnology implications and dangerously confuse even themselves by equating nanotechnology and nanobiotechnology and by not distinguishing science and technology. It seems that they would like to ban nano-everything. It may not be a constructive and realistic objective.

Society indeed *must* be aware of the possible dangers of nanotechnology, microtechnology, semiconductor technology, and so on. Dual-use application of ICs, computers, machinery, and other systems is well known, but even nails or water can cause tremendous harm if not properly used. This book is not devoted to the human trials in biomedical research, genetic manipulation, genetically modified organisms, genome modifications and synthesis, design of nanoweapons, cloning, and other dangerous or questionable developments. The author has been strongly opposed to these developments.

Before we attempt to address societal issues and problems, we need to clarify the subject of discussion. We will discuss microsystems and microtechnology versus nanosystems and nanotechnology.

1.4.1   Microsystems and Microtechnology

Microsystems and microtechnology have been widely accepted without widespread arguments and discussions. It has been easy to accept and appreciate miniaturization (as an evolutionary development) due to outstanding execution (based on well-developed basic physics and matured fabrication technologies). In fact, the transition from mini to micro has been smooth and rapid. For microsystems, the need for miniaturization was an input (necessary condition), and this led to the development of basic research and technology (first and second sufficient conditions) that complemented miniaturization needs, guaranteeing timely and very effective execution.

1.4.2   Nanosystems and Nanotechnology

For nanosystems and nanotechnology, the situation is completely different. Society cannot utilize these systems until the basic theory and fabrication technologies are coherently understood, developed, and matured. Some even question the feasibility of nanotechnology,

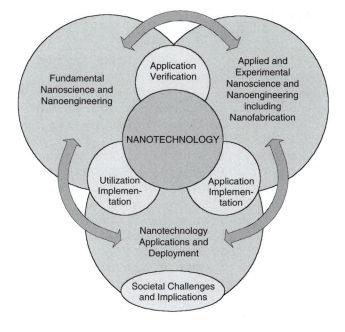

FIGURE 1.4
Nanotechnology and its interactions.

not understanding that fundamental applied and experimental nanotechnology has existed for 100 years and proved to be accurate and feasible. The reader should understand that nanotechnology is a revolutionary paradigm that leads to revolutionary changes, compared with microtechnology (evolutionary approach). The nanoscience and nanoengineering fundamentals, as well as overall changes that result from nanotechnology, have been much more pioneering and innovative, and at the same time disruptive and very risky.

Fully understanding that the author's definition of nanotechnology (provided in Section 1.2) can be disputed, the discussions have been centered on the relevant topics. Correspondingly, justifying and defining nanotechnology as was previously done, we will proceed. Figure 1.4 represents the author's vision of areas and interactions in nanotechnology. It seems that nanotechnology and nanobiotechnology may be viewed as distinct concepts that greatly differ (though they may be related in some areas) and must be dissociated.

1.4.3 Nanoengineering and Nanoscience

Through nanoengineering, one devises, designs, and examines nanoscale structures, devices, and systems whose structures and components exhibit novel physical (electromagnetic, electromechanical, optical, etc.), chemical, electrochemical, and biological properties, phenomena, and effects. These phenomena, properties, and effects are examined and predicted by nanoscience utilizing fundamental laws.

Fundamental and applied nanotechnology (the more accurate terminology is fundamental and applied nanoscience and nanoengineering) has been accepted. There have not been attempts to ban, disregard, or downplay the ideas of Einstein, Planck, and Schrödinger or other theories and discoveries. Yet, to some extent, even natural phenomena and basic physical laws have been controlled by the human reasoning, special interest groups, and emotions, all of which may act in unpredictable and self-centered ways. Here are some examples. Aristotle in 340 B.C. postulated that the Earth, as the center of the universe, is a stationary round sphere and that all planets (including the Sun) orbit the Earth. The post-Aristotelian "scholars" not only attacked and refused to accept Nicholas Copernicus's theory (Copernicus wrote a book titled *De revolutionibus* in 1543 in which he correctly stated that the planets orbit the Sun) but also persuaded the Catholic Church to ban Copernicanism. In 1616 the church declared Copernicanism to be "false and erroneous." In 1623, Galileo Galilei (1564–1642) was allowed by the Catholic Church to write a book and discuss both Aristotle's and Copernicus's theories. This book, titled *Dialog Concerning the Two Chief World Systems*, was published in 1632. Galileo, who accepted and supported Copernicus's theory, was seriously victimized for his book by the Church and later was forced to renounce Copernicus's theory. We can discuss the "evils of Darwinism" and "evils of Copernicanism" and ban them, as well as ban discoveries in the nanodomain (say, renounce atoms or the wave theory of light). One can believe that the theories of Copernicus and Darwin spawned the bloody French Revolution, led to the 1917 Bolshevik/ Communist coup in Russia, and even set the stage for Marxism and Communism (indeed Marx and Lenin acknowledged their indebtedness to Copernicus and Darwin, without whom, they believed, their absurd ideas would not have gained much acceptance). One can say that it may be too bad that nanotechnology was discovered by some critics too late, otherwise the "evils of nanotechnology" could be easily linked to many "orthodox" theories and groups. Most people are familiar with the evolution of the universe and can judge for themselves what is right and what is wrong, but one should be tolerant of other opinions. I hope the reader appreciated these examples from history, and we can now engage in further discussions.

It is important to refer to the Human Genome Project as an example to examine the ethical, legal, bioethical, and social issues. For example, briefly analyze just the availability of genetic information. The societal concerns arising from genetics include justification, implications, fairness, and the right of employers, courts, schools, the military, insurers, adoption agencies, and other entities to use genetic information. Some of the questions that can arise include the following:

- Who should have access to personal genetic information, and how will it be used? Who owns and controls genetic information? (Privacy and confidentiality of genetic information.)
- How does personal genetic information affect an individual and society's perceptions of that individual? (Psychological impact and stigmatization due to genetic differences.)
- Do medical professionals, health-care personnel, employers, insurance companies, and other persons/agencies properly counsel parents about the risks and limitations of genetic technology? How reliable and useful is fetal genetic testing? What are the larger societal issues raised by new reproductive technologies? (Reproductive issues including adequate informed consent for complex and potentially controversial procedures and potentially unreliable methods in the use of genetic information in reproductive decision making.)
- How will genetic tests be evaluated and regulated for accuracy, reliability, utility, and decision making? (Clinical and health-care issues.)
- Should parents have the right to have their children tested for adult-onset diseases? Are genetic tests reliable and interpretable by the medical community? (Uncertainties and difficulties associated with gene tests will appear.)

- Do genes make people or society behave in a particular way? Can people control their own behavior, and can one control other people's behavior? Where is the line between medical treatment and enhancement? (Conceptual and philosophical implications.)

- Who owns genes, DNA, and data? Will patenting DNA sequences limit their accessibility and development results? (Commercialization of products, property rights, intellectual property, and accessibility of data.)

Many liberal arts scholars have been involved in very heated discussions about nanotechnology and nanobiotechnology. These discussions arise mainly around the application and deployment of nanotechnology (frequently mistakenly equating nanotechnology to nanobiotechnology). In particular, the discussions have been frequently concentrated on the implementation of some nanotechnology- and nanobiotechnology-related products. In the first place, it is the author's belief that nanotechnology ≠ nanobiotechnology. Proving that molecule ≠ biomolecule yields nanotechnology ≠ nanobiotechnology. Without a doubt, nanotechnology-based products can and will have both positive and negative effects that should be studied. When we reach the milestone where nanotechnology could lead to the ability to manipulate at the molecular level with atomic precision to devise, design, and fabricate devices and systems that will lead to intelligent functional nanoscale systems, nanotechnology will have a major societal impact. This cannot be neglected. However, we have not approached this stage and may not be there within this century. The statements that "soon there will be molecular manufacturing factories operating at the atomic level manufacturing highly sophisticated and intelligent nanoterminators, nanorobots, nanoscale war machines, and bacterial nanowarriors" are the speculative illusions of romantic dreamers and fantasists. Society should not be scared but rather aware about coming revolutionary technology to make needed national and international regulations as soon as possible and create the required level of protection in advance. Unfortunately, biotechnology and nanobiotechnology may lead to very primitive but very "effective" weapons of mass destruction, and nanobioweaponry is a reality that must be dealt with. Therefore, coordinated international regulations and treaties are urgently needed. However, from the nanotechnology standpoint, this dangerous negative impact of nanobiotechnology may be quite tangential.

Examples of the positive impacts of nanotechnology include medicine and health (nanosystems can sense and diagnose diseases, providing effective treatments); environmental applications (no-pollution technology, for example, by recycling leftover molecules and atoms); low-cost environmentally friendly transportation (lightweight robust materials, enabled by molecular manufacturing, will allow one to design novel affordable and safe propulsion systems and lower the cost), etc. Possible positive impacts of nanobiotechnology are anticipated in medicine (nanobiosystems can sense and rearrange patterns of molecules in the human body, providing the tools needed to provide treatments for various diseases and restore health), biotechnology, pharmaceutics, etc.

The study of ethical implications of nanotechnology needs to address different features of nanotechnology (fundamental, applied, experimental, fabrication, implementation, and deployment) and their likely impacts. There is a need to examine potential consequences of banning or not developing nanotechnology areas (as well as promoting nanotechnology) without speculative and absurd statements that may be made simply to control or influence nanotechnological developments. Among potentially negative effects of nanotechnology, the following can be mentioned:

- Unconventional weapons of mass destruction and terrorism (development of undetectable, low-cost, and highly selective unconventional deadly weapons of mass destruction)

- Undetectable control and monitoring of individuals, security, and human rights violations
- Limited access, availability, and affordability
- Intellectual property obscurity
- Nanotechnology neglect and abuse

It seems that if some countries institute moratoria or bans on overall nanotechnology developments, this will not affect the ability of other nations to develop and deploy nanotechnology. In fact, due to expected tremendous economic and military advantages, a nanotechnology race is already underway and speeding up to the next level. The most polarized debates have been in the United States and some European Union countries. There are many highly developed countries with tremendous scientific, engineering, and technological potential that have significantly advanced nanotechnology without any debate. Recent globalization trends, the end of intellectual emigration, for-profit economies, and other current trends cannot positively contribute to this already quite alarming situation. Furthermore, open exchange should not be anticipated due to highly proprietary research, classified engineering developments, secret technology advances, and so on. It seems that many developed (and undeveloped) countries feel that banning nanotechnology will lead to unilateral disarmament and damage the economy. However, the world is very fragile, and it appears that a serious threat of the possible abuse of nanotechnology and particularly nanobiotechnology is real. The failure to develop nanotechnology as well as failure to develop, sign, and implement collaborative protection mechanisms (through treaties with all countries) will have frightful consequences. Individuals and entities have limited capabilities. Efforts have been made by many scientists,[1-12] but these efforts, unfortunately, have been only partially successful.

Erroneous interpretations of nanotechnology take place with and without emotional bias. Confusion has arisen from misunderstanding rather than from dislike or reluctance for new theories or technologies. Recent discussions about the dangers posed by self-replicating biosystems have created highly charged environments, and rightly so. In particular, if self-replicating weapons systems are developed by any country, nation, agency, entity, or individual, they may become uncontrollable and cause serious unlimited threat to civilization. If related highly specialized nanotechnology research might lead to any nanotechnology-based weapons systems, this research must be stopped and banned due to the severe adverse impact. Obscurity is centered on the confusion between organic/inorganic hybrid and biological (living organisms that can replicate) systems. Though some researchers pursue a doctrine that organic and inorganic systems can be designed as self-replicating, artificially programmable, self-controlled manufacturing systems, it is highly unlikely that this dogma will be supported and demonstrated. How can one create a life using aluminum or silicon atoms?

The development and deployment of nano- and microsystems are critical to the economy and society because nano- and microengineering and science will lead to major breakthroughs in informatics, electronics, computers, medicine, health, manufacturing, transportation, energy, avionics, security, power, energy sources, etc. For example, nano- and microsystems have an important impact on medicine (drug delivery, diagnostics, and imaging), informatics, avionics, and aerospace (actuators and sensors, smart reconfigurable geometry wings and blades, flexible structures, gyroscopes, transducers, and accelerometers), automotive systems, transportation, manufacturing, robotics, safety, etc. Therefore, nano- and microsystems will have tremendous positive (direct and indirect) social and economic impact. New nano- and microsystems will be devised and used in a variety of applications (electronics, medicine, metrology, ecology, etc.). For example, micro- and nanoscale structures, devices, and systems frequently have specific applications such as high-frequency resonators, electromagnetic field and stress sensors, etc. Reported positive and negative impacts of nanotechnology will enhance the responsiveness and accountability of professionals to the ethical implications of discoveries and novel technologies. Due to heated discussions regarding the ethical and societal implications of nanotechnology, one should be prepared for the diverse culture and unsympathetic environment. As society responds to increasing globalization, nanotechnologists will associate with colleagues, peers, and customers from diverse cultures and environments. Leadership skills must be possessed in order to promote and support practices that foster teamwork and integrity in professional and personal development, support understanding, develop vision, enhance culture in the corporate and public worlds, etc. Leadership experience will promote collaboration, effective communication and feedback, conflict management, team development, and ethical decision making. Rapid technological changes and the emerging global marketplace increase opportunities and the need for scientific and technological entrepreneurship. Understanding how to recognize and assess market opportunities and to execute successful business plans will give nanotechnologists the tools they will need to function as valuable team members and leaders in the new environment.

1.5 Conclusions

While the majority of people recognize the benefits that nanotechnology could bring, many people have sound concerns regarding human health, environmental effects, security, privacy, human rights, proliferation in armaments, etc. Some people believe that all developments should be banned until society accurately comprehends the potential effects of nanotechnology. The lack of knowledge has led some participants to call for a moratorium on nano-everything, implying a ban on nanoscience, engineering, and technology (nanotechnology). The author feels that certain areas of nanobiotechnology should be banned and some controversial areas of nanotechnology should be regulated. The mentality that "I am not against nanotechnology but want to be sure of its impact and implications" may not be meaningful because one could, in a similar vein, request a ban on computers (that he or she uses) because computers have been widely used against society and humanity. We do not have any evidence that nanotechnology, if properly and responsibly used, can harm society more than fire or water. It is our hope that all technologies (macro, mini, micro, and nano) will be properly used to benefit people. All countries must urgently develop and sign the treaty to ban any developments of primitive (current threat evolves from nanobiotechnology) and sophisticated (future threat arises from some nanotechnology areas) weapons of mass destruction. It seems that many people do not realize that we are already surrounded by nanotechnology, and biosystems can be considered as a naturally evolved product. Society may desperately need novel technologies to find novel revolutionary solutions addressing health, environmental, economic, energy, and safety concerns, among others. These can be achieved only through the application of new technologies, including nanotechnology. One should not simply criticize as some individuals, groups, or organizations have pursued (sometimes following their own self-interest), but should try to educate society. This will possibly lead the nano-educated critics to become nano-adapted and more accepting, focusing their efforts on real danger. It seems that the inconsistent manner in which nanotechnology frequently has been presented demonstrates a lack of faith in all scientists (from the B.C. era to the 20th century, not to mention this century) and the science on which nanotechnology is based. It is well known that far-reaching scientific, engineering, and technological developments that were initially oriented to achieve and maintain military superiority have been later gradually focused on peaceful uses. Some examples are the Internet, computers, nuclear energy, transportation, agriculture, and medicine. Using semi-conductor devices and integrated circuits, important improvements have been made in communications, consumer electronics, and computers (not only for military uses). These developments have advanced and enhanced the quality of human life at unprecedented rates and created a thriving world economy. Every new scientific, engineering, and techno-logical breakthrough has been perceived as both a triumph and a failure. Nanotechnology, as a cutting-edge revolutionary endeavor, has arrived. The emergence of nanotechnology is pervasive and irreversible. Though societal, ethical, and environmental effects are hard to accurately foresee, ultimately nanoengineering and science are critical to our world. I believe that further developments in nanoengineering should change people's lives for the better. The coherent evaluation of implications of current trends in nanotechnology is a very difficult task that can be considered to be quite ambiguous due to a great number of uncertainties. All who read this book should be aware that the abuse of nanotechnology and nanobiotechnology can lead to the elimination of all living organisms on our planet or severe unrecoverable effects. Please be responsible and accountable by protecting not only your own life but the lives of others as well.

Homework Problems

1. Explain the difference between nano- and microengineering, science, and technology.
2. Provide examples where nano- and microscience, engineering, and technology have been utilized.
3. Where are nano- and microscience, engineering, and technology, as well as nan-otechnology, envisioned to be utilized and commercialized?
4. For any technical paper on nanotechnology (or microtechnology) areas that appeal to you, address and discuss scientific and societal issues as well as potential positive and negative effects of the results reported.

References

1. Drexler, K.E., *Nanosystems: Molecular Machinery, Manufacturing, and Computations*, Wiley-Interscience, New York, 1992.
2. Drexler, K.E., Molecular engineering: An approach to the development of general capabilities for molecular manipulation, *Proc. Natl. Acad. Sci. U.S.A.*, 78, 5275–5278, 1981.
3. Feynman, R.P., There's plenty of room at the bottom: An invitation to enter a new field of physics, in *Handbook of Nanoscience, Engineering, and Technology*, Goddard, W., Brenner, D., Lyshevski, S.E., and Iafrate, G., Eds., CRC Press, Boca Raton, FL, 2003, pp. 1.1–1.9.
4. *Handbook of Nanoscience, Engineering, and Technology*, Goddard, W.A., Brenner, D.W., Lyshevski, S.E., and Iafrate, G.J., Eds., CRC Press, Boca Raton, FL, 2002.
5. Hess, K., Room at the bottom, plenty of tyranny at the top, in *Handbook of Nanoscience, Engineering, and Technology*, Goddard, W., Brenner, D., Lyshevski, S.E., and Iafrate, G., Eds., CRC Press, Boca Raton, FL, 2003, pp. 2.1–2.7.
6. Lane, N., The grand challenges of nanotechnology, *J. Nanopart. Res.*, 3, 99–103, 2001.
7. Ratner, M.A. and Ratner, D., *Nanotechnology: A Gentle Introduction to the Next Big Idea*, Prentice Hall PTR, Upper Saddle River, NJ, 2003.

8. Roco, M.C., Worldwide trends in nanotechnology, *Proc. Semiconductor Conf.*, 1, 3–12, 2001.

9. Roco, M.C. and Bainbridge, W.S., Eds., *Societal Implications of Nanoscience and Nanotechnology*, Kluwer Academic Publishers, Dordrecht, The Netherlands, 2001.

10. *Small Wonders, Endless Frontiers: A Review of the National Nanotechnology Initiative*, National Research Council, Washington, DC, 2002.

11. Smalley, R.E., Of chemistry, love and nanobots—How soon will we see the nanometer-scale robots envisaged by K. Eric Drexler and other molecular nanotechnologists? The simple answer is never, *Sci. Am.*, 285, 76–77, 2001.

12. Drexler, K.E. and Smalley, R.E., Nanotechnology: Drexler and Smalley make the case for and against 'molecular assemblers,' *Chem. Eng. News*, 81, 37–42, 2003.

2

Nano- and Microscale Systems, Devices, and Structures

2.1 Sizing Features: From Micro- to Nanoscale, and from Nano- to Stringoscale

Before being engaged in discussions of nano- and microscale systems, devices, and structures, let us provide some discussions regarding the size of these systems, devices, structures, and elements. The diameter of a human hair is approximately 0.00001 m, i.e., 10 μm. Proteins vary in size from 1 to 50 nm. Now we consider atoms. The radius of an atom is not a precisely defined value because the electron probability density determines the size of an atom. The radii are difficult to define experimentally, and different experiments give different values. One may define the radius as the spacing between the atoms in a uniform media (for example, crystal structure). However, due to different valence states, distinct radii result. It is common to define (estimate) the size of an atom by the radius of the electron orbital. As we emphasized, the size of an atom depends on how the measurement was made or how the calculations were performed. Correspondingly, different values have been reported even for the simplest hydrogen atom. Therefore, the term *atomic radius* is not predominantly useful, although it is widely used. We use the atomic radius from the sizing assessment. The diameter of the helium atom, reported in the literature, is from 53 pm (53×10^{-12} m) to 79 pm (79×10^{-12} m or 0.79 Å (100 pm = 1 Å). For the hydrogen atom, the covalent radii acknowledged in the literature vary from 32 to 37 pm, and the van der Waals radius is reported to be 120 pm. It is interesting that the empirical atomic radius for H is reported to be 25 pm. Using the orbital, it will be documented that the atomic radius for helium is 0.0529 nm.

The largest atoms have the sixth shell (elements start with number 55 and higher in the Mendeleyev periodic table of the elements). The first element is cesium ($_{55}$Cs), and it has $6s^1$. The empirical atomic radius reported in the literature is 260 pm, while the calculated atomic radius is 298 pm. Some sources report 3.34 Å for the Cs atomic radius. Using the orbital, the higher value for the atomic radius of cesium is obtained. All these data document the imprecise definitions and data available. However, it is not very important, and our goal is to understand the scale of things.

2.1.1 Mendeleyev's Periodic Table of Elements and Electronic Configurations

In the early 19th century, chemists noted that chemical and physical properties of elements showed similarities. Correspondingly, they rightly postulated that the elements could be formed into groups. In 1817, Dobereiner showed that elements came in groups of three, and in 1863, A. E. Béguyer de Chancourtois created a list of the elements arranged by increasing atomic weight. In 1868, Dmitry Mendeleyev, a chemistry professor (St. Petersburg University, Russia), arranged elements into seven columns, corresponding to various chemical and physical properties. Mendeleyev's periodic table of elements is a vital tool to examine all chemical and physical properties. Some typical elemental properties are atomic weight (Dalton), density (specific gravity), melting and boiling points, electron negativity, electron affinity, first and second ionization potentials, covalent radius, atomic radii, thermal conductivity, common valence state, crystal structure, etc. These properties are related to its electronic configuration. Each of the elements in the periodic table is arranged in order of ascending atomic number (proton number) and can be represented in the following way:

$$_{\text{atomic number}} \text{ELEMENT}$$

The first elements are $_1$H, $_2$He, $_3$Li, $_4$Be, and $_5$B. Their electron configurations, respectively, are $1s^1$, $1s^2$, $1s^2\,2s^1$, $1s^2\,2s^2$, and $1s^2\,2s^2\,p^1$. These electronic configurations gradually complicate. For example, for $_{29}$Cu, we have $1s^2\,2s^2\,2p^6\,3s^2\,3p^6\,3d^{10}\,4s^1$. When two electrons occupy the same orbital, they have different spins (Pauli exclusion principle), and the pairing raises the energy. A half-filled subshell and a fully filled subshell lower the energy, ensuring stability.

Atomic radii vary among elements in the same group and among elements in the same period (within the same period, atomic radii decrease as the atomic number increases). The highest atomic orbitals occupied by electrons largely determine the properties of the elements. The periodic table can be studied utilizing *s*-, *p*-, *d*-, and *f*-blocks. The filling order of atomic orbitals is as follows:

1s

2s 2p

3s 3p

4s 3d 4p

5s 4d 5p

6s 4f 5d 6p

7s 5f 6d 7p

The *s*- and *p*-blocks of elements are called main group elements. The *d*-block elements are called transition elements. The *f*-block elements are called the inner transition elements. The *s*-, *p*-, and *d*-block elements are in the main body of the periodic table, whereas the *f*-block elements are placed below the main body. Media are classified as diamagnetic, paramagnetic, and ferromagnetic. Diamagnetism is a weak interaction of all material with a magnetic field. Correspondingly, diamagnetic materials (water, nitrogen, copper, etc.) are slightly repelled by a magnetic field and not attracted by such a magnetic field. If a material is attracted into a magnetic field, the material is said to be paramagnetic. A paramagnetic material has unpaired electrons in the molecular or atomic orbitals. When atoms of an element have unpaired electrons, the element will be paramagnetic. For example, the electronic configuration of $_{27}$Co is $1s^2\,2s^2\,2p^6\,3s^2\,3p^6\,3d^7\,4s^2$, and electrons on the $3d$ shell are unpaired. Iron ($_{26}$Fe) with electronic configuration $1s^2\,2s^2\,2p^6\,3s^2\,3p^6\,3d^6\,4s^2$ is a ferromagnetic material that is strongly attracted by magnets. The electronic configuration of the iron ion Fe^{+2} is $1s^2\,2s^2\,2p^6\,3s^2\,3p^6\,3d^6$.

How does one use the electronic configuration to estimate atomic radii? For atoms, the radii of the first (K) 1s, second (L) 2s, third (M) 3s, fourth (N) 4s, fifth (O) 5s, and sixth (P) 6s orbitals are as follows:

$$0.529 \text{ Å}$$
$$0.529 \text{ Å} \times (2)2 = 2.116 \text{ Å}$$
$$0.529 \text{ Å} \times (3)2 = 4.716 \text{ Å}$$
$$0.529 \text{ Å} \times (4)2 = 8.464 \text{ Å}$$
$$0.529 \text{ Å} \times (5)2 = 13.225 \text{ Å}$$
$$0.529 \text{ Å} \times (6)2 = 33.856 \text{ Å}$$

Using these values, the atomic radii can be derived for the following elements:

H $1s^1$
He $1s^2$
Li $1s^2\,2s^1$
Be $1s^2\,2s^2$
B $1s^2\,2s^2\,2p^1$
C $1s^2\,2s^2\,2p^2$

...

$_{26}$Fe $1s^2\,2s^2\,2p^6\,3s^2\,3p^6\,3d^6\,4s^2$
$_{27}$Co $1s^2\,2s^2\,2p^6\,3s^2\,3p^6\,3d^7\,4s^2$

...

$_{62}$Sm $1s^2\,2s^2\,2p^6\,3s^2\,3p^6\,3d^{10}\,4s^2\,4p^6\,4d^{10}\,4f^6\,5s^2\,5p^6\,6s^2$
$_{63}$Eu $1s^2\,2s^2\,2p^6\,3s^2\,3p^6\,3d^{10}\,4s^2\,4p^6\,4d^{10}\,4f^7\,5s^2\,5p^6\,6s^2$

Thus, the atomic radius of the hydrogen atom is 0.0529 nm. Fig. 2.1 illustrates the atom image and three-dimensional molecular assembly. Both are in the nanometer scale.

In nanoscale, one encounters difficulties using the meter as a unit because of the enormous number of zeros. Correspondingly, other units for the length are used, e.g., picometers, nanometers, and micrometers. We recall that

$$1 \text{ pm} = 1 \times 10^{-12} \text{ m}, \quad 1 \text{ nm} = 1 \times 10^{-9} \text{ m}, \quad \text{and} \quad 1\ \mu\text{m} = 1 \times 10^{-6} \text{ m}$$

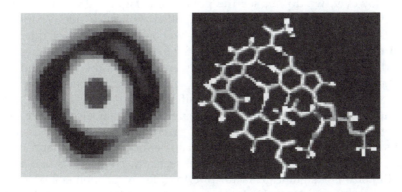

FIGURE 2.1
Image of the hydrogen atom (atomic radius is 0.0529 nm) and molecular assembly (1 nm).

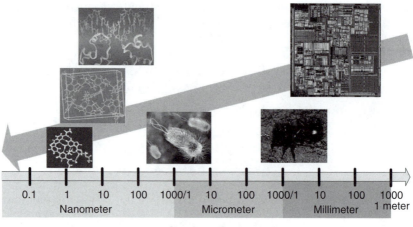

Size (logarithmic scale)

FIGURE 2.2 (See color insert following page 332)
Revolutionary progress from micro- to nanoelectronics with feature sizing: molecular assembly (1 nm), three-dimensional functional nano-ICs topology with doped carbon complex ($2 \times 2 \times 2$-nm cube), nanobio-ICs (10 nm), *E. coli* bacteria (2 μm) and ant (5 mm) which integrate nanobiocircuitry, and 1.5×1.5-cm 478-pin Intel® Pentium® processor with millions of transistors.

Furthermore,

$$1 \text{ nm} = 10 \text{ Å}, \quad \text{and} \quad 1 \text{ Å} = 1 \times 10^{-10} \text{ m}$$

The scale of things was reported in Figure 1.1, and this data provides the reader with the associated features of organic, inorganic, and biological nano-, micro-, and minisystems. In this book, among other topics, nanoelectronics will be emphasized. In additional to organic and inorganic microelectronics, nanobioelectronics exist in enormous variety. Figure 2.2 depicts the revolutionary progress from micro- to nanoelectronics utilizing distinct paradigms, e.g., organic-based carbon-centered three-dimensional nanoelectronics.

Have you seen pollen? Maybe not, but many people suffer from allergy due to pollen. The size of a pollen grain is easy to measure, and pollen grains of forget-me-not species (myosotis) are from 2 to 5 μm, whereas some squash (*Cucurbita pepo*) pollen grains can be up to 200 μm. For wind-pollinated plants, the sizes vary from 10 to 12 μm (*Urtica dioica*) to 125 μm (abies). The size of an average airborne pollen grain is between 10 and 60 μm in diameter. However, we do not see the pollen, and Fig. 2.3 illustrates the microscope image of ragweed pollen. If one cannot see pollen without a microscope, can we see quantum dots without a microscope? The answer is no. The size of quantum dots, as documented in Fig. 2.3, can be in the range of 1 to 3 nm, i.e., 10,000 times smaller than pollen. This difference in size is almost the same as between the moon and a zeppelin.

Nano- and microsystems studied in this book, including nano- and microelectromechanical systems (NEMS and MEMS), should be classified and defined. In Section 2.2 the coherent definitions and classification for MEMS, NEMS, micro- and nanoelectromechanical devices, as well as for micro- and nanostructures, are given. One can try to classify nano- and microsystems simply by size. For example, one can define (postulate) that nano- and microsystems must be within 100 nm and 100 μm, respectively. Though it is undeniable that the system (device) size is a significant factor due to different phenomena observed in nano- and microscale as well as distinct physics (not to mention different fabrication technologies),[1] the author seriously doubts the exclusive size-based classification approach. For example, why not "define" the size of nanosystems to be 1 nm or 10 nm? To classify nano- and

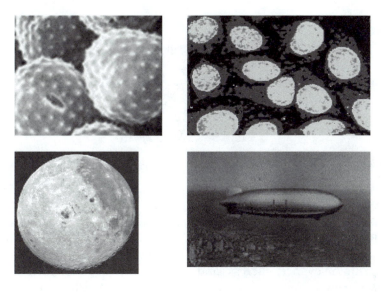

FIGURE 2.3
Ragweed pollen (20 μm) and quantum dots (2 nm) as well as the moon (3476 km) and a zeppelin (350 m).

microscale systems and devices, the designer rather should examine the basic physics, cornerstone laws to be applied, major phenomena utilized, and fabrication processes used. It should be emphasized that there exist a limited number of man-made micro- and nano-systems in the range of 100 μm and 100 nm (we do not consider nano- and microbiosystems). The reader also must not be confused with micro- and nanoscale systems and structures; thin films can be easily fabricated within a few nanometers in thickness (sputtering, evaporation, and other deposition processes) and within a couple of micrometers in width and length. However, there are enormous differences between structure and device as well as between device and system as the reader will realize (some readers may understand this already). To date, only a limited number of 100-μm micro-devices have been fabricated (we are not considering semiconductor nano-devices such as nanodiodes, nanocapacitors, and nanotransistors that have been scaled down to 50 nm and less). Furthermore, 5-mm packaged micro-accelerometers (likely miniaccelerometers) are considered to be MEMS. It should be emphasized that nanodevices and nanosubsystems can be a part of MEMS.

In the late 1950s, Dr. Richard Feynman, who largely promoted nanoengineering and science, offered a $1000 award to the first person who could build an electrical motor (motion device) "smaller than 1/64th of an inch." William McLellan demonstrated his 4-mm-diameter minimotor in the early 1960s. For the past 40 years, we have been progressing to microscale electromechanical motion devices. These days, many companies mass-fabricate 2-mm high-performance minimachines using the MEMS fabrication technology, and different prototypes of 100-μm micromotors have been designed, analyzed, fabricated, tested, and characterized.

Let us compare the fabricated 50-μm silicon-technology micromotor (which cannot be seen without a microscope) with the *Escherichia coli* (*E. coli*) 45- to 50-nm-diameter nanobiomotor made from proteins.[2] Figure 2.4 illustrates the rotor of a micro scale motion device. The sizing difference of the silicon-technology micromotor and the *E. coli* bacteria nanobiometer is 1000 times. Though one can fabricate extremely high-cost 10-μm micromotors using the most advanced fabrication technologies, it is highly unlikely that these micro- and nanometer motors will be mass-fabricated within many decades from now. Welcoming the reader to the micro- and nanodomains, it should be emphasized that one cannot simply scale down

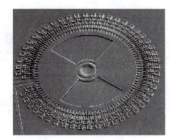

FIGURE 2.4
Silicon-technology micromotor (50 μm).

systems because the basic physics of mini-, micro- and nanoscale systems and devices can be different. To date, the operating principles of *E. coli* nanobiomotors are unknown.

One of the major goals of this book is to shape nanoengineering and science into a paradigm with the relevance, success, and vitality of modern engineering and science. Nano- and microengineering, which are based on highly interdisciplinary enterprises of science (physics, chemistry, biology, mathematics, etc.) and engineering, have enormous potential for revolutionizing and enhancing different emerging technologies. Nano- and microscience, engineering, and technologies have accomplished phenomenal growth over the past few years due to rapid advances in theoretical developments, experimental results (using state-of-the-art measurement and instrumentation hardware), high-performance computer-aided design software, and computationally efficient environments. Recent fundamental and applied research and developments in nanoelectromechanics, nanoelectronics, informatics, and biomimetics have notably contributed to the current progress. Leading-edge basic research, novel technologies, software, and hardware are integrated to devise new systems and study existing systems. These prominent trends provide researchers, engineers, and students with the needed concurrent design of integrated complex nano- and microsystems. The synergy of engineering, science, and technology is essential to attaining the goals and objectives. It becomes increasingly difficult to perform analysis and design of nanoscale systems, subsystems, devices, and structures without the unified theme and multidisciplinary synergy due to integration of new phenomena, complex processes, and compatibility. Therefore, nanodomain paradigms must be further developed and applied. Nanoscience and nanoengineering are based on fundamental theory, engineering practice, and leading-edge technologies in fabrication of nanoscale systems, subsystems, devices, and structures that have dimensions of nanometers.

2.1.2 Nanoengineering and Nanoscience

Nanoengineering deals with devising (synthesis), design, analysis, and optimization of nanoscale structures, devices, and systems that exhibit and utilize novel physical (electromagnetic, electromechanical, electrochemical, optical, etc.), chemical, electrochemical, and biological properties, phenomena, and effects that can be discovered, examined, and predicted by nanoscience (fundamental laws).

Using the pure sizing classification (which, in general, has serious deficiencies), the dimension of nanosystems and their components (nanostructures) is from 10^{-10} m (atom/molecule size) to 10^{-7} m, that is, from 0.1 to 100 nm. To fabricate nanostructures, nanotechnology is applied. In contrast, the dimension of microsystems and their components (devices and microstructures) is from 100 nm to the millimeter range. Conventional microelectronics technologies (specifically, nanolithography-enhanced CMOS technology and processes) are usually used to fabricate semiconductor microelectronic devices, ICs, MEMS, and microsystems. Studying nanoscale systems, one concentrates on the atomic and molecular levels in fabrication, design, analysis, optimization, integration, synthesis, etc. Reducing the dimension of systems leads to the application of novel materials (fullerenes, carbon nanotubes, molecular wires, molecular clusters, biomolecules, etc.) that can be accomplished utilizing newly emerging fabrication technologies. The problems to be solved range from high-yield mass production, assembling, and self-organization to devising novel enhanced-functionality high-performance nano- and microsystems. For example, nano- and microscale switches, logic gates, actuators, and sensors should be devised (synthesized), designed, examined, optimized, fabricated, tested, and characterized.

All living biological systems function as a result of atomic and molecular interactions. Different biological and organic systems, subsystems, and structures are built. The molecular building blocks (proteins, nucleic acids, lipids, carbohydrates, DNA, and RNA) are examined to perform biomimicking and prototyping to devise novel nano- and microsystems. The devised nano- and microsystems must have the desired functionality, integrity, properties, and characteristics. Analytical and numerical methods must be developed in order to analyze the dynamics, three-dimensional geometry, interaction, bonding, and other features of atoms and molecules. In order to accomplish this analysis, complex electromagnetic, mechanical, and other physical and chemical phenomena, effects, and properties must be studied.

Nano- and microsystems will be widely used in medicine and health. Among possible applications are drug synthesis, drug delivery, nanosurgery, nanotherapy, diagnostics, actuation, sensing, diagnosis, prevention, nonrejectable artificial organs and implants, biocompatible materials and structures, etc. For example, with drug delivery, the therapeutic potential will be enormously enhanced due to direct effective delivery of new types of drugs to the specified body sites. The molecular building blocks of DNA-based structures, proteins, nucleic acids, and lipids, as well as carbohydrates and their nonbiological mimics, are examples of materials that possess unique properties determined by their size, folding, and patterns at the nano- and microscale. Significant progress has been made, and new analytical tools must be developed that are capable of characterizing the chemical, electrical, electromagnetic, and mechanical properties of cells including processes such as cell division, information propagation, locomotion, actuation, etc. A defining feature of nanoscale systems and devices is that the behavior and properties of materials and media may differ in fundamental ways from that observed at the macro- and microscales. For example, at nanoscale, quantum effect, nanoscale interface and interaction (interatomic and interfacial forces, chemical bonds, chemical recognition and matching, etc.) can be utilized. There exist microdevices with operating principles similar to nanodevices, as will be illustrated in this book.

Nanosystems can lead to revolutionary increases in computing, processing, and information capabilities, as well as novel functionality. The semiconductor industry has demonstrated scaling of microelectronic, magnetic, and optical devices to nanometer size, leading to better performance (speed, robustness, execution time, power consumption, information propagation, etc.) and higher density. Moore's law (exponential increase in computing power) may be ensured for this decade, and the computing power will double every 18 months. Ambitious research programs have been performed in nanoelectronics and nano-ICs because, to the best of the author's knowledge, there are serious difficulties in expanding the CMOS technology beyond the roadmap horizon of 2015. The International Technology Roadmap for Semiconductors predicts that within 10 years, the physical gate lengths will be less than 10 nm. Physical limits and secondary effects (quantum phenomena, interface, and interference) will significantly limit fabrication technologies and performance. Advances in nanofabrication have demonstrated fabrication of devices at the atomic level, and there is a great potential for improving the properties and functionality of nanostructured materials (carbon nanotubes, quantum dots, molecular wires, etc.). These nanostructures can be used in novel electronic devices, sensors, communications systems, actuators, propulsion systems, etc.

2.1.3   Smaller Than Nano-, Pico-, and Femtoscale—Reality or Fantasy?

The size of the hydrogen atom is estimated to be 0.1 nm. Electrons and quarks are much smaller and are measured in picometers. What is beyond pico- and femtoscale? Planck units have been widely used. To find the numerical values for Planck units, we use the universal gravitational constant G, Planck's constant h ($h = 6.62606876 \times 10^{-34}$ J · sec) and the speed of light c ($c = 299{,}792{,}458$ m/sec). The modified Planck's constant is $\hbar = \frac{h}{2\pi} = 1.054571596 \times 10^{-34}$ J · sec.

Some units may be familiar to the reader. For example, the Planck energy is found to be $E_p = \sqrt{\frac{\hbar c^5}{G}} = 1.22 \times 10^{19}$ GeV , and the Planck mass is $m_p = \sqrt{\frac{\hbar c}{G}} = 2.1767 \times 10^{-8}$ kg. Here, the Newtonian constant of gravitation is $G = 6.673 \times 10^{-11}$ m³kg⁻¹sec⁻². However, some Planck units may not be familiar. For example, among possible unfamiliar Planck units, we introduce the *Planck area*, which is given as $\frac{G\hbar}{c^3} = 2.612 \times 10^{-70}$ m². One can provide an explanation for the nature of the Planck area. We multiply the Planck force and area quantity ($\hbar \times c$) and then divide the results by the Planck force (c^4/G). This gives $\frac{G\hbar}{c^3}$. This area is the square of the Planck length.

A Planck area is the area enclosed by a square that has the side length equal to the Planck length λ_p. The Plank length is defined as the length scale on which the quantized nature of gravity should become evident, and $\lambda_p = \sqrt{\frac{Gh}{c^3}} = 4.05096 \times 10^{-35}$ m.

Using the modified Planck's constant, we obtain $\lambda_p = \sqrt{\frac{G\hbar}{c^3}} = 1.616 \times 10^{-35}$ m.

The Planck time is the Planck length divided by the speed of light, i.e.,

$$t_p = \frac{\lambda_p}{c} = \sqrt{\frac{G\hbar}{c^5}} = 5.3906 \times 10^{-44} \text{ sec}$$

In 1993 the Dutch theoretical physicist G. Hooft proposed the *holographic principle*, which postulates that the information contained in some region of space can be represented as a hologram that gives the bounded region of space that contains at *most* one degree of freedom per Planck unit of area ($\lambda_p = 1.616 \times 10^{-35}$ m).

In this book, we will utilize the so-called *standard model* (particles are considered to be points moving through space and coherently represented by mass, electric charge, interaction, spin, etc.), see Fig. 2.5. The standard model was developed within a quantum field theory

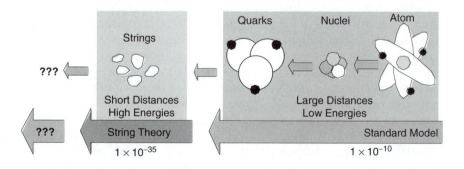

FIGURE 2.5
Visualization of the standard model, string theory, and unknown frontiers.

paradigm that is consistent with quantum mechanics and the special theory of relativity. Electromagnetism and the strong and weak nuclear forces are integrated. However, the gravity (fourth interaction) from Einstein's general relativity theory does not fit into quantum field theory. In fact, the quantum field theory and general relativity cannot be integrated. To overcome these limits, *string theory* has been developed. In string theory, instead of particles, one utilizes a fundamental building block called a *string* that can be closed (loop) or open. As the string moves through time, it traces out a tube (closed string) or a sheet (open string). The string is free to vibrate, and different vibration modes of the string represent the different particle types, since different modes are seen as different masses or spins. For example, modes of vibration (*notes*) may represent electrons or photons. String theory provides a consistent unified method that integrates electromagnetism, strong and weak nuclear forces, and gravity. The particles known in nature are classified according to their spin into bosons (integer spin) or fermions (odd half-integer spin). Examining the forces, we emphasize that the photon carries electromagnetic force, the gluon carries the strong nuclear force, and the graviton carries the gravitational force. Recently, *superstrings* and *heterotic string* theories have been introduced, and many other concepts have emerged. The unified paradigm is called the *M-theory*. We usually apply three dimensions of space and one of time. String theory utilizes 10-dimensional space and time. These extra dimensions, which can confuse the reader, could be unobservable, but compact dimensions result from mathematical deviations.

One should understand the difference between an atom and a string. Compared to a string, an atom is enormous. Dividing the diameter of the hydrogen atom (1.058×10^{-10} m) by the size of a string (1.6161×10^{-35} m) gives us 6.45×10^{24}. The Sun is 1,391,000 km in diameter, while the Earth is 12,742 km in diameter. Let us compare the Sun and a hydrogen atom. The ratio of the diameter of the Sun to the diameter of a hydrogen atom is 1.31×10^{19}. Hence, the Sun–atom ratio is much smaller than the atom–string ratio. Now we can grasp the string's dimension and size. Pluto has the largest orbit radius, and Pluto orbits the Sun with an orbit diameter of 1.184×10^{13} m. Dividing Pluto's orbit diameter by the diameter of the hydrogen atom (1.058×10^{-10} m), we have 1.12×10^{23}. This is still not enough, but the ratio is finally only 57 times smaller than that of the hydrogen atom to the string.

2.2 MEMS and NEMS Definitions

Micro- and nanotechnology have been identified as the most promising technologies of this century because of their potential for making affordable enhanced-functionality high-performance micro- and nanoscale systems and devices. We classify micro- and nanotechnology as evolutionary and revolutionary technologies, respectively.

MEMS, as a batch-fabricated integrated microsystem, integrates motion and motionless microdevices (actuators, sensors, transducers, communication devices, etc.), radiating energy microdevices (antennas, microstructures with windings), microscale driving/sensing circuitry, controlling/processing integrated circuits (ICs), and energy sources. Microsystems can be examined utilizing the biomimetics paradigm examining micro- and nanobiosystems. These micro- and nanobiosystems, which are a vital part of all living biosystems, exist in nature in enormous variety. Many microsystems have components and architectures similar to those in microbiosystems. Among the components, actuators, sensors, circuitry, and power sources should be emphasized. Figure 2.6 illustrates the functional block diagram of MEMS and NEMS (energy source is not illustrated) that compose different devices.

Let us formulate definitions in order to define micro- and nanoscale systems, devices, and structures. MEMS is defined by the author as follows. The MEMS is the batch-fabricated

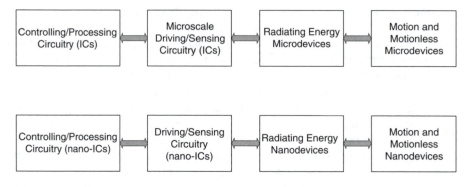

FIGURE 2.6
Functional block diagram of MEMS (batch-fabricated integrated microsystem) and NEMS.

integrated microsystem (motion, motionless, radiating energy, energy sources, and optical microdevices/microstructures, driving/sensing circuitry, and controlling/processing ICs) that has the following characteristics:

- Converts physical stimuli, events, and parameters to electrical, mechanical, and optical signals and vice versa
- Performs actuation, sensing, and other functions
- Comprises control (intelligence, decision making, evolutionary learning, adaptation, self-organization, etc.), diagnostics, signal processing, and data acquisition features
- Comprises microscale features of electromechanical, electromagnetic, electronic, electro-optical, optical, electrochemical, and biological components (structures, devices, and subsystems), architectures, and operating principles that are the basis for the synthesis, operation, design, analysis, and fabrication of the MEMS

MEMS are composed and built using microscale subsystems, devices, and structures. From the aforementioned definition, the MEMS, as the batch-fabricated integrated microsystem, can integrate the following components (devices or subsystems):

- Motion and motionless microdevices
- Radiating energy microdevices
- Energy sources
- Optical microdevices
- Driving/sensing circuitry; controlling/processing ICs

There are many microsystems. For example, microoptoelectromechanical systems (MOEMS) compose a class of microsystems. There are some specific topics for MOEMS (microoptics, photonics, and others) covered in specialized textbooks and manuscripts. In general, the enormous diversity of microsystems is commonly abbreviated as MEMS.

The following classification defines the microdevice.

The microdevice is the batch-fabricated integrated microscale motion, motionless (stationary), electromagnetic, radiating energy, or optical microscale device that has the following characteristics:

- Converts physical stimuli, events, and parameters to electrical, mechanical, and optical signals and vice versa

- Performs actuation, sensing, and other functions
- Comprises microscale features of electromechanical, electromagnetic, electronic, electro-optical, optical, electrochemical, and biological structures, topologies, and operating principles that are the basis for the synthesis, operation, design, analysis, and fabrication of microdevices.

Finally, we present the definition of the microstructure.

The microstructure is the batch-fabricated microscale electromechanical, electromagnetic, mechanical, electro-optical, optical, or electrochemical composite microstructure (that can integrate simple microstructures) that is a functional component of the microdevice and serves to attain the desired microdevice's operating features (e.g., convert physical stimuli, events, and parameters to electrical, mechanical, and optical signals and vice versa, as well as perform actuation, sensing, and other functions). Microscale features of electromechanical, electromagnetic, mechanical, electro-optical, optical, electrochemical, or biological structure (geometry, topology, operating principles, etc.) are the basis of the synthesis, operation, design, analysis, and fabrication of the microstructure.

To date, simple MEMS (accelerometers, micromachines with controlling and signal-processing ICs, system-on-chip sensors, etc.) and motion microdevices smaller than 50 μm have been synthesized, designed, fabricated, tested, and characterized. Having defined MEMS, we provide the definition of NEMS.

The NEMS is the organic, inorganic, or hybrid nanotechnology-based fabricated integrated nanosystem (motion, motionless, radiating energy, energy sources, and optical nanodevices/nanostructures, driving/sensing circuitry, and controlling/processing nano-ICs) that has the following characteristics:

- Converts physical stimuli, events, and parameters to electrical, mechanical, and optical signals and vice versa
- Performs actuation, sensing, and other functions
- Comprises control (intelligence, decision-making, evolutionary learning, adaptation, self-organization, etc.), diagnostics, signal processing, and data acquisition features
- Comprises nanoscale features of electromechanical, electrochemical, electromagnetic, electronic, electro-optical, optical, and biological components (structures, devices, and subsystems), architectures, and operating principles that are the basis for the synthesis, operation, design, analysis, and fabrication of the NEMS

To the best of the author's knowledge, organic, inorganic, and hybrid NEMS have not been fabricated and tested to date (nanobiosystems are not considered to be NEMS). There exist many nanobiosystems (e.g., bacteria that are hundreds of nanometers in length) that exhibit high functionality and perform the aforementioned functions—locomotion, energy conversion, actuation, sensing, control, intelligence, adaptation, etc. For example, the next section covers the *E. coli* bacterium, which integrates a number of nanosubsystems (actuation, sensing, controlling, communication, locomotion, etc.) and nanodevices. This nanobiosystem can be defined as bio-NEMS. However, unified terminology does not exist, and consequently some obscurity results.

It is evident that NEMS include nanodevices and nanostructures.

A nanodevice is a nanotechnology-based fabricated integrated nanoscale motion, motionless (stationary), electromagnetic, radiating energy, or optical nanoscale device that has the following characteristics:

- Converts physical stimuli, events, and parameters to electrical, mechanical, and optical signals and vice versa

- Performs actuation, sensing, and other functions
- Comprises nanoscale features of electromechanical, electromagnetic, electronic, optical, and biological structures, topologies, and operating principles that are the basis for the synthesis, operation, design, analysis, and fabrication of nanodevices

The nanostructure is the nanotechnology-based fabricated nanoscale electromechanical, electromagnetic, mechanical, or optical composite nanostructure (that can integrate simple nanostructures) that is a functional component of the nanodevice and serves to attain the desired nanodevice's operating features (e.g., convert physical stimuli, events, and parameters to electrical, mechanical, and optical signals and vice versa, as well as perform actuation, sensing, and other functions). Nanoscale features of electromechanical, electromagnetic, mechanical, optical, or biological structures (geometry, topology, operating principles, etc.) are the basis for the synthesis, operation, design, analysis, and fabrication of nanostructures.

In general, nanoscale systems and devices require new fabrication approaches (nano-technology-based fabrication) compared with MEMS. Nanofabrication paradigms are fundamentally different compared with microfabrication. Microstructures are typically formed by top-down techniques (sequential patterning, deposition, and etching). The fabrication of nanostructures can be based on bottom-up self-assembly and synthesis (processes where structures are built up from atomic or molecular-scale units and blocks in larger and increasingly complex nanoscale structures and devices). These processes are involved and observed in biological systems. Combinations of top-down (lithography, deposition, and etching) and bottom-up (self-assembly synthesis) technologies and processes also can be utilized.

Compared with nano- and microstructures, which are elementary building blocks of nano- and microsystems, the scope of nano- and microsystems and nano- and microdevices has been further expanded to include devising novel systems and devices, developing novel paradigms and theories, system-level integration, high-fidelity modeling, data-intensive analysis, adaptive control, optimization, intelligence, decision making, computer-aided design, fabrication, and implementation. These have been done through coherent fundamental developments, novel physics, biomimetics, and prototyping that allow one to expand the existing devised and man-made systems and devices. We will discuss nanobiosystems (bacteria), nanodevices (see nanobiomotor as illustrated in Figure 2.4), and nanostructures.

This book examines MEMS, NEMS, and micro- and nanoscale motion devices, as well as corresponding circuitry to control micro- and nanosystems. Correspondingly, micro- and nanobiosystems and devices are studied. In 1962 Professor Heinz A. Lowenstam discovered magnetite (Fe_3O_4) biomineralization in the teeth of chitons (mollusks of the class Polyplacophora), demonstrating that living organisms are able to precipitate the mineral magnetite. The next critical finding was the discovery by Richard Blakemore in 1975 of magnetotactic bacteria. Distinct 3-billion-year-evolved magnetotactic bacteria contain magnetosomes (magnetic mineral particles) enclosed in the protein-based membranes. In most cases, the magnetosomes are arranged in a chain or chains fixed within the cell. In many magnetotactic bacteria, the magnetosome mineral particles are either 30- to 100-nm magnetite (Fe_3O_4) or, in marine and sulfidic environments, greigite (Fe_3S_4). These nanoscale permanent magnets sense the magnetic field, and bacteria swim (migrate) along the magnetic field lines. The magnetosome chains are usually oriented so that a 111 crystallographic axis of each particle lies along the chain direction, while the chained greigite particles are usually oriented so that a 100 crystallographic axis of each particle is oriented along the chain direction. Whether the magnetic mineral particles are magnetite or greigite, the chain of magnetosome particles constitutes a permanent magnetic dipole fixed within the bacterium. Therefore, magnetotactic bacteria have

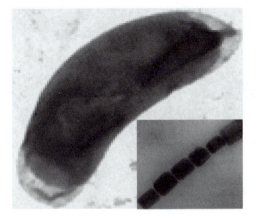

FIGURE 2.7
Magnetotactic bacterium and image of a chain of 60- to 100-nm-diameter cylindrical magnetosome mineral magnetic particles (rectangular, octahedral, prismatic, and other shapes of 30- to 100-nm magnetosome particles exist).

two magnetic poles, depending on the orientation of the magnetic dipole within the cell. The poles can be remagnetized by a magnetic pulse that is greater than the coercive force of the chain of particles. The magnetosome particles are uniformly magnetized, forming permanent magnetic domains. All particles are arranged along the chain axis such that the crystallographic magnetic easy axes are also aligned. The size specificity and crystallographic orientation of the chain assembly is optimally desig-ned for magnetotaxis in the geomagnetic field. Magnetosome particles occur in at least three different crystal forms. The simplest form, found in *M. magnetotacticum*, is cubo-octahedral, which preserves the cubic crystal symmetry of magnetite. A second type, found in coccoid and vibrioid strains, is an elongated hexagonal prism with the axis of elongation parallel to the 111 crystal direction. A third type, observed in some uncultured cells, is an elongated cubo-octahedral form producing cylindrical, bullet-shaped, tear-drop, and arrowhead particles (see Fig. 2.7). The growth (fabrication) mechanisms for these forms are unknown, but particle shapes may be related to anisotropic ion flux through the magnetosome membrane or from constraints imposed by the surrounding membrane structure. Whereas the cubo-octahedral form is common in inorganic magnetites, the prevalence of elongated hexagonal forms in magnetosomes appears to be a unique feature of the biomineralization process.

The iron oxide (Fe_3O_4) can be easily fabricated with 99% purity, and the size of particles is from 15 to 30 nm. The morphology is spherical, and the density is 5 g/cm^3. Figure 2.8 illustrates the fabricated Mo (99% purity, 85-nm size, and 10 g/cm^3 density), Fe (99% purity, 25-nm size, and 8 g/cm^3 density), Co (99.8% purity, 27-nm size, and 9 g/cm^3 density), and Cu (99.9% purity, 25-nm size, and 9 g/cm^3 density) nanoparticles and their assemblies that can be used in electrostatic and electromagnetic NEMS, MEMS, and nano- and microdevices. The reported Mo, Fe, and Co nanoparticles can be magnetized making the nanomagnet arrays, while Cu nanoassemblies can be used as the nanowires and nanointerconnects.

Different materials can be used. For example, carbon nanotubes and carbon fibers can be applied in nanoscale mechanical structures and electronics. Figure 2.9 illustrates carbon nanotubes (1.4 to 5 nm in diameter) and carbon fibers (96% purity with 4% catalyst metals, 50 nm diameter, and up to 100 μm long). The chiral single-wall carbon nanotube with the chiral vector $C_h = \{n1, n2\}$ can be represented as illustrated in Figure 2.9 for $C_h = \{5, 0\}$.

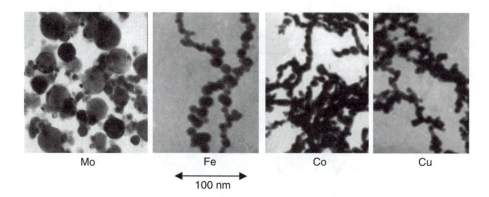

FIGURE 2.8
Mo, Fe, Co, and Cu nanoparticles and their nanoassemblies, respectively.

These carbon nanotubes are nanostructures that can be integrated into nanodevices (nano-resonator, nanoswitch, etc.).

We have explicitly introduced and defined nano- and microscale systems, devices, and structures and have given coherent definitions for systems, devices, and structures. There are significant differences between micro- and nanosystems. Consider electromechanical systems and electronics. In general, electromechanical systems can be classified as conventional electromechanical systems, microelectromechanical systems (MEMS), and nanoelectromechanical systems (NEMS), while nanoelectronics (nanoelectronic systems) can be classified as microelectronics, nanotechnology-enhanced microelectronics, and nanoelectronics.

In Chapter 1, we defined nanoelectronics as a fundamental/applied/experimental research that envisions and strives to devise (discover) new functional and high-performance atomic- and molecular-scale structures, devices, and systems (discovering, understanding, and utilizing novel phenomena and effects) and fabricate them by developing novel nanotechnologies and processes for supercomplex high-yield high-performance nano-ICs that will revolutionize electronics and information technologies. Distinct electromechanical and electronics systems are presented in Fig. 2.10, emphasizing the differences and relationships and assessing fundamental and fabrication differences.

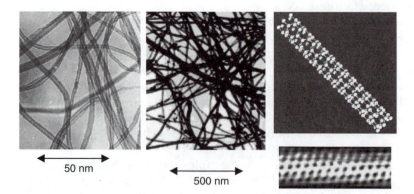

FIGURE 2.9
Carbon nanotubes, carbon nanofabrics, and carbon nanotube.

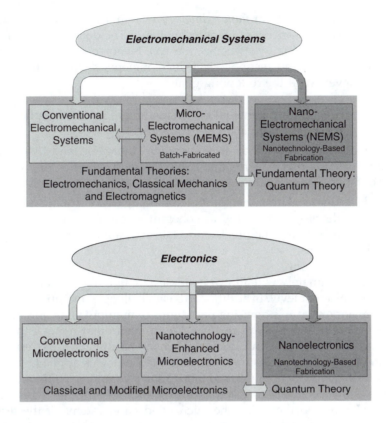

FIGURE 2.10
Electromechanical and electronic systems (conventional, microscale, and nanoscale).

The question to be discussed now is how to attain the synthesis, analysis, and design of electromechanical systems. It is evident that the basic physics laws and fabrication technologies for conventional, micro-, and nanoelectromechanical systems are different (see Figure 2.10). The operational principles and basic foundations of conventional electromechanical systems and MEMS can be identical or similar, while nanosystems require the application of quantum theory (single-electron devices, quantum wires, quantum dots, etc.) and utilize novel phenomena at nanoscale. Many attempts have been made to examine some classes of nanosystems using classical mechanics and electromagnetism. Many methods are shown to be applicable to coherently modify classical mechanics and electromagnetism. Thus, the designer applies classical mechanics (Lagrange equations of motion, the Hamiltonian concept, the Timoshenko paradigm, etc.) as well as electromagnetism (Maxwell's equations) to study conventional electromechanical systems, MEMS, and some NEMS. However, in general, quantum theory must be applied to nanosystems. Figure 2.10 illustrates the fundamental theories used to study the phenomena in conventional, micro-, and nanosystems.

It was emphasized that technologies to fabricate MEMS and NEMS are different. When we defined MEMS and NEMS, we used the terms *batch-fabricated* and *nanotechnology-based fabrication*. The differences between these technologies have been explicitly defined. For example, nanotechnology-based fabrication can include organic and inorganic synthesis, molecular chemistry, and DNA engineering, while the batch-fabricated (micromachining) technologies do not integrate these techniques. However, some processes, techniques, and materials used in micromachining and nanotechnology-based fabrication can be complementary. The term *nanotechnology* was first used by N. Taniguchi in his 1974 paper "On the Basic Concept of Nanotechnology." The definition of nanotechnology can be stated in the following way (see Section 1.2). Nanotechnology is the combination of research, engineering, and technological developments at the atomic, molecular, or macromolecular levels with the following goals:

- To provide a coherent fundamental understanding of interactive phenomena and effects in materials and complexes at the nanoscale
- To devise, design, and implement new organic, inorganic, and hybrid nanoscale structures, devices, and systems that exhibit novel properties due to their basic physics or fabrication technologies, and to utilize these novel phenomena and effects
- To provide a coherent fundamental understanding of controlling matter at nanoscale and to develop paradigms to coherently control and manipulate at the molecular level with atomic precision to fabricate advanced and novel complex devices and systems

Nanotechnology ultimately will allow one to design and fabricate nano- and microsystems built at the molecular level. In addition to basic challenges and unsolved fundamental problems, the major obstacle is devising paradigms to manipulate materials at the atomic level in order to build systems and devices. We can state that nano- and microsystems differ due to different physics and distinct phenomena utilized that result in different basic operating principles and distinct fabrication technologies applied.

Distinct physics, novel phenomena, and fabrication technologies ultimately result in the size difference. However, the sizing is a necessary, but not sufficient, condition to attain coherent classification. Furthermore, the micro- and nanosystems, if the terminology is properly used, immediately define their (or their components') sizing features, e.g., micro- and nanoscale.

Here are some definitions that can be stated as a coproduct of the above discussions:

- The microsystem (device) is the batch-fabricated integrated system (device) that is examined using conventional mechanics and electromagnetics.
- The nanosystem (device) is the nanotechnology-based fabricated integrated system (device) that is examined using quantum mechanics and (if justifiable) enhanced conventional mechanics and electromagnetics.

Having defined and classified MEMS and NEMS, we attack other problems. One of the major problems is how to devise (synthesize), design, analyze, and fabricate these systems and devices. That is, how do we progress from synthesis to fabrication of MEMS, NEMS, and micro- and nanodevices? The following section addresses this problem.

2.3   Introduction to Taxonomy of Nano- and Microsystem Synthesis and Design

For nano- and microscale systems and devices, the designer must study the taxonomy of devising (synthesis), which is relevant to cognitive study, classification, and synthesis of any systems. Devising of nano- and microsystems is an evolutionary process of discovering and examining evolving architectures (topologies) utilizing basic physical laws and studying possible systems evolutions based on synergetic integration of nanoscale structures and subsystems in the unified functional core. The ability to devise and optimize nano- and microsystems to a large extent depends on the basic physical principles comprehended and applied. Design, analysis, and optimization tasks can be performed only as a coherent functional nano- and microsystem is devised (synthesized). High-level hierarchy and abstraction, computational efficiency, adaptability, functionality, integrity, compliance, robustness, flexibility, prototype capability, clarity, interactivity, decision-making capacity, and intelligence of the CAD design meaningfully complement design, analysis, and optimization. It is likely that the synthesis taxonomy, fundamental physical laws, and applied experimental results in conjunction with coherent CAD will allow one to devise, prototype, design, model, analyze, and optimize nanosystems. The synergetic quantitative synthesis and symbolic descriptions can be efficiently used in searching and evaluating possible organizations, architectures, configurations, topologies, geometries, and other descriptive features providing the evolutionary features. Specifically, biosystems provide a proof-of-concept principle for highly integrated multifunctional organic, inorganic, and hybrid nano- and microsystems. The engineering biomimetics paradigm provides a meaningful conceptual tool to understand how biological and man-made systems coherently perform their functions, tasks, roles, and missions. One may cognitively examine microbiosystems with the coherent synthesis of microsystems, and then concentrate on design, analysis, and optimization. The *E. coli* bacterium (1 μm diameter, 2 μm length, and 1 pg weight), shown in Figure 2.11, has a plasma membrane, cell wall, and capsule that contains the cytoplasm and nucleoid.[2] From all viewpoints (locomotion, sensing, actuation, decision-making, and others), this bacterium achieves remarkable levels of efficiency, survivability, adaptation, and robustness. For example, the control and propulsion exhibited by this simple bacterium have not been achieved in any man-made nano-, micro-, mini-, or conventional underwater vehicles including most advanced torpedoes and submarines. Advanced conventional torpedoes (not considering the supercavitating rocket-propelled Shkval torpedo) achieve a maximum speed of 20 m/sec, and speed is a function of the vehicle length. The bacterium is propelled with a maximum speed of 20 μm/sec. That is, the speed–length ratio is 10. Only the most advanced torpedoes can maintain this ratio for a short time. Other examples reported in Figure 2.11 describe ants and butterflies. Micro air vehicles cannot achieve the agility, controllability, and maneuverability exhibited by butterflies.

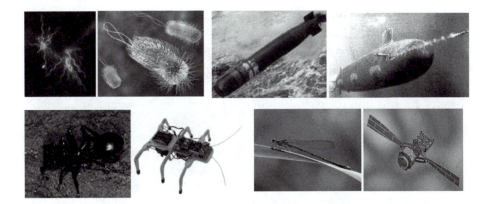

FIGURE 2.11
Fluorescent flagellar filaments of *E. coli* bacteria, bacteria and torpedo images, ant and microrobot (with infrared emitters and receivers), and dragonfly and micro air vehicle (UC Berkeley).

There is a need for system-level synthesis, design, analysis, and optimization. For example, nanobiomotors and nanobiosensors can be studied as complex nanosystems to attain computing, control, signal processing, diagnostics, health monitoring, etc. To devise, design, analyze, and optimize novel high-performance organic, inorganic, and hybrid nano- and microsystems, the designer synthesizes advanced architectures, discovers advanced topologies, examines new operating principles to attain superior functionality, applies cornerstone physical laws and fabrication technologies, integrates computer-aided design, etc. Synthesis and implementation of novel coherent architectures with actuators/sensors, electronics, controllers, and other subsystems (for example, from the circuitry viewpoint to attain communication, memory hierarchies, and multiple parallelism to guarantee high-performance computing, robust signal processing, and real-time decision making) play a critical role.

It should be emphasized that biosystems perform the following functions:

- Evolutionary control, adaptive self-organization, precise assembly of complex atomic compositions, and architecture reconfigurations of very complex and multifunctional structures. For example, simple proteins and complex protein complexes, which integrate millions of atoms, become dysfunctional by the change of a single atom.

- Actuation, sensing, communication (information propagation), computing, and control. For example, biosystems can sense the evolving environment, perform decision making, and deliver the measured information to control subsystems for the appropriate response and action. Correspondingly, synergy of actuation (locomotion), sensing, communication, computing, and control tasks are involved.

- Self-diagnostics, health monitoring, reconfiguration, and repairing. For example, biosystems can sense and identify damage, perform adaptation, execute robust decision making, and repair (reconfigure) itself to the functional state and operational level.

- Adaptation, reconfiguration, optimization, tuning, and switching based on evolutionary learning in response to short, long, reversible, and irreversible environmental changes including rapidly evolving dynamic adverse environments.

- Robust self-assembling and self-organization. In particular, thousands of individual components and compounds are precisely defined in complex functional and operational structures.

- Building highly hierarchical multifunctional complex structures and subsystems.

- Patterning and prototyping, e.g., templates of DNA or RNA are used as blueprints for structures, devices, and systems.

- Designing low-entropy structures increasing the entire system entropy. For example, protein folding, which is a highly ordered arrangement of the amino acids, is driven by the maximization of the entropy.

- Reproducing and mimicking the similar structures, subsystems, and systems.

Although general and systematic approaches may be developed to analyze and optimize nano- and microsystems and radiating energy, optical, and electronic nanodevices, it is unlikely that general CAD devising tools based on abstract concepts can be effectively applied for complex nano- and microsystems. This is due to a great variety of possible solutions, architectures, subsystem device structure organizations, and physical (quantum, electromagnetic, optical, mechanical, and chemical) principles and phenomena. However, by restricting the number of possible solutions based on the specifications and requirements, the application-specific synthesis can be performed using informatics theory, artificial intelligence, knowledge-based libraries, and expert techniques. For example, nanomachines and nanosensors can be synthesized. Correspondingly, analysis, optimization, and design of nano- and microsystems can be accomplished utilizing CAD. However, these CADs have not been developed, and it seems that very limited attempts even to address this problem have been performed to date.

The cognitive conceptual design of nano- and microsystems may integrate synthesis through biomimetics with consecutive design, analysis, optimization, and verification tasks. Currently, there do not exist CAD environments that can discover novel architectures (not to mention coherent synthesis), perform functional synthesis, and carry out high-fidelity modeling, data-intensive analysis, robust design, or optimization for even simple nano- and microsystems. Further synergetic efforts are needed to perform integrated optoelectromechanical analysis and design. The ultimate goal is to progress beyond the existing three-dimensional solid-model representation of nano- and microsystems (solid models have a limited degree of merit and sometime are needed just to portray the system and visualize fabrication tasks). In particular, one should devise, design, analyze, and optimize nano- and microsystems using intelligent discovery paradigms, heterogeneous synthesis, coherent libraries, basic physics, data-intensive analysis, data mining, assessment analysis tools, etc.

The top-level specifications are used to devise novel nano- and microsystems and then to devise electronic/electromechanical/electrooptomechanical/optical/electrochemical or other nanoscale devices and structures. High-level integrated heterogeneous synthesis must be carried out using intelligent databases (libraries) of structures, components, devices, and subsystems. The current design tools allow one to perform the steady-state analysis and design of a limited number of nanostructures and devices. Some CAD tools were extended to simulate nanodevices using linear and nonlinear differential equations. For example, three-dimensional modeling and simulations of thin films, membranes, cantilever beams, and carbon nanotubes were performed. These modeling efforts have frequently used a large number of assumptions and simplifications. However, synthesis and classification of nano- and microscale systems, devices (transducers, sensors, actuators, and electronic nanodevices), and three-dimensional ICs, as well as coherent design, analysis, and optimization problems, have not been solved and many problems have not even been addressed. The reusability and leverage of the extensive CAD developed for nanostructures are limited due to the critical need to devise and design novel systems. Thus, heterogeneous CAD must be developed, and this task is a far-reaching, long-term task.

Let us discuss and study what problems and tasks are associated. Nano- and microsystems, which integrate structures, devices and systems, encompasses four distinct domains of synthesis:

- Devising (synthesis)
- Modeling, analysis, simulation, and design
- Optimization and refining
- Fabrication and testing

A coherent design map is obtained and reported in Fig. 2.12. The evolutionary synthesis of nano- and microsystems is represented as a bidirectional X-flow map. This map illustrates the heterogeneous synthesis flow from devising (synthesis) to modeling/analysis/simulation/design, from modeling/analysis/simulation/design to optimization/refining, and finally from optimization/refining to fabrication/testing as sequential evolutionary processes. The devising hierarchy level is a most critical one, and the designer synthesizes novel systems utilizing specific phenomena and effects.

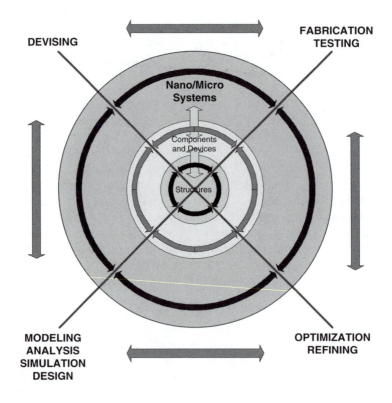

FIGURE 2.12
X-flow map with four domains.

To attain the desired degree of abstraction in the synthesis of nano- and microsystems, one should apply this X-flow map to devise novel nano- and microsystems that integrate nanoscale structures and components (devices and subsystems). The failure to verify the design feasibility for components in the early phases results in the failure to design high-performance or even functional nano- and microsystems and leads to redesign. The interaction between the four domains allows one to guarantee highly interactive bidirectional *top-down* and *bottom-up* design by applying low-level data to high-level design and using the high-level requirements in devising and designing low-level components. The X-flow map ensures the desired level of abstraction, hierarchy, modularity, locality, integrity, and other important features. For nano- and microsystems that can be devised within an indefinite number of operating principles (basic physics), architectures, and topologies, the taxonomy of the evolutionary synthesis is a critical issue. We propose to represent the synthesis taxonomy using the X-taxonomy map, which guarantees multiple abstraction levels (see Fig. 2.13).

This book emphasizes various fabrication aspects. In fact, nano- and microsystems, which comprise structures, devices, and systems, must be fabricated, tested, and characterized. The following block diagram, as shown in Fig. 2.14, illustrates the flowchart of the different tasks involved, from basic physics to verification. It should be emphasized that nano- and microsystems include subsystems, devices, and structures (see Figure 2.14).

Biomimetics, biomimicking, and prototyping are very important paradigms to devise novel nano- and microsystems. Let us briefly discuss the molecular basis of energy transduction. For example, bioenergetics has been studied in the past 20 years due to the availability of structures for some protein complexes catalyzing the reactions of the major pathways. These studies have provided new insights in the bioenergetic mechanisms,

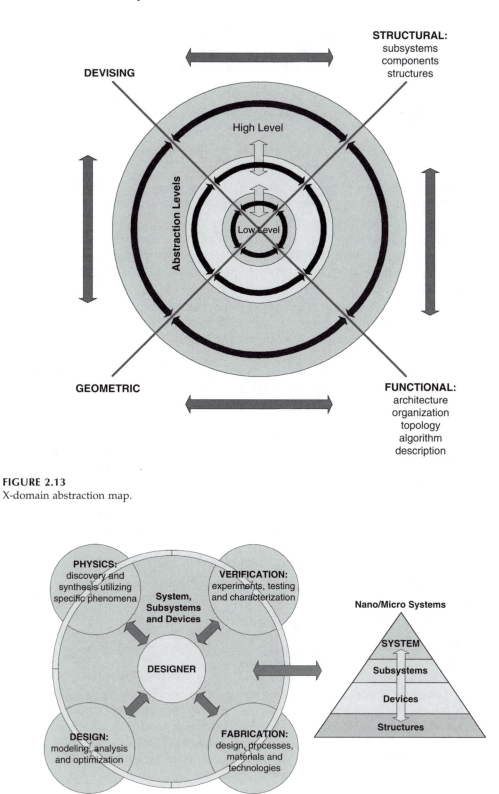

FIGURE 2.13
X-domain abstraction map.

FIGURE 2.14
Tasks flowchart and the system components.

identified some molecular devices, and provided possibilities for understanding processes fundamental to energy conversion (physics of the coupling energy transfer mechanisms through which energy transduction occurs). In particular, the following results have been obtained:

- Photosynthesis has led to an enhancement of electron transfer theory and spectroscopy.
- Structures of light-harvesting complex have resulted in excitation transfer processes and excitation energy funneling.
- ATP complexes have led to the possibility of designing nanoscale radial-topology nanobiomotors, nanobiomachines, and nanobiopropulsors.
- Respiratory complexes have resulted in further study of electron transfer and proton pumping coupling (bc_1 complex has revealed a mechanism of electron transfer over distance, while cytochrome oxidases are showing a mechanism for coupling electron transfer to local protonic potential through deep proton wells into the catalytic core).
- Structures of bacteriorhodopsin have provided the coherent dynamic representation of a photomolecular pump (protons activated through photoisomerization are switched between proton conduction pathways in and out of the protein).

Many species of bacterium move around their aqueous environment using flagella, which is the protruding helical filament driven by the rotating molecular motor. This bionanomotor–flagella complex provides the propulsion thrust for cells to swim. Let us examine the *E. coli* bacteria. Biological (bacterial) nanomotors convert chemical energy into electrical energy, and electrical energy into mechanical energy. The bionanomotor uses the proton or sodium gradient maintained across the cell's inner membrane as the energy source. The motion is due to the downhill transport of ions. The research in complex chemoelectromechanical energy conversion allows one to understand the torque generation, energy conversion, bearing, and sensing/feedback/control mechanisms. With the ultimate goal being to devise novel organic, inorganic, and hybrid micro- and nanomachines through biomimicking and prototyping, one can invent the following:

- Unique radial and axial micro- and nanomachine topologies
- Electrostatic- and electromagnetic-based actuation mechanisms
- Noncontact electrostatic bearing
- Novel sensing/feedback/control mechanisms
- Advanced excitation concepts
- New micro- and nanomachine configurations

Biomimetic systems are man-made systems based on biological principles or on biologically inspired building blocks integrated as the systems structures, devices, and subsystems. These developments benefit greatly from adopting strategies and architectures from the biological world. Based on biological principles, bio-inspired systems and materials are currently being formed by self-assembling and other patterning methods. Artificial organic, inorganic, and hybrid nanomaterials can be introduced into cells with diagnostics features as well as active (smart) structures.

Let us study nanobiomotors that can be utilized in devising (synthesis) and design of new high-performance micro- and nanomachines with fundamentally new organization, topologies, and operating principles, enhanced functionality, superior capabilities, and

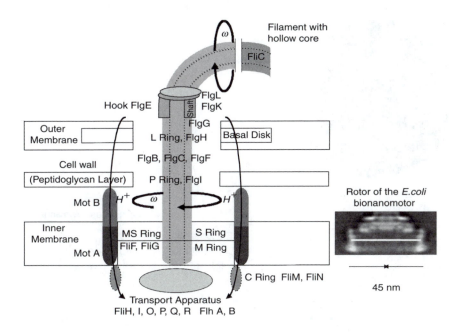

FIGURE 2.15
E. coli nanobiomotor: nanobiomotor/coupling/flagella complex with different proteins and rings and 45-nm-diameter rotor image.

enhanced operating envelopes. In particular, consider the nanobiomotor of *E. coli* bacteria (see Figure 2.15). The flagella (rotated by nanobiomotors) are used for propulsion. The bacterium is propelled with a maximum speed of 20 μm/sec by flagellar filaments. This filament is driven by a 45-nm rotor of the nanobiomotor embedded in the cell wall. The cytoplasmic membrane forms a stator. This nanobiomotor integrates more than 20 proteins and operates as a result of the axial *protonomotive* force resulting from the proton flux.

The rated nanobiomotor parameters were estimated as follows: angular velocity, 20 rad/sec; torque, 1×10^{-16} N · m; efficiency, 50%. It was reported in the literature that the nanobiomotor has three *switch* proteins (FliG, FliM, and FliN) that control the torque, angular velocity, and direction of rotation. Berg[2] found that FliG interacts with FliM, FliM interacts with itself, and FliM interacts with FliN. The flagellum, flexible joint (proximal hook), and nanobiomotor are shown in Fig. 2.15.[2,3] The nanobiomotor has two major parts: a stator (connected to the cell wall/peptidoglycan) and a rotor (connected to the flagellar filament through flexible joint). The *switching* proteins (without any speculations on how they interact or operate because this is not relevant to the presented discussions and may not be comprehended within many decades) can be viewed as three-dimensional (3D) nanobio-ICs.

The protonomotive force in the *E. coli* bionanomotors likely is axial (it should be emphasized that the protonomotive force can be radial as well to attain the rotation). Through biomimicking, one can state that two possible machine (rotational motion device) topologies are radial and axial.

In particular, the flux in the radial (or axial) direction interacts with the time-varying axial (or radial) field, and the rotating torque (to actuate the motor) can be produced. Radial topology nanobiomotors, powered by ATP, can be synthesized from DNA and RNA protein complexes. Figure 2.16 illustrates six strands of RNA complexes and a DNA strand serving as a rotor. The cornerstone idea is that due to ATP reaction, the RNA strands (stator) can develop the force rotating the DNA (rotor). Theoretically, the F_1 gamma-subunit rotates with respect to the a_3b_3-hexamer, and the rotor of the nanobiomotors should turn.

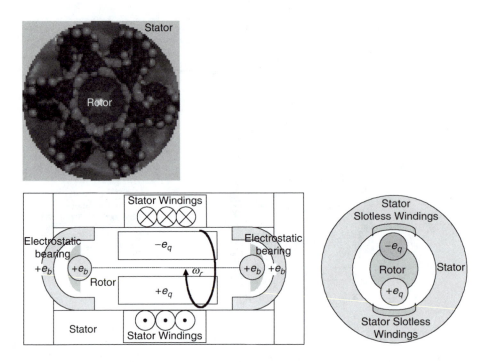

FIGURE 2.16
Radial topology DNA-RNA ATP-powered hypothetical nanobiomotor with nanorotor (DNA) and stator (six RNA). Radial topology machine with electrostatic noncontact bearings (poles are $+e_q$ and $-e_q$, and electrostatic bearing formed by $+e_b$).

Some experiments are reported for the bacterial virus phi 29 that infects *Bacillus subtilis*. In this nanobiomotor, RNAs bind to the connector serving as the building blocks for hexamer assembly of the DNA translocation nanobiomotor. Using the radial topology of the described hypothetical DNA-RNA nanobiomotor, the cylindrical micro- or nanoscale machine with permanent-magnet poles on the rotor is shown in Figure 2.16. The electrostatic noncontact bearings allow one to maximize efficiency and reliability, improve ruggedness and robustness, minimize cost and maintenance, decrease size and weight, optimize packaging and integrity, etc.

The advantage of radial topology is that the net radial force on the rotor is zero. The disadvantages are that it is difficult to fabricate and assemble these machines with nano- and microstructures (stator with deposited windings and rotor with deposited magnets), and the air gap is not adjustable.

Analyzing the *E. coli* nanobiomotor, the nano- and microscale machines (rotational transducers or motion devices) with axial flux topology are devised. The synthesized axial machine is illustrated in Fig. 2.17. The advantages of axial topology are as follows:

- The machines are affordable and easy to fabricate and assemble because permanent magnets are flat (permanent-magnet thin films can be deposited using matured microfabrication technologies, processes, materials, and chemicals reported in Chapters 4 and 10 of this book).
- There are no strict shape/geometry and sizing requirements imposed on the magnets.
- There is no rotor back ferromagnetic material required (silicon or silicon carbide techniques and processes can be straightforwardly applied to fabricate at least micromachines).

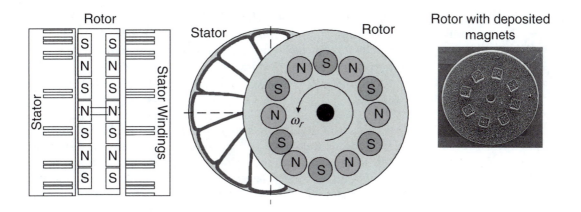

FIGURE 2.17
Axial permanent-magnet synchronous nanomachine.

- The air gap can be adjusted.
- It is easy to fabricate planar windings including nanowires on the flat stator.

Utilizing the axial topology and endless electromagnetic system, one can synthesize permanent-magnet synchronous micro- and nanomachines as documented in Figure 2.17. This nanomachine has well-defined topological analogy compared with the *E. coli* nanobio-motor. It must be emphasized that the documented motion nanodevice can be fabricated, and a prototype of the micromachine with a 50-μm rotor is illustrated in Figure 2.17. The planar segmented nanomagnet array, as evident from Figure 2.17, can be deposited as thin-film nanomagnets.

The reader will find out how these machines with radial and axial topologies operate in Chapter 6, which provides coherent coverage. Therefore, just a brief introductory description, based on basic physics, is reported in this section. The stationary magnetic field is established by the permanent magnets, and stators and rotors can be fabricated using surface micromachining and high-aspect-ratio technologies. The following is a sequential (step-by-step) procedure in the design of nano- and microdevices, with emphasis on the discussed micro- and nanoscale machines:

- Devise machines researching operating principles, topologies, configurations, geometry, electromagnetic systems (closed-ended, open-ended, or integrated), interior or exterior rotor, etc.
- Define application and environmental requirements as well as specify performance specifications.
- Perform electromagnetic, energy conversion, mechanical, thermal, and sizing estimates.
- Define technologies, techniques, and processes to fabricate, assemble, and package micro- and nanostructures (e.g., stator with deposited windings, rotor with permanent magnets, and bearing).
- Based on data-intensive analysis, determine air gap, select permanent magnets and materials, and perform thorough electromagnetic, mechanical, vibroacoustic, and thermodynamic design with performance analysis and outcome prediction.

- Modify and refine the design optimizing machine performance.
- Design control laws to control micromachines and implement these controllers using ICs interfacing machines (this task can be broken down into many subtasks and problems related to control laws design, optimization, analysis, sensing, circuitry design, IC topologies, IC fabrication, micromachine–IC integration, interfacing, networking, etc.).

The reported results clearly illustrate that before being engaged in the fabrication, there is a critical need to address fundamental synthesis, design, analysis, and optimization issues that are common to all micro- and nanomachines. In fact, machine performance and its applicability directly depend on synthesized topologies, configurations, electromagnetic systems, etc. Optimization methods applied, simulation software used, control laws designed, ICs applied, and machine–IC integration are also key factors.

Systematic synthesis and modeling allow the designer to devise novel phenomena and new operating principles guaranteeing synthesis of superior machines with enhanced functionality and operability. To design high-performance nano- and microsystems, fundamental, applied, and experimental research must be performed to further develop the synergetic basic theories. Fundamental electromagnetic and mechanical laws include quantum mechanics, Maxwell's equations, nonlinear mechanics, energy conversion, thermodynamics, vibroacoustics, etc. It was shown that the design of nano- and microsystems is not a simple task because complex electromagnetic, mechanical, thermodynamic, and vibroacoustic problems must be examined in the time domain to solve partial differential equations. Advanced CAD, interactive software with application-specific toolboxes, robust methods, and novel computational algorithms must be used. Computer-aided design of nano- and microsystems offers the following advantages:

- Calculation and thorough evaluation of a large number of options with data-intensive heterogeneous performance analysis and outcome prediction
- Knowledge-based intelligent synthesis and evolutionary design, which allow one to define optimal solutions with minimal effort, time, cost, but with a highest degree of reliability, confidence, and accuracy
- Concurrent nonlinear electromagnetic, mechanical, thermodynamic, and vibroacoustic analysis to attain superior performance of nano- and microsystems while avoiding costly and time-consuming fabrication and testing
- Possibility of solving complex partial differential equations in the time domain by integrating systems patterns with nonlinear material characteristics
- Development of robust, accurate, interactive, and efficient rapid design and prototyping environments and tools that have innumerable features to help the user set up the problem and obtain the engineering parameters

Through structural synthesis, the designer devises nano- and microsystems that must be modeled, analyzed, simulated, and optimized. Synthesis, design, and optimization guarantee superior performance capabilities. As shown in Fig. 2.18, when devising and developing novel micro- and nanomachines, one maximizes efficiency, reliability, power and torque densities, ruggedness, robustness, durability, survivability, compactness, simplicity, controllability, and accuracy. The application of the structural synthesis paradigm leads to the minimization of cost, maintenance, size, weight, volume, and losses. Packaging and integrity are optimized using the structural synthesis concept.

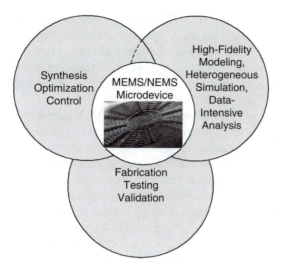

FIGURE 2.18
Tasks in design of high-performance micro- and nanoscale systems and devices.

Nanotechnologies drastically change the fabrication and manufacturing of structures, devices, and systems through the following means:

- Predictable properties and characteristics of nanocomposites, compounds, clusters, and materials (e.g., low weight and high strength; thermal stability; low volume and size; extremely high power, torque, force, charge, and current densities; specified thermal conductivity and resistivity; robustness)
- Design cycle, cost, and maintenance reduction
- Improved accuracy, precision, reliability, functionality, robustness, and durability
- Higher efficiency, safety, capability, flexibility, integrity, supportability, affordability, survivability, and redundancy
- Environmental compatibility, integrity, and flexibility

For nano- and microsystems, many engineering problems can be formulated, attacked, and solved utilizing distinct nanoscience and engineering paradigms that deal with benchmarking and emerging problems in integrated electrical, mechanical, and computer engineering, science, and technology. Many of these problems have not been addressed and solved. Frequently, the existing solutions cannot be treated as the optimal one. This reflects obvious trends in fundamental, applied, and experimental research in response to long-standing unsolved problems, as well as engineering and technological enterprise and entreaties of steady evolutionary demands. Nanoscience and engineering focus on the multidisciplinary synergy, integrated design, analysis, optimization, biomimicking, and virtual prototyping of high-performance nano- and microsystems, system intelligence, learning, adaptation, decision making, and control through the use of advanced hardware and computationally efficient heterogeneous software. Integrated multidisciplinary features and the need for synergetic paradigms approach quickly. The structural complexity of nano- and microsystems has been increased drastically due to specifications imposed on systems, devices, hardware and software advancements, and stringent *achievable* performance requirements. To meet the demands of rising systems complexity, performance specifications, and intelligence, the fundamental theory must be further expanded.

In particular, in addition to devising subsystems, devices, and structures, there are other issues that must be addressed in view of the constantly evolving nature of nano- and microsystems (e.g., synthesis, analysis, design, modeling, simulation, optimization, complexity, intelligence, decision making, diagnostics, fabrication, packaging, etc.). Competitive *optimum-performance* nano- and microsystems can be designed only by applying the advanced hardware and software concepts.

The most challenging problems in systems design are basic physics, topology/architecture/ configuration synthesis, system integration, fabrication technologies, and selection of affordable hardware and software (analytic and numerical methods, computation algorithms, tools, and environments to perform control, sensing, execution, emulation, information flow, data acquisition, simulation, visualization, virtual prototyping, and evaluation). As was emphasized, attempts to design state-of-the-art high-performance nano- and microsystems and to guarantee the integrated design can be pursued through analysis of complex patterns and paradigms of evolutionary developed biological systems. Even at the device level, nano- and microdevices must be devised first, and structural synthesis must be performed integrating modeling, analysis, optimization, and design problems.

As was emphasized and illustrated, synergetic multidisciplinary research must be carried out. For example, rotational and translational MEMS, which integrate motion microdevices (transducers), radiating energy microdevices (antennas), communication microdevices, driving/sensing ICs, and controlling/processing ICs, are widely used. High-performance MEMS must be devised, designed, analyzed, and optimized. The need for innovative methods to perform structural synthesis, comprehensive design, high-fidelity modeling, data-intensive analysis, heterogeneous simulation, and optimization has facilitated theoretical developments within the overall scope of engineering and science. To attack and solve the problems mentioned, far-reaching multidisciplinary research must be performed. A MEMS synthesis paradigm reported in this book (Chapter 5) allows one to synthesize, classify, and optimize rotational and translational motion microdevices based on electromagnetic and geometric design. Microsystems can be synthesized using a number of different operating principles, topologies, configurations, and operating features. Optimal electromagnetic, mechanical, optical, vibroacoustic, and thermal design must be performed to attain superior performance. In MEMS, the issues of operational variability, viability, controllability, topology synthesis, packaging, and electromagnetic design have consistently been the most basic problems. One of the main goals of this book is to present a general conceptual framework with practical examples that expand up to systems design and fabrication. Therefore, innovative research and manageable developments in synthesis and design of rotational and translational MEMS are reported. The MEMS classifier and synthesis concepts, which are based on the classification paradigm reported in Chapter 5, play a central role. These benchmarking results are applied to design high-performance MEMS.

Explicit distinctions must be made between various MEMS configurations, possible operating principles, and topologies. High-fidelity modeling, lumped-parameter modeling, and data-intensive analysis with outcome prediction can be performed using the developed nonlinear mathematical models that allow the designer to analyze, simulate, and assess the MEMS performance in the time domain. Mathematical modeling, analysis, design, and fabrication can be performed only after the designer devises (synthesizes) MEMS. For example, as the MEMS is synthesized based on the electromagnetic features, its performance and controllability are researched, control algorithms are developed, and functionality is assessed. Microsystems are regulated using radiating-energy microdevices. Furthermore, control laws are implemented using ICs. Thus, the designer derives the MEMS configuration integrating microdevices and then performs modeling, simulation, optimization, and assessment analysis with outcome prediction. Modeling and control of nano- and microsystems are covered in Chapters 6, 9, and 10.

Chapters 4 and 10 discuss the fabrication technologies, techniques, processes, and materials. However, the designer realizes that before being engaged in the fabrication (processes developments, sequential steps integration, materials and chemical selection) the MEMS must be devised. Electromagnetic and electrostatic MEMS, which are widely used in various sensing and actuation applications, must be devised, designed, modeled, analyzed, simulated, optimized, and verified. Furthermore, these rotational and translational motion microdevices need to be controlled. Chapter 5 introduces the synthesis paradigm to perform the structural synthesis of MEMS based on electromagnetic features. As motion microdevices are devised, modeling, analysis, simulation, control, optimization, and validation are emphasized. Chapter 6 reports the results in modeling, analysis, and simulation of microelectromechanical motion devices. Quantum theory, carbon nanotubes, and nanoelectronics are covered in Chapters 7 and 8. Control topics for MEMS are covered in Chapter 9. Chapter 10 documents the case studies in design, analysis, and fabrication of MEMS.

Finally, a general comment by the author regarding nano- and microsystems is stated. In general, it seems that MEMS has progressed as an evolutionary development in the miniaturization and integration of miniscale electromechanical (optoelectromechanical or biooptoelectromechanical) systems and ICs. There are fundamental differences between nano- and microsystems because nanosystems can be viewed as a revolutionary breakthrough. For example, electromagnetic microactuators and quantum nanoactuators operate utilizing distinct physics, different phenomena are involved, and distinct fabrication technologies must be used to fabricate these devices. The examples reported in this chapter illustrate that for many biosystems, the basic operating principles are unknown, and we have limited knowledge regarding their functionality. Instead of the traditional evolutionary engineering and science efforts to reduce size and power while enhancing performance and functionality of diverse MEMS, synthesis and design of NEMS have been utilized to radically reduce the scale, maximizing performance of nanosystems by employing novel basic principles and fundamental concepts. It is the author's belief that significant milestones and benchmarking results will be achieved in the observable future, and we will be able to synthesize, design, and test affordable NEMS.

2.4 Introduction to Design and Optimization of Nano- and Microsystems in the Behavioral Domain

It was emphasized that recent trends in engineering and science have increased the emphasis on coherent synthesis, analysis, design, optimization, and control of advanced nano- and microsystems. The synthesis and design processes are evolutionary in nature. They start with a given set of requirements and specifications on functionality, operability, environment, robustness, performance, compliance, integrity, sizing, etc. High-level functional physical phenomena-based synthesis is performed first in order to start design at the system-, subsystem-, component-, device-, and structure-levels. Using synthesized (devised) high-performance subsystems, components, devices, and structures that utilize explicit and well-defined physics, the initial design can be performed, and MEMS/NEMS are then evaluated and tested (through modeling-simulation or experiments) against the requirements. The fabricated nano- and microsystems can be experimentally tested and characterized. Many requirements and specifications are assigned in the behavioral domain, e.g., bandwidth, settling time, overshoot, etc. If requirements and specifications are not met, the designer revises and refines the system architecture, integrates new components, synthesizes novel devices, and performs optimization. Different alternative solutions are sought and examined. The synthesis and design taxonomy was reported in Section 2.3. This section emphasizes the design of nano- and microsystems in the behavioral domain. That is, the steady-state and dynamic performances are addressed and discussed. For example, what is the speed of the *E. coli* bacteria and its nanobiomotor, what are the acceleration capabilities of this bacteria and motion microdevice, and how much time is needed to achieve the desired speed?

Analysis and design of nano- and microsystems in the behavioral domain lead to the performance analysis, transient dynamics evaluation, behavioral studies under disturbances, etc. At each level of the design hierarchy, the system and device performance in the behavioral domain is used to evaluate, optimize, and refine the synthesis and design processes as well as to devise new solutions in the sequence of evolutionary steps. Each level of the design hierarchy corresponds to a particular abstraction level and has the specified set of evolutionary activities and design tools that support the design at this level. The coherent synthesis and design require the application of a wide spectrum of paradigms, methods, computational environments, fabrication technologies, etc. It is the author's goal to cover manageable topics, advanced methods, novel paradigms, and promising leading-edge technologies to design nano- and microsystems that will achieve optimal performance in the behavioral domain. The conceptual view of the synthesis and design was introduced in order to set the objectives, goals, and problems encountered, as well as to illustrate the need for synergetic optimization of transient dynamics. The generic synthesis and design issues are covered in Chapter 5. Chapters 6 and 7 are devoted to the high-fidelity and lumped-parameter modeling of nano- and microsystems. Control of NEMS and MEMS is covered in Chapter 9. The case studies are reported in Chapter 10. The development, simulation, and control of mathematical models are covered with the ultimate objective to examine the system dynamics in the behavioral domain. This section introduces the reader to the problem under consideration.

Requirements and specifications integrated within sequential synthesis, analysis/modeling, optimization, and fabrication must be examined by researching system dynamics, i.e., accomplishing the data-intensive analysis in the behavioral (time) domain. Different criteria are used to synthesize and design nano- and microsystems due to different basic physics, behavior, physical properties, operating principles, and performance criteria imposed. Studying nano- and microsystems in the behavioral domain (which allows one to attain analysis of both steady-state responses and transient dynamics), it should be emphasized that the level of the system hierarchy must be defined and studied. For example, there may be no need to study the very fast behavior of millions of nanotransistors of nano-IC subsystems. The existing nano-ICs ensure the transient dynamics within picosecond range. Due to the "slow" (microseconds) end-to-end microsystem behavior, in many cases electromechanical dynamics should be evaluated instead. That is, the steady-state end-to-end (input-output) nano-IC responses can be sufficient from the system-level viewpoints. Therefore, nano-ICs can be designed as the stand-alone MEMS nanoelectronic components. However, nano-ICs must guarantee MEMS operating and functional features (for example, control of electromagnetic or electrostatic microdevices, input-output interface, analog-to-digital and digital-to-analog conversion, filtering, data processing and acquisition, communication, etc.). The system architecture and component dynamics are illustrated in Fig. 2.19.

Hence, to attain the system behavioral analysis, due to the complexity of nano- and microsystems, one may examine only the input-output steady-state responses of the fast subsystems and devices (for example, nano-ICs) and the "slow" transient dynamics that can be due to the *torsional-mechanical* dynamics. The reader may recall that the torsional-mechanical dynamics can be described by Newton's second law of motion. It must be emphasized that the microsecond-range "slow" dynamics in reality are extremely fast (for

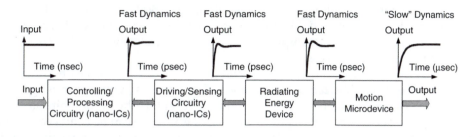

FIGURE 2.19

Functional block diagram of MEMS with the fast (psec) and "slow" (msec) transient dynamics of the components.

example, our car accelerates within 10 seconds, that is, 1,000,000 times slower than the "slow" dynamics of the majority of MEMS and motion microdevices). The nano- and microsystem dynamics will be examined and optimized in the behavioral domain examining the steady-state responses and dynamic transient dynamics. The design flow concerning the optimization of nano- and microsystems in the behavioral domain is illustrated in Fig. 2.20. It is evident that synthesis of nano- and microsystems is a first step to carry out. Mathematical model developments, heterogeneous simulation, data-intensive analysis, and control are the sequential steps to be performed.

The automated synthesis and optimization can be applied to implement the design flow introduced. The synthesis and design start from the specification of requirements, progressively proceeding to perform functional design and dynamic optimization in the behavioral domain for the system and all its components. These are gradually refined through a series of sequential synthesis steps. Specifications typically include the performance requirements derived from desired system functionality, complexity, operating

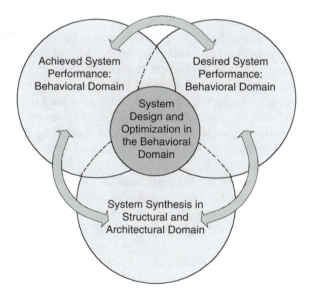

FIGURE 2.20

Design and optimization of nano- and microsystems in the behavioral domain.

envelope, affordability, reliability, and other requirements. Both top-down and bottom-up approaches should be combined to design high-performance nano- and microsystems augmenting hierarchy, integrity, complexity, regularity, modularity, compliance, and completeness in the synthesis process.

Even though the basic foundations have been developed in the literature, some urgent areas have not been emphasized and examined. The systems synthesis must guarantee an eventual consensus between behavioral and structural (synthesis) domains, as well as ensure descriptive and integrative features in the design. These can be achieved by applying the evolutionary synthesis that will allow one to extend and augment the results of basic physics (quantum and classical mechanics, electromagnetism, microelectronics, informatics, and control theories), as well as to apply advanced integrated hardware and software. To acquire and expand the engineering/science/technology core of nano- and microsystems, there is the need to augment interdisciplinary areas as well as to link and place the synergetic perspectives integrating hardware (electronics, actuators, sensors, etc.) with system intelligence, control, decision making, signal processing, information processing, data acquisition, etc. New multidisciplinary developments are needed, and this book introduces and coherently covers most important aspects of modern nano- and microengineering, science, and technology.

The need for innovative integrated methods to perform data-intensive analysis, high-fidelity modeling, and design of nano- and microsystems in the behavioral domain has facilitated theoretical developments within the overall spectrum of engineering and science. This book provides one with viable tools to perform synthesis as well as to accomplish modeling, analysis, optimization, and control in the time domain. The integrated design, analysis, optimization, and virtual prototyping of high-performance nano- and microsystems can be addressed, researched, and solved through the use of advanced theory, state-of-the-art hardware, novel technologies, and leading-edge software. Many problems can be formulated, attacked, and solved using nano- and microengineering. Nano- and microengineering, as the integrated paradigms in the synthesis, design, analysis, and optimization of high-performance MEMS/NEMS, deal with benchmarking and emerging problems in integrated electrical, mechanical, and computer engineering, science, and technologies. In addition, sensing, communication, and computing must be addressed. Integrated multidisciplinary features approach quickly, and the synergetic nano- and microengineering and science are taking place. Novel CAD tools and environments are required to support nano- and microsystem synthesis, analysis, simulation, design, optimization, and fabrication. Efforts have been devoted to attain the specified steady-state and dynamic performance to meet the criteria and requirements imposed. Currently, MEMS are largely designed, optimized, and analyzed using available software packages based on the linear and steady-state analysis. However, highly detailed heterogeneous nonlinear electromagnetic and mechanical modeling must be performed to design high-performance MEMS/NEMS and coherently examine them in the behavioral domain. Therefore, the research is concentrated on high-fidelity mathematical modeling, data-intensive analysis, nonlinear simulations, and control (design of control algorithms to attain the desired performance). The reported synthesis, modeling, analysis, simulation, optimization, and control concepts, tools, and paradigms ensure a cost-effective solution and can be used to attain rapid prototyping of high-performance state-of-the-art nano- and microsystems.

Modeling, simulation, analysis, virtual prototyping, and visualization are critical and urgently important aspects for developing and prototyping advanced nano- and microsystems. As a flexible high-performance modeling and design environment, MATLAB® has become a standard cost-effective tool. Competition has prompted cost and product cycle reductions. To speed up design with assessment analysis, facilitate enormous gains in productivity and creativity, attain intelligence, accomplish evolutionary learning, integrate control and signal processing, accelerate prototyping features, visualize the results, perform data acquisition, and ensure data-intensive analysis with outcome prediction, the MATLAB environment is used. In MATLAB, the following commonly used toolboxes can be applied: SIMULINK®, Optimization, Symbolic Math, Partial Differential Equations, Neural Netwoks, Control System, and other application-specific toolboxes. The MATLAB environment offers a rich set of capabilities to efficiently solve a variety of complex analysis, modeling, simulation, control, and optimization problems. A wide array of nano- and microsystems can be modeled, simulated, analyzed, and optimized. The nano- and microscale components, devices, and structures can be designed and simulated using other environments. For example, the VHDL (Very High-Speed Integrated Circuit Hardware Description Language) and SPICE (Simulation Program with Integrated Circuit Emphasis) environments are used to design, simulate, and analyze ICs. Currently, other software tools and packages are becoming available for nano-ICs.

Homework Problems

1. Use the periodic table and literature sources to derive the size of Si, Fe, and Co atoms (one may use chemistry and physics books, as well as handbooks, for example, *Handbook of Chemistry and Physics*, 83rd ed., Lide, D.R., Ed., CRC Press, Boca Raton, FL, 2002).

2. What is the difference between MEMS and NEMS? Why are MEMS and NEMS, as well as micro- and nanodevices, difficult (or impossible) to classify using merely their size?

3. Develop definitions for nano- and microscale subsystems. Briefly discuss the difference between nano- and microsubsystems.

4. Using the course syllabus, identify what topics and areas will be emphasized in the course, e.g., synthesis, design, analysis, or fabrication.

5. Using the taxonomy of the MEMS/NEMS design, choose a specific MEMS or NEMS (propulsion system, motion device, sensor, etc.) and explicitly identify problems and tasks that need to be addressed, studied, and resolved in order to design and implement the examined system or device.

6. Explain why nano- and microsystems must be examined in the behavioral domain. Using Newtonian mechanics, illustrate that classical mechanics leads to the differential equations to model systems. *Hint*: Use Newton's second law for the translational and rotational motions and recall that the linear and angular accelerations are the second derivative of the linear and angular displacements.

7. Using the Web or library resources, find an appealing paper from the latest issues of the suggested journals (*Journal of Microelectromechanical Systems, Journal of Micromechanics and Microengineering, Micromachine Devices, Sensor and Actuators A: Physical, Nanotechnology, IEEE Transactions on Magnetics, Physics Review, Physics Review Letters, MST News, Sensors, Biosensors and Bioelectronics, Nanobiology, IEEE Spectrum, Science*, and *Nature*; the journals are listed in the order suggested by the author) or conference proceedings (MEMS, Nanotechnology,

ASME meetings, etc.) in which you are most interested. Write a report in which you address the following:

Explain why this paper is appealing to you

7.1 Describe the significance of the results from engineering, scientific, and technological viewpoints

7.2 Explain how the results reported may benefit (or harm) society and improve (or degrade) our lives and nature[4]

7.3 Discuss any questionable and debatable issues (scientific and societal) that could result due to the results documented

7.4 Describe the basic physics, concepts, methods, techniques, approaches, and technologies applied and covered in the paper

7.5 Explain how you utilized the knowledge and material covered in other courses (calculus, physics, chemistry, biology, engineering, etc.) to assess and comprehend the results documented

7.6 Identify and discuss the novelty of the results

7.7 Try to propose and develop alternative possible solutions or directions

7.8 Assess the overall quality of the paper (discuss writing style, readability, clarity, significance, soundness, technical merit, references, etc.)

References

1. Goddard, W.A., Brenner, D.W., Lyshevski, S.E., and Iafrate, G.J., Eds., *Handbook of Nanoscience, Engineering, and Technology*, CRC Press, Boca Raton, FL, 2002.
2. Berg, H.C., The rotary motor of bacterial flagella, *J. Annu. Rev. Biochem.*, 72, 19–54, 2003.
3. Lyshevski, S.E., *MEMS and NEMS: Systems, Devices, and Structures*, CRC Press, Boca Raton, FL, 2002.
4. Bryzek, J., Impact of MEMS technology on society, *Sensors Actuators A: Phys.*, 56, 1–9, 1996.

3

Nano- and Microsystems: Classification and Consideration

3.1 Biomimetics, Biological Analogies, and Design of NEMS and MEMS

3.1.1 Biomimetics Fundamentals

Even though the basic foundations for nano- and microsystems have been developed, some very important areas have not been emphasized and researched thoroughly. There exist organic, inorganic, and hybrid nano- and microsystems. Novel organic and inorganic systems can be designed utilizing the phenomena and effects observed in biosystems. To be able to utilize these phenomena and to fabricate systems, one must first understand these phenomena.

It is evident that systems synthesis must guarantee an eventual consensus and coherence among basic phenomena and their utilization, system architecture, behavioral and structural (synthesis) domains coherence, as well as ensure descriptive and integrative features in the design. These issues were emphasized in Chapter 2. These problems can be tackled by examining the development of evolutionary biological systems. Figure 3.1 illustrates radial and axial topology motion devices that have biological analogies and can be prototyped as documented below.

To acquire and expand the engineering-science-technology core, we must supplement the interdisciplinary areas as well as link together and place the synergetic perspectives, integrating hardware (machines, actuators, sensors, circuits, etc.) with system intelligence, control, decision-making, signal processing, information propagation, data acquisition, etc. New multidisciplinary developments are needed. Biomimetics, as a meaningful paradigm in the design of micro- and nanosystems, was introduced to tackle, integrate, and solve a great variety of emerging problems. Through biomimetics, fundamentals of engineering and science can be utilized with the ultimate objective of guaranteeing the synergistic combination of systems design, basic physics, fundamental theory, precision engineering, microelectronics, microfabrication, and informatics in design, analysis, and optimization of nano- and microsystems. For example, as reported in Chapter 2, micro- and nanoscale machines must be designed and integrated with the corresponding radiating energy devices, controlling and signal processing ICs, input-output devices, etc. In biological systems, the principles of matching and compliance are the general design principles that require that the system architectures be synthesized integrating all subsystems and components. The matching conditions, functionality, operationability, and systems compliance must be determined and guaranteed. For example, the actuators–sensors–antennas–ICs compliance and operating functionality must be satisfied. It is evident that the systems devised must be controlled. These controllers should be designed and implemented.

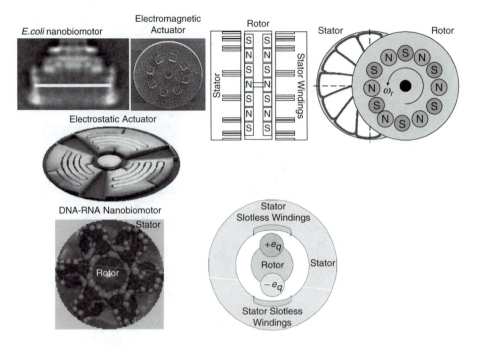

FIGURE 3.1
Axial topology machines: *E. coli* nanobiomotor and micromotors (electromagnetic and electrosytatic). Radial topology machines: DNA-RNA nanobiomotor and micromachine.

Control of biosystems is performed and executed by high-performance, multifunctional, robust, and adaptive three-dimensional (3D) nanobiocircuitry. Thus control, information propagation, interface, and other functions in NEMS and MEMS could be performed (implemented) by 3D nano-ICs, but these 3D nano-ICs have not yet been designed, and fundaments have been barely studied. Research into systems control, 3D nano-ICs, nano-electronics, decision making and other problems aims to find methods for designing intelligent systems using biomimetics.

Let us provide a definition for biomimetics: Biomimetics is the coherent abstraction in devising prototyping in the design of organic, inorganic, and hybrid nano- and microsystems using biological analogies through bioinformatics, prototyping, bioarchitectronics, modeling, analysis, and optimization.

Complex nano- and microscale biosystems exist in nature in enormous variety and sophistication. One can apply complex biological patents to devise, analyze, and examine distinct systems. Biosystems cannot be blindly copied due to the fact that many complex phenomena and effects have not yet been comprehended and understood. Thus one cannot utilize these phenomena. For example, nanosystem architectures have not been coherently examined, control mechanisms are unknown, functionality of 3D nano-biocircuitry is unknown, etc. The attention is concentrated on examining existing biosystems, examining the unknown phenomena, devising novel control paradigms, exploring new operating principles, examining functionality of different subsystems and devices, exploring compliance, researching novel topologies, studying advanced architectures, characterizing distinct complexes and materials, etc. Different nano- and microsystems and devices can be devised and designed through biomimetics based upon basic physical laws and bioprototyping. Our ultimate objective is to provide the focused study and application of biomimetics in systematic design.

3.1.2 Biomimetics for NEMS and MEMS

Significant research efforts have been dedicated to examining nanosystems. The fundamental, applied, and experimental research has been supported to understand and control molecular-scale processes to attain the controlled synthesis and direct self-assembly of functional structures and components into functional nanoscale devices and systems.[1-4] These novel molecular-scale devices and systems can be used in nanopropulsion, computing, information processing, wireless communication, sensing, actuation, nanopower, etc. Integration of engineering and science provides unique opportunities in the synthesis and design of high-performance systems that mimic complex biosystems. In science, biomimetics is the mimicking of biosystems architecture, functionality, design, assembly, and other basic features. For example, fish and bacteria propulsion mechanisms, based upon undulating actuation, can be mimicked. This type of propulsion, which is different from conventional propeller-based marine vehicles, theoretically leads to superior propulsion systems. However, not all biosystems, biodevices, biomachines, and biocircuitry can be prototyped due to limited knowledge and inadequate understanding of the basic physics at micro- and nanoscale as well as fabrication technological deficiencies. As a result, one may focus attention on the engineering biomimetics. For many problems, engineering biomimetics will provide the basics to attain synthesis, design, prototyping, and assembly of micro- and nanosystems with the highest degree of efficiency, integrity, adaptability, functionality, compatibility, and optimality.

The basic physics, complex phenomena, topology, design, and operation of majority nanobiosystems are unsolved, long-standing problems that may be unanswered in the foreseeable future. Despite these unfavorable facts, novel nanosystems and nanodevices can be synthesized and designed applying biomimetics. There is a need for system-level synthesis, design, analysis, control, and optimization. To devise, design, analyze, and optimize novel high-performance nanosystems, the designer synthesizes advanced architectures, discovers advanced topologies, examines new operating principles to attain superior functionality, applies cornerstone physical laws and fabrication technologies, integrates CAD, etc. Synthesis and implementation of novel coherent architectures, with actuators/sensors, electronics, and controllers, and other subsystems (to attain communication, memory hierarchies, and multiple parallelism to guarantee high-performance computing, robust signal processing, and real-time decision making) play a critical role. Biological systems perform the following basic tasks:

- Patterning and prototyping. (Templates of DNA or RNA are used as blueprints for structures.)
- Robust system-level self-assembling and self-organization. (Thousands of individual components and compounds are precisely defined in complex functional and operational systems.)
- Adaptive health-monitoring (self-diagnostics), reconfiguration, and repairing. (For example, a biosystem can sense and identify damage, perform adaptation, execute optimal decisions, and repair itself to the functional state and operational level.)
- Precise assembly of the atomic composition and architecture reconfiguration of very complex and multifunctional structures. (For example, proteins and protein complexes, which integrate millions of atoms, become dysfunctional by the change of a single atom.)
- Building highly hierarchical multifunctional complex subsystems and devices.
- Reproducing and mimicking the similar structures, subsystems, and systems.

- Coherent and efficient actuation, sensing, communication (information propagation), computing, and control. (For example, biosystems can sense the evolving environment, perform decision making, and deliver the measured information to control and actuate subsystems for the appropriate response and action; also, synergy of actuation, sensing, communication, computing, and control tasks are involved.)

- Evolutionary real-time reconfigurable control and adaptive decision making (for example, adaptation, reconfiguration, optimization, tuning, and switching based upon evolutionary learning in response to short, long, reversible, and irreversible environmental changes including rapidly evolving dynamic adverse environment).

- Evolutionary designing low entropy structures increasing the entire system entropy. (For example, protein folding, which is a highly ordered arrangement of the amino acids, is driven by the maximization of the entropy.)

Biosystems provide a proof-of-concept principle for highly integrated multifunctional NEMS and MEMS. The engineering biomimetics paradigm provides a meaningful conceptual tool to understand how biological and man-made systems coherently perform their functions, tasks, roles, and missions. The design of high-performance systems implies the subsystems, components, devices, and structures synthesis, design, and developments. Among a large variety of issues, the following problems must be resolved:

- Synthesis, characterization, and design of micro- and nanoscale transducers, actuators, and sensors according to their applications and overall systems requirements

- Design of high-performance radiating energy, electronic, and optical devices

- Integration of actuators with sensors and circuitry

- Robust control and diagnostic

- Wireless communication

- Affordable and high-yield fabrication technologies and techniques

- Specific CAD software and tools to solve the above mentioned problems

Many reported problems can be solved applying well-defined, sequential, fundamental, applied, and experimental research. It was documented that synthesis, modeling, analysis, and simulation are the sequential activities. The synthesis starts with the discovery of new or application of existing physical principles, examination of novel phenomena and effects, analysis of specifications imposed on the behavior, study of the system performance, preliminary modeling and simulation, and assessment of the available experimental results. Heterogeneous simulation and analysis start with the model developments based upon the system devised. The designer mimics, studies, analyzes, evaluates, and assesses the systems behavior using state, performance, control, events, disturbance, decision making, and other variables. Thus fundamental, applied, and experimental research and engineering developments are needed.

In general, the coordinated behavior, motion, visualization, sensing, actuation, decision making, memory, learning, and other functions performed by living organisms are the results of the electromagnetic and electrochemical transmission of information, energy conversion, and other mechanisms that are based on distinct phenomena. One cubic centimeter of the brain contains millions of nerve cells, and these cells communicate with thousands of neurons creating high-performance data-processing (communication) networks by means of high-density data-transmission channels. The information from the brain to the muscles is transmitted within milliseconds, so that baseball, football, basketball, and tennis players, for example, can anticipate and calculate the speed and velocity of the ball, analyze the situation, make a decision based upon the prediction, and respond in time (e.g., run or jump, processing speed, throw or hit the ball, etc.). The reader can imagine the computing power, processing speed the performance of information propagation channels and sensors-actuators, needed to sense the information, transmit the data, perform calculations, transmit the signals, and accurately execute the coordinated optimum response (action) within 0.2 seconds.

3.1.3 Biomimetics, Nano-ICs, and Nanocomputer Architectronics

Nano-ICs and nanocomputers can be developed thorough biomimetics examining 3D nano-biocircuits. It is possible that even bioelectrochemical nanocomputers will be designed to store and process information utilizing complex electrochemical interactions and changes. Bioelectrochemical nanobiocircuits that store, process, and compute exist in nature in enormous variety and complexity evolving through thousands of years of evolution. The development and prototyping of bioelectrochemical nanocomputer and three-dimensional circuits have progressed through engineering bioinformatic and biomimetic paradigms.

Possible basic concepts in the development of nanocomputers are listed below. Mechanical "computers" have a very rich history that can be traced back thousands of years. While creative theories and machines have been developed and demonstrated, the feasibility of mechanical nanocomputers is questioned by many researchers due to needed controlled mechanical nanocomponents and unsolved fabrication, assembling, packaging, and other difficulties. These mechanical nanocomputers unlikely have any bioanalogies.

Many facets of nanocomputer device and design benefit from biomimetics. Chemical nanocomputational structures can be designed based on the processing information by making or breaking chemical bonds and storing the information in the resulting chemicals. In contrast, in quantum nanocomputers, the information can be represented by a quantum state (e.g., the spin of the atom can be controlled by the electromagnetic field). Nanocomputers can be designed utilizing electronic nanodevices. (Current computers, fabricated using planar two-dimensional ICs, have been used and advanced for the last 30 years.) Molecular and single-electron transistors, quantum dots, molecular logics, and other electronic nanodevices can be used as the basic elements. The nanoswitches, logic gates, and registers can be fabricated on the scale of a single molecule. These concepts have analogies in biosystems.

The so-called quantum dots can be viewed as boxes that hold the discrete number of electrons that are changed by applying the electromagnetic field. The quantum dots are arranged in the quantum dot cells. Consider the quantum dot cells that have five dots and two quantum dots with electrons. Two different states are illustrated in Figure 3.2. (The shaded dots contain the electron, while the white dots do not contain the electron.) The quantum dots can be used to synthesize the logic nanodevices.

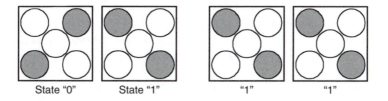

State "0"   State "1"   "1"   "1"

FIGURE 3.2
Quantum dots with states "0" and "1", and "1 1" configuration.

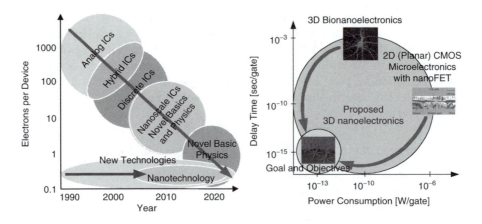

FIGURE 3.3
Moore's first law and revolutionary step toward 3-D nano-ICs.

First, second, third, and fourth generations of computers emerged, and tremendous progress has been achieved. The Intel® Pentium® processors (up to 4 GHz) were built using advanced Intel® NetBurst™ microarchitecture. These processors ensure high-performance processing, and are fabricated using 0.13 micron technology. Currently, 65- and 90-nm technologies are emerging and are applied to fabricate high-yield, high-performance planar (two-dimensional) ICs with billions of transistors on a single 1-cm² die. However, further progress is needed, and novel, rather revolutionary developments are emerging.[3,5,6] The future generation of computers will be built using emerging nano-ICs. Synthesis, integration, and implementation of new affordable high-yield nano-ICs are critical to meet Moore's first law. Figure 3.3 illustrates Moore's first law. The reported data and foreseen trends in power-based analysis can be viewed as controversial and subject to adjustments, but the major trends and tendencies are obvious. It is critical to develop entirely new concepts in design of ICs, and, in particular, 3D nano-ICs as documented in Figure 3.3. Chapter 8 provides an extensive detailed discussion of new trends in nano-IC design and introduces the reader to nanoelectronics as a far-reaching frontier.

Nanotechnology and fundamental progress will lead to 3D nano-ICs. This will result in synthesis and design of nanocomputers with novel computer architectures, topologies, and organizations. These nano-ICs and nanocomputers will guarantee a superior level of overall performance. In particular, compared with the most advanced existing computers, in nanocomputers the execution time, switching frequency, bandwidth, and size will be decreased by the order of millions, while the memory capacity will be increased by the order of millions. However, significant challenges must be overcome particularly in the synthesis and design of multiterminal electronic nanodevices suited for 3D nano-ICs, logic design, architecture synthesis, etc. Many problems (novel architectures, advanced organizations, robust adaptive topologies, high-fidelity modeling, data-intensive analysis, heterogeneous simulations, optimization, reconfiguration, self-organization, robustness, utilization, etc.) must be addressed, researched, and solved. Many of the aforementioned problems have not even been addressed yet. Due to tremendous challenges, much effort must be focused to solve these problems.

The theory of computing, computer architecture, information processing, and networking are the study of efficient robust processing and communication, modeling, analysis, optimization, adaptive networks, architecture, and organization synthesis, as well as other problems of hardware and software design. One can address and study fundamental problems in

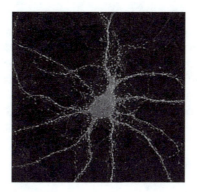

FIGURE 3.4
Vertebrate neuron (soma, axon with synaptic terminals, dendrites, and synapses).

nanocomputer architecture synthesis, design, and optimization applying three-dimensional organization and utilizing multiterminal electronic nanodevices, as well as multithreading, error recovery, massively parallel computing organization, shared memory parallelism, message passing parallelism, etc. These features have been observed in biological systems. Nanocomputer fundamentals, operation, and functionality can be devised, examined, and prototyped based on principles and observations of neuroscience and biomimetics.

Let us examine communication and information processing between nerve cells. Neuronal cells have a large number of synapses. A typical nerve cell in the human brain has thousands of synapses that establish communication and information processing. The communication and processing are not fixed, but rather constantly change and adapt. Neurons and their subsystems function in the hierarchically distributed robust adaptive network manner. During information transfer, some synapses are selectively triggered to release specific neurotransmitters, while other synapses remain passive. Neurons consist of a cell body with a nucleus (soma), an axon (which transmits information from the soma), and dendrites (which transmit information to the soma); see Figure 3.4. It becomes possible to design 3D nano-ICs using biomimetic analogies. For example, the complex organic and inorganic dendrite-like trees can be implemented using carbon-based technology. (Fullerenes and branched carbon nanotube networks ensure robust ballistic switching and amplification behavior, which is desired in nano-ICs).

Nerve cells (neurons) collect and transmit information about the internal state of the organism and the external environment, process and evaluate this information, as well as perform learning, diagnostics, adaptation, evolutionary decision making, coordination, and control. For example, sensory neurons transmit information received from sensory receptors to the interneurons of the central nervous system, while motor neurons transmit the information from the interneurons (located within the central nervous system) to motor (effector) cells. The information between transmitting and receiving neurons is transferred along synapses, for example, from synaptic terminals on the transmitting axon to synaptic vesicles containing neurotransmitter molecules or ions to the synaptic gap to dendrites of the receiving neuron to postsynaptic membrane channels to neuron cytoplasm. Thus synaptic terminals (axonal endings of transmitting neuron) transfer information to the dendrites of the receiving neuron through the synaptic gap (analogy of wireless communication). Neurons receive and transmit information by means of complex electrochemical phenomena. Synapses connect neurons, creating networked information buses. Information processing is based on electrochemical phenomena. Unfortunately, many phenomena are still not well understood.

The common theory states that a signal propagates along a neuronal axon and synaptic terminals as ionic current. The ionic gradient across the axonal plasma membrane is a function of so-called membrane permeability, and the permeability coefficients for various cations (K^+, Na^+, and Ca^{2+}) and anions (Cl^-) are known. The model of this process has a solid mathematical fundamental, in particular, the Lorenz equation, which describes chemoelectromechanical phenomena as $\nabla \cdot \mathbf{A} = -\frac{\partial V}{\partial t}$. This equation allows one to explicitly express the magnetic vector potential $\mathbf{A}$ as a function of the scalar electric potential V.

The vector potential wave equation to be solved is given as

$$-\nabla^2 \mathbf{A} + \mu\sigma \frac{\partial \mathbf{A}}{\partial t} + \mu\varepsilon \frac{\partial^2 \mathbf{A}}{\partial t^2} = -\mu\sigma\nabla V$$

The following chapters will discuss these two equations. In addition to the narrative nature, experiments, visualization, chemical reactions, and other important descriptions, there exist the solid mathematical and physical fundamentals that allow the designer to coherently examine (analyze, simulate, optimize, control, etc.) complex processes and phenomena from engineering standpoints. In general, biomimetics can be synergistically formulated using modern engineering and science.

From the accepted theoretical fundamentals that are reported in the majority of biochemistry and biophysics textbooks, the following information is documented. The electric potential may vary in the range of tens of mV. However, taking into account that the membrane thickness is 50 Å the electric field can be up to 20,000,000 V/m. The K^+ and Na^+ channels (gates) open and close, varying the potential. The open time varies from ~0.5 to ~1 msec, and the Na^+ ions diffuse into the cell at the rate of ~6000 ions/msec. The axon can transmit impulses with a maximum frequency of ~200 Hz, and the propagation speed is from 1 to 100 m/sec. Since the nerve impulses have almost the same amplitude, it can be hypothesized that the data is transmitted using the pulse-width-modulation mechanism. However, other theories exist, as will be discussed in Chapter 8.

The potential at the presynaptic membrane triggers the release of neurotransmitters (synaptic vesicles contain neurotransmitter molecules or ions) into the synaptic gap. For example, well-studied acetylcholine (ACh): An ~400-Å synaptic vesicle contains ~100,000 ACh molecules. The released ACh molecules trigger the opening of the Ca^{2+} channels on the postsynaptic membrane of the dendrite of the receiving neuron. The nanotransmitters cross the fluid-filled synaptic cleft (20–30 nm) and bind to the corresponding receptors on the postsynaptic membrane, changing the membrane potential. There exist excitatory and inhibitory synapses with selective neurotransmitters triggered by distinct receptors. Anion channels are inhibitory, while cation channels are excitatory. The neuron response depends on the specific receptor and neurotransmitter identity. Hence, the precise selectivity and control are established. The receiving neurons transmit the electric current (signal) to the soma. A neuron can have thousands of postsynaptic potentials to simultaneously process. Signals (information) propagate from one neuron to another due to electrochemical information processing, and neurons form hierarchically interconnected adaptive reconfigurable networks.

In addition to information transfer, neurons of the central nervous system perform information processing and computing. Each brain neurotransmitter works within a widely spread but specific brain region that has different effect and function, and has a distinctive physiological role. There are more than 60 human neurotransmitters that have been identified within the following four classes:

- Cholines (ACh being the most important one)
- Biogenic amines (serotonin, histamine and catecholamines)

- Amino acids (glutamate and aspartate being the excitatory transmitters, and gamma-aminobutyric acid [GABA], glycine, and taurine being inhibitory neurotransmitters)
- Neuropeptides (more than 50 of which are involved in modulation, transmission, and processing)

It is possible to map the complex electrochemical information transfer performed by the sensory, motor, and interneurons by two stream-oriented, input-output operators. In particular, we have receive–process (compute)–send.

The neurons maintain communication using *receive* and *send* messages. The execution of these messages is due to electrochemical processes. The data is transferred between the brain (processor) and sensor/receptor (I/O devices) using a communication bus. The processing brain neurons perform computing, and though the specific mechanism of the data processing is under research, it is commonly reported that neurons process the data in a hybrid form. Two triggering states in neurons are mapped to be *off* (closed) and *on* (open), for example, 0 and 1. Therefore, a string of 0s and 1s result, providing analogy to a computer. However, it is possible that there exist intermediate states (due to the rate control) or quantum states. This results in tremendous advantages for devising novel computationally efficient and robust processing paradigms. The organization and topology of adaptive networked information (data) transfer and processing are of great interest with regard to the examination of information propagation (transfer) and parallel processing (computing).

3.1.4 Biomimetics and Nervous Systems

Following the accepted theory, this section introduces the neuroscience concepts that are related to the scope of the book. (The author will report some advanced hypotheses in the end of Chapter 8, but it is not the aim of this book to report the current status of neuroscience. It should be noted that heated debates have recently emerged regarding the cornerstone fundamentals of neuroscience.) The human central nervous system, which includes the brain and spinal cord, serves as the link between the sensors (sensor receptors) and motors composing the peripheral nervous system (effector, muscle, and gland cells). The nervous system has the following major functions: sensing, integration, decision making (processing, computing, and adaptation) and motoring (actuation). The human brain consists of the hindbrain (which controls homeostasis and coordinates movement), midbrain (which receives, integrates, and processes sensory information), and forebrain (where neural processing and integration of information, image processing, short- and long-term memories, learning functions, adaptation, decision making and motor command development take place). The peripheral nervous system consists of the sensory system. Sensory neurons transmit information from internal and external environment to the central nervous system, and motor neurons carry information from the brain or spinal cord to effectors. The sensory system supplies information from sensory receptors to the central nervous system, and the motor nervous system feeds signals (commands) from the central nervous system to muscles (effectors) and glands. The spinal cord mediates reflexes that integrate sensor inputs and motor outputs, and, through the spinal cord, the neurons carry information to and from the brain. The transmission of information along neurons is a very complex process. The membrane potential for a nontransmitting neuron is due to the unequal distribution of ions (sodium and potassium) across the membrane. The resting potential is maintained due to the differential ion permeability and the so-called Na^+–K^+ pump. The stimulus changes the membrane permeability, and ions can

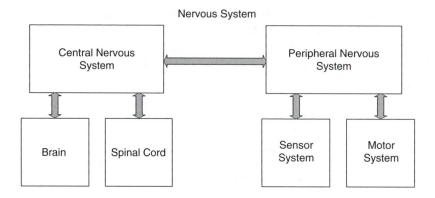

FIGURE 3.5
Vertebrate nervous system: High-level functional diagram.

depolarize or hyperpolarize the membrane's resting potential. This potential (voltage) change is proportional to the strength of the stimulus. The stimulus is transmitted due to the axon mechanism. The nervous system is illustrated in Figure 3.5.

All biological systems have biosensors as documented in Figure 3.5. Biosensors utilize biochemical reactions and sense chemicals, electromagnetic fields, specific compounds, materials, radiation, etc. Many biosensors have been developed to measure and sense various compounds, chemicals, or stimuli. For example, a biosensor can be an immobilized enzyme or cell that monitors specific changes in the microenvironment. Microbial sensors are applied to the industrial process to measure organic compounds. These microbial sensors consist of immobilized whole cells and an oxygen probe used for determination of substrates and products. (Concentration of compounds is determined from microbial respiration activity that can be directly measured by an oxygen probe). Immobilized microorganisms and an electrode, which are the major components of the microbial sensors, sense organic compounds. (Concentration of compounds is indirectly determined from electroactive metabolites such as proton, carbon dioxide, hydrogen, acid, and others measured by the electrode.)

There is a great diversity in the organization of different nervous systems. The cnidarian (*hydra*) nerve net is an organized system of simple nerves (with no central control) that performs elementary tasks (for example, swimming in jellyfishes). Echinoderms have a central nerve ring with radial nerves (for example, the central and radial nerves with nerve net in sea stars). Planarians have small brains that send information through two or more nerve trunks, as illustrated in Figure 3.6.

Let us very briefly examine the jellyfish. Jellyfishes (made up of 95% water, 3–4% salts, and 1–2% protein) have been on the earth for over 650 million years, and they have no heart, bones, brain, or eyes. A network of nerve cells allows them to move and react to food, danger, light, cold, etc. Sensors around the bell rim provide information as to whether they are heading up or down, to the light or away from it. Using the jet propulsion system, jellyfishes can swim. Jellyfishes are very efficient predators, killing their pray by stinging it. Thus actuation (propulsion using pulsing muscle within the bell and stinging cells, cnidocytes, which contain tiny harpoons called nematocyst triggered by the contact) and sensing mechanisms are in place even in the simplest invertebrate.

Living organisms, which consist of atoms, molecules, molecular structures (assemblies), and molecular systems, must be studied at the nanoscale using the corresponding theories. In fact, sugars, amino acids, hormones, and DNA are nanometers in size. In general,

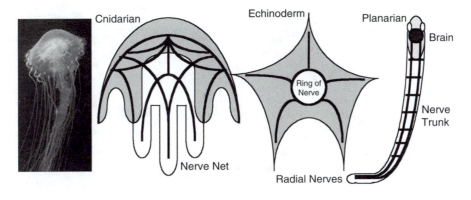

FIGURE 3.6
Overview of invertebrate nervous systems.

membranes that separate one cell from another or one subcellular organelle from another are larger structures. However, they are also built using atoms and molecules. These nanoscale molecular structures and systems perform different tasks and functions. For example, potassium and sodium ions generate nerve impulses.

The ability of organisms to function in a particular way depends on the presence, absence, concentration, location, interaction, and architectures (configurations) of these nanoscale structures and systems. Biotechnology is the synergy of nanoscience, engineering, and technology (nanofabrication) to complement and enhance fundamental research and applied developments. For example, one can apply complex biological processes to devise and fabricate nanosystems. In particular, fabrication and biomimicking can be performed by researching structures, architectures, and biological materials of biomolecules, cells, tissues, membranes, biomotors, etc.

In general, living organisms and their systems, components, and devices provide a proof-of-concept principle for highly integrated multifunctional man-made MEMS and NEMS. Micro- and nanoengineering, science, and technology will provide paradigms, concepts, methods, and tools to understand how living systems perform their functions, tasks, and roles.

3.2 Micro- and Nanoelectromechanical Systems: Scaling Laws and Mathematical Modeling

When advancing synthesis, analysis, and design toward the molecular level with the ultimate goal to design, fabricate, and test nanocomputers, nanomanipulators, and other nano- and microsystems that may have nanocomputers as the core components, the designer faces a great number of unsolved problems and challenges. Through biosystems analogy and biomimicking, a great variety of man-made electromechanical systems have been designed, made, and implemented. As conventional electromechanical systems, many micro- and nanotransducers, actuators, sensors, and other devices are controlled by changing the electromagnetic field. To analyze, design, develop, and deploy novel NEMS and MEMS, the designer must perform synthesis, apply basic theories, integrate the latest advances in available fabrication technologies, utilize nano-ICs, etc. Novel systems

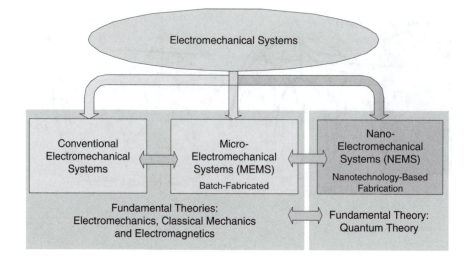

FIGURE 3.7
Electromechanical systems with distinct basic theories and fabrication technologies.

(with multiprocessors, memory hierarchies, and multiple parallelism in order to guarantee high-performance computing and real-time decision making), new actuators, advanced sensors, and novel nano-ICs, as well as other subsystems and components, play a critical role in advancing the research, development, and implementation of micro- and nanoelectromechanical systems. In this book we report synthesis, modeling, analysis, simulation, design, optimization, and fabrication aspects for some NEMS and MEMS. Electromechanical systems, as shown in Figure 3.7, can be classified as conventional electromechanical systems, MEMS, and NEMS.

A great variety of electromechanical systems have been designed. Many MEMS (and some NEMS) can be prototyped from conventional electromechanical systems. The operational principles and basic foundations of conventional electromechanical systems and MEMS can be identical, while NEMS can utilize phenomena that exist and can be applied only at nanoscale. Frequently the designer applies the classical mechanics as well as electromagnetics (Maxwell's equations) to study conventional electromechanical systems and MEMS. In contrast, quantum theory is usually applied to NEMS. Quantum theory and classical mechanics/electromagnetics are closely related. Figure 3.7 also documents the difference between systems from the fabrication viewpoint. As complex multifunctional systems, MEMS and NEMS can integrate different structures, devices, and subsystems as their components. The research in integration and optimization of these subsystems and devices has not been performed. Through this book we will study different MEMS architectures, and fundamental and applied theoretical concepts will be developed and documented in order to devise and design the next generation of superior high-performance systems. This section concentrates on the application of scaling laws to design MEMS. Microsystems can integrate a large variety of components and devices that should be examined.

It is well known that the properties of thin-film materials widely used in nano- and microsystems are different from the same bulk material. For example, thin-film Cu resistivity (conductor), Co and Ni saturation magnetization (magnets), Fe permeability (ferromagnetic), and other properties vary. These differences are due to the distinct processes used to fabricate thin-film and bulk materials as well as effects at the nanoscale (nonuniformity, grain fluctuations, contamination, surface characteristics, quantization, etc.) that significantly influence the material properties.

3.2.1 Mechanical Systems

The mass is proportional to the volume of an object, and if the linear dimension of an object is reduced by a factor of x, the volume (and mass) of this object is reduced by a factor of x^3 (this is a reason why insects survive a fall from a great height without damage and can lift objects many times their size). Thus the basic physical laws apply. When a mechanical structure (cantilever beam or membrane) is scaled down by a factor x, its mechanical stiffness k, $k = \frac{1}{4} w E \frac{t^3}{l^3}$, is scaled down by a factor of x. Here, w, t, and l are the beam width, thickness, and length, respectively; E is the elastic modulus. Thus the mechanical strength of an object is reduced much more slowly (factor of x) than the developed inertial force (factor of x^3). Microstructures can also sustain high accelerations without damage. For example, MEMS accelerometers measure acceleration of more than 100,000 g.

3.2.2 Fluidic Systems

The Reynolds number (Re), which characterizes the flow turbulence, is a function of the system geometry. In particular, $\text{Re} = \frac{\rho v}{\mu} d$, where ρ is the density, v is the characteristic velocity, μ is the fluid viscosity, and d is the diameter. It is well known that Re < 2000 corresponds to laminar flow, while Re > 4000 corresponds to turbulent flow. However, the experiments show that turbulent and chaotic fluid flow is observed in most macroscopic systems, while the fluid flow in microsystems is usually laminar. The dimensions and Reynolds number of the fluidic system are scaled down by x.

3.2.3 Chemical (Biological) Systems

The scaling of chemical and biological systems is extremely difficult due to distinct fundamental physics of complex optoelectrochemomechanical processes. Biomimetics and bioinformatics, which are covered in this book, provide the meaningful paradigms by which to scale and prototype some biological systems, components, and devices. In general, a coherent analysis must be performed and carried out.

3.2.4 Thermal Systems

As the linear dimensions of an object are reduced by x, the thermal mass of an object (i.e., the thermal capacity times the volume) scales down by x^3, while the rate of heat transfer is scaled down by the factor x^2. These scaling factors are important to carry out for various microfabrication processes. For nanostructures, quantum theory should be applied.

3.2.5 Electromagnetic Systems

A great number of sections of this book are devoted to electromagnetic-based micro- and nanoscale actuators and sensors. Comprehensive theories that should be applied are reported using Maxwell's equations, classical electromechanics, energy conversion principle, functional density concept, Schrödinger equations, etc. Here, introductory and simplified equations that cover only preliminary energy analysis are discussed. In the electromagnetic-based actuators and sensors the energy is stored in the airgap. The stored electric and magnetic volume energy densities are found as

$$\rho_{We} = \frac{1}{2} \varepsilon E^2 \quad \text{for electric (electrostatic) transducers}$$

$$\rho_{Wm} = \frac{1}{2} \frac{B^2}{\mu} = \frac{1}{2} \mu H^2 \quad \text{for magnetic (electromagnetic) transducers}$$

where ε is the permittivity ($\varepsilon_0 = 8.85 \times 10^{-12}$ F/m); E is the electric field intensity; μ is the permeability, $\mu = \mu_r \mu_0$ ($\mu_0 = 4\pi \times 10^{-7}$ T-m/A); and B and H are the magnetic field intensity and density. Electrostatic and electromagnetic actuators are shown in Figure 3.1.

We conclude that the maximum energy density of electrostatic actuators is limited by the maximum field (voltage) that can be applied before electrostatic breakdown occurs. In mini- and microstructures, the maximum electric field is bounded, resulting in the maximum energy density. For example, in 100×100-μm to -mm size structures with a few μm airgap, $E < 3 \times 10^6$ V/m, which gives the maxim energy density: $\rho_{We} < 40$ J/m^3. In contrast, for electromagnetic actuators, the maximum energy density is limited by saturation flux density B_{sat} (usually $B_{sat} < 2.4$ T) and material permeability (the relative permeability μ_r varies from 100 to 300,000). Thus the resulting magnetic energy density is 200,000 J/m^3 or higher. We conclude that $\rho_{We} \ll \rho_{Wm}$. Thus the electromagnetic transducers can store at least 5000 times more energy than electrostatic transducers.

In microscale devices, condition $\rho_{We} \ll \rho_{Wm}$ is always met. For nano- and microsize airgaps, fewer ionization collisions occur, and a larger electric field can be applied before a cascade electrostatic breakdown occurs. These short airgaps result in larger maximum voltage applied. The breakdown voltage is a nonlinear function of the airgap and structure dimensions. Thus for electrostatic microtransducers with less than 1 μm airgaps, larger voltages can be supplied, resulting in larger electric fields E and energy densities. However, the ratio $\frac{\rho_{Wm}}{\rho_{We}} > 1000$ is guaranteed even for the most favorable microscale transducers dimensions even if the soft micromagnets (Fe, Ni, or NiFe) with low B_{sat} are used. The application of hard magnets, which can be micromachined, ensures $\frac{\rho_{Wm}}{\rho_{We}} > 5000$.

The saturation flux density B_{sat}, maximum relative permeability μ_r, electrical resistivity, and Curie temperature (T_C) for some soft magnetic materials are reported in Table 3.1. The details and comprehensive data for magnetic materials are provided in Chapter 10.

These brief discussions of the scaling laws are provided to offer preliminary information. The complexity of phenomena and effects in nano- and microsystems require one to pursue new fundamental and applied research as well as engineering and technological developments. In addition, there is a critical need for coordination across a broad range of mathematical methods and software tools. Design of advanced nano- and microdevices, synthesis of optimized (balanced) architectures, development of new programming languages and compilers, performance and debugging tools, operating system and resource management, high-fidelity visualization and data-representation systems, and design of high-performance networks are needed. This leads to the hardware-software design and codesign problems. To support nano- and microsystems, the attention should be concentrated on new algorithms and data structures, advanced system software, distributed access to very large data archives, sophisticated data mining, visualization techniques, advanced data-intensive analysis, etc. In addition, actuators (we compare electrostatic and electromagnetic transducers), advanced processors, and multiprocessors are required to achieve sustained capability with required functionally and adaptability of *usable* systems.

TABLE 3.1

Characteristics of Some Soft Magnetic Materials

Material	B_{sat} [T]	μ_r	$\rho_e \times 10^{-6}$, ohm-m	T_C, [°C]
Silicon iron	2	5000	0.25–0.55	800
Iron	2.1	5500	0.1	760
Hiperco 27	2.36	2800	0.58	925
Molybdenum 4-79 permalloy	0.8	400,000	0.55	454
Ferrites	0.2–0.5	150–10,000	0.1×10^6–5×10^6	140–480

Though scaling laws and factors can provide some useful insight, high-fidelity modeling, heterogeneous simulation, and data-intensive analysis must be performed. The fundamental and applied research in MEMS and NEMS has been dramatically affected by the emergence of high-performance computing. Analysis and simulation have significant effects. However, the problems in analysis, modeling, and simulation of MEMS and NEMS that involve coherent analysis of molecular dynamics cannot be solved because the quantum theory cannot be effectively applied even to simple stand-alone molecules or the simplest nanostructures. (A 10-nm^3 nanostructure can have thousands of molecules.) There are a number of very challenging problems in which advanced theory and high-end computing are required to advance the basic theory and engineering practice. The multidisciplinary fundamentals of nano- and microsystems must be developed to guarantee the possibility to synthesize, analyze, and fabricate high-performance systems with desired (specified) performance characteristics and desired level of confidence in the accuracy of the results. Synergetic fundamental and applied research will dramatically shorten the time and cost of systems development for medical, biomedical, aerospace, automotive, electronic, and manufacturing applications.

The importance of mathematical model developments and numerical analysis has been emphasized. Numerical simulation enhances and complements, but does not substitute, fundamental research. Furthermore, heterogeneous simulations are based on reliable fundamental studies (mathematical model developments, data-intensive analysis, efficient computational methods and algorithms, etc.). These simulations must be validated through experiments. It is evident that heterogeneous simulations lead to understanding of system performance, as well as reduce the time and cost of deriving and leveraging the MEMS/NEMS from concept to systems. Fundamental and applied research is the core of modeling and simulation. Thus focused efforts must be concentrated on high-fidelity modeling and data-intensive analysis.

To comprehensively study MEMS and NEMS, advanced modeling and computational tools are required primarily for 3D+ (three-dimensional geometry dynamics in time domain) data-intensive analysis in order to study the end-to-end dynamic behavior. The heterogeneous mathematical models of systems, augmented with efficient computational algorithms, terascale computers, and advanced software, will play a major role in stimulating the synthesis and design of nano- and microsystems from biomimicking and virtual prototyping standpoints. There are three broad categories of problems for which new algorithms and computational methods are significant:

- Problems for which basic fundamental theories are developed, but the complexity of solutions is beyond the range of current and near-future computing technologies. For example, the conceptually simple quantum mechanics and molecular dynamics cannot be straightforwardly applied even for simple nanoscale structures. In contrast, it will be illustrated that it is possible to perform robust predictive simulations of molecular-scale behavior for micro- and nanoscale structures that contain millions of molecules using feasible paradigms.

- Problems for which fundamental theories are not completely developed to justify direct simulations, but basic research can be adva-nced or developed through numerical results.

- Problems for which the developed advanced modeling and simulation methods will produce major advances and will have a major impact. For example, 3D+ data-intensive analysis and study of end-to-end behavior of nano- and microstructures and devices.

High-fidelity modeling and massive data-intensive computational simulations (mathematical models development within contemporary intelligent libraries and databases/archives, intelligent experimental data manipulation and storage, heterogeneous data grouping and correlation, visualization, data mining and interpretation, etc.) offer the promise of developing and understanding the mechanisms, phenomena, and processes in order to design high-performance systems at nanoscale. Predictive model-based simulations require terascale computing and an unprecedented level of integration between engineering and science. These modeling and simulation developments will lead to new fundamental results. To model and simulate nano- and microsystems, one may augment modern quantum mechanics, electromagnetics, and electromechanics. Our goal is to further enhance and cover distinct theories that can be applied.

One can perform steady-state and dynamic (transient) analysis. While steady-state analysis is important, and the structural optimization to comprehend the actuators/sensors–antennas–ICs design can be performed, systems must be analyzed in the time domain. One of our goals is to develop and introduce basic fundamental conceptual theories in order to study and examine the interactions between actuation and sensing, computing and communication, and other processes and phenomena in MEMS and NEMS. Using different paradigms, modern theory can be enhanced and applied to nanostructures and devices in order to predict the performance through analytic analysis and numerical simulations. Mathematical models of nodes can be developed, and both single molecules and molecular complexes can be studied and examined. It is critical to perform this research to determine a number of parameters to make accurate performance evaluation and to analyze the phenomena performing simulations and compare experimental, modeling, and simulation results.

Current advances and developments in mathematical modeling and simulation of complex phenomena are increasingly dependent on new approaches to heterogeneously model, compute, visualize, and validate the results. This research is needed to clarify, correlate, define, and describe the limits between the numerical results and the qualitative-quantitative analytic analysis in order to understand the basic features. Though the heterogeneous simulations of nano- and microsystems require terascale computing that will be available, high-performance software environments such as MATLAB can be effectively utilized. The computational limitations and inability to develop explicit mathematical models (some nonlinear phenomena cannot be comprehended, fitted, described, and precisely mapped) focus advanced studies on the basic research in robust modeling and simulation under uncertainties. Modeling and design are critical to advance and foster the theoretical and engineering enterprises. We focus our research on developments and enhancements of different theories in order to model, simulate, and design novel MEMS and NEMS. For example, nano- and microscale actuators and sensors should be modeled and analyzed in 3D+ (three-dimensional geometry dynamics in time domain). Rigorous methods for quantifying uncertainties for robust analysis also should be developed. However, these problems are beyond the scope of this book. Uncertainties result due to the fact that it is impossible to explicitly comprehend the complex interacted processes in nanosystems (actuators, sensors, smart structures, antennas, nano-ICs, data movement, storage, and management across multilevel memory hierarchies, archives, networks, and periphery), accurately solve high-order nonlinear partial differential equations, precisely model structural and environmental changes, measure all performance variables and states, etc.

To analyze and design MEMS and NEMS, we will develop tractable analytical mathematical models. There are a number of areas where the advances must be made in order to realize the promises and benefits of modern theoretical developments. For example, to perform high-fidelity 3D+ modeling and data-intensive analysis of actuators and sensors, advanced analytical and numerical methods and algorithms must be used. Novel algorithms in geometry and mesh generation, data assimilation, dynamic adaptive mesh refinement, computationally efficient techniques, and robust methods can be implemented in the MATLAB environment. There are a number of fundamental and computational problems that have not been addressed, formulated, and solved due to the complexity of MEMS and NEMS (e.g., large-scale hybrid models, limited ability to generate and visualize the massive amount of data, computational complexity of the existing methods, etc.). Other problems include nonlinearities and uncertainties that imply fundamental limits to our ability to accurately formulate, set up, and solve analysis and design problems with the desired accuracy. Therefore, one should develop rigorous methods and algorithms for quantifying and modeling uncertainties, 3D+ geometry and mesh generation techniques, and methods for adaptive robust modeling and simulations under uncertainties. A broad class of fundamental and applied problems ranging from fundamental theories (quantum mechanics, electromagnetics, electromechanics, thermodynamics, structural synthesis, optimization, optimized architecture design, control, modeling, analysis, etc.) to numerical computing (to enable the major progress in design and virtual prototyping through the heterogeneous simulations, data-intensive analysis, and visualization) will be addressed and studied in this book. Due to the obvious limitations and the limited scope of this book, a great number of problems and phenomena cannot be addressed and discussed (among them, robust modeling, analysis of large-scale MEMS and NEMS, etc.). However, using the results reported, these problems can be approached, examined, and advanced.

3.3 MEMS Examples and MEMS Architectures

3.3.1 MEMS Examples

Different MEMS and NEMS must be designed depending upon the specifications, requirements, objectives, and applications. Electromechanical, optoelectromechanical, and optochemoelectromechanical micro- and nanosystems have been developed. These MEMS and NEMS have been designed for different applications (for example, actuation versus sensing, energy source versus wireless communication, etc.). Correspondingly, these MEMS and NEMS have distinct architectures and configurations. In general, it is very difficult to make the matching comparison. The application requirements and specifications are applied. For example, bandwidth, electromagnetic interference, temperature, vibration, and radiation can be the factors used to make the preferred selection.

Consider micro- and nanoscale actuators. The actuator size is determined by the force or torque densities, which are the functions of the materials used and size (volume). That is, the size is determined by the force or torque requirements and materials used (for example, conductor resistivity, maximum torque density, saturation magnetization, B–H curve, permeability, etc.). In contrast, for sensors, the size of MEMS and NEMS is defined by what should be sensed (voltage versus size, bioagent versus electromagnetic field, molecule versus temperature), sensitivity, and output requirements (output signal must have the specified power, distortion, and noise to guarantee the interface). Although distinct MEMS and NEMS have some common features, there are many differences. In addition to different fundamental theories used to study MEMS and NEMS, distinct physical phenomena and effects, and different fabrication technologies are applied. For example, MEMS are usually fabricated using complementary metal oxide semiconductor (CMOS) technology, micromachining, and high-aspect-ratio (LIGA) technologies and processes. Therefore, MEMS leverage conventional microelectronics techniques, processes, and materials.

Electromagnetic-based MEMS integrate motion microstructures or microtransducers (actuators and sensors) controlled by ICs using radiating energy microdevices. Thus microstructures/microtransducers, radiating energy devices, and ICs must be integrated.

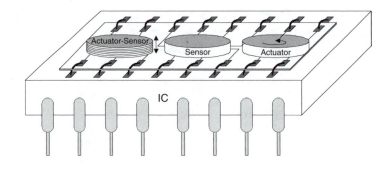

FIGURE 3.8
Flip-chip monolithic MEMS with actuators and sensors.

The direct chip attaching technique was developed and widely deployed. In particular, flip-chip MEMS assembly replaces wire banding to connect ICs with micro- and nanoscale actuators and sensors. The use of the flip-chip technique allows one to eliminate parasitic resistance, capacitance, and inductance. This results in improvements of performance characteristics. In addition, the flip-chip assembly offers advantages in the implementation of advanced flexible packaging, improving reliability and survivability, reduces weight and size, etc. The flip-chip assembly involves attaching microtransducers (actuators and sensors) directly to ICs. For example, the microtransducers can be mounted face down with bumps on the pads that form electrical and mechanical joints to the IC's substrate. Figure 3.8 illustrates flip-chip MEMS.

Integrated MEMS (a single chip that can be mass-produced at low cost using the CMOS, micromachining, LIGA, and other technologies) can integrate

- *N* nodes of microtransducers (actuators/sensors and smart structures)
- ICs and radiating energy devices (antennas)
- Optical and other devices to attain the wireless communication features
- Processor and memories
- Interconnection networks (communication buses)
- Input-output (IO) devices

Different architectures and configurations can be synthesized, and these problems are discussed and covered in this book. One uses MEMS and NEMS to control complex systems, processes, and phenomena. In order to control systems, many performance and decision-making variables (states, outputs, events, etc.) must be measured. Thus in addition to actuation and sensing (performed by microtransducers integrated with ICs and radiating energy devices), computational, communication, networking, signal processing, and other functions must be performed. One can represent a high-level functional block diagram of the dynamic system–MEMS configuration as illustrated in Figure 3.9.

Actuators actuate dynamic systems. These actuators respond to command stimuli (control signals) and develop torque and force. There are a great number of biological (e.g., nanobiomotor, jellyfish, human eye, locomotion system) and man-made actuators. Biological actuators are based on chemical, electromagnetic, and mechanical phenomena and processes. Man-made actuators (electromagnetic, electrostatic, hydraulic, fluidic, thermal, acoustic, and other motors) are devices that receive signals or stimulus (electromagnetic field, stress, pressure, thermo-, acoustic, etc.) and respond with torque or force. Consider flight vehicles: Aircraft, spacecrafts, missiles, and interceptors are controlled by displacing

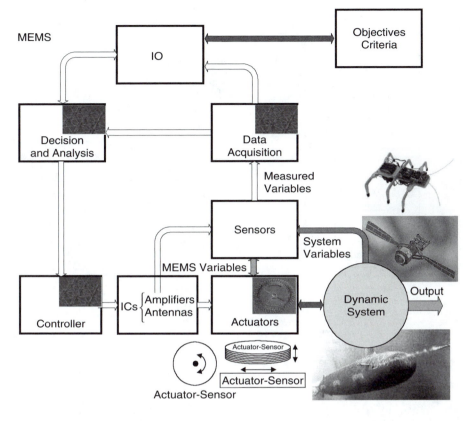

FIGURE 3.9
High-level functional block diagram of the large-scale MEMS and their interaction with distinct dynamic systems.

the control surfaces as well as by changing the control surface and wing geometry. For example, ailerons, elevators, elevons, canards, fins, flaps, rudders, stabilizers, and tips of advanced aircraft can be controlled by micro- and miniscale actuators using MEMS-based smart actuator technology. This actuator technology is uniquely suitable in flight actuator applications. Figure 3.10 illustrates the aircraft where translational and rotational actuators are used to actuate control surfaces, as well as to change the wing and control surface geometry. The application of microtransducers allows one to attain the active aerodynamic flows control minimizing the drag (improving the fuel consumption and increasing the velocity). In addition, maneuverability, controllability, agility, stability, and reconfigurability of flight vehicles are significantly improved, expanding the flight envelope. It must be emphasized that although the force and torque density of microtransducers is usually the same as the conventional miniscale actuators, due to small dimension, the stand-alone microtransducer develops smaller force or torque. However, integrated in the large-scale multinode arrays, microtransducers (controlled by the hierarchically distributed systems) can develop the desired force and actuate control surfaces. Sensors are devices that receive and respond to signals or stimulus. For example, the aerodynamic loads (which flight vehicles experience during the flight), vibrations, temperature, pressure, velocity, acceleration, noise, and radiation can be measured by micro- and nanoscale sensors. (See Figure 3.10.) It should be emphasized that there are many other sensors that can be applied to measure the electromagnetic interference, displacement, orientation, position, voltages, currents, resistance, and other physical variables of interest.

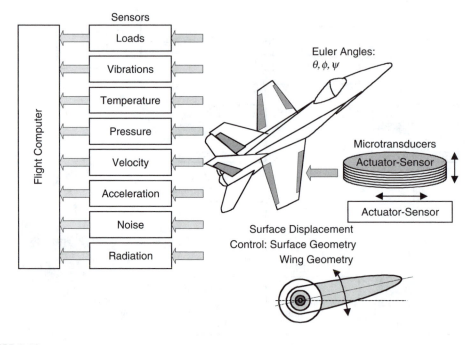

FIGURE 3.10
Application of micro- and nanoscale actuators and sensors in aircraft.

Usually, several energy conversion processes are involved to produce electric, electromagnetic, or mechanical output sensor signals. The conversion of energy is our particular interest. Using energy-based analysis, the general theoretical fundamentals will be studied. The major developments in MEMS and NEMS have been fabrication-technology driven. The applied research has been performed mainly to manufacture structures and devices, as well as to analyze some performance characteristics for specific applications. Extensive research has been done in micro- and nanostructures. In particular, ICs and thin-film sensors have been thoroughly researched, and computer-aided design environments are available to examine them. In addition, mini- and microscale smart structures and nanocomposites have been studied, and feasible fabrication techniques, materials, and processes have been developed. Recently carbon nanotubes were discovered, and molecular wires and molecular transistors were built. However, to our best knowledge, micro- and nanodevices, MEMS and NEMS, have not been comprehensively studied at the system level. In addition, limited efforts have been concentrated to develop the fundamental theory. In this book, we will apply the quantum theory, charge density concept, advanced electromechanical theory, Maxwell's equations, and other cornerstone methods in order to model nanostructures and micro- and nanodevices (antennae, actuators, sensors, etc.). A large variety of nano- and microdevices with different operating features are modeled and simulated. To perform high-fidelity integrated 3D+ modeling and data-intensive analysis with postprocessing and animation, partial and ordinary nonlinear differential equations are solved.

3.3.2 Nanostructures: Giant Magnetoresistance and Multilayered Nanostructure

Nanoscale structures and devices are devised and designed using novel phenomena and effects observed at the nanoscale. In this section we will briefly illustrate the application of nanostructures in the sensor applications. The giant magneto resistive effect was independently discovered by Peter Gruenberg (Germany) and Albert Fert (France) in the late 1980s. They experimentally observed large resistance changes in materials comprising thin layers of various metals. Two thin films (magnetic layers with nanometer thickness) with in-plane magnetization that are antiferromagnetically coupled (magnetizations are in opposite directions in the absence of an external magnetic field) can be set into ferromagnetic alignment (magnetizations are in the same direction) applying external magnetic field. A large change in the perpendicular resistance of the multilayered nanostructures results due to these changes. This effect, known as giant magneto resistance (GMR), has enormous application features including magnetoresistive recording heads and sensors. The antiferromagnetic coupling in multilayered Fe/Cr thin films has initiated intensive studies of thin multilayered nanostructures consisting of ferromagnetic and nonferromagnetic thin films. The magnetic coupling is changed by the nonferromagnetic materials (called spacers), e.g., Cr, Cu, and Re thin films. The GMR phenomenon is associated with the spin-dependent scattering of the conduction electrons and changes in the relative band structure. The electrical resistance is due to scattering of electrons within a material. Depending on the magnetic direction, a single-domain magnetic material will scatter electrons with up or down spin differently. When the magnetic layers in the multilayered GMR effect-based nanostructures are aligned antiparallel, the resistance is high because up electrons that are not scattered in one layer can be scattered in the other. When the layers are aligned in parallel, all up electrons will not significantly contribute (not scattered) regardless of which layer they pass through, leading to a lower resistance.

Due to the fact that the saturation field of the Fe/Cr and Co/Cu thin films is large, soft magnetic multilayered nanostructures, which exhibit the GMR phenomenon, are used. For example, the *permalloy* $Ni_{80\%}Fe_{20\%}$ and copper (fabricated by sputtering in the computerized deposition systems with two independent sputtering sources) multilayered nanostructures have been studied. For different thickness of the permalloy and copper thin films, the saturation fields is from 0.3 to 15 A/m, and the GMR effect leads up to 20% variations of resistance. The dependencies of the GMR effect on the thin films' (for example, permalloy and copper) thickness and geometry results in possibility to shape the multilayered structure's properties for specific requirements and different applications. Using Co in the permalloy-Cu multilayers improves the relative field sensitivity and guarantees 0.1% accuracy. The dependence of the antiferromagnetic coupling on the Cu thickness is very strong, and the first and second maxima are at 0.85- and 2-nm thickness of copper. The GMR effect is maximized (22% of resistance changes) using 39 permalloy-Cu multilayers with 0.8-nm and 2-nm thickness, respectively.

Multilayered nanostructures, fabricated using Co/Cu, CoCu/Cu, CoFe/Cu, and other more complex thin-film structures can be applied to enhance the operating envelopes. As an example, the nanostructure that integrates Ta (5 nm)/$Ni_{80\%}Fe_{20\%}$ (2 nm)/$Ir_{20\%}Mn_{80\%}$ (10 nm)/$Co_{90\%}Fe_{10\%}$ (2 nm)/Ru (1 nm)/$Co_{90\%}Fe_{10\%}$ (2 nm)/Cu (3 nm)/$Co_{90\%}Fe_{10\%}$ (1 nm)/$Ni_{80\%}Fe_{20\%}$ (5 nm)/Ta (5 nm) thin film can be fabricated, and the coefficient of the thermal expansion must be matched, because the single deposition process is desired. One concludes that multilayered nanostructures, which comprise alternating ferromagnetic and nonferromagnetic thin films (each film can be within a few atomic layers thick), exhibit novel functional capabilities due to new phenomena and effects. The discussed GMR phenomena arise from quantum confinement of electrons in spin-dependent potential wells provided by the ferromagnet-spacer layer boundaries. Layers of 3d ferromagnet transition metals are indirectly magnetically coupled via spacer layers comprising many nonferromagnetic 3d, 4d, and 5d transition metals. The magnetic coupling occurs between ferromagnetic and antiferromagnetic layers as the function of thickness of the spacer layer, and the strength varies with the spacer d-band filling. The period of the coupling is related to the electronic structure of the spacer thin film and can be changed by varying the composition of the spacer layer

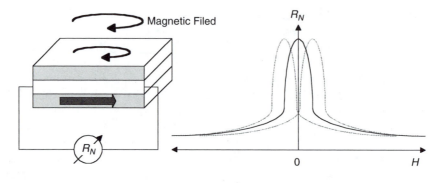

FIGURE 3.11
Normalized resistance in the multilayered nanostructure as functions of the applied magnetic field.

or varying its crystallographic orientation. The resistance of metallic multilayered nanostructures depends on the magnetic arrangement of the magnetic moments of thin layers. Thus the resistance varies as the function of the external magnetic field. The multilayered nanostructures (sensors) display much larger magnetoresistance than any metals or alloys. It was emphasized that the origin of the GMR phenomena derives from spin-dependent scattering of the conduction carriers within the magnetic layers or at the boundaries of the magnetic layers. (Spin-dependent scattering at the ferromagnet-spacer layer interfaces is dominant.) Giant magnetoresistance can be used to detect magnetic fields, and, therefore, is used to read the state of magnetic bits in advanced magnetic disk drives.

Figure 3.11 illustrates the normalized resistance R_N in the multilayered nanostructure with hysteresis (dashed line) and without hysteresis (solid line) as functions of the applied magnetic field.

3.3.3 Integration of Microactuators and ICs

MEMS integrate microassembled devices (electromechanical, electromagnetic, and electronic microsystems and devices on a single chip) that have electrical, electromagnetic, electronic, and mechanical components. To fabricate MEMS, modified microelectronics technologies, techniques, processes, and materials are used. Actuation and sensing cannot be viewed as the peripheral function in many applications. Integrated sensors-actuators (motion microtransducers) with ICs compose an essential class of MEMS that has been widely studied. Micro- and nanoscale actuators were reported already in this book. The reader may wonder if microtransducers have been fabricated. To date, high-performance radial and axial topology transducers with 2-mm diameter have been fabricated by a number of companies (Faulhaber, Namiki, Maxon, etc.) using MEMS technology. In particular, these motors (the reader should decide whether we should call 2-mm motors mini- or micromotors taking into account the fact that accelerometers that are much bigger than the considered motors are called microaccelerometers) have up to $1:100,000$ 2-mm diameter planetary gearhead, because the angular velocity is up to 100,000 rpm (revolutions-per-minute). These motors must be controlled using ICs already available. For example, one can use dual power operational amplifiers (e.g., Motorola TCA0372, DW Suffix plastic package case 751G, DP2 Suffix plastic package case 648, DP1 Suffix plastic package case 626, or others), as monolithic ICs can be used to control DC micromotors or microstructures, as shown in Figure 3.12.

Simply scaling down conventional electromechanical motion devices and augmenting them with available ICs has not met the need, and fundamental theory and fabrication processes have been developed beyond component replacement and miniaturization.

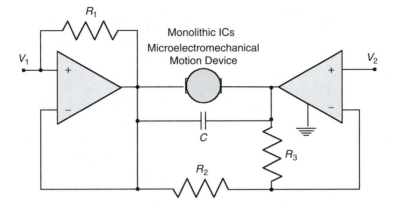

FIGURE 3.12
Application of monolithic IC to control micromotor (MEMS motion microdevice).

Only recently has it become possible to develop high-yield fabrication techniques and fabricate high-performance microelectromechanical motion devices at low cost. There is a critical demand for continuous fundamental, applied, and technological improvements in order to integrate motion microdevices and ICs. The general lack of theory to augment nonlinear electromagnetics, mechanics, signal processing, and control is known. Therefore, synergetic multidisciplinary activities are required. These problems must be addressed and solved through focused efforts. The set of long-range goals that challenge the synthesis, analysis, design, development, fabrication, commercialization, and deployment of high-performance MEMS are

- Devising novel microscale transducers, sensors, and actuators
- Novel sensing, actuation, communication, and networking mechanisms
- Integration and compliance of sensors, actuators, ICs, and radiating energy devices
- Advanced high-performance MEMS architectures, topologies, and configurations
- High-fidelity modeling, data-intensive analysis, heterogeneous design, and optimization
- Computer-aided design and development of computationally efficient environments
- MEMS applications, commercialization, and deployment
- Advanced materials, processes, techniques, and technologies (for example, low resistivity conductors, high-permeability ferromagnetic materials, high flux density micromagnets, etc.)
- Fabrication, packaging, microassembly, testing, and characterization

Significant progress in the application and enhancing of CMOS (including biCMOS) technology enables the industry to fabricate microscale actuators and sensors with the corresponding ICs. This guarantees significant breakthrough and confidence. The field of MEMS has been driven by the rapid global progress in ICs, VLSI, solid-state devices, materials, microprocessors, memories, and DSPs that has revolutionized instrumentation, control, and systems design philosophy. In addition, this progress has facilitated explosive growth in data processing and communications in high-performance systems.

In microelectronics, many emerging problems deal with nonelectric effects, phenomena, and processes (thermal and structural analysis and optimization, stress, ruggedness, and packaging). It was emphasized that ICs are the necessary components to perform control, conversion, data acquisition, networking, and interfacing. For example, control signals (voltage or currents) are computed, converted, modulated, and fed to actuators and radiating energy devices by ICs. It is evident that MEMS have found applications in a wide array of microscale devices (accelerometers, pressure sensors, gyroscopes, pumps, valves, and optical interconnects) due to an extremely high level of integration between electromechanical, microelectronic, optical, and mechanical components with low cost and maintenance, accuracy, efficiency, reliability, robustness, ruggedness, and survivability.

3.3.4 Microelectromechanical Systems Definitions

Though MEMS and NEMS are defined and classified in Chapters 1 and 2, additional examples, further details, and expanded discussions have been presented already in this chapter. Section 3.3.3 introduced motion microdevices (which have microwindings to form the radiating energy microstructure) and ICs. Now let us refresh the definition for MEMS and discuss some additional topics. Microengineering aims to integrate motion microdevices (actuators and sensors) with radiating energy microdevices (antennae, microstructures with windings), microscale driving/sensing circuitry, controlling/processing ICs, and optoelectronic devices. As was mentioned even in the first edition of this book, MEMS are batch-fabricated microscale devices (ICs and motion microstructures) that convert physical parameters to electrical signals and vice versa, and in addition, microscale features of mechanical and electrical components, architectures, structures, and parameters are important elements of their operation and design.

As was illustrated, MEMS integrate motion microstructures, microdevices, ICs, radiating energy, and communication devices on a single chip or on a hybrid chip. The scope of MEMS has been further expanded by devising novel paradigms, synthesizing new architectures, performing system-level synergetic integration, high-fidelity modeling, heterogeneous simulation, data-intensive analysis, control, optimization, fabrication, and implementation.

We defined MEMS as the batch-fabricated integrated microscale system (motion, motionless, radiating energy, energy sources, and optical microdevices/microstructures, driving/sensing circuitry, and controlling/processing ICs) that

- Converts physical stimuli, events, and parameters to electrical, mechanical, and optical signals and vice versa
- Performs actuation, sensing, and other functions
- Comprises control (intelligence, decision-making, evolutionary learning, adaptation, self-organization, etc.), diagnostics, signal processing, and data acquisition features, and microscale features of electromechanical, electronic, electro-optical, optical, electrochemical, and biological components (structures, devices, and subsystems), architectures, and operating principles are basics of the MEMS operation, design, analysis, and fabrication.

MEMS comprise and are built using microscale subsystems, devices, and structures. Recall that in Chapter 2 the following important definition was made for microdevice:

The microdevice is the batch-fabricated integrated microscale motion, motionless (stationary), electromagnetic, radiating energy, or optical microscale device that

- Converts physical stimuli, events, and parameters to electrical, mechanical, and optical signals and vice versa

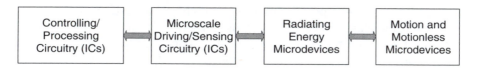

FIGURE 3.13
Functional block-diagram of MEMS.

- Performs actuation, sensing, and other functions, and microscale features of electromechanical, electronic, electro-optical, optical, electrochemical and biological structures, topologies, and operating principles are basics of the microdevice operation, design, analysis, and fabrication.

These MEMS, which comprise microdevices and microstructures as illustrated in Figure 3.13, must be fabricated. The introduction to fabrication processes is given in section 3.4. It must be emphasized that though the fabrication technologies are very important, these technologies largely depend on the infrastructure and equipment available. Furthermore, it is almost impossible to cover all possible scenarios and infrastructure settings. Thus the basics and fundamentals are reported thoroughly emphasizing basic equipment and processes. For example, e-beam masking tool, CVD (LPCVD and PECVD), PVD (sputtering systems), dry etching hardware, lithography tools (at least *g*- and *i*-line), steppers, SEM and metrology systems, wet etching and plating tools, packaging hardware, and some other equipment are absolutely necessary to fabricate microsystems, devices, and structures.

3.3.5 MEMS and NEMS Architectures

Networked MEMS that integrate multiprocessors are of far greater complexity than standalone single-node MEMS (microdevices) commonly used today. Large-scale MEMS and NEMS can integrate

- Thousands of nodes of high-performance stand-alone transducers (actuators and sensors) controlled by ICs and antennae (radiating energy devices can be on-chip and out-of-chip to control transducers)
- High-performance superscalar multiprocessors
- Multilevel memory and storage hierarchies with different latencies (thousands of secondary and tertiary storage devices supporting data-related tasks)
- Interconnected, distributed, heterogeneous databases
- High-performance wireless communication networks (robust, adaptive intelligent networks)

It was shown that even the simplest microdevice (for example, stand-alone microactuator) usually could not function alone. For example, at least the ICs used to control microactuator or microsensor must be used to measure the performance variables (e.g., position, displacement, velocity, force, torque, electromagnetic field, current, voltage, etc.). In addition, the energy microsource should provide the energy needed. A large variety of micro- and nanoscale structures, devices, and systems have been widely used, and a worldwide market for MEMS and NEMS and their applications will drastically increase in the near future. The differences in micro- and nanosystems were emphasized.

Different specifications are imposed on MEMS and NEMS depending on their applications and operating principles. For example, using carbon nanotubes as the nanowires and computing fabrics, the current density and carbon nanotube functionality is defined by the media properties (e.g., resistivity, thermal conductivity, I-V curve, electron transmission, etc.). For example, the maximum current is defined by the diameter and the number of the carbon nanotube layers. These carbon nanotubes can be integrated to form nano-ICs that can perform distinct functions (arithmetic, logics, memory, etc.). Interconnect and control of motion or motionless micro- and nanodevices can be accomplished by carbon nanotube nanowires and nano-ICs. Different processes and techniques have been applied to manufacture MEMS and NEMS, and the properties of nanostructures can be controlled and changed. Microsystems have been mainly fabricated using surface micromachining (silicon-based technology as modification of conventional microelectronics CMOS), LIGA, and LIGA-like technologies.

To deploy and commercialize MEMS and NEMS, a spectrum of problems must be solved, and a portfolio of software design tools needs to be developed using multidisciplinary concepts. In recent years, much attention has been given to MEMS fabrication, synthesis, modeling, analysis, and optimization. It is evident that MEMS and NEMS can be studied with different levels of detail and comprehensiveness, and different application-specific architectures should be synthesized and optimized. The majority of research papers study micro- and nanoscale structures, actuators-sensors, or ICs. That is, MEMS and NEMS components are examined. A great number of publications have been devoted to the carbon nanotubes (nanostructures that can be integrated in nanoelectronic devices). While the results for different MEMS and NEMS components are extremely important and manageable, systems-level research must be performed, because the specifications are imposed on the systems, not on the individual elements, structures, and subsystems. Thus MEMS and NEMS must be developed and studied to attain the comprehensiveness of the analysis and design.

The actuators are controlled by changing the voltage or current (using ICs) or by regulating the electromagnetic field using radiating energy devices. The ICs and antennae (MEMS components) are regulated using controllers that can include central processor and memories (as core), IO devices, etc. Micro- and nanoscale sensors are also integrated as components of MEMS and NEMS. For example, using molecular wires, carbon nanotubes, or optical devices, one feeds the information to the IO devices of the nanoprocessor. That is, MEMS and NEMS integrate a large number of structures, devices, and subsystems that must be studied. As a result, the designer cannot consider MEMS and NEMS as six-degree-of-freedom structures using conventional mechanics (the linear or angular displacement is a function of the applied or developed force or torque) completely ignoring the problems of how these forces or torques are generated and regulated, how components interact, how communication is performed, how energy and signal conversions are achieved, how communication is performed, etc.

In this book we illustrate how to integrate and study the basic components of MEMS and NEMS. The synthesis, design, modeling, simulation, analysis, optimization, and prototyping of MEMS and NEMS must be approached using advanced theories. Even though a wide range of nanoscale structures and devices (e.g., molecular diodes and transistors, transducers, switches, logics) can be fabricated with atomic precision, comprehensive systems analysis must be performed before the designer embarks in costly fabrication. Through optimization of architecture, synthesis, structural optimization of components (transducers, ICs, and antennae), modeling, simulation, analysis as well as visualization, rapid evaluation, and prototyping with assessment analysis can be performed. This facilitates cost-effective solutions reducing the design cycle as well as guaranteeing design of high-performance MEMS and NEMS that satisfy the requirements and specifications.

Large-scale integrated MEMS and NEMS may integrate N high-performance single nodes that include

- N transducers (actuators/sensors, smart structures, and other motion devices)
- Radiating energy devices
- Optical devices
- Communication devices
- Processors and memories
- Interconnected networks (communication buses)
- Driving and sensing ICs
- Controlling and processing ICs
- Input-output (IO) devices

Different MEMS configurations can be synthesized, and diverse architectures can be implemented based upon requirements and specifications imposed. For example, linear, star, ring, and hypercube large-scale micro- and nanosystems architectures are illustrated in Figure 3.14.

More complex architectures can be designed, and the hypercube-connected-cycle node configuration is illustrated in Figure 3.15.

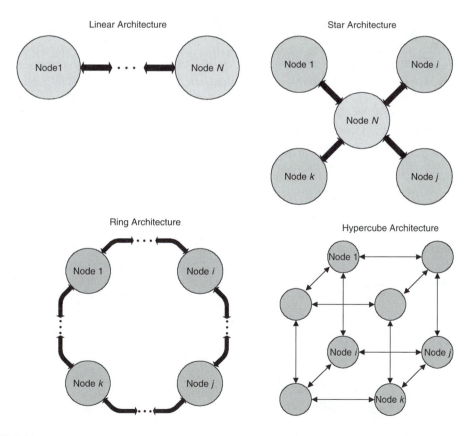

FIGURE 3.14
Linear, star, ring, and hypercube architectures.

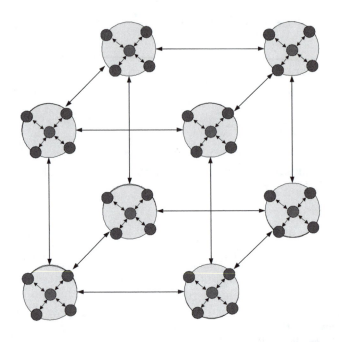

FIGURE 3.15
Hypercube-connected-cycle node architecture.

The nodes (stand-alone MEMS or NEMS) can be synthesized, and the elementary node can be simply actuator or sensor controlled by the ICs. The elementary MEMS or NEMS can be controlled by the external electromagnetic field (that is, ICs or antennae are not a part of the device). As an alternative, each node can integrate actuators, sensors, microstructures, antennae, ICs, processors (with computing, data acquisition, decision-making, and other capabilities), controlling and signal processing ICs, memories, IO devices, communication buses, etc. Figure 3.16 illustrates exclusive and elementary nodes.

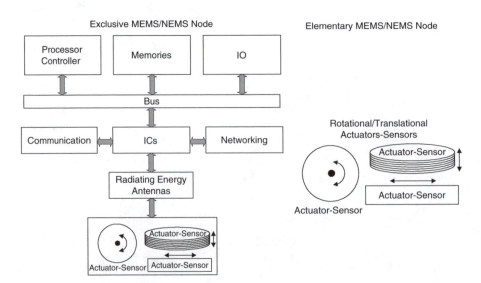

FIGURE 3.16
Micro- and nanosystems nodes.

MEMS and NEMS are used to control, operate, and guarantee functionality for different physical systems, for example, immune system, drug delivery, propeller, wing, relay, lock, car, aircraft, missiles, etc. To illustrate the basic components and their integration, a high-level functional block diagram was shown in Figure 3.9. The desired flight path of aircraft (maneuvering and landing) is maintained by displacing the control surfaces (ailerons, elevators, canards, flaps, rudders, fins, and stabilizers) and/or changing the control surface and wing geometry. The application of the MEMS to actuate the control surfaces was reported in Figure 3.10. It should be emphasized that the digital signal-level signals are generated by the flight computer, and these digital signals are converted into the desired voltages or currents fed to the microactuators by ICs. These signal-level control signals can be converted in the electromagnetic flux intensity to displace the actuators by radiating energy devices. It is also important that microtransducers can be used as sensors. As an example, the loads on the aircraft structures (airframe, wings, etc.) during the flight can be sensed.

Sensing and actuation cannot be viewed as the peripheral function in many applications. Integrated actuators/sensors–radiating energy microdevices–ICs compose the major class of MEMS and NEMS. Due to the use of CMOS technologies in fabricating microscale actuators and sensors, MEMS leverage microelectronics in important additional areas that revolutionize the application capabilities of microsystems. Only recently has it become possible to manufacture affordable and reliable microdevices. However, there is a critical demand for continuous fundamental and applied research, engineering developments, and technological improvements. Therefore, multidisciplinary activities are required. The general lack of synergetic theory to augment actuation, sensing, signal processing, and control is known. Therefore, these issues must be addressed through focused efforts. The set of long-range goals has been emphasized, and the challenges facing the development of MEMS and NEMS are

- Synthesis, modeling, analysis, optimization, and design
- Novel high-performance micro- and nanoscale transducers (actuators and sensors)
- New actuation and sensing mechanisms
- Sensors-actuators-ICs integration and MEMS/NEMS configurations
- Sensing-communication-computing-control-actuation in MEMS and NEMS
- Advanced technologies, techniques, processes, and materials
- Packaging, microassembly, metrology, characterization, and testing
- MEMS implementations and applications

Significant progress in the application of CMOS technology enables the industry to fabricate microscale actuators and sensors with the corresponding ICs. This guarantees significant breakthroughs. The field of MEMS has been driven by the rapid global progress in ICs, VLSI, solid-state devices, microprocessors, memories, and DSPs that have revolutionized instrumentation and control. In addition, this progress has facilitated explosive growth in data processing and communications in high-performance systems. In microelectronics, many emerging problems deal with nonelectric phenomena and processes (optics, thermal and structural analysis and optimization, vibroacoustic, packaging, etc.). However, these nonelectric phenomena frequently result due to electromagnetic effects.

It has been emphasized that ICs are the necessary component to perform control, amplification, signal conditioning, signal processing, computing, information processing, data acquisition, decision making, etc. For example, control signals (voltage or currents) are computed, converted, modulated, filtered, amplified, and fed to actuators, antennae, or microwindings. It is evident that MEMS will found application due to an extremely high integration level of electromechanical ICs components with low cost and maintenance, high performance, accuracy, reliability, and ruggedness. The manufacturability issues were addressed. It was shown that one can design and manufacture individually fabricated devices. However, these devices and subsystems may be impractical if high cost or poor performance result.

Piezoactuators and permanent-magnet technology have been used widely, and rotating and linear microtransducers (actuators and sensors) were designed. For example, piezo-active materials are used in ultrasonic motors. Frequently, conventional concepts of the electric machinery theory (rotational and linear direct-current, induction, and synchronous machine) are used to design and analyze MEMS-based microtransducers. The use of piezoactuators is possible as a consequence of the discovery of advanced materials in sheet and thin-film forms, especially lead zirconate titanate (PZT) and polyvinylidene fluoride. The deposition of thin films allows piezo-based micromachines to become a promising candidate for microactuation and sensing. These microtransducers can be fabricated using a deep x-ray lithography and electrodeposition processes.

To fabricate nanoscale structures, devices, and systems, molecular manufacturing methods and techniques must be developed. Self- and positional-assembly concepts are the preferable techniques compared with individually fabricated molecular structures. To perform self- and positional assembly, complementary pairs (CP) and molecular building blocks (MBB) should be designed. These CP and MBB, which can be built from a few to thousands of atoms, can be examined and designed using the biomolecular analogy. The nucleic acids consist of two major classes of molecules, e.g., DNA and RNA. Deoxyribonucleic acid (DNA) and ribonucleic acid (RNA) are large and very complex organic molecules that are composed of carbon, oxygen, hydrogen, nitrogen and phosphorus. The structural units of DNA and RNA are nucleotides. Each nucleotide consists of three components (nitrogen-base, pentose, and phosphate) joined by dehydration synthesis. The double-helix DNA was discovered by Rosalind Franklin and Maurice Wilkins (King's College, London) and reported by Watson and Crick (Cambridge University) in 1953. Deoxyribonucleic acid (long double-stranded polymer with a double chain of nucleotides held together by hydrogen bonds between the bases), as the genetic material, performs two fundamental roles. It replicates (identically reproduces) itself before a cell divides, and provides pattern for protein synthesis directing the growth and development of all living organisms according to the information DNA carries. Different DNA architectures provide the mechanism for the replication of genes. Specific pairing of nitrogenous bases obeys base-pairing rules and determines the combinations of nitrogenous bases that form the rungs of the double helix. In contrast, RNA performs the protein synthesis using the DNA information. The four DNA nitrogenous bases are A (adenine), G (guanine), C (cytosine), and T (thymine). The ladder-like DNA molecule is formed due to hydrogen bonds between the bases that are paired in the interior of the double helix (the base pairs are 0.34 nm apart and there are 10 pairs per turn of the helix). Two backbones (sugar and phosphate molecules) form the uprights of the DNA molecule, while the joined bases form the rungs. Figure 3.17 illustrates the hydrogen bonding of the bases. In particular, A bonds to T, G bonds to C (details are documented in Section 5.4.1), and the complementary base sequences result.

In RNA molecules (single strands of nucleotides), the complementary bases are A–U (uracil), and G–C. The complementary base bonding of DNA and RNA molecules gives one the idea of possible sticky-ended assembling (through complementary pairing) in nanoscale structures and devices with the desired level of specificity, architecture, topology, organization, and reconfigurability. In structural assembling and design, the key element is the ability of CP or MBB (atoms or molecules) to associate with each other (recognize and identify other atoms or molecules by means of specific base pairing relationships). It was emphasized that in DNA, A (adenine) bonds to T (thymine), and

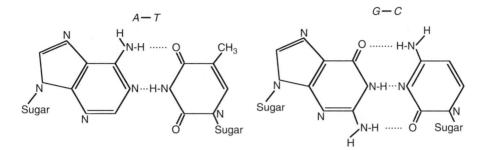

FIGURE 3.17
Deoxyribonucleic acid pairing due to hydrogen bonds.

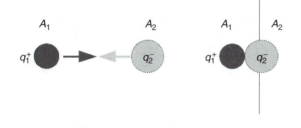

FIGURE 3.18
Sticky-ended electrostatic complementary pair A_1–A_2.

G (guanine) bonds to C (cytosine). Using this idea, one can design CP such as A_1–A_2, B_1–B_2, C_1–C_2, D_1–D_2, E_1–E_2, M_1–M_2, N_1–N_2, etc. That is, A_1 pairs with A_2, B_1 pairs with B_2, and so on. This complementary pairing can be studied using electromagnetics (Coulomb's law) and chemistry (chemical bonding, for example, hydrogen bonds in DNA between nitro-genous bases A and T, and G and C). Figure 3.18 shows how two nanoscale elements with sticky ends form a complementary pair. In particular, + is the sticky end and – is its complement. Thus the electrostatic complementary pair A_1–A_2 results.

An example of assembling a ring micro- or nanostructure is illustrated in Figure 3.19. Using the sticky-ended, segmented (asymmetric), electrostatic CP, self-assembling of nanostructure is performed in the two-dimensional (x–y) plane. It is evident that three-dimensional structures can be formed through self-assembling, developing, electrostatic CP.

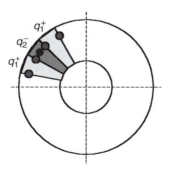

FIGURE 3.19
Two-dimensional ring self-assembling.

There are several advantages to using sticky-ended electrostatic CP. In particular, the ability to recognize (identify) the complementary pair is reliably predicted, and the possibility to form stiff, strong, and robust structures is attained. Self-assembled complex nanostructures can be fabricated using subsegments to form the branched junctions. This concept is well-defined electromagnetically (Coulomb's law) and geometrically (branching connectivity). Using these subsegments, nanostructures with the desired geometry (e.g., cubes, octahedrons, spheres, cones, etc.) can be fabricated. Furthermore, the geometry of nanostructures can be controlled sequentially introducing the CP and pairing MBB. It must be emphasized that it is possible to generate quadrilateral self-assembled nanostructures using different CP. For example, electrostatic and chemical CP can be employed. Single- and double-stranded structures can be generated and linked in the desired topological and architectural manners. The self-assembling must be controlled during the fabrication cycle, and CP and MBB that can be paired and topologically/architecturally bonded must be added in the desired sequence. For example, polyhedral and octahedral synthesis can be made when building elements (CP or MBB) are topologically or geometrically specified. The connectivity of nanostructures determines the minimum number of linkages that flank the branched junctions. The synthesis of complex three-dimensional nanostructures is the design of topology, and the structures are characterized by their branching and linkaging.

3.3.5.1   *Linkage Groups in Molecular Building Blocks*

Hydrogen bonds, which are weak, hold DNA and RNA strands. Strong bonds are desirable to form stiff, strong, and robust micro- and nanostructures. Using polymer chemistry, functional multimonomer groups can be designed. Polymers made from monomers with two linkage groups do not exhibit desired stiffness and strength. Tetrahedral MBB structures with four linkage groups lead to stiff and robust structures. Polymers are made from monomers, and each monomer reacts with two other monomers to form linear chains. Synthetic and organic polymers (large molecules) are nylon and dacron (synthetic), and proteins and RNA, respectively.

There are two feasible assembling techniques: self-assembly and positional assembly. Self-assembling is widely used at the molecular scale, e.g., DNA and RNA. Positional assembling is used in manufacturing and microelectronic manufacturing. The current difficulties implementing positional assembly at the nanoscale with the same flexibility and integrity that is achieved in microelectronic fabrication limits the range of nanostructures that can be made. Therefore, the efforts are focused on developments of MBB, as applied to manufacturing nanostructures, with the ultimate goal to guarantee

- Affordable (low-cost) mass production
- High-yield, scalability and reliability
- Simplicity, predictability, and controllability of synthesis and fabrication
- High-performance, repeatability, and similarity of characteristics
- Stiffness, strength, and robustness
- Tolerance to contaminants

It is possible to select and synthesize MBB that satisfy requirements and specifications (nonflammability, nontoxicity, pressure, temperatures, stiffness, strength, robustness, resistivity, permeability, permittivity, etc.). Molecular building blocks are characterized by the number of linkage groups and bonds they have. The linkage groups and bonds that can be used to connect MBB are dipolar bonds (weak), hydrogen bonds (weak), transition metal complexes bonds (weak), and amide and ester linkages (weak and strong).

It must be emphasized that large MBB can be made from elementary MBB. There is a need to synthesize robust three-dimensional structures. Molecular building blocks can form planar structures that are strong, stiff, and robust in-plane, but weak and compliant in the third dimension. This problem can be resolved by forming tubular structures. It is difficult to form three-dimensional structures using MBB with two linkage groups. Molecular building blocks with three linkage groups form planar structures, which are strong, stiff, and robust in-plane but bend easily. This plane can be rolled into tubular structures to guarantee stiffness. In contrast, MBB with four, five, six, and 12 linkage groups form the strong, stiff, and robust three-dimensional structures needed to synthesize robust micro- and nanostructures. Molecular building blocks with L linkage groups are paired forming *L*-pair structures, and planar and nonplanar (three-dimensional) micro- and nanostructures result. These MBB can have in-plane linkage groups and out-of-plane linkage groups that are normal to the plane. For example, hexagonal sheets are formed using three in-plane linkage groups (MBB is a single carbon atom in a sheet of graphite) with adjacent sheets formed using two out-of-plane linkage groups. These structures have the hexagonal symmetry.

Carbon-based technology has been widely examined. We already briefly covered carbon nanotubes. Let us focus our attention on another meaningful carbon family. C_{60}, fullerene, named after architect Buckminister Fuller, who designed a geodesic dome with the same fundamental symmetry, is the roundest and the largest symmetrical known molecule. This C_{60} molecule consists of 60 carbon atoms arranged in a series of interlocking hexagons and pentagons, forming a structure that looks similar to a soccer ball. C_{60}, which forms truncated icosahedron (consisting of 12 pentagons and 20 hexagons with hollow spheroids, see Figure 3.20) was discovered in 1985 by Professor Sir Harry Kroto (U.K.), Professor Richard E. Smalley, and Professor Robert F. Curl, Jr. (who were jointly awarded the 1996 Nobel prize for chemistry). Fullerenes have cavities, and let us estimate the size of the cavity at the center of C_{60}. How big are fullerenes? The diameter of fullerene molecules is comparable to that of the double helix of DNA. Specifically, the diameter of C_{60} is approximately 7 Å, measured from the nucleus of one carbon to the nucleus of the carbon on the opposite side. If we take into account that the van der Waals diameter of the carbon atoms is 3.4 Å, we obtain a cavity with a diameter equal to 3.6 Å. The complexes can be formed by layers of hexagonally packed C_{60}, each of them connected to six neighbors by 2/2 cycloaddition of double bonds. An equilibrium intermolecular distance is 9.2 Å. Rhombohedral C_{60} solid has higher energy than free C_{60} molecules by 2.1 eV/molecule, and the barrier for dissociation to free C_{60} molecules is approximately 1.6 eV/molecule. It is important that the lowest energy conformation of this new phase of solid C_{60} is a semiconductor, but defects in the intermolecular bonding pattern lead to semimetal properties.

It is known that buckyballs (C_{60}, C_{70}, C_{76}, C_{80}, C_{84}, C_{102}, etc.), studied in the literature as MBB, are formed with six functional groups. Molecular building blocks with six linkage groups can be connected together in the cubic structure. These six linkage groups correspond

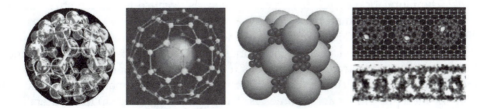

FIGURE 3.20
C_{60}, endohedral C_{60}, doped C_{60} complex, and single-wall carbon nanotube with filled buckyballs.

to six sides of the cube or rhomb. Thus for the studied case, MBB with six linkage groups form solid three-dimensional structures as cubes and rhomboids. Molecular building blocks with six in-plane linkage groups form strong planar structures. Robust, strong, and stiff cubic or hexagonal closed-packed crystal structures are formed using 12 linkage groups. Molecular building blocks synthesized and applied should guarantee the desirable performance characteristics (stiffness, strength, robustness, resistivity, permeability, permittivity, coefficient of thermal extension, etc.), affordability, integrity, and manufacturability. It is evident that stiffness, strength, and robustness are predetermined by bonds (weak and strong), while resistivity, permeability, and permittivity are the functions of MBB of electron transport in media. Carbon-based assemblies exhibit remarkable electromechanical and electronic properties. For example, buckyballs can be used as highly robust mechanical structures to fabricate complex assemblies for micro- and nanoscale motion devices, and the same doped fullerenes exhibit semiconductor properties at low temperature. Furthermore, one can place C_{82} buckyballs containing gadolinium atoms inside carbon nanotubes. Gadolinium and other metal atoms were known to change the electronic structure of C_{xx} buckyballs affecting the electronic properties of carbon nanotubes. For example, single-wall carbon nanotubes filled with gadolinium-doped C_{80} buckyballs (GdC_{80}) self-lined inside carbon nanotubes with 1-nm displacement (after evaporation process) change electronic properties at room temperature. Figure 3.20 illustrates C_{60}, endohedral C_{60}, C_{60} complex, and a 1.4-nm diameter single-wall carbon nanotube with filled buckyballs.

3.4   Introduction to Microfabrication and Micromachining

The manufacturability issues in MEMS and NEMS must be addressed. One can design and manufacture individually fabricated microscale devices and structures, but they may be impractical due to very high cost or low performance. Therefore, high-yield enhanced CMOS and other fabrication technologies (e.g., surface micromachining and LIGA) are used. Microsystems at least have electromechanical motion microdevices and ICs. Microfabricated smart multifunctional materials are used to make microscale actuators, sensors, pumps, valves, optical switches, and other devices. Hundreds of millions of transistors on a chip are currently fabricated by the microelectronic industry, and enormous progress in scaling down transistor dimensions to nanoscale has been made. However, MEMS operational capabilities are measured by the functionality, system-on-a-chip integration, cost, performance, efficiency, size, reliability, robustness, compliance, survivability, intelligence, life time, and other criteria. There is a number of challenges in MEMS fabrication because conventional CMOS technology must be modified and integration strategies (to integrate motion structures and ICs) need to be developed. What should be fabricated first: ICs or mechanical micromachined structures? The fabrication of ICs first faces challenges because to reduce stress in silicon thin films (multifunctional materials to make motion microstructures), a high-temperature annealing (usually in the 1000°C range) is needed for several hours. The aluminum IC's interconnect will be destroyed (melted), and tungsten can be used instead for interconnected metallization. This process leads to difficulties for commercially manufactured MEMS due to high cost and low reproducibility. Let us illustrate the basic steps making use of the analog devices fabrication process for accelerometers. The ICs are fabricated up to the metallization step. Then mechanical structures are made using high-temperature annealing. (Suspended microstructure is fabricated before metallization.) Finally ICs are interconnected. This allows the manufacturer to use low-cost, high-yield conventional aluminum interconnects.

The third option is to fabricate mechanical structures and then ICs. To overcome step coverage, stringer, and topography problems, motion mechanical microstructures can be fabricated in the bottoms of the etched shallow trenches (packaged directly) of the silicon substrate. These trenches are filled with a sacrificial silicon dioxide, and the silicon is planarized through chemical-mechanical polishing.

The motion mechanical microstructures can be protected (sensor applications such as accelerometers, gyroscopes, etc.) and unprotected (interactive environment actuator and sensors). Therefore, MEMS can be packaged, encased, and encapsulated in a clean, hermetically sealed package. However, some elements can be unprotected to allow the direct interaction and active interfacing with the environment. This creates challenges in packaging. It is important to develop novel electromechanical motion microstructures and microdevices (sticky multi-layers, thin films, magnetoelectronic, electrostatic, quantum-effect–based devices, etc.) and examine their properties. Microfabrication of very large-scale integrated circuits (VLSI), microscale structures and devices, and optoelectronics must be addressed for MEMS. Fabrication processes include lithography, film growth, diffusion, implantation, deposition, etching, metallization, planarization, packaging, etc. Complete microfabrication processes with integrated sequential processes are of great importance. These issues are covered in this book.

It was emphasized that microsensors sense physical variables, and microactuators control (actuate) real-world systems. These microactuators are regulated and controlled using ICs. In addition, ICs also perform computations, input-output interfacing, networking, signal conditioning, signal processing, filtering, decision making, and other functions. For example, in microaccelerometers, the motion microstructure displaces. Then using the measured capacitance difference, the acceleration is calculated. In microaccelerometers, computing, signal conditioning, filtering, input-output interfacing, and data acquisition are performed by ICs. In the ADL-series accelerometers, these ICs are built using thousands of transistors.

Microelectromechanical systems contain microscale subsystems, devices, and structures designed and manufactured using different technologies, techniques, processes, and materials. A single silicon substrate can be used to fabricate integrated microscale actuators, sensors, and ICs using the modified CMOS fabrication technologies, techniques, and processes. These MEMS must be assembled, interfaced, and packaged. Different micro-fabrication techniques for MEMS components and subsystems exist to attain simplicity, compactness, reliability, and superior performance. As was illustrated, MEMS typically integrate the following components: microscale actuators (actuate systems), microscale sensors (detect and measure changes of the physical variables and stimuli), radiating energy and communication microdevices, ICs (signal processing, data acquisition, signal conversion, interfacing, communication, and control), optical microdevices, etc. Micro-actuators are needed to develop force or torque (mechanical variable). Typical examples are microdrives, moving mirrors, pumps, servos, and valves. Actuation can be achieved using electromagnetic (electrostatic, magnetic, piezoelectric, optoelectric, etc.), hydraulic, vibration, and thermal effects. This book covers electromagnetic and electrostatic micro-actuators. The so-called electrostatic comb drives (surface micromachined motion microdevices) have been widely used. These drives have movable and stationary plates or fingers. When voltage is applied, a force is developed between two plates, and the motion results. Due to the limited power, force, and torque densities of electrostatic actuators, magnetic actuators (mostly permanent magnet based) have been designed and fabricated. A wide variety of microscale magnetic actuators have been optimized and characterized. These actuators guarantee superior power, force, and torque densities. The difficulties associated with microwinding (coil) and micromagnet fabrication are a common problem. However, these challenges have already been overcome. The choice of magnetic core materials and permanent magnets is also a critical issue. Electromagnetic actuators typically can be fabricated through micromachining technologies using copper, aluminum, and other conducting materials to fabricate microcoils. Nickel, nickel-iron, chrome, and other thin-film alloys utilized as magnetic and ferromagnetic materials. (See Chapter 4 and Chapter 10 for details.)

Piezoelectric microactuators have found wide application due to their simplicity and ruggedness. (Force is generated if one applies the voltage across a film of piezoelectric material). If the voltage is applied, the silicon membrane with PZT thin film deforms. Thus the PZT-silicon–based membrane structures can be used in pumps as movable diaphragms. The disadvantage of PZT actuators is small deflection, which is up to 0.2% of the total length.

Microsensors are devices that convert one physical variable (quantity) to another. For example, electromagnetic phenomena can be converted to mechanical or optical output. There is a number of different types of microscale sensors designed utilizing MEMS technology. For example, microscale thermosensors were designed and fabricated using the thermoelectric effect (resistivity varies with temperature). Extremely low-cost thermoresistors (thermistors) were fabricated on the silicon wafer, and ICs were built on the same substrate. The thermistor resistivity is a highly nonlinear function of the temperature, and compensating circuitry is used to take into account the nonlinear effects. Microelectromagnetic sensors measure electromagnetic fields, e.g., the Hall-effect sensors. Optical sensors can be fabricated on crystals that exhibit a magneto-optic effect, e.g., optical fibers. In contrast, the quantum effect sensors can sense extremely weak electromagnetic fields. Silicon-fabricated piezoresistors (silicon doped with impurities to make it *n*- or *p*-type) belong to the class of electromechanical sensors. When the force is applied to the piezoelectric resistor, the charge (voltage) induced is proportional to the applied force. Zinc oxide and lead zirconate titanate (PZT, $PbZrTiO_3$), which can be easily deposited applying well-developed techniques, are used as piezoelectric crystals.

In this book the microscale accelerometers and gyroscopes, as well as microtransducers, are studied. Accelerometers and gyroscopes are based on the capacitive sensing mechanism. In two parallel conducting plates, separated by an insulating material, the capacitance between the plates is a function of distance between plates (capacitance is inversely proportional to the distance). Thus, measuring the capacitance, the distance can be easily calculated. In accelerometers and gyroscopes, the proof mass (or rotor) is suspended on silicon springs. Another important class of sensors is the thin-film pressure sensors. Thin-film membranes are the basic components of pressure sensors. The deformation of the membrane is usually sensed by resistors, piezoresistors or capacitive microsensors.

We have illustrated the critical need for physical- and system-level synergetic concepts in micro- and nanosystems synthesis, analysis, design, and optimization. Advances in physical-level multidisciplinary research have tremendously expanded the horizon of MEMS and NEMS. For example, magnetic-based (magnetoelectronic) memories have been thoroughly studied. (Magnetoelectronic devices can be grouped into three categories based on the physics of their operation: all-metal spin transistors and valves, hybrid ferromagnetic semiconductor structures, and magnetic tunnel junctions.) Writing and reading the cell data are based on different physical mechanisms. It was demonstrated that low cost, high densities, low power, high reliability, and speed (write/read cycle) memories result. As the physical-level synthesis, analysis, and design are performed at the microstructure/microdevice level, the system-level synthesis, analysis, design and optimization must be accomplished because the design of the integrated large-scale MEMS and NEMS is required. This section introduces the reader to microfabrication techniques and technologies, reporting the basic processes and steps to make microstructures. Different MEMS fabrication technologies are developed and applied.[7-11] In particular, micromachining and high-aspect-ratio technologies will be reported in Chapter 4 and Chapter 10. However, let us focus our attention on the basic overview to facilitate the understanding of microfabrication basics. The MEMS microfabrication technology has been adopted from microelectronics. For example, many microscale structures, devices, and systems are fabricated on silicon wafers, and the basic processes, steps, materials, and equipment are same as in the fabrication of ICs. In addition, the microstructures are usually made from thin films patterned using photolithographic processes and deposited applying methods developed for ICs. It is evident that some processes and materials used in MEMS fabrication are different from conventional microelectronic technologies. In fact, it is important to attain the compliance, compatibility, and integrity in fabrication of mechanical and microelectronic components in order to make MEMS. The basic processes, which should be emphasized first introducing the MEMS microfabrication technology, are (1) deposition (depositing thin films of materials on substrates), (2) photolithography (applying patterned masks on top of the thin films), and (3) etching (etching thin films selectively to the photomask). The importance of these processes is due to the fact that the MEMS fabrication must be based on well-defined sequential sequence of these processes to form microscale structures, devices, and systems.

3.4.1 Thin-Film Deposition

Deposition can be based on chemical reaction (chemical vapor deposition, electrodeposition, epitaxy, and thermal oxidation) as well as physical reaction (physical vapor deposition and casting). The chemical deposition processes make solid materials directly from chemical reactions in gas or liquid or with the substrate material. In contrast the materials can be positioned on the substrate using the physical deposition processes.

In the case of the chemical vapor deposition (CVD), the substrate is placed inside a reactor to which gases are supplied. Chemical reactions occur between the source gases. The product of these reactions is a solid material that condenses on the surfaces inside the reactor.

Two widely applied CVD techniques are

- Low-pressure chemical vapor deposition (LPCVD)
- Plasma-enhanced chemical vapor deposition (PECVD)

Through LPCVD, thin film layers with excellent uniformity of thickness and material characteristics can be made. The major drawbacks of the LPCVD are high deposition temperature 600°C or higher) and slow deposition rate. The PECVD allows one to perform the deposition at low temperature 3000°C due to the energy supplied to the gas molecules by the plasma in the reactor. However, the uniformity and characteristics of thin films fabricated through PECVD do not match the quality fabricated using LPCVD. Furthermore, the PECVD systems usually deposit materials on one side of wafers, while the LPCVD systems deposit thin films on both sides of wafers.

Electrodeposition (electroplating and electroless plating) processes are used to electrochemically deposit conductive materials. To perform electroplating, the substrate is placed in a liquid solution (electrolyte), the electric potential is formed between a substrate and an electrode (external power supply is used), and due to a chemical redox process, a thin layer of material on the substrate is formed (deposited). Thus electrodeposition is applied to fabricate thin films (thickness from 0.1 μm to more than 100 μm) of different metals (copper, gold, iron, nickel, and others). In case of the electroless plating process, complex chemical solutions are used. The deposition occurs spontaneously on any surface that forms a sufficiently high electrochemical potential with the solution. Electroless plating does not require electric potential and contact to the substrate. However, the electroless plating process is difficult to control to attain the uniformity and other desired deposit characteristics.

The so-called half reactions for Al, Cr, Fe, Ni, Cu, and Ag electroplating with half-cell potentials at 25°C are

$$Al^{3+} + 3e^- \equiv Al, \quad \mathscr{E} = -1.662 \text{ V},$$

$$Cr^{3+} + 3e^- \equiv Cr, \quad \mathscr{E} = -0.744 \text{ V},$$

$$Fe^{2+} + 2e^- \equiv Fe, \quad \mathscr{E} = -0.44 \text{ V},$$

$$Ni^{2+} + 2e^- \equiv Ni, \quad \mathscr{E} = -0.25 \text{ V},$$

$$Cu^{2+} + 2e^- \equiv Cu, \quad \mathscr{E} = 0.337 \text{ V},$$

$$Ag^+ + e^- \equiv Ag, \quad \mathscr{E} = 0.799 \text{ V}.$$

(The reader can easily find these reactions, detailed descriptions, and calculations in any freshman Chemistry book.)

If the substrate is an ordered semiconductor crystal (silicon), using the epitaxy process, one fabricates the microstructures with the same crystallographic orientation. (For amorphous-polycrystalline substrates, the amorphous or polycrystalline thin films can be made.) To attain epitaxial growth, a number of gases are introduced in a heated reactor where only the substrate is heated, and high deposition rate is achieved making films with thickness grater than 100 μm. Usually epitaxy is used to fabricate silicon structures on the insulator substrates. Therefore, this technique is primarily used for deposition of silicon to make ICs and insulated structures.

Thermal oxidation is the deposition process usually applied to form thin films for electrical insulation. Thermal oxidation is based upon oxidation of the substrate surface in an oxygen-rich atmosphere at high temperature (800°C to 1,100°C). Thermal oxidation is limited to materials that can be oxidized, and this process can only form films that are oxides of the material, e.g., to form silicon dioxide on a silicon substrate.

Using physical vapor deposition (PVD) processes, materials are released from sources and transferred to the substrates. Usually by using PVD it is difficult to attain good thin-film characteristics. However, PVD allows one to fabricate affordable microstructures. The basic PVD methods are evaporation and sputtering. Applying evaporation techniques, one places the substrate in a vacuum chamber, in which a source of the material to be deposited is located. The source material is heated to the point where it starts to evaporate. Due to the vacuum, the molecules evaporate freely in the chamber, and they subsequently condense on all surfaces. The heating methods applied are electron-beam (an electron beam is directed at the source material causing local heating and evaporation) and resistive (material source is heated electrically using high current to make the material evaporate) evaporation. Different methods should be used. For example, aluminum is difficult to evaporate using resistive heating, while tungsten deposition can be straightforwardly performed through resistive evaporation. In contrast to evaporation, sputtering is a technique in which the material is released from the source at lower temperature than evaporation. The substrate is placed in a vacuum chamber with the source material, and an inert gas (argon) is introduced at low pressure. A gas plasma is struck using a power source, causing the gas to become ionized. The ions are accelerated toward the surface of the source material, vaporizing atoms of this material. These atoms are condensed on all surfaces including the substrate.

Applying casting, the material (which should be deposited) is dissolved in a solvent. The deposition process is performed through spraying or spinning. Once the solvent is evaporated, a thin film of the material remains on the substrate. Polymer materials can be easily dissolved in organic solvents, and casting is an efficient technique to apply photoresist to substrates (photolithography).

3.4.2 Photolithography

Lithography means the pattern transfer to a photosensitive material by selective exposure of a substrate to a radiation source. Therefore, the MEMS geometry and topography are defined (pattern) through lithographic processes. Lithography is performed as sequential steps, e.g., substrate surface preparation, photoresist deposition, alignment of the mask and wafer, and exposure.

Different photosensitive materials that are affected when exposed to a radiation source can be used. In particular, when photoresist is exposed to a radiation source with a specific wavelength, the chemical resistance of the photoresist changes. When the photoresist is placed in a developer solution after selective exposure, exposed or unexposed regions can be etched away. If the exposed material is etched away by the developer and the unexposed region is resilient, the material is called a positive photoresist. If the exposed material is resilient to the developer and the unexposed region is etched away, it is called a negative photoresist. To fabricate MEMS, the patterns for different sequential lithographic steps, associated to a particular microstructure, must be aligned. Therefore, the first pattern transferred to a wafer usually includes a set of alignment marks with high precision features that are needed as the reference when positioning subsequent patterns with respect to the first pattern. The alignment marks can be integrated and used in other patterns. Each pattern layer must have a precise alignment feature so that it may be registered to the rest of the layers. The alignment is needed to fabricate the desired MEMS. The alignment location, specifications, and size vary with the type of the alignment and lithographic equipment. Two alignment marks are used to align the mask and wafer, one alignment mark is sufficient to align the mask and wafer in the x- and y-axis, but two spaced marks are needed for rotational alignment.

The exposure characteristics must be examined to achieve accurate pattern transfer from the mask to the photosensitive layer taking into account the wavelength of the radiation source and the dose required to achieve the needed photoresist properties change based upon the photoresist sensitivity. The exposure dose required per unit volume of photoresist is available from the photoresist's data library. However, the highly reflective layer under the photoresist may result in higher exposure dose, and the photoresist thickness must be also counted. Thus the secondary effects (reflectiveness, thickness, interference, nonuniformity, flatness, roughness, etc.) affect the pattern transfer and should be examined.

The sequential lithography steps are as follows:

- Preparation of the wafer surface (dehydrate the substrate to attain photoresist adhesion)
- Photoresist deposition
- Substrate coating
- Soft baking
- Alignment (align pattern on mask to the specified microstructure topography and geometry)
- Exposure with the optimal dosage
- Postexposure bake
- Photoresist develop (selective removal of photoresist after exposure)
- Hard bake (drive off remaining solvent form the photoresist)
- Descum (remove photoresist scum)

3.4.3   Etching

To fabricate microstructures with the ultimate goal of making MEMS, it is necessary to etch the deposited thin films as well as etch the substrate itself. Two classes of etching processes are wet etching (material can be dissolved as it immersed in a chemical solution) and dry etching (material can be sputtered or dissolved using reactive ions or vapor phase etchant). Though wet etching is a very simple etching technique, in general, complex topographies and geometries cannot be made. In addition, materials exhibit *anisotropic* and *isotropic* phenomena. This leads to the application of anisotropic etching with different etch rates in different material directions. The anisotropic and isotropic etching effects can be beneficial to fabricate the desired microstructures. However, complex topography, accurate sizing, processes flexibility, and other requirements usually lead to the application of dry etching.

Reactive ion etching (RIE), deep reactive ion etching (DRIE), sputter etching, and vapor phase etching are the techniques used to perform dry etching.

Using the RIE, the substrate is placed inside a reactor in which several gases are introduced. A plasma is struck in the gas mixture using a power source, resulting in releasing ions from the gas molecules. These ions are accelerated toward the surface of the material being etched forming gaseous material (the chemical part of reactive ion etching). The physical processes of the RIE are similar to the sputtering deposition processes. In particular, if the ions have high energy, they can knock atoms out of the material to be etched without a chemical reaction. It is a very complex task to develop the processes that balance chemical and physical etching because many parameters must be adjusted and regulated to attain the desired etching characteristics. Furthermore, the secondary effects must be examined.

Sputter etching is performed through the ion bombardment. To perform the etching through the vapor phase etching, the substrate to be etched is placed inside a chamber, into which one or more gases are introduced. The material to be etched is dissolved at the surface in a chemical reaction with the gas molecules. The most common vapor phase etching techniques are silicon dioxide etching (using hydrogen fluoride) and silicon etching (using xenon diflouride). Different etching techniques are discussed in Chapter 4.

3.4.4   ICs and Microfabrication

Many of the microfabrication techniques and materials used to make MEMS have been used to fabricate ICs for many years. However, microfabrication and micromachining enhance traditional semiconductor industry processes, techniques, and materials. In fact, conventional major IC processes are photolithography, thermal oxidation, dopant diffusion, ion implantation, LPCVD, PECVD, evaporation, sputtering, wet etching, plasma etching, reactive-ion etching, ion milling, and basic materials are silicon, silicon dioxide, silicon nitride, aluminum, and copper.

In microfabrication the following additional processes are used: anisotropic and isotropic etching, anisotropic wet etching of single-crystal silicon, deep reactive-ion etching (DRIE), x-ray lithography, electroplating, low-stress LPCVD, thick-film resist, spin casting, micromolding, batch microassembly, etc. These processes and techniques have been developed to fabricate ferromagnetic and magnetic films (Ni, Fe, Co, Cr, and rare earth alloys), high-temperature materials (SiC and ceramics), piezoelectric films, mechanically robust aluminum alloys, stainless steel, platinum, gold, sheet glass, plastics (PVC and PDMS), etc.

Using photolithography, high-yield ICs and MEMS are fabricated within submicron dimensions (features). The photolithography process begins by selecting a substrate material and assigning the desired geometry using photomask. Usually, single-crystal silicon (Si) wafers are used. The substrate is coated by a photosensitive polymer called a photoresist. A photomask, consisting of a transparent supporting medium with precisely patterned metal, is used to cast a highly detailed shade (draw) on the photoresist. The regions receiving

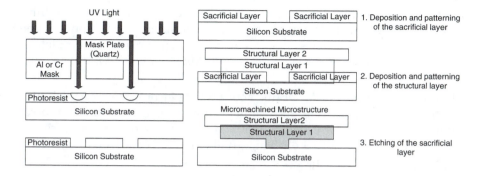

FIGURE 3.21
Surface micromachining process.

an exposure of ultraviolet (UV) light are chemically altered. After exposure the photoresist is immersed in a solution (developer) that chemically removes either the exposed regions (positive process) or the unexposed regions (negative process). This dried photoresist is used as a mask for a subsequent deposition (additive process), or etch out (subtractive process). Finally the photoresist is selectively removed, resulting in a micromachined substrate with the specified geometry. These processes are covered in Chapter 4 and Chapter 10. The reader can also use viable web recourses and books to apply the processes, techniques, and materials that match the design and fabrication infrastructure available.

The processes used to integrate multiple patterned materials in order to fabricate microdevices and MEMS are enormously important. Surface and bulk micromachining have been utilized as the most general methods of MEMS integration.

Surface micromachining is defined as a high-yield process to fabricate MEMS by sequentially depositing, patterning, and etching a sequence of thin films (less than 1 μm to 100 μm).

In Chapter 4 it will be emphasized that one of the most essential processing steps is the selective removal of an underlying film (called sacrificial layer) without attacking an overlying film (called structural layer). Figure 3.21 illustrates the surface micromachining process.

Bulk micromachining is defined as a high-yield process to fabricate MEMS by sequentially depositing, patterning, and etching (anisotropic and isotropic) the bulk of silicon media.

Bulk micromachining differs from surface micromachining because the substrate material (typically single-crystal silicon) is patterned and shaped to form an important functional component of the resulting device. Thus the silicon substrate does not use a rigid mechanical (structural) base, which is typically used in surface micromachining. Exploiting the anisotropic etching characteristics of single-crystal silicon, many highly accurate three-dimensional complexes and assemblies (for example, fluidic and gas channels, pyramidal pits, membranes, cones, vias, nozzles, etc.) have been fabricated.

Homework Problems

1. What could be the difference between MEMS and bioMEMS? (The definition for bioMEMS is not reported in this book. Define bioMEMS and justify your definition.)

2. Why should biosystem-inspired MEMS and NEMS be examined?

3. For a system most appealing to you (CVD, sputtering systems, etcher, lithography tool, stepper, microscope, aircraft, car, computer, missile, power system, ship, submarine, and other systems, you may even consider a complete semiconductor fabrication process), explain how MEMS and NEMS can be utilized in order to:

   a. Improve performance. Be specific about what performance variables and outputs you are considering and estimate how much the performance can be improved. (You may use yield, quality, cost, efficiency, power and force density, acceleration, velocity, fuel efficiency, fuel emission, contamination, etc.)

   b. Enhance functionality. Be precise and specific.

   c. Benefit society. Be explicit in definition of possible benefits.

4. Provide your comments, details, and forecast of the possible application of nanodevices in MEMS. (Try be realistic and avoid speculative or questionable statements.) How can these nanodevices benefit minisystems and MEMS?

5. What are the current drawbacks and challenges in nanoelectronics?

6. Provide examples for two-dimensional (planar) ICs explaining the need for nano-ICs.

7. Provide a very brief description of the major fabrication processes for ICs and MEMS. Why are they important?

8. Using the course syllabus, identify what processes will be emphasized and covered in the course.

9. Explicitly define deposition, photolithography, and etching processes.

10. Using your undergraduate Chemistry or Physics books (you may also use Web site or library resources) write a short report covering:

    a. Electrodeposition of Al, Ag, and Cu (conductors electroplating)

    b. Electrodeposition of Cr, Ni, and Fe (ferromagnetic materials electroplating)

These reactions and calculations are usually covered in the Electrochemistry chapters.

11. Provide the examples when surface and bulk micromachining can be used to fabricate three-dimensional structures and assemblies. Draw these structures and assemblies.

12. Define the processes to use in order to fabricate planar structures that integrate layered thin films of different materials and silicone sphere.

References

1. Drexler, E.K. *Nanosystems: Molecular Machinery, Manufacturing, and Computations*, Wiley-Interscience, New York, 1992.
2. Colbert, D.T. and Smalley, R.E. Past, present, and future of fullerene nanotubes: buckytubes, in *Perspectives of Fullerene Nanotechnology*, Osawa, E., Ed., Kluwer Academic Publishers, Dordrecht, 2002, pp. 3–10.
3. Goddard, W.A., Brenner, D.W., Lyshevski, S.E., and Iafrate, G.J., Eds. *Handbook of Nanoscience, Engineering, and Technology*, CRC Press, Boca Raton, 2002.
4. Seeman, N.C., DNA engineering and its application to nanotechnology, *Nanotechnol.*, 17, 437–443, 1999.

5. Ellenbogen, J.C. and Love, J.C., Architectures for molecular electronic computers, in *Handbook of Nanoscience, Engineering, and Technology*, Goddard, W., Brenner, D., Lyshevski, S., and Iafrate, G., Eds., CRC Press, Boca Raton, 2002, pp. 7.1–7.65.
6. Lyshevski, S.E., Nanocomputers, nanoarchitectronics, and nanoICs, in *Handbook of Electrical Engineering*, Dorf, R., Ed., CRC Press, Boca Raton, 2003.
7. Campbell, S.A., *The Science and Engineering of Microelectronic Fabrication*, Oxford University Press, New York, 2001.
8. Kovacs, G.T.A., *Micromachined Transducers Sourcebook*, McGraw-Hill, Boston, 1998.
9. Lyshevski, S.E., *NEMS and MEMS: Systems, Devices, and Structures*, CRC Press, Boca Raton, 2002.
10. Madou, M.J., *Fundamentals of Microfabrication*, CRC Press, Boca Raton, 2002.
11. Senturia, S.D., *Microsystem Design*, Kluwer Academic Publishers, New York, 2000.

4

Fundamentals of Microfabrication and MEMS Fabrication Technologies

4.1 Introduction and Description of Basic Processes in Microfabrication

It is the author's objective to reach the widest possible range of readers, who are not necessarily experts in microfabrication, as well as to attain the versatility for students and professionals who have been engaged and have experience in microelectronics and MEMS fabrication. Correspondingly, this chapter aims to satisfy the existing growing demands by covering the fabrication processes starting from the basics and coherently progressing to sophisticated fabrication technologies, processes, and materials. Synergetic efforts have been made in this chapter to cover basic fabrication techniques and processes widely used to fabricate microscale systems, devices, and structures. Due to very broad and extensive technologies involved, specialized fabrication-oriented books are written.[1–4] The availability of different fabrication technologies as well as the accessibility of the advanced techniques and processes (which can be utilized depending on the equipment and infrastructure available) allow the author to emphasize the basic concepts and processes. The references are provided for the specialized fabrication processes and materials. Different sources should be used, taking into account the distinct fabrication infrastructure.

The importance of this chapter is the fact that MEMS must be fabricated. Micromachining, as a core MEMS fabrication technology, can be used to produce complex systems, devices, and structures that are micrometers in size. Microelectromechanical systems integrate motion and stationary microstructures (electromechanical components), sensors, actuators, radiating energy devices, energy sources, and microelectronics. These MEMS can be fabricated utilizing different microfabrication technologies, e.g., micromachining (*bulk* and *surface*) or high-aspect-ratio. While the microelectronic components are fabricated using CMOS and bi-CMOS process sequences, the microelectromechanical components can be made using micromachining processes that selectively etch away parts of the silicon wafer or other materials and add (deposit) new structural and sacrificial layers of different materials to form mechanical, electromechanical, and electrooptomechanical devices. The fabrication technologies used in MEMS were developed and reported in books.[1,2] This chapter documents core micromachining and microelectronics techniques and processes. In addition, microfabrication processes and techniques to make microstructures and microtransducers are reported in Chapter 10.

The basic technologies in MEMS fabrication are CMOS and bi-CMOS (to fabricate ICs), micromachining and high-aspect-ratio. Micromachining or high-aspect-ratio processes have been used to fabricate motion and radiating-energy microscale structures and devices.

One of the main goals is to integrate microelectronics with micromachined electromechanical devices in order to produce functional MEMS. To guarantee high-performance, affordability, reliability, and manufacturability, well-developed CMOS-based batch-fabrication processes have been modified and enhanced.[1-4]

In addition, assembling and packaging must be automated. That is, auto- or self-alignment, self-assembly, and other processes should be developed. The MEMS must be protected from mechanical damage and contamination. Thus, MEMS are packaged to protect them from harsh environments, prevent mechanical damage, as well as minimize stresses and vibrations, contamination, electromagnetic interference, etc. Therefore, MEMS are usually sealed. It is impossible to specify a generic MEMS packaging solution. Through input-output interconnects (power, communication, and processing buses) and networking, one delivers the power required, feeds control (command), tests signals, receives the output signals and data, interfaces, makes networks, etc. Robust packages must be designed to minimize electromagnetic interference, noise, vibration, and other undesirable effects. For example, heat generated by MEMS must be dissipated, and the thermal expansion problem must be solved. Conventional MEMS packages are usually ceramic and plastic. In ceramic packages, the die is bonded to a ceramic base, which includes a metal frame and pins for making electric outside connections. Plastic packages are connected in the similar way. However, the plastic package can be molded around the microdevice. Wear tolerance and electromagnetic and thermal insulation, among other problems, are very challenging issues. Different fabrication techniques, processes, and materials must be applied to attain the desired performance, reliability, and cost. Microsystems can be coated directly by thin films of silicon dioxide or silicon nitride, which are deposited using plasma-enhanced chemical vapor deposition. It is possible to deposit (at 700°C to 900°C) carbon (diamond) thin films, which have superior wear capabilities, excellent electric insulation, and superior thermal characteristics. Microelectromechanical systems are connected and interfaced (networked) with other systems and components such as kinematics, communication ports, power buses, sensors, etc.

Bulk and surface micromachining, as well as high-aspect-ratio technologies (LIGA and LIGA-like), are the most developed fabrication methods. Silicon is the primary substrate material used by the microelectronic industry. A single crystal ingot (a solid cylinder hundreds of millimeters in diameter and length) of very high purity silicon is grown, sawed to the desired thickness, and polished using chemical and mechanical polishing techniques. Electromagnetic and mechanical wafer properties depend on the orientation of the crystal growth, concentration, and type of doped impurities. Depending on the silicon substrate, CMOS and bi-CMOS processes are used to manufacture ICs, and the processes are classified as *n-well*, *p-well*, or *twin-well*. The major steps for IC fabrication are diffusion, oxidation, polysilicon gate formations, photolithography, masking, etching, metallization, wire bonding, etc. In addition to conventional IC processing, additional processes are used to make MEMS. There are a number of basic surface silicon micromachining techniques that can be used in order to deposit and pattern thin films (deposited on a silicon wafer) as well as to shape the silicon wafer itself, forming a set of basic microstructures. The basic steps of the silicon micromachining are as follows:

- Lithography
- Deposition of thin films and materials (electroplating, chemical vapor deposition, plasma-enhanced chemical vapor deposition, evaporation, sputtering, spraying, screen printing, etc.)
- Removal of material (patterning) by wet or dry techniques
- Etching (plasma etching, reactive ion etching, laser etching, etc.)

- Doping
- Bonding (fusion, anodic, and other)
- Planarization

4.1.1 Photolithography

To fabricate motion and radiating-energy microstructures and microdevices, the CMOS technology has been modified by developing new processes and applying novel materials. High-resolution photolithography is a technology that is applied to define two-(planar) and three-dimensional shapes (geometry). For example, the geometries of microtransducers and their components (stator, rotor, bearing, coils, etc.) are defined photographically. First, a mask is produced on a glass plate. The silicon wafer is then coated with a polymer that is sensitive to ultraviolet light. This photoresistive layer is called photoresist. Ultraviolet light is shone through the mask onto the photoresist. The positive photoresist becomes softened, and the exposed layer can be removed. There are two types of photoresist—positive and negative. Where the ultraviolet light strikes the positive photoresist, it weakens the polymer. Hence, when the image is developed, the photoresist is rinsed where the light struck it. A high-resolution positive image is needed, and different photolithography processes are developed based on the basic lithography fundamentals. In contrast, if the ultraviolet light strikes negative photoresist, it strengthens the polymer. Therefore, a negative image of the mask results. Different chemical processes are involved to remove the oxide where it is exposed through the openings in the photoresist. When the photoresist is removed, the patterned oxide appears. Alternatively, electron beam lithography can be used.

Photolithography requires the design of photolithography masks, and computer-aided design (CAD) software is available and widely applied to support the photolithography. The photolithography process and a photolithography system are illustrated in Figure 4.1.

Deep-UV lithography processes were developed to decrease the feature sizes of microstructures to 0.1 μm. Different exposure wavelengths λ are used (for example, λ can be 435, 365, 248, 200, 150, 90, and 65 nm). Using the Rayleigh model for image resolution, one finds the expressions for image resolution i_R and depth of focus d_F to be given by the

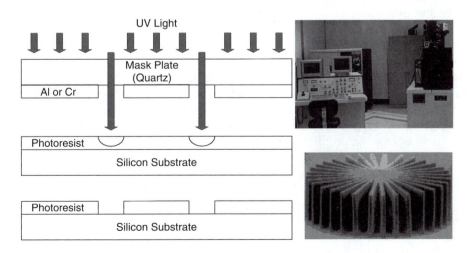

FIGURE 4.1 (See color insert following page 332.)
Photolithography process, computerized photolithography system, and fabricated microstructure.

following formulas:

$$i_R = k_i \frac{\lambda}{N_A}, \quad d_F = k_d \frac{\lambda}{N_A^2}$$

where k_i and k_d are the lithographic process constants, λ is the exposure wavelength, and N_A is the numerical aperture coefficient (for high-numerical aperture, we have $N_A = 0.5\text{--}0.6$).

These formulas indicate that to pattern microstructures with the decreased feature size, the photoresist exposure wavelengths must be decreased and numerical aperture coefficient should be increased. The so-called g- and i-line IBM lithography processes (with wavelengths of 435 nm and 365 nm, respectively) allow one to attain 0.35-μm features. The deep-ultraviolet-light sources (mercury source or excimer lasers) with 248-nm wavelength enable the industry to achieve 0.25-μm resolution.

The changes to short exposure wavelength present both challenges and new highly desired possibilities. Specifically, transparent optical materials with weak absorption as well as organic transparent polymers can serve as the single-layer photoresists. High-purity synthetic fused silica and crystalline calcium fluoride are the practical choices. The ideal optical material should be fully transparent and must remain unaffected after billions of pulses. Fused silica has the absorption coefficient in the range from 0.005 to 0.10 cm^{-1}. Another challenge in the application of 200-nm lithography is the development of robust photoresist processes. The resin photoresists, which are widely used for 365-nm and 248-nm lithography processes, are novolac and polyhydroxystyrene. These photoresists have absorption depths of 30 to 50 nm at 200-nm wavelength. Therefore, resins cannot be used as the single-layer photoresists at $\lambda = 200$ nm. Methacrylates are semitransparent at 200 nm, and these polymers can serve as the 200-nm lithography single-layer photoresists. Acid-catalyzed conversion of t-butyl methacrylate into methacrylic acid provides the chemical underpinning for several versions of these photoresists. This brief description provides the reader with an overview. It is necessary to deeply study different processes in order to fabricate MEMS. Of course, these processes depend on the fabrication facilities available.

Lithography is the process used in ICs and MEMS fabrication to create the patterns defining the ICs', microstructures', and microdevices' features. Different lithography processes are photolithography (as was reported), screen printing, electron-beam lithography, x-ray lithography (high-aspect-ratio technology), etc. Extensions of the currently widely used optical lithography using shorter-wavelength radiation result in the 100-nm minimum feature size. For microstructures and ICs that require less than 100-nm resolution features, the next-generation lithography techniques must be applied. For example, photons and charged particles, which have short wavelengths, can be used. In particular, high-throughput electron-beam lithography is under development. Unlike optical lithography, electron beams are not diffraction-limited, and thus, the ultimate resolution attainable is expanded. Electron-beam lithography has evolved from the early scanning-electron-microscope-type Gaussian beam systems (which expose ICs patterns one pixel at a time) to the massive parallel projection of pixels in electron-projection lithography. This allows one to attain millions of pixels per shot. Figure 4.2 documents a scanning-electron-beam lithography system (IBM VS-2A). The digital pattern generator is based on commercial high-performance RISC processors. The system is capable of creating large area patterns. The system is computerized, and the hardware-software codesign should be accomplished (computer-aided-design and control software must be integrated within the lithography systems).

For last 30 years, optical lithography has been widely used, and the IC features were decreased from 5 μm to 0.1 μm (100 nm). Electron-beam and x-ray lithographies are considered as an alternative solution. The wafer through put with electron-beam lithography is

FIGURE 4.2
Scanning-electron-beam lithography system.

slow, and the electron-beam lithography is considered as complementary to optical lithography. However, optical lithography depends on electron-beam lithography to generate the masks. Because of its intrinsic high resolution, electron-beam lithography is the primary process for 0.1-μm (or smaller resolution) microstructures.

It has been emphasized that the photolithography processes are embedded in IC and MEMS fabrication. The computerized (PC-controlled) integrated stepper platforms (or the so-called step-and-repeat projection aligners) transfer the image of the microstructure or ICs from a master photomask image to a specified area on the wafer surface. The substrate is then moved (stepped), and the image can be exposed once again to another area of the wafer. This process is repeated until the entire wafer is exposed. Different steppers are applied with different capabilities and features, e.g., 65-, 90-, 157-nm, or other application-specific lithography processes, contact or noncontact lithography, single- or multilayer, different wafers (materials, size, 50-, 75-, 100-, 150-, 200-, or 300-mm diameter, thickness, etc.), single- or dual-side alignment (off-axis microscope capable of viewing images on the back of the wafer while exposing features on the front side), prealignment, imaging, versatility, magnetic levitation, etc. The computerized stepper system is illustrated in Figure 4.3.

In photolithography, mask aligners are used to transfer a pattern from a mask to a photoresist on the substrate. In particular, the Karl Suss MJB-3 and MA-6 mask aligners (for 3- and 6-inch wafers, respectively) are commonly used to expose photoresist-coated substrates to ultraviolet light through photo masks (see Figure 4.4). It was emphasized that different photoresists are sensitive to light at different wavelengths. It is important to select the mask aligner with a wavelength optimized with respect to the photoresists used.

The GCA Corporation manufactures a line of photolithography stepper cameras (GCA 6000, 8000, 8500, and others) controlled by computers with different operating systems. The GCA Mann pattern generators, which produce the patterns for ICs and microstructures, consist of a computerized controller, electronics, and cameras. Figure 4.5 illustrates the GCA Mann 3000 photomask pattern generator and GCA 3696 stepper (step-and-repeat) camera, which allow one to achieve micron-resolution features.

Some examples are reported below. Figure 4.6 documents the SEM images of fabricated (1) 5-nm-diameter gold "islands," (2) Cu-Co wires (Co may be magnetized if needed after the fabrication process), (3) 75-nm-wide deposited gold forms current loops (to make antennas or windings), (4) electroplated 100-nm-wide Au for the surface acoustic wave device, and (5) suspended structure for the microdevice with cavity. These structures were fabricated utilizing additional photolithography steps, particularly deposition and etching.

FIGURE 4.3
Computerized stepper system.

4.1.2 Etching

Different microelectromechanical motion devices and microstructures can be designed, and silicon wafers with different crystal orientations can be used to fabricate MEMS. Reactive ion etching (dry-etching process) is commonly applied. Ions are accelerated toward the material to be etched, and the etching reaction is enhanced in the direction of ion traveling. Deep trenches and pits of desired shapes can be etched in a variety of materials including silicon, silicon oxide, and silicon nitride. A combination of dry and wet etching can be sequentially integrated in the fabrication processes to make the desired MEMS.

Metal and alloy thin films can be patterned using the lift-off stenciling technique. As an example, let us document the following simple procedure. A thin film of the assisting material (silicon oxide) is deposited first, and a layer of photoresist is deposited over and patterned. The silicon oxide is then etched to undercut the photoresist. The metal (or any other material) thin film is then deposited on the silicon wafer through the evaporation process. The metal pattern is stenciled through the gaps in the photoresist, which is then removed, lifting off the unwanted metal. The sacrificial layer is then stripped off, leaving the desired metal film pattern.

FIGURE 4.4
Karl Suss MJB-3 and MA-6 mask aligners.

FIGURE 4.5
GCA Mann 3000 photomask pattern generator and GCA 3696 stepper camera.

Isotropic and *anisotropic* wet etching, as well as concentration-dependent etching, are used in bulk silicon micromachining because the microstructures are formed by etching away the bulk of the silicon wafer.

Surface micromachining usually forms the structure in layers of thin films on the surface of the silicon wafer or other substrate. Hence, the surface micromachining process uses thin films of at least two different materials, e.g., structural (usually polysilicon, metals, alloys, etc.) and sacrificial (silicon oxide) material layers. Sacrificial (silicon oxide is deposited on the wafer surface) and structural layers are deposited. Then, the sacrificial material is etched away to release the structure. A variety of different complex motion microstructures with different geometries have been fabricated using the surface micromachining technology.[1,2] Different etching processes and materials are covered in detail in Section 4.2.

Different etching systems are widely used. As an example, the xenon difluoride (gas) etching system is illustrated in Figure 4.7.

4.1.3   Bonding

Micromachined silicon wafers must be bonded together. The anodic (electrostatic) bonding technique is used to bond silicon wafer and glass substrate. The silicon wafer and glass substrate are attached and heated, and an electric field is applied across the joint.

FIGURE 4.6
5-nm gold islands, copper–cobalt wires, gold current loops (0.5 μm and 1 μm diameter), gold antenna for the surface acoustic wave device, and microdevice.

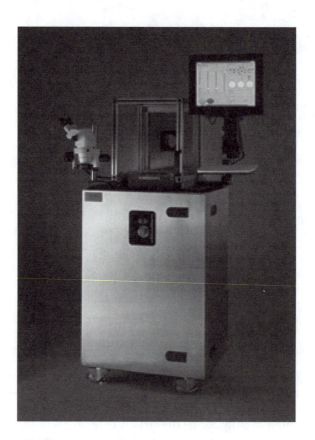

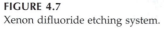

FIGURE 4.7
Xenon difluoride etching system.

This results in extremely strong bonds between the silicon wafer and glass substrate. In contrast, direct silicon bonding is based on applying pressure to bond the silicon wafer and glass substrate. It must be emphasized that to guarantee strong bonds, the silicon wafer and glass substrate surfaces must be flat and clean.

The MEMCAD™ software (current version is 4.6), developed by Microcosm, is widely used to design, model, simulate, characterize, and package MEMS. Using the built-in Microcosm Catapult™ layout editor, augmented with a materials database and components library, three-dimensional solid models of motion microstructures can be developed. Furthermore, customizable packaging is fully supported.

4.1.4   Introduction to MEMS Fabrication and Web Site Resources

The basic fabrication techniques, processes, and materials are covered in this book. The major technologies (micromachining, which is based on the CMOS technology and LIGA) used to fabricate MEMS were emphasized. In particular, surface micromachining is the well-developed, affordable, and high-yield technology that likely will be the dominant method. A simplified surface micromachining process is illustrated in Figure 4.8.

Figure 4.8 demonstrates the major sequential steps of the simplified fabrication of the microstructure through surface micromachining. The description of processes, materials used,

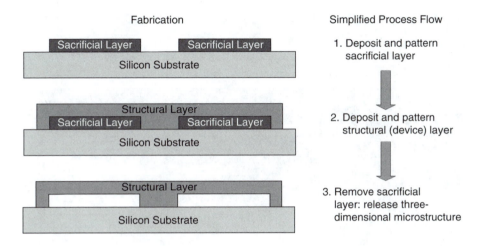

FIGURE 4.8
Simplified surface micromachining process.

and techniques applied are covered in detail in this chapter and Chapter 10. The attractive feature of surface micromachining is the compatibility and compliance of three-dimensional microstructure and IC fabrication. In fact, in addition to the fabrication of microstructures and microdevices, these structures/devices must be controlled. Therefore, the ICs are the important subsystems (components) of MEMS.

The following process in the fabrication of MEMS, along with links to the manufacturer sites, is reported on the MEMS Clearinghouse Web site (see http://mems.isi.edu/): bonding, cleaning, deposition, doping, etching, LIGA, lift-off, lithography, mask making, metrology, packaging, polishing, etc. Throughout this book, the author cannot assume responsibility for the validity and fitness of the techniques, processes, sequential steps, data, chemicals, and materials or for the consequences of their use. Furthermore, the microelectronic (semiconductor) and MEMS manufacturers have different proprietary techniques, processes, and materials. However, for introduction and educational purposes, the references to the MEMS Clearinghouse and university Web sites are quite appropriate. By clicking the corresponding process, the information appears on the screen (see Figure 4.9). In particular, using the Web site http://fab.mems-exchange.org/catalog/ and clicking on deposition, we get a screen like the one shown in Figure 4.9.

Different computerized microscopes are used to attain the visualization, imaging, and examining features. For example, scanning electron microscopes (e.g., Hitachi S-2400, 3000, and 3500 series) are widely used. The Hitachi S-2400 is the general purpose, 25-kV, diffusion-pumped, tungsten-filament-based scanning electron microscope. This scanning electron microscope allows × 300,000 magnification. After the sample loading, high voltage (from 4 to 25 kV) is applied. Focusing is performed manually or automatically, and images can be displayed and zoomed. The 4-nm resolution is achieved, which can be suitable for ICs and microstructures with features on the scale of tens of nanometers. For nanostructures, other microscopes are used. Figure 4.10 documents the Hitachi S-2400 scanning electron microscope.

Different software packages and tools are widely used. For example, L-Edit is widely used in mask making. L-Edit is a computer-aided design tool used to generate photolithography masks for integrated circuits and MEMS devices. Different shapes (rectangles, circles, tori, arcs, polygons, etc.) can be drawn on the layout. Distinct drawing objects are available from the Drawing toolbar. Each object has to belong to a specific mask layer. Each photolithography mask has only one layer. Before drawing any object, the designer must choose the appropriate layer on the Layer Palette in the L-Edit window. One utilizes the

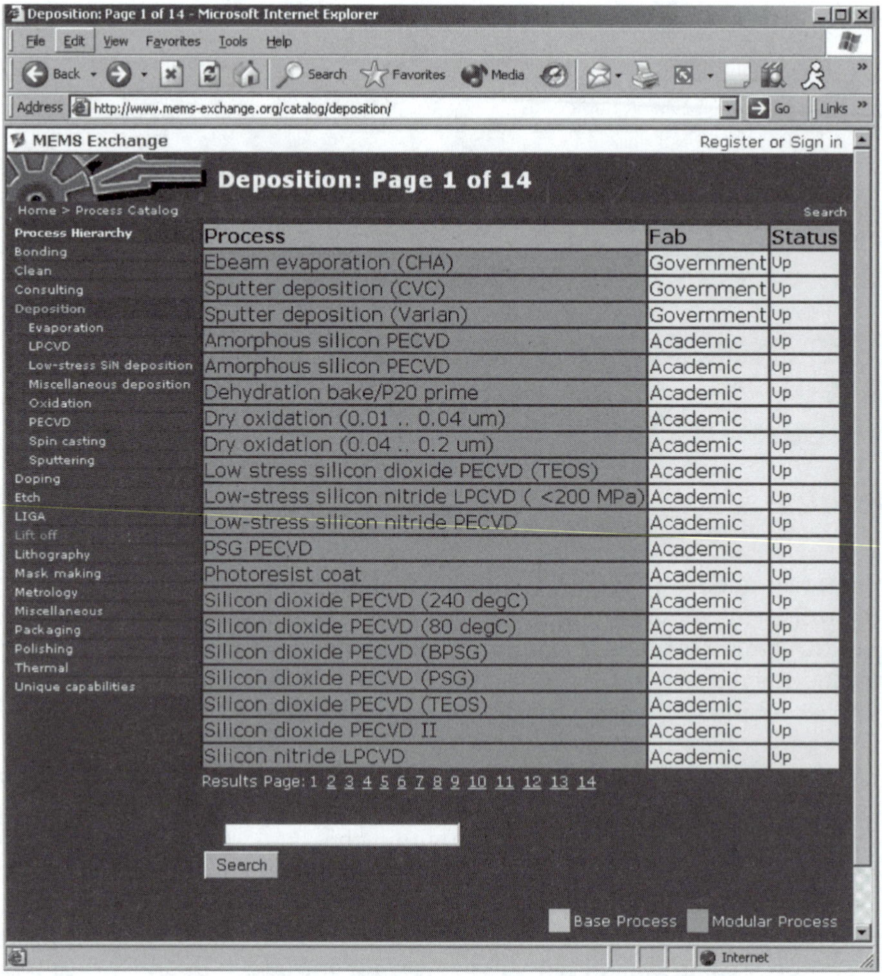

FIGURE 4.9
Anodic bonding description in the MEMS Clearinghouse Web site (http://fab.mems-exchange.org/catalog/).

FIGURE 4.10
Hitachi S-2400 scanning electron microscope.

cells (cells are the basic building blocks of the layout). Cells can contain other cells or just a collection of objects. Instances are references to other cells made on a specific cell. If the reference cell changes, the instances of that cell will also change. Instances are very useful when a pattern repeats throughout the layout. Design rules specify the requirements for MEMS, microdevices, and microstructures. These rules are based on mask misalignment, process variations, materials, etc. The design rules specify the feature size, line width, object spacing and enclosure, etc. To create a mask file, one specifies the explicit information for the mask manufacturer, and the compatibility issues should be accounted for.

4.2 Microfabrication and Micromachining of ICs, Microstructures, and Microdevices

Complementary metal oxide semiconductor (CMOS), high-aspect-ratio, and surface micromachining technologies are key factors for development, implementation, and commercialization of ICs and MEMS.[1,2] For MEMS, micromachining means fabrication of microscale structures and devices controlled by ICs. In general, micromachining has emerged as the extension of CMOS technology. The low-cost, high-yield CMOS technology enables the fabrication of millions of transistors and capacitors on a single chip, and currently, the minimal feature sizing is within the range of tens of nanometers. High-performance signal processing, signal conditioning, interfacing, and control are performed by ICs. Microstructures and microdevices, augmented with ICs, comprise MEMS, which have been widely used in actuator, sensor, and communication applications.

Microfabrication technologies can be categorized as bulk, surface, and high-aspect-ratio (LIGA and LIGA-like) micromachining. The major processes in the fabrication of ICs and microstructures/microdevices are oxidation, diffusion, deposition, patterning, lithography, etching, metallization, planarization, assembling, and packaging.

Complementary metal oxide semiconductors is a well-developed technology for fabrication of ICs and thin films. Thin-film fabrication processes were developed and used for polysilicon, silicon dioxide, silicon nitride, and other materials, e.g., metals, alloys, insulators, etc. For ICs, these thin films are used to build the active and passive circuitry components, as well as attain interconnection. Doping modifies the properties of the media. It was documented that the lithography processes are used to transfer the pattern from the mask (which defines the surface topography and geometry for ICs and microstructures) to the film surface, which is then selectively etched away to remove unwanted thin films, media, and regions to complete the pattern transfer. The number of masks depends on the design complexity, fabrication technology, processes applied, and the desired geometry for the ICs and microstructures. After testing, the wafers are diced, and chips are encapsulated (packaged) as final ICs, microdevices, or MEMS.

The description of the silicon-based fabrication technologies and processes is given below. Crystal growth and slicing are the processes that produce the silicon with the desired (specified) chemical, electromechanical, and thermal characteristics and properties (dimension, orientation, bow, taper, edge contour, surface flatness and scratches, minority carrier lifetime, doping type and concentration, heavy-metal impurity content, electrical resistivity, Young's module, elasticity, Poisson's ratio, stress, thermal conductivity, thermal expansion, etc.). For silicon, the following data is useful:

- Atomic weight is 28.
- Density is 2.33 g/cm^3.

- Melting point is 1415°C.
- Specific heat is 0.7 J/g · K.
- Thermal conductivity is 1.6 W/cm · K at 300°K.
- Coefficient of linear thermal expansion is 0.0000026.
- Intrinsic resistivity is 230,000 ohm · cm.
- DC dielectric constant is 11.9.

To design and fabricate MEMS, the material compatibility issues must be addressed and analyzed. For example, the thermal conductivity and thermal expansion phenomena must be studied, and the appropriate materials with the closest possible coefficients of thermal conductivity and expansion must be chosen. However, the desired electromagnetic and mechanical properties (permeability, resistivity, strength, elasticity, etc.) must also be attained (see Chapter 8 for details). In fact, electromagnetic and mechanical characteristics significantly influence MEMS performance. Therefore, the trade-offs must be examined with the ultimate goal to optimize the overall performance of MEMS. Thus, it is evident that even the simplest microstructures (silicon with deposited thin films) must be the topic of mechanical, electromagnetic, and thermal analysis. Furthermore, though the designer tries to match the mechanical and thermal phenomena and effects and attempts to use magnetic materials with good electromagnetic characteristics, the feasibility and affordability of the fabrication processes are also studied. The mechanical properties of different materials are documented in Table 4.1.

The mechanical properties of the following bulk and thin-film materials, which are commonly used in ICs and MEMS fabrication, are given on the MEMS Clearinghouse Web site. In particular, one can consult the Material Database at http://www.mems-net.org/material/. The materials are aluminum (Al), amorphous carbon (a-C:H), amorphous hydrogenated silicon (a-Si:H), amorphous silicon (a-Si), amorphous silicon dioxide (a-SiO$_2$), antimony (Sb), arsenic (As), barium titanate (BaTiO$_3$), beryllium oxide (BeO), bismuth (Bi), boron (B), boron carbide (B$_4$C), boron nitride (BN), cadmium (Cd), carbon (C), carbon nitride (CN$_x$), chromium boride (CrB$_2$), chromium carbide (Cr$_3$C$_2$), chromium (Cr), chromium nitride, chromium oxide (Cr$_2$O$_3$), cobalt (Co), copper (Cu), copper molybdenum (CuMo), diamond (C), diamond-like carbon (DLC), fullerite (C$_{60}$ and C$_{70}$), gallium arsenate (GaAs), gallium (Ga), germanium (Ge), glass (7059 and SiO$_2$), gold (Au), graphite (C), indium (In), iron (Fe), lead (Pb), lead zirconate titanate (PZT), lithium (Li), magnesium (Mg), molybdenum (Mo), molybdenum silicide (MoSi$_2$), mullite (3Al$_2$O$_3$2SiO$_2$), nickel (Ni), niobium oxide (Nb$_2$O$_5$), nitride coatings, oxide-induced layer in polysilicon, palladium (Pd), phosphorous bronze metal, phosphorous (P),

TABLE 4.1

Mechanical Properties of Materials

Materials	Strength (10^9 N/m²)	Hardness (kg/mm²)	Young's Modulus (10^9 Pa)	Density (g/cm³)	Thermal Conductivity (W/cm · K)	Thermal Expansion (10^{-6} 1/K)
Si	7	850	190	2.3	1.6	2.6
SiC	21	2500	710	3.1	3.5	3.3
Diamond	54	6700	980	3.5	19	1
SiO$_2$	8.3	800	70	2.5	0.015	0.54
Si$_3$N$_4$	15	3500	380	3.1	0.2	0.8
Fe	13	420	200	7.8	0.8	12
Al	0.2	135	70	2.8	2.4	25
Mo	2	280	340	10	1.4	5.2
W	3.9	500	400	19	1.8	4.6

Dupont polyimide (PI 2611D), piezoelectric sheet, piezoresistors (diffused), plastic, platinum (Pt), polyimide, polyimide hinges, polysilicon, poly vinylidene fluoride, porosilicon, sapphire, silicon carbide (SiC), silicon dioxide (SiO_2), silicon nitride hydrogen (SiN_xH_y), silicon nitride (Si_xH_y), silicon oxide (SiO_x), silicon (Si), sillimanite ($Al_2O_3 SiO_2$), silver (Ag), sodium silicate ($Na_2O : SiO_2$), stainless steel, tellurium (Te), thallium (Tl), tin (Sn), titanium aluminum (Ti_3Al and $TiAl$), titanium boride (TiB_2), titanium carbide (TiC), titanium nickel (NiTi), titanium nitride (TiN and TiN_x), titanium oxide (TiO_2), titanium (Ti), tungsten carbide (WC), tungsten silicide (WSi_2), tungsten (W), zinc oxide (ZnO), zinc (Zn), zirconium oxide (ZrO_2), and zircon ($SiO_2 ZrO_2$).

Emphasizing the mechanical and thermal matching (compliance) and the need for optimization of electromagnetic properties (covered in Chapter 10), we describe the major fabrication processes.

4.2.1 Oxidation

Oxidation of silicon wafers is used for passivation of the silicon surface (the formation of a chemically, electronically, and electromechanically stable surface), diffusion, ion implantation, making dielectric films, and interfacing substrate and other materials (for example, chemical material and biosensors).

Silicon, exposed to the air at 25°C, is covered by a 20-Å (2-nm) layer of silicon oxide. Thicker silicon oxide (SiO_2) layers can be grown at elevated temperatures in dry or wet oxygen environments. The wet and dry reactions are as follows:

$$Si + 2H_2O \rightarrow SiO_2 + H_2$$

$$Si + O_2 \rightarrow SiO_2$$

The oxide growth rate is expressed by the following formula:

$$\frac{dx}{dt} = \frac{1}{N} k_o c_o$$

where N is the number of molecules of oxidant per unit volume of oxide, and k_o and c_o are the oxidation rate constant and concentration of oxidant, which are nonlinear functions of the oxide thickness, temperature, oxidant used, crystal orientation, diffusivity, pressure, etc.

If the temperature is constant, the relationship between the thickness of oxide and time is parabolic. The rate of growth is a nonlinear function of the oxygen pressure and the crystal orientation.

During the oxidation process, the dopant concentration profile in the Si—SiO_2 structure is redistributed due to nonuniformity of the equilibrium concentration of the impurity in silicon and silicon dioxide, and the impurity segregation coefficient (the ratio of the equilibrium concentration of the impurity in Si to that in SiO_2) is used. For boron-doped silicon, the boron in Si near the interface is depleted, while for phosphorous, there is a buildup at the interface. High concentrations of the standard dopant atoms will increase the oxidation rate within a certain temperature range.[1-4]

4.2.2 Photolithography

Photolithography (lithography) is the process used to transfer the mask pattern (desired pattern, surface topography, and geometry) to a layer of radiation- or light-sensitive material (photoresist), which is used to transfer the pattern to the films or substrates through etching processes.

Ultraviolet and radiation (optical, x-ray, electron beam, and ion beam) lithography processes are used in fabrication of ICs, motion, and radiating-energy microstructures and microdevices. The major steps in lithography are the fabrication of masks (pattern/topography generation) and transfer of the pattern to the wafer (see Figure 4.1).

The description of photolithography was documented in Section 4.1. Positive and negative photoresists are applied. The pattern on the positive photoresist after development is the same as that on the mask, while on the negative photoresist it is reversed. Important photoresist characteristics are resolution, sensitivity, etch resistance, thermal stability, adhesion, viscosity, flash point, toxicity rate, etc. The photoresist processing includes dehydration baking and priming, coating, soft baking, exposure, development, inspection, post bake (UV hardening), etc. Then, the specified pattern is transferred to the wafer through etching, after which the photoresist is stripped by strong acid solutions (H_2SO_4), acid–oxidant solutions ($H_2SO_4 + Cr_2O_3$), organic solvents, alkaline strippers, oxygen plasma, or gaseous chemical reactants. Using wet and dry stripping, the photoresist must be removed without damaging silicon structures. The photolithographic cycle is completed.

In surface micromachining, an alternative solution to etching (in order to transfer the patterns from photoresist to thin films) is the lift-off process, which is an additive process. In lift-off, the photoresist is first patterned, the thin film to be patterned is deposited, and then the photoresist is dissolved. The photoresist acts as a sacrificial material under thin-film regions to be removed.

4.2.3   Etching

Etching is used to delineate patterns, remove surface damage, clean the surface, remove contaminations, and fabricate two-(planar) and three-dimensional microstructures. Wet chemical etching and dry etching (sputtering, ion beam milling, reactive ion etching, plasma etching, etc.) are used to etch semiconductors, conductors (metals), alloys, and insulators (silicon oxide, silicon nitride, etc.). Different etchants are used in micromachining and microelectronics.

Wet and dry etchants, commonly used in micromachining and ICs, are listed in Table 4.2 and Table 4.3.[3,4]

A large number of dry-etching processes such as physical etching (sputtering and ion milling), chemical plasma etching, and the combinations of physical and chemical etching (reactive ion etching and reactive ion-beam etching) are used, and recipes are available. Dry-etching processes are based on plasmas. Plasmas are fully or partially ionized gas molecules and neutral atoms/molecules sustained by the applied electromagnetic field. Usually, less than 0.1% of the gas is ionized, and the concentration of electrons is much lower than the concentration of gas molecules. The electron temperature is greater than 10,000°K, although the gas thermal temperature is within the range from 50 to 100°C. Plasma etching processes involve highly reactive particles in a relatively cold medium. Adjusting process parameters controls the particles energies and gas temperature. The gas and flow rate, excitation power, frequency, reactor configuration, and pumping determine the electron density and distribution, gas density, and residence time defining the reactivity. These and wafer parameters (temperature and surface potential) define the surface interaction and etching characteristics.[1–10]

The most important properties of dry-etching processes are feature size control, wall profile, selectivity (the ratio of etch rates of the layer/material to be etched and the layer/material to be kept), controllability, in-wafer and interwafer uniformities, defects, impurity, throughput, radiation damage to dielectrics, etc.

TABLE 4.2

Wet Etchants

Materials	Liquid Etchant and Etch Rate
Polysilicon	6 ml HF, 100 ml HNO_3, 40 ml H_2O, 8000 Å/min, smooth edges
	1 ml HF, 26 ml HNO_3, 33 ml CH_3COOH, 1500 Å/min
Phosphorous-doped	Buffered hydrofluoric acid (BHF)
silicon dioxide (PSG)	28 ml HF, 170 ml H_2O, and 113 g NH_4F, 5000 Å/min
	1 ml BHF and 7 ml H_2O, 800 Å/min
Silicon nitride (Si_3N_4)	Hydrofluoric acid (HF)
	140 Å/min CVD at 1100°C
	750 Å/min CVD at 900°C
	1000 Å/min CVD at 800°C
Silicon dioxide (SiO_2)	Buffered hydrofluoric acid (BHF)
	28 ml HF, 170 ml H_2O, and 113 g NH_4F, 1000–2500 Å/min
	1 ml BHF and 7 ml H_2O, 700–900 Å/min
Aluminum (Al)	4 ml H_3PO_4, 1 ml HNO_3, 4 ml CH_3COOH, 1 ml H_2O, 350 Å/min
	16–19 ml H_3PO_4, 1 ml HNO_3, 0–4 ml H_2O, 1500–2400 Å/min
Gold (Au)	3 ml HCl, 1 ml HNO_3, 25–50 μm/min
	4 g KI, 1 g I_2, 40 ml H_2O, 0.5–1 μm/min
Chromium (Cr)	1 ml HCl, 1 ml glycerine, 800 Å/min (need depassivation)
	1 ml HCl, 9 ml saturated $CeSO_4$ solution, 800 Å/min (need depassivation)
	1 ml (1 g NaOH in 2 ml H_2O), 3 ml (1 g $K_3Fe[CN]_6$ in 3 ml H_2O), 250–100 Å/min (photoresist mask)
Tungsten (W)	34 g KH_2PO_4, 13.4 g KOH, 33 g $K_3Fe(CN)_6$, and H_2O to make 1 liter, 1600 Å/min (photoresist mask)

4.2.4 Doping

Doping processes are used to selectively dope the substrate to produce either n- or p-type regions. These doped regions are used to fabricate passive and active circuitry components, form etch-stop layers (a very important feature in buck and surface micromachining), and produce conductive silicon-based micromechanical devices. Diffusion is achieved by placing wafers in a high-temperature furnace and passing a carrier gas that contains the desired dopant. For silicon, boron is the most common p-type dopant (acceptors), and arsenic and phosphorous are n-type dopants (donors). The dopant sources may be solid, liquid, or gaseous. Nitrogen is usually used as the carrier gas. Two major steps in diffusion are predeposition

TABLE 4.3

Dry Etchants

Materials	Etchant (Gas) and Etch Rate
Silicon dioxide (SiO_2) Phosphorous-doped silicon dioxide (PSG)	$CF_4 + H_2$, C_2F_6, C_3F_8, or CHF_3, 500–800 Å/min
Silicon	$SF_6 + Cl_2$, 1000–5000 Å/min
(single-crystal and polycrystalline)	CF_4, CF_4O_2, CF_3Cl, SF_6Cl, Cl_2+H_2, $C_2ClF_5O_2$, SF_6O_2, SiF_4O_2, NF_3, $C_2Cl_3F_5$, or CCl_4He
Silicon nitride (Si_3N_4)	CF_4O_2, CF_4+H_2, C_2F_6, or C_3F_8, SF_6He
Polysilicon	Cl_2, 500–900 Å/min
Aluminum (Al)	BCl_3, CCl_4, $SiCl_4$, BCl_3Cl_2, CCl_4Cl_2, or $SiCl_4Cl_2$
Gold (Au)	$C_2Cl_2F_4$ or Cl_2
Tungsten (W)	CF_4, CF_4O_2, C_2F_6, or SF_6
Al, Al-Si, Al-Cu	$BCl_3 + Cl_2$, 500 Å/min

(impurity atoms are transported from the source to the wafer surface and diffused into the wafer; the number of atoms that enter the wafer surface is limited by the solid solubility of the dopant in the wafer) and drive-in (deposited wafer is heated in a diffusion furnace with an oxidizing or inert gas to redistribute the dopant in the wafer to reach a desired doping depth and uniformity). After deposition, the wafer has a thin, highly doped oxide layer on the silicon, and this oxide layer is removed by hydrofluoric acid.

The diffusion theory is based on Fick's first and second laws. Assuming constant diffusion coefficients at the process temperature, we have

$$J = -D \frac{\partial C(t, x)}{\partial x}$$

$$\frac{\partial C(t, z)}{\partial t} = D \frac{\partial^2 C(t, z)}{\partial z^2} = D \nabla^2 C$$

where J is the net flux, D is the diffusion coefficient, C is the impurity concentration, and z is the depth.

For predeposition diffusion, the surface concentration is constant, and the diffusion profile is described as a complementary error function. Using the following boundary conditions,

$$C(0, z) = 0, \ C(t, 0) = C_0, \quad \text{and} \quad C(t, 4) = 0$$

we have the expression for the impurity concentration:

$$C(t, z) = C_0 erfc^{\frac{z}{2\sqrt{Dt}}}, \quad t > 0$$

For drive-in, the total dopant in the wafer is constant, and the diffusion profile obeys a Gaussian function. These descriptions are valid for low doping concentrations.

For high doping concentration, the diffusion coefficient is a nonlinear function of concentration and other parameters. Using Ohm's law, we have

$$J = -D \frac{\partial C}{\partial z} + \mu C E, \quad E = -\eta \frac{kT}{qC} \frac{dC}{dz}$$

where μ is the mobility, $\mu = \frac{q}{kT}$; E is the electric field; and η is the screening factor that varies from 0 to 1. Thus, one has $J = -D(1+\eta)\frac{dC}{dz}$.

Complex mathematical models, given in the form of nonlinear partial differential equations, were derived to describe the diffusion profiles for boron, arsenic, phosphorous, and antimony at high concentrations. Silicon dioxide is usually used as the diffusion mask for silicon wafers. Ion implantation, as a well-developed technique for introducing impurity atoms into a wafer below the surface by bombarding it with a beam of energetic impurity ions, is widely used.

4.2.5   Metallization

Metallization is the formation of metal films for interconnections, ohmic contacts, rectifying metal-semiconductor contacts, and protection (e.g., attenuation of electromagnetic interference and radiation). Metal thin films can be deposited on the surface by vacuum evaporation (deposition of single-element conductors, resistors, and dielectrics), sputtering, chemical vapor deposition, plating, and electroplating. Electroplating will be covered in Chapter 10. Sputtering is the deposition of compound materials and refractive metals by removal of the surface atoms or molecular fragments from a solid cathode (target) by bombarding it with positive ions from an inert gas (argon), and these atoms or molecular fragments are deposited on the substrate to form a thin film.

4.2.6   Deposition

Atmospheric pressure and low-pressure chemical vapor deposition (APCVD and LPCVD) are used to deposit metals, alloys, dielectrics, silicon, polysilicon, and other semiconductor, conductor, and insulator materials and compounds. The chemical reactants for the desired thin film are introduced into the CVD chamber in the vapor phase. The reactant gases then pyrochemically react at the heated surface of the wafer to form the desired thin film. Epitaxial growth, as the CVD process, allows one to grow a single crystalline layer upon a single crystalline substrate. Homoepitaxy is the growth of the same type of material on the substrate (e.g., p+ silicon etch-stop layer on an n-type substrate for layer formation). Heteroepitaxy is the growth of one material on a substrate that is a different type of medium. Silicon homoepitaxy is used in bulk micromachining to form the etch-stop layers. Plasma-enhanced chemical vapor deposition (PECVD) uses RF-induced plasma to provide additional energy to the reaction. The major advantage of PECVD is that it allows one to deposit thin films at lower temperatures compared with conventional CVD.

4.2.7   MEMS Assembling and Packaging

Assembling and packaging of MEMS includes microstructure and die inspection, separation, attachment, wire bonding, and packaging or encapsulation. For robust packaging, one must match the thermal expansion coefficients for the microstructures to minimize mechanical stresses. The connections and the package must provide actuation and sensing capabilities, high-quality sealing, robustness, protection, noise immunity, high-fidelity path for signals to and from the chip, reliable input-output interconnection, etc. Application-specific considerations include operation in harsh adversarial environments (contaminates, electromagnetic interference, humidity, noise, radiation, shocks, temperature, vibration, etc.), hybrid multichip packages, direct exposure of portions of the MEMS to outside stimuli (e.g., light, gas, pressure, vibration, temperature, and radiation), etc.

Bonding, assembling, and packaging are processes of great importance. In MEMS, different bonding techniques are used to assemble individually micromachined (fabricated) structures to form microdevices or complex microstructures, as well as integrate microstructures/microdevices with controlling ICs bonding an entire wafer or individual dies. Silicon direct bonding is used to bond a pair of silicon wafers together directly (face-to-face), while anodic bonding is used to bond silicon to glass.[1] In silicon direct bonding, the polished sides of two silicon wafers are connected face-to-face, and the wafer pair is annealed at high temperature. During annealing, the bonds are formed between the wafers (these bonds can be as strong as bulk silicon). The process is carried out through the following steps: (1) wafers cleaning in a strong oxidizing solution (organic clean, HF dip, and ionic clean), which results in the hydrophilic wafer surface; (2) wafers rinsing in water and drying; and (3) wafers squeezing (face-to-face), which results in the wafers sticking together due to hydrogen bonding of hydroxyl groups and van der Waals forces on the surface of the wafers. The bonded pairs are then annealed at high temperature to guarantee strong bonds in an inert environment (nitrogen) for approximately 1 hour.

The silicon direct-bonding technique is a simple process that does not require special equipment other than a cleaning station and an oxidation furnace. The requirements imposed on wafers are smoothness, flatness, and minimal possible level of contaminants. The quality of the wafer bonding is determined by inspection and testing using infrared illumination. Silicon-, silicon dioxide–, and silicon nitride–coated surfaces can be used for bonding. By contacting in a controlled ambient, microdevices can be sealed in a gas or vacuum environment.

Anodic (electrostatic) bonding is used to bond silicon to glass. The glass can be in the form of a plate or wafer, or of a thin film between two silicon wafers. Anodic bonding is performed at lower temperatures (450°C or less), bonding metallized microdevices. In anodic bonding, the silicon wafer is placed on a heated plate, the glass plate is placed on top of the silicon wafer, and a high negative voltage is applied to the glass. As the glass is heated, positive sodium ions become mobile and drift toward the negative electrode. A depletion region is formed in the glass at the silicon interface, resulting in a high electric field at the silicon–glass interface. This field forces the silicon and glass into intimate contact and bonds oxygen atoms from the glass with silicon in the silicon wafer, leading to permanent hermetically sealed bonds. Anodic bonding of two silicon wafers can be formed by coating two wafer surfaces using sputtered glasses. Anodic bonding requires smooth bonding surfaces. This requirement is not so critical compared with the fusion bonding process because the high electrostatic forces pull small gaps into contact. It is important to select glass with a thermal expansion coefficient that matches silicon. The difference in the thermal expansion coefficients of the glass and silicon will result in stress between the bonded pair after cooling to room temperature. Corning 1729 and Pyrex 7740 are widely used.

4.3   MEMS Fabrication Technologies

Microelectromechanics integrates fundamental theories (electromagnetism, micromechanics, and microelectronics), engineering practice, and fabrication technologies. Using fundamental research, MEMS can be devised, designed, and optimized. In addition to theoretical fundamentals, affordable (low-cost) high-yield fabrication technologies are required in order to make three-dimensional microscale structures, devices, and MEMS. Micromachining and high-aspect-ratio are key fabrication technologies for MEMS. Microelectromechanical system fabrication technologies fall into three broad categories: bulk micromachining, surface micromachining, and LIGA (LIGA-like) technologies.

Different fabrication techniques and processes are available. In general, the fabrication processes, techniques, and materials depend on the available facilities and equipment. High-yield proprietary integrated CMOS-based MEMS-oriented industrial fabrication technologies are developed by the leading MEMS manufacturers. It is the author's goal that this book will be used for the MEMS, microengineering, and nanoengineering courses, which may integrate laboratories and experiments. However, due to different facilities, equipment, infrastructure, distinct course structures, and different number of credit hours allocated, it is difficult to focus and provide detailed comprehensive coverage of all possible MEMS fabrication processes including packaging, calibration, and testing. There exist an infinite number of different developments, experiments, demonstrations, and laboratories in MEMS, microdevices, microstructures, and microelectronics. Excellent Web sites (Case Western University, Georgia Institute of Technology, Massachusetts Institute of Technology, Stanford University, Universities of California at Berkeley, Los Angeles, and Santa Barbara,

University of Wisconsin Madison, etc.) support the possible fabrication technologies, equipment, facilities, and MEMS/NEMS infrastructure developments.

General information and specific fabrication technologies, techniques, and processes are available at http://microlab.berkeley.edu/ (Microlab home page of the University of California at Berkeley). This information can be very useful and applicable. In particular, the following topics are covered: processes, mask making, CMOS baseline, etc.

The Web site http://www-snf.stanford.edu/ (Stanford Nanofabrication Facility) provides process information for specific equipment. In addition, general fabrication processes and techniques are covered with useful links—in particular, e-beam, microscopy, mask making, optical photolithography, chemical vapor deposition, annealing, oxidation, doping, etching, and other standard processes and tools.

It is evident that a great deal of additional important information regarding the basic fabrication techniques, processes, and materials can be obtained from the aforementioned as well as other Web sites. The Web sites listed are given as the possible sources of additional data from educational standpoints because the high-technology industry has developed proprietary technique and processes to guarantee affordability and high yield. The fitness, applicability, and reliability depend on the equipment and infrastructure, as well as MEMS needed to be made.

4.3.1 Bulk Micromachining

Bulk and surface micromachining are based on the modified CMOS technology with specifically designed micromachining processes. Bulk micromachining was developed more than 30 years ago to fabricate three-dimensional microstructures.[11] Bulk micromachining of silicon uses wet- and dry-etching techniques in conjunction with etch masks and etch-stop layers to develop microstructures from silicon substrates. Bulk micromachining of silicon is an affordable, high-yield, and well-developed technology. Microstructures are fabricated by etching areas of the silicon substrates, thereby releasing the desired three-dimensional microstructures.

The *anisotropic* and *isotropic* wet-etching processes, as well as concentration-dependent etching techniques, are widely used in bulk micromachining. The microstructures are formed by etching away the bulk of the silicon wafer. Bulk machining with crystallographic and dopant-dependent etch processes, when combined with wafer-to-wafer bonding, produces complex three-dimensional microstructures with the desired geometry. One fabricates microstructures by etching deeply into the silicon wafer. There are several ways to etch silicon wafers.

Anisotropic etching uses etchants (usually potassium hydroxide KOH, sodium hydroxide NaOH, H_2N_4, and ethylene-diamine-pyrocatecol EDP) that etch different crystallographic directions at different etch rates. Certain crystallographic planes (stop-planes) etch very slowly. Through anisotropic etching, three-dimensional structures (cones, pyramids, cubes, and channels into the surface of the silicon wafer) are fabricated. In contrast, *isotropic* etching etches all directions in the silicon wafer at the same (or nearly the same) etch rate. Therefore, hemisphere and cylinder structures can be made. Deep reactive ion etching uses plasma to etch straight-walled structures (cubical, rectangular, triangular, etc.). For example, the Si structure with etched 2-μm holes (to be filled, for example, with ferromagnetic material to be magnetized or high-permeability material to ensure different reluctance) is illustrated in Figure 4.11.

In bulk micromachining, wet- and dry-etching processes are widely used. Wet etching is the process of removing material by immersing the wafer in a liquid bath of the chemical etchant. Wet etchants are categorized as *isotropic* etchants (which attack the material being etched at the same rate in all directions) and *anisotropic* etchants (which attack the material

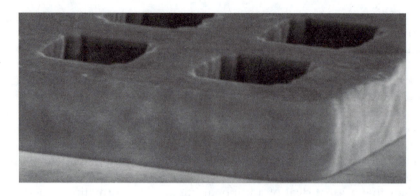

FIGURE 4.11
Si with 2-μm etched holes.

or silicon wafer at different rates in different directions, and therefore shapes/geometry can be precisely controlled). In other words, *isotropic* etching has a uniform etch rate at all orientations, while for *anisotropic* etching, the etch rate depends on crystal orientation. Some etchants attack silicon at different rates depending on the concentration of the impurities in the silicon (concentration-dependent etching). *Isotropic* etchants are available for silicon, silicon oxide, silicon nitride, polysilicon, gold, aluminum, and other commonly used materials. Since *isotropic* etchants attack the material at the same rate in all directions, they remove material horizontally under the etch mask (undercutting) at the same rate as they etch through the material. Hydrofluoric acid etches the silicon oxide faster than silicon. *Anisotropic* etchants, which etch different crystal planes at different rates, are widely used, and the most popular *anisotropic* etchant is potassium hydroxide (KOH) because it is the safest one to use. The application of concentration-dependent etching is illustrated below.

High levels of boron (p-type dopant) will reduce the rate at which the doped silicon is etched in the KOH system by several orders of magnitude, stopping the etching of the boron-rich silicon (as described above, the boron impurities are doped into the silicon by diffusion). Let us illustrate the technique. A thick silicon oxide mask is formed over the silicon wafer and patterned to expose the surface of the silicon wafer where the boron is to be doped. The silicon wafer is then placed in a furnace in contact with a boron diffusion source. Over a period of time, boron atoms migrate into the silicon wafer. As the boron diffusion is completed, the oxide mask is stripped off. A second mask can be deposited and patterned before the wafer is immersed in the KOH system that etches the silicon that is not protected by the mask etching around the boron-doped silicon.

The available anisotropic etchants of silicon (ethylene-diamine-pyrocatecol, potassium hydroxide, and hydrazine) etch single-crystal silicon along the crystal planes. From CMOS technology, etch masks and etch-stop techniques are available, which can be used in conjunction with silicon anisotropic etchants to selectively prevent regions of silicon from being etched. Therefore, microstructures are fabricated on a silicon substrate by combining etch masks and etch-stop patterns with different anisotropic etchants.

Wet etching of silicon is used for shaping and polishing, as well as for characterizing structural and compositional features. The fundamental etch reactions are electrochemical. Oxidation/reduction is followed by dissolution of the oxidation products. The etching process is either reaction-rate limited (etching process depends on the chemical reaction rate) or diffusion-limited (etching process depends on the transport of etchant by diffusion to or from the surface through the liquid). Diffusion-controlled processes have lower activation energies than reaction-rate-controlled processes. Therefore, diffusion-controlled processes are robust (insensitive) to temperature variations. However, diffusion-controlled processes are affected by agitation, which increases the supply of reactant material to the semiconductor surface, increasing the etch rate. Changes in the etching conditions and parameters (temperature, etchant components, their molarity, and proportions) change the rate-limiting process. The supply of minority carriers to the semiconductor surface limits the dissolution rate in etching reactions that result in a depletion of electrons or holes. Creation of electron-hole pairs on the surface (by illumination or by application of electric currents) or providing generation sites increases the etch rate. Additional factors that determine the rate of etching of crystalline semiconductors include orientation, type and concentration of doping atoms, lattice defects, and surface structure.

One of the most important microfabrication features is the etching directionality. If the etch rate in the x and y directions is equal to that in the z direction, the etch process is said to be *isotropic* (nondirectional). The etching of single-crystal silicon, polycrystalline, and amorphous silicon in HF, BHF, HNO_3, or CH_3COOH etchants (which form the so-called HNA etchant system) results in *isotropic* process. For etch processes that are *anisotropic* or *directional*, the etch rate in the z direction is higher than the lateral (x or y direction) etch rate. An example of this etch profile is the etching of (100) single-crystal silicon in the KOH/ water or ethylene-diamine-pyrocatechol/water (EDP) etchants. *Vertical anisotropic* etching is the directional etching in which the lateral etch rate is zero (for example, etching of the (110) single-crystal silicon in the KOH system, or etching the silicon substrate by ion bombardment–assisted plasma etching, e.g., reactive ion etching or ion beam milling).

Isotropic etching in liquid reagents is the most widely used process for removal of damaged surfaces, creating structures in single-crystal slices, and patterning single-crystal or polycrystalline semiconductor films.

For isotropic etching of silicon, the most commonly used etchants are mixtures of hydrofluoric (HF) and nitric (HNO_3) acids in water or acetic acid (CH_3COOH). In this so-called HNA etchant system, after the hole injection and OH^- attachment to the silicon to form $Si(OH)_2$, hydrogen is released to form SiO_2. Hydrofluoric acid is used to dissolve SiO_2 to form water-soluble H_2SiF_6. The reaction is

$$Si + HNO_3 + 6HF \rightarrow H_2SiF_6 + H_2NO_2 + H_2O + H_2$$

Water can be used as a diluent for this etchant. However, acetic acid (CH_3COOH) is preferred because it controls the dissociation of the nitric acid and preserves the oxidizing power of HNO_3 for a wide range of dilution (i.e., it acts as a buffer). Thus, the oxidizing power of the etchant remains almost constant. The HF-HNO_3 system was examined by Judy, Muller, and Zappe.[12] Figure 4.12 shows the results in the form of isoetch curves for various constituents by weight.

It should be noted here that normally available acid concentrations are 49% and 70% for HF and HNO_3, respectively. Either water (dashed-line curves) or acetic acid (solid-line curves) is used as the diluent in this system. At high HF and low HNO_3 concentrations, the etch rate is controlled by the concentration of HNO_3. Etching tends to be difficult to initiate, with the actual onset of etching highly variable. In addition, it results in relatively unstable silicon surfaces that proceed to slowly grow a layer of SiO_2 over a period of time. The etch is limited by the rate of the oxidation-reduction reaction, and therefore, it tends to be orientation dependent. At low HF and high HNO_3 concentrations, the etch rate is controlled by the ability of HF to remove the SiO_2 as it is formed. These etches are self-passivating in that the surface is covered with a 30- to 50-Å layer of SiO_2. The primary limit on the etch rate is the rate of removal of the silicon complexes by diffusion. The etching process in this region is *isotropic* and acts as the polishing etching. The etch system in the HF : HNO_3 = 1 : 1 range is initially insensitive to the addition of diluent when the percentage of diluent is less than 10%. From 10–30%, the etch rate decreases with the

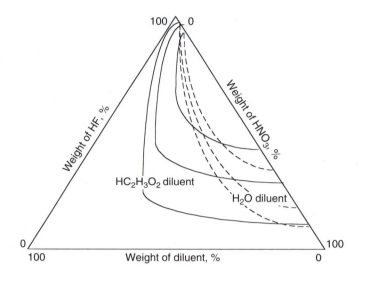

FIGURE 4.12
Isotropic etching curves for the silicon using the HF-HNO₃-diluent etchant system.

addition of diluent. If diluent is greater than 30%, even small changes of diluent cause large changes in the etch rate.

Anisotropic etchants of silicon, such as EDP, KOH, and hydrazine, are orientation dependent. That is, they etch the different crystal orientations with different etch rates. Anisotropic etchants etch the (100) and (110) silicon crystal planes faster than the (111) crystal planes. For example, the etch rates are 500 : 1 for (100) versus (111) orientations, respectively.[13] Silicon dioxide, silicon nitride, and metallic thin films (chromium and gold) provide good etch masks for typical silicon anisotropic etchants. These films are used to mask areas of silicon that must be protected from etching and to define the initial geometry of the regions to be etched.

Two techniques have been widely used in conjunction with silicon anisotropic etching to guarantee the etch-stop. Heavily boron-doped silicon (so-called p⁺ etch-stop) is effective in stopping the etch. The pn-junction technique can be used to stop etching when one side of a reverse-biased junction diode is etched away. Anisotropic etchants for silicon are usually alkaline solutions used at elevated temperatures. For isotropic etchants, two main reactions are oxidation of the silicon, followed by dissolution of the hydrated silica. The commonly used oxidant is H_2O in aqueous alkaline systems (NaOH or KOH),[14-16] cesium hydroxide,[17] hydrazine and EDP,[18] quaternary ammonium hydroxides,[19] or sodium silicates.[20] The most commonly used anisotropic etchants (etchant systems) for silicon are EDP with water, KOH with water or isopropyl, NaOH with water, and H_2N_4 with water or isopropyl (the etch rate of mask varies from 1 Å/min to 20 Å/min). The KOH-water etching system exhibits much higher (110) to (111) etch ratios than the EDP system. The etch rate ratios of the 100-, 110-, and 111-plane in the EDP are 50, 30, and 1,[21] while in the KOH system, the rate ratios are 100, 600, and 1.[16] Therefore, the KOH system is used for groove etching in 110-plane silicon wafers. The differences in etch ratios permit deep, high-aspect-ratio grooves with minimal undercutting of the mask. A disadvantage of the KOH system is that silicon dioxide (SiO_2) is etched at a rate that limits its use as a mask in many applications. For microstructures requiring long etching times, the silicon nitride (Si_3N_4) is the preferred masking material for the KOH system, while if the EDP system is used, masks can be made applying a variety of materials (for example, SiO_2, Si_3N_4, Cr, or Au).

The etching process is a charge-transfer mechanism, and etch rates depend on dopant type and concentration. Highly doped materials may exhibit high etch rate due to the greater availability of mobile carriers. This occurs in the HNA etching system ($HF : HNO_3 : CH_3COOH$ or $H_2O = 2 : 3 : 8$), where typical etch rates are 1–3 μm/min at p or n concentrations.[22] The anisotropic etchants (EDP and KOH) exhibit a different preferential etching behavior. Silicon heavily doped with boron reduces the etch rate in the range from 5 to 100 times when etching in the KOH system, and by 250 times when etching in the EDP system. Thus, the etch rate is a function of boron concentration, and the etch-stops formed by the p^+ technique are less than 10 μm thick. The electrochemical etch-stop process, which does not require heavy doping and guarantees the possibility to create thicker etch-stop layers because the etch-stop layer can be grown epitaxially, is widely used.

A widely implemented dry-etching process in micromachining applications is reactive ion etching. In this process, ions are accelerated toward the material to be etched, and the etching reaction is enhanced in the direction of travel of the ions. Reactive ion etching is an anisotropic etching process. Deep trenches and pits (up to a few tens of microns) of the specified shape with vertical walls can be etched in a variety of commonly used materials, e.g., silicon, polysilicon, silicon oxide, and silicon nitride. Compared with anisotropic wet etching, dry etching is not limited by the crystal planes in the silicon. Figure 4.13 illustrates the anisotropic-etched 400-μm-deep and 20-μm-wide grooves (in 110-silicon), and three-dimensional silicon structures are made using reactive ion etching.

4.3.2 Surface Micromachining

Different techniques and processes for depositing and patterning thin films are used to produce complex microstructures and microdevices on the surface of silicon wafers (surface silicon micromachining) or on the surface of other substrates. Surface micromachining technology allows one to fabricate the structure as layers of thin films. This technology guarantees the fabrication of three-dimensional microdevices with high accuracy, and surface micromachining can be called a thin-film technology. Each thin film is usually limited to a thickness up to 5 μm, which leads to fabrication of high-performance planar-type microscale structures and devices. The advantage of surface micromachining is the use of standard CMOS fabrication processes and facilities, as well as compliance with ICs. Therefore, this technology is widely used to manufacture microscale actuators and sensors (microdevices).

Surface micromachining has become the major fabrication technology in recent years because it allows one to fabricate complex three-dimensional microscale structures

 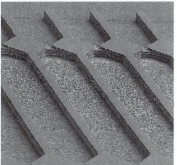

FIGURE 4.13
Isotropic and reactive ion etching of silicon.

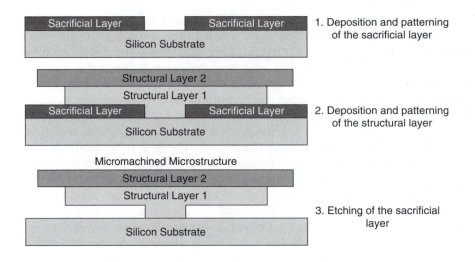

FIGURE 4.14
Surface micromachining.

and devices. Surface micromachining with single-crystal silicon, polysilicon, silicon nitride, silicon oxide, and silicon dioxide (as structural and sacrificial materials that are deposited and etched), as well as metals and alloys, is widely used to fabricate thin micromechanical structures and devices on the surface of a silicon wafer.

This affordable, low-cost, high-yield technology is integrated with electromechanical micro-structures—IC fabrication processes guaranteeing the needed microstructure—IC fabrication compatibility. Surface micromachining is based on the application of sacrificial (temporary) layers that are used to maintain subsequent layers and are removed to reveal (release) fabricated microstructures. This technology was first demonstrated for ICs and applied to fabricate motion microstructures in the 1980s. On the surface of a silicon wafer, thin layers of structural and sacrificial materials are deposited and patterned. Then, the sacrificial material is removed, and the microelectromechanical structure or device is fabricated.

Figure 4.14 illustrates a typical process sequence of the surface micromachining fabrication technology.

Usually, the sacrificial layer is made using silicon dioxide (SiO_2), phosphorous- doped silicon dioxide (PSG), or silicon nitride (Si_3N_4). The structural layers are then typically formed with polysilicon, metals, and alloys. The sacrificial layer is removed. In particular, after fabrication of the surface microstructures and microdevices (micromachines), the silicon wafer can be wet bulk etched to form cavities below the surface components, which allows a wider range of desired motion for the device.

The wet etching can be done using the following:

- Hydrofluoric acid (HF)
- Buffered hydrofluoric acid (BHF)
- Potassium hydroxide (KOH)
- Ethylene-diamene-pyrocatecol (EDP)
- Tetramethylammonium hydroxide (TMAH)
- Sodium hydroxide (NaOH)

Surface micromachining has been widely used in the commercial fabrication of MEMS and microdevices (microtransducers, actuators and sensors such as rotational/translational microservos, accelerometers, gyroscopes, etc.) and microstructures (gears, flip-chip electrostatic actuators, membranes, mirrors, etc.). As was emphasized, surface micromachining means the fabrication of micromechanical structures and devices by deposition and etching of structural and sacrificial layers (thin films). Simple microstructures (beams, gears, membranes, etc.) and complex microdevices (actuators, motors, and sensors) are fabricated on top of a silicon substrate. The most important attractive features of the surface micromachining technology are the small microstructure dimensions and the opportunity to integrate micromechanics, microelectronics (ICs), and optics on the same chip. Using IC-compatible batch processing, affordable, low-cost, high-yield microstructure fabrication is achieved for high-volume applications. For example, to fabricate microscale gears (microgear train), a sacrificial silicon dioxide layer is deposited on the wafer and patterned. Then, a structural layer of polysilicon is deposited and patterned. This polysilicon layer becomes the structural element of the microgear. Other layers are then deposited and patterned, creating the rest of the microstructure (microscale gears). Etching in hydrofluoric or buffered hydrofluoric acid removes the sacrificial layers, releasing the microgear.[1]

There are three key challenges in the fabrication of microstructures using surface micromachining: control and minimization of stress and stress gradient in the structural layer to avoid bending or buckling of the released microstructure; high selectivity of the sacrificial layer etchant to structural layers and silicon substrate; and avoidance of stiction of the released (suspended) microstructure to the substrate. By choosing appropriate deposition and doping parameters, the stress and stress gradient in thin films can be controlled and optimized (minimized). The sacrificial layers can be etched with high selectivity against the structural layer and silicon substrate using hydrofluoric or buffered hydrofluoric acid. Two methods to prevent stiction are commonly used: (1) application of gaseous hydrofluoric acid and control of temperature using the substrate heater, and (2) supercritical phase transition of carbon dioxide above the critical point (73 bar and 31°C). After etching the sacrificial layer and rinsing, the rinsing liquid is exchanged by liquid carbon dioxide, which is subsequently transferred in the supercritical state; thus, the transition from liquid to gaseous phase is avoided, and capillary forces do not occur.

The sacrificial layers are removed by lateral etching. This selective etching can be performed using hydrofluoric acid, which etches SiO_2 but not single-crystal and polycrystalline silicon. Alternatively, the KOH etching system with polysilicon as the sacrificial layer and silicon nitride as the cover material can be used. If 100-oriented silicon is used, substrate etching will terminate on the 111-plane. The deposition is performed using low-pressure chemical vapor deposition (LPCVD) from pure silane. The requirements on the deposition rate, thickness, and stress controls lead one to an analysis of mechanical properties and film morphology. The morphological range is controlled by deposition and nucleation conditions. Calculations based on single-crystal data and texture functions indicate that fine-grained, randomly oriented films are needed to attain isotropic mechanical properties. This requirement restricts the film growth conditions to the amorphous-polycrystalline boundary or, for LPCVD silicon, to the temperature region from 575 to 610°C. The optimization of mechanical properties of thin films is achieved via flow, pressure, and temperature control. This requires fast measurement and control methods.

The primary issue in the deposited thin films is the control of the built-in strain. For example, the maximum membrane deflection d for a loaded plate that is subject to built-up strain is a nonlinear function of the pressure (p), strain (e), length (l), thickness (h), Young's module (E), and Poisson ratio (ρ).

In particular, one has the following formula:

$$d = \frac{l^4(1+p^2)\rho}{Eh^2} f(e)$$

If the strain field is compressive ($e < 0$), bending and buckling occur. If the field is tensile, membrane deflections will be reduced for a given pressure. Compressive polysilicon can be converted to tensile polysilicon via annealing, which converts thin films to the fine-grained form (this involves a volume contraction, which causes the tensile field). The comprehensive database for the mechanical properties of the deposited thin films was developed in order to achieve computer-aided design and manufacturing capabilities. Polysilicon surface micromachining technology, as applied to the microscale sensors, is described by Guckel.[23]

As was shown, the beam-design surface micromachining process employs thin films of two different materials (polysilicon is typically used as the structural material and silicon oxide as the sacrificial material). These materials are deposited and patterned. The sacrificial material is etched away to release three-dimensional structure. Complex microstructures have many layers, and fabrication complexity continues to increase.

A simple surface micromachined cantilever beam is shown in Figure 4.13. A sacrificial layer of silicon oxide is deposited on the surface of the silicon wafer. Two layers of polysilicon and ferromagnetic alloy are then deposited and patterned using dry etching. The wafer is wet etched to remove the silicon oxide layer under the beam and release the beam, which is attached to the wafer by the anchor.

Surface micromachining is an additive fabrication technique that uses modified CMOS technology and materials (e.g., doped and undoped single-crystal silicon and polysilicon, silicon nitride, silicon oxide, and silicon dioxide for the electrical and mechanical microstructures, and aluminum alloys for the metal connections) and involves the building of a microstructure or microdevice on top of the surface of a supporting substrate. This technique complies with other CMOS technologies to fabricate ICs on a substrate.

To fabricate high-performance mechanical microstructures and microdevices using silicon and other materials, the internal stresses of thin films must be controlled. It is desirable to grow/deposit the polysilicon, silicon nitride, silicon dioxide, metal, alloy, and insulator thin films within minimum time. However, the high deposition speed results in high internal stress in thin films, and this highly compressive internal stress leads to bending and buckling effects. Thus, the thin-film deposition process should be controlled and optimized in order to minimize or eliminate the internal stress. For example, the stress of a polysilicon thin film can be controlled by doping it with boron, phosphorus, or arsenic. However, doped polysilicon films are rough and interfere with ICs. The stress in polysilicon can be controlled by annealing (annealing the polysilicon after deposition at elevated temperatures changes thin films to be stress-free or tensile). The annealing temperature sets the films' final stress. Using this method, ICs can be embedded into polysilicon films through selective doping, and hydrofluoric acid will not change the mechanical properties of the material. The stress of a silicon nitride film can be controlled by regulating the deposition temperature and the silicon/nitride ratio. The stress of a silicon dioxide thin film can be controlled and minimized by changing the deposition temperature and post-annealing. It is difficult to control the stress in silicon dioxide accurately. Therefore, silicon dioxide is usually not used as the structural material. Silicon dioxide is used for electric insulation or as a sacrificial layer under the polysilicon structural layer. Sacrificial layers are temporary layers that will be selectively removed later to allow partial or complete release of the structures. Silicon nitride may also be used for electronic insulation and as a sacrificial layer.

4.3.2.1 Example Process

The design and fabrication of motion microstructures start with the microstructure synthesis, identification of the microstructure functionality, specifications, and performance. Let us develop the fabrication flow to fabricate the thin membrane. The polysilicon membrane can be fabricated by oxidizing a silicon substrate, patterning the silicon dioxide, deposition and patterning of polysilicon over the silicon dioxide, and removal of the silicon dioxide. To attain the actuation features, the NiFe thin-film alloy (magnetic material) should then be deposited.

The major fabrication steps and processes for thin membranes are given in Table 4.4.

Conventional CMOS processes and materials were used to develop the fabrication flow (steps) in order to fabricate thin-film membranes. Therefore, CMOS fabrication facilities can be converted to fabricate microstructures, microdevices, and MEMS. Figure 4.15 illustrates the application of surface micromachining technology to fabricate the polysilicon thin-film membrane on the silicon substrate.

In electromagnetic microstructures and microdevices, metals, alloys, ferro-magnetic materials, magnets, and wires (windings) must be deposited. Different processes are used.[1,24] Using electron-beam lithography (a photo-graphic process that uses an electron microscope to project an image of the required structures onto a silicon substrate coated with a photosensitive resist layer), the process of fabrication of micromagnets on the silicon substrate is illustrated in Figure 4.16. Development removes the photoresist, which has been exposed to the electron beam. A ferromagnetic metal or alloy is then deposited, followed by lift-off of the unwanted material. This process allows one to make microstructures within nanometer-dimension features.

The trade-offs in performance and efficiency must be studied by researching different materials. For example, 95% (Ni) and 5% (Fe) to 80% (Ni) and 20% (Fe) nickel–iron alloys lead to the saturational magnetization in the 2-T range,[25] and the permeability can be greater than 2000. In addition, narrow (soft) and medium *B-H* characteristics are achieved. The saturation magnetization and the *B-H* curves for $Ni_{x\%}Fe_{100-x\%}$ thin films are illustrated in Figure 4.17. As emphasized in Chapter 8, the magnetic characteristics depend on other factors including thickness.

The piezoelectric effect has been used in translational (cantilever, membrane, etc.) and rotational microtransducers (actuators and sensors). The piezoelectric thin films are controlled by the electromagnetic field. For example, when voltage is applied to the piezoelectric film, the film expands or contracts. Typical piezoelectric thin films being used in microactuators are zinc oxide (ZnO), lead zirconate titanate (PZT), polyvinylidene difluoride (PVDF), and lead magnesium niobate (PMN). The linear microactuators fabricated using shape memory alloys (TiNi)[26,27] have found limited application compared with the piezoelectric thin films.

For MEMS, there is a critical need to develop the fabrication technologies that are compatible with the silicon-based microelectronic fabrication technologies, processes, and materials used to manufacture ICs.

The electromechanical design of microtransducers (actuators and sensors), which are called microdevices, is divided into the following steps: devising operating principles based on electromagnetic phenomena and effects, analysis of electromagnetic/mechanical features (ruggedness, elasticity, friction, vibration, thermodynamics, etc.), modeling, simulation, optimization, fabrication, testing, validation, and performance analysis.

Microtransducers have stationary and rotating members (stator and rotor) and radiating-energy microdevices. Different direct-current, induction, and synchronous microtransducers were fabricated and tested. Many microtransducers were designed using permanent magnets. Surface micromachining technology was used to fabricate rotational micromachines with a minimal rotor outer radius of 50 μm, an air gap of 1 μm, and a bearing clearance of 0.2 μm.[1,28–32] In particular, heavily phosphorous-doped polysilicon was used to fabricate rotors and stators, and silicon nitride was used for electrical insulation. The maximal angular velocity achieved for the permanent-magnet stepper micromotor was 1000 rad/sec.

TABLE 4.4

Major Fabrication Steps and Processes

Process Steps	Description
Step 1: Grow silicon dioxide	Silicon dioxide is grown thermally on a silicon substrate. For example, growth can be performed in a water vapor ambient at 1000°C for 1 hour. The silicon surfaces will be covered by 0.5 to 1 μm of silicon dioxide (thermal oxide thickness is limited to a few microns due to the diffusion of water vapor through silicon dioxide). Silicon dioxide can be deposited without modifying the surface of the substrate, but this process is slow to minimize the thin-film stress. Silicon nitride may also be deposited, and its thickness is limited to 4–5 μm.
Step 2: Photoresist	A photoresist (photosensitive material) is applied to the surface of the silicon dioxide. This can be done by spin coating the photoresist suspended in a solvent. The result after spinning and driving off the solvent is a photoresist with thickness from 0.2 to 2 μm. The photoresist is then soft baked to drive off the solvents inside.
Step 3: Photolithography exposure and development	The photoresist is exposed to ultraviolet light patterned by a photolithography mask (photomask). This photomask blocks the light and defines the pattern to guarantee the desired surface topography. Photomasks are usually made using fused silica, and optical transparency at the exposure wavelength, flatness, and thermal expansion coefficient must be met. On one surface of the glass (or quartz), an opaque layer is patterned (usually a chromium layer hundreds of Å thick). A photomask is generated based on the desired form of the polysilicon membrane. The surface topography is specified by the mask. The photoresist is developed next. The exposed areas are removed in the developer. In a positive photoresist, the light will decrease the molecular weight of the photoresist, and the developer will selectively remove (etch) the lower-molecular-weight material.
Step 4: Etch silicon dioxide	The silicon dioxide is etched. The remaining photoresist will be used as a *hard mask* that protects sections of the silicon dioxide. The photoresist is removed by wet etching (hydrofluoric acid, sulfuric acid, and hydrogen peroxide) or dry etching (oxygen plasma). The result is a silicon dioxide thin film on the silicon substrate.
Step 5: Deposit polysilicon	Polysilicon thin film is deposited over the silicon dioxide. For example, polysilicon can be deposited in the LPCVD system at 600°C in a silane (SiH_4) ambient. The typical deposition rate is 65–80 Å/min to minimize the internal stress and prevent bending and buckling (polysilicon thin film must be stress-free or have a tensile internal stress). The thickness of the thin film is up to 4 μm.
Step 6: Photoresist	Photoresist is applied to the polysilicon thin film, and the planarization must be done. The patterned silicon dioxide thin film changes the topology of the substrate surface. It is difficult to apply a uniform coat of photoresist over a surface with different heights. This results in photoresist film with different thicknesses and nonuniformity, and corners and edges of the patterns may not be covered. For 1 μm (or less) height, this problem is not significant, but for thicker films and multiple layers, replanarization is required.
Step 7: Photolithography exposure and development	A photomask containing the desired topography (form) of the polysilicon membrane is aligned to the silicon dioxide membrane. Alignment accuracy (tolerances) can be done within the nanometer range, and the accuracy depends on the size of the microstructure features.
Step 8: Etch polysilicon	The polysilicon thin film is etched with the photoresist protecting the desired polysilicon membrane form. It is difficult to find a wet etch for polysilicon that does not attack photoresist. Therefore, dry etching through plasma etching can be applied. Selectivity of the plasma between polysilicon and silicon dioxide is not a concern because the silicon dioxide will be removed later. Therefore, the polysilicon can be overetched by etching it longer than needed. This results in higher yield.

TABLE 4.4

Continued

Process Steps	Description
Step 9: Remove photoresist	The photoresist protecting the polysilicon membrane is removed.
Step 10: Deposit NiFe	NiFe thin film is deposited.
Step 11: Remove silicon dioxide and release the thin-film membrane	The silicon dioxide is removed by wet etching (hydrofluoric or buffered hydrofluoric acid) because plasma etching cannot easily remove the silicon dioxide in the confined space under the polysilicon thin film. Hydrofluoric acid does not attack pure silicon. Hence, the polysilicon membrane and silicon substrate will not be etched. After the silicon dioxide is removed, the polysilicon membrane is formed (released). This membrane can bend down and stick to the surface of the substrate during drying after the wet etch. To avoid this, a rough polysilicon, which does not stick, can be used. Another solution is to fabricate the polysilicon membrane with internal stress so that the polysilicon membrane is bent (curved) up during drying. Both solutions lead to the specific mechanical properties of the polysilicon membrane, which might not be optimal from the operating requirements standpoint. Therefore, alternatives are sought, and, in general, it is possible to fabricate the polysilicon membrane with no stress.

The cross section of the slotless synchronous microtransducer (motor and generator) fabricated on the silicon substrate with polysilicon stator with deposited windings, polysilicon rotor with deposited permanent magnets, and bearing is illustrated in Figure 4.18.

Silicon dioxide, formed by thermal oxidation of silicon, can be used as the insulating layer (SiO_2 is commonly applied as mask and sacrificial material). The driving/sensing and controlling/processing ICs control the brushless micromotor.

To fabricate microtransducers and ICs on a single- or double-sided chip (the application of the double-sided technique significantly enhances the performance), similar fabrication technologies and processes must be used, and the compatibility issues should be addressed and resolved. The surface micromachining processes were integrated with the CMOS technology (e.g., similar materials, lithography, etching, and other techniques). The analysis of the microtransducer feature size clearly indicates that the feature size of the microstructures and ICs is on the same order (for example, the microwindings must be deposited, microbearings and microcavities made, etc.). Furthermore, the surface micromachining processes are application-specific and strongly affected by the microtransducer devised.

To fabricate the integrated MEMS, post-, mixed-, and pre-CMOS/micromachining techniques can be applied. In a post-CMOS/micromachining technique, the ICs are passivated to protect them from the surface micromachining processes. The aluminum metallization is replaced by the tungsten (which has low resistivity and has a thermal extension coefficient matching the silicon thermal extension coefficient) metallization in order to raise the post-CMOS temperature higher than 450°C, but the diffusion barrier (formed by TiN or $TiSi_2$) must be used to avoid the WSi_2 formation at 600°C as well as adhesion and contact layer for the tungsten metallization. However, due to hillock formation in the tungsten during annealing, high contact resistance, and performance degradation due to the heavily doped structural and sacrificial layers, the mixed-CMOS/micromachining technique is used. In particular, performing the processes in sequence to fabricate ICs and microstructures/microdevices, the performance can be optimized, and for the standard bipolar CMOS (bi-CMOS) technology, the minimal modifications are required. The pre-CMOS/micromachining technique allows one to fabricate microstructures/microdevices before ICs. However, due to the vertical three-dimensional microstructure features, passivation, oxidation, step coverage, and interconnection cause difficulties.[1] Microstructures can be fabricated in trenches

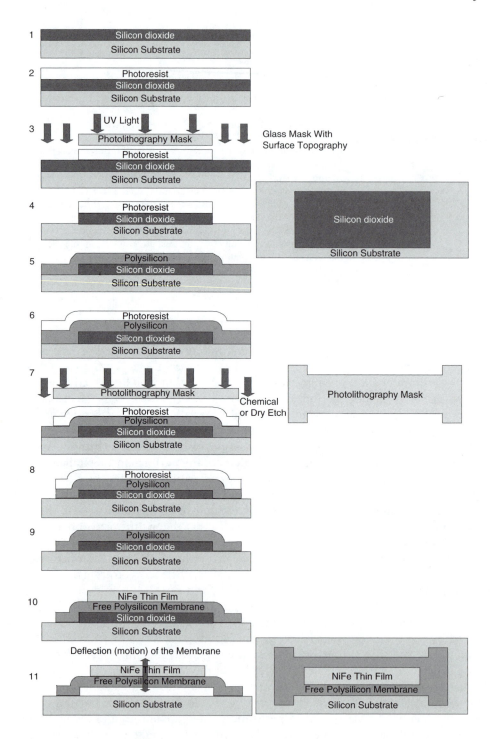

FIGURE 4.15 (See color insert.)
Micromachining fabrication of the polysilicon thin-film membrane.

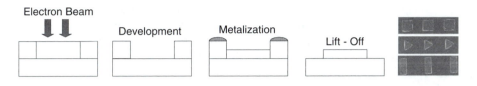

FIGURE 4.16
Electron beam lithography in micromagnet fabrications.

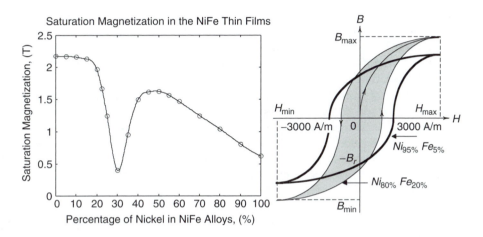

FIGURE 4.17
Saturation magnetization and the *B-H* curves for $Ni_{x\%}Fe_{100-x\%}$ thin films.

etched in the silicon epilayer, and then the trenches are filled with the silicon oxide, planarized (polished), and sealed. After completing these steps, ICs are fabricated using conventional CMOS technology, and additional steps are integrated to expose and release the embedded microstructures/microdevices.[1]

4.3.3 High-Aspect-Ratio (LIGA and LIGA-Like) Technology

There is a critical need to develop the fabrication technologies allowing one to fabricate high-aspect-ratio microstructures and microdevices. The LIGA process, which denotes lithography, galvanoforming, and molding (in German, *Lithografie*, *Galvanik*, and

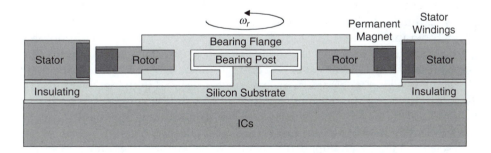

FIGURE 4.18
Cross-section schematics for slotless permanent-magnet brushless microtransducer with ICs.

*A*bformung), is capable of producing three-dimensional microstructures of a few milli-meters high with an aspect ratio (depth versus lateral dimension) of more than 100. This ratio can be achieved only through bulk micromachining using wet anisotropic etching.

The LIGA technology is based on x-ray lithography, which guarantees shorter wave-length (from few to 10 Å, which leads to negligible diffraction effects) and larger depth of focus compared with optical lithography. The ability to fabricate microstructures and microdevices in the centimeter range is particularly important in actuator applications since the specifications are imposed on the rated force and torque developed by the microdevices. Due to the limited force and torque densities, the designer faces the need to increase the actuator's dimensions. The LIGA and LIGA-like processes are based on deep x-ray lithography and electroplating of metal and alloy structures, allowing one to achieve structural heights in the centimeter range.[32–35] This type of processing expands the material base significantly and allows the fabrication of new high-performance elec-tromechanical microtransducers.

In translational (linear) and rotational microactuators, the electromagnetic force and torque depend on the change in energy that is stored in the active volume and the energy density of the material.[36] In particular, the expression for the coenergy is used to derive the electro-magnetic force and torque. High-performance actuators with maximized active volumes and minimized surface areas have been designed and fabricated using LIGA and LIGA-like technologies.[1] In these processes, a substrate with a plating base is covered with a thick photoresist (thickness can be in the centimeter range). The photoresist is cured and exposed by x-rays from a synchrotron source (x-ray lithography).

Photoresist strain, which is due to adhesion, causes well-known difficulties. This problem is solved by combining surface micromachining, patterning the sacrificial layers under the plating base, and optimizing the processes (details are reported in Chapter 10). The achievable structural height of LIGA or LIGA-like fabricated structures is defined mainly by the photoresist process-ing. The photoresist procedures (based on solvent bonding of polymethyl methacrylate [PMMA] and subsequent mechanical height adjustments) have been optimized and used to produce low-strain photoresist layers with thicknesses from 50 μm to the centimeter range. Large-area exposures of photoresist with thicknesses up to 10 cm have been achieved with x-ray masks. After electroplating, replanarization can be made through precision polishing. PMMA positive resists are based on special grades of polymethyl methacrylate designed to provide high contrast, high resolution for e-beam, deep UV (220–250 nm), and x-ray lithographic processes. Standard PMMA resists have 495,000 and 950,000 molecular weights in a wide range of film thicknesses formulated in chlorobenzene or the safer solvent anisole. Copolymer resists are based on a mixture of PMMA and methacrylic acid (typically 10%). Copolymer MAA can be used in combination with PMMA in bilayer lift-off resist processes where independent control of size and shape of each resist layer is needed. Standard copolymer resists are formu-lated in the safer solvent ethyl lactate and are available in a wide range of film thicknesses.

Figure 4.19 illustrates the basic sequential processes (steps) in LIGA technology. Here, the x-ray lithography is used to produce patterns in very thick layers of photoresist. The x-rays from a synchrotron source are shone through a special mask onto a thick photoresist layer (sensitive to x-rays) that covers a conductive substrate (step 1). This photoresist is then developed (step 2). The pattern formed is electroplated with metal (step 3). The metal structures produced can be the final product; however, it is common to produce a metal mold (step 4). This mold can then be filled with a suitable media (e.g., metal, alloy, polymer) as shown in step 5. The final structure is released (step 6).

The described LIGA technology (frequently referred to as the high-aspect-ratio tech-nique) allows one to fabricate microstructures with small lateral dimensions compared with thickness. Thick and narrow microstructures guarantee high ruggedness in the direc-tion perpendicular to the substrate and compliance in the lateral directions. For actuators,

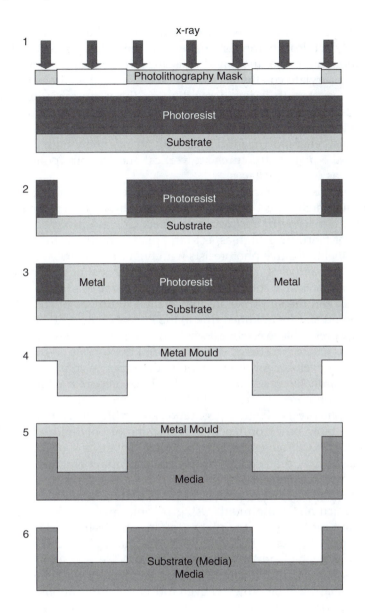

FIGURE 4.19 (See color insert.)
LIGA fabrication technology.

high-aspect-ratio technology offers the possibility to fabricate high-torque and high-force-density microtransducers. As was emphasized, high-intensity, low-divergence, and hard x-rays are used as the exposure source for the lithography.

For exposure wavelengths , the image resolution and the depth of focus are (see Section 4.1)

$$i_R = k_i \frac{\lambda}{N_A} \quad \text{and} \quad d_F = k_d \frac{\lambda}{N_A^2}$$

Due to short exposure wavelengths, the desired feature size is achieved. These x-rays are usually produced by a synchrotron radiation source.[1,32–35] Polymethyl methacrylate

(PMMA) and polylactides are used as the x-ray resists because PMMA (Plexiglas™ or Lucite™) and polylactides photoresists have high sensitivity to x-rays, thermal stability, desired absorption, high resolution, and resistance to chemical, ion, and plasma etching. Polyglycidyl-methacrylate-co-ethylacrylate (PGMA) is used as the negative x-ray resist.

The exposure wavelength varies depending on the x-ray radiation source used. For example, the 0.2-nm x-ray wavelength allows one to transfer the pattern from the high-contrast x-ray mask into the photoresist layer with a thickness of a few centimeters so that the photoresist relief may be fabricated with an extremely high depth-to-width ratio. The sidewalls of the plated structures are vertical and smooth (polished), and therefore they can be used as optical surfaces.

Photolithography using commercially available positive photoresists and near-UV light sources can produce high-aspect-ratio plating molds. Although, in comparison to LIGA, this technique is limited in terms of thickness and aspect ratio, positive photoresists can provide a simple means of fabricating high-aspect-ratio plating molds with conventional photolithography equipment. Positive photoresists with high transparency and high viscosity can be used to achieve coatings of 20–80 μm thick. Multiple coatings are needed to obtain the thicker layers of photoresist. If contrast printing is used, edge bead removal and good contact between the mask and the substrate are important. Conditions of softbake, exposure, and development should also be modified due to the large thickness of the photoresist. Longer softbake times are preferable to remove the solvent from the photoresist, and a high exposure is necessary. In particular, the energy density needed to expose a 30-μm-thick photoresist is 1500 mJ/cm^2. Long development time is required in order to completely remove the resist of exposed area. Hardbake conditions must be optimized because while hardbake improves the adhesion of the photoresist and chemical resistance to the plating solution, it also causes distortion of the photoresist. High aspect ratios (10 or higher) can be obtained for 20-μm thin films. Dry etching, based on reactive ion etching of polyimides to form high-aspect-ratio molds, is used. Electromagnetically controlled dry etching of fluorinated polyimides with Ti or Al masks has been used for deep etching to attain high aspect ratio, good mask selectivity, and smooth sidewalls.

A critical part of the high-aspect-ratio process is plating to form the metallic electromechanical microstructures in the mold. Using plating, metal is deposited from ions in a solution following the shape of the plating mold. This is the additive process, and the thickness of the plated metal can be large since the plating rate can be high. A variety of metals (Al, Au, Cu, Fe, Ni, and W) and alloys (NiCo, NiFe, and NiSi) can be deposited or codeposited. It is important that roughness (smoothness) of the reflective metal surfaces with the desired shape can be achieved even for optical applications. Electroplating (well known from chemistry and covered in Chapter 10) and electroless plating (in which reduction of the metal ions occurs by the chemical reaction between a reducing agent and metal ion on a properly activated substrate) are the commonly used plating processes. The metal seed layer can be deposited and removed from the substrate or sacrificial layer. The plating rate and the grain size are controlled by the current density, temperature, duty cycle, etc. (see Chapter 10).

The UV-LIGA microfabrication technology that is based on the use of SU-8 photoresist is a straightforward method to fabricate MEMS. The sacrificial molds obtained through photolithography are utilized for the batch-fabrication of electrodeposited components. Figure 4.20 illustrates the image of a 5-mm minidevice fabricated by the Swiss UV-LIGA Company with a maximal aspect ratio of 10, a minimal line width of 10 μm, and lateral tolerances of ±2 μm. Microstructures can be fabricated, and Figure 4.20 documents 50-μm-high microspears.

SU-8 (formulated in GBL) and SU-8 2000 (formulated in cyclopentanone) are chemically amplified, epoxy-based negative resists. The thin films can be fabricated with

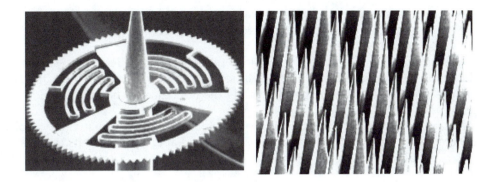

FIGURE 4.20
Swiss UV-LIGA Company and microspears fabricated using LIGA technology.

thicknesses from less than 1 μm to more than 200 μm. SU-8 and SU-8 2000 resists have high functionality and high optical transparency and are sensitive to near-UV (350–400 nm) radiation. High-aspect-ratio structures with straight sidewalls can be formed by contact-proximity or projection printing. Cured SU-8 is highly resistant to solvents, acids, and bases and has excellent thermal stability.

Usually, fabrication of MEMS is done using copper, iron, nickel, alloys, and other materials through electrodeposition, electroless plating, sputtering, and other deposition processes on the selected areas of the silicon substrates— for example, deposition of the copper microwindings and magnetic NiFe alloy thin films. The electroless plating can be conducted on a chemically activated silicon substrate without deposition on the photoresist mold.[1] The optimization of plating conditions (changing the current density, temperature, pH, waveforms, duty cycle, forward and reverse current, etc.) is critical to obtain the desired characteristics, e.g., resistivity, conductivity, smooth surfaces, deposition rates, etc.

Homework Problems

1. Describe the difference between *bulk* and *surface* micromachining. How can bulk and surface micromachining be used to fabricate the cantilever beam?
2. List the basic steps of silicon micromachining.

Answer:
- Lithography
- Deposition of thin films and materials (electroplating, chemical vapor deposition, plasma-enhanced chemical vapor deposition, evaporation, sputtering, spraying, screen printing, etc.)
- Removal of material (patterning) by wet or dry techniques
- Etching (plasma etching, reactive ion etching, laser etching, etc.)
- Doping
- Bonding (fusion, anodic, and other)
- Planarization

3. Why is photolithography important in MEMS fabrication?

Answer:
High-resolution photolithography is a technology that is applied to define two-(planar) and three-dimensional shapes (geometry). Photolithography is a core process used in ICs and MEMS fabrication to create the patterns defining the ICs', microstructures', and microdevices' features.

4. Calculate the image resolution and depth of focus for a 90-nm nanolithography process assuming that the lithographic process and aperture constants (k_i, k_d, and N_A) are equal to 0.5.

Answer:
The expressions for the image resolution i_R and the depth of focus d_F are

$$i_R = k_i \frac{\lambda}{N_A} \quad \text{and} \quad d_F = k_d \frac{\lambda}{N_A^2}$$

Thus, $i_R = 65$ nm and $d_F = 130$ nm.

5. Describe high-aspect-ratio technologies (LIGA and LIGA-like) and how they can be used to fabricate a microactuator.

References

1. Madou, M., *Fundamentals of Microfabrication*, CRC Press, Boca Raton, FL, 2002.
2. Campbell, S.A., *The Science and Engineering of Microelectronic Fabrication*, Oxford University Press, New York, 2001.
3. Ghandi, S., *VLSI Fabrication Principles*, John Wiley, New York, 1983.
4. Wolfe, S. and Tauber, R., *Silicon Processing for the VLSI Era*, Lattice Press, Sunset Beach, CA, 1986.
5. Beadle, W., Tsai, J., and Plummer, R., Eds., *Quick Reference Manual for Silicon IC Technology*, John Wiley, New York, 1985.
6. Colclaser, R., *Microelectronics—Processing and Device Design*, John Wiley, New York, 1980.
7. Mead, C. and Conway, L., *Introduction to VLSI Systems*, Addison-Wesley, Reading, MA, 1980.
8. Reinhard, D., *Introduction to Integrated Circuit Engineering*, Houghton Mifflin, Boston, MA, 1987.
9. Sze, S.M., *Semiconductor Devices—Physics and Technology*, John Wiley, New York, 1985.
10. Sze, S.M., *VLSI Technology*, McGraw-Hill, New York, 1983.
11. Petersen, K.E., Silicon as a mechanical material, *IEEE Proc.*, 420–457, 1982.
12. Robins, H. and Schwartz, B., Chemical etching of silicon, the system HF, HNO_3, H_2O and $HC_2H_3O_2$, *J. Electrochem. Soc.*, 114, 108–112, 1960.
13. Kendall, D.L., On etching very narrow grooves in silicon, *J. Appl. Phys. Lett.*, 26, 195–201, 1975.
14. Zwicker, W.K. and Kurtz, S.K., Anisotropic etching of silicon using electrochemical displacement reactions, in *Semiconductor Silicon*, Huff, H.R. and Burgess, R.R., Eds., Electrochemical Society Press, Princeton, NJ, 1973.
15. Price, J.B., Anisotropic etching of silicon with $KOH-H_2O$-isopropyl alcohol, in *Semiconductor Silicon*, Huff, H.R. and Burgess, R.R., Eds., Electrochemical Society Press, Princeton, NJ, 1973, pp. 338–353.
16. Bean, K.E., Anisotropic etching of silicon, *IEEE Trans. Electron. Dev.*, ED-25, 1185–1993, 1978.
17. Clark, L.D., Lund, J.L., and Edell, D.J., Cesium hydroxide (CsOH): A useful etchant for micromachining silicon, *Tech. Digest of IEEE Solid-State Sensor and Actuator Workshop*, Hilton Head, SC, 5–8, 1988.

18. Declercq, M.J., DeMoor, J.P., and Lambert, J.P., A comparative study of three anisotropic etchants for silicon, *Electrochem. Soc. Ext. Abstr.*, 75, 446–448, 1975.
19. Asano, M., Dho, T., and Muraoka, H., Application of choline in semiconductor technology, *Electrochem. Soc. Ext. Abstr.*, 76, 911–916, 1976.
20. Pugacz-Muraszkiewicz, I.J. and Hammond, B.R., Applications of silicates to the detection of flaws in glassy passivation films deposited on silicon substrate, *J. Vac. Sci. Technol.*, 14, 49–55, 1977.
21. Seidel, H., The mechanism of anisotropic silicon etching and its relevance for micromachining, *Digest of Tech. Papers, Transducers 87, Intl. Conf. Solid-State Sensors and Actuators*, 120–125, 1987.
22. Huraoka, H., Ohhashi, T., and Sumitomo, T., Controlled preferential etching technology, in *Semiconductor Silicon*, Huff, H.R. and Burgess, R.R., Eds., Electro-chemical Society Press, Princeton, NJ, 1973.
23. Guckel, H., Surface micromachined physical sensors, *Sensors Mater.*, 4, 251–264, 1993.
24. Ahn, C.H., Kim, Y.J., and Allen, M.G., A planar variable reluctance magnetic micromotor with fully integrated stator and wrapped coil, *Proc. IEEE Micro Electro Mech. Syst. Workshop*, Fort Lauderdale, FL, 1–6, 1993.
25. Judy, J.W., Muller, R.S., and Zappe, H.H., Magnetic microactuation of poly-silicon flexible structure, *J. Microelectromech. Syst.*, 4, 162–169, 1995.
26. Walker, J.A., Gabriel, K.J., and Mehregany, M., Thin-film processing of TiNi shape memory alloy, *Sensors Actuators*, A21–A23, 1990.
27. Johnson, A.D., Vacuum deposited TiNi shape memory film: Characterization and application in microdevices, *J. Micromech. Microeng.*, 34, 1991.
28. Guckel, H., Skrobis, K.J., Christenson, T.R., Klein, J., Han, S., Choi, B., Lovell, E.G., and Chapman, T.W., Fabrication and testing of the planar magnetic micromotor, *J. Micromech. Microeng.*, 1, 135–138, 1991.
29. Mehregany, M. and Tai, Y.C., Surface micromachined mechanisms and micro-motors, *J. Micromech. Microeng.*, 1, 73–85, 1992.
30. Omar, M.P., Mehregany, M., and Mullen, R.L., Modeling of electric and fluid fields in silicon microactuators, *Int. J. Appl. Electromagn. Mater.*, 3, 249–252, 1993.
31. Bart, S.F., Mehregany, M., Tavrow, L.S., Lang, J.H., and Senturia, S.D., Electric micromotor dynamics, *Trans. Electron. Devices*, 39, 566–575, 1992.
32. Becker, E.W., Ehrfeld, W., Hagmann, P., Maner, A., and Mynchmeyer, D., Fabrication of microstructures with high aspect ratios and great structural heights by synchrotron radiation lithography, galvanoformung, and plastic moulding (LIGA process), *Microelectron. Eng.*, 4, 35–56, 1986.
33. Guckel, H., Christenson, T.R., Skrobis, K.J., Klein, J., and Karnowsky, M., Design and testing of planar magnetic micromotors fabricated by deep x-ray lithography and electroplating, *Technical Digest of International Conference on Solid-State Sensors and Actuators, Transducers 93*, Yokohama, Japan, 60–64, 1993.
34. Guckel, H., Skrobis, K.J., Christenson, T.R., and Klein, J., Micromechanics for actuators via deep x-ray lithography, *Proc. SPIE Symp. Microlithog.*, San Jose, CA, 39–47, 1994.
35. Guckel, H., Christenson, T.R., Klein, J., Earles, T., and Massoud-Ansari, S., Microelectromagnetic actuators based on deep x-ray lithography, *Proc. Int. Symp. Microsyst. Intelligent Mater. Robots*, Sendai, Japan, 1995.
36. Lyshevski, S.E., *Nano- and Microelectromechanical Systems: Fundamentals of Nano- and Microengineering*, CRC Press, Boca Raton, FL, 2000.
37. Judy, J.W., Microelectromechanical systems (MEMS): Fabrication, design and applications, *J. Smart Mater. Struct.*, 10, 1115–1134, 2001.

5

Devising and Synthesis of NEMS and MEMS

5.1   Motion Nano- and Microdevices: Synthesis and Classification

New advances in nanostructures, nano- and microscale motion and radiating-energy devices, driving/sensing and controlling/processing ICs, and fabrication technologies provide enabling benefits and capabilities to design and fabricate NEMS and MEMS. Critical issues are to design high-performance systems that satisfy the specified criteria and requirements. For example, functionality, performance, compatibility, integrity, compliance, power, thermal management, and other factors are considered. While enabling technologies have been developed to accomplish nanoscale-featured fabrication and novel enabling multifunctional materials have been emerged, a spectrum of challenging problems in devising high-performance systems remains. There are several key focus areas to be examined. This chapter covers synthesis (discovery and devising) and classification issues. The reported results have direct impact on design, analysis, optimization, fabrication, testing, and characterization. In fact, before engaging in design, fabrication, and other tasks, NEMS and MEMS must be synthesized.

The designer can synthesize NEMS and MEMS by examining the basic physics, devising and prototyping of new operational principles (that can be due nanoscale size), synthesizing high-performance motion and radiating-energy microdevices, mimicking nano- and microscale driving/sensing circuitry (including three-dimensional circuits), synthesizing topologies and architectures for controlling/processing ICs, etc.

A step-by-step procedure in the design of NEMS and MEMS is as follows:

*   Define application and environmental requirements.
*   Specify performance specifications.
*   Devise NEMS and MEMS that include motion/motionless microstructures and microdevices, radiating-energy microdevices, micro-scale driving/sensing circuitry, controlling/processing ICs, energy sources, etc.
*   Develop the affordable high-yield fabrication process (for example, micromachining and high-aspect-ratio technologies compatible with CMOS technologies for MEMS).
*   Perform electromagnetic, energy conversion, mechanical, thermal, vibroacoustic, sizing, and other estimates.
*   Perform heterogeneous electromagnetic, optical, mechanical, thermal, vibroacoustic, electrochemical, and other designs with performance analysis and outcome prediction.
*   Verify, modify, and refine design with ultimate goals to optimize the performance.

Different nano- and microscale systems have been covered in previous chapters. Due to the variety of possible systems (basic physics, topologies, architectures, etc.), it is virtually impossible to cover the synthesis of generic NEMS and MEMS. Correspondingly, the major emphasis is focused on electromagnetic systems. Specifically, we address and examine the synthesis of motion nano- and microdevices stressing electromagnetic systems, shape/geometry synthesis, optimization, etc. The proposed concept allows the designer to devise novel high-performance NEMS and MEMS and then optimize them. The Synthesis and Classification Solver reported directly leverages high-fidelity modeling, allowing the designer to attain physical and behavioral (steady-state and transient) data-intensive analysis, heterogeneous simulations, optimization, performance assessment, outcome prediction, etc.

The taxonomy of NEMS and MEMS synthesis was emphasized in Chapter 2. One of the most important domains of the bidirectional X-flow map in the NEMS/MEMS design is devising (synthesis) (see Figure 5.1). The interaction between four domains allows one to ensure highly interactive bidirectional *top-down* and *bottom-up* design applying low-level data to high-level design and using the high-level requirements to devise/design low-level components. The X-domain abstraction map is reported in Figure 5.1. It was also emphasized that MEMS and NEMS include subsystems, devices, and structures as shown in Figure 5.1.

The *E. coli* bacteria axial-flux nanobiomotor, as illustrated in Figure 5.2, can be prototyped in order to synthesize electromagnetic axial-flux nano- and micromotors.

The possible topologies of nanomotors can be radial, axial, and integrated. Radial topology nanobiomotors, powered by ATP, can be synthesized from DNA and RNA protein complexes with, for example, six strands of RNA complexes and a DNA strand serving as a rotor. In this nanomotor, the RNA strands (stator) can develop the force rotating the DNA (rotor) due to ATP reaction (in theory, F_1 gamma-subunit rotates with respect to the a_3b_3-hexamer, and the rotor of nanobiomotors should turn). Using the radial flux topology of the described hypothetical DNA-RNA nanobiomotor, the cylindrical micro- or nanoscale machine with permanent-magnet poles on the rotor is shown in Figure 5.3.

In this section, the design and optimization of motion microdevices is reported. To illustrate the procedure, we consider two-phase permanent-magnet synchronous slotless microtransducers as shown in Figure 5.4.

It is evident that the electromagnetic system is *endless*, and different geometries can be utilized as shown in Figure 5.4. In contrast, in translational (linear) synchronous microtransducers, the *open-ended* electromagnetic system results. For the classified microelectromechanical motion devices, qualitative and quantitative comprehensive analysis must be performed.

Motion microdevice geometry and electromagnetic systems must be integrated into the synthesis, analysis, design, and optimization patterns. Motion microdevices can have the plate, spherical, toroidal, conical, cylindrical, and asymmetrical geometry. Using these distinct geometry and diverse electromagnetic systems, we propose to classify MEMS. This idea is extremely useful in the study of existing MEMS as well as in the synthesis of an infinite number of innovative motion microdevices. In particular, using the possible geometry and electromagnetic systems (*endless*, *open-ended*, and *integrated*), novel high-performance MEMS can be synthesized.

The basic electromagnetic microtransducers (microdevices) under our consideration are induction, synchronous, rotational, and translational (linear). That is, microdevices are classified using a type classifier

$$Y = \{y : y \in Y\}$$

Motion microdevices are categorized using a geometric classifier (plate P, spherical S, toroidal T, conical N, cylindrical C, or asymmetrical A geometry) and an electromagnetic system

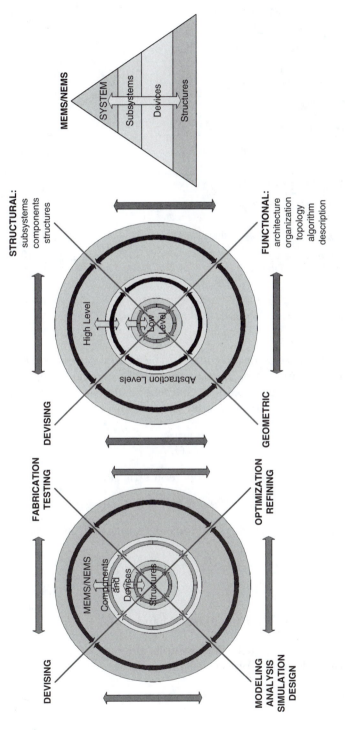

FIGURE 5.1
X-domain synthesis flow map and abstraction map.

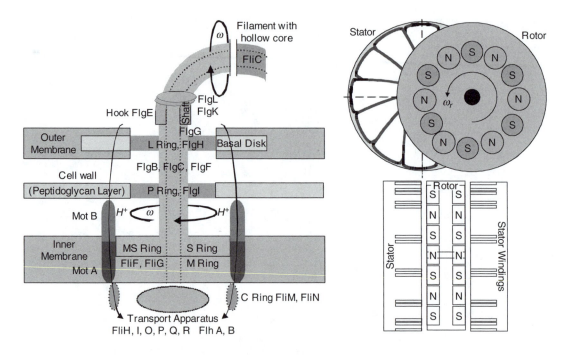

FIGURE 5.2
E. coli axial-flux nanobiomotor and electromagnetic axial-flux nanomotor.

classifier (*endless E*, *open-ended O*, or *integrated I*). The microdevice classifier, documented in Table 5.1, is partitioned into three horizontal and six vertical strips and contains 18 sections, each identified by ordered pairs of characters, such as (*E*, *P*) or (*O*, *C*).

In each ordered pair, the first entry is a letter chosen from the bounded electromagnetic system set

$$M = \{E, O, I\}$$

The second entry is a letter chosen from the geometric set

$$G = \{P, S, T, N, C, A\}$$

That is, for electromagnetic microdevices, the electromagnetic system/geometric set is

$$M \times G = \{(E, F), (E, S), (E, T), \ldots, (I, N), (I, C), (I, A)\}$$

In general, we have

$$M \times G = \{(m, g) : m \in M \text{ and } g \in G\}$$

Other categorizations can be applied. For example, single-, two-, three-, and multiphase microdevices are classified using a phase classifier

$$H = \{h : h \in H\}$$

Therefore, we have

$$Y \times M \times G \times H = \{(y, m, g, h) : y \in Y, m \in M, g \in G \text{ and } h \in H\}$$

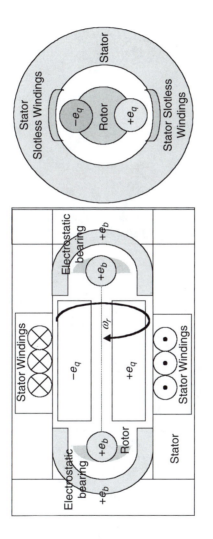

FIGURE 5.3

Radial topology machine with electrostatic noncontact bearings (poles are $+e_q$ and $-e_q$ and electrostatic bearing formed by $+e_b$).

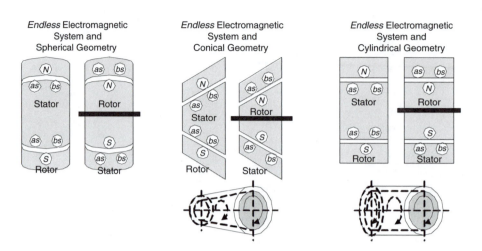

FIGURE 5.4

Permanent-magnet synchronous microtransducers with *endless* electromagnetic system and different geometry.

TABLE 5.1

Classification of Electromagnetic Microdevices (Microtransducers) Using the
Electromagnetic System–Geometry Classifier

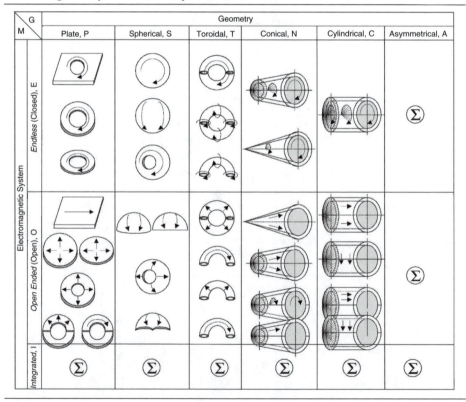

Topology (radial or axial), permanent magnets shaping (strip, arc, disk, rectangular, triangular, etc.), thin-film permanent-magnet characteristics (*B-H* demagnetization curve, energy product, hysteresis minor loop, etc.), commutation, emf distribution, cooling, power, torque, size, torque-speed characteristics, packaging, and other distinct features are easily classified. Permanent-magnet stepper micromotors, fabricated and tested in the mid 1990s and covered in this book, are two-phase synchronous micromotors.

Hence, the devised electromagnetic microdevices (microtransducers) are classified by an *N*-tuple as microdevice type, electromagnetic system, geometry, topology, phase, winding, connection, cooling, fabrication, materials, packaging, etc.

To solve a large variety of problems in modeling, analysis, performance prediction, optimization, control, and fabrication, MEMS must be devised (synthesized) first. Neural networks or generic algorithms can be efficiently used. Neural networks and generic algorithms have evolved to the mature concepts that allow the designer to perform reliable analysis, design, and optimization. Qualitative reasoning in the synthesis, classification, and optimization of MEMS is based on artificial intelligence, and the ultimate goal is to analyze, model, and optimize qualitative models of MEMS when knowledge, processes, and phenomena are not precisely known due to uncertainties. For example, micromachined motion microstructures properties and characteristics (charge density, thermal noise, mass, geometry, etc.) are not precisely known, nonuniform, and varying. It is well known that qualitative models and classifiers are more reliable compared with traditional models if there is a need to perform qualitative analysis, classification, design, optimization, and prediction. Quantitative analysis, classification, and design use a wide range of physical laws and mathematical methods to guarantee validity and robustness using partially available quantitative information.

Synthesis and performance optimization can be based on the knowledge domain. Qualitative representations and compositional (three-dimensional geometric) modeling are used to create control knowledge (existing knowledge, modeling and analysis assumptions, specific plans and requirements domains, task domain and preferences) for solving a wide range of problems through evolving decision making. The solving architectures are based on qualitative reusable fundamental domains (physical laws and phenomena). Qualitative reasoning must be applied to solve complex physics problems in MEMS, as well as to perform engineering analysis and design.

Emphasizing the heuristic concept for choosing the initial domain of solutions, the knowledge domain is available to efficiently and flexibly map all essential phenomena, effects, characteristics, and performances. In fact, the classification table, documented in Table 5.1, ensures classification, modeling, synthesis, and optimization in qualitative and quantitative knowledge domains carrying out analytical and numerical analysis of MEMS. To avoid excessive computations, high-performance (optimal) structures and devices can be found using qualitative analysis and design. That is, qualitative representations and compositional structural modeling can be used to create control knowledge in order to solve fundamental and engineering problems efficiently. The Synthesis and Classification Solver, which gives knowledge domain using compositional structural classification, modeling, analysis, and synthesis, was developed applying qualitative representations. This Synthesis and Classification Solver can integrate modeling and analysis assumptions, expertise, structures, knowledge-based libraries, and preferences that are used in constraining the search (initial structural domain).

Synthesis, classification, and structural optimization are given in terms of qualitative representations and compositional modeling. This guarantees explicit domain due to the application of fundamental concepts. The Synthesis and Classification Solver can be verified by solving problems analytically and numerically. Heuristic synthesis strategies and knowledge regarding physical principles must be augmented for designing micro- and

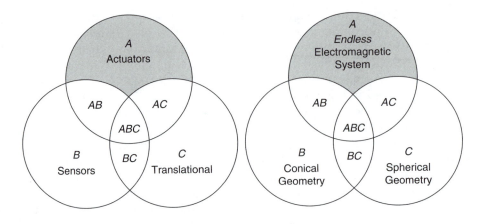

FIGURE 5.5
Venn diagram, $p = 3$: The closed curves are circles, and eight regions are labeled with the interiors that are included in each intersection. The eighth region is the outside, corresponding to the empty set.

nanostructures as well as micro- and nanodevices. Through qualitative analysis, classification, and design, one constrains the search domain. The solutions can be automatically generated, and the synthesized MEMS performance characteristics and end-to-end behavior can be predicted through mathematical modeling, simulations, and analysis. Existing knowledge, specific plan and requirement domains, task domains, preferences, and logical relations make it possible to reason about the modeling and analysis assumptions explicitly, which is necessary to successfully solve fundamental and engineering problems.

The Venn diagram provides a way to represent information about different MEMS topologies, configurations, and architectures. One can use regions labeled with capital letters to represent sets and use lowercase letters to represent elements. By constructing a diagram that represents some initial sets, the designer can deduce other important relations. The basic conventional form of the Venn diagram is three intersecting circles as shown in Figure 5.5. In this diagram, each of the circles represents a set of elements that have some common property or characteristic. Let A stand for actuators, B stand for sensors, and C stand for translational motion microstructure. Then, the region ABC represents actuators and sensors that are synthesized as translational motion microstructure, while the BC maps sensors that are the translational motion microstructures (e.g., *i*MEMS accelerometer, which will be studied in this chapter).

It was illustrated that microtransducers can be designed using the *endless* electromagnetic system and conical, spherical, and conical-spherical geometry (see Figure 5.4). The corresponding Venn diagram is illustrated in Figure 5.5.

Let $A = \{a_1, a_2, \ldots, a_{p-1}, a_p\}$ is the collection of simple closed curves in the *x-y* plane. The collection A is said to be an independent family if the intersection of $b_1, b_2, \ldots, b_{p-1}, b_p$ is nonempty, where each b_i is either $int(a_i)$ (the interior of a_i) or is $ext(a_i)$ (the exterior of a_i). If, in addition, each such intersection is connected, then A is a p-Venn diagram, where p is the number of curves in the diagram.

We consider sets A and B in $\mathbb{R}^n$ and $\mathbb{R}^m$. If $a = \{a_1, a_2, \ldots, a_{n-1}, a_n\}$ and $b = \{b_1, b_2, \ldots, b_{m-1}, b_m\}$ are elements of A and B, then one writes $a \in A$ and $b \in B$. If a and b are not elements of A and B, we write $a \notin A$ and $b \notin B$. The union of two sets A and B is denoted as $A \cup B$, and the intersection is denoted by $A \cap B$. The common set operations are the union, intersection, and complement. The union of A and B is the set of all elements that are either in A or B (or both). Thus, $A \cup B = \{x \mid x \in A \text{ or } x \in B\}$. The union in the Venn diagram is shown in Figure 5.6. The intersection of A and B is the set of all elements that

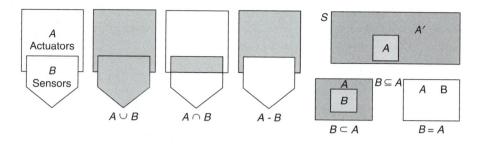

FIGURE 5.6
Two sets (A and B), their union ($A \cup B$), intersection ($A \cap B$), and difference ($A - B$).

are common to A and B, i.e., $A \cap B = \{x \mid x \in A \text{ and } x \in B\}$. If A is a subset of S, then A' is the complement of A in S, the set of all elements of S not in A (see Figure 5.6). Figure 5.6 also illustrates the Venn diagram if $B \subseteq A$.

The Cartesian product of two sets, A and B, is the set of all ordered pairs (a, b) with $a \in A$ and $b \in B$, and $A \times B = \{(a, b) \mid a \in A \text{ and } b \in B\}$. Here, $A \times B$ is the set of all ordered pairs whose first component is in A and whose second component is in B. As an example, consider $A = \{a, b\}$ and $B = \{1, 2, 4, 6\}$. Then, we have $A \times B = \{(a, 1), (a, 2), (a, 4), (a, 6), (b, 1), (b, 2), (b, 4), (b, 6)\}$. Let us consider motion MEMS that integrates two electromagnetic systems, e.g., *endless* (E) and *open-ended integrated* (O). Thus, $S = \{E, O\}$. One obtains, $S \times S = \{(E, E), (E, O), (O, E), (O, O)\}$. In other words, if S is the set of outcomes of two electromagnetic systems, then $S \times S$ is the set of outcomes of varying an electromagnetic system twice.

Other example illustrates micromagnets. Consider $A = \{\text{plate, spherical}\}$ and $B = \{\text{Co, NiFe}\}$; then $A \times B = \{(\text{Plate, Co}), (\text{Plate, NiFe}), (\text{Spherical, Co}), (\text{Spherical, NiFe})\}$, which can be viewed as

$$A \times B = \{\text{Plate Co, Plate NiFe, Spherical Co, Spherical NiFe}\}$$

5.1.1 Cardinality

If A is a finite set, then $n(A)$, the number of elements A contains, is called the cardinality of A. If A and B are finite sets, then one has

$$n(A \cup B) = n(A) + n(B) = n(A \cap B)$$

If A and B are disjoint, then

$$n(A \cup B) = n(A) + n(B)$$

If S is a finite universal set, and A is a subset of S, then

$$n(A') = n(S) = n(A) \quad \text{and} \quad n(A) = n(S) = n(A')$$

If A and B are finite sets, then

$$n(A \times B) = n(A)n(B)$$

As an example, we consider a set (S) of 100 distinct magnets. Let F be the set of the soft magnets, H is the set of the hard magnets, and P is the set of magnets that have plate shape (three are soft and the other two are hard magnets). We assign $n(F) = 20$, $n(H) = 10$ and $n(P) = 5$.

Then, $n(F \cup H) = n(F) + n(H) = 20 + 10 = 30$ (note that $F \cap H = \emptyset$),

$$n(F \cup P) = n(F) + n(P) - n(F \cap P) = 20 + 5 - 2 = 23$$

$$n(F') = n(S) - n(F) = 100 - 20 = 80$$

A decision algorithm is a step-by-step procedure that explicitly defines rules or instructions regarding what to do at every step. All decision algorithms require multiple sequential steps, with step 1 having n_1 outcomes, step 2 having n_2 outcomes, and step M having n_M outcomes. The number of outcomes of the algorithm is $n_1 \times n_2 \times \cdots \times n_{M-1} \times n_M$. For example, let the four-step process results in 6, 4, 3, and 3 possible outcomes for the considered steps. Then, the total number of outcomes is $6 \times 4 \times 3 \times 3 = 216$. If a decision algorithm requires a choice among several different alternatives (distinct electromagnetic systems, geometry, magnets, conductors, etc.), then the total number of outcomes is obtained by adding the number of outcomes of each alternative. A permutation of n outcomes taken M at a time is an ordered list of M chosen from a set of n. The number of permutations of n items taken M at a time is given by $P(n,M) = n \times (n-1) \times (n-2) \times \cdots \times (n-M+1) = n!/[(n-M)!]$. A combination of n outcomes taken M at a time is an unordered set of M chosen from n. The number of combinations of n outcomes taken M at a time is

$$C(n,M) = \frac{P(n,M)}{M!} = \frac{n!}{M!(n-M)!} = \frac{n \times (n-1) \times \cdots \times (n-M+1)}{M \times (M-1) \times \cdots \times 2 \times 1}$$

For example, let $n = 8$ and $M = 5$. We have $P(8,5) = 8 \times 7 \times 6 \times 5 \times 4 = 6720$ and

$$C(n,M) = \frac{P(n,M)}{M!} = \frac{8 \times 7 \times 6 \times 5 \times 4}{5 \times 4 \times 3 \times 2 \times 1} = 56$$

5.1.2 Algebra of Sets

A set is a collection of objects (order is not significant and multiplicity is usually ignored) called the elements of the set. Symbols are used widely in the algebra of sets. One can have finite and infinite sets, for example, $\{1, 2, 3, 4, \ldots, 9999\}$ and $\{1, 2, 3, 4, \ldots\}$, respectively.

If a is an element of set A, we have $a \in A$. For example, $c \in \{a\ b\ c\ d\ e\}$.

If a is not an element of set A, one writes $a \notin A$. For example, $v \notin \{a\ b\ c\ d\ e\}$.

If a set A contains only the single element a, it is denoted as $\{a\}$.

The null set (set does not contain any elements) is denoted as $\emptyset$.

Two sets A and B are equal ($A = B$) if $a \in A$ iff $a \in B$. For example,
$\{1, 2, 3, 4, 5\} = \{3, 2, 1, 5, 4\}$.

If $a \in A$ implies that $a \in B$, then A is a subset of B, and $A \subset B$.

The symbols $\subset$ and $\subseteq$ describe a proper and an improper subset.

$B \subset A$ means that B is a proper subset of A, i.e., $B \subseteq A$, but $B \neq A$. For example, $\{1, 2, 3\} \subset \{1, 2, 3, 4\}$ or $\{a, b, c\} \subset \{a, b, c, d\}$.

$B \subseteq A$ means that B is a subset of A, and every element of B is also an element of A. For example, $\{1, 2, 3, 4\} \subseteq \{1, 2, 3, 4\}$ and $\{a, b, c, d\} \subseteq \{a, b, c, d\}$. Other examples are $\{1, 2, 3\} \subseteq \{1, 2, 3, 4\}$ and $\{a, b, c\} \subseteq \{a, b, c, d\}$.

For example, if $A \subset B$ and $B \subset A$, then A is called an improper subset of B, $A = B$ (if there exists element b in B which is not in A, then A is a proper subset of B).

If the set of all elements under consideration make up the universal set U, then $A \subset U$.

The set A' is the complement of set A if it is made up of all the elements of U which are not elements of A. For each set A there exists a unique set A' such that $A \cup A' = U$ and $A \cap A' = \emptyset$. Furthermore, $(A')' = A$.

Two operations on sets are union $\cup$ and intersection $\cap$.

For example, an element $a \in A \cup B$ iff $a \in A$ *or* $a \in B$.

In contrast, an element $a \in A \cap B$ iff $a \in A$ *and* $a \in B$.

Using $\cup$ and $\cap$ operators, we have the following well-known algebra of sets laws:

- Closure: There is a unique set $A \cup B$ which is a subset of U, and there is a unique set $A \cap B$ which is a subset of U.

- Commutative: $A \cup B = B \cup A$ and $A \cap B = B \cap A$.

- Associative: $(A \cup B) \cup C = A \cup (B \cup C)$ and $(A \cap B) \cap C = A \cap (B \cap C)$.

- Distributive: $A \cup (B \cap C) = (A \cup B) \cap (A \cup C)$ and $A \cap (B \cup C) = (A \cap B) \cup (A \cap C)$. Using the index set Λ, $\lambda \in \Lambda$, one has $A \cup (\bigcap_{\lambda \in \Lambda} B_\lambda) = \bigcap_{\lambda \in \Lambda}(A \cup B_\lambda)$ and $A \cap (\bigcup_{\lambda \in \Lambda} B_\lambda) = \bigcup_{\lambda \in \Lambda}(A \cap B_\lambda)$.

- Idempotent: $A \cup A = A$ and $A \cap A = A$.

- Identity: $A \cup \emptyset = A$ and $A \cap U = A$.

- DeMorgan's: $(A \cup B)' = A' \cap B'$ and $(A \cap B)' = A' \cup B'$.

- U and $\emptyset$ laws: $U \cup A = U$, $U \cap A = A$, $\emptyset \cup A = A$ and $\emptyset \cap A = \emptyset$.

Additional rules and properties of the complement are as follows:

$$A \subset (A \cup B), \quad (A \cap B) \subset A, \quad A \subset U, \quad \emptyset \subset A$$

If $A \subset B$, then $A \cup B = B$, and if $B \subset A$, then $A \cap B = B$

If the designer performs synthesis, design, or experiments that have one or more outcomes, these possible outcomes become the elements of a *set of outcomes* associated with the synthesis, design, or experiment. For example, if the synthesis results in three distinct electromagnetic systems, the set of outcomes can be written as

$$S = \{endless,\ open\text{-}ended,\ integrated\} \quad \text{or} \quad S = \{E, O, I\}$$

If one examines two distinguishable media (for example, magnets) with, say, four distinct relative permeabilities (say, 10000, 20000, 30000, and 40000), the set of outcomes can be represented as

$$S = \begin{cases} (1,1) & (1,2) & (1,3) & (1,4) \\ (2,1) & (2,2) & (2,3) & (2,4) \\ (3,1) & (3,2) & (3,3) & (3,4) \\ (4,1) & (4,2) & (4,3) & (4,4) \end{cases}$$

For two indistinguishable (identical) media, we have

$$S = \begin{cases} (1,1) & (1,2) & (1,3) & (1,4) \\ & (2,2) & (2,3) & (2,4) \\ & & (3,3) & (3,4) \\ & & & (4,4) \end{cases}$$

5.1.3 Sets and Lattices

A set is simply a collection of elements. For example, a, b, and c can be grouped together as a set, which is expressed as $\{a, b, c\}$ where the curly braces are used to enclose the elements that constitute a set. In addition to the set $\{a, b, c\}$, we define the sets $\{a, b\}$ and $\{d, e, g\}$.

Using the union operation, we have

$$\{a,b,c\} \cup \{a,b\} = \{a,b,c\} \quad \text{and} \quad \{a,b,c\} \cup \{d,e,g\} = \{a,b,c,d,e,g\}$$

while the intersection operation leads us to

$$\{a,b,c\} \cap \{a,b\} = \{a,b\} \quad \text{and} \quad \{a,b,c\} \cap \{d,e,g\} = \varnothing = \{\}$$

where $\{\}$ is the empty (or null) set.

The subset relation can be used to partially order a set of sets. If some set A is a subset of a set B, then these sets are partially ordered with respect to each other. If a set A is not a subset of set B, and B is not a subset of A, then these sets are not ordered with respect to each other. This relation can be used to partially order a set of sets in order to classify MEMS and NEMS using the Synthesis and Classification Solver previously introduced. Sets possess some additional structural, geometrical, and other properties. Additional definitions and properties can be formulated and used applying lattices.

Using a lattice, we have

- $A \subseteq A$ (reflexive law).
- If $A \subseteq B$ and $B \subseteq A$, then $A = B$ (antisymmetric law).
- If $A \subseteq B$ and $B \subseteq C$, then $A \subseteq C$ (transitive law).
- A and B have a unique greatest bound, $A \cap B$. Furthermore, $G = A \cap B$, or G is the greatest lower bound of A and B if: $A \subseteq G$, $B \subseteq G$, and if W is any lower bound of A and B, then $G \subseteq W$.
- A and B have a unique least upper bound, $A \cup B$. Furthermore, $L = A \cup B$, or L is the least upper bound of A and B if: $L \subseteq A$, $L \subseteq B$, and if P is any upper bound of A and B, then $P \subseteq L$.

A lattice is a partially ordered set where for any pair of sets (hypotheses) there is a least upper bound and greatest lower bound. Let our current hypothesis be $H1$ and the current training example be $H2$. If $H2$ is a subset of $H1$, then no change of $H1$ is required. If $H2$ is not a subset of $H1$, then $H1$ must be changed. The minimal generalization of $H1$ is the least upper bound of $H2$ and $H1$, and the minimal specialization of $H1$ is the greatest lower bound of $H2$ and $H1$. Thus, the lattice serves as a map that allows us to locate the current hypothesis $H1$ with reference to the new information $H2$. There exists the correspondence between the algebra of propositional logic and the algebra of sets. We refer to hypotheses as logical expressions, as rules that define a concept, or as subsets of the possible instances constructible from some set of dimensions. Furthermore, union and intersection were the important operators used to define a lattice. In addition, the propositional logic expressions can also be organized into a corresponding lattice to implement the artificial learning.

A general structure S is an ordered pair formed by a set object O and a set of binary relations R such that

$$S = (O, R) = \bigcup_{i=1}^{n} S_i$$

where $O = \{o_1, o_2, \ldots, o_{z-1}, o_z\}$, $\forall o_i \in O$; $R = \{r_1, r_2, \ldots, r_{p-1}, r_p\}$, $\forall r_i \in R$; S_i is the simple structure.

In the set object O we define the input n, output u, and internal a variables. We have $o_i = \{q_1^i, q_2^i, \ldots, q_{g-1}^i, q_g^i\}$, $q_j^i = (n_j^i, u_j^i, a_j^i)$, $q_j^i \in O^3$. Hence, the range of q, as a subset of O, is $R(q)$. Using the input-output structural function, different MEMS and NEMS can be synthesized. The documented general theory of synthesis, classification, and structural optimization, which is built using the algebra of sets, allows the designer to derive relationships and flexibly adapt, fit, and optimize the micro- and nanoscale structures and devices within the sets of given possible solutions.

Using the Synthesis and Classifier Solver, which is given in Table 5.1 in terms of electromagnetic system and geometry, the designer can classify the existing motion microdevices as well as synthesize novel high-performance microdevices. As an example, the spherical, conical, and cylindrical geometries of a two-phase permanent-magnet synchronous microdevice are illustrated in Figure 5.7.

This section documents new results in the MEMS synthesis, which can be used to optimize the microdevice performance. The conical (existing) and spherical-conical (devised) microdevice geometries are illustrated in Figure 5.7. Using the innovative spherical-conical geometry, which is different compared with the existing conical geometry, one increases the active length L_r and average diameter D_r. For radial flux microdevices, the electromagnetic torque T_e is proportional to the squared rotor diameter and axial length. In particular,

$$T_e = k_T D_r^2 L_r$$

where k_T is the constant.

From the above relationship, it is evident that the spherical-conical micromotors develop higher electromagnetic torque compared with the conventional design. In addition, improved cooling, reduced undesirable torques components, and increased ruggedness and robustness contribute to the viability of the proposed solution. Thus, using the synthesis (classifier) paradigm, novel microdevices with superior performance can be devised.

The cross section of the slotless radial-topology micromotor, fabricated on the silicon substrate with polysilicon stator (with deposited windings), polysilicon rotor (with deposited permanent magnets), and contact bearing, is illustrated in Figure 5.8. The fabrication of this micromotor and the processes were reported in Chapter 4.

The analysis of the *Escherichia coli (E. coli)* nanobiomotor, which was illustrated in Figure 5.2, leads the designer to the axial-topology micro- and nanoscale transducers. As was emphasized, the designer can devise novel high-performance micro- and nanotransducers (micro- and nanomachines) through biomimicking and prototyping. The distinguished beneficial features of the devised transducers are as follows:

- Unique radial and axial topologies
- Electrostatic- and electromagnetic-based actuation mechanisms
- Different electromagnetic systems
- Noncontact electrostatic bearings

Two distinct transducer topologies are radial and axial. The magnetic flux in the radial (or axial) direction interacts with the time-varying axial (or radial) electromagnetic field with ultimate goal to produce the electromagnetic torque. Using the radial flux topology, the cylindrical transducer with permanent-magnet poles on the rotor was illustrated in Figure 5.3.

The major advantages of the radial topology are that high torque and power densities can be achieved, and the net radial force on the rotor is zero. The disadvantages are that it is difficult to fabricate and assemble micro- and nanostructures (stator and rotor), and the air gap is not adjustable.

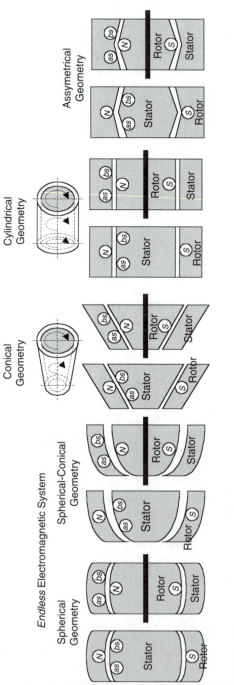

FIGURE 5.7

Two-phase permanent-magnet synchronous microdevice with *endless* electromagnetic system and distinct microtransducer geometry.

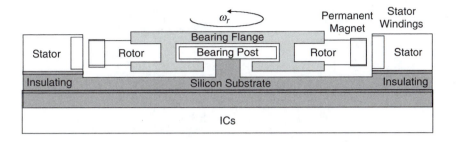

FIGURE 5.8
Cross-section schematics for slotless radial-topology permanent-magnet brushless micromotor (microtransducer) with ICs.

The advantages of the axial topology transducers, as shown in Figure 5.2, are the possibility to affordably fabricate because permanent magnets have flat surfaces (thin-film permanent magnets can be used), there are no strict shape requirements, there is no rotor back ferromagnetic material required (silicon can be used), the air gap can be adjusted, and it is easy to lay out (implant and deposit) nano- and microscale wires to make the windings on the flat stator (silicon).

The stationary magnetic field is established by the permanent magnets, and stators and rotors can be fabricated using micromachining and high-aspect-ratio technologies. Slotless stator windings can be deposited on the silicon as the implanted nanowires.

Here is step-by-step procedure for micro- and nanotransducer design:

- Use the Synthesis and Classifier Concept to devise novel transducers by researching operational principles, topologies, configurations, geometry, electromagnetic systems (closed, open, or integrated), interior, exterior, etc.
- Define application and environmental requirements as well as specify performance specifications.
- Perform electromagnetic, energy conversion, electromechanical, and sizing/dimension estimates.
- Define and design technologies, techniques, and processes to fabricate micro- and nanostructures (e.g., stator, rotor, bearing, windings, etc.), and assemble/integrate them in the transducer (device).
- Select materials and chemicals (substrate, insulators, conductors, permanent magnets, etc.).
- Perform thorough electromagnetic, mechanical, thermodynamic and vibroacoustic design with performance analysis and outcome prediction.
- Test and examine the designed micro- or nanotransducer.
- Modify and refine the design.
- Optimize the overall micro- or nanotransducer performance.

A variety of paradigms can be applied utilizing the set theory reported. For example, nonlinear and linear programming, optimization, entropy analysis, probability theory, Bayesian analysis, and other concepts enhance and complement the previous analysis. As an illustration, let us briefly recall the probability fundamentals. The conditional probability $P(e\,|\,k)$ is the probability of the event e, given the event k. Correspondingly, one can examine failures, uncertainties and other characteristics within the bidirectional synthesis

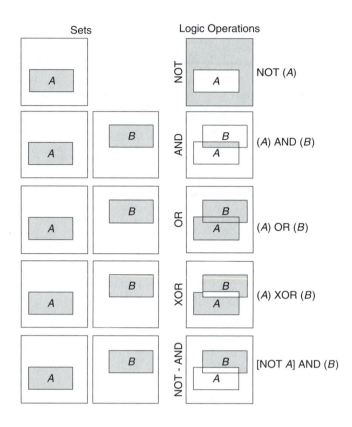

FIGURE 5.9
Logic operations with sets.

taxonomy in NEMS and MEMS design. In general, the conditional probability is found using the following equation:

$$P(e \mid k) = \frac{P(e \cap k)}{P(k)}$$

If all outcomes are equally likely, then $P(e \mid k) = n(e \cap k)/n(k)$, where $n(k)$ is the number of outcomes in the event k. The experimental probability is calculated as $P(e \mid k) = f(e \cap k)/f(k)$, where $f(k)$ is the frequency of the event k.

Having illustrated different electromagnetic systems, geometries, and other distinct features of nano- and microscale motion devices, it is straightforward to illustrate that the studied devices can be described utilizing the mathematical morphology that integrates algebra of sets and logics. For example, AND, OR, NOT, and XOR (exclusive OR) logic operations can be utilized. Though logic operations are restricted to binary variables, these logic operations have a one-to-one correspondence with the set operations. Figure 5.9 demonstrates how one can straightforwardly apply the logic operations in NEMS and MEMS classification.

5.2 Microaccelerometers as Microelectromechanical Microdevices

Different MEMS have been discussed, and it was emphasized that MEMS can be used as actuators, sensors, and transducers (actuators-sensors). Due to the limited torque and force densities, microdevices usually cannot develop high torque and force, and multinode

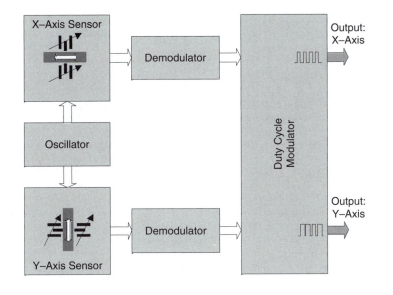

FIGURE 5.10
Functional block diagram of the ADXL202 accelerometer.

cooperative microdevices are used. In contrast, these characteristics (power, torque, and force densities) are not critical in sensor applications. Therefore, MEMS are widely used as microscale sensors. Signal-level signals, measured by sensors, are fed to analog, digital, or hybrid ICs, and sensor design, signal processing, signal conditioning, and interfacing are extremely important in engineering practice.

Smart integrated sensors are the sensors in which in addition to sensing the physical variables, data acquisition, filtering, data storage, communication, interfacing, and networking are embedded. Thus, while the primary component is the sensing element (microstructure), multifunctional integration of sensors and ICs is the current demand. High-performance accelerometers, manufactured by Analog Devices using integrated microelectro-mechanical system technology (*i*MEMS), which is based on CMOS processes, are studied in this section. In addition, the application of smart integrated sensors is briefly discussed.

We study the dual-axis, surface-micromachined ADXL202 accelerometer (manufactured on a single monolithic silicon chip), which combines highly accurate acceleration sensing motion microstructure (proof mass) and signal-processing electronics (signal conditioning ICs). As documented in the Analog Devices catalog data, this accelerometer measures dynamic positive and negative acceleration (vibration) as well as static acceleration (force of gravity). The functional block diagram of the ADXL202 accelerometer with two digital outputs (ratio of pulse width to period is proportional to the acceleration) is illustrated in Figure 5.10.

Polysilicon surface-micromachined sensor motion microstructure is fabricated on the silicon wafer by depositing polysilicon on the sacrificial oxide layer, which is then etched away leaving the suspended proof mass. Polysilicon springs suspend this proof mass over the surface of the wafer. The deflection of the proof mass is measured using the capacitance difference (see Figure 5.11).

The proof mass (1.3 μm, 2 μm thick) has movable plates, which are shown in Figure 5.11. The free-space (air) capacitances C_1 and C_2 (capacitances between the movable plate and two stationary outer plates) are functions of the corresponding displacements x_1 and x_2. The parallel-plate capacitance is proportional to the overlapping area between the plates (125 μm $\times$ 2 μm) and the displacement (up to 1.3 μm).

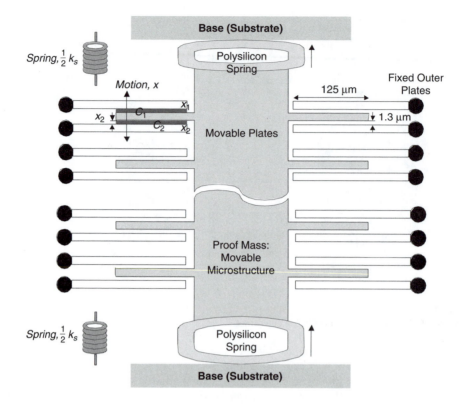

FIGURE 5.11
Accelerometer structure: proof mass, polysilicon springs, and sensing elements.

Neglecting the fringing effects (nonuniform distribution near the edges), the parallel-plate capacitance is

$$C = \varepsilon \frac{A}{d} = \varepsilon_A \frac{1}{d}$$

where ε is the permittivity, A is the overlapping area, d is the displacement between plates, and $\varepsilon_A = \varepsilon A$.

If the acceleration is zero, the capacitances C_1 and C_2 are equal because $x_1 = x_2$ (in ADXL202 accelerometer, $x_1 = x_2 = 1.3 \ \mu m$).

Thus, one has

$$C_1 = C_2$$

where $C_1 = \varepsilon_A (1/x_1)$ and $C_2 = \varepsilon_A (1/x_2)$.

The proof mass (movable microstructure) displacement x results due to acceleration. If $x \neq 0$, we have the following expressions for capacitances:

$$C_1 = \varepsilon_A \frac{1}{x_1 + x} \quad \text{and} \quad C_2 = \varepsilon_A \frac{1}{x_2 - x} = \varepsilon_A \frac{1}{x_1 - x}$$

The capacitance difference is found to be

$$\Delta C = C_1 - C_2 = 2\varepsilon_A \frac{x}{x^2 - x_1^2}$$

Measuring ΔC, one finds the displacement x by solving the nonlinear algebraic equation

$$\Delta C x^2 - 2\varepsilon_A x - \Delta C x_1^2 = 0$$

This equation can be simplified. For small displacements, the term $\Delta C x^2$ is negligible. Thus, $\Delta C x^2$ can be omitted. Then, from

$$x \approx -\frac{x_1^2}{2\varepsilon_A} \Delta C$$

one concludes that the displacement is approximately proportional to the capacitance difference ΔC.

For an ideal spring, according to Hook's law, the spring exhibits a restoring force F_s which is proportional to the displacement x. Thus,

$$F_s = k_s x$$

where k_s is the polysilicon spring constant (Figure 5.11 illustrates that two springs are used).

From Newton's second law of motion, neglecting the air friction (which is negligibly small), the following differential equation results:

$$ma = m\frac{d^2 x}{dt^2} = k_s x$$

Thus, the displacement due to the acceleration is

$$x = \frac{m}{k_s} a$$

while the acceleration, as a function of the displacement, is

$$a = \frac{k_s}{m} x$$

Then, making use of the measured ΔC, the acceleration is found to be

$$a = -\frac{k_s x_1^2}{2m\varepsilon_A} \Delta C$$

Making use of Newton's second law of motion, we have

$$ma = m\frac{d^2 x}{dt^2} = \underset{\text{spring force}}{f_s(x)}$$

where $f_s(x)$ is the spring restoring force, which is a nonlinear function of the displacement, $f_s(x) = k_{s1} x + k_{s2} x^2 + k_{s3} x^3$, and k_{s1}, k_{s2}, and k_{s3} are the spring constants.

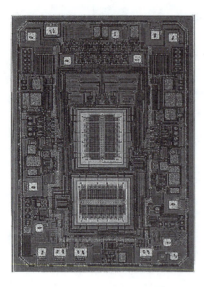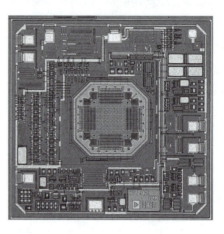

FIGURE 5.12 (See color insert following page 332.)
ADXL202 and ADXL250 accelerometers: proof mass with fingers and ICs (courtesy of Analog Devices).

Therefore, the nonlinear equation is

$$ma = k_{s1}x + k_{s2}x^2 + k_{s3}x^3$$

The expression for the acceleration is found to be

$$a = \frac{1}{m}(k_{s1}x + k_{s2}x^2 + k_{s3}x^3)$$

where $x \approx -(x_1^2/2\varepsilon_A)\Delta C$.

This equation can be used to calculate the acceleration a using the measured capacitance difference ΔC. Two proof masses (motion microstructures) can be placed orthogonally to measure the accelerations in the x- and y-axis (ADXL250), and the movable plates can be mounted along the sides (ADXL202). The signal conditioning, filtering, computing, and input-output interface are performed by the ICs.

Figure 5.12 documents the ADXL202 and ADXL250 accelerometers, which integrate the microscale microstructure (moving masses, springs, etc.) and ICs.

Responding to acceleration, the proof mass moves due to the mass of the movable microstructure (m) along the x- and y-axes relative to the stationary member (accelerometer). The motion of the proof mass is constrained, and the polysilicon springs hold the movable microstructure (beam). Assuming that the polysilicon springs and the proof mass obey Hook's and Newton's laws, it was shown that the acceleration is calculated as

$$a = \frac{k_s}{m}x$$

The fixed outer plates are excited by two square wave 1-MHz signals of equal magnitude that are 180 degrees out of phase from each other. When the movable plates are centered between the fixed outer plates, we have $x_1 = x_2$. Thus, the capacitance difference ΔC and the output signal are zero. If the proof mass (movable microstructure) is displaced due to the acceleration, we have $\Delta C \neq 0$. Thus, the capacitance is not balanced, and the amplitude of the output voltage is a function of (proportional to) the displacement of the proof mass x.

Phase demodulation is used to determine the sign (positive or negative) of acceleration. The AC signal is amplified by buffer amplifier and demodulated by a synchronous synchronized demodulator. The output of the demodulator drives the high-resolution duty cycle modulator. In particular, the filtered signal is converted to a PWM signal by the 14-bit duty cycle modulator. The zero acceleration produces 50% duty cycle. The PWM output fundamental period can be set from 0.5 to 10 msec.

The Analog Devices data for different *i*MEMS accelerometers ADXL202/ADXL210 and ADXL150/ADXL250 are available on the Analog Devices Web site.

There is a wide range of industrial systems where smart integrated sensors are used. For example, accelerometers can be used for the following:

- Vibroacoustic sensing, detection, and diagnostics
- Active vibration and acoustic control
- Situation awareness
- Health and structural integrity monitoring
- Internal navigation systems
- Earthquake-actuated safety systems
- Seismic monitoring and detection

For example, current activities in analysis, design, and optimization of flexible structures (aircraft, missiles, manipulators, robots, spacecraft, underwater vehicles, etc.) are driven by requirements and standards that must be guaranteed. The vibration, structural integrity, and structural behavior must be addressed and studied. For example, fundamental, applied, and experimental research in aeroelasticity and structural dynamics is conducted to obtain a fundamental understanding of the basic phenomena involved in flutter, force and control responses, vibration, and control. Through optimization of aeroelastic characteristics as well as applying passive and active vibration control, the designer minimizes vibration and noise, and current research integrates development of aeroelastic models and diagnostics to predict stalled/whirl flutter, force and control responses, unsteady flight, aerodynamic flow, etc. Vibration control is a very challenging problem because the designer must take into account complex interactive physical phenomena (elastic theory, structural and continuum mechanics, radiation and transduction, wave propagation, chaos, etc.). Thus, it is necessary to accurately measure the vibration, and the accelerometers, which allow one to measure the acceleration in the micro-g range, are used. The application of the MEMS-based accelerometers ensures small size, low cost, ruggedness, hermeticity, reliability, and flexible interfacing with microcontrollers, microprocessors, and DSPs.

High-accuracy low-noise accelerometers can be used to measure velocity and position. This provides the backup in the case of the GPS system failures or for reckoning applications (the initial coordinates and speed are assumed to be known). The acceleration, velocity, and position in the x-y plane are found using integration. In particular,

$$v_x(t) = \int_{t_0}^{t_f} a_x(t)dt, \quad v_y(t) = \int_{t_0}^{t_f} a_y(t)dt \quad \text{and} \quad x_x(t) = \int_{t_0}^{t_f} v_x(t)dt, \quad x_y(t) = \int_{t_0}^{t_f} v_y(t)dt$$

Microgyroscopes have been designed, fabricated, and deployed using tech-nology similar to *i*MEMS accelerometers. Using the difference capacitance (between the movable rotor and stationary stator plates), the angular acceleration is measured. The butterfly-shaped polysilicon rotor is suspended above the substrate, and Figure 5.13 shows the microgyroscope.

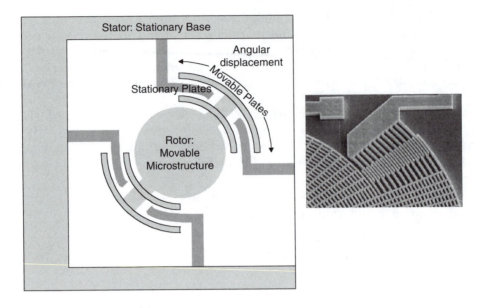

FIGURE 5.13
Angular microgyroscope structure.

5.3 Optimization with Application to Synthesis and Classification Solver

Utilizing set theory, as integrated with other mathematical concepts, one can explicitly describe MEMS and NEMS. Sections 5.1 and 5.2 document distinct nano- and microsystems that can be described and expressed. Other important problems are involved, including optimization. The algorithmic concept in the synthesis, classification, and structural and performance optimization starts by selecting an initial set of competing configurations and solutions (electromagnetic system, geometry, topology, architecture, etc.) for a particular problem using specifications and requirements imposed. The solutions can be generated randomly from the entire domain. However, as was emphasized earlier, available information and accessible knowledge can be readily used in order to formulate the partial (specific) domain (classifier subset). Intelligent knowledge-based libraries and adaptive decision-making algorithms can be applied. The solutions are evaluated for their efficiency and fitness. Performance and regret functionals can be designed to integrate weighted-cost integrands (terms). Linear and nonlinear optimization (linear and nonlinear programming) allow one to find optimal solutions solving, for example, the constrained optimization problem. The maxima or minima can be found using the gradient-based search. Alternatively, the evolutionary algorithms can be used, and the performance functionals are applied to compare and rank the competing solutions. The analysis and evaluation of candidate solutions are very complex problems due to infinite number of possible solutions (it is very difficult or impossible to find solutions randomly from the entire classifier domain of all possible solutions). Thus, the solutions should be examined in the partial domain (subset) of most efficient, feasible, and suitable solutions. This will allow one to define the partial classifier domain of solutions generation to efficiently solve practical problems in MEMS and NEMS design and optimization.

The following should be performed to simplify the search and optimize the algorithm to solve a wide variety of synthesis, classification, and structural optimization problems:

- Formulate and apply rules and criteria for solution sustaining based on performance analysis, assessments, and outcomes.
- Synthesize performance and regret functionals with constraints imposed.
- Develop and generate the partial classifier domain (subset). Select solution representations.
- Download available data; initialize and execute solutions.
- Analyze and compare distinct solutions and derive optimal candidates.
- Develop the fabrication processes to fabricate MEMS or NEMS.
- Refine the design and find the optimal solution.

Using the classifier developed, the designer can synthesize novel high-performance nano-, micro-, and miniscale devices (actuators and sensors). As an example, the synthesis of a two-phase permanent-magnet synchronous microtransducer with *endless* electromagnetic system was performed using distinct geometry. The spherical, spherical-conical, conical, cylindrical, and asymmetrical geometries of the synthesized actuator/sensor are documented in Figure 5.7. The radial- and axial-topology nano- and microtransducers were devised and analyzed (see Figure 5.2 and Figure 5.3).

5.3.1 Illustrative Examples of the MATLAB Application

Let us illustrate the application of the MATLAB. For example, a simple nonlinear function

$$z(x,y) = \frac{\sin\sqrt{2x^2 + 2y^2 + \varepsilon}}{\sqrt{2x^2 + 2y^2 + \varepsilon}}$$

is calculated and plotted as illustrated in Figure 5.14. We assign $\varepsilon = 1 \times 10^{-2}$ within $-10 \le x \le 10$ and $-10 \le y \le 10$. In particular, calculations and three-dimensional plotting are performed running the following statements:

```
>> [x,y]=meshgrid([-10:0.1:10]);
xy=sqrt(2*x.^2+2*y.^2) +1e-2;z=sin(xy)./xy; plot3(x,y,z)
```

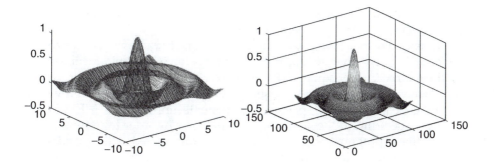

FIGURE 5.14
Three-dimensional plots of $z(x,y) = \frac{\sin\sqrt{2x^2 + 2y^2 + \varepsilon}}{\sqrt{2x^2 + 2y^2 + \varepsilon}}$.

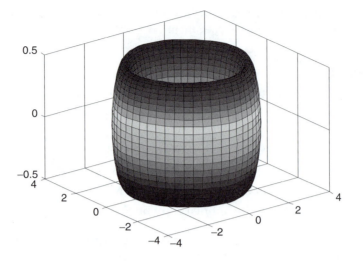

FIGURE 5.15
Surface as calculated and plotted in the MATLAB environment.

```
[x,y]=meshgrid([-10:0.2:10]);
xy=sqrt(2*x.^2+2*y.^2) +1e-2;z=sin(xy)./xy; mesh(z)
```

The tensor-product splines allow one to calculate and plot surfaces. For example, Figure 5.15 illustrates the results of running the following statement:

```
x=0:4; y=-2:2; R=3; r=2;
v(1,:,:)=[R (r+R)/2 r (r+R)/2 R].'*[1 0 -1 0 1];
v(2,:,:)=[R (r+R)/2 r (r+R)/2 R].'*[0 1 0 -1 0];
v(3,:,:)=[0 (R-r)/2 0 (r-R)/2 0].'*[1 1 1 1 1];
sur=csape({x,y},v,'periodic'); fnplt(sur);
```

5.3.2 Linear Programming

Linear programming is the problem that can be formulated and expressed in the so-called standard form as

$$\text{minimize } cx \text{ subject to } Ax = b, x \geq 0$$

where x is the vector of variables to be solved, A is the matrix of known coefficients, and c and b are the vectors of known coefficients.

The term cx is called the objective function, and $Ax = b$, $x \geq 0$ are the constraints.

All these entities must have the appropriate dimensions. It should be emphasized that in general, the matrix A is not square. Therefore, one cannot solve the linear programming problems by using inverse matrix A^{-1}. Usually A has more columns than rows, and $Ax = b$ can be defined based on the specific requirements imposed on MEMS and NEMS enabling a great spectrum in the choosing of variables x in order to minimize cx.

Although all linear programming problems can be formulated in the standard form (all variables are nonnegative), in practice it may be necessary to integrate the constraints and bounds, that is, $x_{min} \leq x \leq x_{max}$ (for example, the electromagnetic field intensity, charge density, material properties, velocity, and other physical quantities are bounded). This allows one to bound the variations of variables within explicit upper or lower bounds, although this implies the limit because problems may have no finite solution. In addition, the constraints can be imposed on Ax. That is, $b_{min} \leq Ax \leq b_{max}$.

It is evident that the user needs to integrate the inequality constraints in order to solve the specific practical problems in the synthesis of MEMS and NEMS. In fact, the importance of linear programming derives by its straightforward applications and by the existence of well-developed general-purpose techniques and computationally efficient software for finding optimal solutions. Simplex methods, introduced 50 years ago, use the *basic* solutions computed by fixing the variables at their bounds to reduce the constraints $Ax = b$ to a square system in order to solve it for unique values of the remaining variables. The *basic* solutions give extreme boundary points of the feasible region defined by $Ax = b$, $x \geq 0$. Therefore, the simplex method is based on moving from one point to another along the edges of the boundary. In contrast, barrier (interior-point) methods utilize points within the interior of the feasible region. The integer linear programming requires that some or all variables are integers. Widely used general-purpose techniques for solving integer linear programming use the solutions to a series of linear programming problems to manage the search for integer solutions and to prove the optimality.

Most linear programming problems can be straightforwardly solved using the available robust computationally efficient software. In fact, the problems with thousands variables and constraints are treated the same as the small-dimensional one. Problems having hundreds of thousands of variables and constraints are tractable and can be solved using different computational environments.

Modern linear programming and optimization software comes in two related but different tools:

- *Algorithmic codes*, which allow one to find the optimal solutions to specific linear problems using a compact listing of the variables as input as providing the compact listing of optimal solution values and related information as outputs
- *Modeling systems*, which allow one to formulate the problems and analyze their solutions using the descriptions of linear programs in a natural and convenient form as inputs allowing the solution output to be viewed in similar terms through automatic conversion to the forms required by algorithmic codes

The collection of statement forms for the input is often called a *modeling language*.

Large-scale linear programming algorithmic codes rely on general-structure sparse matrix techniques and other developed refinements. In additional to a variety of codes available, specialized toolboxes are applied. For example, in the MATLAB environment, the Optimization Toolbox can be very effectively used.

5.3.3 Nonlinear Programming

In contrast to the linear programming, *nonlinear programming* is a problem that can be formulated as

$$\text{minimize } F(x) \text{ subject to } g_i(x) = 0, \, h_j(x) \geq 0$$

$$\text{for } i = 1, \ldots, n, \, n \geq 0, \, j = n+1, \ldots, m, \, m \geq n$$

Other formulations of the nonlinear programming can be set. It is evident that one minimizes the scalar-valued function F of several variables (x is the vector) subject to one or more other functions that serve to limit or define the values of these variables. Here, $F(x)$ is the objective function (criterion), while the other functions are called the constraints. If maximization is needed, one multiplies $F(x)$ by -1.

Nonlinear programming is a much more difficult problem compared with linear programming. As a result, special cases have been studied. The solution is found if the constraints $g_i(x)$ and $h_j(x)$ are linear (the linearly constrained optimization problem). If the objective function $F(x)$ is quadratic, the problem is called quadratic programming. One of the greatest challenges in nonlinear programming is the issues associated with the local optima where the requirements imposed on the derivatives of the functions are satisfied. Algorithms that overcome this difficulty are called the global optimization algorithms, and the corresponding techniques are available.

To solve the nonlinear programming problems, specific codes are used because, in general, globally optimal solutions are sought. The nonlinear optimization can be performed in MATLAB using the Optimization Toolbox. The MAPLE and MATHEMATICA (Global Optimization Toolbox) environments are also available to perform nonlinear optimization.

The analysis of electromagnetic micro- and nanoscale structures and transducers will be covered in the following chapters. However, even with the incomplete background, let us illustrate the application of nonlinear programming. It was documented that the micromotors develop the electromagnetic torque estimated as

$$T_e = k_T D_r^2 L_r$$

That is, T_e is proportional to the squared rotor diameter and axial length. Here, k_T is the constant. It is frequently desired to maximize the electromagnetic torque in order to attain high torque density within the allowed micromotor dimension. That is, using the rotor diameter and axial length as the state variables, the nonlinear programming problem results. In general, D_r and L_r are related due to electromagnetic features and fabrication processes. It will be illustrated that the simple formulation reported is the idealization of complex electromagnetic phenomena and effects in micro- and nanoscale transducers. In general, the nonlinear programming in design and optimization of MEMS and NEMS must be formulated using Maxwell's equations or other comprehensive mathematical models that describe complex phenomena and effects with minimum level of simplifications and assumptions.

Example 5.1   APPLICATION OF MATLAB AND OPTIMIZATION TOOLBOX

The illustrated results are given without specific NEMS and MEMS applications. However, the specifications, objective, and regret functionals are given in the multidimensional space using the system variables. Correspondingly, nonlinear functions result. Frequently nonlinear functions should be minimized. In this illustrative example, we will utilize the MATLAB Optimization Toolbox.

Consider the axial topology microdevice as illustrated in Figure 5.16. The vibration is a function of the placement of permanent magnets. Specifically, the designer can place four permanent magnets with respect to the microwindings in order to minimize vibration and noise (the magnets must be placed also to maximize torque and power densities, minimize heat, etc.). The vibration minimization problem will be solved using unconstrained and constrained optimization methods.

Let the magnitude of vibration, as a function of permanent magnets' position in the x-y plane, be $f(x,y) = f(x_1, x_2) = 4x_1^2 + 4x_2^2 + 10\sin x_1 + 20\cos x_2$. It should be emphasized that the placement of the permanent magnets is symmetrical, and correspondingly we consider only one either S or N magnet.

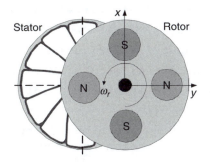

FIGURE 5.16
Axial-topology micromotor.

We will utilize different built-in MATLAB functions to solve the problem as given in the script reported below:

```
% assign the nonlinear function to be minimized
%                   2     2
% f(x1,x2) = 4x1 + 4x2 + 10sin(x1) + 20cos(x2)
% calculate and plot this function
[x1,x2]=meshgrid([-10:.1:10]);
f=(4*x1.^2+4*x2.^2+10*sin(x1)+20*cos(x2));
plot3(x1,x2,f)
% input f(x1,x2) utilizing the inline MATLAB function
fun=inline('(4*x(1)^2+4*x(2)^2+10*sin(x(1))+20*cos(x(2)))');
% initialize the possible solution as [x10 x20] = [1 5]x0=[1 5];
% Finding a minimum to the function: unconstrained
% optimization problem
% Unconstrained minimization problem solution using the
% optimset MATLAB
% function with the Large-Scale algorithm
options = optimset('LargeScale','off');
% calculating minimum of f (x and fvalue) using the
% fminunc MATLAB function
[x, fvalue, exitflag, output] = fminunc(fun, x0, options);
% MATLAB "optimizer" solver found a solution (minimum
% of f) at the following x1 and x2 with fvalue
x, fvalue
% substitute the minimum values in f(x) and verify
% the solution at derived x1 and x2
f=4*x(1)^2+4*x(2)^2+10*sin(x(1))+20*cos(x(2));
disp('x(1), x(2), f'); x(1), x(2), f
% total number of function evaluations was output.funcCount
disp('End of the unconstrained optimization')
```

The results are displayed in the MATLAB command Window as

```
>> optimization1
[x1,x2]=meshgrid([-10:.1:10]);
```

```
f=(4*x1.^2+4*x2.^2+10*sin(x1)+20*cos(x2));
plot3(x1,x2,f)
% input f(x1,x2) utilizing the inline MATLAB function
fun=inline('(4*x(1)^2+4*x(2)^2+10*sin(x(1))+20*cos(x(2)))');
% initialize the possible solution as [x10 x20] = [1 5]
x0=[1 5];
% Finding a minimum to the function: unconstrained
% optimization problem
% Unconstrained minimization problem solution using the
% optimset MATLAB function
% with the Large-Scale algorithm
options = optimset('LargeScale','off');
% calculating minimum of f (x and fvalue) using the
% fminunc MATLAB function
[x, fvalue, exitflag, output] = fminunc(fun, x0, options);
Optimization terminated successfully:
   Current search direction is a descent direction, and magnitude of
   directional derivative in search direction less than
   2*options.TolFun
% MATLAB "optimizer" solver found a solution (minimum
% of f) at the following x1 and x2 with fvalue
x, fvalue
x =
   -0.8371      2.1253
fvalue =
      2.9131
% substitute the minimum values in f(x) and verify the
% solution at derived x1 and x2
f=4*x(1)^2+4*x(2)^2+10*sin(x(1))+20*cos(x(2));
disp('x(1), x(2), f'); x(1), x(2), f
x(1), x(2), f
ans =
   -0.8371
ans =
    2.1253
f =
    2.9131
%  total number of function evaluations was
output.funcCount
ans =
    33
disp('End of the unconstrained optimization')
```

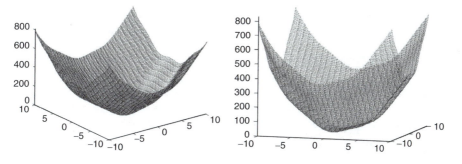

FIGURE 5.17
Three-dimensional plots of $f(x_1, x_2) = 4x_1^2 + 4x_2^2 + 10\sin x_1 + 20\cos x_2$.

```
End of the unconstrained optimization
>>
```

The nonlinear function was minimized. The minimum value of $f(x_1, x_2)$ is 2.9131 at $x_1 = -0.8371$ and $x_2 = 2.1253$. Figure 5.17 documents the plot and illustrates the validity of numerical results. Correspondingly, we conclude that the unconstrained minimization problem was solved, and the optimal placement of a permanent magnet is –0.8371 and 2.1253 in the *x-y* plane. The placement of all other magnets is evident due to the symmetric magnet configuration.

In general, the constrained optimization problem must be solved. It is evident that the designer cannot place the windings, magnets, bearings, and other components without any geometrical restrictions. For example, one cannot place asymmetrically permanent magnets. We assume that the magnitude of vibration is given as positioning of permanent magnets in the *x-y* plane as $f(x, y) = f(x_1, x_2) = 4x_1^2 + 4x_2^2 + 10\sin x_1 + 20\cos x_2$. However, the magnet placement is constrained as $x \geq 0$, $y \geq 0$, $-xy \leq -10$, and $xy - x - y \leq -1.5$. To solve the constrained optimization problem with the specified constraints, the MATLAB file is developed and reported below (we apply the MATLAB Optimization Toolbox, which allows the designer to attain a highly effective and cost-effective solution). Utilizing the `fmincon` MATLAB function to solve the constrained optimization problem, the nonconstrained optimization problem can be solved as well. As illustrated below, the solution is found to be 0 and 2.1253. The derived results are different from the unconstrained optimization solution (–0.8371 and 2.1253) due to the fact that by imposing the lower bounds [0 0] we specify that the solution must be positive.

```
>>
x=fmincon(inline('4*x(1)^2+4*x(2)^2+10*sin(x(1))+20*...
cos(x(2))'),[1;1],[],[],[],[],[0 0])
Warning: Large-scale (trust region) method does not currently
solve this type of problem,
switching to medium-scale (line search).
Optimization terminated successfully:
  Magnitude of directional derivative in search direction
  less than 2*options.TolFun and maximum constraint violation
  is less than options.TolCon
```

```
Active Constraints:

    1

x =

        0

    2.1253
```

Having demonstrated the uncomplicated application of MATLAB, we report more general constrained optimization problem. The m-file is given below.

```
% assign the nonlinear function to be minimized
%                 2    2
% f(x1,x2) = 4x1 + 4x2 + 10sin(x1) + 20cos(x2)
%      subject to
%      1.5 + x1x2 - x1 - x2 <= 0
%      -x1x2 <= 10
%      x1 > = 0, x2 >= 0
% calculate and plot this function
[x1,x2]=meshgrid([-10:.1:10]);
f=(4*x1.^2+4*x2.^2+10*sin(x1)+20*cos(x2));
plot3(x1,x2,f)
% input the object function f(x1,x2) utilizing the
% inline MATLAB function
f=inline('(4*x(1)^2+4*x(2)^2+10*sin(x(1))+20*cos(x(2)))'); f
% input the constraints utilizing the build-in confun
% MATLAB function
% (Optimization Toolbox)
type confun
% initialize the possible solution as [x10 x20] = [-5 5]
x0=[-5 5];
% Finding a minimum to the function: constrained
% optimization problem
% Constrained minimization problem solution using the
% fmincon MATLAB function
% The syntaxes is X = FMINCON(FUN, X0, [],[],[],[], LB,
% UB, OPTIONS, NONLCON)
% with lower and upper bounds (LB and UB, respectively)
% That is, the FMINCON MATLAB solver solves problems of the form:
% min F(X)   subject to:   A*X  <= B, Aeq*X  = Beq (linear
% constraints)
%    X                         C(X) <= 0, Ceq(X) = 0    (nonlinear
% constraints)
%                             LB <= X <= UB
% lower and upper bounds
LB=zeros(1,2); % Lower bounds X >= 0
UB=[];         % No upper bounds
```

```
% constrained (bounded) optimization solution
[x,fvalue,exitflag,output]=fmincon(f,x0,[],[],[],[],LB,UB,'confun',[]);
% [] indicates that no data is set
% MATLAB solver founds a solution (minimum of f) at the
% following x1 and x2 with fvalue
x, fvalue
% The constraint values at the solution are:
[c, ceq]=confun(x)
% substitute the minimum values in f(x) and verify the
% solution at derived x1 and x2
F=exp(x(1))*(4*x(1)^2+4*x(2)^2+10*sin(x(1))+20*cos(x(2)));
disp('x(1), x(2), F'); x(1), x(2), F
% total number of function evaluations was
% output.funcCount
disp('End of the constrained optimization')
```

The results, displayed in the MATLAB Command Window, are

```
>> optimization2
f =
     Inline function:
     f(x) = (4*x(1)^2+4*x(2)^2+10*sin(x(1))+20*cos(x(2)))
function [c, ceq] = confun(x)
% Nonlinear inequality constraints:
%   Copyright 1990-2002 The MathWorks, Inc.
c = [1.5 + x(1)*x(2) - x(1) - x(2);
     -x(1)*x(2) - 10];
% no nonlinear equality constraints:
ceq = [];
Warning: Large-scale (trust region) method does not currently
solve this type of problem,
switching to medium-scale (line search).
Optimization terminated successfully:
  Magnitude of directional derivative in search direction
  less than 2*options.TolFun and maximum constraint violation
  is less than options.TolCon
Active Constraints:
     1
x =
         0    2.1253
fvalue =
    7.5372
c =
  -0.6253
```

```
     -10.0000
 ceq =
      []
 x(1), x(2), F
 ans =
      0
 ans =
     2.1253
 F =
     7.5372
 ans =
     27
End of the constrained optimization
```

The nonlinear function was minimized. The minimum value of $f(x_1, x_2)$ is 7.5372 (2.9131 was obtained without constraints) at $x_1 = 0$ and $x_2 = 2.1253$ (in the unbounded case, we received $x_1 = -0.8371$ and $x_2 = 2.1253$). The total number of functions evaluated is 27. Unconstrained and constrained minimization problems were solved, and the optimal placement of permanent magnets was obtained in the *x-y* plane.

5.4   Nanoengineering Bioinformatics and Its Application

This section sometimes may use advanced mathematics and biology fundamentals. This should not frustrate the reader who has a limited knowledge and background in engineering or science (the level of difficulty is junior-level undergraduate engineering and science). The readers with deficient backgrounds may skip the mathematical deviations.

5.4.1   Introduction and Definitions

The application of biomimetics and biological analogies in the design of NEMS and MEMS were covered in Section 3.1. In this section, we examine engineering bioinformatics as a meaningful paradigm to complement modern nanotechnology as well as synthesize novel NEMS and MEMS. The author defined engineering bioinformatics as a coherent abstraction in devising complex micro- and nanoscale systems and devices utilizing biological systems analogies as well as science and engineering fundamentals.

It should be emphasized that design, optimization, analysis, and other tasks (prototyping, scaling, simulation, visualization, fabrication, etc.) can be performed as systems or devices were devised. Current far-reaching research in genomics, biophysics, informatics, and bioengineering have been focused on understanding the fundamental theory and integrated processes that utilize the information contained in the genome into functioning biological entities. Further developments are essential for characterizing the emergence of biological phenotypes from underlying hierarchies and complex patterns. It was shown that nano- and microscale biological systems exist in nature in enormous variety and sophistication. Complex biological patterns can be utilized in order to devise and examine distinct systems. One may not be able blindly copy biosystems due to the fact that many complex phenomena and effects have not been comprehended (or even examined). Furthermore, in general, system architectures and functionalities have not been fully comprehended. Typical examples include a variety of unsolved problems to comprehend the simplest *E. coli* and *Salmonella typhimurium* bacteria that have been studied for many decades. From an engineering viewpoint directly related to the book coverage, these bacteria, among a variety of subsystems, integrate (1) three-dimensional nanobiocircuitry (computing, processing, networking and controlling nanobioelectronics) and (2) nanobiomotors and nanobiosensors.

Correspondingly, attention should be concentrated on devising novel paradigms in systematic synthesis through bioinformatics with the ultimate objective to synthesize and design distinct systems and devices applying engineering bioinformatics. This will allow one to derive new operating principles examining functionality of different subsystems, research novel structures, study advanced architectures, examine different topologies, and characterize distinct systems, subsystems, and devices reaching benchmarking *nanosystematics* and *nanoachitectronics* paradigms. In this section, we examine complex patterns in biosystems because superior systems can be devised through engineering bioinformatics. Our ultimate objective is to provide the focused study of engineering bioinformatics and systematic design. These problems are far-reaching frontiers of modern nanoscience and nanoengineering. Synergetic paradigms should be derived by researching biosystems and coherently examining distinct nanostructures, complexes, and subsystems.

Significant research efforts have been dedicated to examining biosystems and their subsystems, devices, and components, e.g., organic three-dimensional net-working, processing, computing bioelectronics, biosensors, biomotors,[1-5] etc. These biosystems are made from proteins (organic molecules). If numerous "bio-NEMS" and "bio-MEMS" exist in nature, can one devise, design, and fabricate bio-inspired organic, inorganic, or hybrid NEMS/MEMS? In general, the answer is yes, and MEMS fabricated using bulk and surface micromachining were reported. Typical examples are axial micromachines, DNA-based nanocircuits and nanobiosensors, micropumps, etc. The fundamental, applied, and experimental research has been progressed to understand and control molecular-scale processes to attain the controlled synthesis and direct self-assembly of structures and components into functional nano- and microscale devices and systems. These novel molecular-scale devices and systems can be used in nanoelectronics, computing, information processing, wireless communication, sensing, actuation, propulsion, energy sources, biometrics, etc. Despite of the importance of bioinformatics and efforts to apply it to attack unsolved problems, limited progress has been achieved to date. This section examines engineering bioinformatics to approach and solve different problems in systems synthesis, analysis, and design. Bioinformatics is viewed as a coherent abstraction in devising, prototyping, design, optimization, and analysis of complex systems that will allow one to coherently apply nano- and microtechnology.

5.4.2 Basic Bioscience Fundamentals

All biosystems (living organisms) share the same genetic code paradigm. The amino acid sequences are programmed by a gene. We will also discuss the role of deoxyribonucleic acid (DNA) and ribonucleic acid (RNA). Genes (DNA) and their products (proteins) are the genetic hereditary material that organisms inherit. A DNA molecule is very long (can be several millimeters) and consists of thousands of genes (thousands or millions of base pairs holding two chains together) that occupy a specific (exact) position along the single molecule. One DNA molecule represents a large number of genes, each one a particular segment of helix. The gene can be defined as a DNA sequence coding for a specific polypeptide chain. This definition needs clarification. Most eukaryotic genes contain noncoding regions (these noncoding regions are called introns, while the coding regions are called exons).

Therefore, large portions of these genes do not have corresponding segments in polypeptides. Promoters and other regulatory regions of DNA are also included within the gene boundaries. In addition, DNA that codes for RNA (rRNA, tRNA, and snRNA) can also be considered. These genes do not have polypeptide products, and, correspondingly, a gene can be defined as a region of DNA as required for the production of an RNA molecule (distinct definitions for a gene are given for different situations and other definitions for genes are used).

The information content of DNA (genetic material) is in the form of specific sequences of nucleotides (nucleic acids) along the DNA strands. The DNA inherited by an organism results in specific traits by dictating the synthesis of particular proteins. Proteins are the links between genotype and phenotype. Nucleic acids and proteins have specific sequences of monomers that comprise information, and this information from genes to proteins is provided in the linguistic form. Note that the accurate statement is *one gene, one polypeptide*. However, most proteins consist of single polypeptide. Following the common terminology, we will refer to proteins as the gene product.

In nucleic acids, the monomers are four types of nucleotides that differ in their nitrogen content. Genes are typically hundreds or thousands of nucleotides long, and each gene has a specific sequence of bases. A protein also has monomers arranged in a particular linear order, but its monomers are 20 amino acids. Correspondingly, nucleic acids and proteins contain information in two distinct forms. The so-called transcription and translation processes (steps) are involved. Transcription is the synthesis of RNA under the direction of DNA. Both nucleic acids use the same "language," and the information is simply copied (translated) from one molecule to other. A gene's unique sequence of DNA nucleotides provides a template for assembling a unique sequence of RNA nucleotides. The resulting RNA molecule (called the messenger RNA and denoted as mRNA) is a transcript of the gene's protein-building instructions. Thus, the function of mRNA is to transcribe a genetic message from DNA to the protein-synthesis machinery of the cell. Translation is the concrete synthesis of a polypeptide that occurs under the direction of mRNA. The "language" is changed. The cell must translate the base sequence of an mRNA molecule into the amino acid sequence of a polypeptide. The sites of translation are ribosomes, i.e., complex particles with many enzymes and other agents that facilitate the orderly linking of amino acids into polypeptide chains. Though there exist differences in how transcription and translation are organized for prokaryotes and eukaryotes, genes program protein synthesis by sending genetic messages in the form of RNA. The sequence chain is DNA $\rightarrow$ RNA $\rightarrow$ protein.

It is important that the instructions for assembling amino acids into a specific order are encoded in the "language" of DNA, and this genetic code has been broken—four nucleotides to specify 20 amino acids. Triplets of bases code for amino acids, and these triplets can be considered as the "units" of information or coding instructions. In particular, three consecutive bases specify an amino acid, and there are 64 (4^3) possible code words. Correspondingly, the information is stored and coded as a *triple code*. In other words, the genetic instructions for a polypeptide chain are stored and written in the DNA as a series of three-nucleotide words (symbols) called *codons*. As was emphasized, a cell cannot directly translate a gene's codons into amino acids. The intermediate transcription step (gene determines the codon sequence of an mRNA molecule) is in place. For each gene, only one of two DNA strands is transcribed, i.e., the coding strand of the gene is used. The noncoding strand serves as a template for making a new coding strand when the DNA replicates. The RNA bases are assembled on the template according to the base-pairing rules, and, hence, an mRNA molecule is complementary, rather than identical, to its DNA template. For example, if the codon strand of a gene has a codon GCG, the codon at the corresponding position along the mRNA molecule will be CGC. During translation, the sequence of codons along a genetic message (mRNA molecule) is decoded (translated) into a sequence of amino acids resulting in a polypeptide chain. It takes $3N$ nucleotide strands to code N amino acids.

As was emphasized, nucleic acids are polymers of monomers called nucleotides. Each nucleotide is itself composed of three parts (nitrogenous base is joined to a pentose that is bonded to a phosphate group). The DNA molecules consist of two polynucleotide chains (strands) that spiral around forming a double helix. These polynucleotide chains are held together by hydrogen bonds between the paired bases. DNA is a linear double-stranded polymer of four nitrogenous bases, i.e.,

- Deoxyadenosine monophosphate or adenine (A)
- Deoxythymidine monophosphate or thymine (T) in DNA, and uracil (U) in RNA
- Deoxyguanosine monophosphate or guanine (G)
- Deoxycytidine monophosphate or cytosine (C)

Each nucleotide integrates a nitrogenous base joined to a pentose (five-carbon sugar) which is bonded to a phosphate group as demonstrated in Figure 5.18. There are two families of nitrogenous bases: pyrimidines (C, T, and U) and purines (A and G). A pyrimidine is characterized by a hexagon ring of carbon and hydrogen atoms. In purines, a pentagon ring is

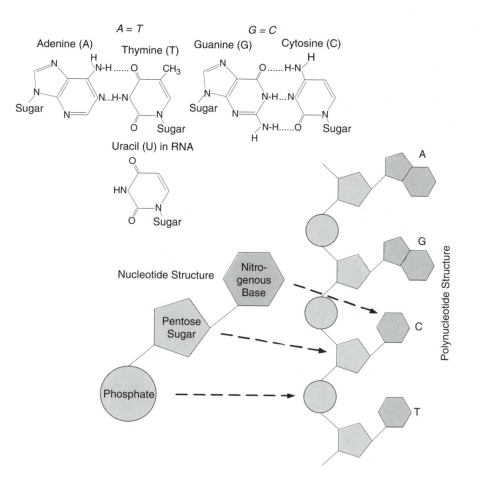

FIGURE 5.18

Nucleotides (monomers of nucleic acids composed of three smaller molecular building blocks: a nitrogenous base (purine or pyrimidine), pentose sugar, and phosphate group) and polynucleotides (phosphate group bonded to the sugar of the next nucleotide, and the polymer has a sugar-phosphate backbone).

fused to the pyrimidine ring (see Figure 5.18). The monomers are joined by covalent bonds between the phosphate of one nucleotide and the sugar of the next monomer. The backbone with a repeating pattern of sugar-phosphate-sugar-phosphate results. The linear order of bases encoded in a gene specifies the amino acid sequence of a protein specifying protein's functionality. The DNA molecule consists of two polynucleotide chains that spiral around forming a double helix. Two sugar-phosphate backbones are on the outside of the helix, and the nitrogenous bases are paired in the interior of the helix. Two polynucleotide chains (strands) are held together by the hydrogen bonds between the paired bases. Most DNA molecules are long (millimeters in length) with up to millions of base pairs holding two chains. The base-pairing rule (A always pairs with T, A=T, while G always pairs with C, G≡C) specifies that two strands of the double helix are complementary. If a stretch of a strand has the base sequence AATTGGCC, then the base-pairing rule dictates that the same stretch of the other strand must have the sequence TTAACCGG. As a cell prepares to divide, two strands of each gene separate, guaranteeing precise copying (each strand serves as a template to order nucleotides into a new complementary strand).

Figure 5.19 represents the sugar-phosphate backbone of two DNA strands. These two strands are held together by hydrogen bonds between the nitrogenous bases that are paired in the interior of the 2-nm-diameter double helix. The base pairs are 0.34 nm apart, and there are 10 pairs per helix turn.

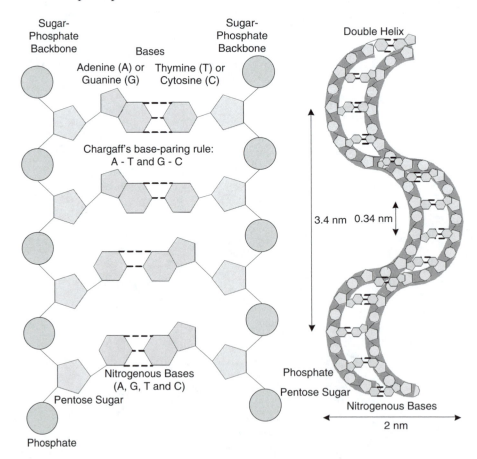

FIGURE 5.19 (See color insert.)
Two DNA strands are held together by hydrogen bonds between the bases that are paired in the interior of the double helix (base pairs are 0.34 nm apart and there are 10 pairs per helix turn).

Proteins are large biomolecules that occur in every living organism. All proteins are chemically similar since all are made from amino acids linked together in long chains. The amino acids can be considered as building blocks with nitrogenous bases as the units to build these building blocks. Amino acids are organic molecules containing both carboxyl and amino groups. Cells build their proteins from 20 amino acids. These 20 amino acids are specified by codon sequences. The following abbreviations are used: Ala (alanine), Arg (arginine), Asn (asparagine), Asp (aspartic), Cys (cysteine), Gln (glutamine), Glu (glutamic), Gly (glycine), His (histidine), Ile (isoleucine), Lys (lysine), Leu (leucine), Met (methionine), Phe (phenylalanine), Pro (praline), Ser (serine), Thr (threonine), Trp (tryptophan), Tyr (tyrosine), and Val (valine).

Each amino acid has a hydrogen atom, a carboxyl group, and an amino acid group bonded to the alpha (α) carbon. It is important to note that all 20 amino acids that make up proteins differ only in what they are attached to by the fourth bond to the α carbon. The amino acids are grouped according to the properties of the side chains (R-group). Physical and chemical properties of the side chain define the unique characteristics of amino acids. Amino acids are classified in the groups based on the chemical and structural properties of their side chains and polarity, e.g., nonpolar (hydrophobic), polar (hydrophilic), electrically charged, etc. Twenty protei-nogenic amino acids, as well as their structures and properties, are reported in Table 5.2.

The amino acid backbone determines the primary sequence of a protein, but the nature of the side chains determines the protein's properties. Amino acid side chains can be polar, nonpolar, or neutral. Polar side chains tend to be present on the surface of a protein where they can interact with the aqueous environment found in cells. Nonpolar amino acids tend to reside within the center of the protein where they can interact with similar nonpolar neighbors (this can create a hydrophobic region within an enzyme where chemical reactions can be conducted in a nonpolar atmosphere). There are different methods to classify amino acids—for example, by their structure, size, charge, hydrophobicity, etc. The charge residue is normally found on the surface of the protein (interact with water and bind to other molecules) and is seldom buried in the interior of a folded protein. Arginine, histidine, and lysine are positively charged, while aspartic and glutamic are negatively charged amino acids. One can compare amino acids by differences in R-groups (alanine with other acids) and similarities (aspartate with asparagines or leucine with isoleucine). By grouping amino acids, based on their charges, hydrophobicity, and polarity, Table 5.3 differentiates the amino acids providing pI (pH at the isoelectric point).

Amino acids can be also classified using the distinct optical properties. A tetrahedral carbon atom with four distinct constituents is said to be chiral. The amino acid that does not exhibit chirality is glycine because its R-group is a hydrogen atom. Chirality describes the handedness of a molecule, which is observable by the ability of a molecule to rotate the plane of polarized light either to the right (dextrorotatory) or to the left (levorotatory). All of the amino acids in proteins are levorotary-α-amino acids (dextrorotary amino acids have never been found in proteins, although they exist in nature and are found in polypeptide antibiotics). The aromatic R-groups in amino acids absorb ultraviolet light, and the maximum of absorbance is observed at 280 nm. The ability of proteins to absorb ultraviolet light is mainly due to the presence of tryptophan, which strongly absorbs ultraviolet light.

A polymer of many amino acids linked by peptide bonds is a polypeptide chain. At one end of the chain is a free amino group, and at the opposite end, there is a free carboxyl group. Hence, the chain has a polarity with an N-terminus (nitrogen of the amino group) and a C-terminus (carbon of the carboxyl group). A protein consists of one or more twisted, wound, and folded polypeptide chains forming a micromolecule (protein) with a defined three-dimensional shape (geometry). This three-dimensional shape is frequently called conformation. A protein's functionality depends on its geometry, which is function of specific

TABLE 5.2

Amino Acids

Amino Acid (molecular weight)	Symbol Abbreviation	Molecular Formula	Structure	Properties
Alanine (89.09)	Ala, A	$C_3H_7NO_2$		aliphatic hydrophobic neutral
Arginine (174.2)	Arg, R	$C_6H_{14}N_4O_2$		polar (strongly) hydrophilic positive charged (+)
Asparagine (132.12)	Asn, N	$C_4H_8N_2O_3$		polar hydrophilic neutral
Aspartic acid (aspartate) (133.1)	Asp, D	$C_4H_7NO_4$		polar hydrophilic negative charged (−)
Cysteine (121.15)	Cys, C	$C_3H_7NO_2S$		polar (weakly) hydrophilic neutral
Glutamine (146.15)	Gln, Q	$C_5H_9NO_4$		polar hydrophilic neutral
Glutamic acid (glutamate) (147.13)	Glu, E	$C_5H_{10}N_2O_3$		polar hydrophilic negative charged (−)
Glycine (75.07)	Gly, G	$C_2H_5NO_2$		aliphatic neutral
Histidine (155.16)	His, H	$C_6H_9N_3O_2$		aromatic polar (strongly) hydrophilic positive charged (+)
Isoleucine (131.17)	Ile, I	$C_6H_{13}NO_2$		aliphatic hydrophobic neutral
Leucine (131.17)	Leu, L	$C_6H_{13}NO_2$		aliphatic hydrophobic neutral

TABLE 5.2

(*Continued*)

Amino Acid (molecular weight)	Symbol Abbreviation	Molecular Formula	Structure	Properties
Lysine (146.21)	Lys, K	$C_6H_{14}N_2O_2$		polar (strongly) hydrophilic positive charged (+)
Methionine (149.21)	Met, M	$C_5H_{11}NO_2S$		hydrophobic neutral
Phenylalanine (165.19)	Phe, F	$C_9H_{11}NO_2$		aromatic hydrophobic neutral
Praline (115.13)	Pro, P	$C_5H_9NO_2$		hydrophobic neutral
Serine (105.09)	Ser, S	$C_3H_7NO_3$		polar hydrophobic neutral
Threonine (119.12)	Thr, T	$C_4H_9NO_3$		polar hydrophobic neutral
Tryptophan (204.23)	Trp, W	$C_{11}H_{12}N_2O_2$		aromatic hydrophobic neutral
Tyrosine (181.19)	Tyr, Y	$C_9H_{11}NO_3$		aromatic polar hydrophobic
Valine (117.15)	Val, V	$C_5H_{11}NO_2$		aliphatic hydrophobic neutral

linear sequence of the amino acids that make up the polypeptide chain. Some proteins are spherical (globular) while others are fibrous.

The structure of a protein is defined by the sequence of amino acids. Table 5.4 documents resulting amino acids as a function of nucleotides (bases) in the codon (note that some exceptions to this standard table have been discovered and reported). The three bases of an mRNA codon are designated as the first, second, and third bases. The codon AUG stands not only for the methionine amino acid (Met), but also indicates the "start" mark. Three of the total 64 (4^3) codons are "stop" marks (termination codon indicates the end of a genetic message). Hence, four amino acids are "start" and "stop" ones that carry out start and end signals of genetic messages.

TABLE 5.3

Classification of Amino Acids Using Charges, Hydrophobicity, and Polarity

Nonpolar and Hydrophobic		Negatively Charged (Acidic Amino Acids), Polar, and Hydrophilic		No Charge (Nonacidic Amino Acids), Polar, and Hydrophilic		Positively Charged (Nonacidic Amino Acids), Polar, and Hydrophilic	
Amino acid	pI	Amino acid	pI	Amino acid	pI	Amino acid	pI
Phenylalanine (phe)	5.48	Aspartic (asp)	2.77	Cysteine (cys)	5.07	Histidine (his)	7.59
Methionine (met)	5.74	Glutamic (glu)	3.22	Asparagine (asn)	5.41	Lysine (lys)	9.74
Tryptophan (trp)	5.89			Threonine (thr)	5.6	Arginine (arg)	10.76
Valine (val)	5.96			Glutamine (gln)	5.65		
Leucine (leu)	5.98			Tyrosine (tyr)	5.66		
Alanine (ala)	6.00			Serine (ser)	5.68		
Isoleucine (ile)	6.02			Glycine (gly)	5.97		
Proline (pro)	6.30						

Triplets of bases are the units of uniform length that code all amino acids. Each arrangement of three consecutive bases specifies an amino acid, and there are 4^3 (64) possible arrangements, which are represented in the form of 4^3 (64) code words. The flow of information from gene to protein is based on a triplet code. That is, the genetic instructions are written (and stored) in the DNA as a series of three-nucleotide words called codons. This genetic code is a universal one and shared by all biosystems (from bacteria to humans). The linguistics and fundamental mathematics of this highly descriptive universal "language" have been gradually comprehended. Correspondingly, these findings can be applied to the man-made systems and will significantly positively contribute to biotechnology, medicine, etc.

The protein synthesis allows one to coherently analyze the synthesis taxonomy. As we discussed, genetic instructions from DNA are written and coded in three-nucleotide units

TABLE 5.4

Nucleotides (Bases) in Codon and Resulting Amino Acids

Codon Base First (1st)	Second U	C	(2nd) base A	G	Codon Base Third (3rd)
U	Phe (UUU)	Ser	Tyr	Cys	U
	Phe (UUC)	Ser	Tyr	Cys	C
	Leu (UUA)	Ser	Stop	Stop	A
	Leu (UUG)	Ser	Stop	Trp	G
C	Leu (CUU)	Pro	His	Arg	U
	Leu (CUC)	Pro	His	Arg	C
	Leu (CUA)	Pro	Gln	Arg	A
	Leu (CUG)	Pro	Gln	Arg	G
A	Ile	Thr	Asn	Ser	U
	Ile	Thr	Asn	Ser	C
	Ile	Thr	Lys	Arg	A
	Met–Start	Thr	Lys	Arg	G
G	Val	Ala	Asp	Gly (GGT)	U
	Val	Ala	Asp	Gly (GGC)	C
	Val	Ala	Glu	Gly (GGA)	A
	Val	Ala	Glu	Gly (GGG)	G

(codons) that indirectly precisely specify particular amino acids. The genetic code and genetic "language" can be viewed as a generic and universal one because they are shared by all living organisms as distinct as bacteria and humans. Having introduced the genetic linguistics, it is clear that these codes culminate in the assembly of extremely complex functional systems. This can be related from the description of organic and inorganic NEMS/ MEMS to their fabrication. The assembly and fabrication are very different. It is unlikely that the coherence, power, efficiency, robustness, accuracy, and other features observed in biosystem assembly can be prototyped and achieved in the fabricated NEMS/MEMS. The reported synthesis philosophy highlights the need for further developments of engineering bioinformatics, including coherent descriptive languages (to attain bidirectional synthesis and design taxonomy as were reported in Sections 2.3 and 5.1) including CAD. In fact, both nucleic acids and proteins are informational polymers assembled from linear sequences of nucleotides and amino acids, respectively. Messenger RNA (as the information carrier from the coding strand of a gene) is the intermediate in the flow of information from DNA to proteins. Can one utilize and apply this paradigm? Likely yes, but there exist a great number of unsolved problems that may not be resolved in the observable future.

We should carefully use the word *language*. In fact, though one can utilize linguistics using abbreviations for nitrogenous bases A, U, T, G, and C) and 20 amino acids (from alanine to valine), complex biochemical synthesis is involved. Hence, the linguistic descriptions must be coherent within biochemical processes. Correspondingly, the "language" cannot be mistakenly considered as, for example, Boolean algebra or binary arithmetic where the designer without biological-based restrictions performs, say, logic or circuit design.

In biosystems, information processing, verification, assembling, and other processes are completed within so-called transcription and translation (see Figure 5.20). Transcription results in nucleotide-to-nucleotide transfer of information from DNA to RNA. RNA synthesis on a DNA template is catalyzed by RNA polymerase. Promoters (specific nucleotides sequences flanking the start of a gene) signal the initiation of mRNA synthesis. Transcription factors (proteins) help RNA polymerase recognize promoter sequences and bind to the RNA. Transcription continues until the RNA polymerase reaches the termination (stop) sequence of nucleotides on the DNA template. As the mRNA peels away, the

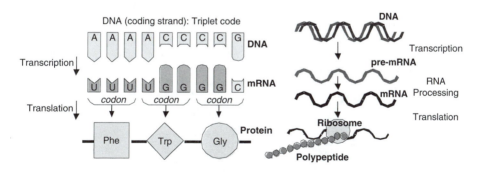

FIGURE 5.20
For each gene, one strand of DNA serves as the coding strand. The sequence of bases specifies the protein (with certain amino acid sequence). During transcription, the DNA coding strand provides a template for synthesis of mRNA (molecule of complementary sequence using the base-pairing rule). During translation, a sequence of base triplets (analog to three-letter code), called codon, explicitly specifies amino acids to be made at the corresponding position along a growing complex protein chain. In a eukaryotic cell, two main steps of protein synthesis (transcription and translation) occur in nucleus (transcription) and cytoplasm (translation); mRNA is translocated from nucleus to cytoplasm through pores in the nuclear envelope (RNA is first synthesized as pre-mRNA, which is processed by enzymes before leaving the nucleus as mRNA). In contrast, in a prokaryotic cell (no nucleus), mRNA is produced without processing.

DNA double helix reforms. Translation results in the informational transfer from RNA nucleotides to polypeptide amino acids (transfer RNA interprets the genetic code during translation, and each kind of tRNA brings a specific amino acid to ribosomes). Transfer RNA molecules pick up specific amino acids and line up by means of their anticodon triplets at complementary codon sites on the mRNA molecule. The binding of a specific amino acid to its particular tRNA is a precise ATP process catalyzed by aminoacryl-tRNA synthetase enzymes. Ribosomes control the coupling of tRNA to mRNA codons. They provide a site for the binding of mRNA, as well as P and A sites for holding adjacent tRNA as amino acids are linked in the growing polypeptide chain. There are three major stages: initiation (integrates mRNA with tRNA with the attached first amino acid), elongation (polypeptide chain is completed, adding amino acids attached to its tRNA by binding and translocation tRNA and mRNA along the ribosome), and termination (termination codons cause the protein release freeing the polypeptide chain and dislocation of the ribosome subunits). Several ribosomes can read a single mRNA, forming polyribosome clusters. Complex proteins usually undertake one or several changes during and after translation that affect their three-dimensional structures influencing cells. This leads to the cell transitional dynamics.

It is important to emphasize that genes can be transcribed and translated after they are transplanted from one species to another. For example, bacteria can be programmed by the insertion of a humane gene to synthesize the protein insulin.

Proteins are utilized to build the structures, information storage and processing, actuation, sensing, and other "components" in all living organisms. As a protein sequence is defined, the efforts have been directed to examine protein structure and protein functionality. These efforts have not been even nearly completed. The protein structure (three-dimensional geometry) is due to folding of a peptide chain as well as multiple peptide chains. Proteins are the most structurally complex molecules known. Amino acid bonds determine the folding (alpha helix resulting in helix-loop-helix, beta-pleated sheet, random conformations, and others). However, most proteins go through several intermediate states to form stable structures and conformation. There are many unsolved problems in protein folding. Figure 5.21 illustrates a protein folding. The folded molecule contains the four chains of the ElbB (enhancing lycopene biosynthesis) gene of *E. coli*.[4–6] We use the protein data bank source 1OY1[7] and ElbB gene involved in isoprenoid biosynthesis. All four chains are generated from identical protein sequences. The image was generated from the protein database using the protein explorer.[8,9]

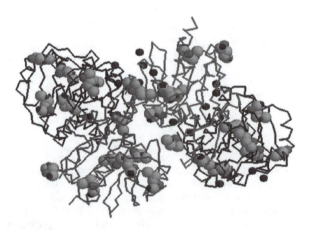

FIGURE 5.21 (See color insert.)
ElbB gene geometry composed of four similar protein sequence geometries.

In biosystems, there are thousands of different kind of proteins, each with specific structure and function—for example, (1) structural proteins, (2) storage (store amino acids) proteins, (3) transport proteins, (4) receptor (sensor) proteins, (5) contractile (actuator) proteins, (6) defensive (protect and combat diseases) proteins, (7) hormonal proteins, (8) enzymatic proteins, and many other functional proteins. The conceptual functional categories for protein can be explicitly defined.

Engineering bioinformatics has become increasingly database driven due to large-scale functional genomics and proteomics. A significant challenge is integrating and utilizing databases to coherently examine large-scale different data. The single-gene research (functionality, folding, interaction, etc.) has progressed to a large population of proteins emphasizing descriptive genomics and functional proteomics. Proteomics includes protein identification, as well their characterization and functionality. The databases have been developed—e.g., the SCOP, CATH, and FSSP databases classify proteins based on structural similarity; Pfam and ProtoMap identify families of proteins based on sequence homology; and PartList and GeneCensus examine the occurrence of protein families in various genomes. The large-scale genomics and proteomics are the forefront of not only biological and genomic research, but also engineering and technology developments. Machine learning methods (clustering, Bayesian networks, decision trees, neural networks, etc.) can be used to discover trends and patterns in the large-scale data. Genome sequences for different organisms are available. Even human genome sequences are accessible; in particular, (1) GenBank, DDBJ, and EMBL provide nucleic acid sequences; (2) PIR and SWISS-PROT report protein sequences; and (3) Protein Data Bank offers three-dimensional protein structures. In addition to sequence and structure databases, efforts have been directed toward various functionality aspects. Correspondingly, integrated data-intensive large-scale analysis and heterogeneous intelligent data mining are essential. There is a need to develop novel paradigms that will allow one to integrate genomic data from different databases in a common framework. A general problem is to integrate the large-scale diverse genomic information in the viable taxonomies or categories. Currently, the majority of methods are based on statistical analysis employing unsupervised learning, self-organization, classification, hierarchical clustering, etc. For example, a clustering method establishes multitiered partitioning of the data sets. Using the Pearson correlation coefficient $r_{ij} = (1/N - 1)X_i \cdot X_j$, given as a dot product of two "profiles" X_i and X_j, the similarity between genes (or groups of genes) is obtained. Furthermore, the measurement expression ratio profile is found using the average x_{av} and the standard deviation σ_x, i.e.,

$$X(k) = \frac{x(k) - x_{av}}{\sigma_x}$$

The aggregation of proteomic data from multiple sources must be performed to identify and predict various protein properties, functionality, and features.

The DNA sequences of several human pathogens are known. To achieve reasonable accuracy and high-quality continuous sequences, each base pair was sequenced many times (8 to 10 times). As a result, 90–93% of the euchromatin sequence has an error rate of less than one base in 10,000 bases.[10] However, different sequencing technologies, mathematical methods, distinct procedures, and measurement techniques have been used. Correspondingly, it is very difficult to estimate the accuracy, and there are many gaps and unknown strings of bases in the large-scale genomic sequence data. There are differences even in the count of genes. For example, the public human genome database reports 31,780 genes (2.693 billion bases sequenced). These include 15,000 known genes and 17,000 predicted genes. However, it is estimated that there can be less than 20,000 actual genes. Some predicted genes can be "pseudogenes" (noncoding) or fragments of real genes leading to predictions that there could be only 7000 real genes. For example, Celera reported 39,114 genes (2.654 billion bases sequenced) advising that 12,000 genes are "weak." Correspondingly, it is very difficult to identify the disease-associated genes. *Escherichia coli* bacteria is one of the most thoroughly studied living organisms. It is also remarkably diverse because some *E. coli* are harmless, whereas other distinct genotypes (extraintestinal, enteropathogenic, enterohemorrhagic, enteroinvasive, enterotoxigenic, and enteroaggregative *E. coli*) cause morbidity and mortality as human intestinal pathogens. Distinct *E. coli* genome sequences were reported and compared, e.g., MG1655, CFT073, EDL933, and other strains. For different strains (MG1655 and CFT073), the genome length is 4,639,221 and 5,231,428 base pairs, specifying 4293 and 5533 protein-coding genes (less than 50% of proteins are known).[6] After sequencing, one should examine genes and proteins. For example, after identifying the critical genes involved in virulence mechanisms and host response, one can design effective detectors, vaccines, and treatments. Furthermore, the studies of multigene pathways (cluster of genes) for multifunctional disorders are extremely important. This analysis can provide the following sequential analysis: (1) examine and characterize changes in genes and proteins of distinct *E. coli* strains; (2) identify protein functionality (study structural, receptor, defensive, and virulence genes and proteins); (3) compare and examine genes and protein profiles; and (4) integrate the experimental and computational data into the particular data-intensive analysis methods. Different methods, reported below, have been applied to attain analysis, comparison, and data recognition. However, traditional (conventional) approaches may not be well suited, and novel accurate efficient information-theoretic methods must be devised to attain data-intensive robust analysis and heterogeneous data mining with decision making. In particular, there is a need to examine genes that are unique and analogous to same and other species.

Different statistical techniques have been applied to attain global and local sequence comparisons.[11–14] However, under even the simplest random models and scoring systems, the distribution of optimal global alignment scores is unknown. Monte Carlo experiments potentially can provide some promising results for specific scoring systems and sequence compositions, but these results cannot be generalized. Compared with global alignments, statistics for the scores of local alignments, particularly for local alignments lacking gaps, is well posed. A local alignment without gaps consists simply of a pair of equal-length segments, one from each of the two sequences being compared. For example, in the BLAST program, the database search can be performed utilizing high-scoring segment pairs (HSPs). To analyze the score probability, a model of random sequences is applied. For proteins, the simplest model chooses the amino acid residues in a sequence independently, with specific background probabilities for the various residues, and the expected score for aligning a random pair of amino acid is required to be negative. For sequence (with lengths m and n), the HSP score statistics are characterized by the scaling parameters K and λ. The expected number of HSPs with scores of at least S is given as $E = mnKe^{-\lambda S}$. One obtains the E-value for the score S. However, the length of sequence changes E and sound methods to find the scaling positive parameters K and λ have not been reported. The score is normalized as $S' = (\lambda S - \ln K)/\ln 2$ to obtain the so-called bit score S'. The E-value is $E = mn2^{-\sigma S'}$. The number of random HSPs with score greater than or equal to S is described by a Poisson distribution. For example, the probability of finding exactly A HSPs with score $\geq S$ is given by $e^{-E}E^A/A!$. The probability of finding at least one such HSP is $P = 1 - e^{-E}$. This is the P-value associated with the score S. In BLAST, the E-value is used to compare two proteins of lengths m and n. To assess the significance of an alignment that arises from the comparison of a protein of length m to a database containing many different proteins of varying lengths, one view is that all proteins in the database are *a priori* equally likely to be related to the query. This implies that a low E-value for an alignment involving a short database sequence should carry the same weight as a low E-value for an alignment involving a long database sequence. To calculate a "database search" E-value, one multiplies the pairwise-compared E-value by the number of sequences in the database, for example the FASTA protein comparison programs. As was emphasized, the approaches applied to date have a sound theoretical foundation only for local alignments that are not permitted to have gaps, short sequences, or "estimation" of K and λ. Different amino acid substitution scores $S_{ij} = (1/\lambda)\ln(q_{ij}/p_i p_j)$ are reported. Here, q_{ij} is the target frequency; p_i and p_j are the background frequencies for the various residues. The target frequencies and the corresponding substitution matrix may be calculated for any given evolutionary distance. However, this method has serious deficiencies, and there have been efforts to develop novel methods. For example, utilizing the log-odds matrices, multiple alignments of distantly related protein regions were examined. While we have discussed substitution matrices in the context of protein sequence comparison, the main challenge is to perform the DNA sequence comparison. These DNA sequences contain coding information to be examined. Special attention must be given to the all regions (low, medium, and high complexity). The BLAST program filters low-complexity regions from proteins before executing a database search. Due to the application of vague mathematical methods, low-complexity regions lead to unsolved difficulties in sequence similarity searching (high scores result for sequences that are not related, existing match cannot be found, etc.). However, complete genomes must be analyzed by performing data-intensive analysis coherently utilizing all available information for the sequenced genes ensuring complete topologies and sequence preservation. Correspondingly, new sound methods that are not based on assumed hypotheses and simplifications must be developed and demonstrated. Recent results further expand the statistical methods developing information-enhanced procedures to perform large-scale analysis. One may intend to overcome the existing formidable challenges achieving data-intensive analysis and coherent data mining by developing entirely new methods through a fundamental understanding of complex and unique phenomena in the frequency domain. For example, concepts that are based on the spectra analysis utilizing entropy can be applied.

Despite the progress, it appears that novel methods are needed. In fact, statistics-based methods test *a priori* hypotheses against the data with a great number of assumptions and simplifications. Correspondingly, novel information-theoretic methods based on robust analysis must be developed to be performed in the frequency domain utilizing the most advanced array-based matrix methods in order to attain systematic analysis. These methods will discover new (previously unknown or hidden) patterns of the large-scale data sets, examine (interpret, represent, and map) these complex patterns, and create and utilize intelligent libraries. The key question is how to distinguish, detect, and recognize protein functionality under uncertainties. The methods must guarantee the following:

- Homology search and genes detection with superior accuracy and robustness under uncertainties (distinct genes such as cancer and others have been accurately detected with 100% accuracy)
- Accurate and robust analysis and decision making
- Superior computationally efficiency and soundness
- Information extraction and information retrieval
- Correlation between large-scale datasets from multiple databases
- Detection of potential homologs in the databases

The detailed analysis of molecular mechanisms (binding, protein-to-protein interactions, and molecular recognition) is the next task in enhancing our understanding of biosystems.

5.4.3   Bioinformatics and Its Applications to *Escherichia coli* Bacteria

The *E. coli* and *Salmonella typhimurium* bacteria are two of the best-studied microorganisms with sequenced genomes.[6] Information for each *E. coli* gene (the EcoGene12 release includes 4293 genes with 706 predicted or confirmed gene start sites for the MG1655 strain[6]) is organized into separate *gene pages*. The lengths of the genome sequences are different. For example, the *E. coli* MG1655 and CFT073 strains are 4,639,221 and 5,231,428 base-pair strains. For MG1655 these 4,639,221 base pairs (A, C, G, and T) specify 4293 genes. Though there are 717 proteins whose N-terminal amino acids have been verified by sequencing, only 50% of proteins are known. The problem to locate 4293 genes (each of which starts with a ribosome binding site) from 4,639,221 possibilities is extremely difficult, and, using information theory, the number of choices is $\log_2(4639221/4293) = 10$ bits.

5.4.4   Sequential, Fourier Transform and Autocorrelation Analysis in Genome Analysis

Our goal is to examine a sequence similarity for quaternary sequences. One can measure the sum over the length of the sequences of alphabetic similarities at all positions. Alphabetic similarities are symmetrically defined on the Cartesian square of the alphabet. These similarities equal zero whenever the two elements differ, and, in contrast to the Hamming similarity, the reported alphabetic similarities take individual values whenever two elements are identical. Hence, lower and upper bounds can be derived.

Let $\mathbf{A} = \{A, C, G, T\}$; this is the *symbolic quaternary alphabet*. This symbolic alphabet can be represented numerically as $\mathbf{A} = \{0\ 1\ 2\ 3\}$ or $\mathbf{A} = \{1 + j, -1 + j, 1 - j, -1 - j\}$.

The arbitrary pairs of quaternary N-sequences (words of length N) are

$$x = (x_1, x_2, \ldots x_{N-1}, x_N),\ x_i \in \mathbf{A}$$

$$y = (y_1, y_2, \ldots y_{N-1}, y_N),\ y_i \in \mathbf{A}$$

For a pair (x, y) of quaternary words, the similarity is defined as

$$S(x,y) = \sum_{i=1}^{N} s(x_i, y_i)$$

The gene, protein, or genome alphabetic similarity can be expressed as

$$s(x_i, y_i) = \begin{cases} 1 & \text{if } x = y = A \\ 2 & \text{if } x = y = T \\ 3 & \text{if } x = y = G \\ 4 & \text{if } x = y = C \end{cases} \quad \text{or} \quad s(x_i, y_i) = \begin{cases} 1+j & \text{if } x = y = A \\ 1-j & \text{if } x = y = T \\ -1+j & \text{if } x = y = G \\ -1-j & \text{if } x = y = C \end{cases}$$

Thus, one can consider two *complementary* pairs of symbols. If $x = y$, then $S(x,x)$ corresponds to the *self similarity* of x. If $x \neq y$, then $S(x,y)$ represents the *cross-similarity* of pair (x,y). We denote by $D(x,y)$ the Hamming distance between x and y, i.e., the number of positions in which words x and y are different. The DNA code (quaternary code) or amino acid code similarity can be examined. In fact, genes and proteins are represented by finite sequences (*strands*) of A, C, G, and T nucleotides.

Example 5.2

Let $N = 12$, $x = $ (A A A C G T T A C C T A), and $y = $ (A C G T G A A A T C G G). To perform the analysis, let us, for example for two quaternary DNA sequences, assign 0 and 3 for C and G, and 1 and 2 for A and T. The cross-similarity of (x,y) is $S(x, y) = 1 + 0 + 0 + 0 + 3 + 0 + 0 + 1 + 0 + 4 + 0 + 0 = 9$, and the self-similarity of x $S(x, x)$ can be derived.

The Fourier transform offers superior computational advantages, accuracy, versatility, and coherence in order to examine complex genomes composed of millions of experimentally obtained nucleotides (bases). The frequency analysis of DNA and amino acid sequences is an important complement of experimental studies in order to identify protein-coding genes in genomic DNA and perform analysis, synthesis, and design. Furthermore, the Fourier transformation can identify the protein-coding genes, define structural and functional characteristics, analyze the data, identify patterns in gene sequences, etc. The sequence of genes in *E. coli* is available with satisfactory accuracy. In particular, the strings of nitrogenous bases A, C, G, and T or amino acids Ala, Arg, . . . , Tyr, and Val have been found. For example, the FliG gene sequence is given below (genomic address of FliG: 2012902 bp left end, and 2013897 bp right end, 996 length)[6]:

```
ATGAGTAACCTGACAGGCACCGATAAAAGCGTCATCCTGCTGATGACCATTGGCGAAGACC
GGGCGGCAGAGGTGTTCAAGCACCTCTCCCAGCGTGAAGTACAAACCCTGAGCGCTGCAA
TGGCGAACGTCACGCAGATCTCCAACAAGCAGCTAACCGATGTGCTGGCGGAGTTTGAGC
AAGAAGCTGAACAGTTTGCCGCACTGAATATCAACGCCAACGATTATCTGCGCTCGGTATTG
GTCAAAGCTCTGGGTGAAGAACGTGCCGCCAGCCTGCTGGAAGATATTCTCGAAACTCGCG
ATACCGCCAGCGGTATTGAAACGCTCAACTTTATGGAGCCACAGAGCGCCGCCGATCTGAT
TCGCGATGAGCATCCGCAAATTATCGCCACCATTCTGGTGCATCTGAAGCGCGCCCAAGCC
GCCGATATTCTGGCGTTGTTCGATGAACGTCTGCGCCACGACGTGATGTTGCGTATCGCCAC
CTTTGGCGGCGTGCAGCCAGCCGCGCTGGCGGAGCTGACCGAAGTACTGAATGGCTTGCTC
GACGGTCAGAATCTCAAGCGCAGCAAAATGGGCGGCGTGAGAACGGCAGCCGAAATTATC
AACCTGATGAAAACTCAGCAGGAAGAAGCCGTTATTACCCCGTGCGTGAATTCGACGGCGA
GCTGGCGCAGAAAATCATCGACGAGATGTTCCTGTTCGAGAATCTGGTGGATGTCGACGAT
CGCAGCATTCAGCGTCTGTTGCAGGAAGTGGATTCCGAATCGCTGTTGATCGCGCTGAAAG
GAGCCGAGCAGCCACTGCGCGAGAAATTCTTGCGCAATATGTCGCAGCGTGCCGCCGATAT
TCTGCGCGACGATCTCGCCAACCGTGGTCCGGTGCGTCTGTCGCAGGTGGAAAACGAACA
GAAAGCGATTCTGCTGATTGTGCGCCGCCTTGCCGAAACTGGCGAGATGGTAATTGGCAGC
GGCGAGGATACCTATGTCTGA
```

The FliM and FliN proteins are as follows[6]:

Genomic address of FliM: 2018109 bp left end, and 2019113 bp right end, 1005 length
```
ATGGGCGATAGTATTCTTTCTCAAGCTGAAATTGATGCGCTGTTGAATGGTGACAGCGAAGT
CAAAGACGAACCGACAGCCAGTGTTAGCGGCGAAAGTGACATTCGTCCGTACGATCCGAA
TACCCAACGACGGGTTGTGCGCGAACGTTTGCAGGCGCTGGAAATCATTAATGAGCGCTTT
GCCCGCCATTTTCGTATGGGGCTGTTCAACCTGCTGCGTCGTAGCCCGGATATAACCGTCGG
GGCCATCCGCATTCAGCCGTACCATGAATTTGCCCGCAACCTGCCGGTGCCGACCAACCTG
AACCTTATCCATCTGAAACCGCTGCGCGGCACTGGGCTGGTGGTGTTCTCACCGAGTCTGGT
GTTTATCGCCGTGGATAACCTGTTTGGCGGCGATGGACGCTTCCCGACCAAAGTGGAAGGT
CGCGAGTTTACCCATACCGAACAGCGCGTCATCAACCGCATGTTGAAACTGGCGCTTGAAG
GCTATAGCGACGCCTGGAAGGCGATTAATCCGCTGGAAGTTGAGTACGTGCGTTCGGAAAT
GCAGGTGAAATTTACCAATATCACCACCTCGCCGAACGACATTGTGGTTAACACGCCGTTCC
ATGTGGAGATTGGCAACCTGACCGGCGAATTTAATATCTGCCTGCCATTCAGCATGATCGAG
CCGCTACGGGAATTGTTGGTTAACCCGCCGCTGGAAAACTCGCGTAATGAAGATCAGAACT
GGCGCGATAACCTGGTGCGCCAGGTGCAGCATTCACAGCTGGAGCTGGTCGCCAACTTTGC
CGATATCTCGCTACGCCTGTCGCAGATTTTAAAACTGAACCCCGGCGACGTCCTGCCGATAG
AAAAACCCGATCGCATCATCGCCCATGTTGACGGCGTCCCGGTGCTGACCAGTCAGTATGG
CACCCTCAACGGTCAGTATGCGTTACGGATAGAACATTTGATTAACCCGATTTTAAATTCTCT
GAACGAGGAACAGCCCAAATGA
```

Genomic address of FliN: 2019110 bp left end, and 2019523 bp right end, 414 length
```
ATGAGTGACATGAATAATCCGGCCGATGACAACAACGGCGCAATGGACGATCTGTGGGCTG
AAGCGTTGAGCGAACAAAAATCAACCAGCAGCAAAAGCGCTGCCGAGACGGTGTTCCAGC
AATTTGGCGGTGGTGATGTCAGCGGAACGTTGCAGGATATCGACCTGATTATGGATATTCCGG
```

TCAAGCTGACCGTCGAGCTGGGCCGTACGCGGATGACCATCAAAGAGCTGTTGCGTCTGAC
GCAAGGGTCCGTCGTGGCGCTGGACGGTCTGGCGGGCGAACCACTGGATATTCTGATCAAC
GGTTATTTAATCGCCCAGGGCGAAGTGGTGGTCGTTGCCGATAAATATGGCGTGCGGATCAC
CGATATCATTACTCCGTCTGAGCGAATGCGCCGCCTGAGCCGTTAG

Our goal is to apply the mathematical fundamentals to attain analytical, numerical, pattern, visual, and interactive analysis. Consider a sequence of nitrogenous bases A, T, C, and G. We can assign the number a to the character A, the number t to the character T, the number c to the character C, and the number g to the character G. These (a, t, c, and g) can be complex numbers. There exists a numerical sequence resulting from a character string of length N. In particular, we have

$$x[n] = au_A[n] + tu_T[n] + cu_C[n] + gu_G[n], \quad n = 0, 1, 2, \ldots, N - 1$$

where $u_A[n]$, $u_T[n]$, $u_C[n]$, and $u_G[n]$ are the binary indicator sequences (take the value of either 1 or 0 at location n depending on whether the corresponding character exists or not at location n); N is the length of the sequence.

For amino acids, we have the following expression for the amino acid sequence:

$$x[n] = A_{la}u_{Ala}[n] + A_{rg}u_{Arg}[n] + \cdots + T_{yr}u_{Tyr}[n] + V_{al}u_{Val}[n], \quad n = 0, 1, 2, \ldots, N - 1$$

Using the amino acids alphabet (utilizing the common amino acid codes), we have

$$\mathbf{A} = \{\text{Ala}, \text{Arg}, \ldots, \text{Tyr}, \text{Val}\} \quad \text{or}$$

$$\mathbf{A} = \{\text{A, R, N, D, C, Q, E, G, H, I, L, K, M, F, P, S, T, W, Y, V}\}$$

Thus, the amino acid sequence is

$$x[n] = au_a[n] + ru_r[n] + \cdots + yu_y[n] + vu_v[n], \quad n = 0, 1, 2, \ldots, N - 1$$

We obtain the linguistic strings that coherently represent DNA and amino acids. Four-dimensional Fourier transform (for DNA) and 20-dimensional Fourier transform for amino acids can be derived. In particular, the discrete Fourier transform of a sequence $x[n]$ of length N is

$$X[k] = \sum_{n=0}^{N-1} x[n]e^{-j\frac{2\pi}{N}kn}, \quad k = 0, 1, 2, \ldots, N - 1$$

This Fourier transform provides a measure of the frequency content at frequency k that corresponds to a period of N/k samples. The resulting sequences $U_A[k]$, $U_T[k]$, $U_C[k]$, and $U_G[k]$ are the discrete Fourier transforms of the binary indicator sequences $u_A[n]$, $u_T[n]$, $u_C[n]$, and $u_G[n]$. In particular,

$$U_A[k] = \sum_{n=0}^{N-1} u_A[n]e^{-j\frac{2\pi}{N}kn}, \quad U_T[k] = \sum_{n=0}^{N-1} u_T[n]e^{-j\frac{2\pi}{N}kn}, \quad U_C[k] = \sum_{n=0}^{N-1} u_C[n]e^{-j\frac{2\pi}{N}kn},$$

$$U_G[k] = \sum_{n=0}^{N-1} u_G[n]e^{-j\frac{2\pi}{N}kn}, \quad k = 0, 1, 2, \ldots, N - 1$$

If we assign numerical values a, t, c, and g, then

$$X[k] = aU_A[k] + tU_T[k] + cU_C[k] + gU_G[k], \quad k = 0, 1, 2, \ldots, N - 1$$

In general, DNA character strings lead to the sequences $U_A[k]$, $U_T[k]$, $U_C[k]$, and $U_G[k]$ resulting in four-dimensional representation of the frequency spectrum, with

$$U_A[k] + U_T[k] + U_C[k] + U_G[k] = \begin{cases} 0, & k \neq 0 \\ N, & k = 0 \end{cases}$$

The total power spectral content of the DNA character string at the frequency k is

$$S[k] = |U_A[k]|^2 + |U_T[k]|^2 + |U_C[k]|^2 + |U_G[k]|^2$$

For the amino acids, the frequency spectra and power analysis are identical to those reported for DNA. Though the analysis can be accomplished examining multidimensional Fourier transforms, the high dimensionality problem can be resolved by assigning the numerical values to the nitrogenous bases and amino acids. This concept allows one to obtain well-defined one-dimensional Fourier transforms. An example of such a spectrum is given in Figure 5.22, which reports the frequency spectrum of amino acids of the gene FliG.[9]

The autocorrelation analysis can be performed. The deterministic autocorrelation sequence $r_{xx}[n]$ of a sequence $x[n]$ is given as

$$r_{xx}[n] = \sum_{k=-\infty}^{\infty} x[k]x[n+k], \quad n = 0, 1, 2, \ldots, N-1$$

where $x[n]$ is a sequence of either nitrogenous bases or amino acids.

The autocorrelation sequence measures the dependence of values of the sequence at different positions in the sequence. A finite random sequence has an autocorrelation sequence of all zeros with the exception of a single large value at zero. We examine the "randomness" of the studied protein sequence applying the autocorrelation analysis.[9] Figure 5.23 and Figure 5.24 show the autocorrelation sequences of the nitrogenous bases and amino acid sequences of the ElbB and FliG genes.

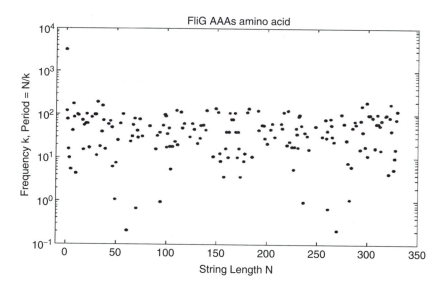

FIGURE 5.22
Fourier transform of FliG amino acids.

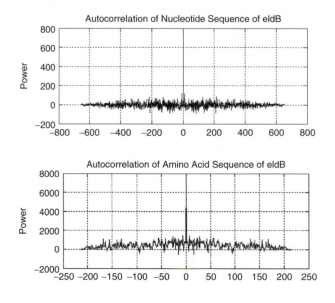

FIGURE 5.23

Autocorrelation of the nucleotide and amino acid sequence for ElbB gene.

The results, documented in Figure 5.23 and Figure 5.24, indicate that the amino acid sequence can be used for analysis. The recently found ElbB gene (which may be involved in isoprenoid biosynthesis) and FliG gene (so-called nanobiomotor *switch*) indicate that there exist specific templates that must be examined. Large-scale DNA and amino acid sequences can be examined using statistical methods. These sequences have different

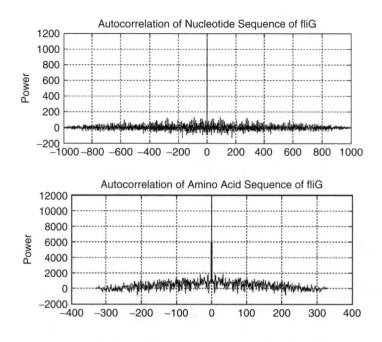

FIGURE 5.24

Autocorrelation of the nucleotide and amino acid sequence for FliG gene.

probabilistic characteristics, not to mention that different genes have distinct characteristics, coding, and uncoding regions that can be distinguished. However, in general, the entropy analysis must be used to overcome well-known difficulties in the application of statistical methods[11-14] to the large-scale genomic data.

Example 5.3

FliG and FliN proteins have attracted attention because mutant phenotypes suggest that these *switch* proteins are needed not only for nanobiomotor assembly but also for torque generation and control. Perform the correlation analysis.

SOLUTION

The cross-correlation analysis is achieved using the equation

$$r_{xy}[n] = \sum_{k=-\infty}^{\infty} x[k]y[n+k]$$

Figure 5.25 reports the cross-correlation of FliG and FliN proteins. The correlation coefficients is given as

$$\rho_{xy}[n] = \frac{r_{xy}[n]}{\sqrt{r_x[0]r_y[0]}}$$

The derived $\rho_{xy}[n]$ provides a normalized measure of correlation. If $\rho_{xy} > 0$, then the sequences are positively correlated, while if $\rho_{xy} = 0$, the sequences are uncorrelated. In addition, $\rho_{xy}[n]$ validates the results. In the studied case, $\rho_{xy} = 0$.

5.4.5 Entropy Analysis

The fundamental question to be examined is how much information is needed to describe the patterns. Messenger RNA (mRNA) carries information specifying amino acid sequences of proteins from DNA to ribosomes. Transfer RNA (tRNA) translates mRNA nucleotide sequences into protein amino acid sequences. Finally, ribosomal RNA (rRNA)

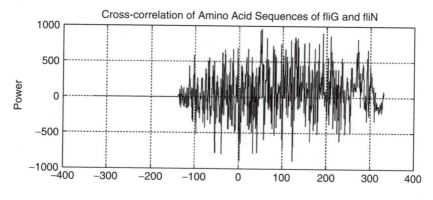

FIGURE 5.25
Cross-correlation of the FliG and FliN *switch* proteins.

and small nuclear RNA (snRNA) play structural and enzymatic roles. In addition to a binding site for mRNA, each ribosome has two binding sites for tRNA. The P site (peptidyl-tRNA site) holds the tRNA carrying the growing polypeptide chain, while the A site (aminoacyl-tRNA site) holds the tRNA carrying the next amino acid to be added to the chain. The ribosome holds the tRNA and mRNA molecules close together, while proteins catalyze the transfer of an amino acid to the carboxyl end of the growing polypeptide chain. The mRNA is moved through the ribosome in the $5` \rightarrow 3`$ direction only (ribosome and mRNA move relative to each other unidirectionally, codon by codon). RNA polymerases (enzymes) add nucleotides to the $3`$ end of the growing polymer, and the RNA molecule elongates in its $5` \rightarrow 3`$ direction. Specific sequences of nucleotides along the DNA mark the initiation and termination sites where transcription of a gene begins and ends.

Before binding, the ribosome's binding sites have four possibilities and do not distinguish among them. Thus, each finger is uncertain by $\log_2 4 = 2$ bits. After binding, the uncertainty at each finger is lower, e.g., $\log_2 1 = 0$ bits. The uncertainty after binding for each finger (Shannon entropy of position l) is

$$H(l) = -\sum_{b \in A} f(b,l) \log_2 f(b,l)$$

where $\mathbf{A}$ is the cardinality of the M-letter alphabet (four-letter of the quaternary DNA code alphabet, $\mathbf{A} = \{A, C, G, T\}$ or 20-letter for the amino acid alphabet $\mathbf{A} = \{Ala, Arg, \ldots, Tyr, Val\}$), and $f(b,l)$ is the frequency of base b at position l.

It should be emphasized that for DNA the maximum uncertainty at any given position is

$$\log_2 \mathbf{A} = 2$$

For amino acids, the alphabet is

$$\mathbf{A} = \{Ala, Arg, \ldots, Tyr, Val\}$$

or

$$\mathbf{A} = \{A, R, N, D, C, Q, E, G, H, I, L, K, M, F, P, S, T, W, Y, V\}$$

Therefore, for amino acids, the maximum entropy at any given position is

$$\log_2 \mathbf{A} = 4.32 \text{ bits}$$

Using the Shannon entropy $H(l)$, one derives the information at every position in the site as

$$R(l) = \log_2 \mathbf{A} - \left(-\sum_{b \in A} f(b,l) \log_2 f(b,l) \right)$$

The total amount of pattern in ribosome binding sites is found by adding the information from each position, e.g., $R_{\Sigma}(l) = \sum_l R(l)$ bits per site. For *E. coli* one finds 10 ± 0.4 bits per site. There is enough pattern at ribosome binding sites for them to be found in the genetic material of the cell (there is no excess and no shortage of patterns).

Let us apply entropy paradigm to the *E. coli* genome. Our ultimate goal is to utilize fundamental mathematical methods to identify interesting sections of a genome. Examining genomes, it is well known that DNA strings have so-called low- and high-complexity regions. By making use of entropy, one can consistently and accurately detect these low- and high-complexity regions. In general, entropy depends on the probability model attributed to the source, and entropy measures *unexpectedness*. Repetitions have low entropy. Thus, if the DNA or amino acid code is redundant with multiple encodings with the same symbol, a random sequence is recognized by another observer using complementary pairs. Probability models can be built upon coding sequences for peptides and proteins encoded using codons (sequences of nitrogenous bases triples). One can examine codons, using a specific genome length, and estimate the probability of occurrence of each codon. The entropy of this sequence is then

$$H = \sum p_i \log_2 \frac{1}{p_i}$$

where p_i is the probability of the specific codon.

The probabilities p should be found. The most direct means of doing this is to calculate the frequency at which each of the codons appears in a given "window." This "window" can be smaller than the entire genome but large enough to capture representative statistics for a given region. To identify the different and important sections of a genome, one can slide this window over the genome and estimate the entropy within it. However, instead of calculating the entropy of all 4^3 (64) possible codons, the entropy of amino acids is calculated. These 64 codons map 20 amino acids as well as "start" and "stop" sequences. The choice of the size of the window and by how much it should slide is not well posed. There is no mathematical basis for determining the window and step sizes. As will be illustrated, a window size directly influences the entropy analysis.

We first consider the case of a large window size. The probabilities calculated in this window will be closest to a normal distribution, and the entropies will be near the maximum. If one uses a small window size, then the distribution of amino acids will be less varied, and the entropy will be lower. The results are reported in Figure 5.26. In particular, two different entropy calculations are performed. The first utilizes a fixed window size of 3^8 (6561) nitrogenous bases and a step size of 3^4 (81) bases. The second

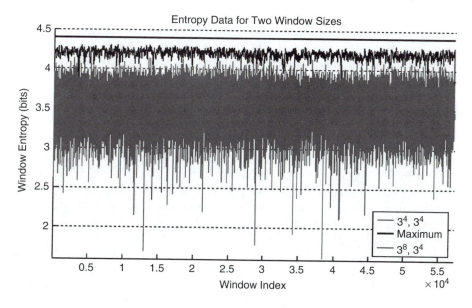

FIGURE 5.26
Entropy for two window sizes compared to maximum entropy.

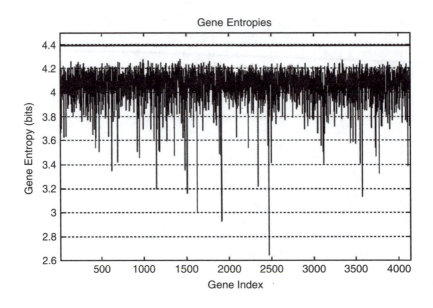

FIGURE 5.27
Entropy of the entire *E. coli* genome.

set of entropies is calculated using a window size of 3^4 (81) nitrogenous bases and a step size of 3^4 (81) bases. This last window represents the case of nonoverlapping probability measurements. The maximum entropy is plotted as a reference in Figure 5.26. The first set of entropies (the large overlapping window size) was found to be in the range of 4.20 bits with a standard deviation of 0.0459 bits. The second set was centered at 3.51 bits with a standard deviation of 0.218 bits. It is evident that the resulting entropies are dependent on the chosen window and step sizes. Therefore, it is important that these quantities have a sound basis if any meaningful information is to be extracted. The EcoGene database[6] provides the needed genomic data. The database defines specific start and stop positions for genes within the entire *E. coli* genome. We use the sizes of various genes as variable window sizes and step the window one gene at a time. Figure 5.27 presents the entropy of verified gene sequences given in the EcoGene database for all 4134 genes. This set is sorted by the natural sequencing order given in the genome. The minimum and maximum gene entropies are 2.65 and 4.27 bits, respectively, with an average at 4.06 bits and a standard deviation of 0.1212 bits.

This section approaches important problems in analysis and design of nanosystems applying the engineering bioinformatics fundamentals. Distinct genes have been examined to prove the periodicity and patterns in their design. The Fourier transform, correlation, and entropy analysis were performed to illustrate that the template patterns should be examined. This analysis provides a viable method in synthesis, pattern recognition, functionality analysis, optimization, redundancy, prototyping, etc. The engineering bioinformatics paradigm should be complemented by the accurate protein sequencing, protein functionality analysis, protein characterization, etc. Unfortunately, these problems have not been solved yet. Correspondingly, one may be unable to fully utilize elegant and powerful mathematic tools and advanced software that allow us to examine all genomes by performing aggregative data-intensive analysis and coherent data mining.

References

1. Lyshevski, S.E., *MEMS and NEMS: Systems, Devices, and Structures*, CRC Press, Boca Raton, FL, 2002.
2. Madigan, M., Martinko, J., and Parker, J., *Biology of Microorganisms*, Prentice Hall, Englewood Cliffs, NJ, 1997.
3. Neidhardt, F., Ingraham, J., and Schaechter, M., *Physiology of the Bacterial Cell: A Molecular Approach*, Sinauer, Sunderland, MA, 1990.
4. Berg, H.C., The rotary motor of bacterial flagella, *J. Annu. Rev. Biochem.*, 72, 19–54, 2003.
5. Welch, R.A., Burland, V., Plunkett, G., Redford, P., Roesch, P., Rasko, D., Buckels, E.L., Liou, S.R., Boutin, A., Hackett, J., Stroud, D., Mayhew, G.F., Rose, D.J., Zhou, S., Schwartz, D.C., Perna, N.T., Mobley, H.L.T., Donnenberg, M.S., and Blattner, F.R., Extensive mosaic structure revealed by the complete genome sequence of uropathogenic *Escherichia coli*, *Microbiology*, 99, 17020–17024, 2002.
6. Rudd, K.E., EcoGene: A genome sequence database for *Escherichia coli* K-12, *Nucleic Acids Res.*, 28, 60–64, 2000.
7. *Protein Data Bank*, retrieved in 2003 from http://www.rcsb.org/pdb/
8. Martz, E., *Protein Explorer*, retrieved in 2003 from http://www.proteinexplorer.org
9. Lyshevski, S.E., Krueger, F.A., and Theodorou, E., Nanoengineering bioinformatics: Nanotechnology paradigm and its applications, *Proc. IEEE Conf. Nanotechnol.*, San Francisco, CA, pp. 896–899, 2003.
10. Bork, P. and Copley, R., Filling in the gap, *Nature*, 409, 818–820, 2001.
11. Altschul, S.F., Madden, T.L., Schäffer, A.A., Zhang, J., Zhang, Z., Miller, W., and Lipman, D.J., Gapped BLAST and PSI-BLAST: A new generation of protein database search programs, *Nucleic Acids Res.*, 25, 3389–3402, 1997.
12. Bertone, P. and Gerstein, M., Integrative data mining: The new direction in bioinformatics, *IEEE Eng. Med. Biol.*, 4, 33–40, 2001.
13. Lusman, G.G. and Lancet, D., Visualizing large-scale genomic sequences, *IEEE Eng. Med. Biol.*, 4, 49–54, 2001.
14. Myers, E.W. and Miller, W., Optimal alignments in linear space, *Comput. Appl. Biosci.*, 4, 11–17, 1988.

6

Modeling of Micro- and Nanoscale Electromechanical Systems and Devices

6.1 Introduction to Modeling, Analysis, and Simulation

As micro- and nanosystems are synthesized, they must be analyzed, modeled, simulated, optimized, etc. The design and analysis of micro- and nanoelectromechanical systems, devices, and structures is not a simple task because electromagnetic, mechanical, thermodynamic, vibroacoustic, and other phenomena and effects must be studied in the time domain. The modeling fundamentals, e.g., the application of basic physics laws to derive equations that coherently and accurately describe system behavior, are reported in books covering quantum mechanics, electromagnetics, electromechanics, optics, solid-state physics, thermodynamics, vibroacoustics, etc. This chapter is written for readers who may lack a strong background, progressing to different areas of physical modeling of major classes of MEMS and NEMS including PZT-based [1]. We have already discussed that it is virtually unfeasible to develop a generic approach to synthesize, design, and model micro- and nanosystems for example studied in [2, 3]. Though it is impossible to cover application- and phenomena-specific MEMS and NEMS, the key electromagnetic and electromechanical classes are covered. In particular, we will develop mathematical models in the form of differential equations and solve them. In addition, computer-aided design is addressed. Computer-aided design of systems, devices, and structures is valuable due to

- Calculation and thorough evaluation of a large number of options with performance analysis and outcome prediction
- Knowledge-based intelligent synthesis and evolutionary design that allow one to define an optimal solution with minimal effort, time, cost, reliability, and accuracy
- Data-intensive nonlinear electromagnetic and mechanical analysis to attain superior performance of systems, devices, and structures while avoiding costly and time-consuming fabrication and testing
- The possibility of solving complex partial differential equations in the time domain integrating nonlinear media characteristics
- Feasibility of developing robust and accurate rapid design software tools that have innumerable features to assist the user to set the problem up and to obtain the engineering parameters
- Synergy between high-performance software and hardware developments.

Through synthesis, classification, and structural optimization, as reported in Chapter 5, the designer devises micro- and nanoscale systems and devices that must be modeled, simulated, analyzed, and optimized. Heterogeneous synthesis, design, and optimization

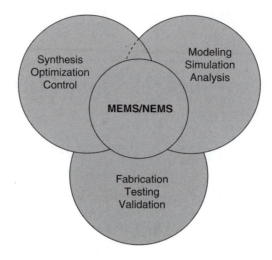

FIGURE 6.1
Design of high-performance MEMS and NEMS.

guarantee superior performance capabilities. Devising and developing novel MEMS and NEMS, one can maximize efficiency, reliability, power, and torque densities, ruggedness, robustness, durability, survivability, compactness, simplicity, controllability, and accuracy. These features are accomplished while minimizing cost, maintenance, size, weight, volume, and losses, as well as optimizing packaging and integrity. (See Figure 6.1.)

Micro- and nanotechnologies drastically change the fabrication and manufacturing of materials, devices, and systems through

- Predictable properties of nanocomposites and materials (e.g., light weight, high strength, thermal stability, low volume and size, extremely high power, torque, force, charge and current densities, specified thermal conductivity and resistivity, etc.)
- Virtual prototyping (design cycle, cost, and maintenance reduction)
- Improved accuracy, precision, reliability, and durability
- High degree of efficiency and capability
- Guaranteed flexibility, integrity, and adaptability
- Attained supportability and affordability
- Survivability, redundancy, and safety
- Improved stability and robustness
- Environmental competitiveness.

Using fabrication technologies (reported in Chapter 4 and Chapter 10), MEMS and NEMS devised using the synthesis and classifier solver can be fabricated. As was reported in Chapter 5, synthesis, classification, and structural optimization are based on the consideration and synthesis of the electromagnetic system, analysis of the magnetomotive force, geometrical and topological designs, biomimicking, and optimization. For example, different rotational (radial and axial) and translational motion microdevices (transducers) are classified using endless (closed), open-ended (open), and integrated electromagnetic systems. Our goal is to approach and solve a wide range of practical problems encountered in nonlinear design, modeling, analysis, control, and optimization of micro- and nanoscale electromechanical systems, devices, and structures. Those devices are controlled utilizing antennas [4–9].

When studying MEMS and NEMS, the emphasis is placed on [5]

- Devising and design of high-performance systems by discovering innovative motion and radiating energy devices, micro- and nanoscale driving/sensing circuitry, and controlling/signal processing ICs
- Optimization and analysis of rotational and translation motion devices
- Development of high-performance signal processing and controlling ICs
- Derivation of high-fidelity mathematical models with minimum level of simplifications and assumptions in the time domain
- Attaining data-intensive analysis capabilities using complete mathematical models and robust computationally effective methods and analytical/numerical algorithms to solve nonlinear equations
- Design of optimal robust control architectures and algorithms
- Design of intelligent systems through self-adaptation, self-organization, evolutionary learning, decision making, and intelligence
- Development of advanced software and hardware to attain the highest degree of synergy, intelligence, integration, efficiency, and performance.

In this chapter our goal is to perform nonlinear modeling, analysis, and simulation. To attain these objectives, we apply the synthesis paradigm (synthesis and classification solver), develop nonlinear mathematical models to model complex electromagnetic-mechanical (electromechanical) dynamics, perform optimization, design closed-loop control systems, and perform data-intensive analysis in the time domain.

For MEMS and NEMS, devices and structures, many engineering problems can be formulated, approached, and solved using micro- and nanoelectromechanics. These paradigms deal with benchmarking and emerging problems in integrated electrical-mechanical-computer engineering, science, and technology. Many of these problems have not been approached and solved, and in general, the existing solutions cannot be treated as optimal. This reflects obvious trends in synergetic fundamental, applied, and experimental research in response to long-standing unsolved problems, engineering and technological enterprise.

Micro- and nanoelectromechanics focus on the integrated design, data-intensive analysis, heterogeneous simulation, optimization, and virtual prototyping of high-performance MEMS and NEMS. In addition other important problems are system intelligence, evolutionary learning, adaptation, decision making, and control through the use of advanced hardware and leading-edge software.

Integrated multidisciplinary features have been addressed. The structural complexity of micro- and nanoscale systems and devices has been increased drastically due to hardware and software advancements as well as stringent imposed achievable performance requirements. Answering the demands of rising systems complexity, performance specifications, and intelligence, fundamental theory must be further expanded. In addition to devising subsystems, devices, and structures, there are other issues that must be addressed and solved in view of the constantly evolving nature of MEMS and NEMS (e.g., complexity, intelligence, decision making, diagnostics, packaging, etc.). Competitive optimum-performance MEMS and NEMS must be designed within the advanced hardware and software concepts.

One of the most challenging problems in MEMS and NEMS systems design is devising novel high-performance motion and radiating energy devices, architecture/configuration synthesis, system integration, and optimization. Other issues are related to selection of hardware and software (environments, tools and computation algorithms to perform control, sensing, execution, emulation, information flow, data acquisition, simulation, visualization, virtual prototyping, and evaluation). As was emphasized, the attempts to design state-of-the-art high-performance MEMS and NEMS and to guarantee integrated design can be pursued through analysis of complex patterns and paradigms of evolutionary developed biological systems.

6.2   Basic Electromagnetics with Applications to MEMS and NEMS

To study MEMS and NEMS, micro- and nanoscale devices and structures, ICs, and antennae, one applies electromagnetic field theory and mechanics. Electric force holds atoms and molecules together. Electromagnetics plays a central role in molecular biology. For example, two DNA chains wrap around one another in the shape of a double helix. These two strands are held together by electrostatic forces. Electric force is responsible for energy-transforming processes in all living organisms (metabolism). Electromagnetism is used to study protein synthesis and structure, nervous systems, etc.

Electrostatic interaction was investigated by Charles Coulomb.

For charges q_1 and q_2, separated by a distance x in free space, the magnitude of the electric force is

$$F = \frac{|q_1 q_2|}{4\pi\varepsilon_0 x^2}$$

where ε_0 is the permittivity of free space, $\varepsilon_0 = 8.85 \times 10^{-12}$ F/m or C^2/N-m^2, and

$$\frac{1}{4\pi\varepsilon_0} = 9 \times 10^9 \text{ N-m}^2/\text{C}$$

The unit for the force is the Newton [N], while the charges are given in coulombs [C]. The force is the vector. Therefore, in general, we have

$$\vec{F} = \frac{q_1 q_2}{4\pi\varepsilon_0 x^2}\vec{a}_x$$

where $\vec{a}_x$ is the unit vector that is directed along the line joining these two charges.

The capacity, elegance, and uniformity of electromagnetics arise from a sequence of fundamental laws linked together and necessary to the study the field quantities.

We denote the vector of electric flux density as $\vec{D}$ [F/m] and the vector of electric field intensity as $\vec{E}$ [V/m or N/C]. Using Gauss's law, the total electric flux Φ [C] through a closed surface is found to be equal to the total force charge enclosed by the surface, that is,

$$\Phi = \oint_s \vec{D} \cdot d\vec{s} = Q_s, \quad \vec{D} = \varepsilon\vec{E}$$

where $d\vec{s}$ is the vector surface area, $d\vec{s} = ds\vec{a}_n$; $\vec{a}_n$ is the unit vector that is normal to the surface; ε is the permittivity of the medium; Q_s is the total charge enclosed by the surface.

Ohm's law for circuits is $V = ir$. However, for a medium, Ohm's law relates the volume charge density $\vec{J}$ and electric field intensity $\vec{E}$ using conductivity σ. In particular,

$$\vec{J} = \sigma\vec{E}$$

where σ is the conductivity [A/V-m] for copper, $\sigma = 5.8 \times 10^7$, and for aluminum, $\sigma = 3.5 \times 10^7$.

Here one may emphasize that the application of Ohm's law to the Ampere-Maxwell equation

$$\nabla \times \vec{B} = \mu \vec{J} + \mu \varepsilon \frac{\partial \vec{E}}{\partial t}, \quad \vec{J} = \sigma \vec{E}$$

results in the wave equation

$$\nabla^2 \vec{B} = \mu \varepsilon \frac{\partial^2 \vec{B}}{\partial t^2} + \mu \sigma \frac{\partial \vec{B}}{\partial t}$$

This equation can be solved. For example, for the one-dimensional equation

$$\frac{\partial^2 B_x}{\partial x^2} = \mu \varepsilon \frac{\partial^2 B_x}{\partial t^2} + \mu \sigma \frac{\partial B_x}{\partial t}$$

the solution is $B_x(t, x) = B_{x0} e^{i(kx - \omega t)}$. Taking note of the derivatives

$$\frac{\partial^2 B_x}{\partial x^2} = -k^2 B_x, \quad \frac{\partial B_x}{\partial t} = -i\omega B_x, \quad \text{and} \quad \frac{\partial^2 B_x}{\partial t^2} = -\omega^2 B_x$$

one finds $k^2 = \mu \varepsilon \omega^2 + \mu \sigma i \omega$.

The derived partial differential equations can be solved. There exist a variety of numerical and analytical methods to solve partial differential equations. For example, the three-dimensional wave equation (for $\sigma \approx 0$)

$$\nabla^2 B(t, x, y, z) = \mu \varepsilon \frac{\partial^2 B(t, x, y, z)}{\partial t^2}$$

with boundary conditions of $B = 0$ when $x = 0$, $x = a$, $y = 0$, $y = b$, $z = 0$, and $z = d$ has the following solution:

$$B(t, x, y, z) = \sin\left(n\pi \frac{x}{a} \right) \sin\left(m\pi \frac{y}{b} \right) \sin\left(l\pi \frac{z}{d} \right) e^{i\omega t}$$

where integers n, m, and l are the so-called mode numbers, and the lowest frequency mode is the (1 1 1) mode.

Taking note of the derivatives

$$\nabla^2 B = -\left[\left(n\pi \frac{x}{a} \right)^2 + \left(m\pi \frac{y}{b} \right)^2 + \left(l\pi \frac{z}{d} \right)^2 \right] B \quad \text{and} \quad \frac{\partial^2 B}{\partial t^2} = -\omega^2 B$$

one obtains the frequency of oscillation:

$$\omega = \pi \sqrt{\frac{1}{\mu \varepsilon}} \sqrt{\left(\frac{n}{a} \right)^2 + \left(\frac{m}{b} \right)^2 + \left(\frac{l}{d} \right)^2}$$

It is important to emphasize that the equation of the electric field in the conducting medium is

$$\nabla^2 \vec{E} = \mu \varepsilon \frac{\partial^2 \vec{E}}{\partial t^2} + \mu \sigma \frac{\partial \vec{E}}{\partial t}$$

with the solution for one-dimensional case $E_x(t,x) = E_{x0}e^{i(kx-\omega t)}$. Taking note of $k^2 = \mu\varepsilon\omega^2 + \mu\sigma i\omega$, we have $\mathrm{Re}(k) = \omega\sqrt{\frac{1}{2}\mu\varepsilon}\sqrt{\sqrt{1+(\frac{\sigma}{\omega\varepsilon})^2}+1}$ and

$$\mathrm{Im}(k) = \omega\sqrt{\tfrac{1}{2}\mu\varepsilon}\ \sqrt{\sqrt{1+\left(\frac{\sigma}{\omega\varepsilon}\right)^2}-1}\ .$$

In vacuum, $\sigma = 0$, and one has $\mathrm{Re}(k) = \omega\sqrt{\mu_0\varepsilon_0}$ and $\mathrm{Im}(k) = 0$. In a perfect conductor, the charges and currents are all at the surface. In a real conductor, the surface currents penetrate a finite distance (so-called skin depth) which is found as

$$d_{\text{skin}} = \frac{1}{\mathrm{Im}(k)} = \left(\omega\sqrt{\tfrac{1}{2}\mu\varepsilon}\sqrt{\sqrt{1+\left(\frac{\sigma}{\omega\varepsilon}\right)^2}-1}\right)^{-1}$$

As the frequency increases, the skin depth decreases. At high frequencies (microwaves), the attenuation becomes significant, and thus the resistance increases with frequency. The reader can examine the real and imaginary parts for high conductivity and examine the mathematics and physics of high-conductivity media. The ratio $\sigma/\varepsilon\omega$ defines whether the conductivity is large or small.

Consider the current density equation

$$\nabla\cdot\vec{J} = -\frac{\partial\rho}{\partial t}$$

Using Ohm's law, we have $\nabla\cdot(\sigma\vec{E}) = -\partial\rho/\partial t$. Making note of the Gauss law $\nabla\cdot\vec{E} = \rho/\varepsilon$ and assuming constant conductivity, one finds

$$\frac{\partial\rho}{\partial t} = -\frac{\sigma}{\varepsilon}\rho$$

with the solution in the form $\rho(t,x,y,z) = \rho_0(x,y,z)e^{-\frac{\sigma}{\varepsilon}t}$. Correspondingly the time constant is σ/ε. However, to obtained the dimensionless time, one has $\sigma/\varepsilon\omega$.

The current i is proportional to the potential difference, and the resistivity ρ of the conductor is the ratio between the electric field $\vec{E}$ and the current density $\vec{J}$. Thus,

$$\rho = \frac{\vec{E}}{\vec{J}}$$

The resistance r of the conductor is related to the resistivity and conductivity by the following formulas

$$r = \frac{\rho l}{A} \quad\text{and}\quad r = \frac{l}{\sigma A}$$

where l is the length and A is the cross-sectional area.

It is important to emphasize that the parameters vary. Let us illustrate this using the wire. The resistances of the wire vary due to heating. In fact, the resistivity depends on temperature T [°C], and we have

$$\rho(T) = \rho_0[1 + \alpha_{p1}(T - T_0) + \alpha_{p2}(T - T_0)^2 + \cdots]$$

where α_{p1} and α_{p2} are the coefficients.

As an example, over the small temperature range (up to 160°C) for copper at $T_0 = 20$°C, we have $\rho(T) = 1.7\times10^{-8}[1 + 0.0039(T - 20)]$.

The basic principles of electromagnetic theory are reviewed below with illustrative examples.

The total magnetic flux through the surface is given by

$$\Phi = \int \vec{B} \cdot d\vec{s}$$

where $\vec{B}$ is the magnetic flux density.

The Ampere circuital law is

$$\int_l \vec{B} \cdot d\vec{l} = \mu_0 \int_s \vec{J} \cdot d\vec{s}$$

where μ_o is the permeability of free space, $\mu_o = 4\pi \times 10^{-7}$ H/m or T-m/A.

For the filamentary current, Ampere's law connects the magnetic flux with the algebraic sum of the enclosed (linked) currents (net current) i_n and

$$\oint_l \vec{B} \cdot d\vec{l} = \mu_o i_n$$

The time-varying magnetic field produces the electromotive force (emf), denoted as $\mathcal{E}$, which induces the current in the closed circuit. Faraday's law relates the emf (induced voltage due to conductor motion in the magnetic field) to the rate of change of the magnetic flux Φ penetrating the loop. In approaching the analysis of electromechanical energy conversion, Lenz's law should be used to find the direction of emf and the current induced. In particular, the emf is in such a direction as to produce a current whose flux, if added to the original flux, would reduce the magnitude of the emf. According to Faraday's law, the induced emf in a closed-loop circuit is defined in terms of the rate of change of the magnetic flux Φ. One has

$$\mathcal{E} = \oint_l \vec{E}(t) \cdot d\vec{l} = -\frac{d}{dt} \int_s \vec{B}(t) \cdot d\vec{s} = -N \frac{d\Phi}{dt} = -\frac{d\psi}{dt}$$

where N is the number of turns and ψ denotes the flux linkages.

This formula represents the Faraday law of induction, and the induced emf (induced voltage) given by

$$\mathcal{E} = -\frac{d\psi}{dt} = -N \frac{d\Phi}{dt}$$

is our particular interest.

The current flows in an opposite direction to the flux linkages. The electromotive force (energy-per-unit-charge quantity) represents a magnitude of the potential difference V in a circuit carrying a current. We have

$$V = -ir + \mathcal{E} = -ir - \frac{d\psi}{dt}$$

The unit for the emf is volts.

Kirchhoff's voltage law states that around a closed path in an electric circuit, the algebraic sum of the emf is equal to the algebraic sum of the voltage drop across the resistance.

Another formulation is that the algebraic sum of the voltages around any closed path in a circuit is zero.

Kirchhoff's current law states that the algebraic sum of the currents at any node in a circuit is zero.

The magnetomotive force (mmf) is the line integral of the time-varying magnetic field intensity $\vec{H}(t)$. That is,

$$mmf = \oint_l \vec{H}(t) \cdot d\vec{l}$$

One concludes that the induced mmf is the sum of the induced current and the rate of change of the flux penetrating the surface bounded by the contour. To show that, we apply Stoke's theorem to find the integral form of Ampere's law (second Maxwell's equation), as given by

$$\oint_l \vec{H}(t) \cdot d\vec{l} = \int_s \vec{J}(t) \cdot d\vec{s} + \int_s \frac{d\vec{D}(t)}{dt} d\vec{s}$$

where $\vec{J}(t)$ is the time-varying current density vector.

The unit for the magnetomotive force is amperes or ampere-turns.

The duality of the emf and mmf can be observed using the following two equations given in terms of the electric and magnetic field intensity vectors

$$\mathscr{E} = \oint_l \vec{E}(t) \cdot d\vec{l}$$

$$mmf = \oint_l \vec{H}(t) \cdot d\vec{l}$$

The inductance (the ratio of the total flux linkages to the current which they link, $L = N\Phi/i$) and reluctance (the ratio of the mmf to the total flux, $\mathfrak{R} = mmf/\Phi$) are used to find emf and mmf.

Using the following equation for the self-inductance $L = \psi/i$, we have

$$\mathscr{E} = -\frac{d\psi}{dt} = -\frac{d(Li)}{dt} = -L\frac{di}{dt} - i\frac{dL}{dt}$$

And, if $L = $ const, one obtains $\mathscr{E} = -L(di/dt)$.

That is, the self-inductance is the magnitude of the self-induced emf per unit rate of change of current.

Solenoids usually integrate movable (plunger) and stationary (fixed) members made from high-permeability ferromagnetic materials. The windings wound with a helical pattern. These solenoids, as electromechanical devices, convert electrical energy to mechanical energy or vice versa. Performance of solenoids is strongly affected by the electromagnetic system, materials, mechanical geometry, magnetic permeability, winging resistance, inductance, friction, etc. The plunger moves with respect to the stationary member as shown in Figure 6.2. When the voltage is applied to the winding, current flows in the winding, and the electromagnetic force is developed causing the plunger to move. When the applied voltage becomes zero, the plunger may resume its original position due to the spring. Furthermore the undesirable phenomena such as residual magnetism and friction must be minimized. Different materials for the central guide (nonmagnetic sleeve) and plunger coating (plating) should be chosen to attain minimum friction and minimize wear. Glass-filled nylon and brass for the guide and silver, copper, aluminum, tungsten, platinum

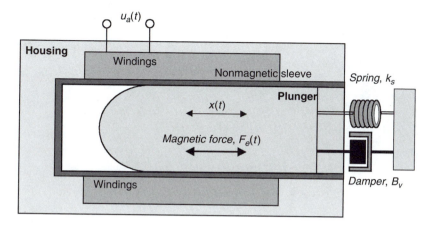

FIGURE 6.2
Schematic of a microsolenoid.

plating, or other low friction coatings for the plunger are possible candidates. The friction coefficients of lubricated (solid film and oil) and unlubricated for different possible candidates are tungsten on tungsten 0.04–0.1 and 0.3; copper on copper 0.04–0.08 and 1.2–1.5; aluminum on aluminum 0.04–0.12 and 1; platinum on platinum 0.04–0.25 and 1.2; titanium on titanium 0.04–0.1 and 0.6.

It is evident that the design, analysis, and optimization of solenoids require application of basic physics, electromechanics, and fabrication technologies. In many cases it is essential to make tradeoffs among a variety of mechanical, electrical, thermal, acoustical, and other physical properties.

Example 6.1

Find the self-inductances of a nanosolenoid (assuming that it can be fabricated) with air-core and filled-core ($\mu = 1000\mu_0$). The solenoid has 10 turns ($N = 10$), the length is 100 nm ($l = 1 \times 10^{-7}$ m), and the uniform circular cross-sectional area is 4×10^{-17} m^2 ($A = 4 \times 10^{-17}$ m^2).

SOLUTION

The magnetic field inside a solenoid is $B = (\mu_0 Ni)/l$. From $\mathscr{E} = -N(d\Phi/dt) = -L(di/dt)$ and applying $\Phi = BA = (\mu_0 NiA)/l$, one obtains the following expression for the self-inductance:

$$L = \frac{\mu_0 N^2 A}{l}$$

Then, for the solenoid with air-core, one obtains $L = 1.6\pi \times 10^{-14}$ H.

If the solenoid is filled with a magnetic material, we have

$$L = \frac{\mu N^2 A}{l}$$

and $L = 1.6\pi \times 10^{-11}$ H.

Example 6.2

Derive a formula for the self-inductance of a torroidal solenoid that has a rectangular cross section ($2a \times b$) and mean radius r.

SOLUTION

The magnetic flux through a cross section is

$$\Phi = \int_{r-a}^{r+a} B b dr = \int_{r-a}^{r+a} \frac{\mu N i}{2\pi r} b dr = \frac{\mu N i b}{2\pi} \int_{r-a}^{r+a} \frac{1}{r} dr = \frac{\mu N i b}{2\pi} \ln\left(\frac{r+a}{r-a}\right)$$

Thus,

$$L = \frac{N\Phi}{i} = \frac{\mu N^2 b}{2\pi} \ln\left(\frac{r+a}{r-a}\right)$$

It is well-known that the electromagnetic torque $\vec{T}$ [N-m] in a current loop is:

$$\vec{T} = \vec{M} \times \vec{B}$$

where $\vec{M}$ denotes the magnetic moment.

Let us examine the torque-energy relations in micro- and nanoscale actuators. Our goal is to study the electromagnetic field energy. It is known that the energy stored in the capacitor is $\frac{1}{2}CV^2$, while the energy stored in the inductor is $\frac{1}{2}Li^2$. Observe that the energy in the capacitor is stored in the electric field between plates, while the energy in the inductor is stored in the magnetic field within the coils.

Let us find the expressions for energies stored in electrostatic and magnetic fields in terms of field quantities. The total potential energy stored in the electrostatic field is found using the potential difference V. We have

$$W_e = \frac{1}{2} \int_v \rho_v V dv \text{ [J]}$$

where ρ_v is the volume charge density [C/m³], $\rho_v = \vec{\nabla} \cdot \vec{D}$, and $\vec{\nabla}$ is the curl operator.

Thus, the potential energy W_e should be found using the amount of work that is required to position the charge in the electrostatic field. In particular, the work is found as the product of the charge and the potential.

Considering the region with a continuous charge distribution (ρ_v = const), each charge is replaced by $\rho_v dv$, and hence the following equation results:

$$W_e = \frac{1}{2} \int_v \rho_v V dv$$

In the Gauss form, by taking note of $\rho_v = \vec{\nabla} \cdot \vec{D}$ and making use of $\vec{E} = -\vec{\nabla} V$, one obtains the following expression for the energy stored in the electrostatic field:

$$W_e = \frac{1}{2} \int_v \vec{D} \cdot \vec{E} dv$$

and the electrostatic volume energy density is $\frac{1}{2} \vec{D} \cdot \vec{E}$ [J/m³].

For a linear isotropic medium we have

$$W_e = \frac{1}{2} \int_v \varepsilon |\vec{E}|^2 \, dv = \frac{1}{2} \int_v \frac{1}{\varepsilon} |\vec{D}|^2 \, dv$$

The electric field $\vec{E}(x,y,z)$ is found using the scalar electrostatic potential function $V(x,y,z)$ as

$$\vec{E}(x,y,z) = -\vec{\nabla} V(x,y,z)$$

In the cylindrical and spherical coordinate systems, we have

$$\vec{E}(r,\phi,z) = -\vec{\nabla} V(r,\phi,z) \quad \text{and} \quad \vec{E}(r,\theta,\phi) = -\vec{\nabla} V(r,\theta,\phi)$$

Using $W_e = \frac{1}{2} \int_v \rho_v V dv$, the potential energy that is stored in the electric field between two surfaces (for example, in a capacitor) is found to be

$$W_e = \frac{1}{2} QV = \frac{1}{2} CV^2$$

Using the principle of virtual work, for the lossless conservative system, the differential change of the electrostatic energy dW_e is equal to the differential change of mechanical energy dW_{mec}. That is,

$$dW_e = dW_{mec}$$

For translational motion, one has

$$dW_{mec} = \vec{F}_e \cdot d\vec{l}$$

where $d\vec{l}$ is the differential displacement.

One obtains $dW_e = \vec{\nabla} W_e \cdot d\vec{l}$.

Hence, the force is the gradient of the stored electrostatic energy,

$$\vec{F}_e = \vec{\nabla} W_e$$

In the Cartesian coordinates, we have

$$F_{ex} = \frac{\partial W_e}{\partial x}, \quad F_{ey} = \frac{\partial W_e}{\partial y}, \quad \text{and} \quad F_{ez} = \frac{\partial W_e}{\partial z}$$

Example 6.3

Consider the capacitor (the plates have area A and they are separated by x), which is charged to a voltage V. The permittivity of the dielectric is ε. Find the stored electrostatic energy and the force F_{ex} in the x direction.

SOLUTION

Neglecting the fringing effect at the edges, one concludes that the electric field is uniform, and $E = V/x$. Therefore,

$$W_e = \frac{1}{2} \int_v \varepsilon |\vec{E}|^2 \, dv = \frac{1}{2} \int_v \varepsilon \left(\frac{V}{x} \right)^2 dv = \frac{1}{2} \varepsilon \frac{V^2}{x^2} Ax = \frac{1}{2} \varepsilon \frac{A}{x} V^2 = \frac{1}{2} C(x) V^2$$

Thus, the force is

$$F_{ex} = \frac{\partial W_e}{\partial x} = \frac{\partial\left(\frac{1}{2}C(x)V^2\right)}{\partial x} = \frac{1}{2}V^2\frac{\partial C(x)}{\partial x}$$

Taking note that $C = \varepsilon(A/x)$, one finds

$$F_{ex} = \frac{1}{2}V^2\frac{\partial C(x)}{\partial x} = -\frac{1}{2}\varepsilon A V^2 \frac{1}{x^2}$$

To find the stored energy in the magnetostatic field in terms of field quantities, the following formula is used:

$$W_m = \frac{1}{2}\int_v \vec{B}\cdot\vec{H}dv$$

The magnetic volume energy density is $\frac{1}{2}\vec{B}\cdot\vec{H}$ [J/m³].

Using $\vec{B} = \mu\vec{H}$, one obtains two alternative formulas:

$$W_m = \frac{1}{2}\int_v \mu\left|\vec{H}\right|^2 dv = \frac{1}{2}\int_v \frac{\left|\vec{B}\right|^2}{\mu}dv$$

To show how the energy concept studied is applied to electromechanical devices, we find the energy stored in inductors. To approach this problem, we substitute

$$\vec{B} = \vec{\nabla}\times\vec{A}$$

Using the following vector identity

$$\vec{H}\cdot\vec{\nabla}\times\vec{A} = \vec{\nabla}\cdot(\vec{A}\times\vec{H}) + \vec{A}\cdot\vec{\nabla}\times\vec{H}$$

one obtains

$$W_m = \frac{1}{2}\int_v \vec{B}\cdot\vec{H}dv = \frac{1}{2}\int_v \vec{\nabla}\cdot(\vec{A}\times\vec{H})dv + \frac{1}{2}\int_v \vec{A}\cdot\vec{\nabla}\times\vec{H}dv = \frac{1}{2}\int_s (\vec{A}\times\vec{H})\cdot d\vec{s} + \frac{1}{2}\int_v \vec{A}\cdot\vec{J}dv = \frac{1}{2}\int_v \vec{A}\cdot\vec{J}dv$$

Using the general expression for the vector magnetic potential $\vec{A}(\vec{r})$ [Wb/m], as given by

$$\vec{A}(\vec{r}) = \frac{\mu_0}{4\pi}\int_{v_A} \frac{\vec{J}(\vec{r}_A)}{x}dv_J, \quad \vec{\nabla}\cdot\vec{A} = 0$$

we have

$$W_m = \frac{\mu}{8\pi}\int_v \int_{v_J} \frac{\vec{J}(\vec{r}_A)\cdot\vec{J}(\vec{r})}{x}dv_J dv$$

Here, v_J is the volume of the medium where $\vec{J}$ exists.

The general formula for the self-inductance $i = j$ and the mutual inductance $i \neq j$ of loops i and j is

$$L_{ij} = \frac{N_i\Phi_{ij}}{i_j} = \frac{\psi_{ij}}{i_j}$$

where ψ_{ij} is the flux linkage through the ith coil due to the current in the jth coil; i_j is the current in the jth coil.

The Neumann formula is applied to find the mutual inductance. We have

$$L_{ij} = L_{ji} = \frac{\mu}{4\pi} \oint_{l_i} \oint_{l_j} \frac{d\vec{l}_j \cdot d\vec{l}_i}{x_{ij}}, \quad i \neq j$$

Then, using

$$W_m = \frac{\mu}{8\pi} \int_v \int_{v_j} \frac{\vec{J}(\vec{r}_A) \cdot \vec{J}(\vec{r})}{x} dv_j dv$$

one obtains

$$W_m = \frac{\mu}{8\pi} \int_{l_i} \int_{l_j} \frac{i_j d\vec{l}_j \cdot i_i d\vec{l}_i}{x_{ij}}$$

Hence, the energy stored in the magnetic field is found to be

$$W_m = \tfrac{1}{2} i_i L_{ij} i_j$$

As an example, the energy, stored in the inductor is $W_m = \tfrac{1}{2} L i^2$.

The differential change in the stored magnetic energy should be found.

Using

$$\frac{dW_m}{dt} = \tfrac{1}{2}\left(L_{ij} i_j \frac{di_i}{dt} + L_{ij} i_i \frac{di_j}{dt} + i_i i_j \frac{dL_{ij}}{dt} \right)$$

we have

$$dW_m = \tfrac{1}{2}(L_{ij} i_j di_i + L_{ij} i_i di_j + i_i i_j dL_{ij})$$

For translational motion, the differential change in the mechanical energy is expressed by

$$dW_{mec} = \vec{F}_m \cdot d\vec{l}$$

Assuming that the system is conservative (for lossless systems $dW_{mec} = dW_m$), in the rectangular coordinate system we obtain the following equation:

$$dW_m = \frac{\partial W_m}{\partial x} dx + \frac{\partial W_m}{\partial y} dy + \frac{\partial W_m}{\partial z} dz = \vec{\nabla} W_m \cdot d\vec{l}$$

Hence, the force is the gradient of the stored magnetic energy, and

$$\vec{F}_m = \vec{\nabla} W_m$$

In the x-y-z coordinate system for the translational motion, we have

$$F_{mx} = \frac{\partial W_m}{\partial x}, \quad F_{my} = \frac{\partial W_m}{\partial y}, \quad \text{and} \quad F_{mz} = \frac{\partial W_m}{\partial z}$$

For the rotational motion, the torque should be used. Using the differential change in the mechanical energy as a function of the angular displacement θ, the following formula results if the rigid body (nano- or microactuator) is constrained to rotate about the z-axis:

$$dW_{mec} = T_e d\theta$$

where T_e is the z-component of the electromagnetic torque.

Assuming that the system is lossless, one obtains the following expression for the electromagnetic torque:

$$T_e = \frac{\partial W_m}{\partial \theta}$$

Example 6.4

Calculate the magnetic energy of the torroidal microsolenoid if the self-inductance is $L = 2 \times 10^{-11}$ H when the current is $i = 1 \times 10^{-6}$ A.

SOLUTION

The stored field energy is $W_m = \frac{1}{2}Li^2$. Hence, $W_m = 1 \times 10^{-23}$ J.

Example 6.5

Derive the expression for the electromagnetic force developed by the microelectromagnet with the cross-sectional area A. The current $i_a(t)$ in N coils (microwindings) produces the constant flux Φ_m. (See Figure 6.3.)

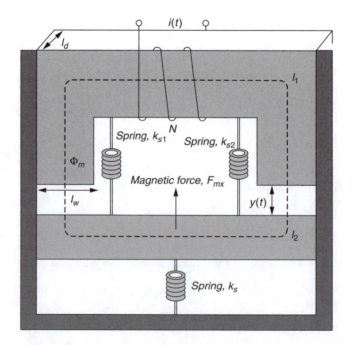

FIGURE 6.3
Microelectromagnet with springs.

SOLUTION

We assume that the flux is constant. Taking into account the fact that the displacement (the virtual displacement is denoted as dy) changes only the magnetic energy stored in the air gaps, from

$$W_m = \tfrac{1}{2}\int_v \mu \left|\vec{H}\right|^2 dv = \tfrac{1}{2}\int_v \frac{\left|\vec{B}\right|^2}{\mu}\,dv$$

we have

$$dW_m = dW_{m\,air\,gap} = 2\frac{B^2}{2\mu_0}A\,dy = \frac{\Phi_m^2}{\mu_0 A}\,dy$$

Thus, if Φ_m = const (the current is constant), one concludes that the increase of the air gap (dy) leads to an increase of the stored magnetic energy. Using $F_{mx} = \partial W_m/\partial x$, one finds the expression for the force in the following form:

$$\vec{F}_{mx} = -\vec{a}_y \frac{\Phi_m^2}{\mu_0 A}$$

The result indicates that the force tends to reduce the air-gap length, and the movable member is attached to the springs, which develop three spring forces in addition to the electromagnetic force.

The fringing effect should be integrated in high-fidelity modeling. The air-gap reluctance (two air gaps are in series) is expressed as

$$\Re_g = \frac{2x}{\mu_0\left(k_{g1}l_w l_d + k_{g2}x^2\right)}$$

where k_{g1} and k_{g2} are the nonlinear functions of the ferromagnetic material, l_d/l_w ratio, B-H curve, load, etc.

The reluctances of the ferromagnetic materials of stationary and movable members (microstructures) $\Re_1$ and $\Re_2$ are found as

$$\Re_1 = \frac{l_1}{\mu_0\mu_1 A} = \frac{l_1}{\mu_0\mu_1 l_w l_d} \quad \text{and} \quad \Re_2 = \frac{l_2}{\mu_0\mu_2 A} = \frac{l_2}{\mu_0\mu_2 l_w l_d}$$

The inductance is expressed as

$$L(x) = \frac{N^2}{\Re_g(x) + \Re_1 + \Re_2}$$

The electromagnetic torque is found as

$$F_{mx} = \tfrac{1}{2}i^2\frac{dL(x)}{dx} = \tfrac{1}{2}i^2\frac{d\left(\dfrac{N^2}{\Re_g(x)+\Re_1+\Re_2}\right)}{dx} = \tfrac{1}{2}i^2\frac{d\left(\dfrac{N^2}{\dfrac{2}{\mu_0}\left(\dfrac{l_w l_d}{x}+k_{g1}l_w+k_{g2}l_d+k_{g3}x\right)^{-1}+\Re_1+\Re_2}\right)}{dx}$$

The reader is advised to perform the differentiation to derive the expression for the electromagnetic force. (Note: A similar differentiation is performed in Example 6.6.)

In micro- and nanoscale electromechanical motion devices, the coupling (magnetic interaction) between windings that are carrying currents is represented by their mutual inductances. In fact, the current in each winding causes the magnetic field in other windings. The mutually induced emf is characterized by the mutual inductance, which is a function of the position x or the angular displacement θ. By applying the expressions for the coenergy $W_c[i, L(x)]$ (translational motion) or $W_c[i, L(\theta)]$ (rotational motion), the developed electromagnetic torque can be found as

$$T_e(i, x) = \frac{\partial W_c[i, L(x)]}{\partial x}$$

and

$$T_e(i, x) = \frac{\partial W_c[i, L(\theta)]}{\partial \theta}$$

Example 6.6

Consider the microactuator (microelectromagnet) that has N turns. (See Figure 6.4.) The distance between the stationary and movable members is denoted as $x(t)$. The mean lengths of the stationary and movable members are l_1 and l_2, and the cross-sectional area is A. Neglecting the leakage flux, find the force exerted on the movable member if the time-varying current $i_a(t)$ is supplied (one can feed sinusoidal current to the winding). The permeabilities of stationary and movable members are μ_1 and μ_2.

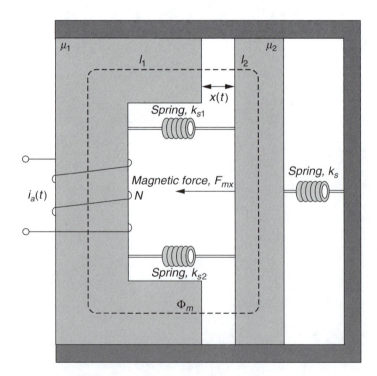

FIGURE 6.4
Schematic of a microelectromagnet.

SOLUTION

The electromagnetic force is

$$F_{mx} = \frac{\partial W_m}{\partial x}$$

where $W_m = \frac{1}{2} L i_a^2(t)$.

The magnetizing inductance is

$$L = \frac{N\Phi}{i_a(t)} = \frac{\psi}{i_a(t)}$$

where the magnetic flux is found as

$$\Phi = \frac{N i_a(t)}{\Re_1 + \Re_x + \Re_x + \Re_2}$$

The reluctances of the ferromagnetic materials of stationary and movable members $\Re_1$ and $\Re_2$, as well as the reluctance of the air gap $\Re_x$, are found as

$$\Re_1 = \frac{l_1}{\mu_0 \mu_1 A}, \quad \Re_2 = \frac{l_2}{\mu_0 \mu_2 A}, \quad \text{and} \quad \Re_x = \frac{x(t)}{\mu_0 A}$$

and the magnetic circuit analog with the reluctances of the various paths is illustrated in Figure 6.5.

By making use of the reluctances in the movable and stationary members and air gap, one obtains the following formula for the flux linkages:

$$\psi = N\Phi = \frac{N^2 i_a(t)}{\dfrac{l_1}{\mu_0 \mu_1 A} + \dfrac{2x(t)}{\mu_0 A} + \dfrac{l_2}{\mu_0 \mu_2 A}}$$

and the magnetizing inductance is a nonlinear function of the displacement.

Thus, we have

$$L(x) = \frac{N^2}{\dfrac{l_1}{\mu_0 \mu_1 A} + \dfrac{2x(t)}{\mu_0 A} + \dfrac{l_2}{\mu_0 \mu_2 A}} = \frac{N^2 \mu_0 \mu_1 \mu_2 A}{\mu_2 l_1 + \mu_1 \mu_2 2x(t) + \mu_1 l_2}$$

Using

$$F_{mx} = \frac{\partial W_m}{\partial x} = \frac{1}{2} \frac{\partial \left(L(x(t)) i_a^2(t) \right)}{\partial x}$$

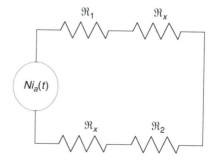

FIGURE 6.5
Magnetic Circuit analog.

the force in the x direction is found to be

$$F_{mx} = -\frac{N^2 \mu_0 \mu_1^2 \mu_2^2 A i_a^2}{[\mu_2 l_1 + \mu_1 \mu_2 2x(t) + \mu_1 l_2]^2}$$

Differential equations must be developed to model the microelectromagnet studied. Using Newton's second law of motion, one obtains following nonlinear differential equations to model, analyze, and simulate the microelectromagnet dynamics:

$$\frac{dx}{dt} = v$$

$$\frac{dv}{dt} = \frac{1}{m}\left(\frac{N^2 \mu_0 \mu_1^2 \mu_2^2 A i_a^2}{-[\mu_2 l_1 + \mu_1 \mu_2 2x(t) + \mu_1 l_2]^2} - k_s x + k_{s1} x + k_{s2} x\right)$$

This set of two differential equations gives the lumped-parameter model if microactuator. It is evident that nonlinear differential equations are found. These equations accurately model microactuators and, therefore, can be used for analysis, simulation, and design.

Example 6.7

Two microcoils have mutual inductance 0.00005 H ($L_{12} = 0.00005$ H). The current in the first microcoil is $i_1 = \sqrt{\sin 4t}$. Find the induced emf in the second microcoil.

SOLUTION

The induced emf is given as

$$\mathcal{E}_2 = L_{12}\frac{di_1}{dt}$$

By using the power rule for the time-varying current in the first coil $i_1 = \sqrt{\sin 4t}$, we have

$$\frac{di_1}{dt} = \frac{2\cos 4t}{\sqrt{\sin 4t}}$$

Hence, one obtains

$$\mathcal{E}_2 = \frac{0.0001\cos 4t}{\sqrt{\sin 4t}}$$

Example 6.8

Figure 6.6 illustrates a microactuator (microscale electromechanical device) with a stationary member and movable plunger. Our goal is to derive the differential equations to perform modeling. That is, the ultimate objective is to illustrate the lumped-parameter mathematical model development for microactuators.

SOLUTION

Applying Newton's second law of motion, one finds the differential equation to model the translational motion. In particular,

$$F(t) = B_v \frac{dx}{dt} + (k_s x - k_{s1} x) + F_e(t) = m\frac{d^2 x}{dt^2}$$

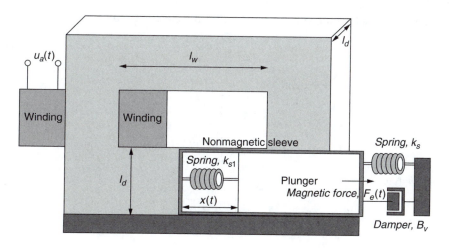

FIGURE 6.6
Schematic of a microactuator.

where x denotes the displacement of a plunger (air gap); m is the mass of a movable plunger; B_v is the viscous friction coefficient; k_s and k_{s1} are the spring constants; and $F_e(t)$ is the magnetic force, which is given as

$$F_e(i,x) = \frac{\partial W_c(i,x)}{\partial x}$$

The restoring/stretching force exerted by the springs is

$$F_s = k_s x - k_{s1} x$$

Assuming that the magnetic system is linear, the coenergy is expressed as

$$W_c(i,x) = \tfrac{1}{2} L(x) i^2$$

The electromagnetic force is

$$F_e(i,x) = \tfrac{1}{2} i^2 \frac{dL(x)}{dx}$$

The inductance can be found by using the following formula:

$$L(x) = \frac{N^2}{\Re_f + \Re_g} = \frac{N^2 \mu_f \mu_0 A_f A_g}{A_g l_f + A_f \mu_f (x + 2d)}$$

where $\Re_f$ and $\Re_g$ are the reluctances of the ferromagnetic material and air gap; A_f and A_g are the associated cross-sectional areas; and l_f and $(x + 2d)$ are the lengths of the magnetic material and the air gap, respectively.

Hence,

$$\frac{dL}{dx} = -\frac{N^2 \mu_f^2 \mu_0 A_f^2 A_g}{[A_g l_f + A_f \mu_f (x + 2d)]^2}$$

Using Kirchhoff's law, the voltage equation for the electric circuit is

$$u_a = ri + \frac{d\psi}{dt}$$

where the flux linkage ψ is expressed as $\psi = L(x)i$.

One obtains

$$u_a = ri + L(x)\frac{di}{dt} + i\frac{dL(x)}{dx}\frac{dx}{dt}$$

Thus, we have

$$\frac{di}{dt} = -\frac{r}{L(x)}i + \frac{1}{L(x)}\frac{N^2\mu_f^2\mu_0 A_f^2 A_g}{[A_g l_f + A_f \mu_f(x+2d)]^2}iv + \frac{1}{L(x)}u_a$$

Augmenting this equation with the second-order differential equation for the mechanical system, three first-order nonlinear differential equations are found as

$$\frac{di}{dt} = -\frac{r[A_g l_f - A_f \mu_f(x-2d)]}{N^2\mu_f\mu_0 A_f A_g}i + \frac{\mu_f A_f}{A_g l_f + A_f \mu_f(x+2d)}iv + \frac{A_g l_f - A_f \mu_f(x-2d)}{N^2\mu_f\mu_0 A_f A_g}u_a$$

$$\frac{dx}{dt} = v$$

$$\frac{dv}{dt} = -\frac{N^2\mu_f^2\mu_0 A_f^2 A_g}{2m[A_g l_f + A_f \mu_f(x+2d)]^2}i^2 - \frac{1}{m}(k_s x - k_{s1}x) - \frac{B_v}{m}v$$

It must be emphasized that the fringing effect should be integrated in high-fidelity modeling and analysis. The air gap reluctance is expressed as

$$\mathfrak{R}_g = \frac{x - 2d}{\mu_0(k_{g1}l_w l_d + k_{g2}x^2)}$$

where k_{g1} and k_{g2} are the nonlinear functions of the ferromagnetic material, l_d/l_w ratio, B-H curve, load, etc.

Therefore, the inductance can be found as $L(x) = N^2/(\mathfrak{R}_f + \mathfrak{R}_g)$, and the accurate mathematical model can be straightforwardly derived using the procedure reported.

Having documented various electromagnetic microdevices, the electrostatic microdevice is examined in Example 6.9.

Example 6.9

Rotational electrostatic micromotors have been widely examined in the literature. For example, in References 19 and 20. The cross-sectional view of the electrostatic motor is shown in Figure 6.7. For example, the motor can be built on a layer of 1 μm silicon nitride, which is deposited on a layer of 100 nm silicon dioxide on a silicon wafer. The silicon nitride layer is used to prevent electrical breakdown between the motor and the silicon substrate. The thin oxide layer below the nitride layer is used to reduce the stress between the silicon nitride and the substrate. Below the polysilicon rotor, there is a thin (say 300 nm) layer of polysilicon, which is needed to shield the electric field between the rotor and the substrate (so that there will be no vertical electrostatic forces on the rotor). The rotor is supported by a flanged hub.

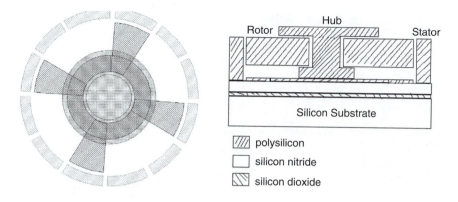

FIGURE 6.7
Top and cross-sectional view of the electrostatic micromotor.

(There is a need to minimize the frictional forces between the rotor and substrate). Derive the differential equations to perform modeling. That is, find the lumped-parameter mathematical model and estimate the major coefficients of differential equations. Make your conclusions regarding the benefits and drawbacks of the electrostatic micromotors.

SOLUTION

As the voltage V is applied to the parallel conducting rotor and stator plates, the charge is $Q = CV$, where C is the capacitance, $C = \varepsilon(A/g) = \varepsilon(WL/g)$; A is the area of the plate; ε is the permittivity of the media between the plates (in vacuum, $\varepsilon_0 = 8.85 \times 10^{-12}$ C²/N-m² = 8.85×10^{-12} F/m); g is the gap distance between the plates; and W and L are the width and length of the plates, respectively.

The energy associated with electrical potential is $W = \frac{1}{2}CV^2$.

Hence, the electrostatic force is found as

$$F_{el} = \frac{\partial W}{\partial g} = -\frac{1}{2}\frac{\varepsilon WL}{g^2}V^2$$

The tangential force due to misalignment is

$$F_t = \frac{\partial W}{\partial x} = \frac{1}{2}\frac{\varepsilon}{g}\frac{\partial(WL)}{\partial x}V^2$$

where x is the direction in which misalignment occurs. If the misalignment occurs in the width direction,

$$F_{t,w} = \frac{\partial W}{\partial x} = \frac{1}{2}\frac{\varepsilon L}{g}V^2$$

The capacitance of a cylindrical capacitor must be found. The voltage between the cylinders can be obtained by integrating the electric field. The electric field at a distance r from a conducting cylinder has only a radial component denoted as E_r. We have, $E_r = \rho/(2\pi\varepsilon r)$, where ρ is the linear charge density, and $Q = \rho L$. Hence, the potential difference is found as

$$\Delta V = V_a - V_b = \int_a^b \vec{E} \cdot d\vec{l} = \int_a^b E_r \cdot dr = \frac{\rho}{2\pi\varepsilon}\int_{r_1}^{r_2}\frac{1}{r}dr = \frac{\rho}{2\pi\varepsilon}\ln\frac{r_2}{r_1}$$

Thus,

$$C = \frac{Q}{\Delta V} = \frac{2\pi\varepsilon L}{\ln\frac{r_2}{r_1}}$$

The capacitance per unit length is

$$\frac{C}{L} = \frac{\rho}{\Delta V} = \frac{2\pi\varepsilon}{\ln\frac{r_2}{r_1}}$$

Therefore, using the stator-rotor electrodes (plates) overlap, for the studied rotational electrostatic micromotor, the capacitance can be expressed as a function of the angular displacement. In particular,

$$C(\theta_r) = N\frac{2\pi\varepsilon}{\ln\frac{r_2}{r_1}}\theta_r$$

where N is the number of overlapping stator-rotor electrodes (plates); r_1 and r_2 are the radii of the rotor and stator electrodes, respectively.

Other expressions for capacitances can be found. For example, accounting for the second-order effects, more detailed expressions for the capacitance can be derived. Our goal is to demonstrate the basic features, and thus the simple expression $C(\theta_r) = N\frac{2\pi\varepsilon}{\ln\frac{r_2}{r_1}}\theta_r$ is used.

The electrostatic torque developed is found as

$$T_{el} = \frac{1}{2}\frac{\partial C(\theta_r)}{\partial\theta_r}V^2 = N\frac{\pi\varepsilon}{\ln\frac{r_2}{r_1}}V^2$$

The torsional-mechanical equations of motion are found using Newton's law. We have

$$\frac{d\omega_r}{dt} = \frac{1}{J}(T_{el} - B_m\omega - T_L) = \frac{1}{J}\left(N\frac{\pi\varepsilon}{\ln\frac{r_2}{r_1}}V^2 - B_m\omega_r - T_L\right)$$

$$\frac{d\theta_r}{dt} = \omega_r$$

These differential equations describe the dynamics of the studied electrostatic micromotor. The parameters of the fabricated micromotor can be estimated using the micromotor size (to obtain N, r_1 and r_2). The fabrication technologies and processes significantly influence the micromotor dimensions and coefficients. For example, one can have $N = 12$, $r_1 = 100$ μm and $r_2 = 105$ μm. Using this sizing data, as well as the density of materials, the moment of inertia J can be found.

It must be emphasized that motors rotate only if $T_{el} > T_L$. Thus, taking note of the loads, the rated electrostatic torque must be examined to guarantee $T_{el} > T_L$. This results in the specific micromotor dimensions and voltage applied. In fact, one uses the equation

$$T_{el} = N\frac{\pi\varepsilon}{\ln\frac{r_2}{r_1}}V^2$$

to examine the rated electrostatic torque developed.

Though fabrication of electrostatic micromotors may appear to be simpler than electromagnetic motion devices, the fact that the high voltage must be supplied to the stator plates and the rotor must be grounded (voltage may be supplied to the rotor plates) significantly limits the application of the electrostatic rotational micromachines. It also must be emphasized that the torque density of electrostatic micromachines is significantly lower than the torque density of electromagnetic motion microdevices. The electromagnetic micromotors can be also straightforwardly fabricated and assembled using axial windings and premade permanent magnets.

6.3 Model Developments of Micro- and Nanoactuators Using Electromagnetics

Electromagnetic theory and mechanics form the basis for the development of MEMS and even NEMS models. The electrostatic and magnetostatic equations in linear isotropic media are found using the vectors of the electric field intensity $\vec{E}$, electric flux density $\vec{D}$, magnetic field intensity $\vec{H}$, and magnetic flux density $\vec{B}$. In addition one uses the constitutive equations

$$\vec{D} = \varepsilon \vec{E} \quad \text{and} \quad \vec{B} = \mu \vec{H}$$

where ε is the permittivity and μ is the permeability.

The basic equations are given in Table 6.1.

In the static (time-invariant) fields, electric and magnetic field vectors form separate and independent pairs. That is, $\vec{E}$ and $\vec{D}$ are not related to $\vec{H}$ and $\vec{B}$, and vice versa. However, in reality, the electric and magnetic fields are time-varying, and the changes of magnetic field influence the electric field, and vice versa.

The partial differential equations are found by using Maxwell's equations. In particular, four of Maxwell's equations in the differential form for time-varying fields are

$$\nabla \times \vec{E}(x,y,z,t) = -\mu \frac{\partial \vec{H}(x,y,z,t)}{\partial t} \quad \text{(Faraday's law)}$$

$$\nabla \times \vec{H}(x,y,z,t) = \sigma \vec{E}(x,y,z,t) + \vec{J}(x,y,z,t) = \sigma \vec{E}(x,y,z,t) + \varepsilon \frac{\partial \vec{E}(x,y,z,t)}{\partial t}$$

$$\nabla \cdot \vec{E}(x,y,z,t) = \frac{\rho_v(x,y,z,t)}{\varepsilon} \quad \text{(Gauss's law)}$$

$$\nabla \cdot \vec{H}(x,y,z,t) = 0$$

where $\vec{E}$ is the electric field intensity, and using the permittivity ε, the electric flux density is $\vec{D} = \varepsilon \vec{E}$; $\vec{H}$ is the magnetic field intensity, and using the permeability μ, the magnetic

TABLE 6.1

Fundamental Equations of Electrostatic and Magnetostatic Fields in Media

	Electrostatic Model	MagnetostaticModel
Governing equations	$\nabla \times \vec{E}(x,y,z,t) = 0$	$\nabla \times \vec{H}(x,y,z,t) = 0$
	$\nabla \cdot \vec{E}(x,y,z,t) = \dfrac{\rho_v(x,y,z,t)}{\varepsilon}$	$\nabla \cdot \vec{H}(x,y,z,t) = 0$
Constitutive equations	$\vec{D} = \varepsilon \vec{E}$	$\vec{B} = \mu \vec{H}$

flux density is $\vec{B} = \mu\vec{H}$; $\vec{J}$ is the current density, and using the conductivity σ, we have $\vec{J} = \sigma\vec{E}$; ρ_v is the volume charge density; the total electric flux through a closed surface is $\Phi = \oint_s \vec{D} \cdot d\vec{s} = \oint_v \rho_v dv = Q$ (Gauss's law); and the magnetic flux crossing surface is $\Phi = \oint_s \vec{B} \cdot d\vec{s}$.
The second equation

$$\nabla \times \vec{H}(x,y,z,t) = \sigma\vec{E}(x,y,z,t) + \vec{J}(x,y,z,t) = \sigma\vec{E}(x,y,z,t) + \varepsilon\frac{\partial\vec{E}(x,y,z,t)}{\partial t}$$

was derived by Maxwell adding the term

$$\vec{J}_d(x,y,z,t) = \varepsilon\frac{\partial\vec{E}(x,y,z,t)}{\partial t}$$

to the Ampere law.

The constitutive (auxiliary) equations are given using the permittivity ε, permeability tensor μ, and conductivity σ. In particular, one has

$$\vec{D} = \varepsilon\vec{E} \quad \text{or} \quad \vec{D} = \varepsilon\vec{E} + \vec{P}$$

$$\vec{B} = \mu\vec{H} \quad \text{or} \quad \vec{B} = \mu(\vec{H} + \vec{M})$$

$$\vec{J} = \sigma\vec{E} \quad \text{or} \quad \vec{J} = \rho_v\vec{v}$$

Maxwell's equations can be solved using the boundary conditions on the field vectors. In two-region media, we have

$$\vec{a}_N \times (\vec{E}_2 - \vec{E}_1) = 0$$

$$\vec{a}_N \times (\vec{H}_2 - \vec{H}_1) = \vec{J}_s$$

$$\vec{a}_N \cdot (\vec{D}_2 - \vec{D}_1) = \rho_s$$

$$\vec{a}_N \cdot (\vec{B}_2 - \vec{B}_1) = 0$$

where $\vec{J}_s$ is the surface current density vector; $\vec{a}_N$ is the surface normal unit vector at the boundary from region 2 into region 1; and ρ_s is the surface charge density.

The constitutive relations that describe media can be integrated with Maxwell's equations, which relate the fields in order to find two partial differential equations. Using the electric and magnetic field intensities $\vec{E}$ and $\vec{H}$ to model electromagnetic fields in MEMS, one has

$$\nabla \times (\nabla \times \vec{E}) = \nabla(\nabla \cdot \vec{E}) - \nabla^2\vec{E} = -\mu\frac{\partial\vec{J}}{\partial t} - \mu\frac{\partial^2\vec{D}}{\partial t^2} = -\mu\sigma\frac{\partial\vec{E}}{\partial t} - \mu\varepsilon\frac{\partial^2\vec{E}}{\partial t^2}$$

$$\nabla \times (\nabla \times \vec{H}) = \nabla(\nabla \cdot \vec{H}) - \nabla^2\vec{H} = -\mu\sigma\frac{\partial\vec{H}}{\partial t} - \mu\varepsilon\frac{\partial^2\vec{H}}{\partial t^2}$$

The following pair of homogeneous and inhomogeneous wave equations is equivalent to four Maxwell's equations and constitutive relations:

$$\nabla^2\vec{E} - \mu\sigma\frac{\partial\vec{E}}{\partial t} - \mu\varepsilon\frac{\partial^2\vec{E}}{\partial t^2} = \nabla\left(\frac{\rho_v}{\varepsilon}\right)$$

$$\nabla^2\vec{H} - \mu\sigma\frac{\partial\vec{H}}{\partial t} - \mu\varepsilon\frac{\partial^2\vec{H}}{\partial t^2} = 0$$

For some cases, these two equations can be solved independently. It must be emphasized that it is not always possible to use the boundary conditions using only $\vec{E}$ and $\vec{H}$. Thus, the problem cannot always be simplified to two electromagnetic field vectors. Therefore, the electric scalar and magnetic vector potentials are used. Denoting the magnetic vector potential as $\vec{A}$ and the electric scalar potential as V, we have

$$\nabla \times \vec{A} = \vec{B} = \mu \vec{H} \quad \text{and} \quad \vec{E} = -\frac{\partial \vec{A}}{\partial t} - \nabla V$$

The electromagnetic field is derived from the potentials. Using the Lorentz equation

$$\nabla \cdot \vec{A} = -\frac{\partial V}{\partial t}$$

the inhomogeneous vector potential wave equation to be solved is

$$-\nabla^2 \vec{A} + \mu\sigma \frac{\partial \vec{A}}{\partial t} + \mu\varepsilon \frac{\partial^2 \vec{A}}{\partial t^2} = -\mu\sigma \nabla V$$

To model motion microdevices, the mechanical equations must be used, and Newton's second law is usually applied to derive the equations of motion.

Using the volume charge density ρ_v, the Lorenz force, which relates the electromagnetic and mechanical phenomena, is found as

$$\vec{F} = \rho_v(\vec{E} + \vec{v} \times \vec{B}) = \rho_v \vec{E} + \vec{J} \times \vec{B}$$

The Lorenz force law is

$$\vec{F} = \frac{d\vec{p}}{dt} = q(\vec{E} + \vec{v} \times \vec{B}) = q\left[-\nabla V - \frac{\partial \vec{A}}{\partial t} + \vec{v} \times (\nabla \times \vec{A})\right]$$

where for the canonical momentum $\vec{p}$ we have $d\vec{p}/dt = -\nabla\Pi$.

That is, the Maxwell equations illustrate how charges produce the electromagnetic fields, and the Lorenz force law demonstrates how the electromagnetic fields affect charges.

The energy per unit time per unit area transported by the electromagnetic fields is called the Poynting vector:

$$\vec{S} = \frac{1}{\mu}(\vec{E} \times \vec{B})$$

The electromagnetic force can be found by applying the Maxwell stress tensor method. This concept employs a volume integral to obtain the stored energy, and stress at all points of a bounding surface can be determined. The sum of local stresses gives the net force. In particular, the electromagnetic stress is

$$\vec{F} = \int_v \rho_v(\vec{E} + \vec{v} \times \vec{B})dv = \int_v (\rho_v \vec{E} + \vec{J} \times \vec{B})dv = \frac{1}{\mu}\oint_s \vec{\vec{T}} \cdot d\vec{s} - \varepsilon\mu \frac{d}{dt}\int_v \vec{S}dv$$

The force per unit volume is

$$\vec{F}_u = \rho_v \vec{E} + \vec{J} \times \vec{B} = \nabla \cdot \vec{\vec{T}} - \varepsilon\mu \frac{\partial \vec{S}}{\partial t}$$

The electromagnetic stress energy tensor $\vec{\vec{T}}$ (the second Maxwell stress tensor) is

$$(\vec{a} \cdot \vec{\vec{T}})_j = \sum_{i=x,y,z} a_i T_{ij}$$

We have

$$(\nabla \cdot \vec{\vec{T}})_j = \varepsilon\left[(\nabla \cdot \vec{E})E_j + (\vec{E} \cdot \nabla)E_j - \tfrac{1}{2}\nabla_j E^2\right] + \frac{1}{\mu}\left[(\nabla \cdot \vec{B})B_j + (\vec{B} \cdot \nabla)B_j - \tfrac{1}{2}\nabla_j B^2\right]$$

or

$$T_{ij} = \varepsilon\left(E_i E_j - \tfrac{1}{2}\delta_{ij}E^2\right) + \frac{1}{\mu}\left(B_i B_j - \tfrac{1}{2}\delta_{ij}B^2\right)$$

where i and j are the indexes that refer to the coordinates x, y, and z (the stress tensor $\vec{\vec{T}}$ has nine components: T_{xx}, T_{xy}, T_{xz}, ..., T_{zy} and T_{zz}); δ_{ij} is the Kronecker delta-function, which is defined to be 1 if the indexes are the same and 0 otherwise, $\delta_{xx} = \delta_{yy} = \delta_{zz} = 1$ and $\delta_{xy} = \delta_{xz} = \delta_{yz} = 0$.

The electromagnetic torque developed by motion microstructures is found using the electromagnetic field. The electromagnetic stress tensor is given as

$$T_s = T_s^E + T_s^M = \begin{bmatrix} E_1 D_1 - \tfrac{1}{2}E_j D_j & E_1 D_2 & E_1 D_3 \\ E_2 D_1 & E_2 D_2 - \tfrac{1}{2}E_j D_j & E_2 D_3 \\ E_3 D_1 & E_3 D_2 & E_3 D_3 - \tfrac{1}{2}E_j D_j \end{bmatrix}$$

$$+ \begin{bmatrix} B_1 H_1 - \tfrac{1}{2}B_j H_j & B_1 H_2 & B_1 H_3 \\ B_2 H_1 & B_2 H_2 - \tfrac{1}{2}B_j H_j & B_2 H_3 \\ B_3 H_1 & B_3 H_2 & B_3 H_3 - \tfrac{1}{2}B_j H_j \end{bmatrix}$$

For the Cartesian, cylindrical, and spherical coordinate systems, which can be used to develop the mathematical model, we have

$$E_x = E_1, E_y = E_2, E_z = E_3, D_x = D_1, D_y = D_2, D_z = D_3$$

$$H_x = H_1, H_y = H_2, H_z = H_3, B_x = B_1, B_y = B_2, B_z = B_3$$

$$E_r = E_1, E_\theta = E_2, E_z = E_3, D_r = D_1, D_\theta = D_2, D_z = D_3$$

$$H_r = H_1, H_\theta = H_2, H_z = H_3, B_r = B_1, B_\theta = B_2, B_z = B_3$$

$$E_\rho = E_1, E_\theta = E_2, E_\phi = E_3, D_\rho = D_1, D_\theta = D_2, D_\phi = D_3$$

$$H_\rho = H_1, H_\theta = H_2, H_\phi = H_3, B_\rho = B_1, B_\theta = B_2, B_\phi = B_3$$

Maxwell's equations can be solved using the MATLAB environment (for example, using the Partial Differential Equations Toolbox).

The electromotive and magnetomotive forces are found as

$$emf = \oint_l \vec{E} \cdot d\vec{l} = \underbrace{\oint_l (\vec{v} \times \vec{B}) \cdot d\vec{l}}_{\text{motional induction (generation)}} - \underbrace{\oint_s \frac{\partial \vec{B}}{\partial t} d\vec{s}}_{\text{transformer induction}}$$

and

$$mmf = \oint_l \vec{H} \cdot d\vec{l} = \oint_s \vec{J} \cdot d\vec{s} + \oint_s \frac{\partial \vec{D}}{\partial t} d\vec{s}$$

The motional emf is a function of the velocity and the magnetic flux density, while the electromotive force induced in a stationary closed circuit is equal to the negative rate of increase of the magnetic flux (transformer induction).

We introduced the vector magnetic potential $\vec{A}$. Using the equation

$$\vec{B} = \nabla \times \vec{A}$$

one has the following nonhomogeneous vector wave equation

$$\nabla^2 \times \vec{A} - \mu\varepsilon \frac{\partial^2 \vec{A}}{\partial t^2} = -\mu\vec{J}$$

and the solution gives the waves traveling with the velocity $1/\sqrt{\mu\varepsilon}$.

6.3.1 Lumped-Parameter Mathematical Models of MEMS

The problems of modeling, analysis, and control of MEMS are very important in many applications. A mathematical model is a mathematical description (in the form of equations) of MEMS, which integrate motion microdevices (microscale actuators and sensors), radiating energy microdevices, microscale driving/sensing circuitry, and controlling/signal processing ICs. The purpose of the model development is to understand and comprehend the phenomena, as well as to analyze the end-to-end behavior and study the system performance.

To model MEMS, advanced analysis methods are required to accurately cope with the involved highly complex physical phenomena, effects, and processes. The need for high-fidelity analysis, computationally efficient algorithms, and simulation-time reduction increases significantly for complex microdevices, restricting the application of Maxwell's equations to a small number of solvable problems. As was illustrated in the previous section, nonlinear electromagnetic and energy conversion phenomena are described by partial differential equations. The application of Maxwell's equations contradicts the need for computationaly tractable analysis and performance evaluation capabilities with outcome prediction within overall modeling domains for simulation and analysis of MEMS. Hence, other modeling and analysis methods must be applied. The lumped mathematical models, described by ordinary differential equations, can be used. The process of mathematical modeling and model development is given below.

The first step is to formulate the modeling problem:

* Examine and analyze MEMS using a multilevel hierarchy concept: Develop multivariable input-output subsystem (device) pairs, e.g., motion microstructures or microdevices (microscale actuators and sensors)-radiating energy microdevices-microscale circuitry- ICs-controller-input-output devices.
* Understand and comprehend the MEMS structure and system configuration.
* Gather (collect) the data and information.
* Develop input-output variable pairs, identify the independent and dependent control, disturbance, output, reference (command), state and performance variables, and events.

- Make accurate assumptions, and simplify the problem to make the studied MEMS mathematically tractable. Mathematical models are the idealization of physical phenomena. These mathematical models are never absolutely accurate. Even comprehensive models simplify the reality to allow the designer to perform the thorough analysis and accurate predictions of the system performance. Even using Maxwell's equations, one makes assumptions in order to solve these partial differential equations.

The second step is to derive equations that relate the variables and events:

- Define and specify the basic laws (Kirchhoff's, Lagrange's, Maxwell's, Newton's, and others) to be used to obtain the equations of motion. Mathematical models of electromagnetic, electronic, and mechanical microscale subsystems can be found and augmented to derive mathematical models of MEMS using defined variables and events.
- Derive mathematical models.

The third step is simulation, analysis, and validation:

- Identify the numerical and analytic methods to be used in analysis and simulations.
- Analytically or numerically solve the mathematical equations (e.g., differential or difference equations, nonlinear equations, etc.).
- Using information variables (measured or observed) and events, synthesize the fitting and mismatch functionals.
- Verify the results through the comprehensive comparison of the solution (model input-state-output-event mapping sets) with the experimental data (experimental input-state-output-event mapping sets).
- Calculate the fitting and mismatch functionals.
- Examine the analytical and numerical data against new experimental data and evidence.

If matching with the desired accuracy is not guaranteed, the mathematical model of MEMS must be refined, and the designer must start the cycle again.

Electromagnetic theory and classical mechanics form the basis for the development of mathematical models of MEMS. It was illustrated that MEMS can be modeled using Maxwell's equations and torsional-mechanical equations of motion. Forces and torques are found using the Maxwell stress tensor. Nonlinear partial differential equations result. However, for modeling, simulation, analysis, design, control, and optimization, the lumped-parameter mathematical models as given by ordinary differential equations can be derived and used.

Consider the rotational microstructure (bar magnet, current loop, and microsolenoid) in a uniform magnetic field. (See Figure 6.8.) The microstructure rotates if the electromagnetic torque is developed. Therefore, the electromagnetic field must be studied to find the electromagnetic torque developed.

The torque tends to align the magnetic moment $\vec{m}$ with $\vec{B}$, and it is well known that

$$\vec{T} = \vec{m} \times \vec{B}$$

For a magnetic bar with the length l, the pole strength is Q.
The magnetic moment is given as $m = Ql$, and the force is $F = QB$.

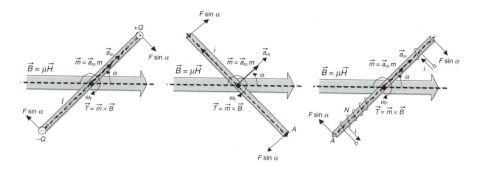

FIGURE 6.8
Clockwise rotation of a microstructure (magnetic bar, current loop, and solenoid).

The electromagnetic torque is found to be $T = 2F\frac{1}{2}l\sin\alpha = QlB\sin\alpha = mB\sin\alpha$.

Using the vector notations, one obtains $\vec{T} = \vec{m} \times \vec{B} = \vec{a}_m m \times \vec{B} = Ql\vec{a}_m \times \vec{B}$, where $\vec{a}_m$ is the unit vector in the magnetic moment direction.

For a current loop with the cross-sectional area A, the torque is found as $\vec{T} = \vec{m} \times \vec{B} = \vec{a}_m m \times \vec{B} = iA\vec{a}_m \times \vec{B}$.

For a solenoid with N turns, one obtains $\vec{T} = \vec{m} \times \vec{B} = \vec{a}_m m \times \vec{B} = iAN\vec{a}_m \times \vec{B}$.

The straightforward application of Newton's second law for the rotational motion gives

$$\sum \vec{T}_\Sigma = J\frac{d\omega_r}{dt}$$

where $\Sigma\vec{T}_\Sigma$ is the net torque; ω_r is the angular velocity; and J is the equivalent moment of inertia.

The transient evolution of the angular displacement θ_r is modeled as

$$\frac{d\theta_r}{dt} = \omega_r$$

Augmenting the equations for the electromagnetic torque (found in terms of the magnetic field variables $\vec{m}$ and $\vec{B}$) and the torsional-mechanical dynamics (the state variables are the angular velocity and displacement), the mathematical model of micro- and nanoscale rotational actuators results.

The energy is stored in the magnetic field, and media are classified as diamagnetic, paramagnetic, ferromagnetic, antiferromagnetic, and superparamagnetic. Using the magnetic susceptibility χ_m, the magnetization is expressed as $\vec{M} = \chi_m \vec{H}$.

Magnetization curves should be studied, and the permeability is used. The magnetic field density B lags behind the magnetic flux intensity H, and this phenomenon is called hysteresis. Thus, the *B-H* magnetization curves must be studied.

The per-unit volume magnetic energy stored is $\oint_B H dB$. The density of the energy stored in the magnetic field is $\frac{1}{2}\vec{B}\cdot\vec{H}$. We have the expression for the total energy stored in the magnetic field as $\frac{1}{2}\int_v \vec{B}\cdot\vec{H}dv$.

For translational motion, Newton's second law states that the net force acting on the object is related to its acceleration as $\Sigma\vec{F} = m\vec{a}$.

In the *x-y-z* coordinate system, one obtains

$$\sum F_x = ma_x, \quad \sum F_y = ma_y, \quad \text{and} \quad \sum F_z = ma_z$$

The force is the gradient of the stored magnetic energy. That is, $\vec{F}_m = \vec{\nabla} W_m$. Hence, in the x-y-z directions, we have

$$F_{mx} = \frac{\partial W_m}{\partial x}, \quad F_{my} = \frac{\partial W_m}{\partial y}, \quad \text{and} \quad F_{mz} = \frac{\partial W_m}{\partial z}$$

where the stored magnetic energy is found using the volume current density $\vec{J}$. In particular,

$$W_m = \frac{\mu}{8\pi} \int_v \int_{v_A} \frac{\vec{J}(\vec{r}_A) \cdot \vec{J}(\vec{r})}{R} dv_A dv$$

Applying the field quantities, we have

$$W_m = \frac{1}{2} \int_v \vec{A} \cdot \vec{J} dv = \frac{1}{2} \int_v \vec{B} \cdot \vec{H} dv$$

The magnetic energy density is

$$w_m = \frac{1}{2} \vec{A} \cdot \vec{J} = \frac{1}{2} \vec{B} \cdot \vec{H}$$

Using Newton's second law and the derived expression for stored magnetic energy, we have nine highly coupled nonlinear differential equations for the x-y-z translational motion of microactuator. In particular,

$$\frac{dF_{xyz}}{dt} = f_F(F_{xyz}, v_{xyz}, x_{xyz}, H)$$

$$\frac{dv_{xyz}}{dt} = f_v(F_{xyz}, v_{xyz}, x_{xyz}, F_{Lxyz})$$

$$\frac{dx_{xyz}}{dt} = f_x(v_{xyz}, x_{xyz})$$

where F_{xyz} is the forces developed; v_{xyz} and x_{xyz} are the linear velocities and positions; and F_{Lxyz} is the load forces.

The expressions for energies stored in electrostatic and magnetic fields in terms of field quantities should be derived. The total potential energy stored in the electrostatic field is obtained using the potential difference V as

$$W_e = \frac{1}{2} \int_v \rho_v V dv$$

where the volume charge density is found as $\rho_v = \vec{\nabla} \cdot \vec{D}$, where $\vec{\nabla}$ is the curl operator.

In the Gauss form, using $\rho_v = \vec{\nabla} \cdot \vec{D}$ and making use of $\vec{E} = -\vec{\nabla} V$, one obtains the following expression for the energy stored in the electrostatic field: $W_e = \frac{1}{2} \int_v \vec{D} \cdot \vec{E} dv$. Thus, the electrostatic volume energy density is $\frac{1}{2} \vec{D} \cdot \vec{E}$. For a linear isotropic medium, one finds

$$W_e = \frac{1}{2} \int_v \varepsilon |\vec{E}|^2 dv = \frac{1}{2} \int_v \frac{1}{\varepsilon} |\vec{D}|^2 dv$$

The electric field $\vec{E}(x,y,z)$ is found using the scalar electrostatic potential function $V(x,y,z)$ as

$$\vec{E}(x,y,z) = -\vec{\nabla} V(x,y,z)$$

In the cylindrical and spherical coordinate systems, we have

$$\vec{E}\,(r,\phi,z) = -\vec{\nabla}V(r,\phi,z) \quad \text{and} \quad \vec{E}(r,\theta,\phi) = -\vec{\nabla}V(r,\theta,\phi).$$

Using the principle of virtual work, for the lossless conservative micro- and nanoelectromechanical systems, the differential change of the electrostatic energy dW_e is equal to the differential change of mechanical energy dW_{mec}, $dW_e = dW_{mec}$. For translational motion, $dW_{mec} = \vec{F}_e \cdot d\vec{l}$, where $d\vec{l}$ is the differential displacement.

One obtains

$$dW_e = \vec{\nabla}W_e \cdot d\vec{l}$$

Hence the force is the gradient of the stored electrostatic energy, $\vec{F}_e = \vec{\nabla}W_e$. In the Cartesian coordinates, we have

$$F_{ex} = \frac{\partial W_e}{\partial x}, \quad F_{ey} = \frac{\partial W_e}{\partial y}, \quad \text{and} \quad F_{ez} = \frac{\partial W_e}{\partial z}$$

6.3.2 Energy Conversion in NEMS and MEMS

Energy conversion takes place in nano- and microscale electromechanical motion devices (actuators and sensors, smart structures), antennae, and ICs. We study electromechanical motion devices that convert electrical energy (more precisely electromagnetic energy) to mechanical energy and vice versa (conversion of mechanical energy to electromagnetic energy). Fundamental principles of energy conversion, applicable to micro- and nanoelectromechanical motion devices, were studied to provide basic foundations. Using the principle of conservation of energy we can formulate the following: For lossless micro- and nanoelectromechanical motion devices (in the conservative system no energy is lost through friction, heat, or other irreversible energy conversion), the sum of the instantaneous kinetic and potential energies of the system remains constant.

The energy conversion is represented in Figure 6.9.

The general equation that describes the energy conversion is given as

$$\underset{\text{Electrical Energy Input}}{\mathbf{E}_E} - \underset{\text{Ohmic Losses}}{\mathbf{L}_E} - \underset{\text{Magnetic Losses}}{\mathbf{L}_M} = \underset{\text{Mechanical Energy}}{\mathbf{E}_M} + \underset{\text{Friction Losses}}{\mathbf{L}_E} + \underset{\text{Stored Energy}}{\mathbf{L}_S}$$

For conservative (lossless) energy conversion, one can write

$$\underset{\text{Change in Electrical Energy Input}}{\Delta\mathbf{W}_E} = \underset{\text{Change in Mechanical Energy}}{\Delta\mathbf{W}_M} + \underset{\text{Change in Electromagnetic Energy}}{\Delta\mathbf{W}_m}$$

The total energy stored in the magnetic field is

$$W_m = \tfrac{1}{2}\int_v \vec{B} \cdot \vec{H}dv$$

where $\vec{B}$ and $\vec{H}$ are related using the permeability μ as $\vec{B} = \mu\vec{H}$.

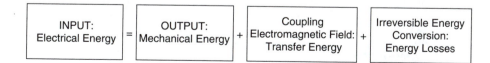

FIGURE 6.9
Energy transfer in micro- and nanoelectromechanical systems.

The material becomes magnetized in response to the external field $\vec{H}$, and the dimensionless magnetic susceptibility χ_m or relative permeability μ_r are used. We have

$$\vec{B} = \mu\vec{H} = \mu_0(1+\chi_m)\vec{H} = \mu_0\mu_r\vec{H} = \mu\vec{H}$$

Based upon the value of the magnetic susceptibility χ_m, the materials are classified as

- Diamagnetic, $\chi_m \approx -1 \times 10^{-5}$ ($\chi_m = -9.5 \times 10^{-6}$ for copper, $\chi_m = -3.2 \times 10^{-5}$ for gold, and $\chi_m = -2.6 \times 10^{-5}$ for silver)
- Paramagnetic, $\chi_m \approx 1 \times 10^{-4}$ ($\chi_m = 1.4 \times 10^{-3}$ for Fe_2O_3, and $\chi_m = 1.7 \times 10^{-3}$ for Cr_2O_3)
- Ferromagnetic, $|\chi_m| \gg 1$ (iron, nickel, cobalt, neodymium-iron-boron, and samarium-cobalt permanent magnets, etc.)

Table 6.2 provides additional data for some basic materials utilized in MEMS.

The magnetization behavior of the ferromagnetic materials is mapped by the magnetization curve, where H is the externally applied magnetic field, and B is total magnetic flux density in the medium. Typical B-H curves for hard and soft ferromagnetic materials are given in Figure 6.10.

The B versus H curve allows one to establish the energy analysis. Assume that initially $B_0 = 0$ and $H_0 = 0$. Let H increase from $H_0 = 0$ to H_{max}. Then B increases from $B_0 = 0$ until the maximum value of B (denoted as B_{max}) is reached. If H then decreases to H_{min}, B decreases to B_{min} through the remnant value B_r (the so-called residual magnetic flux density) along the different curve (See Figure 6.10.) For variations of H, $H \in [H_{min} \; H_{max}]$, B changes within the hysteresis loop, and $B \in [B_{min} \; B_{max}]$.

In the per-unit volume, the applied field energy is

$$W_F = \oint_B H dB$$

TABLE 6.2

Relative Permeability of Some Diamagnetic, Paramagnetic, Ferromagnetic, and Ferrimagnetic Materials

Material	Maximum Achievable Relative Permeability, μ_r
Diamagnetic	
Silver	0.9999736
Copper	0.9999906
Paramagnetic	
Aluminum	1.000021
Tungsten	1.00008
Platinum	1.0003
Manganese	1.001
Ferromagnetic	
Purified iron (99.96% Fe)	280,000
Electric steel (99.6% Fe)	5000
Permalloy ($Ni_{78.5\%}Fe_{21.5\%}$)	70,000
Superpermalloy ($Ni_{79\%}Fe_{15\%}Mo_{5\%}Mn_{0.5\%}$)	1,000,000
Ferrimagnetic	
Nickel-zinc ferrite	600–1000
Manganese-zinc ferrite	700–1500

This data for the bulk materials is with maximum achievable μ_r. The permeability strongly depends on the materials, processes, and fabrication technologies, and degrades with the size due to bulk and surface micromachining processes.

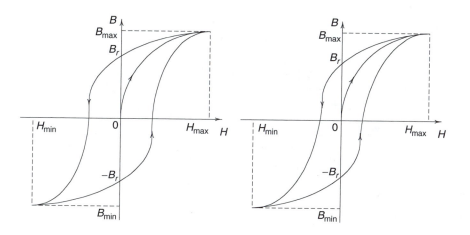

FIGURE 6.10
B-H curves for hard and soft ferromagnetic materials.

while the stored energy is expressed as

$$W_c = \oint_H BdH$$

In the volume v, we have the following expressions for the field and stored energy:

$$W_F = v\oint_B HdB$$

and

$$W_c = v\oint_H BdH$$

A complete B versus H loop should be considered, and the equations for field and stored energy represent the areas enclosed by the corresponding curve.

In ferromagnetic materials, time-varying magnetic flux produces core losses that consist of hysteresis losses (due to the hysteresis loop of the *B-H* curve) and the eddy-current losses (proportional to the current frequency and lamination thickness). The area of the hysteresis loop is related to the hysteresis losses. Soft ferromagnetic materials have a narrow hysteresis loop and they are easily magnetized and demagnetized. Therefore, the lower hysteresis losses, compared with hard ferromagnetic materials, result.

Usually the flux linkages are plotted versus the current, because the current and flux linkages are used rather than the flux intensity and flux density. In micro- and nanoelectromechanical motion devices almost all energy is stored in the air gap. Using the fact that the air is a conservative medium, one concludes that the coupling field is lossless.

Figure 6.11 illustrates the nonlinear magnetizing characteristic (normal magnetization curve). The energy stored in the magnetic field is

$$W_F = \oint_\psi id\psi$$

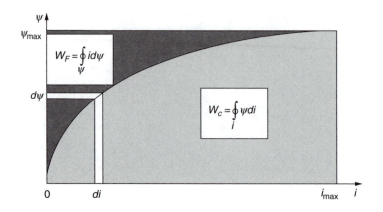

FIGURE 6.11
Magnetization curve and energies.

while the coenergy is found as

$$W_c = \oint_i \psi di$$

The total energy is

$$W_F + W_c = \oint_\psi id\psi + \oint_i \psi di = \psi i$$

The flux linkage is the function of the current i and position x (for translational motion) or angular displacement θ (for rotational motion). That is, $\psi = f(i,x)$ or $\psi = f(i,\theta)$. The current can be found as the nonlinear function of the flux linkages and position or angular displacement. Hence

$$d\psi = \frac{\partial \psi(i,x)}{\partial i}di + \frac{\partial \psi(i,x)}{\partial x}dx, \quad d\psi = \frac{\partial \psi(i,\theta)}{\partial i}di + \frac{\partial \psi(i,\theta)}{\partial \theta}d\theta$$

and

$$di = \frac{\partial i(\psi,x)}{\partial \psi}d\psi + \frac{\partial i(\psi,x)}{\partial x}dx, \quad di = \frac{\partial i(\psi,\theta)}{\partial \psi}d\psi + \frac{\partial i(\psi,\theta)}{\partial \theta}d\theta$$

Therefore, we have

$$W_F = \oint_\psi id\psi = \oint_i i\frac{\partial \psi(i,x)}{\partial i}di + \oint_x i\frac{\partial \psi(i,x)}{\partial x}dx$$

$$W_F = \oint_\psi id\psi = \oint_i i\frac{\partial \psi(i,\theta)}{\partial i}di + \oint_\theta i\frac{\partial \psi(i,\theta)}{\partial \theta}d\theta$$

and

$$W_c = \oint_i \psi di = \oint_\psi \psi \frac{\partial i(\psi,x)}{\partial \psi} d\psi + \oint_x \psi \frac{\partial i(\psi,x)}{\partial x} dx$$

$$W_c = \oint_i \psi di = \oint_\psi \psi \frac{\partial i(\psi,\theta)}{\partial \psi} d\psi + \oint_\theta \psi \frac{\partial i(\psi,\theta)}{\partial \theta} d\theta$$

Assuming that the coupling field is lossless, the differential change in the mechanical energy (which is found using the differential displacement $d\vec{l}$ as $dW_{mec} = \vec{F}_m \cdot d\vec{l}$) is related to the differential change of the coenergy. For displacement dx at constant current, one obtains $dW_{mec} = dW_c$. Hence the electromagnetic force is

$$F_e(i,x) = \frac{\partial W_c(i,x)}{\partial x}$$

For rotational motion, the electromagnetic torque is

$$T_e(i,\theta) = \frac{\partial W_c(i,\theta)}{\partial \theta}$$

Nano- and microscale structures, as well as thin magnetic films, exhibit anisotropy. Consider the anisotropic ferromagnetic element in the Cartesian (rectangular) coordinate systems as shown in Figure 6.12.

The permeability is

$$\mu(x,y,z) = \begin{bmatrix} \mu_{xx} & \mu_{xy} & \mu_{xz} \\ \mu_{yx} & \mu_{yy} & \mu_{yz} \\ \mu_{zx} & \mu_{zy} & \mu_{zz} \end{bmatrix}$$

and therefore

$$\vec{B} = \mu(x,y,z)\vec{H}, \quad \begin{bmatrix} B_x \\ B_y \\ B_z \end{bmatrix} = \begin{bmatrix} \mu_{xx} & \mu_{xy} & \mu_{xz} \\ \mu_{yx} & \mu_{yy} & \mu_{yz} \\ \mu_{zx} & \mu_{zy} & \mu_{zz} \end{bmatrix} \begin{bmatrix} H_x \\ H_y \\ H_z \end{bmatrix}$$

The analysis of anisotropic micro- and nanoscale actuators and sensors can be performed. Some actuators and sensors can be studied assuming that the medium is linear, homogeneous, and isotropic. Unfortunately this assumption is not valid in general.

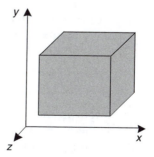

FIGURE 6.12
Material in the *x-y-z* coordinate system.

Control of the microactuators, position and linear velocity, angular displacement and angular velocity, is established by changing electromagnetic field (e.g., E, D, H, or B). In fact, in this chapter it was already shown that the electromagnetic torque and force are derived in the terms of the electromagnetic field quantities. For example, the magnetic field intensity can be considered as a control. Electromagnetic fields are developed by ICs or antennae. Hence the ICs or microantenna dynamics have to be integrated in the MEMS equations of motion. Thus microscale antennae and ICs must be thoroughly considered.

Consider the microactuator controlled by the microantenna. Assume that the linear isotropic medium has permittivity $\varepsilon_0 \varepsilon_m$ and permeability $\mu_0 \mu_m$.

The force is calculated using the stress energy tensor $\bar{\bar{T}}$, which is given in terms of the electromagnetic field as

$$T_{ij} = \varepsilon_0 \varepsilon_m E_i E_j + \mu_0 \mu_m H_i H_j - \tfrac{1}{2}\delta_{ij}(\varepsilon_0 \varepsilon_m E^2 + \mu_0 \mu_m H^2)$$

where δ_{ij} is the Kronecker delta-function, defined as

$$\delta_{ij} = \begin{cases} 1 \text{ if } & i = j \\ 0 \text{ if } & i \neq j \end{cases}$$

The Maxwell's equation for electromagnetic fields can be expressed in the tensor form using the electromagnetic field stress tensor. In particular, the electromagnetic force is found as $\vec{F} = \int_s \bar{\bar{T}} d\vec{s}$.

The results derived can be viewed using the energy analysis, and one has

$$\sum \vec{F}(\vec{r}) = -\nabla\Pi(\vec{r})$$

$$\Pi(\vec{r}) = \frac{\varepsilon_0 \varepsilon_m}{2}\int_s \vec{E}\cdot\vec{E}dv + \frac{1}{2\mu_0\mu_m}\int_s \vec{H}\cdot\vec{H}dv$$

Let us demonstrate how to apply the reported concept in the design of motion microtransducers (micromachines). For preliminary design, it is sufficiently accurate to apply Faraday's or Lenz's laws, which give the electromotive force in terms of the time-varying magnetic field changes. In particular,

$$emf = -\frac{d\psi}{dt} = -\frac{\partial\psi}{\partial t} - \frac{\partial\psi}{\partial\theta_r}\frac{d\theta_r}{dt} = -\frac{\partial\psi}{\partial t} - \frac{\partial\psi}{\partial\theta_r}\omega_r$$

where $\partial\psi/\partial t$ is the transformer term.

The total flux linkage is

$$\psi = \tfrac{1}{4}\pi N_s \Phi_p$$

where N_s is the number of turns and Φ_p is the flux per pole.

For radial topology microtransducers, we have

$$\Phi_p = \frac{\mu i N_s}{P^2 g_e} R_{in\,st}L$$

where i is the current in the phase microwinding (supplied by the IC); $R_{in\,st}$ is the inner stator radius; L is the inductance; P is the number of poles; and g_e is the equivalent gap, which includes the airgap and radial thickness of the permanent magnet.

Denoting the number of turns per phase as N_s, the magnetomotive force is

$$mmf = \frac{iN_s}{P} \cos P\theta_r$$

The simplified expression for the electromagnetic torque for radial topology brushless microtransducers is

$$T = \tfrac{1}{2} P B_{ag} i_s N_s L_r D_r$$

where B_{ag} is the air gap flux density, $B_{ag} = (\mu i N_s / 2 P g_e) \cos P\theta_r$; i_s is the total current; L_r is the active length (rotor axial length); and D_r is the outside rotor diameter.

As was illustrated, the axial topology brushless microtransducers can be designed and fabricated, and the electromagnetic torque is given as $T = k_{ax} B_{ag} i_s N_s D_a^2$, where k_{ax} is the nonlinear coefficient, which is found in terms of active conductors and thin-film permanent magnet length; and D_a is the equivalent diameter, which is a function of windings and permanent-magnet topography.

As the expression for the electromagnetic torque is found, the torsional-mechanical dynamics are used to integrate electromagnetic- and mechanical-based phenomena and derive the complete lumped-parameter mathematical model.

In particular, Newton's second law, as given by

$$\frac{d\omega_r}{dt} = \frac{1}{J} \sum \vec{T}_\Sigma$$

$$\frac{d\theta_r}{dt} = \omega_r$$

is used. Here ω_r and θ_r are the angular velocity and displacement, and $\sum \vec{T}_\Sigma$ is the net torque.

Electromagnetics of NEMS and MEMS must be integrated with torsional-mechanical dynamics. This can be accomplished using distinct methods as will be explicitly demonstrated in the following section.

6.4 Classical Mechanics and Its Application to MEMS

Having demonstrated the application of various modeling concepts, we also present the use of classical mechanics to model distinct electromechanical systems including MEMS. It will be illustrated that the augmented lumped-parameter models can be straightforwardly derived. In many cases, the desired level of accuracy and confidence can be achieved using the ordinary differential equations that can be time-invariant (coefficients of differential equations are constant) or time-varying (coefficients vary). It should be also emphasized that before the attempts to develop high-fidelity mathematical models in order to accomplish data-intensive analysis through heterogeneous simulations, preliminary analysis may be carried out using the basic physics. Classical mechanics provides the basic fundamentals. Our goal is to develop dynamic mathematical models in the form of differential equations that also result in the steady-state solution. But steady-state models do not result in dynamic (behavioral) analysis, and, correspondingly, steady-state analysis has limited capabilities. In fact utilizing computer-aided design tools, one can design, analyze, and evaluate three-dimensional (3D) nanostructures in the steady-state. However, the comprehensive analysis in the time domain needs to be performed, particularly examining devices and systems as well as carrying the system-level approach in their assessment. That is, the designer must study the dynamic evolution of MEMS and NEMS. Quantum mechanics does not allow one to perform numerical analysis for even the simplest MEMS and NEMS in time domain, and even 3D steady-state modeling is restricted to very simple structures under many assumptions. Our goal is to develop a fundamental understanding of basic electromechanical and electromagnetic phenomena in nano- and microscale devices and systems. An addition, the basic theoretical foundations should be developed and used in analysis of MEMS and NEMS from systems standpoints. That is, targeting coherent analysis and searching for generic results, we have departed from steady-state analysis by studying MEMS and NEMS as dynamics systems. This is accomplished using different modeling paradigms.

From modeling, simulation, analysis, and visualization standpoints, MEMS and NEMS are very complex. In fact MEMS and NEMS are modeled using advanced concepts of quantum mechanics, electromagnetic theory, structural dynamics, thermodynamics, thermochemistry, vibroacoustics, etc. It was illustrated that MEMS and NEMS integrate a great number of components (subsystems, devices, and structures), and the mathematical model development is an extremely challenging problem, because the commonly used conventional methods, assumptions, and simplifications may not be applied. For example, Newtonian mechanics is not applicable to the molecular scale. Though Maxwell's equations can be contemporarily applied to study complex electromagnetic phenomena, quantum phenomena lead to distinct analysis paradigms. As a result partial differential equations describe multibody dynamics of multivariable mathematical models. The visualization issues must be addressed to study complex tensor data (tensor fields). Techniques and software for visualizing scalar and vector field data are available to visualize the data in three dimensions. In contrast, techniques to visualize tensor fields are not available due to the complex, multivariate nature of the data and the fact that no commonly used experimental analogy exists. The second-order tensor fields consist of 3×3 matrices defined at each node in a computational grid. Tensor field variables can include stress, viscous stress, rate of strain, and momentum. Tensor variables in conventional structural dynamics include stress and strain. The tensor field can be simplified and visualized as a scalar field. Alternatively the individual vectors that compose the tensor field can be analyzed. However, these simplifications result in the loss of valuable information needed to analyze complex tensor fields. Vector fields can be visualized using streamlines that depict a subset of the data. Hyperstreamlines, as an extension of the streamlines to the second-order tensor fields, provide one with a continuous representation of the tensor field along a three-dimensional path. Due to obvious limitations and scope, this book does not cover the tensor field topologies. The author emphasizes the multidisciplinary nature and complexity of the phenomena in MEMS and NEMS. Thus, the designer, in addition to developing mathematical model, examines visualization concepts to achieve data-intensive analysis.

While some results have been thoroughly studied, many important aspects have not been approached and researched, primarily due to the multidisciplinary nature and complexity of MEMS and NEMS. The major objectives of this book are to study the fundamental theoretical foundations, develop innovative concepts in design and optimization, perform modeling and simulation, as well as solve the motion control problem, and validate the results. In this section, to develop mathematical models, we illustrate that classical mechanics can be effectively utilized. In particular, augmented nano- or microactuator/sensor and circuitry dynamics results. Newtonian and quantum mechanics, and Lagrange's and Hamilton's concepts, as well as other cornerstone theories, are used to model MEMS and NEMS dynamics in the time domain. Taking note of these basic principles and laws, nonlinear mathematical models are found to perform analysis and design.

The control problem is discussed, and control algorithms must be synthesized to attain the desired specifications using a set of requirements imposed on the performance. The designer cannot achieve the optimum achievable performance due to the physical limits imposed. However, different concepts to optimize MEMS and NEMS performance are reported using the mathematical models derived. Nano- and microsystem features must be thoroughly considered when approaching modeling and simulation, analysis, and design. The ability to find mathematical models is a key problem in MEMS and NEMS analysis, optimization, synthesis, control, fabrication, and evaluation. For MEMS, using electromagnetic theory and electromechanics, we develop adequate mathematical models to attain the design objectives. The proposed approach, which augments electromagnetics and electromechanics, allows the designer to solve a broad spectrum of problems compared with finite-element analysis (steady-state modeling paradigm), because an interactive dynamic (behavioral) electromagnetic-mechanical-electronics analysis is performed. This is achieved using the Lagrange equations of motion.

We will study large-scale MEMS and NEMS (in many books actuators and sensors have been primarily studied and analyzed from the fabrication standpoints, not emphasizing analysis and design at the system level), and fundamental theory is covered thoroughly. Applying the theoretical foundations to analyze and regulate in the desired behavioral manner the energy or information flows in dynamic systems, the designer is confronted with the need to find adequate mathematical models of the phenomena. Mathematical models can be found using basic physical concepts. In particular, in electrical, mechanical, fluid, thermal, or vibroacoustic systems, the mechanisms of storing, dissipating, transforming, and transferring energies are analyzed. We use the Lagrange equations of motion, Kirchhoff's and Newton's laws, Maxwell's equations, and quantum theory to illustrate the model developments. MEMS and NEMS integrate many components, subsystems, and devices. One can reduce interconnected systems to simple, idealized subsystems (components). However, this idealization is unpractical in many cases. For example, one cannot study nano- and microscale actuators and sensors without studying subsystems (devices) to be actuated and controlled by these transducers. That is, MEMS and NEMS integrate mechanical and electromechanical motion devices (actuators and sensors), power converters and antennae, processors and IO devices, kinematics, etc. One of our primary objectives is to illustrate how one can develop comprehensive mathematical models using basic principles and laws. Through illustrative examples, differential equations will be found to model dynamic systems.

Based upon the synthesized MEMS and NEMS architectures, to analyze and regulate in the desired manner the energy or information flows, the designer needs to find adequate mathematical models and optimize the performance characteristics through design of control algorithms. Some mathematical models can be found using basic foundations, but mathematical theory to map the dynamics of some processes and system evolution is not yet developed. In this section we mainly study electrical, mechanical, and electromechanical systems. The mechanism of storing, dissipating, transforming, and transferring energies in actuators and sensors that can be devised using a large variety of different paradigms. However, the cornerstone physics in the analysis and modeling of different phenomena has many similarities. In this section we will use the Lagrange equations of motion, as well as Kirchhoff's and Newton's laws, to illustrate the model developments applicable to a large class of nano- and microscale transducers. It has been illustrated that one cannot reduce interconnected systems (MEMS and NEMS) to simple, idealized subsystems (components). For example, one cannot study actuators and smart structures without studying the mechanism to regulate these actuators using ICs and antennae. These ICs and antennae are controlled by the processor that received the information from sensors. Through illustrative examples, differential equations will be found and simulated. Nano- and microelectromechanical systems must be studied using the fundamental laws and basic principles of mechanics and electromagnetics. Let us identify and study these key concepts to illustrate the use of cornerstone principles.

6.4.1   Newtonian Mechanics

6.4.1.1   *Newtonian Mechanics, Energy Analysis, Generalized Coordinates, and Lagrange Equations: Translational Motion*

The study of the motion of systems with the corresponding analysis of forces that cause motion is our interest. The equations of motion for mechanical systems can be found using Newton's second law of motion. In particular, using the position (displacement) vector $\vec{\mathbf{r}}$, the Newton equation in the vector form is given as

$$\sum \vec{F}(t,\vec{\mathbf{r}}) = m\vec{a} \tag{6.1}$$

where $\sum \vec{F}(t,\vec{\mathbf{r}})$ is the vector sum of all forces applied to the body ($\vec{F}$ is the net force); $\vec{a}$ is the vector of acceleration of the body with respect to an inertial reference frame; and m is the mass of the body.

It is worth noting that $m\vec{a}$ represents the magnitude and direction of the applied net force acting on the object. Hence $m\vec{a}$ is not a force. A body is at equilibrium (the object is at rest or is moving with constant speed) if $\sum \vec{F} = 0$, e.g., there is no acceleration.

Using Equation 6.1, in the Cartesian system (*x-y-z* coordinates), we have the mechanical equations of motion

$$\sum \vec{F}(t,\vec{\mathbf{r}}) = m\vec{a} = m\frac{d\vec{\mathbf{r}}^2}{dt^2} = m\begin{bmatrix} \dfrac{d\vec{x}^2}{dt^2} \\[2ex] \dfrac{d\vec{y}^2}{dt^2} \\[2ex] \dfrac{d\vec{z}^2}{dt^2} \end{bmatrix}, \quad \begin{bmatrix} \vec{a}_x \\[2ex] \vec{a}_y \\[2ex] \vec{a}_z \end{bmatrix} = \begin{bmatrix} \dfrac{d\vec{x}^2}{dt^2} \\[2ex] \dfrac{d\vec{y}^2}{dt^2} \\[2ex] \dfrac{d\vec{z}^2}{dt^2} \end{bmatrix}$$

which can be considered as the second-order ordinary differential equations.

Though one can identify these equations as rigid-body mechanical equations of motion, the forces are developed by actuators. For example, the force can be a function of current (electromechanical actuators) or pressure (hydraulic actuators). Thus, Equation 6.1 must be complemented by the circuitry dynamics to achieve the completeness and coherence of modeling, analysis, and design.

In the Cartesian coordinate system, Newton's second law is expressed as

$$\sum F_x = ma_x, \quad \sum F_y = ma_y, \quad \sum F_z = ma_z$$

Newton's second law in terms of the linear momentum, which is found as $\vec{p} = m\vec{v}$, is given by

$$\sum \vec{F} = \frac{d\vec{p}}{dt} = \frac{d(m\vec{v})}{dt}$$

where $\vec{v}$ is the vector of the object velocity.

Thus, the force is equal to the rate of change of the momentum. The object or particle moves uniformly if $d\vec{p}/dt = 0$ (thus, $\vec{p} = const$).

Newton's laws are extended to multibody systems, and the momentum of a system of N particles is the vector sum of the individual momenta. That is, we have

$$\vec{P} = \sum_{i=1}^{N} \vec{p}_i$$

Consider the multibody system of N particles. The position (displacement) is represented by the vector **r**, which in the Cartesian coordinate system has the components x, y, and z. Taking note of the expression for the potential energy $\Pi(\vec{r})$, one has for the conservative mechanical system

$$\sum \vec{F}(\vec{r}) = -\nabla \Pi(\vec{r})$$

Therefore, the work done per unit time is

$$\frac{dW}{dt} = \sum \vec{F}(\vec{r})\frac{d\vec{r}}{dt} = -\nabla\Pi(\vec{r})\frac{d\vec{r}}{dt} = -\frac{d\Pi(\vec{r})}{dt}$$

From Newton's second law one obtains

$$m\vec{a} - \sum \vec{F}(\vec{r}) = m\frac{d^2\vec{r}}{dt^2} - \sum \vec{F}(\vec{r}) = 0$$

Hence, for a conservative system the following equation results:

$$m\frac{d^2\vec{r}}{dt^2} + \nabla\Pi(\vec{r}) = 0$$

For the system of N particles, the equations of motion are

$$m_N\frac{d^2\vec{r}_N}{dt^2} + \nabla\Pi(\vec{r}_N) = 0$$

or

$$m_i\frac{d^2(\vec{x}_i,\vec{y}_i,\vec{z}_i)}{dt^2} + \frac{\partial\Pi(\vec{x}_i,\vec{y}_i,\vec{z}_i)}{\partial(\vec{x}_i,\vec{y}_i,\vec{z}_i)} = 0, \quad i = 1,\ldots,N$$

The total kinetic energy of the particle is $\Gamma = \frac{1}{2}mv^2$.

For N particles, one has the expression for the total kinetic energy

$$\Gamma\left(\frac{d\vec{x}_i}{dt},\frac{d\vec{y}}{dt_i},\frac{d\vec{z}_i}{dt}\right) = \frac{1}{2}\sum_{i=1}^{N} m_i\left(\frac{d\vec{x}_i}{dt},\frac{d\vec{y}}{dt_i},\frac{d\vec{z}_i}{dt}\right)^2$$

Furthermore, we obtain

$$m_i\frac{d(\vec{x}_i,\vec{y}_i,\vec{z}_i)}{dt} = \frac{\partial\Gamma\left(\dfrac{d\vec{x}_i}{dt},\dfrac{d\vec{y}_i}{dt},\dfrac{d\vec{z}_i}{dt}\right)}{\partial\left(\dfrac{d\vec{x}_i}{dt},\dfrac{d\vec{y}_i}{dt},\dfrac{d\vec{z}_i}{dt}\right)}$$

Having documented the Newtonian mechanics and the use of the kinetic and potential energies, let us introduce one of the most general concepts in the modeling of dynamic systems that is based on Lagrangian mechanics.

In contrast to the application of the displacement, velocity, or acceleration, we will utilize the so-called generalized coordinates. In fact, using the generalized coordinates $(q_1, \ldots, q_n)$ and generalized velocities $(dq_1/dt, \ldots, dq_n/dt)$, one finds the total kinetic $\Gamma(q_1, \ldots, q_n, dq_1/dt, \ldots, dq_n/dt)$ and potential $\Pi(q_1, \ldots, q_n)$ energies. Hence, using the expressions for the total kinetic and potential energies, Newton's second law of motion can be given in the following form:

$$\frac{d}{dt}\left(\frac{\partial \Gamma}{\partial \dot{q}_i}\right) + \frac{\partial \Pi}{\partial q_i} = 0$$

That is, the generalized coordinates q_i are used to model multibody systems, and

$$(q_1, \ldots, q_n) = (\vec{x}_1, \vec{y}_1, \vec{z}_1, \ldots, \vec{x}_N, \vec{y}_N, \vec{z}_N)$$

The obtained results are connected to the Lagrange equations of motion, which will be studied later.

Example 6.10

Consider a micropositioning table actuated by a micromotor. How much work is required to accelerate a 2-mg payload ($m = 2$ mg) from $v_0 = 0$ m/sec to $v_f = 1$ m/sec?

SOLUTION

The work needed is calculated as

$$W = \tfrac{1}{2}\left(mv_f^2 - mv_0^2\right) = \tfrac{1}{2} 2 \times 10^{-6} \times 1^2 = 1 \times 10^{-6} \text{ J}$$

Example 6.11

Consider a body of mass m in the x-y coordinate system. The force $\vec{F}_a$ is applied in the x direction. Neglecting Coulomb and static friction, and assuming that the viscous friction force is $F_{fr} = B_v(dx/dt)$, find the equations of motion. Here B_v is the viscous friction coefficient.

SOLUTION

The free-body diagram developed is illustrated in Figure 6.13.

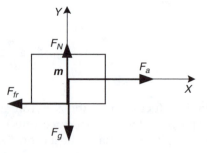

FIGURE 6.13
Free-body diagram.

The sum of the forces, acting in the y direction, is expressed as

$$\sum \vec{F}_Y = \vec{F}_N - \vec{F}_g$$

where $\vec{F}_g = mg$ is the gravitational force acting on the mass m; and $\vec{F}_N$ is the normal force, which is equal and opposite to the gravitational force.

From Equation 6.1, the equation of motion in the y direction is

$$\vec{F}_N - \vec{F}_g = ma_y = m\frac{d^2y}{dt^2}$$

where a_y is the acceleration in the y direction, and $a_y = d^2y/dt^2$.

Making use of $\vec{F}_N = \vec{F}_g$, we have

$$\frac{d^2y}{dt^2} = 0$$

The sum of the forces acting in the x direction is found using the applied force $\vec{F}_a$ and the friction force $\vec{F}_{fr}$. In particular, we have

$$\sum \vec{F}_X = \vec{F}_a - \vec{F}_{fr}$$

The applied force can be time-invariant $\vec{F}_a = const$ or time-varying $\vec{F}_a(t) = f(t, x, y, z)$. For example,

$$\vec{F}_a(t) = x\sin(6t - 4)e^{-0.5t} + \frac{dy}{dt}t^2 + z^3\cos\left(\frac{dx}{dt}t - y^2t^4\right)$$

Using Equation 6.1, the equation motion in the x direction is found to be

$$\vec{F}_a - \vec{F}_{fr} = ma_x = m\frac{d^2x}{dt^2}$$

where a_x is the acceleration in the x direction, $a_x = d^2x/dt^2$, and the velocity in the x direction is $V = dx/dt$.

Assuming that the Coulomb and static friction can be neglected, the friction force, as a function of the viscous friction coefficient B_v and velocity $v = dx/dt$, is given by $F_{fr} = B_v(dx/dt)$.

Hence, one obtains the second-order nonlinear differential equation to model the body dynamics in the x direction:

$$\frac{d^2x}{dt^2} = \frac{1}{m}\left(F_a - B_v\frac{dx}{dt}\right)$$

From the derived equation of motion, a set of two first-order linear differential equations results. In particular,

$$\frac{dx}{dt} = v$$

$$\frac{dv}{dt} = \frac{1}{m}\left(F_a - B_v v\right)$$

The application of the Newtonian mechanics does not necessarily result in the ordinary differential equations. The following examples illustrate that partial differential equations can be found to model structures (elastic membranes and beams). This book also will illustrate how these membranes and beams can be controlled (deflected) using distinct actuation paradigms.

Example 6.12

The elastic membrane is illustrated in Figure 6.14. Derive the mathematical model to model the rectangular membrane vibration. That is, the goal is to study the time varying membrane deflection $d(t, x, y)$ in the x-y plane. The mass of the undeflected membrane per unit area ρ is constant (homogeneous membrane).

SOLUTION

Some assumptions and simplifications can be made. We assume that the membrane is perfectly flexible. For small deflections, the tension T (the force per unit length) is the same at all points in all directions, and suppose that T is constant during the motion. It should be emphasized that because the deflection of the membrane is small, compared with the membrane size ab, the inclination angles are small.

Taking note of these assumptions, the forces acting on the sides are approximated as $F_x = T\Delta x$ and $F_y = T\Delta y$. The membrane is assumed to be perfectly flexible; therefore, forces F_x and F_y are tangential to the membrane.

The horizontal components of the forces are found as the cosine functions of the inclination angles. The horizontal components at the opposite sides (right and left) are equal because angles α and β are small. Thus, the membrane motion in the horizontal direction can be neglected.

The vertical components of the forces are $T\Delta y \sin\beta$ and $-T\Delta y \sin\alpha$.

Using Newton's second law of motion, the net force must be found. We have the following expression:

$$\sum F = T\Delta y(d_x(x+\Delta x, y_1) - d_x(x, y_2)) + T\Delta x(d_y(x_1, y+\Delta y) - d_y(x_2, y))$$

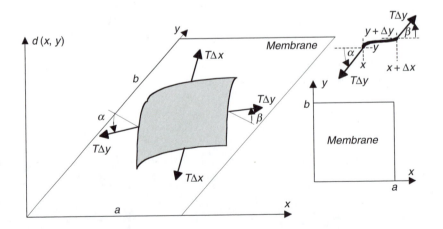

FIGURE 6.14
Vibrating rectangular membrane.

Thus, two-dimensional partial (wave) differential equation is

$$\frac{\partial^2 d(t,x,y)}{\partial t^2} = \frac{T}{\rho}\left(\frac{\partial^2 d(t,x,y)}{\partial x^2} + \frac{\partial^2 d(t,x,y)}{\partial y^2}\right) = \frac{T}{\rho}\nabla^2 d(t,x,y)$$

Using initial and boundary conditions, the solution can be found.

Let the initial conditions be $d(t_0, x, y) = d_0(x, y)$ and

$$\frac{\partial d(t_0, x, y)}{\partial t} = d_1(x, y)$$

Thus, the initial displacement $d_0(x,y)$ and initial velocity $d_1(x,y)$ are given.

Assume that the boundary conditions are $d(t, x_0, y_0) = 0$ and $d(t, x_f, y_f) = 0$.

The solution is found to be

$$d(t,x,y) = \sum_{i=1}^{\infty}\sum_{j=1}^{\infty} d_{ij}(t,x,y) = \sum_{i=1}^{\infty}\sum_{j=1}^{\infty}\left(A_{ij}\cos\lambda_{ij}t + B_{ij}\sin\lambda_{ij}t\right)\sin\frac{i\pi x}{a}\sin\frac{j\pi x}{b}$$

where the characteristic eigenvalues are

$$\lambda_{ij} = \sqrt{\frac{T}{\rho}}\pi\sqrt{\frac{i^2}{a^2} + \frac{j^2}{b^2}}$$

Using initial conditions, the Fourier coefficients are obtained in the form of the double Fourier series. In particular, we have

$$A_{ij} = \frac{4}{ab}\int_0^b\int_0^a d_0(x,y)\sin\frac{i\pi x}{a}\sin\frac{j\pi y}{b}\,dxdy$$

and

$$B_{ij} = \frac{4}{ab\lambda_{ij}}\int_0^b\int_0^a d_1(x,y)\sin\frac{i\pi x}{a}\sin\frac{j\pi y}{b}\,dxdy$$

Example 6.13

Derive the mathematical model of the infinitely long beam on the elastic foundation as shown in Figure 6.15. The load force is the square function. The modulus (the spring stiffness per unit length) of the elastic foundation is k_s.

SOLUTION

Using the Euler beam theory, the deflection $y(x)$ due to the net load force $F(x)$ is modeled by the fourth-order differential equation

$$k_r\frac{d^4y}{dt^4} = F(x)$$

where k_r is the flexural rigidity constant.

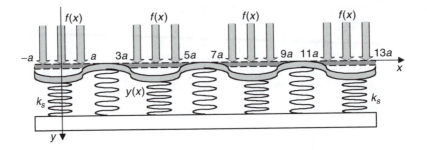

FIGURE 6.15
Beam on elastic foundation under the load force $f(x)$.

Therefore, we have the following differential equation to model the infinite beam under consideration:

$$k_r \frac{d^4 y}{dt^4} + k_s y = f(x)$$

The general homogeneous solution is given by

$$y(x) = e^{\frac{1}{2}\sqrt[4]{k_r}\,x}\left[k_1\,\sin\!\left(\tfrac{1}{2}\sqrt[4]{k_r}\,x\right) + k_2\,\sin\!\left(\tfrac{1}{2}\sqrt[4]{k_r}\,x\right)\right] + e^{-\frac{1}{2}\sqrt[4]{k_r}\,x}\left[k_3\,\sin\!\left(\tfrac{1}{2}\sqrt[4]{k_r}\,x\right) + k_4\,\sin\!\left(\tfrac{1}{2}\sqrt[4]{k_r}\,x\right)\right]$$

where the unknown coefficients k_i can be determined using the initial and boundary conditions.

The boundary-value problem can be relaxed, and the solution can be found in the series form.

The load force is the periodic function, and using the Fourier series we have

$$f(x) = \frac{f_0}{2} + \frac{2f_0}{\pi}\sum_{i=1}^{\infty}\frac{\sin\!\left(\tfrac{1}{2}i\pi\right)}{i}\cos\frac{i\pi x}{2a}$$

The solution of the differential equation

$$k_r \frac{d^4 y}{dt^4} + k_s y = f(x)$$

is found as

$$y(x) = a_0 + \sum_{i=1}^{\infty} a_i \cos\frac{i\pi x}{2a}$$

Differentiating this equation four times gives

$$k_s a_0 = \frac{f_0}{2}$$

and

$$\left(k_r \frac{i^4 \pi^4}{16a^4} + k_s\right) a_i = \frac{2f_0}{\pi}\frac{\sin\!\left(\tfrac{1}{2}i\pi\right)}{i}$$

Thus, the Fourier series coefficients are

$$a_0 = \frac{f_0}{2k_s} \quad \text{and} \quad a_i = \frac{2f_0}{\pi} \frac{\sin(\frac{1}{2}i\pi)}{i\left[k_r\left(\dfrac{i\pi}{2a}\right)^4 + k_s\right]}, \quad i \geq 1$$

Therefore, the solution is given by

$$y(x) = \frac{f_0}{2k_s} + \frac{32f_0 a^4}{\pi} \sum_{i=1}^{\infty} \frac{\sin(\frac{1}{2}i\pi)}{i(k_r i^4 \pi^4 + 16a^4 k_s)} \cos\frac{i\pi x}{2a}$$

The first-order approximation is

$$y(x) \approx \frac{f_0}{2k} + \frac{32f_0 a^4}{\pi(k_r \pi^4 + 16a^4 k_s)} \cos\frac{\pi x}{2a}$$

6.4.1.2 Newtonian Mechanics: Rotational Motion

Having examined the translational motion (linear actuators and sensors, translational kinematics, etc.), it is evident that the majority of devices are rotational. There is no difference in modeling of translational or rotational devices, but the variables used are distinct. In fact, instead of linear displacement and acceleration, angular displacement and acceleration should be used. In addition, for rotational motion, the net torque must be considered.

The rotational analog of Newton's second law for a rigid body is given as

$$\sum \vec{T}(t,\vec{\theta}) = J\vec{\alpha} \tag{6.2}$$

where $\sum\vec{T}$ is the net torque; J is the moment of inertia (rotational inertia); $\vec{\alpha}$ is the angular acceleration vector,

$$\vec{\alpha} = \frac{d}{dt}\frac{d\vec{\theta}}{dt} = \frac{d^2\vec{\theta}}{dt^2} = \frac{d\vec{\omega}}{dt}$$

$\vec{\theta}$ is the angular displacement; and ω denotes the angular velocity.

The angular momentum of the system $\vec{L}_M$ is expressed as

$$\vec{L}_M = \vec{r} \times \vec{p} = \vec{r} \times m\vec{v}$$

and

$$\sum \vec{T} = \frac{d\vec{L}_M}{dt} = \vec{r} \times \vec{F}$$

where $\vec{r}$ is the position vector with respect to the origin.

For the rigid body, rotating around the axis of symmetry, we have

$$\vec{L}_M = J\vec{\omega}$$

For one-dimensional rotational systems, Newton's second law of motion is frequently expressed as

$$M = J\alpha$$

where M is the sum of all moments about the center of mass of a body [N-m]; J is the moment of inertia about its center of mass [kg-m^2]; and α is the angular acceleration of the body [rad/sec^2].

Example 6.14

A micromotor has the equivalent moment of inertia $J = 5 \times 10^{-20}$ kg-m^2. Let the angular velocity of the rotor be $\omega_r = 10t^{1/5}$. Find the angular momentum and the developed electromagnetic torque as functions of time. The load and friction torques are zero.

SOLUTION

The angular momentum is found as $L_M = J\omega_r = 5 \times 10^{-19} t^{1/5}$.

The developed electromagnetic torque is $T_e = dL_M/dt = 1 \times 10^{-19} t^{-4/5}$.

From Newtonian mechanics one concludes that the applied net force or torque plays a key role in quantitatively describing the motion and qualitatively changing the behavioral dynamics. The analysis of motion can be performed using the energy or momentum quantities, which are conserved. The principle of conservation of energy states that energy can be only converted from one form to another.

Kinetic energy is associated with motion, while potential energy is associated with position. The sum of the kinetic (Γ), potential (Π), and dissipated (D) energies is called the total energy of the system (Σ_T), which is conserved, and the total amount of energy remains constant; that is,

$$\Sigma_T = \Gamma + \Pi + D = const$$

For example, consider the translational motion of a body that is attached to an ideal spring that obeys Hooke's law. Neglecting friction, one obtains the following expression for the total energy:

$$\Sigma_T = \Gamma + \Pi = \tfrac{1}{2}(mv^2 + k_s x^2) = const$$

Here the translational kinetic energy is $\Gamma = \tfrac{1}{2}mv^2$; the elastic potential energy is $\Pi = \tfrac{1}{2}k_s x^2$; k_s is the force constant of the spring; and x is the displacement.

For rotating spring, we have

$$\Sigma_T = \Gamma + \Pi = \tfrac{1}{2}(J\omega^2 + k_s \theta^2) = const$$

where the rotational kinetic energy and the elastic potential energy are

$$\Gamma = \tfrac{1}{2}J\omega^2 \quad and \quad \Pi = \tfrac{1}{2}k_s\theta^2$$

The kinetic energy of a rigid body having translational and rotational components of motion is found to be

$$\Gamma = \tfrac{1}{2}(mv^2 + J\omega^2)$$

That is, motion of the rigid body is represented as a combination of translational motion of the center of mass and rotational motion about the axis through the center of mass. The moment of inertia depends upon how the mass is distributed with respect to the axis, and J is different for different axes of rotation. If the body is uniform in density, J can be easily calculated for regularly shaped bodies in terms of their dimensions. For example, a rigid cylinder with mass m (which is uniformly distributed), radius R, and length l has the following horizontal and vertical moments of inertia:

$$J_{horizontal} = \tfrac{1}{2}mR^2 \quad \text{and} \quad J_{vertical} = \tfrac{1}{4}mR^2 + \tfrac{1}{12}ml^2$$

The radius of gyration can be found for irregularly shaped objects, and the moment of inertia can be easily obtained.

In electromechanical motion devices, the force and torque are thoroughly studied. Assuming that the body is rigid and the moment of inertia is constant, one has

$$\vec{T}d\vec{\theta} = J\vec{\alpha}d\vec{\theta} = J\frac{d\vec{\omega}}{dt}d\vec{\theta} = J\frac{d\vec{\theta}}{dt}d\vec{\omega} = J\vec{\omega}d\vec{\omega}$$

The total work, as given by

$$W = \int_{\theta_0}^{\theta_f} \vec{T}d\vec{\theta} = \int_{\omega_0}^{\omega_f} J\vec{\omega}d\vec{\omega} = \tfrac{1}{2}\left(J\vec{\omega}_f^2 - J\vec{\omega}_0^2\right)$$

represents the change of the kinetic energy.

Furthermore,

$$\frac{dW}{dt} = \vec{T}\frac{d\vec{\theta}}{dt} = \vec{T}\times\vec{\omega}$$

and the power is defined by

$$P = \vec{T}\times\vec{\omega}$$

This equation is an analog of $P = \vec{F}\times\vec{v}$, which is applied for translational motion.

Example 6.15

The rated power and angular velocity of a 10-μm micromotor are 0.01 W and 10,000 rad/sec. Calculate the rated electromagnetic torque.

SOLUTION

The electromagnetic torque is

$$T_e = \frac{P}{\omega_r} = \frac{0.01}{10,000} = 1\times10^{-6}\,\text{N-m}$$

Example 6.16

Given a point mass m suspended by a massless unstretchable string of length l (see Figure 6.16), derive the equations of motion for a simple pendulum with negligible friction.

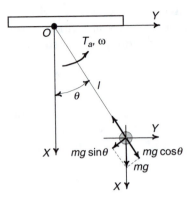

FIGURE 6.16
A simple pendulum.

SOLUTION

The restoring force, which is proportional to $\sin\theta$ and given by $-mg\sin\theta$, is the tangential component of the net force. Therefore, the sum of the moments about the pivot point O is found to be

$$\sum M = -mgl\sin\theta + T_a$$

where T_a is the applied torque and I is the length of the pendulum measured from the point of rotation.

Using Equation 6.2, one obtains the equation of motion

$$J\alpha = J\frac{d^2\theta}{dt^2} = -mgl\sin\theta + T_a$$

where J is the moment of inertia of the mass about the point O.

Hence, the second-order differential equation is found to be

$$\frac{d^2\theta}{dt^2} = \frac{1}{J}(-mgl\sin\theta + T_a)$$

Using the following differential equation for the angular displacement

$$\frac{d\theta}{dt} = \omega$$

one obtains the following set of two first-order differential equations:

$$\frac{d\omega}{dt} = \frac{1}{J}(-mgl\sin\theta + T_a)$$

$$\frac{d\theta}{dt} = \omega$$

The moment of inertia is expressed by $J = ml^2$. Hence, we have the following differential equations to be used in the modeling of a simple pendulum:

$$\frac{d\omega}{dt} = -\frac{g}{l}\sin\theta + \frac{1}{ml^2}T_a$$

$$\frac{d\theta}{dt} = \omega$$

6.4.1.3 *Friction Models in Microelectromechanical Systems*

A thorough consideration of friction is essential for understanding the operation of MEMS, as well as electromechanical systems in general. Friction is a very complex nonlinear phenomenon that is difficult to model. The classical Coulomb friction is a retarding frictional force (for translational motion) or torque (for rotational motion) that changes its sign with the reversal of the direction of motion, and the amplitude of the frictional force or torque are constant. For translational and rotational motions, the Coulomb friction force and torque are

$$F_{Coulomb} = k_{Fc}\,\text{sgn}(v) = k_{Fc}\,\text{sgn}\left(\frac{dx}{dt}\right)$$

$$T_{Coulomb} = k_{Tc}\,\text{sgn}(\omega) = k_{Tc}\,\text{sgn}\left(\frac{d\theta}{dt}\right)$$

where k_{Fc} and k_{Tc} are the Coulomb friction coefficients.

Figure 6.17A illustrates the Coulomb friction.

Viscous friction is a retarding force or torque that is a linear function of linear or angular velocity. The viscous friction force and torque versus linear and angular velocities are shown in Figure 6.17B. The following expressions are commonly used to model the viscous friction:

$$F_{viscous} = B_v v = B_v \frac{dx}{dt} \quad \text{for translational motion}$$

and

$$T_{viscous} = B_m \omega = B_m \frac{d\theta}{dt} \quad \text{for rotational motion}$$

where B_v and B_m are the viscous friction coefficients.

The static friction exists only when the body is stationary and vanishes as motion begins. The static friction is a force F_{static} or torque T_{static}, and we have the following expressions:

$$F_{static} = \pm F_{st}\big|_{v=\frac{dx}{dt}=0}$$

and

$$T_{static} = \pm T_{st}\big|_{\omega=\frac{d\theta}{dt}=0}$$

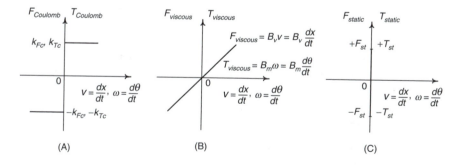

FIGURE 6.17
Functional representations of (A) Coulomb friction; (B) viscous friction; (C) static friction.

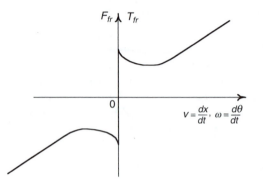

FIGURE 6.18
Friction force and torque are functions of linear and angular velocities.

One concludes that the static friction is a retarding force or torque that tends to prevent the initial translational or rotational motion at beginning. (See Figure 6.17C.)

In general, the friction force and torque are nonlinear functions that must be modeled using frictional memory, presliding conditions, etc. The empirical formulas, commonly used to express F_{static} and T_{static}, are

$$F_{fr} = \left(k_{fr1} - k_{fr2}e^{-k|v|} + k_{fr3}|v|\right)\operatorname{sgn}(v_t) = \left(k_{fr1} - k_{fr2}e^{-k\left|\frac{dx}{dt}\right|} + k_{fr3}\left|\frac{dx}{dt}\right|\right)\operatorname{sgn}\left(\frac{dx}{dt}\right)$$

and

$$T_{fr} = \left(k_{fr1} - k_{fr2}e^{-k|\omega|} + k_{fr3}|\omega|\right)\operatorname{sgn}(\omega) = \left(k_{fr1} - k_{fr2}e^{-k\left|\frac{d\theta}{dt}\right|} + k_{fr3}\left|\frac{d\theta}{dt}\right|\right)\operatorname{sgn}\left(\frac{d\theta}{dt}\right)$$

These F_{static} and T_{static} are shown in Figure 6.18.

Example 6.17   MICROTRANSDUCER MODEL

Figure 6.19 shows a simple electromechanical device (microactuator) with a stationary member and movable plunger. Using Newton's second law, find the lumped-parameter mathematical model in the form of ordinary differential equations.

SOLUTION

As illustrated in the Example 6.8, the mathematical model was derived using Newton's and Kirchhoff's second laws. In particular, using Equation 6.1, $\sum \vec{F}(t,\vec{r}) = m\vec{a}$, we have

$$\sum F(t) = B_v \frac{dx}{dt} + (k_s x - k_{s1} x) + F_e(i, x) = m\frac{d^2x}{dt^2}$$

and taking note of the Kirchhoff second (voltage) law, one obtains

$$u_a = ri + \frac{d\psi}{dt}, \quad \psi = L(x)i$$

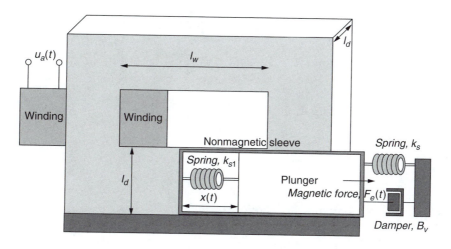

FIGURE 6.19
Schematic of a microactuator.

The net force can be expressed using the electromagnetic force developed by the microactuator

$$F_e(i,x) = \frac{\partial W_c(i,x)}{\partial x}$$

the friction force F_{fr}, and the load force F_L (not shown in Figure 6.19 and thus not used in the differential equations that model the microtransducer dynamics).

Using the expression for the inductance, one finally finds the following nonlinear differential equations for the microtransducer studied:

$$\frac{di}{dt} = -\frac{r[A_g l_f + A_f \mu_f (x+2d)]}{N^2 \mu_f \mu_0 A_f A_g} i + \frac{\mu_f A_f}{A_g l_f + A_f \mu_f (x+2d)} iv + \frac{A_g l_f + A_f \mu_f (x+2d)}{N^2 \mu_f \mu_0 A_f A_g} u_a$$

$$\frac{dx}{dt} = v$$

$$\frac{dv}{dt} = \underbrace{\frac{N^2 \mu_f^2 \mu_0 A_f^2 A_g}{2m[A_g l_f + A_f \mu_f (x+2d)]^2} i^2}_{\text{electromagnetic force developed}} - \frac{1}{m}(k_s x - k_{s1} x) - \frac{1}{m}\underbrace{\left(k_{fr1} - k_{fr2} e^{-k\left|\frac{dx}{dt}\right|} + k_{fr3}\left|\frac{dx}{dt}\right| \right) \text{sgn}\left(\frac{dx}{dt}\right)}_{\text{friction force}}$$

6.4.2 Lagrange Equations of Motion

Electromechanical systems, including MEMS and NEMS, augment mechanical and electronic (microelectronic) components. Therefore, one studies mechanical, electromagnetic, and circuitry transients. It was illustrated that the designer can integrate the torsional-mechanical dynamics and circuitry equations of motion. However, there exist general concepts to model systems. The Lagrange and Hamilton concepts are based on the energy analysis. Using the system variables, one finds the total kinetic, dissipation, and potential energies (which are denoted as Γ, D, and Π).

Taking note of the total

$$\text{kinetic,} \quad \Gamma\left(t, q_1, \ldots, q_n, \frac{dq_1}{dt}, \ldots, \frac{dq_n}{dt}\right)$$

$$\text{dissipation,} \quad D\left(t, q_1, \ldots, q_n, \frac{dq_1}{dt}, \ldots, \frac{dq_n}{dt}\right)$$

and

$$\text{potential,} \quad \Pi(t, q_1, \ldots, q_n)$$

energies, the Lagrange equations of motion are

$$\frac{d}{dt}\left(\frac{\partial \Gamma}{\partial \dot{q}_i}\right) - \frac{\partial \Gamma}{\partial q_i} + \frac{\partial D}{\partial \dot{q}_i} + \frac{\partial \Pi}{\partial q_i} = Q_i \qquad (6.3)$$

Here, q_i and Q_i are the generalized coordinates and the generalized forces (applied forces and disturbances). These generalized coordinates q_i are used to derive expressions for energies $\Gamma(t, q, \ldots, q_n, dq_1/dt, \ldots, dq_n/dt)$, $D(t, q_1, \ldots, q_n, dq_1/dt, \ldots, dq_n/dt)$, and $\Pi(t, q_1, \ldots, q_n)$. Furthermore, the reader should understand that these generalized coordinates are assigned to explicitly and coherently express the energies of the system. Therefore, the linear or angular displacement (for translational and rotational devices) and charges are most commonly used.

Taking into account that for conservative (lossless) systems $D = 0$, we have the following Lagrange's equations of motion:

$$\frac{d}{dt}\left(\frac{\partial \Gamma}{\partial \dot{q}_i}\right) - \frac{\partial \Gamma}{\partial q_i} + \frac{\partial \Pi}{\partial q_i} = Q_i$$

To illustrate the application of the Lagrange equation of motion we consider a familiar example.

Example 6.18   MATHEMATICAL MODEL of a SIMPLE PENDULUM

Derive the mathematical model for a simple pendulum using the Lagrange equations of motion.

SOLUTION

Derivation of the mathematical model for the simple pendulum, shown in Figure 6.16, was performed in Example 6.17 using Newtonian mechanics. For the studied conservative (lossless) system, we have $D = 0$. Thus, the Lagrange equations of motion are

$$\frac{d}{dt}\left(\frac{\partial \Gamma}{\partial \dot{q}_i}\right) - \frac{\partial \Gamma}{\partial q_i} + \frac{\partial \Pi}{\partial q_i} = Q_i$$

The kinetic energy of the pendulum bob is

$$\Gamma = \tfrac{1}{2}m(l\dot{\theta})^2$$

The potential energy is found as

$$\Pi = mgl(1 - \cos\theta)$$

The angular displacement is the generalized coordinate. Thus, $q_i = \theta$.

The generalized force is the torque applied, e.g., $Q_i = T_a$.

One obtains the following expressions for the derivatives:

$$\frac{\partial \Gamma}{\partial \dot{q}_i} = \frac{\partial \Gamma}{\partial \dot{\theta}} = ml^2\dot{\theta}$$

$$\frac{\partial \Gamma}{\partial q_i} = \frac{\partial \Gamma}{\partial \theta} = 0$$

$$\frac{\partial \Pi}{\partial q_i} = \frac{\partial \Pi}{\partial \theta} = mgl\sin\theta$$

Thus, the first term of the Lagrange equation is found to be

$$\frac{d}{dt}\left(\frac{\partial \Gamma}{\partial \dot{\theta}}\right) = ml^2\frac{d^2\theta}{dt^2} + 2ml\frac{dl}{dt}\frac{d\theta}{dt}$$

Assuming that the string is unstretchable, we have

$$\frac{dl}{dt} = 0$$

If this assumption is not valid, one should use the appropriate expression for the length as a function of the generalized coordinates. Assuming that $dl/dt = 0$, we obtain

$$ml^2\frac{d^2\theta}{dt^2} + mgl\sin\theta = T_a$$

Thus, one obtains

$$\frac{d^2\theta}{dt^2} = \frac{1}{ml^2}(-mgl\sin\theta + T_a)$$

Recall that the equation of motion, derived by using Newtonian mechanics, is

$$\frac{d^2\theta}{dt^2} = \frac{1}{J}(-mgl\sin\theta + T_a), \quad \text{where } J = ml^2$$

One concludes that the results are the same, and the equations are

$$\frac{d\omega}{dt} = -\frac{g}{l}\sin\theta + \frac{1}{ml^2}T_a$$

$$\frac{d\theta}{dt} = \omega$$

However, the Lagrange equations of motion provide more general results. We will illustrate that electromechanical systems can be modeled using the Lagrange equations of motion (Newton's laws can be used only to model the torsional-mechanical dynamics), and, as was illustrated, the system parameters can be dependent on generalized coordinate (for example, l can be a function of $q = \theta$).

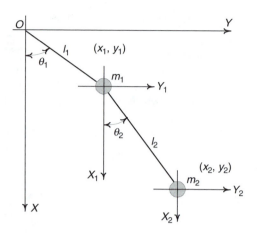

FIGURE 6.20
Double pendulum.

Example 6.19   MATHEMATICAL MODEL of a PENDULUM

Consider a double pendulum of two degrees of freedom with no external forces applied to the system. (See Figure 6.20.) Using the Lagrange equations of motion, derive the differential equations.

SOLUTION

The angular displacement θ_1 and θ_2 are chosen as the independent generalized coordinates. In the x-y plane studied, let (x_1,y_1) and (x_2,y_2) be the rectangular coordinates of m_1 and m_2. Then, we obtain

$$x_1 = l_1 \cos\theta_1, \quad x_2 = l_1 \cos\theta_1 + l_2 \cos\theta_2$$

$$y_1 = l_1 \sin\theta_1, \quad y_2 = l_1 \sin\theta_1 + l_2 \sin\theta_2$$

The total kinetic energy Γ is found to be a nonlinear function of the displacements:

$$\Gamma = \tfrac{1}{2}m_1\left(\dot{x}_1^2 + \dot{y}_1^2\right) + \tfrac{1}{2}m_2\left(\dot{x}_2^2 + \dot{y}_2^2\right) = \tfrac{1}{2}(m_1 + m_2)l_1^2\dot{\theta}_1^2 + m_2 l_1 l_2 \dot{\theta}_1\dot{\theta}_2 \cos(\theta_2 - \theta_1) + \tfrac{1}{2}m_2 l_2^2 \dot{\theta}_2^2$$

Then one obtains

$$\frac{\partial \Gamma}{\partial \theta_1} = m_2 l_1 l_2 \sin(\theta_2 - \theta_1)\dot{\theta}_1\dot{\theta}_2$$

$$\frac{\partial \Gamma}{\partial \dot{\theta}_1} = (m_1 + m_2)l_1^2\dot{\theta}_1 + m_2 l_1 l_2 \cos(\theta_2 - \theta_1)\dot{\theta}_2$$

$$\frac{\partial \Gamma}{\partial \theta_2} = -m_2 l_1 l_2 \sin(\theta_1 - \theta_2)\dot{\theta}_1\dot{\theta}_2$$

$$\frac{\partial \Gamma}{\partial \dot{\theta}_2} = m_2 l_1 l_2 \cos(\theta_2 - \theta_1)\dot{\theta}_1 + m_2 l_1^2\dot{\theta}_2$$

The total potential energy is given by

$$\Pi = m_1 g x_1 + m_2 g x_2 = (m_1 + m_2)g l_1 \cos\theta_1 + m_2 g l_2 \cos\theta_2$$

Hence, the derivatives are given as

$$\frac{\partial \Pi}{\partial \theta_1} = -(m_1 + m_2)gl_1 \sin\theta_1 \quad \text{and} \quad \frac{\partial \Pi}{\partial \theta_2} = -m_2 gl_2 \sin\theta_2$$

The Lagrange equations of motion are

$$\frac{d}{dt}\left(\frac{\partial \Gamma}{\partial \dot{\theta}_1}\right) - \frac{\partial \Gamma}{\partial \theta_1} + \frac{\partial \Pi}{\partial \theta_1} = 0$$

$$\frac{d}{dt}\left(\frac{\partial \Gamma}{\partial \dot{\theta}_2}\right) - \frac{\partial \Gamma}{\partial \theta_2} + \frac{\partial \Pi}{\partial \theta_2} = 0$$

Hence, the dynamic equations of the system are

$$l_1\Big[(m_1 + m_2)l_1\ddot{\theta}_1 + m_2 l_2 \cos(\theta_2 - \theta_1)\ddot{\theta}_2 - m_2 l_2 \sin(\theta_2 - \theta_1)\dot{\theta}_2^2 - m_2 l_2 \sin(\theta_2 - \theta_1)\dot{\theta}_1\dot{\theta}_2$$
$$- (m_1 + m_2)g\sin\theta_1\Big] = 0$$

$$m_2 l_2\Big[l_2\ddot{\theta}_2 + l_1\cos(\theta_2 - \theta_1)\ddot{\theta}_1 + l_1\sin(\theta_2 - \theta_1)\dot{\theta}_1^2 + l_1\sin(\theta_2 - \theta_1)\dot{\theta}_1\dot{\theta}_2 - g\sin\theta_2\Big] = 0$$

It should be emphasized that if the torques T_1 and T_2 are applied to the first and second joints, the following equations of motions result:

$$l_1\Big[(m_1 + m_2)l_1\ddot{\theta}_1 + m_2 l_2 \cos(\theta_2 - \theta_1\ddot{\theta}_2) - m_2 l_2 \sin(\theta_2 - \theta_1)\dot{\theta}_2^2 - m_2 l_2 \sin(\theta_2 - \theta_1)\dot{\theta}_1\dot{\theta}_2$$
$$- (m_1 + m_2)g\sin\theta_1\Big] = T_1$$

$$m_2 l_2\Big[l_2\ddot{\theta}_2 + l_1\cos(\theta_2 - \theta_1)\ddot{\theta}_1 + l_1\sin(\theta_2 - \theta_1)\dot{\theta}_1^2 + l_1\sin(\theta_2 - \theta_1)\dot{\theta}_1\dot{\theta}_2 - g\sin\theta_2\Big] = T_2$$

These torques T_1 and T_2 can be time-varying functions. Furthermore, the load torques, as were illustrated, can be added to the mathematical model derived.

Having illustrated the use of the Lagrange equations of motion to mechanical systems, Example 6.20 demonstrates the use of Lagrangian mechanics to electric circuits.

Example 6.20   MATHEMATICAL MODEL of a CIRCUIT NETWORK

Consider a two-mesh electric circuit, as shown in Figure 6.21. Find the circuitry dynamics.

SOLUTION

We use the electric charges as the generalized coordinates. That is, q_1 and q_2 are the independent generalized coordinates shown in Figure 6.21. Here q_1 is the electric charge in the first loop, and q_2 represents the electric charge in the second loop.

The generalized force (applied voltage u_a), which is applied to the system, is denoted as Q_1.

It must be emphasized that these generalized coordinates are related to the circuitry variables.

In particular, the currents i_1 and i_2 are found in terms of charges as $i_1 = \dot{q}_1$ and $i_2 = \dot{q}_2$.

That is, we have

$$q_1 = \frac{i_1}{s} \quad \text{and} \quad q_2 = \frac{i_2}{s}$$

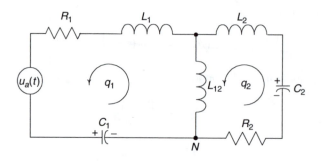

FIGURE 6.21
Two-mesh circuit network.

The generalized force is the applied voltage. Hence, $u_a(t) = Q_1$.

The total magnetic energy (kinetic energy) is expressed by

$$\Gamma = \tfrac{1}{2}L_1\dot{q}_1^2 + \tfrac{1}{2}L_{12}(\dot{q}_1 - \dot{q}_2)^2 + \tfrac{1}{2}L_2\dot{q}_2^2$$

By using this equation for Γ, we have the following expressions for derivatives:

$$\frac{\partial \Gamma}{\partial q_1} = 0, \quad \frac{\partial \Gamma}{\partial \dot{q}_1} = (L_1 + L_{12})\dot{q}_1 - L_{12}\dot{q}_2$$

$$\frac{\partial \Gamma}{\partial q_2} = 0, \quad \frac{\partial \Gamma}{\partial \dot{q}_2} = -L_{12}\dot{q}_1 + (L_2 + L_{12})\dot{q}_2$$

Using the equation for the total electric energy (potential energy)

$$\Pi = \tfrac{1}{2}\frac{q_1^2}{C_1} + \tfrac{1}{2}\frac{q_2^2}{C_2}$$

one finds

$$\frac{\partial \Pi}{\partial q_1} = \frac{q_1}{C_1} \quad \text{and} \quad \frac{\partial \Pi}{\partial q_2} = \frac{q_2}{C_2}$$

The total heat energy dissipated is

$$D = \tfrac{1}{2}R_1\dot{q}_1^2 + \tfrac{1}{2}R_2\dot{q}_2^2$$

Hence,

$$\frac{\partial D}{\partial \dot{q}_1} = R_1\dot{q}_1 \quad \text{and} \quad \frac{\partial D}{\partial \dot{q}_2} = R_2\dot{q}_2$$

The Lagrange equations of motion are expressed using the independent coordinates used. We obtain

$$\frac{d}{dt}\left(\frac{\partial \Gamma}{\partial \dot{q}_1}\right) - \frac{\partial \Gamma}{\partial q_1} + \frac{\partial D}{\partial \dot{q}_1} + \frac{\partial \Pi}{\partial q_1} = Q_1$$

$$\frac{d}{dt}\left(\frac{\partial \Gamma}{\partial \dot{q}_2}\right) - \frac{\partial \Gamma}{\partial q_2} + \frac{\partial D}{\partial \dot{q}_2} + \frac{\partial \Pi}{\partial q_2} = 0$$

Hence, the differential equations for the circuit studied are found to be

$$(L_1 + L_{12})\ddot{q}_1 - L_{12}\ddot{q}_2 + R_1\dot{q}_1 + \frac{q_1}{C_1} = u_a$$

$$-L_{12}\ddot{q}_1 + (L_2 + L_{12})\ddot{q}_2 + R_2\dot{q}_2 + \frac{q_2}{C_2} = 0$$

To perform numerical simulations, the MATLAB environment will be introduced and used in this chapter. Two second-order nonlinear differential equations, which model the circuit dynamics, can be simulated. Specifically, the SIMULINK model can be built using these derived nonlinear differential equations. We have

$$\ddot{q}_1 = \frac{1}{(L_1 + L_{12})}\left(-\frac{q_1}{C_1} - R_1\dot{q}_1 + L_{12}\ddot{q}_2 + u_a\right)$$

and

$$\ddot{q}_2 = \frac{1}{(L_2 + L_{12})}\left(L_{12}\ddot{q}_1 - \frac{q_2}{C_2} - R_2\dot{q}_2\right)$$

The corresponding SIMULINK diagram, built using the above listed equations, is shown in Figure 6.22. It should be emphasized that the currents i_1 and i_2 are expressed in terms of charges as $i_1 = \dot{q}_1$ and $i_2 = \dot{q}_2$. That is $q_1 = i_1/s$ and $q_2 = i_2/s$.

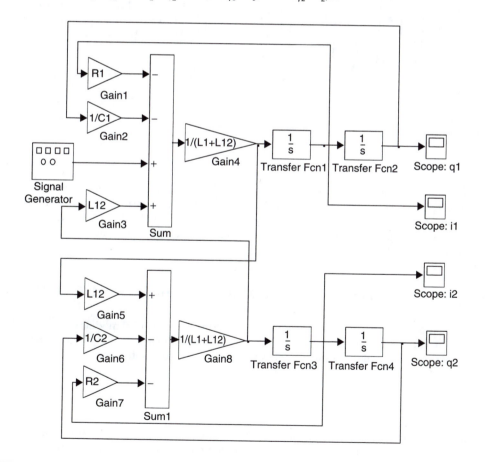

FIGURE 6.22
SIMULINK diagram to perform numerical simulations.

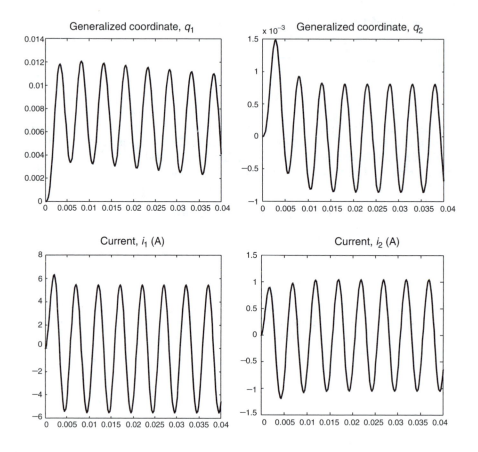

FIGURE 6.23
Circuit dynamics: evolution of the generalized coordinates and currents.

To perform simulations to attain analysis needed, the circuitry parameters are assigned to be $L_1 = 0.01$ H, $L_2 = 0.005$ H, $L_{12} = 0.0025$ H, $C_1 = 0.02$ F, $C_2 = 0.01$ F, $R_1 = 10$ ohm, $R_2 = 5$ ohm, and $u_a = 100\sin(200t)$ V. These parameters are downloaded in the Command Window. The simulation results, which give the time history of $q_1(t)$, $q_2(t)$, $i_1(t)$ and $i_2(t)$, are documented in Figure 6.23.

Example 6.21   MATHEMATICAL MODEL of an ELECTRIC CIRCUIT

Using the Lagrange equations of motion develop the mathematical models for the circuit shown in the Figure 6.24. Prove that the model derived using the Lagrange equations of motion is equivalent to the model developed using Kirchhoff's law.

SOLUTION

Using q_1 and q_2 as the independent generalized coordinates (charges in the first and second loops), the Lagrange equations of motion can be found. Here $i_a = \dot{q}_1$ and $i_L = \dot{q}_2$.

The generalized force applied to the system is denoted as Q_1, and $u_a(t) = Q_1$.

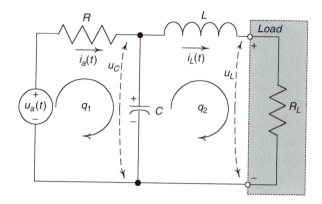

FIGURE 6.24
Electric circuit.

The total kinetic energy is given as

$$\Gamma = \tfrac{1}{2}L\dot{q}_2^2$$

Therefore, we have

$$\frac{\partial \Gamma}{\partial q_1} = 0, \quad \frac{\partial \Gamma}{\partial \dot{q}_1} = 0, \quad \text{and} \quad \frac{d}{dt}\left(\frac{\partial \Gamma}{\partial \dot{q}_1}\right) = 0$$

$$\frac{\partial \Gamma}{\partial q_2} = 0, \quad \frac{\partial \Gamma}{\partial \dot{q}_2} = L\dot{q}_2, \quad \text{and} \quad \frac{d}{dt}\left(\frac{\partial \Gamma}{\partial \dot{q}_2}\right) = L\ddot{q}_2$$

The total potential energy is expressed as

$$\Pi = \tfrac{1}{2}\frac{(q_1 - q_2)^2}{C}$$

Hence, one obtains

$$\frac{\partial \Pi}{\partial q_1} = \frac{q_1 - q_2}{C} \quad \text{and} \quad \frac{\partial \Pi}{\partial q_2} = \frac{-q_1 + q_2}{C}$$

The total dissipated energy is

$$D = \tfrac{1}{2}R\dot{q}_1^2 + \tfrac{1}{2}R_L\dot{q}_2^2$$

Therefore,

$$\frac{\partial D}{\partial \dot{q}_1} = R\dot{q}_1 \quad \text{and} \quad \frac{\partial D}{\partial \dot{q}_2} = R_L\dot{q}_2$$

The Lagrange equations of motion result. In particular, two equations

$$\frac{d}{dt}\left(\frac{\partial \Gamma}{\partial \dot{q}_1}\right) - \frac{\partial \Gamma}{\partial q_1} + \frac{\partial D}{\partial \dot{q}_1} + \frac{\partial \Pi}{\partial q_1} = Q_1$$

and

$$\frac{d}{dt}\left(\frac{\partial \Gamma}{\partial \dot{q}_2}\right) - \frac{\partial \Gamma}{\partial q_2} + \frac{\partial D}{\partial \dot{q}_2} + \frac{\partial \Pi}{\partial q_2} = 0$$

lead one to the following two differential equations:

$$R\dot{q}_1 + \frac{q_1 - q_2}{C} = u_a$$

$$L\ddot{q}_2 + R_L\dot{q}_2 + \frac{-q_1 + q_2}{C} = 0$$

Hence, we have found a set of two differential equations:

$$\dot{q}_1 = \frac{1}{R}\left(\frac{-q_1 + q_2}{C} + u_a\right)$$

$$\ddot{q}_2 = \frac{1}{L}\left(-R_L\dot{q}_2 + \frac{q_1 - q_2}{C}\right)$$

By using Kirchhoff's law, two differential equations result:

$$\frac{du_C}{dt} = \frac{1}{C}\left(-\frac{u_C}{R} - i_L + \frac{u_a(t)}{R}\right)$$

$$\frac{di_L}{dt} = \frac{1}{L}(u_C - R_L i_L)$$

Taking note of $i_a = \dot{q}_1$ and $i_L = \dot{q}_2$, and making use of $C(du_c/dt) = i_a - i_L$, we obtain

$$u_C = \frac{q_1 - q_2}{C}$$

The equivalence of the differential equations derived using the Lagrange equations of motion and Kirchhoff's law is proven.

Example 6.22   MATHEMATICAL MODEL of a BOOST CONVERTER

A high-frequency one-quadrant boost (step-up) dc-dc switching converter is documented in Figure 6.25. Find the mathematical model in the form of differential equations.

SOLUTION

To develop a mathematical model, we will apply Kirchhoff's laws and Lagrange equations of motion. It is evident that the majority are more familiar and get used to Kirchhoff's laws. However, to develop an augmented mathematical model for MEMS, the reader should be used to Lagrange equations of motion.

The differential equations that describe the circuitry dynamics will be developed assuming that the duty ratio (cycle) d_D is 1 and 0. In fact, using the on and off times, one has $d_D = t_{on}/(t_{on} + t_{off})$. Thus, $0 \leq d_D \leq 1$. One controls t_{on} or t_{off}, changing u_s. We consider two cases, e.g., the switch (transistor) is on or off. Then we will augment two sets of differential equation to obtain a general mathematical model to describe the boost converter dynamic and steady-state performance.

When the switch (transistor) S is closed, the diode D is reverse biased. For $d_D = 1$ ($t_{off} = 0$), one obtains the following set of linear differential equations using Kirchhoff's second law:

$$\frac{du_C}{dt} = -\frac{1}{C}i_a$$

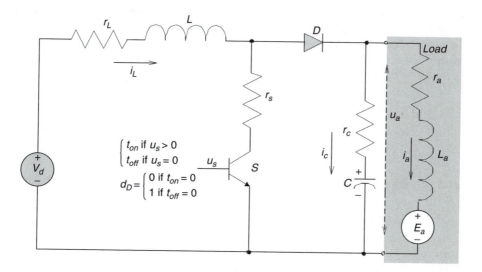

FIGURE 6.25
Boost converter.

$$\frac{di_L}{dt} = \frac{1}{L}(-(r_L + r_s)i_L + V_d)$$

$$\frac{di_a}{dt} = \frac{1}{L_a}(u_C - (r_a + r_c)i_a - E_a)$$

If the switch is open ($d_D = 0$), the diode D is forward biased because the direction of the inductor current i_L does not change instantly. Therefore, one has three linear differential equations:

$$\frac{du_C}{dt} = \frac{1}{C}(i_L - i_a)$$

$$\frac{di_L}{dt} = \frac{1}{L}(-u_C - (r_L + r_c)i_L + r_c i_a + V_d)$$

$$\frac{di_a}{dt} = \frac{1}{L_a}(u_C + r_c i_L - (r_a + r_c)i_a - E_a)$$

Assuming the switching frequency is high, the averaging concept is applied, and we have the resulting set of differential equations:

$$\frac{du_C}{dt} = \frac{1}{C}(i_L - i_a - i_L d_D)$$

$$\frac{di_L}{dt} = \frac{1}{L}(-u_C - (r_L + r_c)i_L + r_c i_a + u_C d_D + (r_c - r_s)i_L d_D - r_c i_a d_D + V_d)$$

$$\frac{di_a}{dt} = \frac{1}{L_a}(u_C + r_c i_L - (r_a + r_c)i_a - r_c i_L d_D - E_a)$$

In these equations, the duty ratio varies as $0 \le d_D \le 1$ in order to regulate the output voltage u_a. Thus, the output voltage is controlled by changing the duty ratio by regulating the on and off time of the switch. For example, if $t_{on} = t_{off}$, then $d_D = 0.5$, and $d_D = t_{on}/(t_{on} + t_{off})$. Considering the duty ratio as the control input, one concludes that a set of nonlinear differential equations result.

Let us illustrate that Lagrange's concept gives the same differential equations. We use the electric charges in the first and the second loops q_1 and q_2 as the independent generalized coordinates. The generalized forces are Q_1 and Q_2. Then we have a set of two equations:

$$\frac{d}{dt}\left(\frac{\partial \Gamma}{\partial \dot{q}_1}\right) - \frac{\partial \Gamma}{\partial q_1} + \frac{\partial D}{\partial \dot{q}_1} + \frac{\partial \Pi}{\partial q_1} = Q_1$$

$$\frac{d}{dt}\left(\frac{\partial \Gamma}{\partial \dot{q}_2}\right) - \frac{\partial \Gamma}{\partial q_2} + \frac{\partial D}{\partial \dot{q}_2} + \frac{\partial \Pi}{\partial q_2} = Q_2$$

For the closed switch, the total kinetic, potential, and dissipated energies are found to be

$$\Gamma = \tfrac{1}{2}\left(L\dot{q}_1^2 + L_a\dot{q}_2^2\right), \quad \Pi = \tfrac{1}{2}\frac{q_2^2}{C}, \quad \text{and} \quad D = \tfrac{1}{2}\left((r_L + r_s)\dot{q}_1^2 + (r_c + r_a)\dot{q}_2^2\right)$$

Assuming that the resistances, inductances, and capacitance are time invariant (constant), one obtains the following expressions for derivatives:

$$\frac{\partial \Gamma}{\partial q_1} = 0, \quad \frac{\partial \Gamma}{\partial q_2} = 0, \quad \frac{\partial \Gamma}{\partial \dot{q}_1} = L\dot{q}_1, \quad \frac{\partial \Gamma}{\partial \dot{q}_2} = L_a\dot{q}_2$$

$$\frac{d}{dt}\left(\frac{\partial \Gamma}{\partial \dot{q}_1}\right) = L\ddot{q}_1, \quad \frac{d}{dt}\left(\frac{\partial \Gamma}{\partial \dot{q}_2}\right) = L_a\ddot{q}_2$$

$$\frac{\partial \Pi}{\partial q_1} = 0, \quad \frac{\partial \Pi}{\partial q_2} = \frac{q_2}{C}$$

$$\frac{\partial D}{\partial \dot{q}_1} = (r_L + r_s)\dot{q}_1, \quad \frac{\partial D}{\partial \dot{q}_2} = (r_c + r_a)\dot{q}_2$$

Therefore,

$$L\ddot{q}_1 + (r_L + r_s)\dot{q}_1 = Q_1$$

$$L_a\ddot{q}_2 + (r_c + r_a)\dot{q}_2 + \frac{1}{C}q_2 = Q_2$$

and thus, if the switch is closed, the differential equations are

$$\ddot{q}_1 = \frac{1}{L}(-(r_L + r_s)\dot{q}_1 + Q_1)$$

$$\ddot{q}_2 = \frac{1}{L_a}\left(-(r_c + r_a)\dot{q}_2 - \frac{1}{C}q_2 + Q_2\right)$$

The total kinetic, potential, and dissipated energies if the switch is open are found to be

$$\Gamma = \tfrac{1}{2}\left(L\dot{q}_1^2 + L_a\dot{q}_2^2\right), \quad \Pi = \tfrac{1}{2}\frac{(q_1 - q_2)^2}{C}, \quad \text{and} \quad D = \tfrac{1}{2}\left(r_L\dot{q}_1^2 + r_c(\dot{q}_1 - \dot{q}_2)^2 + r_a\dot{q}_2^2\right)$$

Thus, for derivatives one has

$$\frac{\partial \Gamma}{\partial q_1} = 0, \quad \frac{\partial \Gamma}{\partial q_2} = 0, \quad \frac{\partial \Gamma}{\partial \dot{q}_1} = L\dot{q}_1, \quad \frac{\partial \Gamma}{\partial \dot{q}_2} = L_a\dot{q}_2$$

$$\frac{d}{dt}\left(\frac{\partial \Gamma}{\partial \dot{q}_1}\right) = L\ddot{q}_1, \quad \frac{d}{dt}\left(\frac{\partial \Gamma}{\partial \dot{q}_2}\right) = L_a\ddot{q}_2$$

$$\frac{\partial \Pi}{\partial q_1} = \frac{q_1 - q_2}{C}, \qquad \frac{\partial \Pi}{\partial q_2} = -\frac{q_1 - q_2}{C}$$

$$\frac{\partial D}{\partial \dot{q}_1} = (r_L + r_c)\dot{q}_1 - r_c \dot{q}_2, \qquad \frac{\partial D}{\partial \dot{q}_2} = -r_c \dot{q}_1 + (r_c + r_a)\dot{q}_2$$

Using

$$L\ddot{q}_1 + (r_L + r_c)\dot{q}_1 - r_c \dot{q}_2 + \frac{q_1 - q_2}{C} = Q_1$$

and

$$L_a \ddot{q}_2 - r_c \dot{q}_1 + (r_c + r_a)\dot{q}_2 - \frac{q_1 - q_2}{C} = Q_2$$

one has the following differential equations for the open switch:

$$\ddot{q}_1 = \frac{1}{L}\left(-(r_L + r_c)\dot{q}_1 + r_c \dot{q}_2 - \frac{q_1 - q_2}{C} + Q_1 \right)$$

$$\ddot{q}_2 = \frac{1}{L_a}\left(r_c \dot{q}_1 - (r_c + r_a)\dot{q}_2 + \frac{q_1 - q_2}{C} + Q_2 \right)$$

It must be emphasized that $i_L = \dot{q}_1$, $i_a = \dot{q}_2$, and $Q_1 = V_d$, $Q_2 = -E_a$.

Taking note of the differential equations when the switch is closed and open, the differential equations in Cauchy's form are found using

$$\frac{dq_1}{dt} = i_L \quad \text{and} \quad \frac{dq_2}{dt} = i_a$$

The duty ratio should be used. The voltage across the capacitor u_c is expressed using the charges q_1 and q_2. When the switch is closed, $u_c = -(q_2/C)$. If the switch is open, $u_C = (q_1 - q_2)/C$. The analysis of the differential equations derived using Kirchhoff's voltage law and the Lagrange equations of motion illustrates that the mathematical models are found using different state variables. In particular, u_C, i_L, i_a and q_1, i_L, q_2, i_a are used. However, the resulting differential equations are the same as one applies the corresponding variable transformations given by

$$\frac{dq_1}{dt} = i_L, \qquad \frac{dq_2}{dt} = i_a, \qquad Q_1 = V_d, \quad \text{and} \quad Q_2 = -E_a$$

Example 6.23 MATHEMATICAL MODEL of a MICROTRANSDUCER in MICROROBOT APPLICATION

Consider a microtransducer with two independently excited stator and rotor windings. (See Figure 6.26.) Derive the differential equations.

SOLUTION

The following notations are used for the microtransducer variables and parameters: i_s and i_r are the currents in the stator and rotor windings; u_s and u_r are the applied voltages to the stator and rotor windings; ω_r and θ_r are the rotor angular velocity and displacement; T_e and T_L are the electromagnetic and load torques; r_s and r_r are the resistances of the stator and rotor windings; L_s and L_r are the self-inductances of the stator and rotor

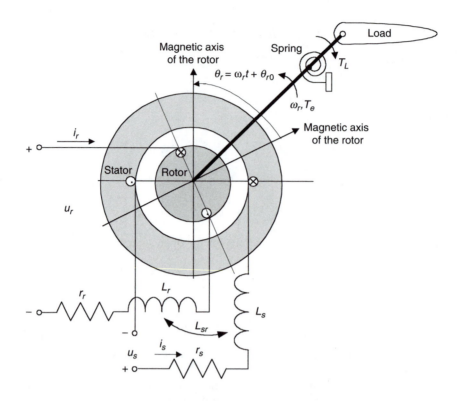

FIGURE 6.26
Microtransducer with stator and rotor windings.

windings; L_{sr} is the mutual inductance of the stator and rotor windings; $\mathfrak{R}_m$ is the reluctance of the magnetizing path; N_s and N_r are the number of turns in the stator and rotor windings; J is the moment of inertia of the rotor and attached load; B_m is the viscous friction coefficient; and k_s is the spring constant.

The magnetic fluxes that cross the air gap produce a force of attraction, and the developed electromagnetic torque T_e is countered by the tortional spring, which causes a counter-clockwise rotation. The load torque T_L should be considered. Our goal is to find a nonlinear mathematical model. In fact, the ability to formulate the modeling problem and find the resulting equations that describe a motion device constitute the most important issues.

By using the Lagrange concept, the independent generalized coordinates must be chosen. We use q_1, q_2, and q_3, where q_1 and q_2 denote the electric charges in the stator and rotor windings, and q_3 represents the rotor angular displacement.

We denote the generalized forces applied to an electromechanical system as Q_1, Q_2, and Q_3, where Q_1 and Q_2 are the applied voltages to the stator and rotor windings, and Q_3 is the load torque.

The first derivative of the generalized coordinates $\dot{q}_1$ and $\dot{q}_2$ represent the stator and rotor currents i_s and i_r, while $\dot{q}_3$ is the angular velocity of the rotor ω_r. We have

$$q_1 = \frac{i_s}{s}, \quad q_2 = \frac{i_r}{s}, \quad q_3 = \theta_r, \quad \dot{q}_1 = i_s, \quad \dot{q}_2 = i_r, \quad \text{and} \quad \dot{q}_3 = \omega_r,$$

$$Q_1 = u_s, \quad Q_2 = u_r, \quad \text{and} \quad Q_3 = -T_L$$

The Lagrange equations are expressed in terms of each independent coordinate, and we have

$$\frac{d}{dt}\left(\frac{\partial \Gamma}{\partial \dot{q}_1}\right) - \frac{\partial \Gamma}{\partial q_1} + \frac{\partial D}{\partial \dot{q}_1} + \frac{\partial \Pi}{\partial q_1} = Q_1$$

$$\frac{d}{dt}\left(\frac{\partial \Gamma}{\partial \dot{q}_2}\right) - \frac{\partial \Gamma}{\partial q_2} + \frac{\partial D}{\partial \dot{q}_2} + \frac{\partial \Pi}{\partial q_2} = Q_2$$

$$\frac{d}{dt}\left(\frac{\partial \Gamma}{\partial \dot{q}_3}\right) - \frac{\partial \Gamma}{\partial q_3} + \frac{\partial D}{\partial \dot{q}_3} + \frac{\partial \Pi}{\partial q_3} = Q_3$$

The total kinetic energy of electrical and mechanical systems is found as a sum of the total magnetic (electrical) Γ_E and mechanical Γ_M energies. The total kinetic energy of the stator and rotor circuitry is given as

$$\Gamma_E = \tfrac{1}{2}L_s \dot{q}_1^2 + L_{sr}\dot{q}_1\dot{q}_2 + \tfrac{1}{2}L_r \dot{q}_2^2$$

The total kinetic energy of the mechanical system, which is a function of the equivalent moment of inertia of the rotor and the payload attached, is expressed by

$$\Gamma_M = \tfrac{1}{2}J\dot{q}_3^2$$

Then we have

$$\Gamma = \Gamma_E + \Gamma_M = \tfrac{1}{2}L_s \dot{q}_1^2 + L_{sr}\dot{q}_1\dot{q}_2 + \tfrac{1}{2}L_r \dot{q}_2^2 + \tfrac{1}{2}J\dot{q}_3^2$$

The mutual inductance is a periodic function of the angular rotor displacement, and

$$L_{sr}(\theta_r) = \frac{N_s N_r}{\Re_m(\theta_r)}$$

The magnetizing reluctance is maximum if the stator and rotor windings are not displaced, and $\Re_m(\theta_r)$ is minimum if the coils are displaced by 90 degrees. Then $L_{sr\,min} \le L_{sr}(\theta_r) \le L_{sr\,max}$, where

$$L_{sr\,max} = \frac{N_s N_r}{\Re_m(90°)} \quad \text{and} \quad L_{sr\,min} = \frac{N_s N_r}{\Re_m(0°)}$$

The mutual inductance can be approximated as a cosine function of the rotor angular displacement. The amplitude of the mutual inductance between the stator and rotor windings is found as

$$L_M = L_{sr\,max} = \frac{N_s N_r}{\Re_m(90°)}$$

Then, in the attempt to derive the analytic expression for the mutual inductance L_{sr} (we should perform differentiation for energies, so the energy terms must be differentiated), we assume that it varies as a cosine function. Therefore,

$$L_{sr}(\theta_r) = L_M \cos\theta_r = L_M \cos q_3$$

One obtains an explicit expression for the total kinetic energy as

$$\Gamma = \tfrac{1}{2}L_s \dot{q}_1^2 + L_M \dot{q}_1\dot{q}_2 \cos q_3 + \tfrac{1}{2}L_r \dot{q}_2^2 + \tfrac{1}{2}J\dot{q}_3^2$$

The following partial derivatives result:

$$\frac{\partial \Gamma}{\partial q_1} = 0, \qquad \frac{\partial \Gamma}{\partial \dot{q}_1} = L_s \dot{q}_1 + L_M \dot{q}_2 \cos q_3$$

$$\frac{\partial \Gamma}{\partial q_2} = 0, \qquad \frac{\partial \Gamma}{\partial \dot{q}_2} = L_M \dot{q}_1 \cos q_3 + L_r \dot{q}_2$$

$$\frac{\partial \Gamma}{\partial q_3} = -L_M \dot{q}_1 \dot{q}_2 \sin q_3, \quad \frac{\partial \Gamma}{\partial \dot{q}_3} = J \dot{q}_3$$

The potential energy of the spring with constant k_s is

$$\Pi = \tfrac{1}{2} k_s q_3^2$$

Therefore, we have $\partial \Pi / \partial q_1 = 0$, $\partial \Pi / \partial q_2 = 0$, and $\partial \Pi / \partial q_3 = k_s q_3$.

The total heat energy dissipated is expressed as

$$D = D_E + D_M$$

where D_E is the heat energy dissipated in the stator and rotor windings, $D_E = \tfrac{1}{2} r_s \dot{q}_1^2 + \tfrac{1}{2} r_r \dot{q}_2^2$; and D_M is the heat energy dissipated by mechanical system, $D_M = \tfrac{1}{2} B_m \dot{q}_3^2$.

Hence, making use of

$$D = \tfrac{1}{2} r_s \dot{q}_1^2 + \tfrac{1}{2} r_r \dot{q}_2^2 + \tfrac{1}{2} B_m \dot{q}_3^2$$

we obtain

$$\frac{\partial D}{\partial \dot{q}_1} = r_s \dot{q}_1, \qquad \frac{\partial D}{\partial \dot{q}_2} = r_r \dot{q}_2, \quad \text{and} \quad \frac{\partial D}{\partial \dot{q}_3} = B_m \dot{q}_3$$

Using the following relationships between the generalized coordinates and state variables

$$q_1 = \frac{i_s}{s}, \qquad q_2 = \frac{i_r}{s}, \qquad q_3 = \theta_r, \qquad \dot{q}_1 = i_s, \qquad \dot{q}_2 = i_r, \qquad \dot{q}_3 = \omega_r,$$

$$Q_1 = u_s, \quad Q_2 = u_r \quad \text{and} \quad Q_3 = -T_L$$

we have three differential equations for the considered microtransducer:

$$L_s \frac{di_s}{dt} + L_M \cos\theta_r \frac{di_r}{dt} - L_M i_r \sin\theta_r \frac{d\theta_r}{dt} + r_s i_s = u_s$$

$$L_r \frac{di_r}{dt} + L_M \cos\theta_r \frac{di_s}{dt} - L_M i_s \sin\theta_r \frac{d\theta_r}{dt} + r_r i_r = u_r$$

$$J \frac{d^2\theta_r}{dt^2} + L_M i_s i_r \sin\theta_r + B_m \frac{d\theta_r}{dt} + k_s \theta_r = -T_L$$

The last equation should be rewritten by making use of the rotor angular velocity:

$$\frac{d\theta_r}{dt} = \omega_r$$

Using the stator and rotor currents, angular velocity, and displacement as the state variables, the nonlinear differential equations in Cauchy's form are

$$\frac{di_s}{dt} = \frac{-r_s L_r i_s - \frac{1}{2} L_M^2 i_s \omega_r \sin 2\theta_r + r_r L_M i_r \cos\theta_r + L_r L_M i_r \omega_r \sin\theta_r + L_r u_s - L_M \cos\theta_r u_r}{L_s L_r - L_M^2 \cos^2\theta_r}$$

$$\frac{di_r}{dt} = \frac{r_s L_M i_s \cos\theta_r + L_s L_M i_s \omega_r \sin\theta_r - r_r L_s i_r - \frac{1}{2} L_M^2 i_r \omega_r \sin 2\theta_r - L_M \cos\theta_r u_s + L_s u_r}{L_s L_r - L_M^2 \cos^2\theta_r}$$

$$\frac{d\omega_r}{dt} = \frac{1}{J}(-L_M i_s i_r \sin\theta_r - B_m \omega_r - k_s \theta_r - T_L)$$

$$\frac{d\theta_r}{dt} = \omega_r$$

The developed nonlinear mathematical model in the form of highly coupled nonlinear differential equations cannot be linearized, and one must model the microtransducer studied using a complete set of four nonlinear differential equations.

6.4.3 Hamilton Equations of Motion

The Hamilton concept allows one to model system dynamics, and the differential equations that describe the system dynamics are found using the generalized momenta p_i: $p_i = \partial L / \partial \dot{q}_i$.

As was emphasized, the generalized coordinates were used in the Lagrange equations of motion to derive the mathematical models.

The Lagrangian function

$$L\left(t, q_1, \ldots, q_n, \frac{dq_1}{dt}, \ldots, \frac{dq_n}{dt}\right)$$

for the conservative systems is the difference between the total kinetic and potential energies. In particular, we can write

$$L\left(t, q_1, \ldots, q_n, \frac{dq_1}{dt}, \ldots, \frac{dq_n}{dt}\right) = \Gamma\left(t, q_1, \ldots, q_n, \frac{dq_1}{dt}, \ldots, \frac{dq_n}{dt}\right) - \Pi(t, q_1, \ldots, q_n)$$

It is evident that

$$L\left(t, q_1, \ldots, q_n, \frac{dq_1}{dt}, \ldots, \frac{dq_n}{dt}\right)$$

is the function of $2n$ independent variables. One has

$$dL = \sum_{i=1}^{n}\left(\frac{\partial L}{\partial q_i}dq_i + \frac{\partial L}{\partial \dot{q}_i}d\dot{q}_i\right) = \sum_{i=1}^{n}(\dot{p}_i dq_i + p_i d\dot{q}_i)$$

We define the Hamiltonian function as

$$H(t, q_1, \ldots, q_n, p_1, \ldots, p_n) = -L\left(t, q_1, \ldots, q_n, \frac{dq_1}{dt}, \ldots, \frac{dq_n}{dt}\right) + \sum_{i=1}^{n}p_i \dot{q}_i$$

$$dH = \sum_{i=1}^{n}(-\dot{p}_i dq_i + \dot{q}_i dp_i)$$

where

$$\sum_{i=1}^{n} p_i \dot{q}_i = \sum_{i=1}^{n} \frac{\partial L}{\partial \dot{q}_i} \dot{q}_i = \sum_{i=1}^{n} \frac{\partial \Gamma}{\partial \dot{q}_i} \dot{q}_i = 2\Gamma$$

Thus, we have

$$H\left(t, q_1, \ldots, q_n, \frac{dq_1}{dt}, \ldots, \frac{dq_n}{dt}\right) = \Gamma\left(t, q_1, \ldots, q_n, \frac{dq_1}{dt}, \ldots, \frac{dq_n}{dt}\right) + \Pi(t, q_1, \ldots, q_n)$$

or

$$H(t, q_1, \ldots, q_n, p_1, \ldots, p_n) = \Gamma(t, q_1, \ldots, q_n, p_1, \ldots, p_n) + \Pi(t, q_1, \ldots, q_n)$$

One concludes that the Hamiltonian function, which is equal to the total energy, is expressed as a function of the generalized coordinates and generalized momenta.

The equations of motion are governed by the following equations

$$\dot{p}_i = -\frac{\partial H}{\partial q_i}$$

$$\dot{q}_i = \frac{\partial H}{\partial p_i}$$

(6.4)

These equations are called the Hamiltonian equations of motion.

It is evident that using Hamiltonian mechanics one obtains the system of $2n$ first-order partial differential equations to model the system dynamics. In contrast, using the Lagrange equations of motion, the system of n second-order differential equations results. However, the derived differential equations are equivalent.

Example 6.24

Consider the harmonic oscillator. The total energy is given as the sum of the kinetic and potential energies, $\Sigma_T = \Gamma + \Pi = \frac{1}{2}(mv^2 + k_s x^2)$. Find the equations of motion using the Lagrange and Hamiltonian concepts.

SOLUTION

The Lagrangian function is

$$L\left(x, \frac{dx}{dt}\right) = \Gamma - \Pi = \frac{1}{2}(mv^2 - k_s x^2) = \frac{1}{2}(m\dot{x}^2 - k_s x^2)$$

Making use of the Lagrange equations of motion,

$$\frac{d}{dt}\frac{\partial L}{\partial \dot{x}} - \frac{\partial L}{\partial x} = 0$$

we have the following equation of motion:

$$m\frac{d^2 x}{dt^2} + k_s x = 0$$

From Newton's second law, the second-order differential equation of motion is given as

$$m\frac{d^2x}{dt^2} + k_s x = 0$$

The Hamiltonian function is expressed as

$$H(x, p) = \Gamma + \Pi = \tfrac{1}{2}(mv^2 - k_s x^2) = \tfrac{1}{2}\left(\frac{1}{m}p^2 - k_s x^2\right)$$

From the Hamiltonian equations of motion,

$$\dot{p}_i = -\frac{\partial H}{\partial q_i} \text{ and } \dot{q}_i = \frac{\partial H}{\partial p_i}$$

as given by Equation 6.4, one obtains

$$\dot{p} = -\frac{\partial H}{\partial x} = -k_s x$$

$$\dot{x} = \dot{q} = \frac{\partial H}{\partial p} = \frac{p}{m}$$

The equivalence of the results and equations of motion is obvious.

6.5 Direct-Current Micromachines

It has been shown that the basic electromagnetic principles and fundamental physical laws are used to design motion micro- and nanostructures. Micro- and nanoengineering leverages from conventional theory of electromechanical motion devices, electromagnetics, integrated circuits, and quantum mechanics. The fabrication of motion microstructures is based upon CMOS- and LIGA-based technologies, and rotational and translational transducers (actuators and sensors) have been manufactured and tested. The major challenge is the difficulty in fabricating brushes and windings for direct-current microdevices (electric micromachines), reliability and ruggedness (due to bearing problems, vibration, heat, and other phenomena), etc. It appears that novel fabrication technologies allow one to overcome many challenges. The most efficient class of microtransducers to be used as MEMS motion microdevices are induction and synchronous. These microtransducers do not have a collector. The stator and rotor windings, as well as permanent-magnets, have been manufactured and tested for synchronous (permanent-magnet synchronous and stepper micromotors) and induction micromachines. Direct-current microtransducers are not the preferable choice. However, these micromachines will be covered first because they are simple from an analysis standpoint, and, in addition, students and engineers are familiar with these transducers. Furthermore, even using the micromachining technologies, mini-/microscale permanent-magnet DC transducers (1–2 mm in diameter and 2–6 mm long) were commercialized and manufactured for pagers, phones, cameras, stages, etc. It must be emphasized that due to the limits imposed on the torque, force, and power densities, the micromachine sizing is defined by the required torque, force, or power needed to be produced. Therefore, though it is possible to design and fabricate micromachines, bigger machines are needed due to specifications imposed.

In this section we develop mathematical models of the permanent-magnet DC micro-machines [5, 10, 11], discuss the operating principles, and give the basic architectures of MEMS with DC micromachines.

The list of basic variables and symbols used in this section is given below:

i_a is the currents in the armature winding.

u_a is the applied voltages to the armature windings.

ω_r and θ_r are the angular velocity and angular displacement of the rotor.

E_a is the electromotive force.

T_e and T_L are the electromagnetic and load torques.

r_a is the resistances of the armature windings.

L_a is the self-inductances of the armature windings.

B_m is the viscous friction coefficient.

J is the equivalent moment of inertia of the rotor and attached load.

Permanent-magnet microtransducers are rotating energy-transfer electromechanical motion devices that convert energy by means of rotational motion. Transducers are the major part of MEMS, and therefore they must be thoroughly studied with the driving ICs. Micro-motors (actuators) convert electrical energy to mechanical energy, while generators (sensors) convert mechanical energy to electrical energy. It is worth mentioning that the same perma-nent-magnet microtransducers can be used as the actuator-motor (if one applies the voltage) or as the sensor-generator. (If one rotates the transducers, the voltage is induced.) Hence, the energy conversion is reversible, and sensors can be operated as actuators and vice versa. That is, permanent-magnet DC microtransducers can be used as the actuators (motors) and sensors (gyroscopes, tachogenerators, etc.). Transducers have stationary and rotating mem-bers, separated by an air gap. The armature winding is placed in the rotor slots and connected to a rotating commutator, which rectifies the induced voltage. (See Figure 6.27.) One supplies the armature voltages to the armature (rotor) windings. The rotor windings and stator permanent magnets are magnetically coupled.

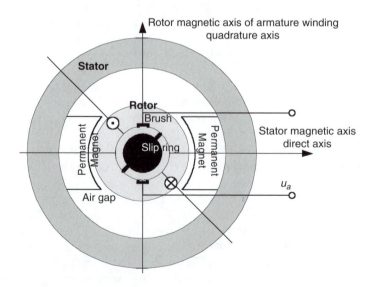

FIGURE 6.27
Permanent-magnet DC microtransducers.

The brushes, which are connected to the armature windings, ride on the commutator. The armature winding consists of identical uniformly distributed coils. The excitation magnetic field is produced by the permanent magnets. Due to the commutator (circular conducting segments), armature windings and permanent magnets produce stationary magnetomotive forces that are displaced by 90 electrical degrees. The armature magnetic force is along the quadrature (rotor) magnetic axis, while the direct axis stands for a permanent-magnet magnetic axis. The electromagnetic torque is produced as a result of the interaction of these stationary magnetomotive forces.

From Kirchhoff's law, one obtains the following steady-state equation for the armature voltage for micromotors:

$$u_a - E_a = i_a r_a$$

(The armature current opposes the induced electromotive force.)

For microgenerators, the armature current is in the same direction as the generated electromotive force, and we have

$$u_a - E_a = -i_a r_a$$

The difference between the applied voltage and the induced electromotive force is the voltage drop across the internal armature resistance r_a. One concludes that transducers rotate at an angular velocity at which the electromotive force generated in the armature winding balances the armature voltage. If a microtransducer operates as a micromotor, the induced electromotive force is less than the voltage applied to the windings. If a microtransducer operates as a generator, the generated (induced) electromotive force is greater than the terminal voltage.

The constant magnetic flux in AC and DC microtransducers are produced by permanent magnets. Microtransducers with permanent-magnet poles are called permanent-magnet microtransducers. The permanent-magnet DC microtransducer was illustrated in Figure 6.27, and a schematic diagram of permanent-magnet DC microtransducers is shown in Figure 6.28.

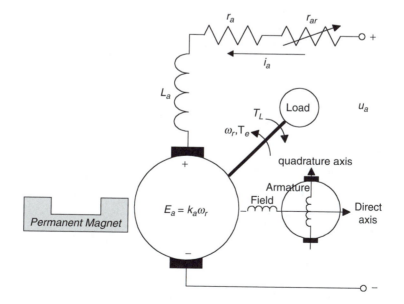

FIGURE 6.28

Schematic diagram of permanent-magnet microtransducers. (Current direction corresponds to the motor operation.)

Using Kirchhoff's voltage law and Newton's second law of motion, the differential equations for permanent-magnet DC microtransducers can be easily derived. Assuming that the susceptibility is constant (in reality, Curie's constant varies as a function of temperature), one supposes that the flux, established by the permanent-magnet poles, is constant. Then, denoting the back emf and torque constants as k_a, we have the following linear differential equations describing the transient behavior of the armature winding and torsional-mechanical dynamics:

$$\frac{di_a}{dt} = -\frac{r_a}{L_a}i_a - \frac{k_a}{L_a}\omega_r + \frac{1}{L_a}u_a$$

$$\frac{d\omega_r}{dt} = \frac{k_a}{J}i_a - \frac{B_m}{J}\omega_r - \frac{1}{J}T_L$$

(6.5)

The lumped-parameter model in the state-space (matrix) form, using Equation 6.5, is found as

$$\begin{bmatrix} \dfrac{di_a}{dt} \\ \dfrac{d\omega_r}{dt} \end{bmatrix} = \begin{bmatrix} -\dfrac{r_a}{L_a} & -\dfrac{k_a}{L_a} \\ \dfrac{k_a}{J} & -\dfrac{B_m}{J} \end{bmatrix} \begin{bmatrix} i_a \\ \omega_r \end{bmatrix} + \begin{bmatrix} \dfrac{1}{L_a} \\ 0 \end{bmatrix} u_a - \begin{bmatrix} 0 \\ \dfrac{1}{J} \end{bmatrix} T_L$$

(6.6)

An *s*-domain block diagram of permanent-magnet DC micromotors is illustrated in Figure 6.29.

The angular velocity can be reversed if the polarity of the applied voltage is changed. (The direction of the field flux cannot be changed.) The steady-state torque-speed characteristic curves obey the following equation:

$$\omega_r = \frac{u_a - r_a i_a}{k_a} = \frac{u_a}{k_a} - \frac{r_a}{k_a^2}T_e$$

and a spectrum of the torque-speed characteristic curves is illustrated in Figure 6.30. The illustrative example (Example 6.26 in the end of this section) as it is demonstrates the application of the results reported.

If the permanent-magnet DC microtransducer is used as the generator, the circuitry dynamics for the resistive load R_L are given as

$$\frac{di_a}{dt} = -\frac{r_a + R_L}{L_a}i_a + \frac{k_a}{L_a}\omega_r$$

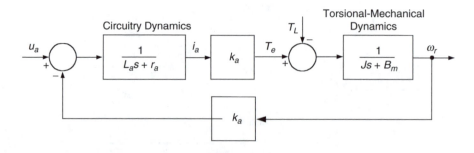

FIGURE 6.29
s-domain block diagram of permanent-magnet DC micromotors.

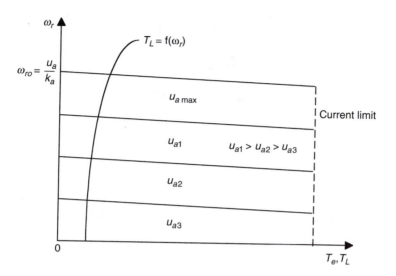

FIGURE 6.30
Torque-speed characteristics for permanent-magnet micromotors.

That is, in the steady-state, the armature current is proportional to the angular velocity, and we have

$$i_a = \frac{k_a}{r_a + R_L} \omega_r$$

Flip-chip MEMS found wide application due to low cost, and well-developed fabrication processes were applied. For example, monolithic dual power operational amplifiers, designed as single-chip ICs, feed DC micromotor to regulate the angular velocity. (See Figure 6.31.)

Motion microdevices (microtransducers—actuators and sensors) are mounted face down with bumps on the pads that form electrical and mechanical joints to the ICs driver. Figure 6.32 illustrates a flip-chip MEMS with permanent-magnet micromotor driven by the MC33030 monolithic ICs driver. Control algorithms are implemented to control the angular velocity of micromotors. The MC33030 integrates on-chip operational amplifier

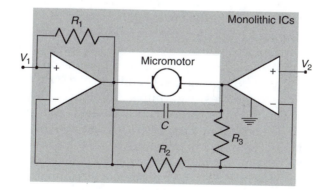

FIGURE 6.31
Application of a monolithic IC to control DC micromotors.

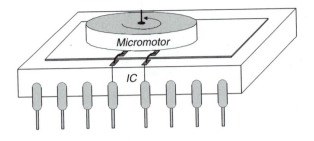

FIGURE 6.32
Flip-chip monolithic MEMS: MC33030 ICs and micromotor.

and comparator, driving and braking logic, PWM four-quadrant converter, etc. The MC33030 data and complete description are given in Reference 4. As in the conventional configurations, the difference between the reference (command) and actual angular velocity or displacement, linear velocity or position, is compared by the error amplifier, and two comparators are used. A pnp differential output power stage provides driving and braking capabilities, and the four-quadrant H-configured power stage guarantees high performance and efficiency. Using the error between the desired (command) and actual angular velocity or displacement, the bipolar voltage u_a is applied to the armature winding. The electromagnetic torque is developed, and micromotor rotates.

6.6   Simulation of MEMS in the MATLAB Environment with Examples

MATLAB® (MATrix LABoratory) is a high-performance interactive data-intensive software environment for high-efficiency engineering and scientific numerical calculations. This environment can be applied to perform heterogeneous simulations and data-intensive analyses of complex MEMS and NEMS, two- and three-dimensional plotting, graphics, visualization, control, optimization, etc.[5, 13–15] MATLAB enables the users to solve a wide spectrum of analytical and numerical problems using matrix-based methods, attain excellent interfacing and interactive capabilities, compile with high-level programming languages, ensure robustness, access and implement state-of-the-art numerical algorithms, guarantee powerful graphical features, etc. [16] Due to high flexibility and versatility, the MATLAB environment has been significantly enhanced and developed during recent years. A broad family of application-specific toolboxes, with a specialized collection of m-files for solving problems, guarantee comprehensiveness and effectiveness. For example, SIMULINK® is a companion graphical mouse-driven interactive environment enhancing MATLAB. A great number of books and MathWorks user manuals in MATLAB, SIMULINK, and different MATLAB toolboxes are available. In addition, the MathWorks, Inc. educational Web sites can be used as references (e.g., http://education.mathworks.com and http://www.mathworks.com). This book is intended to help students and engineers use MATLAB efficiently, showing and demonstrating how MATLAB and SIMULINK can be applied for MEMS and NEMS. The MATLAB environment (MATLAB 6.5, release 13) is covered in this book, and the Web site http://www.mathworks.com/access/helpdesk/help/helpdesk.shtml can assist users to master MATLAB. MATLAB documentation and user manuals (thousands of pages each) are available in the Portable Document Format using the Help Desk. This book focuses on MATLAB applications to micro- and nanosystems, showing one how to solve practical problems using step-by-step instructions.

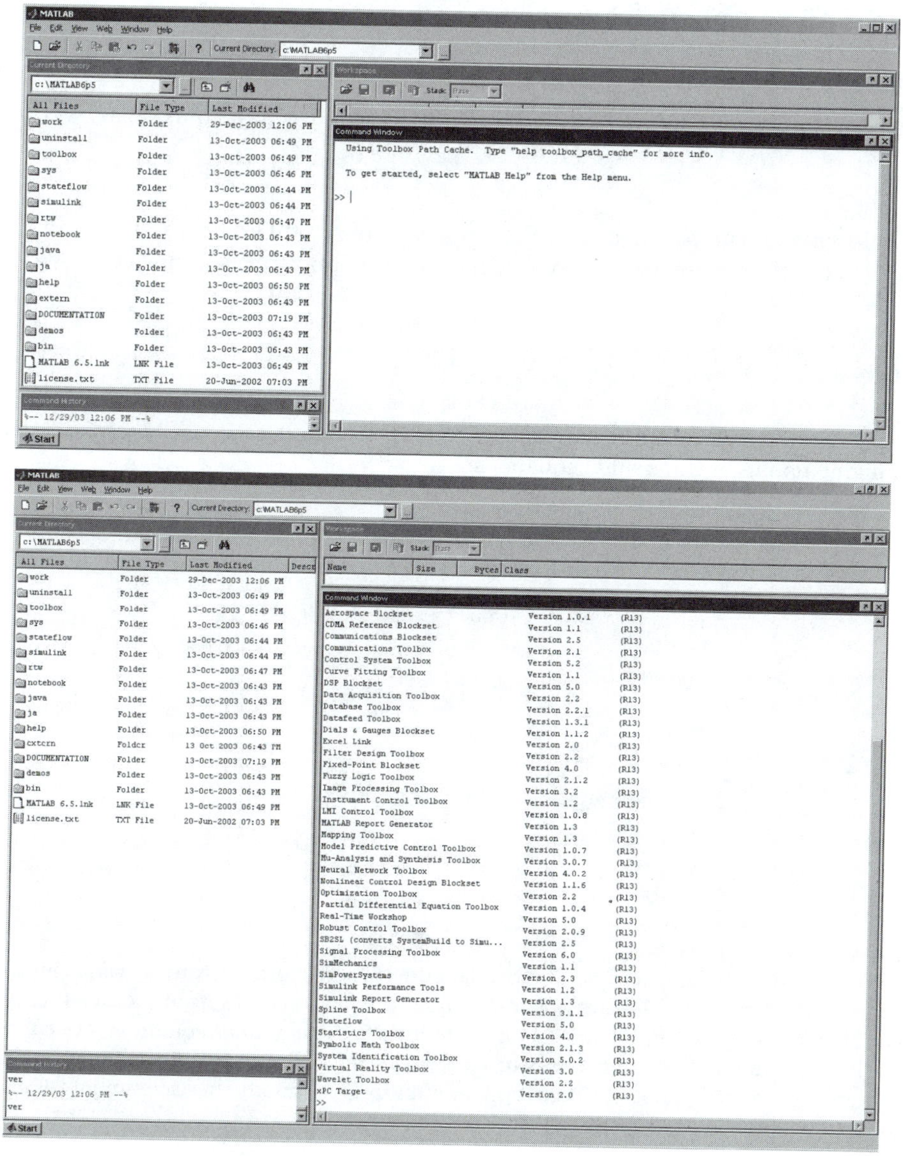

FIGURE 6.33
MATLAB Command window and MATLAB version with toolboxes available.

To start MATLAB, double-click the following MATLAB icon:

MATLAB 6.5 lnk

and the MATLAB Command window with Launch Pad and Command History appear
on the screen. (See Figure 6.33.)

The line

>>

is the MATLAB prompt.

Typing

```
>> a = 1 + 2 + 1 * 3
```

and pressing the Enter (Return) key, we have the value for a. In particular, we have

```
a =
    6
```

The following example illustrates the application of MATLAB to solve the ordinary differential equations to readers who may have some deficiencies in or have never applied MATLAB.

Example 6.25

Using the MATLAB ode45 solver (built-in ode45 function), numerically solve a system of highly nonlinear differential equations:

$$\frac{dx_1(t)}{dt} = -20x_1 + \left| x_2 x_3 \right| + 10x_1 x_2 x_3, \quad x_1(t_0) = x_{10}$$

$$\frac{dx_2(t)}{dt} = -5x_1 x_2 - 10\cos x_1 - \sqrt{\left| x_3 \right|}, \quad x_2(t_0) = x_{20}$$

$$\frac{dx_3(t)}{dt} = -5x_1 x_2 + 50x_2 \cos x_1 - 50x_3, \quad x_3(t_0) = x_{30}$$

The initial conditions are $x_0 = \begin{bmatrix} x_{10} \\ x_{20} \\ x_{30} \end{bmatrix} = \begin{bmatrix} 2 \\ 1 \\ -1 \end{bmatrix}$

SOLUTION

Two m-files (c6_2_1a.m and c6_2_1b.m) are developed in order to numerically simulate this set of three first-order nonlinear differential equations. The evolution of the state variables $x_1(t)$, $x_2(t)$, and $x_3(t)$ must be plotted as the differential equations are solved. To illustrate the transient responses of $x_1(t)$, $x_2(t)$, and $x_3(t)$, the plot function is used. Comments, which are not executed, appear after the % symbol. These comments explain sequential steps in MATLAB scripts.

The MATLAB script with ode45 solver, two-dimensional plotting statements using the plot function, and three-dimensional plotting statements using the plot3 function is listed below. (See c6_2_1a.m)

```
echo on; clear all
tspan=[0 0.7];   % initial and final time
y0=[2 1 -1]';  % initial conditions for state variables
[t,y]=ode45('c6_2_1b',tspan,y0);
% ode45 MATLAB solver using ode45 function
% Plot of the time history found solving three
% differential equations assigned in c6_2_1b.m
plot(t,y(:,1),'--',t,y(:,2),'-',t,y(:,3),':');
% plot the transient dynamics
```

```
xlabel('Time (seconds)'); ylabel('State Variables');
title('Solution of Differential Equations: x1(t), x2(t) and x3(t)');
pause
% 3-D plot usin x1, x2 and x3
plot3(y(:,1),y(:,2),y(:,3));
xlabel('x1'), ylabel('x2'), zlabel('x3')
text(15,-15,10,'x0 Initial'); text(0,0,0,'0 Origin')
v=axis; pause; disp('end')
```

The MATLAB script with the assigned set of differential equations to be numerically solved is (c6_2_1b.m):

```
function yprime = difer(t,y);
a11=-20; a12=1; a13=10; a21=-5; a22=-10; a31=-5; a32=50; a33=-50;
% differential equations coefficients
% Three differential equations (system of three first-order
% differential equations)
yprime=[a11*y(1,:)+a12*abs(y(2,:)*y(3,:))+a13*y(1,:)*y(2,:)* y(3,:);...
a21*y(1,:)*y(2,:)+a22*cos(y(1,:))+sqrt(abs(y(3,:)));...
a31*y(1,:)*y(2,:)+a32*cos(y(1,:))*y(2,:)+a33*y(3,:)];
```

To calculate the transient dynamics and plot the transient dynamics, one types in the Command window

```
>> c6_2_1a
```

and presses the Enter key. The resulting transient behavior (two-dimensional plot) is documented in Figure 6.34. The three-dimensional plot also illustrate the evolution of the state variable and are given in Figure 6.34.

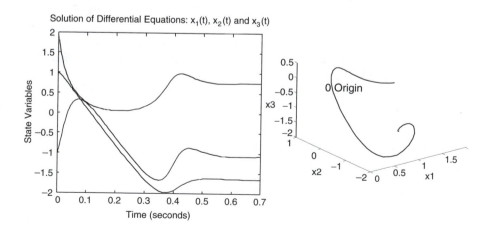

FIGURE 6.34
Evolution of the state variables.

After typing

```
>>   who, z=[t,y]
```

in the Command window, the variables used are displayed, e.g.,

```
Your variables are:
t       tspan   v       y       y0
```

The time evolutions of $x_1(t)$, $x_2(t)$, and $x_3(t)$ were plotted in Figure 6.34. The resulting data, which is displayed in the Command window, is reported below. In particular, we have four columns for time t, as well as for three state variables $x_1(t)$, $x_2(t)$, and $x_3(t)$.

```
z =
         0       2.0000    1.0000   -1.0000
    0.0017       1.9027    0.9915   -0.9643
    0.0034       1.8116    0.9823   -0.9232
    0.0051       1.7263    0.9726   -0.8776
    0.0068       1.6466    0.9622   -0.8283
    0.0107       1.4840    0.9365   -0.7070
    0.0146       1.3442    0.9082   -0.5796
..........
    0.6880       0.7579   -1.6497   -1.0759
    0.6940       0.7583   -1.6495   -1.0751
    0.7000       0.7586   -1.6493   -1.0744
```

Having reported the illustrative example numerically solving highly nonlinear differential equations in MATLAB, we will focus our attention on the numerical simulation of the permanent-magnet DC micromachines.

One can straightforwardly modify the provided files to model permanent-magnet DC micromachines. The next example illustrates the application of state-space modeling and simulation.

The state-space mathematical model of permanent-magnet DC micromotors is given by Equation 6.6. One can apply this state-space description to perform simulations and analysis.

The simulation parameters (final time and initial conditions) as well as micromachine parameters must be assigned. We will perform simulations with zero initial conditions and final time of 0.00015 sec. We set the following micromotor coefficients: $r_a = 200$, $k_a = 0.2$, $L_a = 0.002$, $B_m = 0.00000001$, and $J = 0.00000001$. In the Command window we download these parameters as

```
>> tfinal=0.00015; x1initial=0; x2initial=0;
>> ra=200; ka=0.2; La=0.002; Bm=0.00000001;
>> J=0.000000001;
>> Uaassigned=10;
>> A=[-ra/La -ka/La; ka/J -Bm/J]; B=[1/La; 0]; H=[0 1];
>> D=[0];
```

Taking note of the micromotor parameters and output (angular velocity), we have the following results for the state-space model matrices:

```
>> A, B, H, D
A =

      -100000              -100
     200000000              -10
B =

      500
        0
H =

        0        1
D =

        0
```

That is, we have

$$A = \begin{bmatrix} -100,000 & -100 \\ 200,000,000 & -10 \end{bmatrix}, \quad B = \begin{bmatrix} 500 \\ 0 \end{bmatrix}, \quad H = [0 \quad 1], \quad \text{and} \quad D = [0]$$

In particular, the following MATLAB file solves the simulation problem:

```
>> t=0:.000001:tfinal; u=Uaassigned*ones(size(t));
>> x0=[x1initial x2initial];
>> [y,x]=lsim(A,B,H,D,u,t,x0);
>> plot(t,x(:,1)*1000,t,x(:,2),':');
>> xlabel('Time (seconds)');
>> title('Micromotor Angular Velocity [rad/sec] and Current [mA]');
```

Letting the applied voltage be 10 V ($u_a = 10$ V), the micromotor state variables are plotted in Figure 6.35. The simulation results illustrate that the motor reaches the angular

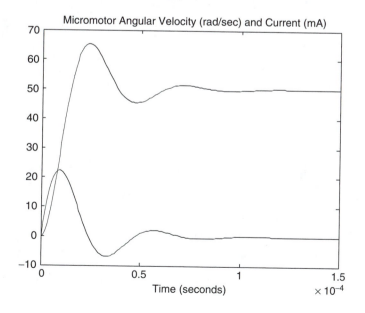

FIGURE 6.35
Motor state variables dynamics: behavior of $i_a(t)$ (solid line) and $\omega_r(t)$ (dashed line).

velocity 50 rad/sec within 0.0001 sec. There is the overshoot for the angular velocity. During the micromotor acceleration, the maximum value of the armature current is 22 mA, and $i_a(t)$ goes to zero because there is no load torque applied and the friction torque is very small.

The simulation of permanent-magnet DC micromachines can also be performed using SIMULINK. SIMULINK (interactive computing package for simulating and analyzing differential equations, mathematical models, and dynamic systems) is a part of the MAT-LAB environment. SIMILINK is a graphical mouse-driven program that allows one to numerically simulate and analyze systems by developing and manipulating blocks and diagrams. To run SIMULINK, one can type in the Command window

>> simulink

and press the Enter key. Alternatively one can click on the SIMULINK icon   .

The problem to be solved is to develop a SIMULINK block diagram. An *s*-domain block diagram for permanent-magnet DC micromotors was developed and illustrated in Figure 6.29. Making use of this *s*-domain block diagram, the corresponding SIMULINK block diagram (mdl model, c6_2_1.mdl) is built and represented in Figure 6.36. The initial conditions,

$$x_0 = \begin{bmatrix} x_{10} \\ x_{20} \end{bmatrix} = \begin{bmatrix} 0 \\ 25 \end{bmatrix}$$

are downloaded. (See "Initial conditions" in the Integrator 2 (angular velocity) block shown in Figure 6.36.) The Signal Generator block is used to set the applied voltage.

As was emphasized, the micromotor parameters can be effectively assigned symbolically using equations rather than numerical values. (This allows us to attain the greatest level of flexibility and adaptability.) Two Gain blocks used are illustrated in Figure 6.36.

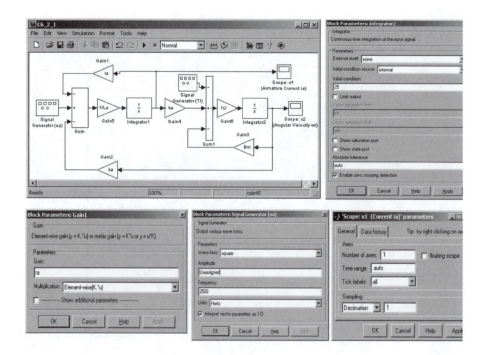

FIGURE 6.36
SIMULINK block diagram to simulate permanent-magnet DC motors (c6_2_1.mdl).

The applied armature voltage is $u_a = 10\text{rect}(2500t)$ V. We assume that there is no load torque; that is, $T_L = 0$. To perform numerical simulations, the micromotor parameters must be downloaded. In particular, we input the coefficients of Equation 6.5. To download the micromotor parameters, in the Command window we type

```
>> ra=200; ka=0.2; La=0.002; Bm=0.00000001;
>> J=0.000000001; Uaassigned=10;
```

The transient responses for two state variables (armature current $x_1 = i_a$ and angular velocity $x_2 = \omega_r$) are illustrated in Figure 6.37 using the scope data. To plot the motor dynamics, one may use the `plot` function. In particular, by making use the stored data arrays $x(:,1)$, $x(:,2)$ and $x1(:,1)$, $x1(:,2)$ we type the following statements:

```
>> plot(x(:,1),x(:,2)); xlabel('Time (sec)');
>> title('Armature Current ia, [A]');
>> plot(x1(:,1),x1(:,2)); xlabel('Time (sec)');
>> title('Velocity wr, [rad/sec]');
```

The resulting plots are documented in Figure 6.37.

As simulations are performed, the analysis can be accomplished. In particular, one analyzes the steady-state and dynamic responses of the state variables, settling time, overshoot, stability, etc. We have demonstrated the application of MATLAB and SIMULINK to

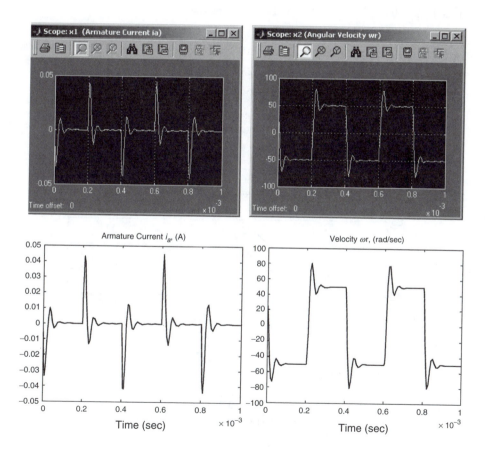

FIGURE 6.37
Permanent-magnet micromotor dynamics.

numerically simulate permanent-magnet DC micromachines. These simulations must be carried out in order to analyze the system performance with the ultimate objective and to optimize the performance to attain the maximum achievable one.

Example 6.26

Calculate and plot the torque-speed characteristic curves for a permanent-magnet DC micromotor studied in this section. Plot the load characteristic if the load is $T_L = f(\omega_r) = 1 \times 10^{-5} + 3 \times 10^{-8} \omega_r^2$ N-m.

SOLUTION

A family of the torque-speed characteristics are found applying equation $\omega_r = \frac{u_a - r_a i_a}{k_a} = \frac{u_a}{k_a} - \frac{r_a}{k_a^2} T_e$. Using different values for the armature voltage u_a in this equation, the steady-state characteristics are found and plotted in Figure 6.38. The load characteristic is also illustrated in Figure 6.38. The MATLAB m-file to calculate and plot the steady-state operating characteristics is given below.

```
% parameters of a permanent-magnet DC micromotor
ra=200; ka=0.2;
Te=0:0.00001:0.0001; % torque in N-m
for ua=1:1:10;        % applied voltage
wr=ua/ka-(ra/ka^2)* Te;
% angular velocity for different voltages
wrl=0:1:50; Tl=1e-5+3e-8*wrl.^2
% load torque at different velocity
```

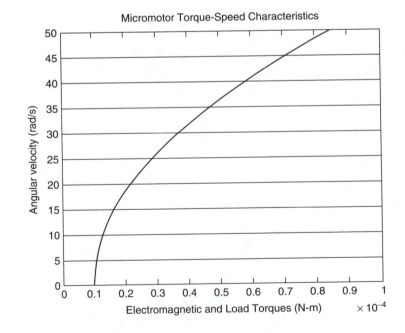

FIGURE 6.38
Torque-speed and load characteristics for permanent-magnet micromotor.

```
plot(Te,wr,'-',Tl,wrl,'-');
title('Micromotor Torque-Speed Characteristics');
xlabel('Electromagnetic and Load Torques (N-m)');
ylabel('Angular velocity (rad/sec)'); hold on;
axis([0, 0.0001, 0, 50]);
end; disp('End')
```

6.7 Induction Micromachines

6.7.1 Introduction and Analogies

The majority of rotating micromachines and microtransducers designed and fabricated to date are synchronous microdevices. Induction micromachines, though they have lower torque and power densities, do not require permanent magnet. Therefore, in this section we cover analysis, modeling, and control issues for induction micromotors that can be used as actuators.

This section studies micro- and nanomachines synthesized utilizing induction electromagnetics. In particular, the rotor currents are induced in the rotor short-circuited nanowindings due to time-varying stator magnetic field as well as motion of the rotor with respect to the stator. The electromagnetic torque results due to the interaction of the time-varying electromagnetic fields.

One of the most challenging problems in micro- and nanomachine synthesis and design is topology synthesis, analysis, and optimization. The synergetic attempts to design micro- and nanomachines have been pursued through analysis of complex patterns and paradigms of evolutionary developed nanobiomachines, e.g., *Escherichia coli, Salmonella typhimurium,* and others. Though promising results have been reported, the basic physics of nanobiomachines is unknown. Thus, distinct micro- and nanomachines have been devised postulating different fundamentals. Though it is possible that the physics of nanobiomachines is based on variable reluctance or nanomagnets, the analysis of rotor structures indicates the possibility of nanowindings in the *E. coli* and *S. typhimurium* rotors. Thus, we study electromagnetic induction-based micro- and nanomachines with the short-circuited nanowindings on the rotor. The current is induced in the rotor nanowindings due to time-varying stator magnetic field as well as motion of the rotor with respect to the stator. This section covers analysis, modeling, and control of two-phase induction micro- and nanomachines that can be fabricated.

Complex three-dimensional organic complexes and assemblies in *E. coli* and *S. typhimurium* bacteria have been covered in previous sections. For example, the 45-nm *E. coli* nanorotor is the so-called MS ring, which consists of FliF and FliG proteins. These proteins' geometry and folding are unknown. We assume that short-circuited nanowindings can be formed by these proteins. It should be emphasized that complex three-dimensional organic circuits (windings) can be engineered. As another example, consider the AAA (ATPases Associated with various cellular Activities) interacting protein superfamily. The AAA protein superfamily is characterized by a highly conserved module of more than 230 amino acid residues including an ATP binding consensus, present in one or two copies in the AAA proteins. The AAA proteins are found in all organisms and are essential for their functionality. Specific attention should be focused on the geometry and folding of different protein complexes and assemblies. Thus, the *E. coli* nanobiomotor and synthesized

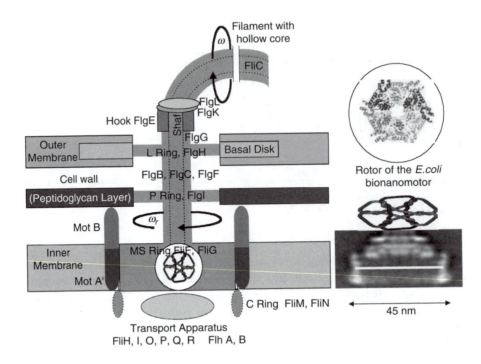

FIGURE 6.39
E. coli nanobiomotor with assumed short-circuited rotor nanobiowindings.

micro- and nanomachines can operate as induction nanomachines, as will be discussed later. (See Figure 6.39.)

The flagella in *E. coli* bacteria, rotated by nanobiomotors, are used for propulsion. The bacterium is propelled with a maximum speed of 20 μm/sec by flagellar filaments. This filament is driven by a 45-nm rotor of the nanobiomotor embedded in the cell wall. The cytoplasmic membrane forms a stator. This nanobiomotor integrates more than 20 proteins and operates as a result of the axial protonomotive force resulting due to the proton flux. The rated nanobiomotor parameters were found to be as follows: Angular velocity is 20 rad/sec, torque is 1×10^{-16} N-m, and efficiency is 50% (estimated). The nanobiomotor has three switch proteins (FliG, FliM, and FliN) that control the torque, angular velocity, and direction of rotation. These proteins are involved in the flagellar assembly. It was found that FliG possibly interacts with FliM, FliM interacts with itself, and FliM interacts with FliN. The flagellum, flexible joint (proximal hook), and nanobiomotor are shown in Figure 6.39. The nanobiomotor has two major parts: a stator (connected to the cell wall—peptidoglycan) and a rotor (connected to the flagellar filament through the flexible joint).

The stator of *E. coli* comprises the so-called MotA and MotB complexes, while the rotor comprises FliF, FliG, FliM, and FliN protein forming the so-called MS and C rings. The shaft is made from proteins FlgB, FlgC, FlgF, and FlgG, while the bearing is built from proteins FlgH and FlgI forming the L and P rings. The MS, P, and L rings each contain many copies of FliF, FlgI, and FlgH proteins, respectively. Some results provide evidence that there are eight stator elements (MotA and MotB complexes), each of which exerts the same force. The torque is developed due to axial flux of protons. (In marine bacteria and bacteria that live at high pH, however, the sodium ions establish the axial flux.) The source of energy is a transmembrane electrical potential or pH gradient. MotA and MotB complexes form a transmembrane channel. The nanobiomotor rotates clockwise or counterclockwise, and the change of the direction results due to proteins FliG, FliM, and FliN.

When the nanobiomotor rotates clockwise, the flagellar filaments work independently, and the cell body moves erratically with little net displacement (bacterium tumbles). When the nanobiomotor rotates counterclockwise, the filaments rotate in parallel in a bundle that propels the cell body forward (bacterium runs). There are multiple copies of proteins that build the flagellum, and, as an example, there are thousands of FliC molecules per helical turn of the filament, which has up to six turns.

Though the *E. coli* nanobiomotor is the most studied one, the analysis of nanobiomotors is far from complete, and there is a significant lack of reliable data and coherent understanding. There are a couple of possible torque production and energy conversion mechanisms that lead to direct analogies of nanobiomotors with electromagnetic micro- and nanomachines:

- Induction electromagnetics—The rotor currents are induced in the rotor windings due to the time-varying stator magnetic field and motion of the rotor with respect to the stator. The torque results due to the interaction of time-varying electromagnetic fields.

- Synchronous electromagnetics—The torque results due to the interaction of time-varying magnetic field established by the stator (rotor) windings and stationary magnetic field established by the rotor (stator) permanent nanomagnets.

- Variable reluctance electromagnetics (synchronous nanomachine)— The torque is produced to minimize the reluctance of the electromagnetic system, that is, the torque is created by the magnetic system in an attempt to align the minimum-reluctance path of the rotor with the time-varying rotating airgap mmf.

With the limited knowledge available to date, we hypothesize that the *E. coli* nanobiomotor operates based on the induction electromagnetics. It can also be concluded with a high degree of confidence that the electromagnetic system is closed and nanomachine. Complex chemoelectromechanical energy conversion is a far-reaching research topic that unlikely will be completed in the near future. It is extremely difficult to comprehend and prototype the torque generation, energy conversion, bearing, sensing–feedback– control, and other mechanisms in nanobiomotors. By making use of the documented assumptions and results, the studied two-phase induction micro- and nanomachine is illustrated in Figure 6.40. It should be emphasized that only micromachines can be fabricated these days. Therefore, this section is focused on induction micromachines. The control variables, to control the angular velocity, are the phase voltages u_{as} and u_{bs}.

6.7.2   Two-Phase Induction Micromotors

A two-phase induction micromotor, shown in Figure 6.40, has two stator and rotor windings. The operation and control of induction micromotors are straightforward. In order to regulate the angular velocity, the magnitude and frequency of the phase voltages u_{as} and u_{bs} are controlled. This section covers control, modeling, analysis, and simulation of induction micromachines.

6.7.2.1   *Control of Induction Micromotors*

The angular velocity of induction micromotors must be controlled, and the torque-speed characteristic curves should be examined. The electromagnetic torque developed by two-phase induction micromotors is given by

$$T_e = -\frac{P}{2}L_{ms}[(i_{as}i'_{ar} + i_{bs}i'_{br})\sin\theta_r + (i_{as}i'_{br} - i_{bs}i'_{ar})\cos\theta_r]$$

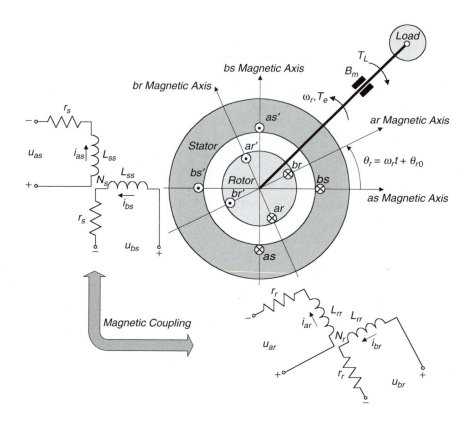

FIGURE 6.40
Two-phase symmetrical induction micromotor.

while for three-phase the electromagnetic torque is

$$T_e = -\frac{P}{2}L_{ms}\left\{\left[i_{as}\left(i'_{ar} - \tfrac{1}{2}i'_{br} - \tfrac{1}{2}i'_{cr}\right) + i_{bs}\left(i'_{br} - \tfrac{1}{2}i'_{ar} - \tfrac{1}{2}i'_{cr}\right) + i_{cs}\left(i'_{cr} - \tfrac{1}{2}i'_{br} - \tfrac{1}{2}i'_{ar}\right)\right]\sin\theta_r \right.$$

$$\left. + \tfrac{\sqrt{3}}{2}\left[i_{as}(i'_{br} - i'_{cr}) + i_{bs}(i'_{cr} - i'_{ar}) + i_{cs}(i'_{ar} - i'_{br})\right]\cos\theta_r \right\}$$

To guarantee the balanced operating condition for two-phase induction micromotors, one supplies the following phase voltages to the stator windings:

$$u_{as}(t) = \sqrt{2}u_M \cos(\omega_f t), \qquad u_{bs}(t) = \sqrt{2}u_M \sin(\omega_f t)$$

and the sinusoidal steady-state phase currents are

$$i_{as}(t) = \sqrt{2}i_M \cos(\omega_f t - \varphi_i), \qquad i_{bs}(t) = \sqrt{2}i_M \sin(\omega_f t - \varphi_i)$$

Here the following notations are used: u_M is the magnitude of the voltages applied to the *as* and *bs* stator windings; i_M is the magnitude of the *as* and *bs* stator currents; ω_f is the angular frequency of the applied phase voltages, $\omega_f = 2\pi f$; f is the frequency of the supplied voltage; and φ_i is the phase difference.

For three-phase induction micromotors, one supplies the following balanced three-phase voltages:

$$u_{as}(t) = \sqrt{2}u_M \cos(\omega_f t), \quad u_{bs}(t) = \sqrt{2}u_M \cos\left(\omega_f t - \tfrac{2}{3}\pi\right), \quad \text{and}$$

$$u_{cs}(t) = \sqrt{2}u_M \cos\left(\omega_f t + \tfrac{2}{3}\pi\right)$$

where the frequency of the applied voltage is $\omega_f = 2\pi f$.

The applied voltage to the micromotor windings cannot exceed the admissible voltage $u_{M\,max}$. That is,

$$u_{M\,min} \le u_M \le u_{M\,max}.$$

The micromotor synchronous angular velocity ω_e is found using the number of poles as

$$\omega_e = \frac{4\pi f}{P}$$

It is evident that the synchronous velocity ω_e can be regulated by changing the frequency f. To regulate the angular velocity, one varies the magnitude of the applied voltages as well as the frequency. The torque-speed characteristic curves of induction micromotors must be thoroughly studied. By performing the transient analysis by solving the derived differential equations, one can find the steady-state curves $\omega_r = \Omega_T(T_e)$ by plotting the angular velocity versus the electromagnetic torque developed.

The following principles are used to control the angular velocity of induction micromotors.

Voltage control: By changing the magnitude u_M of the applied phase voltages to the stator windings, the angular velocity is regulated in the stable operating region. (See Figure 6.41A.) It was emphasized that $u_{M\,min} < u_M < u_{M\,max}$, where $u_{M\,max}$ is the maximum allowed (rated) voltage.

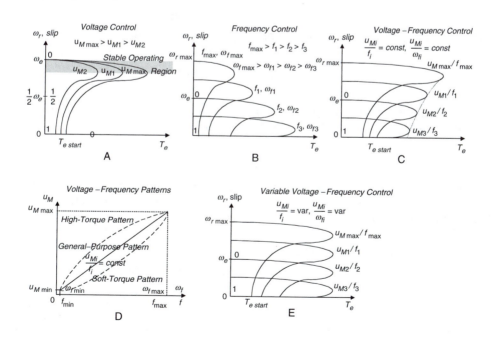

FIGURE 6.41
Torque-speed characteristic curves $\omega_r = \Omega_T(T_e)$: (A) voltage control; (B) frequency control; (C) voltage-frequency control: *constant volts per hertz* control; (D) voltage-frequency patterns; (E) variable voltage-frequency control.

Frequency control: The magnitude of the applied phase voltages is constant $u_M^{constant}$, and the angular velocity is regulated above and below the synchronous angular velocity by changing the frequency of the supplied voltages f. This concept can be clearly demonstrated using the formula $\omega_e = 4\pi f/P$. The torque-speed characteristics for different values of the frequency are shown in Figure 6.41B.

Voltage-frequency control: The angular frequency ω_f is proportional to the frequency of the supplied voltages, $\omega_f = 2\pi f$. To minimize losses, the voltages applied to the stator windings should be regulated if the frequency is changed. In particular, the magnitude of phase voltages can be decreased linearly by decreasing the frequency. That is, to guarantee the constant volts per hertz control, one maintains the following relationship:

$$\frac{u_{Mi}}{f_i} = const \quad \text{or} \quad \frac{u_{Mi}}{\omega_{fi}} = const$$

The corresponding torque-speed characteristics are documented in Figure 6.41C. Regulating the voltage-frequency patterns, one shapes the torque-speed curves. For example, the relation

$$\sqrt{\frac{u_{Mi}}{f_i}} = const$$

can be applied to adjust the magnitude u_M and frequency f of the supplied voltages. To attain the acceleration and settling time specified, and the overshoot and rise time needed, the general purpose (standard), soft- and high-starting torque patterns are implemented based on the requirements and criteria imposed. (See the standard, soft- and high-torque patterns illustrated in Figure 6.41D.)

That is, assigning

$$\omega_f = \varphi(u_M)$$

with domain $u_{M\min} < u_M < u_{M\max}$ and range $\omega_{f\min} < \omega_f < \omega_{f\max}$, one maintains

$$\frac{u_{Mi}}{f_i} = var \quad \text{or} \quad \frac{u_{Mi}}{\omega_{fi}} = var$$

For example, the desired torque-speed characteristics, as documented in Figure 6.41E, can be guaranteed.

6.7.2.2   *Modeling of Induction Micromotors*

To examine and assess a micromachine's performance, mathematical models should be derived. The following variables and symbols are used in this section to develop differential equations that describe machine dynamics:

u_{as}, u_{bs}, and u_{cs} are the phase voltages in the stator windings *as*, *bs*, and *cs*.

u_{qs}, u_{ds}, and u_{os} are the quadrature-, direct-, and zero-axis voltages.

i_{as}, i_{bs}, and i_{cs} are the phase currents in the stator windings *as*, *bs*, and *cs*.

i_{qs}, i_{ds}, and i_{os} are the quadrature-, direct-, and zero-axis stator currents.

ψ_{as}, ψ_{bs}, and ψ_{cs} are the stator flux linkages.

ψ_{qs}, ψ_{ds}, and ψ_{os} are the quadrature-, direct-, and zero-axis stator flux linkages.

u_{ar}, u_{br}, and u_{cr} are the voltages in the rotor windings *ar*, *br*, and *cr*.

u_{qr}, u_{dr}, and u_{or} are the quadrature-, direct-, and zero-axis rotor voltages.

i_{ar}, i_{br}, and i_{cr} are the currents in the rotor windings *ar*, *br*, and *cr*.

i_{qr}, i_{dr}, and i_{or} are the quadrature-, direct-, and zero-axis rotor currents.

ψ_{ar}, ψ_{br}, and ψ_{cr} are the rotor flux linkages.

ψ_{qr}, ψ_{dr}, and ψ_{or} are the quadrature-, direct-, and zero-axis rotor flux linkages.

ω_r and ω_{rm} are the electrical and mechanical angular velocities.

θ_r and θ_{rm} are the electrical and mechanical angular displacements.

T_e is the electromagnetic torque developed by the micromotor.

T_L is the load torque applied.

r_s and r_r are the resistances of the stator and rotor windings.

L_{ss} and L_{rr} are the self-inductances of the stator and rotor windings.

L_{ms} is the stator magnetizing inductance.

L_{ls} and L_{lr} are the stator and rotor leakage inductances.

N_s and N_r are the number of turns of the stator and rotor windings.

P is the number of poles.

B_m is the viscous friction coefficient.

J is the equivalent moment of inertia.

ω and θ are the angular velocity and displacement of the reference frame.

6.7.2.2.1 Modeling of Induction Micromotors Using Kirchhoff's and Newton's Laws

To develop lumped-parameter mathematical models of two-phase induction micromotors, we model the stator and rotor circuitry dynamics. As the control and state variables, we use the voltages applied to the stator (*as* and *bs*) and rotor (*ar* and *br*) windings, as well as the stator and rotor currents and flux linkages.

Using Kirchhoff's voltage law, four differential equations are

$$u_{as} = r_s i_{as} + \frac{d\psi_{as}}{dt}, \quad u_{bs} = r_s i_{bs} + \frac{d\psi_{bs}}{dt}, \quad u_{ar} = r_r i_{ar} + \frac{d\psi_{ar}}{dt}, \quad \text{and} \quad u_{br} = r_r i_{br} + \frac{d\psi_{br}}{dt}$$

Hence, in the state-space form we have

$$\mathbf{u}_{abs} = \mathbf{r}_s \mathbf{i}_{abs} + \frac{d\psi_{abs}}{dt}, \quad \mathbf{u}_{abr} = \mathbf{r}_r \mathbf{i}_{abr} + \frac{d\psi_{abr}}{dt} \tag{6.7}$$

where

$$\mathbf{u}_{abs} = \begin{bmatrix} u_{as} \\ u_{bs} \end{bmatrix}, \quad \mathbf{u}_{abr} = \begin{bmatrix} u_{ar} \\ u_{br} \end{bmatrix}, \quad \mathbf{i}_{abs} = \begin{bmatrix} i_{as} \\ i_{bs} \end{bmatrix}, \quad \mathbf{i}_{abr} = \begin{bmatrix} i_{ar} \\ i_{br} \end{bmatrix}, \quad \boldsymbol{\psi}_{abs} = \begin{bmatrix} \psi_{as} \\ \psi_{bs} \end{bmatrix}, \quad \text{and} \quad \boldsymbol{\psi}_{abr} = \begin{bmatrix} \psi_{ar} \\ \psi_{br} \end{bmatrix}$$

are the phase voltages, currents, and flux linkages;

$$\mathbf{r}_s = \begin{bmatrix} r_s & 0 \\ 0 & r_s \end{bmatrix} \quad \text{and} \quad \mathbf{r}_r = \begin{bmatrix} r_r & 0 \\ 0 & r_r \end{bmatrix}$$

are the matrices of the stator and rotor resistances.

Studying the magnetically coupled micromotor circuits, the following matrix equation for the flux linkages is found

$$\begin{bmatrix} \boldsymbol{\psi}_{abs} \\ \boldsymbol{\psi}_{abr} \end{bmatrix} = \begin{bmatrix} \mathbf{L}_s & \mathbf{L}_{sr} \\ \mathbf{L}_{sr}^T & \mathbf{L}_r \end{bmatrix} \begin{bmatrix} \mathbf{i}_{abs} \\ \mathbf{i}_{abr} \end{bmatrix}$$

where $\mathbf{L}_s$ is the matrix of the stator inductances,

$$\mathbf{L}_s = \begin{bmatrix} L_{ss} & 0 \\ 0 & L_{ss} \end{bmatrix}, \quad L_{ss} = L_{ls} + L_{ms}, \quad L_{ms} = \frac{N_s^2}{\mathfrak{R}_m}$$

$\mathbf{L}_r$ is the matrix of the rotor inductances,

$$\mathbf{L}_r = \begin{bmatrix} L_{rr} & 0 \\ 0 & L_{rr} \end{bmatrix}, \quad L_{rr} = L_{lr} + L_{mr}, \quad L_{mr} = \frac{N_r^2}{\mathfrak{R}_m}$$

and $\mathbf{L}_{sr}$ is the matrix of the stator-rotor mutual inductances,

$$\mathbf{L}_{sr} = \begin{bmatrix} L_{sr}\cos\theta_r & -L_{sr}\sin\theta_r \\ L_{sr}\sin\theta_r & L_{sr}\cos\theta_r \end{bmatrix}, \quad L_{sr} = \frac{N_s N_r}{\mathfrak{R}_m}$$

Using the number of turns in the stator and rotor windings, we have

$$\mathbf{i}'_{abr} = \frac{N_r}{N_s}\mathbf{i}'_{abr}, \quad \mathbf{u}'_{abr} = \frac{N_s}{N_r}\mathbf{u}_{abr}, \quad \text{and} \quad \boldsymbol{\psi}'_{abr} = \frac{N_s}{N_r}\boldsymbol{\psi}_{abr}$$

Then, taking note of the turn ratio, the flux linkages are written in matrix form as

$$\begin{bmatrix} \boldsymbol{\psi}_{abs} \\ \boldsymbol{\psi}'_{abr} \end{bmatrix} = \begin{bmatrix} \mathbf{L}_s & \mathbf{L}'_{sr} \\ \mathbf{L}'_{sr}{}^T & \mathbf{L}'_r \end{bmatrix} \begin{bmatrix} \mathbf{i}_{abs} \\ \mathbf{i}'_{abr} \end{bmatrix} \tag{6.8}$$

where

$$\mathbf{L}'_r = \left(\frac{N_s}{N_r}\right)^2 \mathbf{L}_r = \begin{bmatrix} L'_{rr} & 0 \\ 0 & L'_{rr} \end{bmatrix}$$

$$L'_{rr} = L'_{lr} + L'_{mr}$$

$$\mathbf{L}'_{sr} = \left(\frac{N_s}{N_r}\right)\mathbf{L}_{sr} = L_{ms}\begin{bmatrix} \cos\theta_r & -\sin\theta_r \\ \sin\theta_r & \cos\theta_r \end{bmatrix}$$

$$L_{ms} = \frac{N_s}{N_r}L_{sr}, \quad L'_{mr} = \left(\frac{N_s}{N_r}\right)^2 L_{mr}, \quad L'_{mr} = L_{ms} = \frac{N_s}{N_r}L_{sr}, \quad \text{and} \quad L'_{rr} = L'_{lr} + L_{ms}$$

Substituting the matrices for self- and mutual inductances $\mathbf{L}'_s$, $\mathbf{L}'_r$, and $\mathbf{L}'_{sr}$ in Equation 6.8, one obtains

$$\begin{bmatrix} \psi_{as} \\ \psi_{bs} \\ \psi'_{ar} \\ \psi'_{br} \end{bmatrix} = \begin{bmatrix} L_{ss} & 0 & L_{ms}\cos\theta_r & -L_{ms}\sin\theta_r \\ 0 & L_{ss} & L_{ms}\sin\theta_r & L_{ms}\cos\theta_r \\ L_{ms}\cos\theta_r & L_{ms}\sin\theta_r & L'_{rr} & 0 \\ -L_{ms}\sin\theta_r & L_{ms}\cos\theta_r & 0 & L'_{rr} \end{bmatrix} \begin{bmatrix} i_{as} \\ i_{bs} \\ i'_{ar} \\ i'_{br} \end{bmatrix}$$

Therefore, the circuitry differential equations (Equation 6.7) are rewritten as

$$\mathbf{u}_{abs} = \mathbf{r}_s \mathbf{i}_{abs} + \frac{d\mathbf{\psi}_{abs}}{dt}, \qquad \mathbf{u}'_{abr} = \mathbf{r}'_r \mathbf{i}'_{abr} + \frac{d\mathbf{\psi}'_{abr}}{dt}$$

where

$$\mathbf{r}'_r = \frac{N_s^2}{N_r^2}\mathbf{r}_r = \frac{N_s^2}{N_r^2}\begin{bmatrix} r'_r & 0 \\ 0 & r'_r \end{bmatrix}$$

Assuming that the self- and mutual inductances L_{ss}, L'_{rr}, and L_{ms} are time invariant and using the expressions for the flux linkages, one obtains a set of four nonlinear differential equations to model the circuitry dynamics:

$$L_{ss}\frac{di_{as}}{dt} + L_{ms}\frac{d(i'_{ar}\cos\theta_r)}{dt} - L_{ms}\frac{d(i'_{br}\sin\theta_r)}{dt} = -r_s i_{as} + u_{as}$$

$$L_{ss}\frac{di_{bs}}{dt} + L_{ms}\frac{d(i'_{ar}\sin\theta_r)}{dt} + L_{ms}\frac{d(i'_{br}\cos\theta_r)}{dt} = -r_s i_{bs} + u_{bs}$$

$$L_{ms}\frac{d(i_{as}\cos\theta_r)}{dt} + L_{ms}\frac{d(i_{bs}\sin\theta_r)}{dt} + L'_{rr}\frac{di'_{ar}}{dt} = -r'_r i'_{ar} + u'_{ar}$$

$$-L_{ms}\frac{d(i_{as}\sin\theta_r)}{dt} + L_{ms}\frac{d(i_{bs}\cos\theta_r)}{dt} + L'_{rr}\frac{di'_{br}}{dt} = -r'_r i'_{br} + u'_{br}$$

Cauchy's form of these differential equations is found. In particular, we have the following nonlinear differential equations to model the stator-rotor circuitry dynamics for two-phase induction micromotors:

$$\frac{di_{as}}{dt} = -\frac{L'_{rr}r_s}{L_{ss}L'_{rr} - L_{ms}^2}i_{as} + \frac{L_{ms}^2}{L_{ss}L'_{rr} - L_{ms}^2}i_{bs}\omega_r + \frac{L_{ms}L'_{rr}}{L_{ss}L'_{rr} - L_{ms}^2}i'_{ar}\left(\omega_r\sin\theta_r + \frac{r'_r}{L'_{rr}}\cos\theta_r\right)$$

$$+ \frac{L_{ms}L'_{rr}}{L_{ss}L'_{rr} - L_{ms}^2}i'_{br}\left(\omega_r\cos\theta_r - \frac{r'_r}{L'_{rr}}\sin\theta_r\right) + \frac{L'_{rr}}{L_{ss}L'_{rr} - L_{ms}^2}u_{as}$$

$$- \frac{L_{ms}}{L_{ss}L'_{rr} - L_{ms}^2}\cos\theta_r u'_{ar} + \frac{L_{ms}}{L_{ss}L'_{rr} - L_{ms}^2}\sin\theta_r u'_{br}$$

$$\frac{di_{bs}}{dt} = -\frac{L'_{rr}r_s}{L_{ss}L'_{rr} - L_{ms}^2}i_{bs} - \frac{L_{ms}^2}{L_{ss}L'_{rr} - L_{ms}^2}i_{as}\omega_r - \frac{L_{ms}L'_{rr}}{L_{ss}L'_{rr} - L_{ms}^2}i'_{ar}\left(\omega_r\cos\theta_r - \frac{r'_r}{L'_{rr}}\sin\theta_r\right)$$

$$+ \frac{L_{ms}L'_{rr}}{L_{ss}L'_{rr} - L_{ms}^2}i'_{br}\left(\omega_r\sin\theta_r + \frac{r'_r}{L'_{rr}}\cos\theta_r\right) + \frac{L'_{rr}}{L_{ss}L'_{rr} - L_{ms}^2}u_{bs}$$

$$- \frac{L_{ms}}{L_{ss}L'_{rr} - L_{ms}^2}\sin\theta_r u'_{ar} - \frac{L_{ms}}{L_{ss}L'_{rr} - L_{ms}^2}\cos\theta_r u'_{br}$$

$$\frac{di'_{ar}}{dt} = -\frac{L_{ss}r'_r}{L_{ss}L'_{rr} - L^2_{ms}} i'_{ar} + \frac{L_{ms}L_{ss}}{L_{ss}L'_{rr} - L^2_{ms}} i_{as}\left(\omega_r \sin\theta_r + \frac{r_s}{L_{ss}}\cos\theta_r\right)$$

$$- \frac{L_{ms}L_{ss}}{L_{ss}L'_{rr} - L^2_{ms}} i_{bs}\left(\omega_r \cos\theta_r - \frac{r_s}{L_{ss}}\sin\theta_r\right) - \frac{L^2_{ms}}{L_{ss}L'_{rr} - L^2_{ms}} i'_{br}\omega_r$$

$$- \frac{L_{ms}}{L_{ss}L'_{rr} - L^2_{ms}} \cos\theta_r u_{as} - \frac{L_{ms}}{L_{ss}L'_{rr} - L^2_{ms}} \sin\theta_r u_{bs} + \frac{L_{ss}}{L_{ss}L'_{rr} - L^2_{ms}} u'_{ar}$$

$$\frac{di'_{br}}{dt} = -\frac{L_{ss}r'_r}{L_{ss}L'_{rr} - L^2_{ms}} i'_{br} + \frac{L_{ms}L_{ss}}{L_{ss}L'_{rr} - L^2_{ms}} i_{as}\left(\omega_r \cos\theta_r - \frac{r_s}{L_{ss}}\sin\theta_r\right)$$

$$+ \frac{L_{ms}L_{ss}}{L_{ss}L'_{rr} - L^2_{ms}} i_{bs}\left(\omega_r \sin\theta_r + \frac{r_s}{L_{ss}}\cos\theta_r\right) + \frac{L^2_{ms}}{L_{ss}L'_{rr} - L^2_{ms}} i'_{ar}\omega_r$$

$$++ \frac{L_{ms}}{L_{ss}L'_{rr} - L^2_{ms}} \sin\theta_r u_{as} - \frac{L_{ms}}{L_{ss}L'_{rr} - L^2_{ms}} \cos\theta_r u_{bs} + \frac{L_{ss}}{L_{ss}L'_{rr} - L^2_{ms}} u'_{br} \qquad (6.9)$$

The electrical angular velocity ω_r and displacement θ_r are used in Equation 6.9 as the state variables. Therefore, the torsional-mechanical equation of motion must be incorporated to describe the evolution of ω_r and θ_r. From Newton's second law, we have

$$\sum T = J\frac{d\omega_{rm}}{dt}, \quad \text{where} \quad \sum T = T_e - B_m\omega_{rm} - T_L$$

That is,

$$J\frac{d\omega_{rm}}{dt} = T_e - B_m\omega_{rm} - T_L, \quad \frac{d\theta_{rm}}{dt} = \omega_{rm}$$

The mechanical angular velocity ω_{rm} is expressed by using the electrical angular velocity ω_r and the number of poles P. In particular, $\omega_{rm} = (2/P)\omega_r$.

The mechanical and electrical angular displacements θ_{rm} and θ_r are related as $\theta_{rm} = (2/P)\theta_r$.

Taking note of Newton's second law of motion, one obtains two differential equations as given by

$$\frac{d\omega_r}{dt} = \frac{P}{2J}T_e - \frac{B_m}{J}\omega_r - \frac{P}{2J}T_L$$

$$\frac{d\theta_r}{dt} = \omega_r$$

The electromagnetic torque developed by the induction micromotors must be found. In particular, to find the expression for the electromagnetic torque developed by two-phase induction micromotors, the coenergy $W_c(i_{abs}, i_{abr}, \theta_r)$ is used, and

$$T_e = \frac{P}{2}\frac{\partial W_c(\mathbf{i}_{abs}, \mathbf{i}'_{abr}, \theta_r)}{\partial\theta_r}$$

Assuming that the magnetic system is linear, one has

$$W_c = W_f = \tfrac{1}{2}\mathbf{i}^T_{abs}(\mathbf{L}_s - L_{ls}\mathbf{I})\mathbf{i}_{abs} + \mathbf{i}^T_{abs}\mathbf{L}'_{sr}\mathbf{i}'_{abr} + \tfrac{1}{2}\mathbf{i}'^T_{abr}(\mathbf{L}'_r - L'_{lr}\mathbf{I})\mathbf{i}'_{abr}$$

The self-inductances L_{ss} and L'_{rr}, as well as the leakage inductances L_{ls} and L'_{lr}, are not functions of the angular displacement θ_r, while the following expression for the matrix of stator-rotor mutual inductances $\mathbf{L}'_{sr}$ must be used (assuming pure sinusoidal variations of mutual inductances):

$$\mathbf{L}'_{sr} = L_{ms}\begin{bmatrix} \cos\theta_r & -\sin\theta_r \\ \sin\theta_r & \cos\theta_r \end{bmatrix}$$

Then, for *P*-pole two-phase induction micromotors, the electromagnetic torque is given by

$$T_e = \frac{P}{2}\frac{\partial W_c(\mathbf{i}_{abs}, \mathbf{i}'_{abr}, \theta_r)}{\partial\theta_r} = \frac{P}{2}\mathbf{i}^T_{abs}\frac{\partial \mathbf{L}'_{sr}(\theta_r)}{\partial\theta_r}\mathbf{i}'_{abr}$$

$$= \frac{P}{2}L_{ms}[i_{as} \quad i_{bs}]\begin{bmatrix} -\sin\theta_r & -\cos\theta_r \\ \cos\theta_r & -\sin\theta_r \end{bmatrix}\begin{bmatrix} i'_{ar} \\ i'_{br} \end{bmatrix}$$

$$= -\frac{P}{2}L_{ms}[(i_{as}i'_{ar} + i_{bs}i'_{br})\sin\theta_r + (i_{as}i'_{br} - i_{bs}i'_{ar})\cos\theta_r] \tag{6.10}$$

Using Equation 6.10 for the electromagnetic torque T_e in the torsional-mechanical equations of motion, one obtains

$$\frac{d\omega_r}{dt} = -\frac{P^2}{4J}L_{ms}[(i_{as}i'_{ar} + i_{bs}i'_{br})\sin\theta_r + (i_{as}i'_{br} - i_{bs}i'_{ar})\cos\theta_r] - \frac{B_m}{J}\omega_r - \frac{P}{2J}T_L$$

$$\frac{d\theta_r}{dt} = \omega_r \tag{6.11}$$

These two differential equations must be integrated with the circuitry dynamics. In particular, augmenting Equation 6.9 and Equation 6.11, the following set of highly non-linear differential equations results to model two-phase induction micromachines:

$$\frac{di_{as}}{dt} = -\frac{L'_{rr}r_s}{L_\Sigma}i_{as} + \frac{L^2_{ms}}{L_\Sigma}i_{bs}\omega_r + \frac{L_{ms}L'_{rr}}{L_\Sigma}i'_{ar}\left(\omega_r\sin\theta_r + \frac{r'_r}{L'_{rr}}\cos\theta_r\right)$$

$$+ \frac{L_{ms}L'_{rr}}{L_\Sigma}i'_{br}\left(\omega_r\cos\theta_r - \frac{r'_r}{L'_{rr}}\sin\theta_r\right) + \frac{L'_{rr}}{L_\Sigma}u_{as} - \frac{L_{ms}}{L_\Sigma}\cos\theta_r u'_{ar} + \frac{L_{ms}}{L_\Sigma}\sin\theta_r u'_{br}$$

$$\frac{di_{bs}}{dt} = -\frac{L'_{rr}r_s}{L_\Sigma}i_{bs} - \frac{L^2_{ms}}{L_\Sigma}i_{as}\omega_r - \frac{L_{ms}L'_{rr}}{L_\Sigma}i'_{ar}\left(\omega_r\cos\theta_r - \frac{r'_r}{L'_{rr}}\sin\theta_r\right)$$

$$+ \frac{L_{ms}L'_{rr}}{L_\Sigma}i'_{br}\left(\omega_r\sin\theta_r + \frac{r'_r}{L'_{rr}}\cos\theta_r\right) + \frac{L'_{rr}}{L_\Sigma}u_{bs} - \frac{L_{ms}}{L_\Sigma}\sin\theta_r u'_{ar} - \frac{L_{ms}}{L_\Sigma}\cos\theta_r u'_{br}$$

$$\frac{d'_{ar}}{dt} = -\frac{L_{ss}r'_r}{L_\Sigma}i'_{ar} + \frac{L_{ms}L_{ss}}{L_\Sigma}i_{as}\left(\omega_r\sin\theta_r + \frac{r_s}{L_{ss}}\cos\theta_r\right) - \frac{L_{ms}L_{ss}}{L_\Sigma}i_{bs}\left(\omega_r\cos\theta_r - \frac{r_s}{L_{ss}}\sin\theta_r\right)$$

$$- \frac{L^2_{ms}}{L_\Sigma}i'_{br}\omega_r - \frac{L_{ms}}{L_\Sigma}\cos\theta_r u_{as} - \frac{L_{ms}}{L_\Sigma}\sin\theta_r u_{bs} + \frac{L_{ss}}{L_\Sigma}u'_{ar}$$

$$\frac{di'_{br}}{dt} = -\frac{L_{ss}r'_r}{L_\Sigma}i'_{br} + \frac{L_{ms}L_{ss}}{L_\Sigma}i_{as}\left(\omega_r\cos\theta_r - \frac{r_s}{L_{ss}}\sin\theta_r\right) + \frac{L_{ms}L_{ss}}{L_\Sigma}i_{bs}\left(\omega_r\sin\theta_r + \frac{r_s}{L_{ss}}\cos\theta_r\right)$$

$$+\frac{L_{ms}^2}{L_\Sigma}i'_{ar}\omega_r + \frac{L_{ms}}{L_\Sigma}\sin\theta_r u_{as} - \frac{L_{ms}}{L_\Sigma}\cos\theta_r u_{bs} + \frac{L_{ss}}{L_\Sigma}u'_{br}$$

$$\frac{d\omega_r}{dt} = -\frac{P^2}{4J}L_{ms}[(i_{as}i'_{ar} + i_{bs}i'_{br})\sin\theta_r + (i_{as}i'_{br} - i_{bs}i'_{ar})\cos\theta_r] - \frac{B_m}{J}\omega_r - \frac{P}{2J}T_L$$

$$\frac{d\theta_r}{dt} = \omega_r \tag{6.12}$$

where $L_\Sigma = L_{ss}L'_{rr} - L_{ms}^2$.

These nonlinear differential equations give the lumped-parameter mathematical model of two-phase induction micromotors (microtransducers). In the state-space (matrix) form, a set of six highly coupled nonlinear differential equations (Equation 6.12) is given as

$$\begin{bmatrix} \dfrac{di_{as}}{dt} \\[2mm] \dfrac{di_{bs}}{dt} \\[2mm] \dfrac{di'_{ar}}{dt} \\[2mm] \dfrac{di'_{br}}{dt} \\[2mm] \dfrac{d\omega_r}{dt} \\[2mm] \dfrac{d\theta_r}{dt} \end{bmatrix} = \begin{bmatrix} -\dfrac{L'_{rr}r_s}{L_\Sigma} & 0 & 0 & 0 & 0 & 0 \\[2mm] 0 & -\dfrac{L'_{rr}r_s}{L_\Sigma} & 0 & 0 & 0 & 0 \\[2mm] 0 & 0 & -\dfrac{L'_{ss}r'_r}{L_\Sigma} & 0 & 0 & 0 \\[2mm] 0 & 0 & 0 & -\dfrac{L'_{ss}r'_r}{L_{\Sigma\Sigma}} & 0 & 0 \\[2mm] 0 & 0 & 0 & 0 & -\dfrac{B_m}{J} & 0 \\[2mm] 0 & 0 & 0 & 0 & 1 & 0 \end{bmatrix}\begin{bmatrix} i_{as} \\[2mm] i_{bs} \\[2mm] i'_{ar} \\[2mm] i'_{br} \\[2mm] \omega_r \\[2mm] \theta_r \end{bmatrix}$$

$$+\begin{bmatrix} \dfrac{L_{ms}^2}{L_\Sigma}i_{bs}\omega_r + \dfrac{L_{ms}L'_{rr}}{L_\Sigma}i'_{ar}\left(\omega_r\sin\theta_r + \dfrac{r'_r}{L'_{rr}}\cos\theta_r\right) + \dfrac{L_{ms}L'_{rr}}{L_\Sigma}i'_{br}\left(\omega_r\cos\theta_r - \dfrac{r'_r}{L'_{rr}}\sin\theta_r\right) \\[3mm] -\dfrac{L_{ms}^2}{L_\Sigma}i_{as}\omega_r - \dfrac{L_{ms}L'_{rr}}{L_\Sigma}i'_{ar}\left(\omega_r\cos\theta_r - \dfrac{r'_r}{L'_{rr}}\sin\theta_r\right) + +\dfrac{L_{ms}L'_{rr}}{L_\Sigma}i'_{br}\left(\omega_r\sin\theta_r + \dfrac{r'_r}{L'_{rr}}\cos\theta_r\right) \\[3mm] \dfrac{L_{ms}L_{ss}}{L_\Sigma}i_{as}\left(\omega_r\sin\theta_r + \dfrac{r_s}{L_{ss}}\cos\theta_r\right) - \dfrac{L_{ms}L_{ss}}{L_\Sigma}i_{bs}\left(\omega_r\cos\theta_r - \dfrac{r_s}{L_{ss}}\sin\theta_r\right) - \dfrac{L_{ms}^2}{L_\Sigma}i'_{br}\omega_r \\[3mm] \dfrac{L_{ms}L_{ss}}{L_\Sigma}i_{as}\left(\omega_r\cos\theta_r - \dfrac{r_s}{L_{ss}}\sin\theta_r\right) + \dfrac{L_{ms}L_{ss}}{L_\Sigma}i_{bs}\left(\omega_r\sin\theta_r + \dfrac{r_s}{L_{ss}}\cos\theta_r\right) + \dfrac{L_{ms}^2}{L_\Sigma}i'_{ar}\omega_r \\[3mm] -\dfrac{P^2}{4J}L_{ms}\left[(i_{as}i'_{ar} + i_{bs}i'_{br})\sin\theta_r + (i_{as}i'_{br} + i_{bs}i'_{ar})\cos\theta_r\right] \\[3mm] 0 \end{bmatrix}$$

$$+\begin{bmatrix} \dfrac{L'_{rr}}{L_{\Sigma}} & 0 & 0 & 0 \\[2mm] 0 & \dfrac{L'_{rr}}{L_{\Sigma}} & 0 & 0 \\[2mm] 0 & 0 & \dfrac{L_{ss}}{L_{\Sigma}} & 0 \\[2mm] 0 & 0 & 0 & \dfrac{L_{ss}}{L_{\Sigma}} \\[2mm] 0 & 0 & 0 & 0 \\[2mm] 0 & 0 & 0 & 0 \end{bmatrix}\begin{bmatrix} u_{as} \\[2mm] u_{bs} \\[2mm] u'_{ar} \\[2mm] u'_{br} \end{bmatrix}+\begin{bmatrix} -\dfrac{L_{ms}}{L_{\Sigma}}\cos\theta_r u'_{ar}+\dfrac{L_{ms}}{L_{\Sigma}}\sin\theta_r u'_{br} \\[2mm] -\dfrac{L_{ms}}{L_{\Sigma}}\sin\theta_r u'_{ar}-\dfrac{L_{ms}}{L_{\Sigma}}\cos\theta_r u'_{br} \\[2mm] -\dfrac{L_{ms}}{L_{\Sigma}}\cos\theta_r u_{as}-\dfrac{L_{ms}}{L_{\Sigma}}\sin\theta_r u_{bs} \\[2mm] \dfrac{L_{ms}}{L_{\Sigma}}\sin\theta_r u_{as}-\dfrac{L_{ms}}{L_{\Sigma}}\cos\theta_r u_{bs} \\[2mm] 0 \\[2mm] 0 \end{bmatrix}-\begin{bmatrix} 0 \\[2mm] 0 \\[2mm] 0 \\[2mm] 0 \\[2mm] \dfrac{P}{2J} \\[2mm] 0 \end{bmatrix}T_L \qquad (6.13)$$

6.7.2.2.2 Modeling of Induction Micromotors Using the Lagrange Equations

The mathematical model can be derived using Lagrange's equations for dynamic systems. The generalized independent coordinates are four charges and the rotor angular displacement

$$q_1 = \frac{i_{as}}{s}, \quad q_2 = \frac{i_{bs}}{s}, \quad q_3 = \frac{i'_{ar}}{s}, \quad q_4 = \frac{i'_{br}}{s}, \quad q_5 = \theta_r$$

while the generalized forces are the voltages and load torque:

$$Q_1 = u_{as}, \quad Q_2 = u_{bs}, \quad Q_3 = u'_{ar}, \quad Q_4 = u'_{br}, \quad Q_5 = -T_L$$

Five Lagrange equations are written as

$$\frac{d}{dt}\left(\frac{\partial\Gamma}{\partial\dot{q}_1}\right)-\frac{\partial\Gamma}{\partial q_1}+\frac{\partial D}{\partial\dot{q}_1}+\frac{\partial\Pi}{\partial q_1}=Q_1$$

$$\frac{d}{dt}\left(\frac{\partial\Gamma}{\partial\dot{q}_2}\right)-\frac{\partial\Gamma}{\partial q_2}+\frac{\partial D}{\partial\dot{q}_2}+\frac{\partial\Pi}{\partial q_2}=Q_2$$

$$\frac{d}{dt}\left(\frac{\partial\Gamma}{\partial\dot{q}_3}\right)-\frac{\partial\Gamma}{\partial q_3}+\frac{\partial D}{\partial\dot{q}_3}+\frac{\partial\Pi}{\partial q_3}=Q_3$$

$$\frac{d}{dt}\left(\frac{\partial\Gamma}{\partial\dot{q}_4}\right)-\frac{\partial\Gamma}{\partial q_4}+\frac{\partial D}{\partial\dot{q}_4}+\frac{\partial\Pi}{\partial q_4}=Q_4$$

$$\frac{d}{dt}\left(\frac{\partial\Gamma}{\partial\dot{q}_5}\right)-\frac{\partial\Gamma}{\partial q_5}+\frac{\partial D}{\partial\dot{q}_5}+\frac{\partial\Pi}{\partial q_5}=Q_5$$

The total kinetic, potential, and dissipated energies are expressed as

$$\Gamma = \tfrac{1}{2}L_{ss}\dot{q}_1^2+L_{ms}\dot{q}_1\dot{q}_3\cos q_5-L_{ms}\dot{q}_1\dot{q}_4\sin q_5+\tfrac{1}{2}L_{ss}\dot{q}_2^2+L_{ms}\dot{q}_2\dot{q}_3\sin q_5+L_{ms}\dot{q}_2\dot{q}_4\cos q_5$$

$$+\tfrac{1}{2}L'_{rr}\dot{q}_3^2+\tfrac{1}{2}L'_{rr}\dot{q}_4^2+\tfrac{1}{2}J\dot{q}_5^2$$

$$\Pi = 0$$

$$D = \tfrac{1}{2}\left(r_s\dot{q}_1^2+r_s\dot{q}_2^2+r'_r\dot{q}_3^2+r'_r\dot{q}_4^2+B_m\dot{q}_5^2\right)$$

Thus, the derivatives are found to be

$$\frac{\partial \Gamma}{\partial q_1} = 0, \quad \frac{\partial \Gamma}{\partial \dot{q}_1} = L_{ss}\dot{q}_1 + L_{ms}\dot{q}_3 \cos q_5 - L_{ms}\dot{q}_4 \sin q_5$$

$$\frac{\partial \Gamma}{\partial q_2} = 0, \quad \frac{\partial \Gamma}{\partial \dot{q}_2} = L_{ss}\dot{q}_2 + L_{ms}\dot{q}_3 \sin q_5 + L_{ms}\dot{q}_4 \cos q_5$$

$$\frac{\partial \Gamma}{\partial q_3} = 0, \quad \frac{\partial \Gamma}{\partial \dot{q}_3} = L_{rr}'\dot{q}_3 + L_{ms}\dot{q}_1 \cos q_5 + L_{ms}\dot{q}_2 \sin q_5$$

$$\frac{\partial \Gamma}{\partial q_4} = 0, \quad \frac{\partial \Gamma}{\partial \dot{q}_4} = L_{rr}'\dot{q}_4 - L_{ms}\dot{q}_1 \sin q_5 + L_{ms}\dot{q}_2 \cos q_5$$

$$\frac{\partial \Gamma}{\partial q_5} = -L_{ms}\dot{q}_1\dot{q}_3 \sin q_5 - L_{ms}\dot{q}_1\dot{q}_4 \cos q_5 + L_{ms}\dot{q}_2\dot{q}_3 \cos q_5 - L_{ms}\dot{q}_2\dot{q}_4 \sin q_5$$

$$= -L_{ms}[(\dot{q}_1\dot{q}_3 + \dot{q}_2\dot{q}_4)\sin q_5 + (\dot{q}_1\dot{q}_4 - \dot{q}_2\dot{q}_3)\cos q_5]$$

$$\frac{\partial \Gamma}{\partial \dot{q}_5} = J\dot{q}_5$$

$$\frac{\partial \Pi}{\partial q_1} = 0, \quad \frac{\partial \Pi}{\partial q_2} = 0, \quad \frac{\partial \Pi}{\partial q_3} = 0, \quad \frac{\partial \Pi}{\partial q_4} = 0, \quad \frac{\partial \Pi}{\partial q_5} = 0$$

$$\frac{\partial D}{\partial \dot{q}_1} = r_s\dot{q}_1, \quad \frac{\partial D}{\partial \dot{q}_2} = r_s\dot{q}_2, \quad \frac{\partial D}{\partial \dot{q}_3} = r_r'\dot{q}_3, \quad \frac{\partial D}{\partial \dot{q}_4} = r_r'\dot{q}_4, \quad \frac{\partial D}{\partial \dot{q}_5} = B_m\dot{q}_5$$

Taking note of $\dot{q}_1 = i_{as}$, $\dot{q}_2 = i_{bs}$, $\dot{q}_3 = i_{ar}'$, $\dot{q}_4 = i_{br}'$, and $\dot{q}_5 = \omega_r$, one obtains the following differential equations:

$$L_{ss}\frac{di_{as}}{dt} + L_{ms}\frac{d(i_{ar}'\cos\theta_r)}{dt} - L_{ms}\frac{d(i_{br}'\sin\theta_r)}{dt} + r_s i_{as} = u_{as}$$

$$L_{ss}\frac{di_{bs}}{dt} + L_{ms}\frac{d(i_{ar}'\sin\theta_r)}{dt} + L_{ms}\frac{d(i_{br}'\cos\theta_r)}{dt} + r_s i_{bs} = u_{bs}$$

$$L_{ms}\frac{d(i_{as}\cos\theta_r)}{dt} + L_{ms}\frac{d(i_{bs}\sin\theta_r)}{dt} + L_{rr}'\frac{di_{ar}'}{dt} + r_r'i_{ar}' = u_{ar}'$$

$$-L_{ms}\frac{d(i_{as}\sin\theta_r)}{dt} + L_{ms}\frac{d(i_{bs}\cos\theta_r)}{dt} + L_{rr}'\frac{di_{br}'}{dt} + r_r'i_{br}' = u_{br}'$$

$$J\frac{d^2\theta_r}{dt^2} + L_{ms}[(i_{as}i_{ar}' + i_{bs}i_{br}')\sin\theta_r + (i_{as}i_{br}' - i_{bs}i_{ar}')\cos\theta_r] + B_m\frac{d\theta_r}{dt} = -T_L$$

For *P*-pole induction micromotors, by making use of

$$\frac{d\theta_r}{dt} = \omega_r$$

six differential equations, as found in Equation 6.12, result.

6.7.2.2.3 s-Domain Block Diagram of Two-Phase Induction Micromotors

To perform the analysis of dynamics, to control induction machines, as well as to visualize the results, it is important to develop the s-domain block diagrams. For squirrel-cage induction micromotors, the rotor windings are short-circuited, and hence $u'_{ar} = u'_{br} = 0$. The block diagram is built using Equation 6.12. The resulting s-domain block diagram is shown in Figure 6.42.

6.7.3 Three-Phase Induction Micromotors

Having covered two-phase induction micromachines, let us consider three-phase induction micromotors. Specifically, we will examine modeling, analysis, and simulation issues.

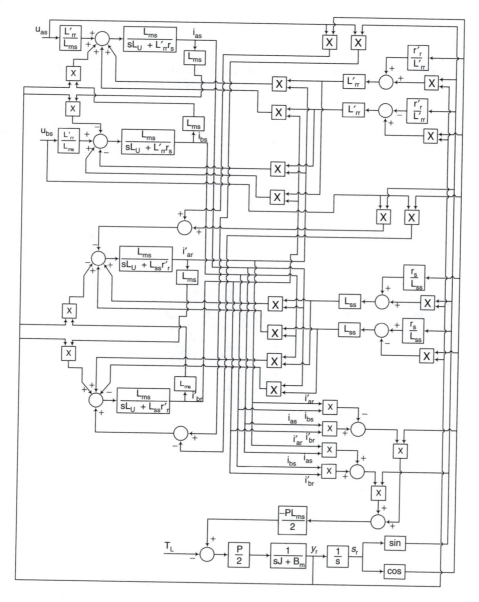

FIGURE 6.42
s-domain block diagram of squirrel-cage induction micromotors.

6.7.3.1 *Dynamics of Induction Micromotors in the Machine Variables*

Our goal is to develop the mathematical model of three-phase induction micromotors, as shown in Figure 6.43, using Kirchhoff's and Newton's second laws.

Studying the magnetically coupled stator and rotor circuitry, Kirchhoff's voltage law relates the *abc* stator and rotor phase voltages, currents, and flux linkages through the set of differential equations.

For magnetically coupled stator and rotor windings, we have

$$u_{as} = r_s i_{as} + \frac{d\psi_{as}}{dt}, \quad u_{bs} = r_s i_{bs} + \frac{d\psi_{bs}}{dt}, \quad u_{cs} = r_s i_{cs} + \frac{d\psi_{cs}}{dt}$$

$$u_{ar} = r_r i_{ar} + \frac{d\psi_{ar}}{dt}, \quad u_{br} = r_r i_{br} + \frac{d\psi_{br}}{dt}, \quad u_{cr} = r_r i_{cr} + \frac{d\psi_{cr}}{dt}$$

(6.14)

It is clear that the *abc* stator and rotor voltages, currents, and flux linkages are used as the variables. In the state-space form, Equation 6.14 is rewritten as

$$\mathbf{u}_{abcs} = \mathbf{r}_s \mathbf{i}_{abcs} + \frac{d\boldsymbol{\psi}_{abcs}}{dt}, \quad \mathbf{u}_{abcr} = \mathbf{r}_r \mathbf{i}_{abcr} + \frac{d\boldsymbol{\psi}_{abcr}}{dt}$$

(6.15)

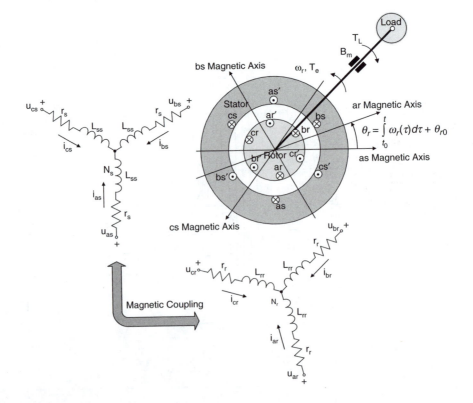

FIGURE 6.43
Three-phase symmetrical induction micromotor.

where the *abc* stator and rotor voltages, currents, and flux linkages are

$$\mathbf{u}_{abcs} = \begin{bmatrix} u_{as} \\ u_{bs} \\ u_{cs} \end{bmatrix}, \quad \mathbf{u}_{abcr} = \begin{bmatrix} u_{ar} \\ u_{br} \\ u_{cr} \end{bmatrix}, \quad \mathbf{i}_{abcs} = \begin{bmatrix} i_{as} \\ i_{bs} \\ i_{cs} \end{bmatrix}, \quad \mathbf{i}_{abcr} = \begin{bmatrix} i_{ar} \\ i_{br} \\ i_{cr} \end{bmatrix}, \quad \mathbf{\psi}_{abcs} = \begin{bmatrix} \psi_{as} \\ \psi_{bs} \\ \psi_{cs} \end{bmatrix}, \quad \text{and} \quad \mathbf{\psi}_{abcr} = \begin{bmatrix} \psi_{ar} \\ \psi_{br} \\ \psi_{cr} \end{bmatrix}$$

In Equation 6.15, the diagonal matrices of the stator and rotor resistances are

$$\mathbf{r}_s = \begin{bmatrix} r_s & 0 & 0 \\ 0 & r_s & 0 \\ 0 & 0 & r_s \end{bmatrix} \quad \text{and} \quad \mathbf{r}_r = \begin{bmatrix} r_r & 0 & 0 \\ 0 & r_r & 0 \\ 0 & 0 & r_r \end{bmatrix}$$

The flux linkages equations must be thoroughly examined, and one has

$$\begin{bmatrix} \mathbf{\psi}_{abcs} \\ \mathbf{\psi}_{abcr} \end{bmatrix} = \begin{bmatrix} \mathbf{L}_s & \mathbf{L}_{sr} \\ \mathbf{L}_{sr}^T & \mathbf{L}_r \end{bmatrix} \begin{bmatrix} \mathbf{i}_{abcs} \\ \mathbf{i}_{abcr} \end{bmatrix} \tag{6.16}$$

where the matrices of self- and mutual inductances $\mathbf{L}_s$, $\mathbf{L}_r$, and $\mathbf{L}_{sr}$ are

$$\mathbf{L}_s = \begin{bmatrix} L_{ls}+L_{ms} & -\tfrac{1}{2}L_{ms} & -\tfrac{1}{2}L_{ms} \\ -\tfrac{1}{2}L_{ms} & L_{ls}+L_{ms} & -\tfrac{1}{2}L_{ms} \\ -\tfrac{1}{2}L_{ms} & -\tfrac{1}{2}L_{ms} & L_{ls}+L_{ms} \end{bmatrix}, \quad \mathbf{L}_r = \begin{bmatrix} L_{lr}+L_{mr} & -\tfrac{1}{2}L_{mr} & -\tfrac{1}{2}L_{mr} \\ -\tfrac{1}{2}L_{mr} & L_{lr}+L_{mr} & -\tfrac{1}{2}L_{mr} \\ -\tfrac{1}{2}L_{mr} & -\tfrac{1}{2}L_{mr} & L_{lr}+L_{mr} \end{bmatrix}$$

$$\mathbf{L}_{sr} = L_{sr}\begin{bmatrix} \cos\theta_r & \cos(\theta_r+\tfrac{2}{3}\pi) & \cos(\theta_r-\tfrac{2}{3}\pi) \\ \cos(\theta_r-\tfrac{2}{3}\pi) & \cos\theta_r & \cos(\theta_r+\tfrac{2}{3}\pi) \\ \cos(\theta_r+\tfrac{2}{3}\pi) & \cos(\theta_r-\tfrac{2}{3}\pi) & \cos\theta_r \end{bmatrix}$$

Using the number of turns N_s and N_r, one finds

$$\mathbf{u}'_{abcr} = \frac{N_s}{N_r}\mathbf{u}_{abcr}, \quad \mathbf{i}'_{abcr} = \frac{N_r}{N_s}\mathbf{i}_{abcr}, \quad \text{and} \quad \mathbf{\psi}'_{abcr} = \frac{N_s}{N_r}\mathbf{\psi}_{abcr}$$

The inductances are expressed as

$$L_{ms} = \frac{N_s}{N_r}L_{sr}, \quad L_{sr} = \frac{N_s N_r}{\Re_m}, \quad \text{and} \quad L_{ms} = \frac{N_s^2}{\Re_m}$$

Then we have

$$\mathbf{L}'_{sr} = \frac{N_s}{N_r}\mathbf{L}_{sr} = L_{ms}\begin{bmatrix} \cos\theta_r & \cos(\theta_r+\tfrac{2}{3}\pi) & \cos(\theta_r-\tfrac{2}{3}\pi) \\ \cos(\theta_r-\tfrac{2}{3}\pi) & \cos\theta_r & \cos(\theta_r+\tfrac{2}{3}\pi) \\ \cos(\theta_r+\tfrac{2}{3}\pi) & \cos(\theta_r-\tfrac{2}{3}\pi) & \cos\theta_r \end{bmatrix}$$

and

$$\mathbf{L}'_r = \frac{N_s^2}{N_r^2}\mathbf{L}_r = \begin{bmatrix} L'_{lr}+L_{ms} & -\tfrac{1}{2}L_{ms} & -\tfrac{1}{2}L_{ms} \\ -\tfrac{1}{2}L_{ms} & L'_{lr}+L_{ms} & -\tfrac{1}{2}L_{ms} \\ -\tfrac{1}{2}L_{ms} & -\tfrac{1}{2}L_{ms} & L'_{lr}+L_{ms} \end{bmatrix}, \quad L'_{lr} = \frac{N_s^2}{N_r^2}L_{lr}$$

From Equation 6.16, one finds

$$\begin{bmatrix} \boldsymbol{\psi}_{abcs} \\ \boldsymbol{\psi}'_{abcr} \end{bmatrix} = \begin{bmatrix} \mathbf{L}_s & \mathbf{L}'_{sr} \\ \mathbf{L}'^{T}_{st} & \mathbf{L}'_r \end{bmatrix} \begin{bmatrix} \mathbf{i}_{abcs} \\ \mathbf{i}'_{abcr} \end{bmatrix} \tag{6.17}$$

Substituting the matrices $\mathbf{L}_s$, $\mathbf{L}'_{sr}$, and $\mathbf{L}'_r$, we have

$$
\begin{bmatrix} \psi_{as} \\ \psi_{bs} \\ \psi_{cs} \\ \psi'_{ar} \\ \psi'_{br} \\ \psi'_{cr} \end{bmatrix} =
\begin{bmatrix}
L_{ls}+L_{ms} & -\tfrac{1}{2}L_{ms} & -\tfrac{1}{2}L_{ms} \\
-\tfrac{1}{2}L_{ms} & L_{ls}+L_{ms} & -\tfrac{1}{2}L_{ms} \\
-\tfrac{1}{2}L_{ms} & -\tfrac{1}{2}L_{ms} & L_{ls}+L_{ms} \\
L_{ms}\cos\theta_r & L_{ms}\cos(\theta_r-\tfrac{2}{3}\pi) & L_{ms}\cos(\theta_r+\tfrac{2}{3}\pi) \\
L_{ms}\cos(\theta_r+\tfrac{2}{3}\pi) & L_{ms}\cos\theta_r & L_{ms}\cos(\theta_r-\tfrac{2}{3}\pi) \\
L_{ms}\cos(\theta_r-\tfrac{2}{3}\pi) & L_{ms}\cos(\theta_r+\tfrac{2}{3}\pi) & L_{ms}\cos\theta_r
\end{bmatrix}
$$

$$
\begin{bmatrix}
L_{ms}\cos\theta_r & L_{ms}\cos(\theta_r+\tfrac{2}{3}\pi) & L_{ms}\cos(\theta_r-\tfrac{2}{3}\pi) \\
L_{ms}\cos(\theta_r-\tfrac{2}{3}\pi) & L_{ms}\cos\theta_r & L_{ms}\cos(\theta_r+\tfrac{2}{3}\pi) \\
L_{ms}\cos(\theta_r+\tfrac{2}{3}\pi) & L_{ms}\cos(\theta_r-\tfrac{2}{3}\pi) & L_{ms}\cos\theta_r \\
L'_{lr}+L_{ms} & -\tfrac{1}{2}L_{ms} & -\tfrac{1}{2}L_{ms} \\
-\tfrac{1}{2}L_{ms} & L'_{lr}+L_{ms} & -\tfrac{1}{2}L_{ms} \\
-\tfrac{1}{2}L_{ms} & -\tfrac{1}{2}L_{ms} & L'_{lr}+L_{ms}
\end{bmatrix}
\begin{bmatrix} i_{as} \\ i_{bs} \\ i_{cs} \\ i'_{ar} \\ i'_{br} \\ i'_{cr} \end{bmatrix}
$$

Using Equation 6.15 and Equation 6.17, one obtains

$$\mathbf{u}_{abcs} = \mathbf{r}_s\mathbf{i}_{abcs} + \frac{d\boldsymbol{\psi}_{abcs}}{dt} = \mathbf{r}_s\mathbf{i}_{abcs} + \mathbf{L}_s\frac{d\mathbf{i}_{abcs}}{dt} + \frac{d(\mathbf{L}'_{sr}\mathbf{i}'_{abcr})}{dt} \tag{6.18}$$

$$\mathbf{u}'_{abcr} = \mathbf{r}'_r\mathbf{i}'_{abcr} + \frac{d\boldsymbol{\psi}'_{abcr}}{dt} = \mathbf{r}'_r\mathbf{i}'_{abcr} + \mathbf{L}'_r\frac{d\mathbf{i}'_{abcr}}{dt} + \frac{d(\mathbf{L}'^{T}_{sr}\mathbf{i}_{abcs})}{dt}$$

where

$$\mathbf{r}'_r = \frac{N_s^2}{N_r^2}\mathbf{r}_r$$

The state-space equations (Equation 6.18) in expanded form using Equation 6.17 are rewritten as

$$u_{as} = r_s i_{as} + (L_{ls}+L_{ms})\frac{di_{as}}{dt} - \frac{1}{2}L_{ms}\frac{di_{bs}}{dt} - \frac{1}{2}L_{ms}\frac{di_{cs}}{dt} + L_{ms}\frac{d(i'_{ar}\cos\theta_r)}{dt} + L_{ms}\frac{d\left(i'_{br}\cos\left(\theta_r+\frac{2\pi}{3}\right)\right)}{dt}$$

$$+ L_{ms}\frac{d\left(i'_{cr}\cos\left(\theta_r-\frac{2\pi}{3}\right)\right)}{dt}$$

$$u_{bs} = r_s i_{bs} - \frac{1}{2} L_{ms} \frac{di_{as}}{dt} + (L_{ls} + L_{ms}) \frac{di_{bs}}{dt} - \frac{1}{2} L_{ms} \frac{di_{cs}}{dt} + L_{ms} \frac{d\left(i'_{ar} \cos\left(\theta_r - \frac{2\pi}{3}\right)\right)}{dt} + L_{ms} \frac{d(i'_{br} \cos\theta_r)}{dt}$$

$$+ L_{ms} \frac{d\left(i'_{cr} \cos\left(\theta_r + \frac{2\pi}{3}\right)\right)}{dt}$$

$$u_{cs} = r_s i_{cs} - \frac{1}{2} L_{ms} \frac{di_{as}}{dt} - \frac{1}{2} L_{ms} \frac{di_{bs}}{dt} + (L_{ls} + L_{ms}) \frac{di_{cs}}{dt} + L_{ms} \frac{d\left(i'_{ar} \cos\left(\theta_r + \frac{2\pi}{3}\right)\right)}{dt}$$

$$+ L_{ms} \frac{d\left(i'_{br} \cos\left(\theta_r - \frac{2\pi}{3}\right)\right)}{dt} + L_{ms} \frac{d(i'_{cr} \cos\theta_r)}{dt}$$

$$u'_{ar} = r'_r i'_{ar} + L_{ms} \frac{d(i_{as} \cos\theta_r)}{dt} + L_{ms} \frac{d\left(i_{bs} \cos\left(\theta_r - \frac{2\pi}{3}\right)\right)}{dt} + L_{ms} \frac{d\left(i_{cs} \cos\left(\theta_r + \frac{2\pi}{3}\right)\right)}{dt} + (L'_{lr} + L_{ms}) \frac{di'_{ar}}{dt}$$

$$- \frac{1}{2} L_{ms} \frac{di'_{br}}{dt} - \frac{1}{2} L_{ms} \frac{di'_{cr}}{dt}$$

$$u'_{br} = r'_r i'_{br} + L_{ms} \frac{d\left(i_{as} \cos\left(\theta_r + \frac{2\pi}{3}\right)\right)}{dt} + L_{ms} \frac{d(i_{bs} \cos\theta_r)}{dt} + L_{ms} \frac{d\left(i_{cs} \cos\left(\theta_r - \frac{2\pi}{3}\right)\right)}{dt} - \frac{1}{2} L_{ms} \frac{di'_{ar}}{dt}$$

$$+ (L'_{lr} + L_{ms}) \frac{di'_{br}}{dt} - \frac{1}{2} L_{ms} \frac{di'_{cr}}{dt}$$

$$u'_{cr} = r'_r i'_{cr} + L_{ms} \frac{d\left(i_{as} \cos\left(\theta_r - \frac{2\pi}{3}\right)\right)}{dt} + L_{ms} \frac{d\left(i_{bs} \cos\left(\theta_r + \frac{2\pi}{3}\right)\right)}{dt} + L_{ms} \frac{d(i_{cs} \cos\theta_r)}{dt} - \frac{1}{2} L_{ms} \frac{di'_{ar}}{dt}$$

$$- \frac{1}{2} L_{ms} \frac{di'_{br}}{dt} + (L'_{lr} + L_{ms}) \frac{di'_{cr}}{dt}$$

Cauchy's form of differential equations, given in the state-space form, is found to be

$$
\begin{bmatrix} \frac{di_{as}}{dt} \\[2mm] \frac{di_{bs}}{dt} \\[2mm] \frac{di_{cs}}{dt} \\[2mm] \frac{di'_{ar}}{dt} \\[2mm] \frac{di'_{br}}{dt} \\[2mm] \frac{di'_{cr}}{dt} \end{bmatrix}
= \frac{1}{L_{\Sigma L}}
\begin{bmatrix}
-r_s L_{\Sigma m} & -\frac{1}{2} r_s L_{ms} & -\frac{1}{2} r_s L_{ms} & 0 & 0 & 0 \\[2mm]
-\frac{1}{2} r_s L_{ms} & -r_s L_{\Sigma m} & -\frac{1}{2} r_s L_{ms} & 0 & 0 & 0 \\[2mm]
-\frac{1}{2} r_s L_{ms} & -\frac{1}{2} r_s L_{ms} & -r_s L_{\Sigma m} & 0 & 0 & 0 \\[2mm]
0 & 0 & 0 & -r_r L_{\Sigma m} & -\frac{1}{2} r_r L_{ms} & -\frac{1}{2} r_r L_{ms} \\[2mm]
0 & 0 & 0 & -\frac{1}{2} r_r L_{ms} & -r_r L_{\Sigma m} & -\frac{1}{2} r_r L_{ms} \\[2mm]
0 & 0 & 0 & -\frac{1}{2} r_r L_{ms} & -\frac{1}{2} r_r L_{ms} & -r_r L_{\Sigma m}
\end{bmatrix}
\begin{bmatrix} i_{as} \\[2mm] i_{bs} \\[2mm] i_{cs} \\[2mm] i'_{ar} \\[2mm] i'_{br} \\[2mm] i'_{cr} \end{bmatrix}
$$

$$+\frac{1}{L_{\Sigma L}}\begin{bmatrix} 0 & 0 & 0 \\ 0 & 0 & 0 \\ 0 & 0 & 0 \\ r_s L_{ms}\cos\theta_r & r_s L_{ms}\cos\left(\theta_r-\tfrac{2}{3}\pi\right) & r_s L_{ms}\cos\left(\theta_r+\tfrac{2}{3}\pi\right) \\ r_s L_{ms}\cos\left(\theta_r+\tfrac{2}{3}\pi\right) & r_s L_{ms}\cos\theta_r & r_s L_{ms}\cos\left(\theta_r-\tfrac{2}{3}\pi\right) \\ r_s L_{ms}\cos\left(\theta_r-\tfrac{2}{3}\pi\right) & r_s L_{ms}\cos\left(\theta_r+\tfrac{2}{3}\pi\right) & r_s L_{ms}\cos\theta_r \end{bmatrix}$$

$$\begin{bmatrix} r_r L_{ms}\cos\theta_r & r_r L_{ms}\cos\left(\theta_r+\tfrac{2}{3}\pi\right) & r_r L_{ms}\cos\left(\theta_r-\tfrac{2}{3}\pi\right) \\ r_r L_{ms}\cos\left(\theta_r-\tfrac{2}{3}\pi\right) & r_r L_{ms}\cos\theta_r & r_r L_{ms}\cos\left(\theta_r+\tfrac{2}{3}\pi\right) \\ r_r L_{ms}\cos\left(\theta_r+\tfrac{2}{3}\pi\right) & r_r L_{ms}\cos\left(\theta_r-\tfrac{2}{3}\pi\right) & r_r L_{ms}\cos\theta_r \\ 0 & 0 & 0 \\ 0 & 0 & 0 \\ 0 & 0 & 0 \end{bmatrix}\begin{bmatrix} i_{as} \\ i_{bs} \\ i_{cs} \\ i'_{ar} \\ i'_{br} \\ i'_{cr} \end{bmatrix}$$

$$+\frac{1}{L_{\Sigma L}}\begin{bmatrix} 0 & 1.3L_{ms}^2\omega_r & -1.3L_{ms}^2\omega_r \\ -1.3L_{ms}^2\omega_r & 0 & 1.3L_{ms}^2\omega_r \\ 1.3L_{ms}^2\omega_r & -1.3L_{ms}^2\omega_r & 0 \\ L_{\Sigma ms}\omega_r\sin\theta_r & L_{\Sigma ms}\omega_r\sin\left(\theta_r-\tfrac{2}{3}\pi\right) & L_{\Sigma ms}\omega_r\sin\left(\theta_r+\tfrac{2}{3}\pi\right) \\ L_{\Sigma ms}\omega_r\sin\left(\theta_r+\tfrac{2}{3}\pi\right) & L_{\Sigma ms}\omega_r\sin\theta_r & L_{\Sigma ms}\omega_r\sin\left(\theta_r-\tfrac{2}{3}\pi\right) \\ L_{\Sigma ms}\omega_r\sin\left(\theta_r-\tfrac{2}{3}\pi\right) & L_{\Sigma ms}\omega_r\sin\left(\theta_r+\tfrac{2}{3}\pi\right) & L_{\Sigma ms}\omega_r\sin\theta_r \end{bmatrix}$$

$$\begin{bmatrix} L_{\Sigma ms}\omega_r\sin\theta_r & L_{\Sigma ms}\omega_r\sin\left(\theta_r+\tfrac{2}{3}\pi\right) & L_{\Sigma ms}\omega_r\sin\left(\theta_r-\tfrac{2}{3}\pi\right) \\ L_{\Sigma ms}\omega_r\sin\left(\theta_r-\tfrac{2}{3}\pi\right) & L_{\Sigma ms}\omega_r\sin\theta_r & L_{\Sigma ms}\omega_r\sin\left(\theta_r+\tfrac{2}{3}\pi\right) \\ L_{\Sigma ms}\omega_r\sin\left(\theta_r+\tfrac{2}{3}\pi\right) & L_{\Sigma ms}\omega_r\sin\left(\theta_r-\tfrac{2}{3}\pi\right) & L_{\Sigma ms}\omega_r\sin\theta_r \\ 0 & -1.3L_{ms}^2\omega_r & 1.3L_{ms}^2\omega_r \\ 1.3L_{ms}^2\omega_r & 0 & -1.3L_{ms}^2\omega_r \\ -1.3L_{ms}^2\omega_r & 1.3L_{ms}^2\omega_r & 0 \end{bmatrix}\begin{bmatrix} i_{as} \\ i_{bs} \\ i_{cs} \\ i'_{ar} \\ i'_{br} \\ i'_{cr} \end{bmatrix}$$

$$+\frac{1}{L_{\Sigma L}}\begin{bmatrix} 2L_{ms}+L'_{lr} & \tfrac{1}{2}L_{ms} & \tfrac{1}{2}L_{ms} \\ \tfrac{1}{2}L_{ms} & 2L_{ms}+L'_{lr} & \tfrac{1}{2}L_{ms} \\ \tfrac{1}{2}L_{ms} & \tfrac{1}{2}L_{ms} & 2L_{ms}+L'_{lr} \\ -L_{ms}\cos\theta_r & -L_{ms}\cos\left(\theta_r-\tfrac{2}{3}\pi\right) & -L_{ms}\cos\left(\theta_r+\tfrac{2}{3}\pi\right) \\ -L_{ms}\cos\left(\theta_r+\tfrac{2}{3}\pi\right) & -L_{ms}\cos\theta_r & -L_{ms}\cos\left(\theta_r-\tfrac{2}{3}\pi\right) \\ -L_{ms}\cos\left(\theta_r-\tfrac{2}{3}\pi\right) & -L_{ms}\cos\left(\theta_r+\tfrac{2}{3}\pi\right) & -L_{ms}\cos\theta_r \end{bmatrix}$$

$$\begin{bmatrix} -L_{ms}\cos\theta_r & -L_{ms}\cos\left(\theta_r+\tfrac{2}{3}\pi\right) & -L_{ms}\cos\left(\theta_r-\tfrac{2}{3}\pi\right) \\ -L_{ms}\cos\left(\theta_r-\tfrac{2}{3}\pi\right) & -L_{ms}\cos\theta_r & -L_{ms}\cos\left(\theta_r+\tfrac{2}{3}\pi\right) \\ -L_{ms}\cos\left(\theta_r+\tfrac{2}{3}\pi\right) & -L_{ms}\cos\left(\theta_r-\tfrac{2}{3}\pi\right) & -L_{ms}\cos\theta_r \\ 2L_{ms}+L'_{lr} & \tfrac{1}{2}L_{ms} & \tfrac{1}{2}L_{ms} \\ \tfrac{1}{2}L_{ms} & 2L_{ms}+L'_{lr} & \tfrac{1}{2}L_{ms} \\ \tfrac{1}{2}L_{ms} & \tfrac{1}{2}L_{ms} & 2L_{ms}+L'_{lr} \end{bmatrix}\begin{bmatrix} u_{as} \\ u_{bs} \\ u_{cs} \\ u'_{ar} \\ u'_{br} \\ u'_{cr} \end{bmatrix}\qquad(6.19)$$

Here the following notations are used:

$$L_{\Sigma L} = (3L_{ms} + L'_{lr})L'_{lr}, \quad L_{\Sigma m} = 2L_{ms} + L'_{lr}, \quad \text{and} \quad L_{\Sigma ms} = \tfrac{3}{2}L_{ms}^2 + L_{ms}L'_{lr}$$

Newton's second law is applied to derive the torsional-mechanical equations, and the expression for the electromagnetic torque must be obtained.

For P-pole three-phase induction machines, as one finds the expression for coenergy $W_c(\mathbf{i}_{abcs}, \mathbf{i}'_{abcr}, \theta_r)$, the electromagnetic torque can be straightforwardly derived as

$$T_e = \frac{P}{2} \frac{\partial W_c(\mathbf{i}_{abcs}, \mathbf{i}'_{abcr}, \theta_r)}{\partial \theta_r}$$

For three-phase induction micromotors we have

$$W_c = W_f = \tfrac{1}{2} \mathbf{i}_{abcs}^T (\mathbf{L}_s - L_{ls}\mathbf{I}) \mathbf{i}_{abcs} + \mathbf{i}_{abcs}^T \mathbf{L}'_{sr}(\theta_r) \mathbf{i}'_{abcr} + \tfrac{1}{2} \mathbf{i}_{abcr}'^T (\mathbf{L}'_r - L'_{lr}\mathbf{I}) \mathbf{i}'_{abcr}$$

Matrices $\mathbf{L}_s$ and $\mathbf{L}'_r$, as well as leakage inductances L_{ls} and L'_{lr}, are not functions of the electrical displacement θ_r. Therefore, we have

$$T_e = \frac{P}{2} \mathbf{i}_{abcs}^T \frac{\partial \mathbf{L}'_{sr}(\theta_r)}{\partial \theta_r} \mathbf{i}'_{abcr}$$

$$= -\frac{P}{2} L_{ms} [i_{as} \quad i_{bs} \quad i_{cs}] \begin{bmatrix} \sin\theta_r & \sin(\theta_r + \tfrac{2}{3}\pi) & \sin(\theta_r - \tfrac{2}{3}\pi) \\ \sin(\theta_r - \tfrac{2}{3}\pi) & \sin\theta_r & \sin(\theta_r + \tfrac{2}{3}\pi) \\ \sin(\theta_r + \tfrac{2}{3}\pi) & \sin(\theta_r - \tfrac{2}{3}\pi) & \sin\theta_r \end{bmatrix} \begin{bmatrix} i'_{ar} \\ i'_{br} \\ i'_{cr} \end{bmatrix}$$

$$= -\frac{P}{2} L_{ms} \left\{ \left[i_{as} \left(i'_{ar} - \tfrac{1}{2} i'_{br} - \tfrac{1}{2} i'_{cr} \right) + i_{bs} \left(i'_{br} - \tfrac{1}{2} i'_{ar} - \tfrac{1}{2} i'_{cr} \right) + i_{cs} \left(i'_{cr} - \tfrac{1}{2} i'_{br} - \tfrac{1}{2} i'_{ar} \right) \right] \sin\theta_r \right.$$

$$\left. + \tfrac{\sqrt{3}}{2} [i_{as}(i'_{br} - i'_{cr}) + i_{bs}(i'_{cr} - i'_{ar}) + i_{cs}(i'_{ar} - i'_{br})] \cos\theta_r \right\} \tag{6.20}$$

Using Newton's second law and Equation 6.20, the torsional-mechanical equations are found to be

$$\frac{d\omega_r}{dt} = \frac{P}{2J} T_e - \frac{B_m}{J} \omega_r - \frac{P}{2J} T_L$$

$$= -\frac{P^2}{4J} L_{ms} \left\{ \left[i_{as} \left(i'_{ar} - \tfrac{1}{2} i'_{br} - \tfrac{1}{2} i'_{cr} \right) + i_{bs} \left(i'_{br} - \tfrac{1}{2} i'_{ar} - \tfrac{1}{2} i'_{cr} \right) + i_{cs} \left(i'_{cr} - \tfrac{1}{2} i'_{br} - \tfrac{1}{2} i'_{ar} \right) \right] \sin\theta_r \right.$$

$$\left. + \tfrac{\sqrt{3}}{2} [i_{as}(i'_{br} - i'_{cr}) + i_{bs}(i'_{cr} - i'_{ar}) + i_{cs}(i'_{ar} - i'_{br})] \cos\theta_r \right\} - \frac{B_m}{J} \omega_r - \frac{P}{2J} T_L \tag{6.21}$$

$$\frac{d\theta_r}{dt} = \omega_r$$

Augmenting Equation 6.19 and Equation 6.21, the mathematical model for three-phase induction micromotors in the machine variables results.

6.7.3.2 Dynamics of Induction Micromotors in the Arbitrary Reference Frame

The *abc* stator and rotor variables must be transformed to the quadrature, direct, and zero quantities. To transform the machine (*abc*) stator voltages, currents, and flux linkages to the quadrature-, direct-, and zero-axis components of stator voltages, currents, and flux linkages, the direct Park transformation is used:

$$\mathbf{u}_{qdos} = \mathbf{K}_s \mathbf{u}_{abcs}, \quad \mathbf{i}_{qdos} = \mathbf{K}_s \mathbf{i}_{abcs}, \quad \mathbf{\psi}_{qdos} = \mathbf{K}_s \mathbf{\psi}_{abcs} \tag{6.22}$$

where the stator transformation matrix $\mathbf{K}_s$ is given by

$$\mathbf{K}_s = \tfrac{2}{3} \begin{bmatrix} \cos\theta & \cos\left(\theta - \tfrac{2}{3}\pi\right) & \cos\left(\theta + \tfrac{2}{3}\pi\right) \\ \sin\theta & \sin\left(\theta - \tfrac{2}{3}\pi\right) & \sin\left(\theta + \tfrac{2}{3}\pi\right) \\ \tfrac{1}{2} & \tfrac{1}{2} & \tfrac{1}{2} \end{bmatrix} \tag{6.23}$$

Here the angular displacement of the reference frame is $\theta = \int_{t_0}^{t} \omega(\tau)d\tau + \theta_0$.

Using the rotor transformations matrix $\mathbf{K}_r$, the quadrature-, direct-, and zero-axis components of rotor voltages, currents, and flux linkages are found by using the *abc* rotor voltages, currents, and flux linkages. In particular,

$$\mathbf{u}'_{qdor} = \mathbf{K}_r \mathbf{u}'_{abcr}, \quad \mathbf{i}'_{qdor} = \mathbf{K}_r \mathbf{i}'_{abcr}, \quad \mathbf{\psi}'_{qdor} = \mathbf{K}_r \mathbf{\psi}'_{abcr} \tag{6.24}$$

where the rotor transformation matrix is

$$\mathbf{K}_r = \tfrac{2}{3} \begin{bmatrix} \cos\left(\theta - \theta_r\right) & \cos\left(\theta - \theta_r - \tfrac{2}{3}\pi\right) & \cos\left(\theta - \theta_r + \tfrac{2}{3}\pi\right) \\ \sin\left(\theta - \theta_r\right) & \sin\left(\theta - \theta_r - \tfrac{2}{3}\pi\right) & \sin\left(\theta - \theta_r + \tfrac{2}{3}\pi\right) \\ \tfrac{1}{2} & \tfrac{1}{2} & \tfrac{1}{2} \end{bmatrix} \tag{6.25}$$

From Equation 6.18,

$$\mathbf{u}_{abcs} = \mathbf{r}_s \mathbf{i}_{abcs} + \frac{d\mathbf{\psi}_{abcs}}{dt}, \quad \mathbf{u}'_{abcr} = \mathbf{r}'_r \mathbf{i}'_{abcr} + \frac{d\mathbf{\psi}'_{abcr}}{dt},$$

by taking note of the inverse Park transformation matrices $\mathbf{K}_s^{-1}$ and $\mathbf{K}_r^{-1}$, we have

$$\mathbf{K}_s^{-1}\mathbf{u}_{qdos} = \mathbf{r}_s \mathbf{K}_s^{-1}\mathbf{i}_{qdos} + \frac{d\left(\mathbf{K}_s^{-1}\mathbf{\psi}_{qdos}\right)}{dt}, \quad \mathbf{K}_r^{-1}\mathbf{u}'_{qdor} = \mathbf{r}'_r \mathbf{K}_r^{-1}\mathbf{i}'_{qdor} + \frac{d\left(\mathbf{K}_r^{-1}\mathbf{\psi}'_{qdor}\right)}{dt} \tag{6.26}$$

Making use of Equation 6.23 and Equation 6.25, one finds inverse matrices $\mathbf{K}_s^{-1}$ and $\mathbf{K}_r^{-1}$. In particular,

$$\mathbf{K}_s^{-1} = \begin{bmatrix} \cos\theta & \sin\theta & 1 \\ \cos\left(\theta - \tfrac{2}{3}\pi\right) & \sin\left(\theta - \tfrac{2}{3}\pi\right) & 1 \\ \cos\left(\theta + \tfrac{2}{3}\pi\right) & \sin\left(\theta + \tfrac{2}{3}\pi\right) & 1 \end{bmatrix} \quad \text{and}$$

$$\mathbf{K}_r^{-1} = \begin{bmatrix} \cos\left(\theta - \theta_r\right) & \sin\left(\theta - \theta_r\right) & 1 \\ \cos\left(\theta - \theta_r - \tfrac{2}{3}\pi\right) & \sin\left(\theta - \theta_r - \tfrac{2}{3}\pi\right) & 1 \\ \cos\left(\theta - \theta_r + \tfrac{2}{3}\pi\right) & \sin\left(\theta - \theta_r + \tfrac{2}{3}\pi\right) & 1 \end{bmatrix}$$

Multiplying the left and right sides of Equation 6.26 by $\mathbf{K}_s$ and $\mathbf{K}_r$, one has

$$\mathbf{u}_{qdos} = \mathbf{K}_s \mathbf{r}_s \mathbf{K}_s^{-1}\mathbf{i}_{qdos} + \mathbf{K}_s \frac{d\mathbf{K}_s^{-1}}{dt}\mathbf{\psi}_{qdos} + \mathbf{K}_s \mathbf{K}_s^{-1}\frac{d\mathbf{\psi}_{qdos}}{dt}$$

$$\mathbf{u}'_{qdor} = \mathbf{K}_r \mathbf{r}'_r \mathbf{K}_r^{-1}\mathbf{i}'_{qdor} + \mathbf{K}_r \frac{d\mathbf{K}_r^{-1}}{dt}\mathbf{\psi}'_{qdor} + \mathbf{K}_r \mathbf{K}_r^{-1}\frac{d\mathbf{\psi}'_{qdor}}{dt} \tag{6.27}$$

The matrices of the stator and rotor resistances $\mathbf{r}_s$ and $\mathbf{r}'_r$ are diagonal, and hence,

$$\mathbf{K}_s \mathbf{r}_s \mathbf{K}_s^{-1} = \mathbf{r}_s \quad \text{and} \quad \mathbf{K}_r \mathbf{r}'_r \mathbf{K}_r^{-1} = \mathbf{r}'_r$$

Performing differentiation, one finds

$$\frac{d\mathbf{K}_s^{-1}}{dt} = \omega \begin{bmatrix} -\sin\theta & \cos\theta & 0 \\ -\sin\left(\theta - \frac{2}{3}\pi\right) & \cos\left(\theta - \frac{2}{3}\pi\right) & 0 \\ -\sin\left(\theta + \frac{2}{3}\pi\right) & \cos\left(\theta + \frac{2}{3}\pi\right) & 0 \end{bmatrix}$$

$$\frac{d\mathbf{K}_r^{-1}}{dt} = (\omega - \omega_r)\begin{bmatrix} -\sin\left(\theta - \theta_r\right) & \cos\left(\theta - \theta_r\right) & 0 \\ -\sin\left(\theta - \theta_r - \frac{2}{3}\pi\right) & \cos\left(\theta - \theta_r - \frac{2}{3}\pi\right) & 0 \\ -\sin\left(\theta - \theta_r + \frac{2}{3}\pi\right) & \cos\left(\theta - \theta_r + \frac{2}{3}\pi\right) & 0 \end{bmatrix}$$

Therefore, we have

$$\mathbf{K}_s \frac{d\mathbf{K}_s^{-1}}{dt} = \omega \begin{bmatrix} 0 & 1 & 0 \\ -1 & 0 & 0 \\ 0 & 0 & 0 \end{bmatrix} \quad \text{and} \quad \mathbf{K}_r \frac{d\mathbf{K}_r^{-1}}{dt} = (\omega - \omega_r)\begin{bmatrix} 0 & 1 & 0 \\ -1 & 0 & 0 \\ 0 & 0 & 0 \end{bmatrix}$$

One obtains the voltage equations for stator and rotor circuits in the arbitrary reference frame when the angular velocity of the reference frame ω is not specified. From Equation 6.27 the following state-space differential equations result:

$$\mathbf{u}_{qdos} = \mathbf{r}_s \mathbf{i}_{qdos} + \begin{bmatrix} 0 & \omega & 0 \\ -\omega & 0 & 0 \\ 0 & 0 & 0 \end{bmatrix}\mathbf{\psi}_{qdos} + \frac{d\mathbf{\psi}_{qdos}}{dt}$$

$$\mathbf{u}'_{qdor} = \mathbf{r}'_r \mathbf{i}'_{qdor} + \begin{bmatrix} 0 & \omega - \omega_r & 0 \\ -\omega + \omega_r & 0 & 0 \\ 0 & 0 & 0 \end{bmatrix}\mathbf{\psi}'_{qdor} + \frac{d\mathbf{\psi}'_{qdor}}{dt}$$

(6.28)

From Equation 6.28, six differential equations in expanded form are found to model the stator and rotor circuitry dynamics. In particular,

$$u_{qs} = r_s i_{qs} + \omega \psi_{ds} + \frac{d\psi_{qs}}{dt}$$

$$u_{ds} = r_s i_{ds} - \omega \psi_{qs} + \frac{d\psi_{ds}}{dt}$$

$$u_{os} = r_s i_{os} + \frac{d\psi_{os}}{dt}$$

(6.29)

$$u'_{qr} = r'_r i'_{qr} + (\omega - \omega_r)\psi'_{dr} + \frac{d\psi'_{dr}}{dt}$$

$$u'_{dr} = r'_r i'_{dr} - (\omega - \omega_r)\psi'_{qr} + \frac{d\psi'_{dr}}{dt}$$

$$u'_{or} = r'_r i'_{or} + \frac{d\psi'_{or}}{dt}$$

Using the matrix equation for flux linkages

$$\begin{bmatrix} \mathbf{\psi}_{abcs} \\ \mathbf{\psi}'_{abcr} \end{bmatrix} = \begin{bmatrix} \mathbf{L}_s & \mathbf{L}'_{sr} \\ \mathbf{L}'^{T}_{sr} & \mathbf{L}'_r \end{bmatrix}\begin{bmatrix} \mathbf{i}_{abcs} \\ \mathbf{i}'_{abcr} \end{bmatrix}$$

we have

$$\boldsymbol{\psi}_{abcs} = \mathbf{L}_s \mathbf{i}_{abcs} + \mathbf{L}'_{sr} \mathbf{i}'_{abcr}, \qquad \boldsymbol{\psi}'_{abcr} = \mathbf{L}'^{T}_{sr} \mathbf{i}_{abcs} + \mathbf{L}'_r \mathbf{i}_{abcr}$$

These equations can be represented using the quadrature, direct, and zero quantities. Employing the Park transformation matrices, one has

$$\mathbf{K}_s^{-1} \boldsymbol{\psi}_{qdos} = \mathbf{L}_s \mathbf{K}_s^{-1} \mathbf{i}_{qdos} + \mathbf{L}'_{sr} \mathbf{K}_r^{-1} \mathbf{i}'_{qdor}$$

$$\mathbf{K}_r^{-1} \boldsymbol{\psi}'_{qdor} = \mathbf{L}'^{T}_{sr} \mathbf{K}_s^{-1} \mathbf{i}_{qdos} + \mathbf{L}'_r \mathbf{K}_r^{-1} \mathbf{i}'_{abcr}$$

Thus, we obtain

$$\boldsymbol{\psi}_{qdos} = \mathbf{K}_s \mathbf{L}_s \mathbf{K}_s^{-1} \mathbf{i}_{qdos} + \mathbf{K}_s \mathbf{L}'_{sr} \mathbf{K}_r^{-1} \mathbf{i}'_{qdor}$$

$$\boldsymbol{\psi}'_{qdor} = \mathbf{K}_r \mathbf{L}'^{T}_{sr} \mathbf{K}_s^{-1} \mathbf{i}_{qdos} + \mathbf{K}_r \mathbf{L}'_r \mathbf{K}_r^{-1} \mathbf{i}'_{abcr}$$

(6.30)

Taking note of the Park transformation matrices and applying the derived expressions for $\mathbf{L}_s$, $\mathbf{L}_{sr}$, and $\mathbf{L}_r$, by multiplying the matrices the following result are found:

$$\mathbf{K}_s \mathbf{L}_s \mathbf{K}_s^{-1} = \begin{bmatrix} L_{ls} + M & 0 & 0 \\ 0 & L_{ls} + M & 0 \\ 0 & 0 & L_{ls} \end{bmatrix}, \qquad \mathbf{K}_s \mathbf{L}'_{sr} \mathbf{K}_r^{-1} = \mathbf{K}_r \mathbf{L}'^{T}_{sr} \mathbf{K}_s^{-1} \begin{bmatrix} M & 0 & 0 \\ 0 & M & 0 \\ 0 & 0 & 0 \end{bmatrix}$$

and

$$\mathbf{K}_r \mathbf{L}'_r \mathbf{K}_r^{-1} = \begin{bmatrix} L'_{lr} + M & 0 & 0 \\ 0 & L'_{lr} + M & 0 \\ 0 & 0 & L'_{lr} \end{bmatrix}, \qquad M = \tfrac{3}{2} L_{ms}$$

In expanded form, the flux linkage equations (Equation 6.30) are

$$\boldsymbol{\psi}_{qs} = L_{lr} i_{qs} + M i_{qs} + M i'_{qr}, \quad \boldsymbol{\psi}_{ds} = L_{ls} i_{ds} + M i_{ds} + M i'_{dr}, \quad \boldsymbol{\psi}_{os} = L_{ls} i_{os}$$

$$\boldsymbol{\psi}'_{qr} = L'_{lr} i'_{qr} + M i_{qs} + M i'_{qr}, \quad \boldsymbol{\psi}'_{dr} = L'_{lr} i'_{dr} + M i_{ds} + M i'_{dr}, \quad \boldsymbol{\psi}'_{or} = L'_{lr} i'_{or}$$

(6.31)

Using the expressions of Equation 6.31 in Equation 6.29, the following differential equations result:

$$u_{qs} = r_s i_{qs} + \omega(L_{ls} i_{ds} + M i_{ds} + M i'_{dr}) + \frac{d(L_{ls} i_{qs} + M i_{qs} + M i'_{qr})}{dt}$$

$$u_{ds} = r_s i_{ds} - \omega(L_{ls} i_{qs} + M i_{qs} + M i'_{qr}) + \frac{d(L_{ls} i_{ds} + M i_{ds} + M i'_{qr})}{dt}$$

$$u_{os} = r_s i_{os} + \frac{d(L_{ls} i_{os})}{dt}$$

$$u'_{qr} = r'_r i'_{qr} + (\omega - \omega_r)(L'_{lr} i'_{dr} + M i_{ds} + M i'_{dr}) + \frac{d(L'_{lr} i'_{qr} + M i_{qs} + M i'_{qr})}{dt}$$

$$u'_{dr} = r'_r i'_{dr} - (\omega - \omega_r)(L'_{lr} i'_{qr} + M i_{qs} + M i'_{qr}) + \frac{d(L'_{lr} i'_{qr} + M i_{ds} + M i'_{qr})}{dt}$$

$$u'_{or} = r'_r i'_{or} + \frac{d(L'_{lr} i'_{or})}{dt}$$

Cauchy's form of differential equations is

$$\frac{di_{qs}}{dt} = \frac{1}{L_{SM}L_{RM} - M^2}[-L_{RM}r_s i_{qs} - (L_{SM}L_{RM} - M^2)\omega i_{ds} + Mr_r' i_{qr}'$$
$$- M(Mi_{ds} + L_{RM}i_{dr}')\omega_r + L_{RM}u_{qs} - Mu_{qr}']$$

$$\frac{di_{ds}}{dt} = \frac{1}{L_{SM}L_{RM} - M^2}[(L_{SM}L_{RM} - M^2)\omega i_{qs} - L_{RM}r_s i_{ds} + Mr_r' i_{dr}'$$
$$+ M(Mi_{qs} + L_{RM}i_{qr}')\omega_r + L_{RM}u_{ds} - Mu_{dr}']$$

$$\frac{di_{os}}{dt} = \frac{1}{L_{ls}}(-r_s i_{os} + u_{os}) \qquad\qquad (6.32)$$

$$\frac{di_{qr}'}{dt} = \frac{1}{L_{SM}L_{RM} - M^2}[Mr_s i_{qs} - L_{SM}r_r' i_{qr}' - (L_{SM}L_{RM} - M^2)\omega i_{dr}'$$
$$+ L_{SM}(Mi_{ds} + L_{RM}i_{dr}')\omega_r - Mu_{qs} + L_{SM}u_{qr}']$$

$$\frac{di_{dr}'}{dt} = \frac{1}{L_{SM}L_{RM} - M^2}[Mr_s i_{ds} + (L_{SM}L_{RM} - M^2)\omega i_{qr}' - L_{SM}r_r' i_{dr}'$$
$$- L_{SM}(Mi_{qs} + L_{RM}i_{qr}')\omega_r - Mu_{ds} + L_{SM}u_{dr}']$$

$$\frac{di_{or}'}{dt} = \frac{1}{L_{lr}'}(-r_r' i_{or}' + u_{or}')$$

where $L_{SM} = L_{ls} + M = L_{ls} + \frac{3}{2}L_{ms}$ and $L_{RM} = L_{lr}' + M = L_{lr}' + \frac{3}{2}L_{ms}$.

One concludes that the nonlinear differential equations are found to describe the stator-rotor circuitry transient behavior of three-phase induction micromotors. To complete the model developments, the torsional-mechanical equations

$$T_e - B_m\omega_{rm} - T_L = J\frac{d\omega_{rm}}{dt}$$
$$\frac{d\theta_{rm}}{dt} = \omega_{rm} \qquad\qquad (6.33)$$

must be used.

The equation for the electromagnetic torque must be obtained in terms of the quadrature- and direct-axis components of stator and rotor currents.

Using the formula for coenergy

$$W_c = \frac{1}{2}\mathbf{i}_{abcs}^T(\mathbf{L}_s - L_{ls}\mathbf{I})\mathbf{i}_{abcs} + \mathbf{i}_{abcs}^T\mathbf{L}_{sr}'(\theta_r)\mathbf{i}_{abcr}' + \frac{1}{2}\mathbf{i}_{abcr}'^T(\mathbf{L}_r' - L_{lr}'\mathbf{I})\mathbf{i}_{abcr}'$$

one finds

$$T_e = \frac{P}{2}\frac{\partial W_c(\mathbf{i}_{abcs}, \mathbf{i}_{abcr}', \theta_r)}{\partial\theta_r} = \frac{P}{2}\mathbf{i}_{abcs}^T\frac{\partial\mathbf{L}_{sr}'(\theta_r)}{\partial\theta_r}\mathbf{i}_{abcr}'$$

Hence, we have

$$T_e = \frac{P}{2}\left(\mathbf{K}_s^{-1}\mathbf{i}_{qdos}\right)^T\frac{\partial\mathbf{L}_{sr}'(\theta_r)}{\partial\theta_r}\mathbf{K}_r^{-1}\mathbf{i}_{qdor}' = \frac{P}{2}\mathbf{i}_{qdos}^T\mathbf{K}_s^{-1^T}\frac{\partial\mathbf{L}_{sr}'(\theta_r)}{\partial\theta_r}\mathbf{K}_r^{-1}\mathbf{i}_{qdor}'$$

By performing multiplication of matrices, the following formula results:

$$T_e = \frac{3P}{4} M(i_{qs} i'_{dr} - i_{ds} i'_{qr})$$ (6.34)

Thus, from Equation 6.33 and Equation 6.34, one has

$$\frac{d\omega_r}{dt} = \frac{3P^2}{8J} M(i_{qs} i'_{dr} - i_{ds} i'_{qr}) - \frac{B_m}{J} \omega_r - \frac{P}{2J} T_L$$

$$\frac{d\theta_r}{dt} = \omega_r$$ (6.35)

Augmenting the circuitry and torsional-mechanical dynamics, as given by Equation 6.32 and Equation 6.35, the model for three-phase induction micromotors in the arbitrary reference frame results. We have a set of eight highly coupled nonlinear differential equations:

$$\frac{di_{qs}}{dt} = \frac{1}{L_{SM}L_{RM} - M^2}[-L_{RM} r_s i_{qs} - (L_{SM}L_{RM} - M^2)\omega i_{ds} + Mr'_r i'_{qr}$$

$$- M(Mi_{ds} + L_{RM}i'_{dr})\omega_r + L_{RM}u_{qs} - Mu'_{qr}]$$

$$\frac{di_{ds}}{dt} = \frac{1}{L_{SM}L_{RM} - M^2}[(L_{SM}L_{RM} - M^2)\omega i_{qs} - L_{RM} r_s i_{ds} + Mr'_r i'_{dr}$$

$$+ M(Mi_{qs} + L_{RM}i'_{qr})\omega_r + L_{RM}u_{ds} - Mu'_{dr}]$$

$$\frac{di_{os}}{dt} = \frac{1}{L_{ls}}(-r_s i_{os} + u_{os})$$

$$\frac{di'_{qr}}{dt} = \frac{1}{L_{SM}L_{RM} - M^2}[Mr_s i_{qs} - L_{SM} r'_r i'_{qr} - (L_{SM}L_{RM} - M^2)\omega i'_{dr}$$

$$+ L_{SM}(Mi_{ds} + L_{RM}i'_{dr})\omega_r - Mu_{qs} + L_{SM}u'_{qr}]$$ (6.36)

$$\frac{di'_{dr}}{dt} = \frac{1}{L_{SM}L_{RM} - M^2}[Mr_s i_{ds} + (L_{SM}L_{RM} - M^2)\omega i'_{dr} - L_{SM} r'_r i'_{dr}$$

$$- L_{SM}(Mi_{qs} + L_{RM}i'_{qr})\omega_r - Mu_{ds} + L_{SM}u'_{dr}]$$

$$\frac{di'_{or}}{dt} = \frac{1}{L'_{lr}}(-r'_r i'_{or} + u'_{or})$$

$$\frac{d\omega_r}{dt} = \frac{3P^2}{8J} M(i_{qs} i'_{dr} - i_{ds} i'_{qr}) - \frac{B_m}{J} \omega_r - \frac{P}{2J} T_L$$

$$\frac{d\theta_r}{dt} = \omega_r$$

The last differential equation in the set (Equation 6.36) can be omitted in the analysis and simulations if induction micromotors are used in microdrives. That is, for electric drives one finds the following state-space equation:

$$
\begin{bmatrix}
\dfrac{di_{qs}}{dt} \\[2mm]
\dfrac{di_{ds}}{dt} \\[2mm]
\dfrac{di_{os}}{dt} \\[2mm]
\dfrac{di_{qr}'}{dt} \\[2mm]
\dfrac{di_{dr}'}{dt} \\[2mm]
\dfrac{di_{or}'}{dt} \\[2mm]
\dfrac{d\omega_r}{dt}
\end{bmatrix}
=
\begin{bmatrix}
-\dfrac{L_{RM}r_s}{L_{SM}L_{RM}-M^2} & -\omega & 0 & \dfrac{Mr_r'}{L_{SM}L_{RM}-M^2} & 0 & 0 & 0 \\[3mm]
\omega & -\dfrac{L_{RM}r_s}{L_{SM}L_{RM}-M^2} & 0 & 0 & \dfrac{Mr_r'}{L_{SM}L_{RM}-M^2} & 0 & 0 \\[3mm]
0 & 0 & -\dfrac{r_s}{L_{ls}} & 0 & 0 & 0 & 0 \\[3mm]
\dfrac{Mr_s}{L_{SM}L_{RM}-M^2} & 0 & 0 & -\dfrac{L_{SM}r_r'}{L_{SM}L_{RM}-M^2} & -\omega & 0 & 0 \\[3mm]
0 & \dfrac{Mr_s}{L_{SM}L_{RM}-M^2} & 0 & \omega & -\dfrac{L_{SM}r_r'}{L_{SM}L_{RM}-M^2} & 0 & 0 \\[3mm]
0 & 0 & 0 & 0 & 0 & -\dfrac{r_r'}{L_{lr}'} & 0 \\[3mm]
0 & 0 & 0 & 0 & 0 & 0 & -\dfrac{B_m}{J}
\end{bmatrix}
$$

$$
\times
\begin{bmatrix}
i_{qs} \\[2mm]
i_{ds} \\[2mm]
i_{os} \\[2mm]
i_{qr}' \\[2mm]
i_{dr}' \\[2mm]
i_{or}' \\[2mm]
\omega_r
\end{bmatrix}
+
\begin{bmatrix}
-\dfrac{M(Mi_{ds}+L_{RM}i_{dr}')\omega\varphi}{L_{SM}L_{RM}-M^2} \\[3mm]
\dfrac{M(Mi_{qs}+L_{RM}i_{qr}')\omega_r}{L_{SM}L_{RM}-M^2} \\[3mm]
0 \\[3mm]
\dfrac{L_{SM}(Mi_{ds}+L_{RM}i_{dr}')\omega_r}{L_{SM}L_{RM}-M^2} \\[3mm]
-\dfrac{L_{SM}(Mi_{qs}+L_{RM}i_{qr}')\omega_r}{L_{SM}L_{RM}-M^2} \\[3mm]
0 \\[3mm]
\dfrac{3P^2}{8J}M(i_{qs}i_{dr}'-i_{ds}i_{qr}')
\end{bmatrix}
$$

$$
+
\begin{bmatrix}
\dfrac{L_{RM}}{L_{SM}L_{RM}-M^2} & 0 & 0 & -\dfrac{M}{L_{SM}L_{RM}-M^2} & 0 & 0 \\[3mm]
0 & \dfrac{L_{RM}}{L_{SM}L_{RM}-M^2} & 0 & 0 & -\dfrac{M}{L_{SM}L_{RM}-M^2} & 0 \\[3mm]
0 & 0 & \dfrac{1}{L_{ls}} & 0 & 0 & 0 \\[3mm]
-\dfrac{M}{L_{SM}L_{RM}-M^2} & 0 & 0 & \dfrac{L_{SM}}{L_{SM}L_{RM}-M^2} & 0 & 0 \\[3mm]
0 & -\dfrac{M}{L_{SM}L_{RM}-M^2} & 0 & 0 & \dfrac{L_{SM}}{L_{SM}L_{RM}-M^2} & 0 \\[3mm]
0 & 0 & 0 & 0 & 0 & \dfrac{1}{L_{lr}'} \\[3mm]
0 & 0 & 0 & 0 & 0 & 0
\end{bmatrix}
\begin{bmatrix}
u_{qs} \\[2mm]
u_{ds} \\[2mm]
u_{os} \\[2mm]
u_{qr}' \\[2mm]
u_{dr}' \\[2mm]
u_{or}'
\end{bmatrix}
-
\begin{bmatrix}
0 \\[2mm]
0 \\[2mm]
0 \\[2mm]
0 \\[2mm]
0 \\[2mm]
0 \\[2mm]
\dfrac{P}{2J}
\end{bmatrix}
T_L
$$

(6.37)

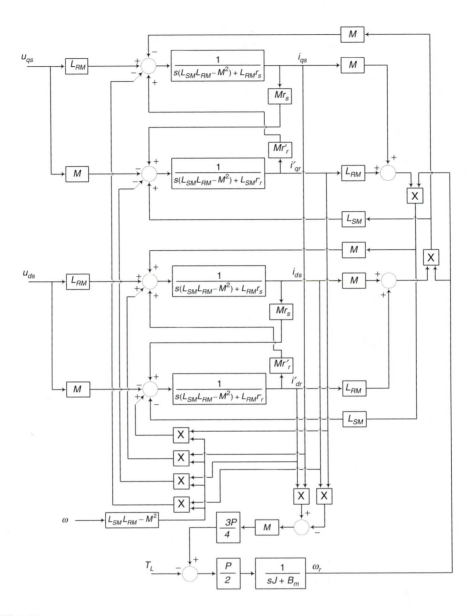

FIGURE 6.44
Block diagram of three-phase squirrel-cage induction micromotors in the *arbitrary* reference frame.

The block diagram for three-phase induction micromotors modeled in the arbitrary reference frame is developed using Equation 6.37. In particular, applying the Laplace operator, one finds the block diagram as shown in Figure 6.44.

Micro- and miniscale induction motors are squirrel-cage motors, and the rotor windings are short-circuited. To guarantee the balanced operating conditions, one supplies the following balanced three-phase voltages:

$$u_{as}(t) = \sqrt{2}u_M \cos(\omega_f t), \quad u_{bs}(t) = \sqrt{2}u_M \cos\left(\omega_f t - \tfrac{2}{3}\pi\right), \quad \text{and}$$

$$u_{cs}(t) = \sqrt{2}u_M \cos\left(\omega_f t + \tfrac{2}{3}\pi\right)$$

where the frequency of the applied voltage is $\omega_f = 2\pi f$.

The quadrature-, direct-, and zero-axis components of stator voltages are obtained by using the stator Park transformation matrix as

$$\mathbf{u}_{qdos} = \mathbf{K}_s \mathbf{u}_{abcs}$$

where

$$\mathbf{K}_s = \frac{2}{3}\begin{bmatrix} \cos\theta & \cos\left(\theta - \frac{2}{3}\pi\right) & \cos\left(\theta + \frac{2}{3}\pi\right) \\ \sin\theta & \sin\left(\theta - \frac{2}{3}\pi\right) & \sin\left(\theta + \frac{2}{3}\pi\right) \\ \frac{1}{2} & \frac{1}{2} & \frac{1}{2} \end{bmatrix}$$

is the Park transformation matrix.

The stationary, rotor, and synchronous reference frames are commonly used. For stationary, rotor, and synchronous reference frames, the reference frame angular velocities are

$$\omega = 0, \quad \omega = \omega_r, \quad \text{and} \quad \omega = \omega_e$$

Hence, the corresponding angular displacement θ results. In particular, for zero initial conditions for stationary, rotor, and synchronous reference frames, one finds

$$\theta = 0, \quad \theta = \theta_r, \quad \text{and} \quad \theta = \theta_e$$

Hence, the quadrature-, direct-, and zero-axis components of voltages can be obtained to guarantee the balance operation of induction micromotors.

Example 6.27 THREE-PHASE INDUCTION MICROMOTORS in the SYNCHRONOUS REFERENCE FRAME

Develop a mathematical model for three-phase induction micromotors in the synchronous reference frame. Develop an *s*-domain block diagram using the differential equations found.

SOLUTION

The most commonly used is the synchronous reference frame. The lumped-parameter mathematical model of three-phase induction micromotors in the synchronous reference frame is found by substituting the frame angular velocity in the differential equations obtained for the arbitrary reference frame (Equation 6.37). Using $\omega = \omega_e$ in Equation 6.37, we have

$$\frac{di_{qs}^e}{dt} = \frac{1}{L_{SM}L_{RM} - M^2}\Big[-L_{RM}r_s i_{qs}^e - (L_{SM}L_{RM} - M^2)\omega_e i_{ds}^e + Mr_r' i_{qr}'^e$$

$$- M\big(Mi_{ds}^e + L_{RM}i_{dr}'^e\big)\omega_r + L_{RM}u_{qs}^e - Mu_{qr}'^e\Big]$$

$$\frac{di_{ds}^e}{dt} = \frac{1}{L_{SM}L_{RM} - M^2}\Big[(L_{SM}L_{RM} - M^2)\omega_e i_{qs}^e - L_{RM}r_s i_{ds}^e + Mr_r' i_{dr}'^e$$

$$+ M\Big(Mi_{qs}^e + L_{RM}i_{qr}'^e\Big)\omega_r + L_{RM}u_{ds}^e - Mu_{dr}'^e\Big]$$

$$\frac{di_{os}^e}{dt} = \frac{1}{L_{ls}}\Big(-r_s i_{os}^e + u_{os}^e\Big)$$

$$\frac{di_{qr}'^e}{dt} = \frac{1}{L_{SM}L_{RM} - M^2}\Big[Mr_s i_{qs}^e - L_{SM}r_r' i_{qr}'^e - (L_{SM}L_{RM} - M^2)\omega_e i_{dr}'^e$$

$$+ L_{SM}\Big(Mi_{ds}^e + L_{RM}i_{dr}'^e\Big)\omega_r - Mu_{qs}^e + L_{SM}u_{qr}'^e\Big]$$

$$\frac{di_{dr}'^e}{dt} = \frac{1}{L_{SM}L_{RM} - M^2}\Big[Mr_s i_{ds}^e + (L_{SM}L_{RM} - M^2)\omega_e i_{qr}'^e - L_{SM}r_r' i_{dr}'^e$$

$$- L_{SM}\Big(Mi_{qs}^e + L_{RM}i_{qr}'^e\Big)\omega_r - Mu_{ds}^e + L_{SM}u_{dr}'^e\Big] \tag{6.38}$$

$$\frac{di_{or}'^e}{dt} = \frac{1}{L_{lr}'}\Big(-r_r' i_{or}'^e + u_{or}'^e\Big)$$

$$\frac{d\omega_r}{dt} = \frac{3P^2}{8J}M\Big(i_{qs}^e i_{dr}'^e - i_{ds}^e i_{qr}'^e\Big) - \frac{B_m}{J}\omega_r - \frac{P}{2J}T_L$$

$$\frac{d\theta_r}{dt} = \omega_r$$

The superscript e denotes the synchronous frame of reference. In the state-space form, using Equation 6.38, we have the following differential equation for electric drives:

$$
\begin{bmatrix} \dfrac{di_{qs}^e}{dt} \\[2ex] \dfrac{di_{ds}^e}{dt} \\[2ex] \dfrac{di_{os}^e}{dt} \\[2ex] \dfrac{di_{qr}'^e}{dt} \\[2ex] \dfrac{di_{dr}'^e}{dt} \\[2ex] \dfrac{di_{or}'^e}{dt} \\[2ex] \dfrac{d\omega_r}{dt} \end{bmatrix} =
\begin{bmatrix}
-\dfrac{L_{RM}r_s}{L_{SM}L_{RM}-M^2} & -\omega_e & 0 & \dfrac{Mr_r'}{L_{SM}L_{RM}-M^2} & 0 & 0 & 0 \\[2ex]
\omega_e & -\dfrac{L_{RM}r_s}{L_{SM}L_{RM}-M^2} & 0 & 0 & \dfrac{Mr_r'}{L_{SM}L_{RM}-M^2} & 0 & 0 \\[2ex]
0 & 0 & -\dfrac{r_s}{L_{ls}} & 0 & 0 & 0 & 0 \\[2ex]
\dfrac{Mr_s}{L_{SM}L_{RM}-M^2} & 0 & 0 & -\dfrac{L_{SM}r_r'}{L_{SM}L_{RM}-M^2} & -\omega_e & 0 & 0 \\[2ex]
0 & \dfrac{Mr_s}{L_{SM}L_{RM}-M^2} & 0 & \omega_e & -\dfrac{L_{SM}r_r'}{L_{SM}L_{RM}-M^2} & 0 & 0 \\[2ex]
0 & 0 & 0 & 0 & 0 & -\dfrac{r_r'}{L_{lr}'} & 0 \\[2ex]
0 & 0 & 0 & 0 & 0 & 0 & -\dfrac{B_m}{J}
\end{bmatrix}
$$

$$\times \begin{bmatrix} i_{qs}^e \\ i_{ds}^e \\ i_{os}^e \\ i_{qr}'^e \\ i_{dr}'^e \\ i_{or}'^e \\ \omega_r \end{bmatrix} + \begin{bmatrix} -\dfrac{M\left(Mi_{ds}^e + L_{RM}i_{dr}'^e\right)\omega_r}{L_{SM}L_{RM} - M^2} \\[3ex] \dfrac{M\left(Mi_{qs}^e + L_{RM}i_{qr}'^e\right)\omega_r}{L_{SM}L_{RM} - M^2} \\[3ex] 0 \\[3ex] \dfrac{L_{SM}\left(Mi_{ds}^e + L_{RM}i_{dr}'^e\right)\omega_r}{L_{SM}L_{RM} - M^2} \\[3ex] -\dfrac{L_{SM}\left(Mi_{qs}^e + L_{RM}i_{qr}'^e\right)\omega_r}{L_{SM}L_{RM} - M^2} \\[3ex] 0 \\[3ex] \dfrac{3P^2}{8J}M\left(i_{qs}^e i_{dr}'^e - i_{ds}^e i_{qr}'^e\right) \end{bmatrix}$$

$$+ \begin{bmatrix} \dfrac{L_{RM}}{L_{SM}L_{RM}-M^2} & 0 & 0 & -\dfrac{M}{L_{SM}L_{RM}-M^2} & 0 & 0 \\[3ex] 0 & \dfrac{L_{RM}}{L_{SM}L_{RM}-M^2} & 0 & 0 & -\dfrac{M}{L_{SM}L_{RM}-M^2} & 0 \\[3ex] 0 & 0 & \dfrac{1}{L_{ls}} & 0 & 0 & 0 \\[3ex] -\dfrac{M}{L_{SM}L_{RM}-M^2} & 0 & 0 & \dfrac{L_{SM}}{L_{SM}L_{RM}-M^2} & 0 & 0 \\[3ex] 0 & -\dfrac{M}{L_{SM}L_{RM}-M^2} & 0 & 0 & \dfrac{L_{SM}}{L_{SM}L_{RM}-M^2} & 0 \\[3ex] 0 & 0 & 0 & 0 & 0 & \dfrac{1}{L'_{lr}} \\[3ex] 0 & 0 & 0 & 0 & 0 & 0 \end{bmatrix} \begin{bmatrix} u_{qs}^e \\ u_{ds}^e \\ u_{os}^e \\ u_{qr}'^e \\ u_{dr}'^e \\ u_{or}'^e \end{bmatrix} - \begin{bmatrix} 0 \\ 0 \\ 0 \\ 0 \\ 0 \\ 0 \\ \dfrac{P}{2J} \end{bmatrix} T_L$$

To guarantee the balanced operation of induction micromotors, the quadrature, direct, and zero voltages u_{qs}^e, u_{ds}^e, and u_{os}^e are found making use of the following relationship:

$$\mathbf{u}_{qdos}^e = \mathbf{K}_s^e \mathbf{u}_{abcs}$$

Taking note that $\theta = \theta_e$, we have

$$\mathbf{K}_s = \tfrac{2}{3}\begin{bmatrix} \cos\theta & \cos\left(\theta - \tfrac{2}{3}\pi\right) & \cos\left(\theta + \tfrac{2}{3}\pi\right) \\[1ex] \sin\theta & \sin\left(\theta - \tfrac{2}{3}\pi\right) & \sin\left(\theta + \tfrac{2}{3}\pi\right) \\[1ex] \tfrac{1}{2} & \tfrac{1}{2} & \tfrac{1}{2} \end{bmatrix}$$

Therefore, one finds

$$\mathbf{K}_s^e = \frac{2}{3} \begin{bmatrix} \cos\theta_e & \cos\left(\theta_e - \frac{2}{3}\pi\right) & \cos\left(\theta_e + \frac{2}{3}\pi\right) \\ \sin\theta_e & \sin\left(\theta_e - \frac{2}{3}\pi\right) & \sin\left(\theta_e + \frac{2}{3}\pi\right) \\ \frac{1}{2} & \frac{1}{2} & \frac{1}{2} \end{bmatrix}$$

Taking note of

$$\begin{bmatrix} u_{qs}^e \\ u_{ds}^e \\ u_{os}^e \end{bmatrix} = \frac{2}{3} \begin{bmatrix} \cos\theta_e & \cos\left(\theta_e - \frac{2}{3}\pi\right) & \cos\left(\theta_e + \frac{2}{3}\pi\right) \\ \sin\theta_e & \sin\left(\theta_e - \frac{2}{3}\pi\right) & \sin\left(\theta_e + \frac{2}{3}\pi\right) \\ \frac{1}{2} & \frac{1}{2} & \frac{1}{2} \end{bmatrix} \begin{bmatrix} u_{as} \\ u_{bs} \\ u_{cs} \end{bmatrix}$$

finally we obtain

$$u_{qs}^e(t) = \frac{2}{3}\left[u_{as}\cos\theta_e + u_{bs}\cos\left(\theta_e - \frac{2}{3}\pi\right) + u_{cs}\cos\left(\theta_e + \frac{2}{3}\pi\right)\right]$$

$$u_{ds}^e(t) = \frac{2}{3}\left[u_{as}\sin\theta_e + u_{bs}\sin\left(\theta_e - \frac{2}{3}\pi\right) + u_{cs}\sin\left(\theta_e + \frac{2}{3}\pi\right)\right]$$

$$u_{os}^e(t) = \frac{1}{3}\left(u_{as} + u_{bs} + u_{cs}\right)$$

Using a balanced three-phase voltage set

$$u_{as}(t) = \sqrt{2}u_M\cos(\omega_f t), \quad u_{bs}(t) = \sqrt{2}u_M\cos\left(\omega_f t - \frac{2}{3}\pi\right), \quad \text{and} \quad u_{cs}(t) = \sqrt{2}u_M\cos\left(\omega_f t + \frac{2}{3}\pi\right)$$

and assuming that the initial displacement of the quadrature magnetic axis is zero, from $\theta_e = \omega_f t$, the following quadrature, direct, and zero stator voltages must be supplied to guarantee the balance operation:

$$u_{qs}^e(t) = \sqrt{2}u_M, \quad u_{ds}^e(t) = 0, \quad u_{os}^e(t) = 0 \qquad (6.39)$$

It should be emphasized that the quadrature-, direct-, and zero-axis components of stator and rotor voltages, currents, and flux linkages have DC form. Furthermore, to control induction micromotors, only the DC quadrature voltage $u_{qs}^e(t)$ is regulated because $u_{ds}^e(t) = 0$ and $u_{os}^e(t) = 0$. The reader should understand that in reality the AC phase voltages are supplied to the phase windings. Therefore, the quadrature-, direct- and zero-axis components of stator voltages, though they can be viewed as the control variables, are not applied to the phase windings. In particular, taking note of $\mathbf{u}_{qdos}^e$, using the rotor displacement (which must be measured or observed), one calculates $\mathbf{u}_{abcs}$ as $\mathbf{u}_{abcs} = (\mathbf{K}_s^e)^{-1}\mathbf{u}_{qdos}^e$ and applies u_{as}, u_{bs}, and u_{cs} to the phase windings.

Using Equation 6.38, the block diagram is developed and documented in Figure 6.45.

6.7.3.3 *Simulation of Induction Micromachines in the MATLAB Environment*

Making use the derived differential equations, the simulation and analysis of induction micromachines can be straightforwardly performed in the MATLAB environment. Let us

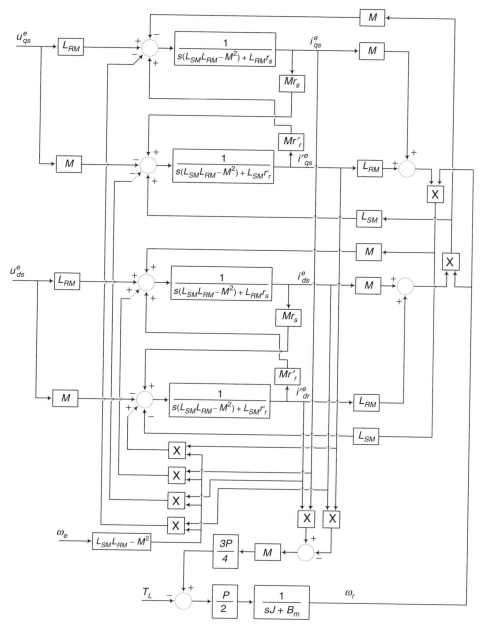

FIGURE 6.45

Block diagram for three-phase squirrel-cage induction micromotors modeled in the synchronous reference frame.

use SIMULINK to simulate induction micromotors. It must be emphasized that Cauchy's and not Cauchy's form of differential equations can be used. The simulation using Cauchy's form is straightforward. Thus we will illustrate how one should use SIMULINK to model not Cauchy's form of differential equations.

As shown in Section 6.7.3.1, the mathematical model of three-phase induction motors is governed by a set of the following nonlinear differential equations (Equation 6.18) and the torsional-mechanical equations of motion (Equation 6.21):

$$u_{as} = r_s i_{as} + (L_{ls} + L_{ms})\frac{di_{as}}{dt} - \frac{1}{2}L_{ms}\frac{di_{bs}}{dt} - \frac{1}{2}L_{ms}\frac{di_{cs}}{dt} + L_{ms}\frac{d(i'_{ar}\cos\theta_r)}{dt}$$

$$+ L_{ms}\frac{d\left(i'_{br}\cos\left(\theta_r + \frac{2\pi}{3}\right)\right)}{dt} + L_{ms}\frac{d\left(i'_{cr}\cos\left(\theta_r - \frac{2\pi}{3}\right)\right)}{dt}$$

$$u_{bs} = r_s i_{bs} - \frac{1}{2}L_{ms}\frac{di_{as}}{dt} + (L_{ls} + L_{ms})\frac{di_{bs}}{dt} - \frac{1}{2}L_{ms}\frac{di_{cs}}{dt} + L_{ms}\frac{d\left(i'_{ar}\cos\left(\theta_r - \frac{2\pi}{3}\right)\right)}{dt}$$

$$+ L_{ms}\frac{d(i'_{br}\cos\theta_r)}{dt} + L_{ms}\frac{d\left(i'_{cr}\cos\left(\theta_r + \frac{2\pi}{3}\right)\right)}{dt}$$

$$u_{cs} = r_s i_{cs} - \frac{1}{2}L_{ms}\frac{di_{as}}{dt} - \frac{1}{2}L_{ms}\frac{di_{bs}}{dt} + (L_{ls} + L_{ms})\frac{di_{cs}}{dt} + L_{ms}\frac{d\left(i'_{ar}\cos\left(\theta_r + \frac{2\pi}{3}\right)\right)}{dt}$$

$$+ L_{ms}\frac{d\left(i'_{br}\cos\left(\theta_r - \frac{2\pi}{3}\right)\right)}{dt} + L_{ms}\frac{d(i'_{cr}\cos\theta_r)}{dt}$$

$$u'_{ar} = r'_r i'_{ar} + L_{ms}\frac{d(i_{as}\cos\theta_r)}{dt} + L_{ms}\frac{d\left(i_{bs}\cos\left(\theta_r - \frac{2\pi}{3}\right)\right)}{dt} + L_{ms}\frac{d\left(i_{cs}\cos\left(\theta_r + \frac{2\pi}{3}\right)\right)}{dt}$$

$$+ (L'_{lr} + L_{ms})\frac{di'_{ar}}{dt} - \frac{1}{2}L_{ms}\frac{di'_{br}}{dt} - \frac{1}{2}L_{ms}\frac{di'_{cr}}{dt}$$

$$u'_{br} = r'_r i'_{br} + L_{ms}\frac{d\left(i_{as}\cos\left(\theta_r + \frac{2\pi}{3}\right)\right)}{dt} + L_{ms}\frac{d(i_{bs}\cos\theta_r)}{dt} + L_{ms}\frac{d\left(i_{cs}\cos\left(\theta_r - \frac{2\pi}{3}\right)\right)}{dt}$$

$$- \frac{1}{2}L_{ms}\frac{di'_{ar}}{dt} + (L'_{lr} + L_{ms})\frac{di'_{br}}{dt} - \frac{1}{2}L_{ms}\frac{di'_{cr}}{dt}$$

$$u'_{cr} = r'_r i'_{cr} + L_{ms}\frac{d\left(i_{as}\cos\left(\theta_r - \frac{2\pi}{3}\right)\right)}{dt} + L_{ms}\frac{d\left(i_{bs}\cos\left(\theta_r + \frac{2\pi}{3}\right)\right)}{dt} + L_{ms}\frac{d(i_{cs}\cos\theta_r)}{dt}$$

$$- \frac{1}{2}L_{ms}\frac{di'_{ar}}{dt} - \frac{1}{2}L_{ms}\frac{di'_{br}}{dt} + (L'_{lr} + L_{ms})\frac{di'_{cr}}{dt}$$

$$\frac{d\omega_r}{dt} = \frac{P^2}{4J}L_{ms}\left\{\left[i_{as}\left(i'_{ar} - \frac{1}{2}i'_{br} - \frac{1}{2}i'_{cr}\right) + i_{bs}\left(i'_{br} - \frac{1}{2}i'_{ar} - \frac{1}{2}i'_{cr}\right) + i_{cs}\left(i'_{cr} - \frac{1}{2}i'_{br} - \frac{1}{2}i'_{ar}\right)\right]\sin\theta_r\right.$$

$$\left. + \frac{\sqrt{3}}{2}[i_{as}(i'_{br} - i'_{cr}) + i_{bs}(i'_{cr} - i'_{ar}) + i_{cs}(i'_{ar} - i'_{br})]\cos\theta_r\right\} - \frac{B_m}{J}\omega_r - \frac{P}{2J}T_L$$

$$\frac{d\theta_r}{dt_r} = \omega$$

Using the differential equations, one must build the SIMULINK model. We rewrite the differential equations as

$$\frac{di_{as}}{dt} = \frac{1}{L_{ls} + L_{ms}}\left[-r_s i_{as} + \tfrac{1}{2}L_{ms}\frac{di_{bs}}{dt} + \tfrac{1}{2}L_{ms}\frac{di_{cs}}{dt} - L_{ms}\frac{d\big(i'_{ar}\cos\theta_r\big)}{dt}\right.$$

$$\left. - L_{ms}\frac{d\big(i'_{br}\cos\big(\theta_r + \frac{2\pi}{3}\big)\big)}{dt} - L_{ms}\frac{d\big(i'_{cr}\cos\big(\theta_r + \frac{2\pi}{3}\big)\big)}{dt} + u_{as}\right],$$

$$\frac{di_{bs}}{dt} = \frac{1}{L_{ls} + L_{ms}}\left[-r_s i_{bs} + \tfrac{1}{2}L_{ms}\frac{di_{as}}{dt} + \tfrac{1}{2}L_{ms}\frac{di_{cs}}{dt} - L_{ms}\frac{d\big(i'_{ar}\cos\big(\theta_r - \frac{2\pi}{3}\big)\big)}{dt}\right.$$

$$\left. - L_{ms}\frac{d(i'_{br}\cos\theta_r)}{dt} - L_{ms}\frac{d\big(i'_{cr}\cos\big(\theta_r + \frac{2\pi}{3}\big)\big)}{dt} + u_{bs}\right],$$

$$\frac{di_{cs}}{dt} = \frac{1}{L_{ls} + L_{ms}}\left[-r_s i_{cs} + \tfrac{1}{2}L_{ms}\frac{di_{as}}{dt} + \tfrac{1}{2}L_{ms}\frac{di_{bs}}{dt} - L_{ms}\frac{d\big(i'_{ar}\cos\big(\theta_r + \frac{2\pi}{3}\big)\big)}{dt}\right.$$

$$\left. - L_{ms}\frac{d\big(i'_{br}\cos\big(\theta_r - \frac{2\pi}{3}\big)\big)}{dt} - L_{ms}\frac{d(i'_{cr}\cos\theta_r)}{dt} + u'_{cs}\right],$$

$$\frac{di'_{ar}}{dt} = \frac{1}{L'_{lr} + L_{ms}}\left[-r'_r i'_{ar} - L_{ms}\frac{d(i_{as}\cos\theta_r)}{dt} - L_{ms}\frac{d\big(i_{bs}\cos\big(\theta_r - \frac{2\pi}{3}\big)\big)}{dt}\right.$$

$$\left. - L_{ms}\frac{d\big(i_{cs}\cos\big(\theta_r + \frac{2\pi}{3}\big)\big)}{dt} + \tfrac{1}{2}L_{ms}\frac{di'_{br}}{dt} + \tfrac{1}{2}L_{ms}\frac{di'_{cr}}{dt} + u_{ar}\right],$$

$$\frac{di'_{br}}{dt} = \frac{1}{L'_{lr} + L_{ms}}\left[-r'_r i'_{br} - L_{ms}\frac{d\big(i_{as}\cos\big(\theta_r + \frac{2\pi}{3}\big)\big)}{dt} - L_{ms}\frac{d(i_{bs}\cos\theta_r)}{dt}\right.$$

$$\left. - L_{ms}\frac{d\big(i_{cs}\cos\big(\theta_r - \frac{2\pi}{3}\big)\big)}{dt} + \tfrac{1}{2}L_{ms}\frac{di'_{ar}}{dt} + \tfrac{1}{2}L_{ms}\frac{di'_{cr}}{dt} + u'_{br}\right]$$

$$\frac{di'_{cr}}{dt} = \frac{1}{L'_{lr} + L_{ms}}\left[-r'_r i'_{cr} - L_{ms}\frac{d\big(i_{as}\cos\big(\theta_r + \frac{2\pi}{3}\big)\big)}{dt} - L_{ms}\frac{d\big(i_{bs}\cos\big(\theta_r + \frac{2\pi}{3}\big)\big)}{dt}\right.$$

$$\left. - L_{ms}\frac{d(i_{cs}\cos\theta_r)}{dt} + \tfrac{1}{2}L_{ms}\frac{di'_{ar}}{dt} + \tfrac{1}{2}L_{ms}\frac{di'_{br}}{dt} + u'_{cr}\right]$$

$$\frac{d\omega_r}{dt} = -\frac{P^2}{4J}L_{ms}\left\{\left[i_{as}\big(i'_{ar} - \tfrac{1}{2}i'_{br} - \tfrac{1}{2}i'_{cr}\big) + i_{bs}\big(i'_{br} - \tfrac{1}{2}i'_{ar} - \tfrac{1}{2}i'_{cr}\big) + i_{cs}\big(i'_{cr} - \tfrac{1}{2}i'_{br} - \tfrac{1}{2}i'_{ar}\big)\right]\sin\theta_r\right.$$

$$\left. + \tfrac{\sqrt{3}}{2}[i_{as}(i'_{br} - i'_{cr}) + i_{bs}(i'_{cr} - i'_{ar}) + i_{cs}(i'_{ar} - i'_{br})]\cos\theta_r\right\} - \frac{B_m}{J}\omega_r \frac{P}{2J}T_L$$

$$\frac{d\theta_r}{dt} = \omega_r$$

To guarantee the balanced operating condition, the following phase voltages should be applied to the induction motor windings:

$$u_{as}(t) = \sqrt{2u_M}\,\cos(\omega_f t), \quad u_{bs}(t) = \sqrt{2u_M}\,\cos\!\left(\omega_f t - \tfrac{2}{3}\pi\right) \text{ and}$$

$$u_{cs}(t) = \sqrt{2u_M}\,\cos\!\left(\omega_f t + \tfrac{2}{3}\pi\right)$$

The SIMULINK block diagram to simulate three-phase induction micromotors is developed and illustrated in Figure 6.46 using the `Derivative, Subsystem` and other blocks. One of the subsystems is reported in Figure 6.47.

Example 6.28   SIMULATION of THREE-PHASE INDUCTION MICROMOTOR

Perform numerical simulation of the micromotor if u_M is 10 V, and the induction micromotor parameters are $r_s = 30$ ohm, $r_r = 20$ ohm, $L_{ms} = 0.025$ H, $L_{ls} = 0.0025$ H, $L_{lr} = 0.025$ H, $J = 0.000000005$ kg-m^2, and $B_m = 0.0000004$ N-m-sec/rad. Repeat simulations assigning $r_s = 300$ ohm and $r_r = 200$ ohm. Examine the results.

SOLUTION

To perform simulations and analysis, the parameters must be used. In general these parameters are available as the micromotor is fabricated. It should be emphasized that these parameters significantly influence the micromotor performance. These parameters are downloaded. In particular, in the Command window, the designer inputs the motor parameters as

```
P=2; Um=10; Rs=30; Rr=20; Lms=0.025; Lls=0.0025; Llr=0.0025;
Bm=0.0000004; J=0.000000005;
```

Using the mdl model developed, nonlinear simulations were performed, and transient dynamics of the stator and rotor currents $i_{as}(t)$, $i_{bs}(t)$, $i_{cs}(t)$, $i'_{ar}(t)$, $i'_{br}(t)$, $i'_{cr}(t)$ as well as the rotor angular velocity $\omega_{rm}(t)$ can be viewed. Figure 6.47 illustrates the transient dynamics for the angular velocity. It should be emphasized that the statement to plot the evolution of $\omega_{rm}(t)$ is

```
plot(wr(:,1),wr(:,2)); xlabel('Time [seconds]');
ylabel('Angular Velocity, wr [rad/sec]');
title('Angular Velocity Dynamics');
```

Induction micromotors can be used as microdrives. The studied micromotor reaches the steady-state angular velocity 377 rad/sec, and the settling time is 0.01 sec. The reader is advised to examine other micromotor variables. For example, the analysis of the motor currents indicates that this micromotor will be unoperational because the currents reach 150 A.

Let us examine the same micromotor if $r_s = 300$ ohm and $r_r = 200$ ohm.

We download the parameters as

```
P=2;  Um=10;  Rs=300;  Rr=200;  Lms=0.025;  Lls=0.0025;  Llr=0.0025;
Bm=0.0000004; J=0.000000005;
```

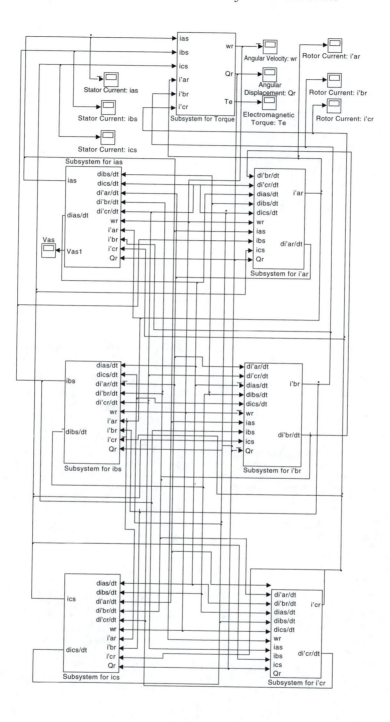

FIGURE 6.46
SIMULINK-block diagram to simulate squirrel-cage induction motors (c6_7_1.mdl). In the developed SIMULINK diagram, different MATLAB blocks from the block library were used.

Nonlinear simulations were performed. Figure 6.48 illustrates the transient dynamics for the angular velocity $\omega_{rm}(t)$, as well as $i_{as}(t)$, $i_{bs}(t)$, and $i_{cs}(t)$. Taking note of the 0.032 A rms value for the stator currents, one concludes that this induction micromotor can be used. However, low speed results, and the rated angular velocity is 10.5 rad/sec.

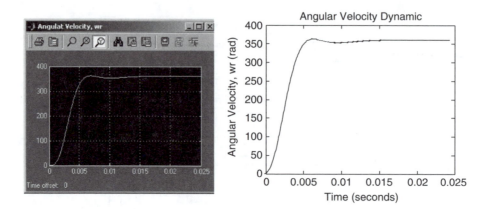

FIGURE 6.47
Transient dynamic for the angular velocity $\omega_{rm}(t)$.

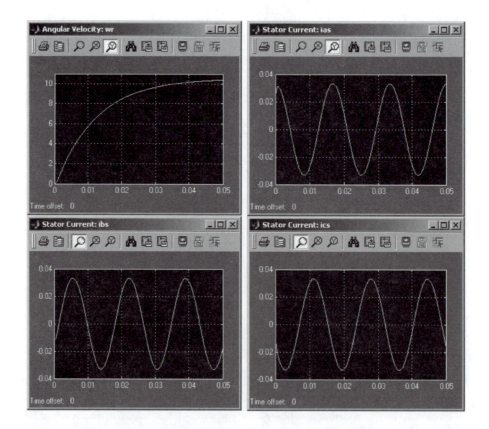

FIGURE 6.48
Transient dynamics of three-phase induction micromotor.

6.8 Synchronous Micromachines

6.8.1 Introduction and Analogies

Rotational and translational nanomachines, controlled by nanoscale integrated circuits (nano-ICs), can be widely used as actuators and sensors. The implications of micro- and nanotechnology to motion nanodevices have received meticulous consideration as technologies to fabricate these nanomachines are being developed. In particular, micromachines that have been fabricated using CMOS and micromachining technologies serve as prototypes and proof-of-concept paradigms. In this section we address and solve a spectrum of problems in the synthesis, analysis, modeling, and control of synchronous machines. All micro- and nanomachines and motion devices must be synthesized before attempts to design and optimize them, because basic physical features, machine topologies, energy conversion, operating principles, and other issues significantly contribute to sequential tasks in analysis, control, optimization, and design. This is of particular significance for electromagnetic motion micro- and nanodevices including reluctance and permanent-magnet synchronous machines.

The benchmarking problems in the synthesis of novel high-performance micro- and nanomachines have challenged researchers for many years. Though some progress has been made in nanobiomotors, many unsolved problems exist in devising micro- and nanomachines. In *E. coli* the filament is driven by a 45-nm rotor of the nanobiomotor embedded in the cell wall. The cytoplasmic membrane forms a stator. This nanobiomotor integrates more than 20 proteins and operates as a result of the axial protonomotive force resulting due to the proton flux. (See Figure 6.49.)

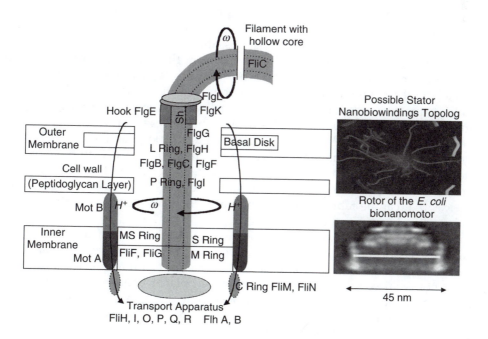

FIGURE 6.49

E. coli nanobiomotor, nanobiomotor–bearing–coupling–flagella complex with different proteins and rings, rotor image, and possible protein-based stator bionanocircuitry geometry.

It was emphasized in the previous section that there are three possible operating principles for electromagnetic nanobiomotors that are based on electromagnetics. In particular,

- Synchronous electromagnetics—The torque results due to the interaction of time-varying magnetic field established by the stator (rotor) windings and stationary magnetic field established by the rotor (stator) permanent nanomagnets;

- Variable reluctance electromagnetics (synchronous nanomachine)— The torque is produced to minimize the reluctance of the electromagnetic system. That is, the torque is created by the magnetic system in attempt to align the minimum-reluctance path of the rotor with the time-varying rotating airgap mmf;

- Induction electromagnetics—The rotor currents are induced in the rotor windings due to the time-varying stator magnetic field and motion of the rotor with respect to the stator. The torque results due to the interaction of time-varying electromagnetic fields.

There is limited knowledge available to date even for the most widely examined *E. coli* nanobiomotor. In Section 6.3 we hypothesized that the *E. coli* nanobiomotor operates based on induction electromagnetics. However, micro- and nanomachines that are based on more feasible synchronous and variable reluctance electromagnetics are examined in this section.

6.8.2 Axial Topology Permanent-Magnet Synchronous Micromachines

6.8.2.1 *Fundamentals of Axial Topology Permanent-Magnet Synchronous Micromachines*

It can be concluded with the highest degree of confidence that the electromagnetic system is closed, and that the nanobiomotor has an axial topology. (Nanobiomotors use the proton or sodium gradient maintained across the cell's inner membrane as the energy source, and the torque is developed due to the axial flux.) Complex chemoelectromechanical energy conversion and other phenomena are topics of far-reaching research that unlikely will be completed soon. It is extremely difficult to comprehend and prototype the torque generation, energy conversion, bearing, sensing-feedback-control, and other mechanisms in nanobiomotors. Despite the limited research and evidence, efficient nanomachines can be synthesized utilizing the axial topology and endless electromagnetic system. Though these nanomachines are different compared with the *E. coli* nanobiomotor, which may not have permanent magnets, a similar topology is utilized. Furthermore we will progress to the well-defined and well-understood inorganic motion nanodevices that can be fabricated utilizing nanotechnology, micromachining technology, and modified CMOS processes.

The advantages of axial topology micro- and nanomachines are feasibility, efficiency, and reliability. Fabrication simplicity results because (1) nanomagnets are flat; (2) there are no strict shape requirements imposed on nano-magnets; (3) rotor-back ferromagnetic material is not required; and (4) it is easy to deposit planar nanowires on the flat stator. Utilizing the axial topology and endless electromagnetic system, we synthesize permanent-magnet synchronous nanomachines. The synthesized nanomachine is reported in Figure 6.50. This nanomachine has well-defined topological analogy compared with the *E. coli* nanobiomotor. It must be emphasized that the documented motion nanodevice can be fabricated, and a prototype of the micromachine rotor is illustrated in Figure 6.50. The planar segmented nanomagnet array, as evident from Figure 6.50, can be deposited as thin-film nanomagnets.

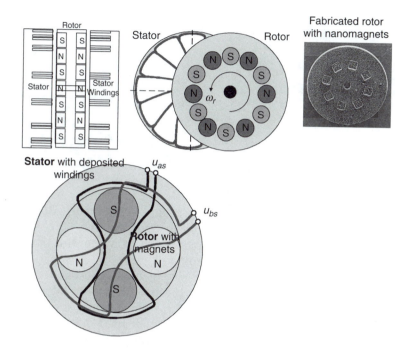

FIGURE 6.50
Axial permanent-magnet synchronous nanomachine.

Single- and two-phase axial permanent-magnet synchronous micromachines are illustrated in Figure 6.50. In particular, for two-phase micromachines, one supplies two phase voltages, u_{as} and u_{bs}, while for a single-phase micromachine, phase voltage u_{as} or u_{bs} is applied.

We now cover the basic physics, modeling, and analysis for axial topology permanent-magnet synchronous micromachines. A good starting point is to consider a current loop in the magnetic field that can be produced by the permanent magnets. Assuming that the magnetic flux is constant through the magnetic plane (current loop), the torque on a planar current loop of any size and shape in the uniform magnetic field is give as

$$\mathbf{T} = i\mathbf{s} \times \mathbf{B} = \mathbf{m} \times \mathbf{B}$$

where i is the current in the loop (winding), and $\mathbf{m}$ is the magnetic dipole moment [A-m²].

The torque on the current loop always tends to turn the loop to align the magnetic field produced by the loop with permanent-magnet magnetic filed causing the resulting electromagnetic torque. For example, for the current loop in a uniform magnetic field with flux density $\mathbf{B} = -0.5\mathbf{a}_y + \mathbf{a}_z$, demonstrated in Figure 6.51, the torque is found to be

$$\mathbf{T} = i\mathbf{s} \times \mathbf{B} = \mathbf{m} \times \mathbf{B} = 1 \times 10^{-3}[(1 \times 10^{-3})(2 \times 10^{-3})\mathbf{a}_z] \times (-0.5\mathbf{a}_y + \mathbf{a}_z) = 1 \times 10^{-9}\mathbf{a}_x \text{ N-m}$$

The electromagnetic force is found as $\mathbf{F} = \oint_l id\mathbf{l} \times \mathbf{B}$.

Thus, as illustrated in Figure 6.50, the interaction of the current (in windings) and magnets will produce the rotating electromagnetic torque. It should be emphasized and shown later that the voltage, applied to the windings (that lead to current), must be a function of the rotor displacement. That is, we need to apply the voltage due to different

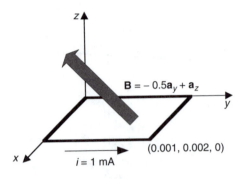

FIGURE 6.51
Rectangular planar loop in a uniform magnetic field.

(south and north) polarity of the deposited nanomagnets. Figure 6.52 documents that the direction of the current in the windings must be changed as a function of the direction of the magnetic flux to develop the electromagnetic torque in one direction achieving the clockwise (or counterclockwise) direction of rotation.

Microscale synchronous transducers can be used as motors and generators. Microgenerators (velocity and position sensors) convert mechanical energy into electrical energy, while micromotors-microactuators convert electrical energy into mechanical energy. A broad spectrum of synchronous microtransducers can be used in MEMS as actuators, for example, electric microdrives, microservos, or microscale power systems. We will develop high-fidelity and lumped-parameter nonlinear mathematical models and perform nonlinear analysis of synchronous permanent-magnet microtransducers.

Distinct electrostatic and electromagnetic micro- and nanomachines can be designed and fabricated using surface micromachining technology. However, all high-performance, high-power, and torque density machines are electromagnetic, and these micro- and nanomachines use permanent micromagnets. Therefore, the issues of fabrication and analysis of micromagnets are of great interest.

Different *soft* and *hard* micromagnets can be fabricated using surface micromachining. For example, Ni, Fe, Co, NiFe, and other magnets can be made. In general four classes of high-performance magnets have been commonly used in electromechanical motion devices: neodymium iron boron ($Nd_2Fe_{14}B$), samarium cobalt (usually Sm_1Co_5 and Sm_2Co_{17}), ceramic (ferrite), and alnico (AlNiCo). The term *soft* is used to describe those magnets that have high saturation magnetization and a low coercivity (narrow *B-H* curve). Another property of these micromagnets is their low magnetostriction. The soft micromagnets have been widely used in magnetic recording heads, and NiFe thin films

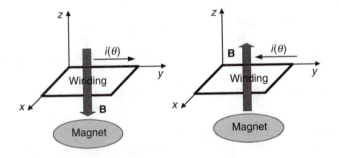

FIGURE 6.52
Rectangular planar loop in a uniform magnetic field.

with the desired properties have been fabricated. Hard magnets have wide *B-H* curves (high coercivity), and, therefore, high-energy storage capacity. These nano- and micromagnets should be used in micromachines in order to attain high force, torque, and power densities. Unfortunately, limited progress has been made in the fabrication of hard micromagnets. Most hard magnets are fabricated using the following processes:

- Metallurgical—Sintering (creating a solid but porous material from a powder), pressure bonding, injection molding, casting, and extruding.
- Vacuum—Evaporation, sputtering, molecular beam epitaxy (MBE), and chemical vapor deposition (CVD).
- Electrochemical—Electroless deposition and electroplating. Many of these processes are incompatible with ICs.

Electroplating has been used to fabricate the Co-based hard micromagnets. The saturation magnetization decreases with increases in the amount of nonmagnetic alloy adding P, As, Sb, Bi, W, Cr, Mo, Pd, Pt, Ni, Fe, Cu, Mn, etc. For example, CoPt and FePt thin films have high coercivities. The Co-based micromagnets have been usually deposited using MBE and sputtering techniques. These processes require high temperatures (up to 700°C) during annealing. Neodymium iron boron micromagnets can be microfabricated using injection molding and pressure bonding methods. Both CoPt and NdFeB hard magnets can be vacuum deposited. However, high temperatures (of either the material or the substrate) need to be achieved to anneal the final deposition. (If it is not annealed, the material acts as a soft magnet.)

The energy density is given as the area enclosed by the *B-H* curve: $w_m = \frac{1}{2}\mathbf{B}\cdot\mathbf{H}$. When nano- or micromagnets are used in micro- and nanomachines, the demagnetization curve (second quadrant of the *B-H* curve) is studied. A basic understanding of the phenomena and operating features of permanent magnets is extremely useful. Permanent magnets store, exchange, and convert energy. In particular, permanent magnets produce stationary magnetic field without external energy sources. The operating point is determined by the permanent-magnet geometry and properties, and the hysteresis minor loop occurs. The second-quadrant *B-H* characteristic is given in Figure 6.53. (Permanent magnets operate on the demagnetization curve of the hysteresis loop.) Figure 6.53 also illustrates the energy product curve for a permanent magnet. The demagnetization and energy product curves are in mutual correspondence.

The operating point is denoted by H_d and B_d. Given an airgap through which the magnetic flux exists in micromachines, according to Ampere's law, we have

$$H_d l_m = H_{ag} l_{ag}$$

where l_m is the length of the magnet; l_{ag} is the length of the air gap parallel to the flux lines; and H_{ag} is the magnetic field intensity in the air gap.

The cross-sectional area of a magnet required to produce a specific flux density in the air gap is

$$A_m = \frac{B_{ag} A_{ag}}{B_d}$$

where A_{ag} is the airgap area, and B_{ag} is the flux density in the airgap.

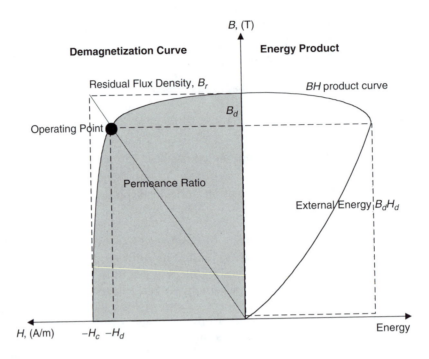

FIGURE 6.53
B-H and energy product curves.

The flux linkages due to the magnet in the air gap are $\psi = N\phi = NB_{ag}A_{ag}$, and the coenergy is given by

$$W = \int_{\psi} i \cdot d\psi = \int_i \psi \cdot di$$

In a lossless system, an informative energy equation for the air gap is

$$Vol_{ag}B_{ag}H_{ag} = A_{ag}l_{ag}\frac{A_m l_m B_d H_d}{A_{ag}l_{ag}} = A_m l_m B_d H_d = \psi i$$

The flux density at position x can be derived, and high-fidelity mathematical models are found. In one-dimensional cases for cylindrical micromagnets (length l_m and radius r) that have near-linear demagnetization curves, the flux density at a distance x can be approximated as

$$B = \frac{B_r}{2}\left(\frac{l_m + x}{\sqrt{r^2 + (l_m + x)^2}} - \frac{x}{\sqrt{r^2 + x^2}}\right)$$

6.8.2.2 *Mathematical Models of Axial Topology Permanent-Magnet Synchronous Micromachines*

Synthesis and prototyping tasks have been reported and performed. As nanomachines are devised (synthesized), the sequential analysis and design problems must be researched.

In this section we develop integrated electromagneticmechanical modeling, analysis, and optimization to comprehensively assess and control electromagnetic and electromechanics in axial micro- and nanomachines. Our goal is to perform lumped-parameter and high-fidelity modeling to achieve the highest degree of confidence in simulation and guarantee accurate analysis. In particular, the problems to be researched are modeling (deviations of equations of motion to model complex phenomena and effects in the time domain), heterogeneous simulation, and data-intensive analysis.

6.8.2.2.1 Lumped-Parameter Modeling and Mathematical Model Development

We apply Kirchhoff's and Newton's laws to derive the equations of motion for axial topology permanent-magnet micro- and nanomotors. Using the axial permanent-magnet synchronous nanomachine documented in Figure 6.50, we assume that this variation of the flux density is sinusoidal, that is,

$$B(\theta_r) = B_{max} \sin^n\left(\tfrac{1}{2} N_m \theta_r\right), \quad n = 1, 3, 5, 7, \ldots$$

where B_{max} is the maximum flux density in the airgap produced by the magnet as viewed from the winding (B_{max} depends on the magnets used, airgap length, temperature, etc.); N_m is the number of magnets; and n is the integer that is a function of the magnet geometry and the waveform of the airgap B.

For example, for $N_m = 2$ and $N_m = 10$, letting $B(\theta_r) = B_{max} \sin(\tfrac{1}{2} N_m \theta_r)$ for $n = 1$ and $B(\theta_r) = B_{max} \sin^3(\tfrac{1}{2} N_m \theta_r)$ for $n = 3$ (common distribution of B), we type the following statements in the MATLAB Command window:

```
th=0:.01:2*pi; Nm=2; B=sin(Nm*th/2); plot(th,B);
xlabel('Rotor Displacement'); ylabel('B');
title('B as a Function on the Rotor Displacement');
```

and

```
th=0:.01:2*pi; Nm=10; B=sin(Nm*th/2).^3; plot(th,B);
xlabel('Rotor Displacement'); ylabel('B');
title('B as a Function on the Rotor Displacement');
```

The resulting MATLAB plots for $B(\theta_r)$ are reported in Figure 6.54.

Using the drawing illustrated in Figure 6.50, the reader can see that the airgap flux density is described as $B(\theta_r) = B_{max} \sin^3(\tfrac{1}{2} N_m \theta_r)$ or $B(\theta_r) = B_{max} \sin^5(\tfrac{1}{2} N_m \theta_r)$, but not as $B(\theta_r) = B_{max}\sin(\tfrac{1}{2} N_m \theta_r)$, due to setting of the permanent magnets with respect to the windings. It should be emphasized that nonuniform magnetization of magnets to attain $B(\theta_r) = B_{max}\sin(\tfrac{1}{2} N_m \theta_r)$ (so-called pure sinusoidal flux linkages distribution) can be achieved only for minimachines, but not for micro- and nanomachines.

The electromagnetic torque developed by single-phase axial topology permanent-magnet synchronous micromotors is found using the expression for the coenergy $W_c(i_{as}, \theta_r)$, which is given as

$$W_c(i_{as}, \theta_r) = NA_{ag} B(\theta_r) i_{as}$$

where N is the number of turns, and A_{ag} is the airgap surface area.

Assuming that the airgap flux density obeys

$$B(\theta_r) = B_{max} \sin\left(\tfrac{1}{2} N_m \theta_r\right)$$

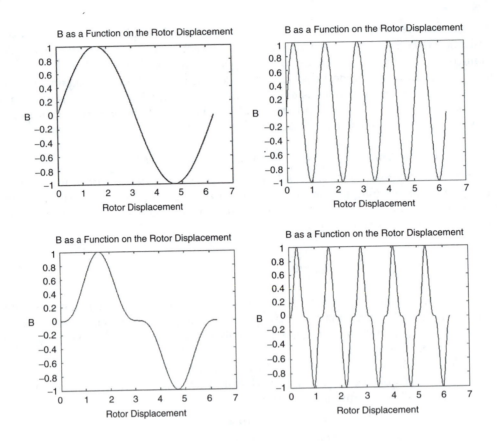

FIGURE 6.54
Plots for $B(\theta_r)$, $N_m = 2$ and $N_m = 10$, $B(\theta_r) = B_{max} \sin(\frac{1}{2} N_m \theta_r)$, and $B(\theta_r) = B_{max} \sin^3(\frac{1}{2} N_m \theta_r)$.

we have

$$W_c(i_{as}, \theta_r) = NA_{ag} B_{max} \sin\left(\frac{1}{2} N_m \theta_r\right) i_{as}$$

and the electromagnetic torque is

$$T_e = \frac{\partial W_c(i_{as}, \theta_r)}{\partial \theta_r} = \frac{\partial\left(NA_{ag} B_{max} \sin\left(\frac{1}{2} N_m \theta_r\right) i_{as}\right)}{\partial \theta_r} = \frac{1}{2} N_m NA_{ag} B_{max} \cos\left(\frac{1}{2} N_m \theta_r\right) i_{as}$$

It is clear that the electromagnetic torque is not developed by the axial synchronous micromotors (microactuator) if one feeds the DC current i_{as} or voltage to the winding. The average value of T_e is not equal to zero if the current is a function of the rotor displacement θ_r. As an illustration, we feed the following current with the magnitude i_M to the micromotor winding:

$$i_{as} = i_M \cos\left(\frac{1}{2} N_m \theta_r\right)$$

Then the electromagnetic torque is

$$T_e = \frac{1}{2} N_m NA_{ag} B_{max} i_M \cos^2\left(\frac{1}{2} N_m \theta_r\right) \neq 0$$

The mathematical model of the single-phase permanent-magnet micromotor is found by using Kirchhoff's and Newton's second laws

$$u_{as} = r_s i_{as} + \frac{d\psi_{as}}{dt} \text{ (circuitry equation)}$$

$$T_e - B_m \omega_r - T_L = J \frac{d^2\theta_r}{dt^2} \text{ (torsional-mechanical equation)}$$

From the flux linkage equation $\psi_{as} = L_{as} i_{as} + NA_{ag} B(\theta_r)$, we have

$$\frac{d\psi_{as}}{dt} = L_{as} \frac{di_{as}}{dt} + NA_{ag} \frac{dB(\theta_r)}{dt} = L_{as} \frac{di_{as}}{dt} + \tfrac{1}{2} N_m NA_{ag} B_{\max} \cos\!\left(\tfrac{1}{2} N_m \theta_r\right)\omega_r$$

Thus, a set of three first-order nonlinear differential equations that models a single-phase axial topology permanent-magnet synchronous micromotor results. In particular,

$$\frac{di_{as}}{dt} = \frac{1}{L_{as}}\left(-r_s i_{as} - \tfrac{1}{2} N_m NA_{ag} B_{\max} \cos\!\left(\tfrac{1}{2} N_m \theta_r\right)\omega_r + u_{as}\right)$$

$$\frac{d\omega_r}{dt} = \frac{1}{J}\left(\tfrac{1}{2} N_m NA_{ag} B_{\max} \cos\!\left(\tfrac{1}{2} N_m \theta_r\right)i_{as} - B_m \omega_r - T_L\right)$$

$$\frac{d\theta_r}{dt} = \omega_r$$

Single-phase axial permanent-magnet synchronous micro- and nanomachines may have a torque ripple. (It will be demonstrated in Example 6.30 that there are no torque ripples for stripe-segmented permanent magnets.) In general, to avoid this torque ripple, one must design two-phase machines. For two-phase machines, the coenergy and electromagnetic torque are given as

$$W_c(i_{as}, \theta_r) = NA_{ag} B_{\max}\left(\sin\!\left(\tfrac{1}{2} N_m \theta_r\right)i_{as} - \cos\!\left(\tfrac{1}{2} N_m \theta_r\right)i_{bs}\right)$$

and the electromagnetic torque is

$$T_e = \frac{\partial W_c(i_{as}, \theta_r)}{\partial \theta_r} = \tfrac{1}{2} N_m NA_{ag} B_{\max}\left(\cos\!\left(\tfrac{1}{2} N_m \theta_r\right)i_{as} + \sin\!\left(\tfrac{1}{2} N_m \theta_r\right)i_{bs}\right)$$

Thus, feeding the phase currents as

$$i_{as} = i_M \cos\!\left(\tfrac{1}{2} N_m \theta_r\right) \quad \text{and} \quad i_{bs} = i_M \sin\!\left(\tfrac{1}{2} N_m \theta_r\right)$$

we obtain

$$T_e = \tfrac{1}{2} N_m NA_{ag} B_{\max} i_M\left(\cos^2\!\left(\tfrac{1}{2} N_m \theta_r\right) + \sin^2\!\left(\tfrac{1}{2} N_m \theta_r\right)\right) = \tfrac{1}{2} N_m NA_{ag} B_{\max} i_M$$

That is, the electromagnetic torque is maximized and there is no torque ripple. This is correct if we approximate the airgap flux as $B(\theta_r) = B_{max} \sin(\frac{1}{2} N_m \theta_r)$. The homework problem is assigned if $B(\theta_r) = B_{max} \sin^3(\frac{1}{2} N_m \theta_r)$, which represents the realistic airgap flux distribution.

We also conclude that to regulate the angular velocity, the phase currents $i_{as}(t)$ and $i_{bs}(t)$ (or voltages), are the functions of the electrical angular displacement θ_r (measured using the Hall-effect microsensors or observed by the controller in the sensorless applications).

Example 6.29

Consider the two-phase axial topology permanent-magnet synchronous micromotor. The airgap flux with respect to the *as* and *bs* windings is given by

$$B_{as}(\theta_r) = B_{max} \sin^5\left(\tfrac{1}{2} N_m \theta_r\right) \quad \text{and} \quad B_{bs}(\theta_r) = B_{max} \cos^5\left(\tfrac{1}{2} N_m \theta_r\right)$$

The micromotor parameters are $N = 100$, $A_{ag} = 0.000001$, $B_{max} = 0.5$, and $N_m = 6$. Assume that the phase currents are $i_{as} = i_M \cos(\tfrac{1}{2} N_m \theta_r)$ and $i_{bs} = i_m \sin(\tfrac{1}{2} N_m \theta_r)$

Derive the expression for the electromagnetic torque and perform calculations and plotting using MATLAB. Study how the electromagnetic torque varies as a function of the rotor displacement. Analyze the results examining the electromagnetic torque variations.

SOLUTION

The coenergy is given as

$$W_c(i_{as}, \theta_r) = N A_{ag} B_{max}\left(\sin^5\left(\tfrac{1}{2} N_m \theta_r\right) i_{as} - \cos^5\left(\tfrac{1}{2} N_m \theta_r\right) i_{bs}\right)$$

One finds the electromagnetic torque performing differentiation

$$T_e = \frac{\partial W_c(i_{as}, \theta_r)}{\partial \theta_r}$$

We utilize the MATLAB environment to solve this problem. The following MATLAB file is developed and used (using the Symbolic Math Toolbox):

```
% In order to use a symbolic variable, create an object of type SYM
x=sym('x');
N=100; Aag=0.000001; Bmax=0.5; Nm=6;
% Input flux density and perform plotting using the EZPLOT function
w1=N*Aag*Bmax*sin(Nm*x/2)^5, ezplot(w1); pause
w2=N*Aag*Bmax*cos(Nm*x/2)^5, ezplot(w2); pause
% Differentiate these function using the DIFF command
d1=diff(w1), d2=diff(w2)
% Assign phase currents
ias=cos(Nm*x/2); ibs=-sin(Nm*x/2);
```

```
% Electromagnetic torque equation
T=d1*ias+d2*ibs, simplify(T), ezplot(T)
```

The results of the calculations are given in the Command window as the following:

```
w1 =
1/20000*sin(3*x)^5
w2 =
1/20000*cos(3*x)^5
d1 =
3/4000*sin(3*x)^4*cos(3*x)
d2 =
-3/4000*cos(3*x)^4*sin(3*x)
T =
3/4000*sin(3*x)^4*cos(3*x)^2+3/4000*cos(3*x)^4*sin(3*x)^2
ans =
3/4000*cos(3*x)^2-3/4000*cos(3*x)^4
```

Having performed the differentiations and calculations, one finds the expression for the electromagnetic torque as

$$T_e = \frac{3}{4000}(\sin^4(3\theta_r)\cos^2(3\theta_r)+\cos^4(3\theta_r)\sin^2(3\theta_r)) \text{ N-m}$$

and the simplified expression is found to be

$$T_e = \frac{3}{4000}(\cos^2(3\theta_r)-\cos^4(3\theta_r)) \text{ N-m}.$$

The plots for the airgap flux density and electromagnetic torque are plotted in Figure 6.55. The electromagnetic torque varies as a sinusoidal-like function of the rotor angular displacement. Therefore, the torque ripple results. This is highly undesirable phenomena due to noise, vibration, degradations, etc. To minimize the torque ripple, one can perform redesign (attaining close to sinusoidal airgap flux distribution) and reshape the phase current.

Example 6.30

Consider a single-phase axial topology permanent-magnet synchronous micromotor with the segmented array of the permanent-magnet strips. (See Figure 6.56.) These magnetic array assemblies are widely used in high-performance hard drives. Figure 6.56 documents a 2 × 2-cm hard drive with miniscale axial topology actuators. Derive the mathematical model in the form of differential equations. Examine how to control this micromachine. That is, derive the expressions for the phase current and voltage to ensure that the electromagnetic torque is developed.

SOLUTION

The magnetic field developed by the permanent magnets is a periodic function, and the period depends on the number of magnets N_m. One can express the variations

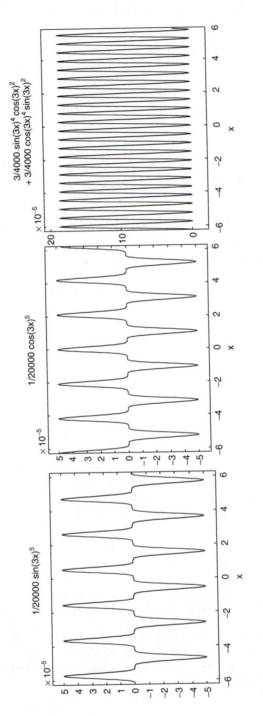

FIGURE 6.55
Airgap flux density and electromagnetic torque.

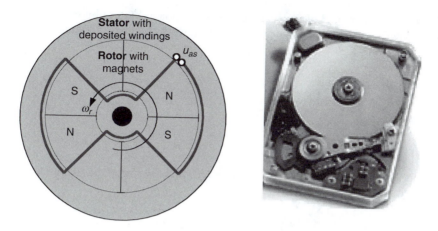

FIGURE 6.56
Axial topology single-phase synchronous micromachine and micro hard drive.

of the magnetic field using continuous and discontinuous functions. For example, using the signum function sign, we have

$$B(\theta_r) = B_{\max} \, \text{sgn}\!\left(\sin\!\left(\tfrac{1}{2} N_m \theta_r\right)\right)$$

For the studied example, $N_m = 4$. Thus, one finds $B(\theta_r) = B_{\max}$ sgn (sin $2\theta_r$).

The value for B_{sat} depends on the magnets used, their geometry, fabrication processes, etc. For the studied minidrive, assuming that one used SmCo magnetic strips with outer diameter 4 mm and thickness 0.5 mm, B_{sat} is from 0.3 to 0.5 T.

The electromagnetic torque, developed by this axial topology permanent-magnet synchronous micromotor is found using the magnetic dipole moment as $T_e = \vec{m} \times \vec{B}$. By taking note of the active coil length l (outer and inner magnetic strip diameter difference), we have

$$T_e = \tfrac{1}{2} N_m N l B_{\max} \, \text{sgn}\!\left(\sin\!\left(\tfrac{1}{2} N_m \theta_r\right)\right) i_{as}$$

The electromagnetic torque is not developed by the axial synchronous micromotors if one feeds the DC current i_{as} or voltage u_{as} to the winding. To develop the electromagnetic torque, the phase current must be a function of the rotor displacement θ_r. For example, if one feeds the following current with the magnitude i_M to the micromotor winding

$$i_{as} = i_M \, \text{sgn}\!\left(\sin\!\left(\tfrac{1}{2} N_m \theta_r\right)\right)$$

the electromagnetic torque is found to be

$$T_e = \tfrac{1}{2} N_m N l B_{\max} i_M$$

To change the direction of rotation, one feeds $i_{as} = -i_M \, \text{sgn}(\sin(\tfrac{1}{2} N_m \theta_r))$, and we obtain

$$T_e = -\tfrac{1}{2} N_m N l B_{\max} i_M$$

The mathematical model of the single-phase micromotor is found by using Kirchhoff's and Newton's second laws. In particular, the following circuitry and torsional-mechanical equations are used:

$$u_{as} = r_s i_{as} + \frac{d\psi_{as}}{dt}$$

$$T_e - B_m \omega_r - T_L = J \frac{d^2\theta_r}{dt^2}$$

From the flux linkage equation $\psi_{as} = L_{as}i_{as} + NA_{ag}B(\theta_r)$, we have the expression for the total derivative

$$\frac{d\psi_{as}}{dt} = L_{as}\frac{di_{as}}{dt} + NA_{ag}\frac{dB(\theta_r)}{dt}$$

Thus, by making use the approximation of $B(\theta_r)$, a set of three first-order nonlinear differential equations that models a single-phase axial topology permanent-magnet synchronous micromotors with segmented magnet arrays results. In particular,

$$\frac{di_{as}}{dt} = \frac{1}{L_{as}}\left(-r_s i_{as} - \underbrace{NA_{ag}\frac{dB(\theta_r)}{dt}}_{\text{back } emf} + u_{as}\right)$$

$$\frac{d\omega_r}{dt} = \frac{1}{J}\left[\frac{1}{2}N_m NlB_{max}\,\text{sgn}\left(\sin\left(\frac{1}{2}N_m\theta_r\right)\right)i_{as} - B_m\omega_r - T_L\right]$$

$$\frac{d\theta_r}{dt} = \omega_r$$

We conclude that to develop the electromagnetic torque, the phase voltage is supplied as a function of the rotor displacement θ_r. In particular, $u_{as} = u_M\,\text{sgn}(\sin(\frac{1}{2}N_m\theta_r))$ and T_e vary by changing the magnitude of the applied voltage u_{as}.

6.8.2.2.2 *High-Fidelity Modeling and Mathematical Model Development*

Electromagnetic field modeling is performed applying Maxwell's equations. The following partial differential equations describe time-varying electromagnetic fields in motion nanodevices that are not based on quantum effects

$$\nabla \times \mathbf{E} = -\mu\frac{\partial \mathbf{H}}{\partial t}, \quad \nabla \times \mathbf{H} = \sigma\mathbf{E} + \mathbf{J}, \quad \nabla \cdot \mathbf{E} = \frac{\rho_v}{\varepsilon}, \quad \nabla \cdot \mathbf{H} = 0$$

where $\mathbf{E}$ and $\mathbf{H}$ are the electric and magnetic field intensities; $\mathbf{J}$ is the current density; ε, μ, and σ are the permittivity, permeability, and conductivity tensors; and ρ_v is the volume charge density.

The Lorenz force, which relates the electromagnetic and mechanical phenomena, is given as

$$\mathbf{F} = \rho_v(\mathbf{E} + \mathbf{v} \times \mathbf{B}) = \rho_v\mathbf{E} + \mathbf{J} \times \mathbf{B}$$

The electromagnetic force is found applying the Maxwell stress tensor. This concept employs a volume integral to obtain the stored energy, and

$$\mathbf{F} = \int_v (\rho_v \mathbf{E} + \mathbf{J} \times \mathbf{B})dv = \frac{1}{\mu} \oint_s \ddot{T}_s \cdot d\mathbf{s}$$

where the electromagnetic stress energy tensor is

$$T_s = T_s^E + T_s^M = \begin{bmatrix} E_1 D_1 - \frac{1}{2} E_j D_j & E_1 D_2 & E_1 D_3 \\ E_2 D_1 & E_2 D_2 - \frac{1}{2} E_j D_j & E_2 D_3 \\ E_3 D_1 & E_3 D_2 & E_3 D_3 - \frac{1}{2} E_j D_j \end{bmatrix}$$

$$+ \begin{bmatrix} B_1 H_1 - \frac{1}{2} B_j H_j & B_1 H_2 & B_1 H_3 \\ B_2 H_1 & B_2 H_2 - \frac{1}{2} B_j H_j & B_2 H_3 \\ B_3 H_1 & B_3 H_2 & B_3 H_3 - \frac{1}{2} B_j H_j \end{bmatrix}$$

For two regions (airgap *ag* and permanent magnets *pm*), we have the airgap and permanent-magnet flux densities as

$$\mathbf{B}_{ag} = \mu_0 \mathbf{H}_{ag}$$

$$\mathbf{B}_{pm} = \mu_0 \mathbf{H}_{pm} + \mathbf{J} = \mu_0 (\mu_r \mathbf{H}_{pm} + \mathbf{M})$$

where $\mathbf{M}$ is the residual magnetization vector, $M = B_r / \mu_0 \mu_r$; B_r is the remanence; and μ_r is the relative recoil permeability.

The negative gradient of the scalar magnetic potential V gives the magnetic field intensity, that is,

$$\mathbf{H} = -\nabla V$$

The scalar magnetic potential satisfies the Laplace equation in free and homogeneous media (with zero current density, that is, $\mathbf{J} = 0$).

For axial topology nanomachines, the cylindrical coordinate system is used. We have

$$\nabla \cdot \mathbf{M} = \frac{1}{r} \frac{\partial (r M_r)}{\partial r} + \frac{1}{r} \frac{\partial M_\phi}{\partial r} + \frac{\partial M_z}{\partial z}$$

Solving the partial differential equation

$$\frac{1 + \chi_t}{r} \frac{\partial}{\partial r} \left(r \frac{\partial V}{\partial r} \right) + \frac{1 + \chi_t}{r^2} \frac{\partial^2 V}{\partial \phi^2} + (1 + \chi) \frac{\partial^2 V_{ag}}{\partial z^2} = \nabla \cdot \mathbf{M}$$

the three-dimensional airgap flux density is found as

$$B_{ag\,z}(r, \phi, z) = \frac{\mu_0 M_0}{1 + \chi} \sum_{i=1}^{\infty} a_i \frac{\sinh \frac{v_i \varepsilon h_r}{r}}{\sinh \frac{v_i \varepsilon (h_r + g_{ag})}{r}} \cosh \frac{v_i \varepsilon z}{r} \sin v_i \phi$$

where χ and χ_t are the reversible susceptibility along the easy and transverse magnetization axes; a_i is the harmonic amplitude coefficient, and for the trapezoidal-wave magnetization, $a_i = \frac{4\sin(2i-1)}{\pi(2i-1)^2}$; and h_r is the rotor thickness,

$$h_r \le z \le g_{ag} + h_r.$$

One-dimensional airgap flux density is found to be

$$B_{ag\,z}(\phi) = \frac{\mu_0 M_0 h_r}{(1+\chi)(h_r + g_{ag})} \sum_{i=1}^{\infty} a_i \sin v_i \phi$$

The maximum flux density in the airgap is

$$B_{\max z} = \frac{\mu_0 M_0 h_r}{(1+\chi)(h_r + g_{ag})} \sum_{i=1}^{\infty} a_i (-1)^{i+1}$$

Using the derived equations for the airgap flux and emf

$$emf = \oint \mathbf{E}d\mathbf{l} = -\int_s \frac{\partial \mathbf{B}}{\partial t} \cdot d\mathbf{s}$$

one finds a three-dimensional electromagnetic model for nanomachine dynamics, torque production, and vibration. It was shown that the electromagnetic torque is

$$\mathbf{T} = \mathbf{m} \times \mathbf{B}$$

The developed equations of motion model electromagnetic and electromechanical behavior of axial topology of synchronous electromagnetic nano- and micromachines.

In general the electromagnetic and mechanical design are based on the application of Maxwell's equations and tensor calculus in order to optimize the complex electromechanical behavior in nanomachines. For example, the nanomachine components (magnets, windings, airgap, etc.) geometry can be optimized to maximize efficiency (η) and robustness as well as minimize torque ripple and losses. In addition to the passive optimization, the active optimization– control problem is formulated and examined. The mathematical formulation of the active optimization–control problem is given as

$$\max_{\mathbf{u} \in U} \min[\eta, \mathbf{T}, p(t,r,\theta,\phi)]$$

where $\mathbf{u}$ is the control vector.

Different control variables can be used. For axial and radial topology synchronous nanomotors, the electro-magnetic field is controlled by varying applied voltage.

The major objectives of this section were to

- Devise and synthesize axial topology permanent-magnet synchronous nanomachines
- Derive lumped-parameter and high-fidelity mathematical models
- Perform electromagnetic and electromechanical analysis
- Address the *active optimization–control* problem

Axial topology permanent-magnet synchronous nano- and micromachines were synthesized. It is very important that these nanomachines can be fabricated expanding the existing technologies. Accurate assessment of machine performance depends on mathematical models used in design and analysis. Therefore, a complete high-fidelity mathematical model of synchronous nanomachines was developed, and electromagnetic phenomena were examined. The major goals were to maximize efficiency, guarantee robustness, reduce vibration, attain affordability, etc.

6.8.3   Radial Topology Single-Phase Synchronous Reluctance Micromotors

6.8.3.1   *Mathematical Model of Synchronous Reluctance Micromotors*

Different nano- and microscale machines, devices, and actuators have been devised and studied. In general the rotational and translational micro- and nanomachines can be classified as synchronous and induction. A step-by-step procedure in the nanoactuator synthesis and design is as follows:

1. Devise micro- and nanomachines researching operational principles, topologies, configurations, geometry, electromagnetic systems and other features.
2. Study electrochemomechanical energy conversion and sensing– feedback–control mechanisms.
3. Define application and environmental requirements.
4. Specify performance specifications imposed.
5. Perform electromagnetic, energy conversion, mechanical, vibroacoustic, and sizing/dimension (stator, rotor, nanomagnets, airgap, winding, etc.) estimates.
6. Define technologies, processes, and materials to fabricate structures (stator, rotor, bearing, post, shaft, etc.) and assemble them as a nanomachine.
7. Perform coherent electromagnetic, mechanical, and thermodynamic design with performance analysis synergetically assessing synthesis, design, and optimization.
8. Modify and refine the design.

In addition to devising micro- and nanomachines, research should examine them coherently. It is important to accurately model, simulate, and analyze very complex electromagnetic, electromechanical, and vibroacoustic phenomena in motion micro- and nanodevices in order to optimize the design, that is, increase efficiency, maximize the torque density, attenuate vibrations and noise, etc. Those problems have not been addressed and solved yet. Therefore, this section focuses on the coherent synthesis and design of radial topology synchronous reluctance micro- and nanomotors.

The first problem to be solved is to devise a micro- or nanomachine. Using even the limited knowledge in nanobiomachines and nanobiostructures, novel nanomachines can be designed. Three-dimensional nanobiocircuitry and nanostructures can be studied researching distinct protein folding and geometry. In addition to the proteins involved to build nanobiomotors [2] (for example, *E. coli* and *S. typhimurium*), complex three-dimensional organic circuits and structures can be engineered [18]. The attachment of specific sticky ends to a DNA branched junction enables the construction of stick assemblies, whose edges are double-stranded DNA. This technology has already been used to make cubes and truncated octahedrons from DNA. Complex arithmetic and logic operations, as well as signal processing, can be performed using these engineered DNA assemblies. For example, theoretically even a biochip computer can be engineered. The protein folding changes the permeability of the media, and this can result in the variable reluctance of the *E. coli* rotor (MS ring that consists of FliF and FliG proteins) as the rotor

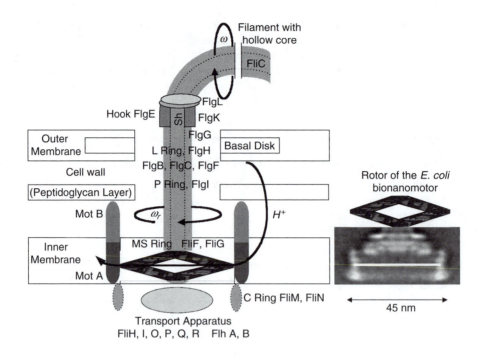

FIGURE 6.57
E. coli nanobiomotor with a possible rotor protein assembly.

rotates. Thus, the *E. coli* nanobiomotor can operate as a synchronous reluctance nano-machine. (See Figure 6.57.)

One can hypothesize that there is a reluctance difference in the closed protonomotive system in the *E. coli* nanobiomotor. Thus, the torque is developed due to change of reluctances as rotor rotates. To illustrate the concept, consider the nanoactuator prototype synthesized using three-dimensional molecular assemblies. For example, triangle, square, pentagon, heptagon, truncated octahedron, cross, star $N_{5,\,6,\,7,...}$, and other geometries of protein assemblies shown in Figure 6.58 result in different reluctances due to the distinct length of the protonomotive flux through the media (protein) as the rotor rotates. Hence the reluctance of the magnetic path varies as the rotor rotates as a function of the rotor angular displacement θ_r. This leads one to the synthesis of the synchronous reluctance nanomachines. In particular, instead of a closed protonomotive system, as in the *E. coli* nanobiomotor, we design a nanomachine with an endless (closed) electromagnetic system that has a different reluctances as the rotor rotates. To illustrate the paradigm, a single-phase reluctance nanoactuator is documented in Figure 6.59. This nanodevice can be fabricated as the nanoactuator reported in Figure 6.60.

Thus, we examine radial topology single-phase reluctance micromotors to study the operation of synchronous microtransducers, analyze important features, and visualize

FIGURE 6.58
Triangle, square, pentagon, heptagon, truncated octahedron, cross, and star $N_{5,6,7,...}$ geometry of protein.

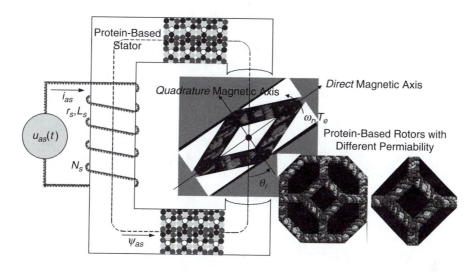

FIGURE 6.59
Single-phase reluctance nanomachine.

mathematical model developments. The quadrature and direct magnetic axes are fixed with the rotor, which rotates with angular velocity ω_r. These magnetic axes rotate with the angular velocity ω. It should be emphasized that under normal operation the angular velocity of synchronous micromachines is equal to the synchronous angular velocity ω_e. Hence, we have $\omega_r = \omega_e$ and $\omega = \omega_r = \omega_e$. The angular displacements of the rotor θ_r and the angular displacement of the quadrature magnetic axis θ are equal, and assuming that the initial conditions are zero,

$$\theta_r = \theta = \int_{t_0}^{t} \omega_r(\tau)d\tau = \int_{t_0}^{t} \omega(\tau)d\tau$$

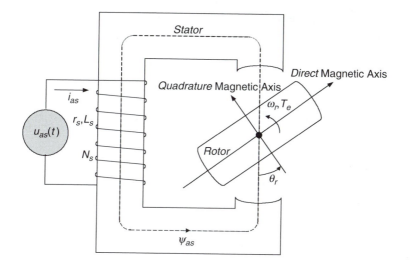

FIGURE 6.60
Radial topology reluctance nanomotor with variable reluctance rotor.

The magnetizing reluctance $\mathfrak{R}_m$ is a function of the rotor angular displacement θ_r. Using the number of turns N_s, the magnetizing inductance is $L_m(\theta_r) = N_s^2 / [\mathfrak{R}_m(\theta_r)]$. This magnetizing inductance varies twice per one revolution of the rotor and has minimum and maximum values. Thus, we have

$$L_{m\min} = \frac{N_s^2}{\mathfrak{R}_{m\max}(\theta_r)}\bigg|_{\theta_r=0,\pi,2\pi,\dots} \quad \text{and} \quad L_{m\max} = \frac{N_s^2}{\mathfrak{R}_{m\min}(\theta_r)}\bigg|_{\theta_r=\frac{1}{2}\pi,\frac{3}{2}\pi,\frac{5}{2}\pi,\dots}$$

Assume that this variation is a sinusoidal function of the rotor angular displacement. Then

$$L_m(\theta_r) = \bar{L}_m - L_{\Delta m}\cos 2\theta_r$$

where $\bar{L}_m$ is the average value of the magnetizing inductance, and $L_{\Delta m}$ is half of the amplitude of the sinusoidal variation of the magnetizing inductance.

The plot for $L_m(\theta_r)$ is documented in Figure 6.61.

The electromagnetic torque developed by single-phase reluctance micromotors is found using the expression for the coenergy $W_c(i_{as}, \theta_r)$. From

$$W_c(i_{as}, \theta_r) = \tfrac{1}{2}(L_{ls} + \bar{L}_m - L_{\Delta m}\cos 2\theta_r)i_{as}^2$$

one finds the electromagnetic torque:

$$T_e = \frac{\partial W_c(i_{as}, \theta_r)}{\partial \theta_r} = \frac{\partial\left(\tfrac{1}{2}i_{as}^2(L_{ls} + \bar{L}_m - L_{\Delta m}\cos 2\theta_r)\right)}{\partial \theta_r} = L_{\Delta m}i_{as}^2 \sin 2\theta_r$$

The electromagnetic torque is not developed by synchronous reluctance micromotors (microactuator) if one feeds the DC current or voltage to the winding. Hence conventional control algorithms cannot be applied, and new methods, which are based on electromagnetic–electromechanical features must be researched. The average value of T_e is not equal to zero if the current is a function of θ_r. As an illustration, we feed the following phase current to the motor winding:

$$i_{as} = i_M \operatorname{Re}\!\left(\sqrt{\sin 2\theta_r}\right).$$

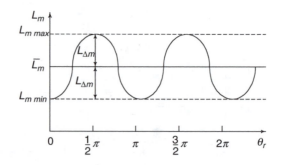

FIGURE 6.61
Magnetizing inductance $L_m(\theta_r)$.

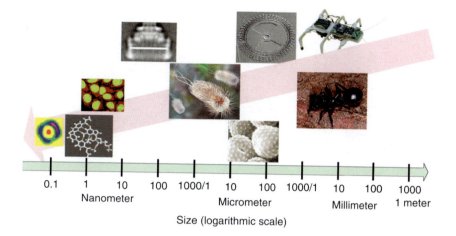

FIGURE 1.1
Scaling common items: hydrogen atom (atomic radius is 0.0529 nm), molecular electronic nanoassembly (1 nm), quantum dots (2 nm), *E. coli* nanobiomotor (45 nm), *E. coli* bacteria (2 μm), ragweed pollen (20 μm), stepper micromotor (50 μm), and ant (5 mm).

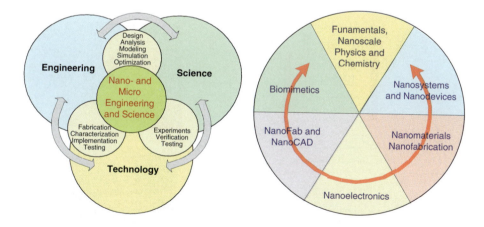

FIGURE 1.3
Representation of nano- and microengineering and science and core areas of nanotechnology.

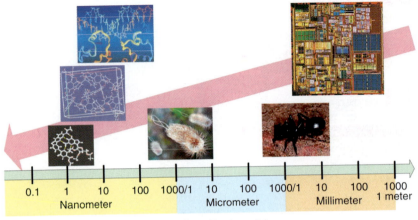

Size (logarithmic scale)

FIGURE 2.2
Revolutionary progress from micro- to nanoelectronics with feature sizing: molecular assembly (1 nm), three-dimensional functional nano-ICs topology with doped carbon complex (2 × 2 × 2-nm cube), nanobio-ICs (10 nm), *E. coli* bacteria (2 μm) and ant (5 mm) which integrate nanobiocircuitry, and 1.5 × 1.5-cm 478-pin Intel® Pentium® processor with millions of transistors.

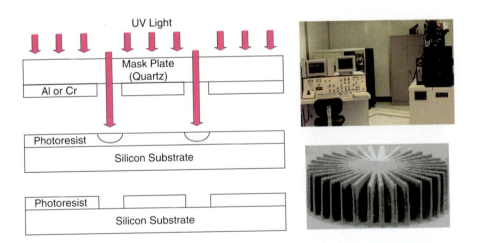

FIGURE 4.1
Photolithography process, computerized photolithography system, and fabricated microstructure.

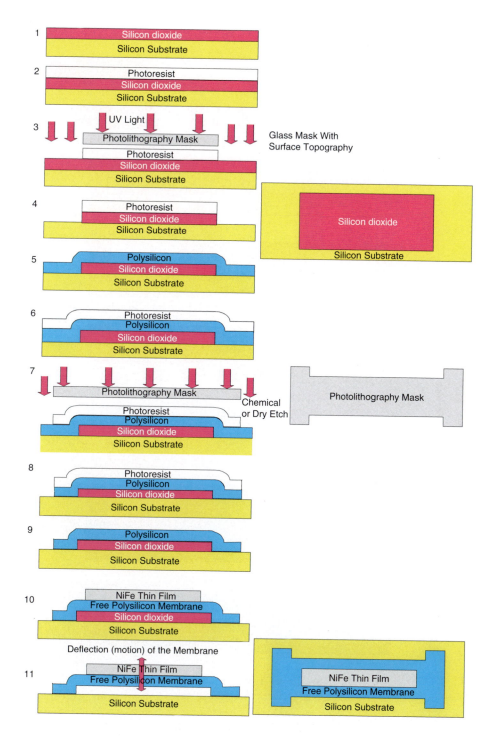

FIGURE 4.15
Micromachining fabrication of the polysilicon thin-film membrane.

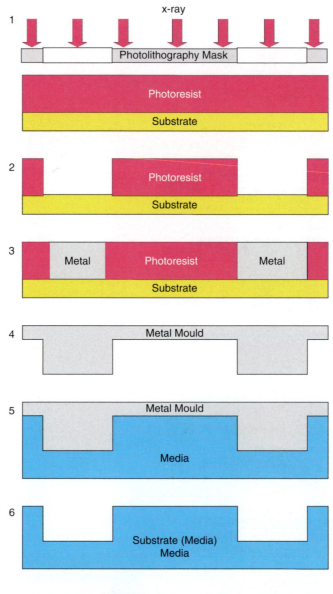

FIGURE 4.19
LIGA fabrication technology.

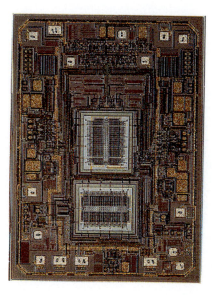
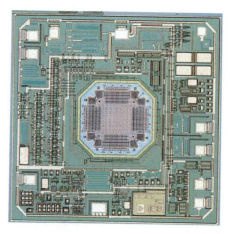

FIGURE 5.12

ADXL202 and ADXL250 accelerometers: proof mass with fingers and ICs (courtesy of Analog Devices).

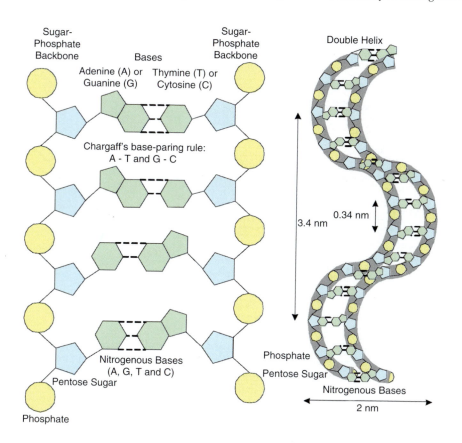

FIGURE 5.19

Two DNA strands are held together by hydrogen bonds between the bases that are paired in the interior of the double helix (base pairs are 0.34 nm apart and there are 10 pairs per helix turn).

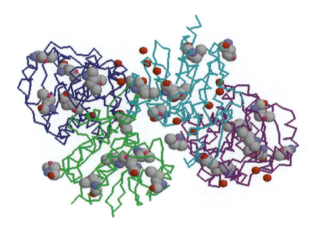

FIGURE 5.21

ElbB gene geometry composed of four similar protein sequence geometries.

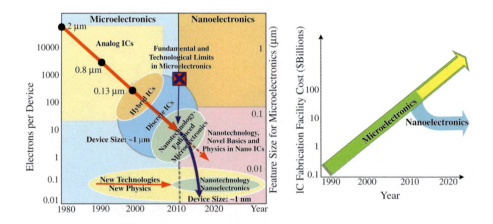

FIGURE 8.1

Projected revolutionary developments in nanoelectronics and nano-ICs compared with microelectronics (Moore's first and second laws for microelectronics are documented with the expected revolutionary changes that will emerge after 2010, drastically affecting the electronics industry and society).

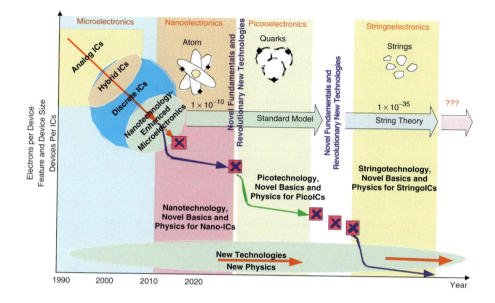

FIGURE 8.2

Projected nano-, pico-, . . . , and stringoelectronics and beyond.

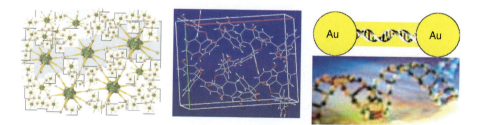

FIGURE 8.3

Three-dimensional nano-ICs made of functional carbon-based complexes, doped carbon complexes in $2 \times 2 \times 2$-nm cube, and double-stranded DNA.

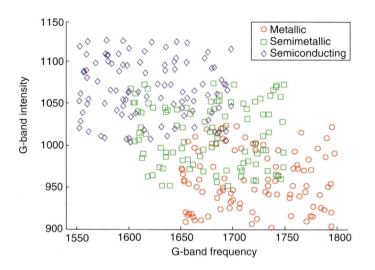

FIGURE 8.16
Metallic, semimetallic, and semiconducting carbon nanotubes.

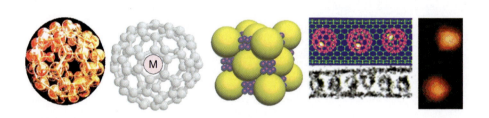

FIGURE 8.17
Empty fullerene C_{60}, endohedral fullerene $M@C_{60}$, functional fullerene complex, single-wall carbon nanotube with fullerenes, and endohedral fullerenes on Si.

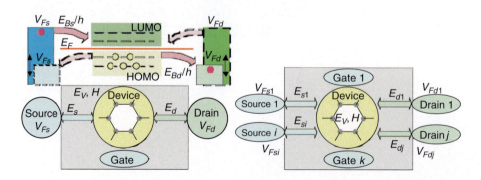

FIGURE 8.26
One-dimensional schematic representation of two- and multichannel electronic nanodevices with Hamiltonian H, $H = H_0 + V_{SC} = H_0 + f_\rho(\rho)$.

Then the electromagnetic torque is

$$T_e = L_{\Delta m} i_{as}^2 \sin 2\theta_r = L_{\Delta m} i_M^2 \left(\text{Re} \sqrt{\sin 2\theta_r} \right)^2 \sin 2\theta_r \neq 0$$

and

$$T_{eav} = \frac{1}{\pi} \int_0^\pi L_{\Delta m} i_{as}^2 \sin 2\theta_r d\theta_r = \tfrac{1}{4} L_{\Delta m} i_M^2$$

The mathematical model of the single-phase reluctance micromotor is found by using Kirchhoff's and Newton's second laws:

$$u_{as} = r_s i_{as} + \frac{d\psi_{as}}{dt} \quad \text{(circuitry equation of motion)}$$

$$T_e - B_m \omega_r - T_L = J \frac{d^2\theta_r}{dt^2} \quad \text{(torsional-mechanical equation of motion)}$$

From $\psi_{as} = (L_{ls} + \bar{L}_m - L_{\Delta m} \cos 2\theta_r) i_{as}$, one obtains a set of three first-order nonlinear differential equations that model single-phase reluctance micromotors. In particular, we have

$$\frac{di_{as}}{dt} = -\frac{r_s}{L_{ls} + \bar{L}_m - L_{\Delta m} \cos 2\theta_r} i_{as} - \frac{2L_{\Delta m}}{L_{ls} + \bar{L}_m - L_{\Delta m} \cos 2\theta_r} i_{as} \omega_r \sin 2\theta_r + \frac{1}{L_{ls} + \bar{L}_m - L_{\Delta m} \cos 2\theta_r} u_{as}$$

$$\frac{d\omega_r}{dt} = \frac{1}{J} \left(L_{\Delta m} i_{as}^2 \sin 2\theta_r - B_m \omega_r - T_L \right)$$

$$\frac{d\theta_r}{dt} = \omega_r$$

6.8.3.2   *Simulation of Reluctance Motors*

Using the derived nonlinear differential equations that describe synchronous reluctance micromotors dynamics,

$$\frac{di_{as}}{dt} = -\frac{r_s}{L_{ls} + \bar{L}_m - L_{\Delta m} \cos 2\theta_r} i_{as} - \frac{2L_{\Delta m}}{L_{ls} + \bar{L}_m - L_{\Delta m} \cos 2\theta_r} i_{as} \omega_r \sin 2\theta_r + \frac{1}{L_{ls} + \bar{L}_m - L_{\Delta m} \cos 2\theta_r} u_{as}$$

$$\frac{d\omega_r}{dt} = \frac{1}{J} \left(L_{\Delta m} i_{as}^2 \sin 2\theta_r - B_m \omega_r - T_L \right)$$

$$\frac{d\theta_r}{dt} = \omega_r$$

one can simulate and analyze these micromotor in SIMULINK by solving the differential equations.

The parameters of the micromotor can be measured as micromachine is fabricated. For illustrative purposes we assign the following parameters:

- $r_s = 1$ ohm, $L_{md} = 0.45$ H, $L_{mq} = 0.03$ H, $L_{ls} = 0.0035$ H, J = 0.000000075 kg-m², and $B_m = 0.0000000003$ N-m-sec/rad.

- The voltage applied to the stator winding is $u_{as} = 10\sin(2\theta_r - 0.62)$.

For no-load scenario ($T_L = 0$ N-m), the micromotor parameters are downloaded as

```
% Synchronous reluctance micromotor parameters
P = 2; rs = 2; Lmd = 0.5; Lmq = 0.035; Lls = 0.0035; J = 0.0000001;
Bm = 0.0000000005;
Lmb = (Lmq + Lmd)/3; Ldm = (Lmd - Lmq)/3;
um = 10; % rms value of the applied voltage
Tl = 0; % load torque
```

We must run this data as an m-file or just type the parameter values in the Command window before running the SIMULINK mdl model. The developed SIMULINK block diagram is documented in Figure 6.62.

The transient responses (micromotor dynamics) for the angular velocity $\omega_r(t)$ is plotted in Figure 6.63. The micromotor reaches 1300 rad/sec within 0.6 sec.

6.8.3.3   *Three-Phase Synchronous Reluctance Micromotors*

The goal of this section is to reinforce the results presented for simple single-phase synchronous reluctance micromotors and demonstrate how one should examine multiple-phase micromachines. We will solve a spectrum of problems in analysis, modeling, and control of three-phase synchronous reluctance micromachines. It has been emphasized that the electromagnetic features must be thoroughly analyzed before attempting to control micromotors. The angular velocity or displacement is regulated by changing the phase voltages applied or phase currents fed to the microwindings. The electromagnetic features significantly restrict the control algorithms to be applied. Depending on the conceptual methods employed to analyze synchronous reluctance micromachines, different control laws can be designed and implemented using ICs. Analysis and control of synchronous reluctance micromotors can be performed using different modeling, analysis, and optimization concepts. Complete lumped-parameter mathematical models of synchronous reluctance micromotors in the machine (*abc*) and in the quadrature-direct-zero (*qd0*) variables should be developed in the form of nonlinear differential equations. In particular, the lumped-parameter mathematical model for circuitry is found using the Kirchhoff's voltage law, and Newtonian mechanics are utilized to derive the torsional-mechanical equations of motion.

The synchronous reluctance micromotor is illustrated in Figure 6.64.

The micromachine parameters are the stator resistance r_s (it is assumed that the phase resistances are equal), the magnetizing inductances in the quadrature and direct axes L_{mq} and L_{md}, the average magnetizing inductance L_m, the leakage inductance L_{ls}, the moment of inertia J, and the viscous friction coefficient B_m.

The circuitry dynamics are modeled as

$$\mathbf{u}_{abcs} = \mathbf{r}_s \mathbf{i}_{abcs} + \frac{d\mathbf{\psi}_{abcs}}{dt}$$

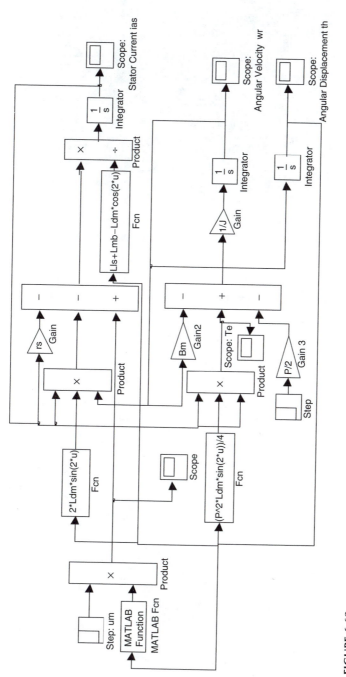

FIGURE 6.62
SIMULINK block diagram for simulation reluctance micromotors (c6_8_1.mdl).

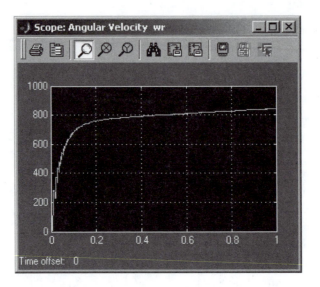

FIGURE 6.63
Micromotor behavior (transient responses) for the angular velocity ω_r.

where u_{as}, u_{bs}, and u_{cs} are the phase voltages; i_{as}, i_{bs}, and i_{cs}, are the phase currents; ψ_{as}, ψ_{bs}, and ψ_{cs} are the flux linkages, $\psi_{abcs} = \mathbf{L}_s \mathbf{i}_{abcs}$;

$$\mathbf{r}_s = \begin{bmatrix} r_s & 0 & 0 \\ 0 & r_s & 0 \\ 0 & 0 & r_s \end{bmatrix}$$

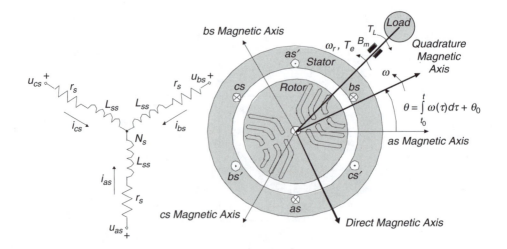

FIGURE 6.64
Three-phase synchronous reluctance micromotor.

$$\mathbf{L}_s = \begin{bmatrix} L_{ls} + \overline{L}_m - L_{\Delta m} \cos(2\theta_r) & -\tfrac{1}{2}\overline{L}_m - L_{\Delta m} \cos 2\left(\theta_r - \tfrac{1}{2}\pi\right) & -\tfrac{1}{2}\overline{L}_m - L_{\Delta m} \cos 2\left(\theta_r - \tfrac{1}{3}\pi\right) \\ -\tfrac{1}{2}\overline{L}_m - L_{\Delta m} \cos 2\left(\theta_r - \tfrac{1}{3}\pi\right) & L_{ls} + \overline{L}_m - L_{\Delta m} \cos 2\left(\theta_r - \tfrac{2}{3}\pi\right) & -\tfrac{1}{2}\overline{L}_m - L_{\Delta m} \cos 2\left(\theta_r + \pi\right) \\ -\tfrac{1}{2}\overline{L}_m - L_{\Delta m} \cos 2\left(\theta_r + \tfrac{1}{3}\pi\right) & -\tfrac{1}{2}\overline{L}_m - L_{\Delta m} \cos 2\left(\theta_r + \pi\right) & L_{ls} + \overline{L}_m - L_{\Delta m} \cos 2\left(\theta_r + \tfrac{2}{3}\pi\right) \end{bmatrix},$$

$$\overline{L}_m = \tfrac{1}{3}(L_{mq} + L_{md}); \quad \text{and} \quad L_{\Delta m} = \tfrac{1}{3}(L_{md} - L_{mq}).$$

The expressions for inductances are nonlinear functions of the electrical angular displacement θ_r. Hence, the torsional-mechanical dynamics must be used. Taking note of Newton's second law of rotational motion, and using ω_r and θ_r (electrical angular velocity and displacement) as the state variables (mechanical variables), one obtains

$$T_e - B_m \frac{2}{P}\omega_r - T_L = J\frac{2}{P}\frac{d\omega_r}{dt}$$

$$\frac{d\theta_r}{dt} = \omega_r$$

where T_e and T_L are the electromagnetic and load torques.

6.8.3.3.1 Torque Production Analysis

Using the coenergy, the electromagnetic torque, which is a nonlinear function of the micromotor variables (phase currents and electrical angular position) and micromotor parameters (number of poles P and inductance $L_{\Delta m}$), is found to be[8]

$$T_e = \frac{P}{2} L_{\Delta m}\Big[i_{as}^2 \sin 2\theta_r + 2i_{as}i_{bs} \sin 2\left(\theta_r - \tfrac{1}{3}\pi\right) + 2i_{as}i_{cs}\sin 2\left(\theta_r + \tfrac{1}{3}\pi\right)$$
$$+ i_{bs}^2 \sin 2\left(\theta_r - \tfrac{2}{3}\pi\right) + 2i_{bs}i_{cs}\sin 2\theta_r + i_{cs}^2 \sin 2\left(\theta_r + \tfrac{2}{3}\pi\right)\Big]$$

6.8.3.3.2 Control of Synchronous Reluctance Micromotors

To control the angular velocity, the electromagnetic torque must be regulated. To maximize the electromagnetic torque, ICs must feed the following phase currents as functions of the angular displacement measuring or observing (sensorless control) the rotor displacement θ_r:

$$i_{as} = \sqrt{2}i_M \sin\left(\theta_r + \tfrac{1}{3}\varphi_i\pi\right), \quad i_{bs} = \sqrt{2}i_M \sin\left(\theta_r - \tfrac{1}{3}(2 - \varphi_i)\pi\right), \quad i_{cs} = \sqrt{2}i_M \sin\left(\theta_r + \tfrac{1}{3}(2 + \varphi_i)\pi\right)$$

Thus, for $\varphi_i = 0.3245$, one obtains $T_e = \sqrt{2}PL_{\Delta m}i_M^2$.

That is, T_e is maximized and controlled by changing the magnitude of the phase currents i_M. Furthermore, in theory, using these equations, there is no torque ripple. In practice, based on the experimental results, and performing the high-fidelity modeling integrating nonlinear electromagnetics using Maxwell's equations, one finds that there exists the torque ripple, which is due to the cogging and fringing effects, eccentricity, bearing, pulse-width modulation, electromagnetic field nonuniformity, nonuniformity of magnetic materials, and other phenomena.

The majority of ICs are designed to control the phase voltages u_{as}, u_{bs}, and u_{cs}. Therefore, the three-phase balance voltage set must be introduced. We have $u_{as} = \sqrt{2}u_M \sin(\theta_r + \frac{1}{3}\varphi_i\pi)$, $u_{bs} = \sqrt{2}u_M \sin(\theta_r - \frac{1}{3}(2-\varphi_i)\pi)$, and $u_{cs} = \sqrt{2}u_M \sin(\theta_r \frac{1}{3}(2+\varphi_i)\pi)$, where u_M is the magnitude of the supplied voltages.

6.8.3.3.3   Lumped-Parameter Mathematical Models

The mathematical model of synchronous reluctance micromotors in the *abc* variables is found to be

$$\frac{di_{as}}{dt} = \frac{1}{L_D}\Big[(r_s i_{as} - u_{as})\big(4L_{ls}^2 + 3\bar{L}_m^2 - 3L_{\Delta m}^2 + 8\bar{L}_m L_{ls} - 4L_{ls}L_{\Delta m}\cos 2\theta_r\big)$$

$$+ (r_s i_{bs} - u_{bs})\big(3\bar{L}_m^2 - 3L_{\Delta m}^2 + 2\bar{L}_m L_{ls} + 4L_{ls}L_{\Delta m}\cos 2(\theta_r - \tfrac{1}{3}\pi)\big)$$

$$+ (r_s i_{cs} - u_{cs})\big(3\bar{L}_m^2 - 3L_{\Delta m}^2 + 2\bar{L}_m L_{ls} + 4L_{ls}L_{\Delta m}\cos 2(\theta_r + \tfrac{1}{3}\pi)\big)$$

$$+ 6\sqrt{3}L_{\Delta m}^2 L_{ls}\omega_r (i_{cs} - i_{bs}) + \big(8L_{\Delta m}L_{ls}^2\omega_r + 12L_{\Delta m}\bar{L}_m L_{ls}\omega_r\big)$$

$$\times \big(\sin 2\theta_r i_{as} + \sin 2(\theta_r - \tfrac{1}{3}\pi)i_{bs} + \sin 2(\theta_r + \tfrac{1}{3}\pi)i_{cs}\big)\Big]$$

$$\frac{di_{bs}}{dt} = \frac{1}{L_D}\Big[(r_s i_{as} - u_{as})\big(3\bar{L}_m^2 - 3L_{\Delta m}^2 + 2\bar{L}_m L_{ls} + 4L_{ls}L_{\Delta m}\cos 2(\theta_r - \tfrac{1}{3}\pi)\big)$$

$$+ (r_s i_{bs} - u_{bs})\big(4L_{ls}^2 + 3\bar{L}_m^2 - 3L_{\Delta m}^2 + 8\bar{L}_m L_{ls} - 4L_{ls}L_{\Delta m}\cos 2(\theta_r + \tfrac{1}{3}\pi)\big)$$

$$+ (r_s i_{cs} - u_{cs})\big(3\bar{L}_m^2 - 3L_{\Delta m}^2 + 2\bar{L}_m L_{ls} + 4L_{ls}L_{\Delta m}\cos 2\theta_r\big)$$

$$+ 6\sqrt{3}L_{\Delta m}^2 L_{ls}\omega_r (i_{as} - i_{cs}) + \big(8L_{\Delta m}L_{ls}^2\omega_r + 12L_{\Delta m}\bar{L}_m L_{ls}\omega_r\big)$$

$$\times \big(\sin 2(\theta_r - \tfrac{1}{3}\pi)i_{as} + \sin 2(\theta_r + \tfrac{1}{3}\pi)i_{bs} + \sin 2\theta_r i_{cs}\big)\Big]$$

$$\frac{di_{cs}}{dt} = \frac{1}{L_D}\Big[(r_s i_{as} - u_{as})\big(3\bar{L}_m^2 - 3L_{\Delta m}^2 + 2\bar{L}_m L_{ls} + 4L_{ls}L_{\Delta m}\cos 2(\theta_r + \tfrac{1}{3}\pi)\big)$$

$$+ (r_s i_{bs} - u_{bs})\big(3\bar{L}_m^2 - 3L_{\Delta m}^2 + 2\bar{L}_m L_{ls} + 4L_{ls}L_{\Delta m}\cos 2\theta_r\big)$$

$$+ (r_s i_{cs} - u_{cs})\big(4L_{ls}^2 + 3\bar{L}_m^2 - 3L_{\Delta m}^2 + 8\bar{L}_m L_{ls} - 4L_{ls}L_{\Delta m}\cos 2(\theta_r - \tfrac{1}{3}\pi)\big)$$

$$+ 6\sqrt{3}L_{\Delta m}^2 L_{ls}\omega_r (i_{bs} - i_{as}) + \big(8L_{\Delta m}L_{ls}^2\omega_r + 12L_{\Delta m}\bar{L}_m L_{ls}\omega_r\big)$$

$$\times \big(\sin 2(\theta_r + \tfrac{1}{3}\pi)i_{as} + \sin 2\theta_r i_{bs} + \sin 2(\theta_r - \tfrac{1}{3}\pi)i_{cs}\big)\Big]$$

$$\frac{d\omega_r}{dt} = \frac{P^2}{4J}L_{\Delta m}\Big(i_{as}^2 \sin 2\theta_r + 2i_{as}i_{bs}\sin 2(\theta_r - \tfrac{1}{3}\pi) + 2i_{as}i_{cs}\sin 2(\theta_r + \tfrac{1}{3}\pi)$$

$$+ i_{bs}^2 \sin 2(\theta_r - \tfrac{2}{3}\pi) + 2i_{bs}i_{cs}\sin 2\theta_r + i_{cs}^2 \sin 2(\theta_r + \tfrac{2}{3}\pi)\Big) - \frac{B_m}{J}\omega_r - \frac{P}{2J}T_L$$

$$\frac{d\theta_r}{dt} = \omega_r$$

In these differential equations, the following notations are used:

$$\bar{L}_m = \tfrac{1}{3}(L_{mq} + L_{md}), \quad L_{\Delta m} = \tfrac{1}{3}(L_{md} - L_{mq}), \quad L_D = L_{ls}\big(9L_{\Delta m}^2 - 4L_{ls}^2 - 12\bar{L}_m L_{ls} - 9\bar{L}_m^2\big)$$

That is, we have a set of five highly coupled nonlinear differential equations. Though analytic solution of these equations is virtually impossible, nonlinear modeling can be straightforwardly performed in MATLAB, which includes the SIMULINK environment.

The mathematical model can be simplified. In particular, in the rotor reference frame, we apply the Park transformation[8]:

$$\mathbf{u}^r_{qd0s} = \mathbf{K}^r_s \mathbf{u}_{abcs}, \quad \mathbf{i}^r_{qd0s} = \mathbf{K}^r_s \mathbf{i}_{abcs}, \quad \boldsymbol{\psi}^r_{qd0s} = \mathbf{K}^r_s \boldsymbol{\psi}_{abcs},$$

$$\mathbf{K}^r_s = \frac{2}{3} \begin{bmatrix} \cos\theta_r & \cos\left(\theta_r - \frac{2}{3}\pi\right) & \cos\left(\theta_r + \frac{2}{3}\pi\right) \\ \sin\theta_r & \sin\left(\theta_r - \frac{2}{3}\pi\right) & \sin\left(\theta_r + \frac{2}{3}\pi\right) \\ \frac{1}{2} & \frac{1}{2} & \frac{1}{2} \end{bmatrix}$$

where u_{qs}, u_{ds}, u_{0s}, i_{qs}, i_{ds}, i_{0s} and ψ_{qs}, ψ_{ds}, ψ_{0s} are the $qd0$ voltages, currents, and flux linkages.

Using the circuitry and torsional-mechanical dynamics, one finds the following nonlinear differential equations to model synchronous reluctance micromotors in the rotor reference frame:

$$\frac{di^r_{qs}}{dt} = -\frac{r_s}{L_{ls} + L_{mq}} i^r_{qs} - \frac{L_{ls} + L_{md}}{L_{ls} + L_{mq}} i^r_{ds}\omega_r + \frac{1}{L_{ls} + L_{mq}} u^r_{qs}$$

$$\frac{di^r_{ds}}{dt} = -\frac{r_s}{L_{ls} + L_{md}} i^r_{ds} + \frac{L_{ls} + L_{mq}}{L_{ls} + L_{md}} i^r_{qs}\omega_r + \frac{1}{L_{ls} + L_{md}} u^r_{ds}$$

$$\frac{di^r_{0s}}{dt} = -\frac{r_s}{L_{ls}} i^r_{0s} + \frac{1}{L_{ls}} u^r_{0s}$$

$$\frac{d\omega_r}{dt} = \frac{3P^2}{8J}(L_{md} - L_{mq})i^r_{qs}i^r_{ds} - \frac{B_m}{J}\omega_r - \frac{P}{2J}T_L$$

$$\frac{d\theta_r}{dt} = \omega_r$$

One can easily observe that this model is much simpler than the lumped-parameter mathematical model derived using the *abc* variables. (However, the angular rotor displacement must be measured to control synchronous micromotors.)

To attain the balanced operation, the quadrature and direct currents and voltages must be derived using the direct Park transformation $\mathbf{i}^r_{qd0s} = \mathbf{K}^r_s \mathbf{i}_{abcs}$, $\mathbf{u}^r_{qd0s} = \mathbf{K}^r_s \mathbf{u}_{abcs}$. The $qd0$ voltages $\mathbf{u}^r_{qs}$, $\mathbf{u}^r_{ds}$, and $\mathbf{u}^r_{0s}$ are found using the three-phase balance voltage set. In particular, we have

$$u^r_{qs} = \sqrt{2}u_M, \quad u^r_{ds} = 0, \quad u^r_{0s} = 0$$

We derived the lumped-parameter mathematical models of three-phase synchronous reluctance micromotors. Based on the differential equations obtained, nonlinear analysis can be performed, and the phase currents and voltages needed to guarantee balanced operating conditions were found. In particular, taking note of electromagnetic features, we found that phase currents and voltages needed to be applied to guarantee the desired operating features. The results reported can be straightforwardly used in nonlinear simulation.

6.8.4 Radial Topology Permanent-Magnet Synchronous Micromachines

To further examine permanent-magnet synchronous machines, we consider the radial topology microdevices. The studied permanent-magnet synchronous microtransducers are brushless micromachines because the excitation flux is produced by permanent magnets deposited on the microrotor. In this section, the following variables and symbols are used:

u_{as}, u_{bs}, and u_{cs} are the phase voltages in the stator windings *as*, *bs*, and *cs*.

u_{qs}, u_{ds}, and u_{0s} are the quadrature-, direct-, and zero-axis stator voltages.

i_{as}, i_{bs}, and i_{cs} are the phase currents in the stator windings *as*, *bs*, and *cs*.

i_{qs}, i_{ds}, and i_{0s} are the quadrature-, direct-, and zero-axis stator currents.

ψ_{as}, ψ_{bs}, and ψ_{cs} are the stator flux linkages.

ψ_{qs}, ψ_{ds}, and ψ_{0s} are the quadrature-, direct-, and zero-axis stator flux linkages.

ψ_m is the magnitude of the flux linkages established by the permanent-magnets.

ω_r and ω_{rm} are the electrical and rotor angular velocities.

θ_r and θ_{rm} are the electrical and rotor angular displacements.

T_e is the electromagnetic torque developed.

T_L is the load torque applied.

B_m is the viscous friction coefficient.

J is the equivalent moment of inertia.

r_s is the resistances of the stator windings.

L_{ss} is the self-inductances of the stator windings.

L_{ms} and L_{ls} are the stator magnetizing and leakage inductances.

L_{mq} and L_{md} are the magnetizing inductances in the quadrature and direct axes.

$\Re_{md}$ and $\Re_{mq}$ are the magnetizing reluctances in the direct and quadrature axes.

N_s is the number of turns of the stator windings.

P is the number of poles.

ω and θ are the angular velocity and displacement of the reference frame.

6.8.4.1 *Mathematical Model of Two-Phase Permanent-Magnet Synchronous Micromotors*

Consider radial topology two-phase permanent-magnet synchronous micromotors. Using Kirchhoff's voltage law, we have the following two equations:

$$u_{as} = r_s i_{as} + \frac{d\psi_{as}}{dt}$$

$$u_{bs} = r_s i_{bs} + \frac{d\psi_{bs}}{dt}$$

where the flux linkages are expressed as

$$\psi_{as} = L_{asas} i_{as} + L_{asbs} i_{bs} + \psi_{asm} \quad \text{and} \quad \psi_{bs} = L_{bsas} i_{as} + L_{bsbs} i_{bs} + \psi_{bsm}$$

Here u_{as} and u_{bs} are the phase voltages in the stator microwindings *as* and *bs*; i_{as} and i_{bs} are the phase currents in the stator microwindings; ψ_{as} and ψ_{bs} are the stator flux linkages;

r_s is the resistances of the stator microwindings; and L_{asas}, L_{asbs}, L_{bsas}, and L_{bsbs} are the mutual inductances.

The flux linkages are periodic functions of the angular displacement (rotor position), and let

$$\psi_{asm} = \psi_m \sin\theta_{rm} \quad \text{and} \quad \psi_{bsm} = -\psi_m \cos\theta_{rm}$$

The self-inductances of the stator windings are found to be

$$L_{ss} = L_{asas} = L_{bsbs} = L_{ls} + \overline{L}_m$$

The stator windings are displaced by 90 electrical degrees.
Hence, the mutual inductances between the stator windings are

$$L_{asbs} = L_{bsas} = 0$$

Thus, we have

$$\psi_{as} = L_{ss}i_{as} + \psi_m \sin\theta_{rm} \quad \text{and} \quad \psi_{bs} = L_{ss}i_{bs} - \psi_m \cos\theta_{rm}$$

Therefore, one finds

$$u_{as} = r_s i_{as} + \frac{d(L_{ss}i_{as} + \psi_m \sin\theta_{rm})}{dt} = r_s i_{as} + L_{ss}\frac{di_{as}}{dt} + \psi_m \omega_{rm} \cos\theta_{rm}$$

$$u_{bs} = r_s i_{bs} + \frac{d(L_{ss}i_{bs} - \psi_m \cos\theta_{rm})}{dt} = r_s i_{bs} + L_{ss}\frac{di_{bs}}{dt} - \psi_m \omega_{rm} \sin\theta_{rm}$$

Using Newton's second law, $T_e - B_m \omega_{rm} - T_L = J d^2\theta_{rm}/dt^2$, we have

$$\frac{d\omega_{rm}}{dt} = \frac{1}{J}(T_e - B_m \omega_{rm} - T_L)$$

$$\frac{d\theta_{rm}}{dt} = \omega_{rm}$$

The expression for the electromagnetic torque developed by permanent-magnet micro-motors can be obtained by using the coenergy

$$W_c = \frac{1}{2}\left(L_{ss}i_{as}^2 + L_{ss}i_{bs}^2\right) + \psi_m i_{as} \sin\theta_{rm} - \psi_m i_{bs} \cos\theta_{rm} + W_{PM}$$

Then one has

$$T_e = \frac{\partial W_c}{\partial\theta_{rm}} = \frac{P\psi_m}{2}[i_{as}\cos\theta_{rm} + i_{bs}\sin\theta_{rm}]$$

Augmenting the circuitry transients with the torsional-mechanical dynamics, one finds the mathematical model of two-phase permanent-magnet micromotors in the following form:

$$\frac{di_{as}}{dt} = -\frac{r_s}{L_{ss}}i_{as} - \frac{\psi_m}{L_{ss}}\omega_{rm}\cos\theta_{rm} + \frac{1}{L_{ss}}u_{as}$$

$$\frac{di_{bs}}{dt} = -\frac{r_s}{L_{ss}}i_{bs} + \frac{\psi_m}{L_{ss}}\omega_{rm}\sin\theta_{rm} + \frac{1}{L_{ss}}u_{bs}$$

$$\frac{d\omega_{rm}}{dt} = \frac{P\psi_m}{2J}(i_{as}\cos\theta_{rm} + i_{bs}\sin\theta_{rm}) - \frac{B_m}{J}\omega_{rm} - \frac{1}{J}T_L$$

$$\frac{d\theta_{rm}}{dt} = \omega_{rm}$$

For two-phase micromotors (assuming the sinusoidal winding distributions and the sinusoidal mmf waveforms), the electromagnetic torque is expressed as

$$T_e = \frac{P\psi_m}{2}(i_{as}\cos\theta_r + i_{bs}\sin\theta_r)$$

Hence, to guarantee the balanced operation, one feeds

$$i_{as} = \sqrt{2}i_M\cos\theta_r \quad \text{and} \quad i_{bs} = \sqrt{2}i_M\sin\theta_r$$

to maximize the electromagnetic torque. In fact, one obtains

$$T_e = \frac{P\psi_m}{2}(i_{as}\cos\theta_r + i_{bs}\sin\theta_r) = \frac{P\psi_m}{2}\sqrt{2}i_M(\cos^2\theta_r + \sin^2\theta_r) = \frac{P\psi_m}{\sqrt{2}}i_M$$

6.8.4.2 Radial Topology Three-Phase Permanent-Magnet Synchronous Micromachines

Three-phase two-pole permanent-magnet synchronous microtransducers (microscale motors and generators) are illustrated in Figure 6.65 and Figure 6.66.

From Kirchhoff's second law, one obtains three differential equations for the *as*, *bs*, and *cs* stator windings. In particular,

$$u_{as} = r_s i_{as} + \frac{d\psi_{as}}{dt}$$

$$u_{bs} = r_s i_{bs} + \frac{d\psi_{bs}}{dt} \tag{6.40}$$

$$u_{cs} = r_s i_{cs} + \frac{d\psi_{cs}}{dt}$$

where the flux linkages are

$$\psi_{as} = L_{asas}i_{as} + L_{asbs}i_{bs} + L_{ascs}i_{cs} + \psi_{asm}, \quad \psi_{bs} = L_{bsas}i_{as} + L_{bsbs}i_{bs} + L_{bscs}i_{as} + \psi_{bsm} \text{ and}$$

$$\psi_{cs} = L_{csas}i_{as} + L_{csbs}i_{bs} + L_{cscs}i_{cs} + \psi_{csm}.$$

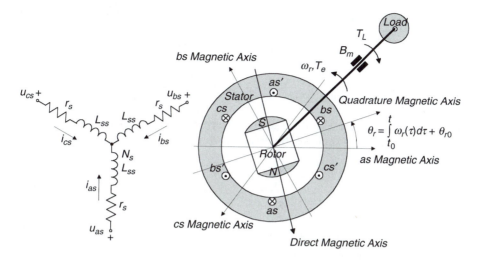

FIGURE 6.65
Two-pole permanent-magnet synchronous micromotor.

From Equation 6.40, one can write

$$\mathbf{u}_{abcs} = \mathbf{r}_s \mathbf{i}_{abcs} + \frac{d\boldsymbol{\psi}_{abcs}}{dt}, \qquad \begin{bmatrix} u_{as} \\ u_{bs} \\ u_{cs} \end{bmatrix} = \begin{bmatrix} r_s & 0 & 0 \\ 0 & r_s & 0 \\ 0 & 0 & r_s \end{bmatrix} \begin{bmatrix} i_{as} \\ i_{bs} \\ i_{cs} \end{bmatrix} + \begin{bmatrix} \dfrac{d\psi_{as}}{dt} \\ \dfrac{d\psi_{bs}}{dt} \\ \dfrac{d\psi_{cs}}{dt} \end{bmatrix}$$

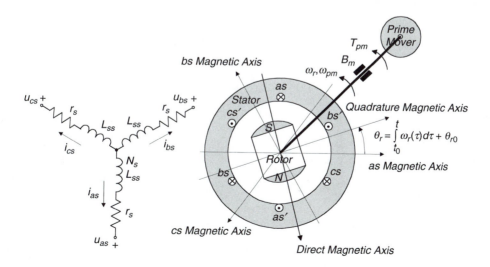

FIGURE 6.66
Three-phase wye-connected synchronous microgenerator.

The flux linkages ψ_{asm}, ψ_{bsm}, and ψ_{csm}, established by the permanent magnet, are periodic functions of θ_r. We assume that ψ_{asm}, ψ_{bsm}, and ψ_{csm} vary obeying the sine law. The stator windings are displaced by 120 electrical degrees. Denoting the magnitude of the flux linkages established by the permanent magnet as ψ_m, one has

$$\psi_{asm} = \psi_m \sin\theta_r, \quad \psi_{bsm} = \psi_m \sin\left(\theta_r - \tfrac{2}{3}\pi\right) \quad \text{and} \quad \psi_{csm} = \psi_m \sin\left(\theta_r + \tfrac{2}{3}\pi\right)$$

Self- and mutual inductances for three-phase permanent-magnet synchronous microtransducers can be derived. In particular, the equations for the magnetizing quadrature and direct inductances are

$$L_{mq} = \frac{N_s^2}{\Re_{mq}} \quad \text{and} \quad L_{md} = \frac{N_s^2}{\Re_{md}}$$

In general the quadrature and direct magnetizing reluctances can be different, and $\Re_{mq} > \Re_{md}$. Hence, we have $L_{mq} < L_{md}$.

The minimum value of L_{asas} occurs periodically at $\theta_r = 0, \pi, 2\pi, \ldots$, while the maximum value of L_{asas} occurs at $\theta_r = \tfrac{1}{2}\pi, \tfrac{3}{2}\pi, \tfrac{5}{2}\pi, \ldots$.

One concludes that the self-inductance $L_{asas}(\theta_r)$, which is bounded as $L_{ls} + L_{mq} \leq L_{asas} \leq L_{ls} + L_{md}$, is a periodic function of (θ_r). Assuming that $L_{asas}(\theta_r)$, varies as a sine function with a DC component, we have

$$L_{asas} = L_{ls} + \overline{L}_m - L_{\Delta m}\cos 2\theta_r$$

where $\overline{L}_m$ is the average value of the magnetizing inductance, and $L_{\Delta m}$ is half the amplitude of the sinusoidal variation of the magnetizing inductance.

The relationships between L_{mq}, L_{md}, and $\overline{L}_m, L_{\Delta m}$, must be found. For three-phase synchronous microtransducers, one obtains

$$L_{mq} = \tfrac{3}{2}(\overline{L}_m - L_{\Delta m}) \quad \text{and} \quad L_{md} = \tfrac{3}{2}(\overline{L}_m + L_{\Delta m})$$

Therefore, $\overline{L}_m = \tfrac{1}{3}(L_{mq} + L_{md})$ and $L_{\Delta m} = \tfrac{1}{3}(L_{md} - L_{mq})$.

Using the expressions for L_{mq} and L_{md}, we have

$$\overline{L}_m = \frac{1}{3}\left(\frac{N_s^2}{\Re_{mq}} + \frac{N_s^2}{\Re_{md}}\right) \quad \text{and} \quad L_{\Delta m} = \frac{1}{3}\left(\frac{N_s^2}{\Re_{md}} - \frac{N_s^2}{\Re_{mq}}\right)$$

Therefore, the following equations for ψ_{as}, ψ_{bs}, and ψ_{cs} result:

$$\psi_{as} = \left(L_{ls} + \overline{L}_m - L_{\Delta m}\cos 2\theta_r\right)i_{as} + \left(-\tfrac{1}{2}\overline{L}_m - L_{\Delta m}\cos 2\left(\theta_r - \tfrac{1}{3}\pi\right)\right)i_{bs}$$

$$+ \left(-\tfrac{1}{2}\overline{L}_m - L_{\Delta m}\cos 2\left(\theta_r + \tfrac{1}{3}\pi\right)\right)i_{cs} + \psi_m \sin\theta_r,$$

$$\psi_{bs} = \left(-\tfrac{1}{2}\overline{L}_m - L_{\Delta m}\cos 2\left(\theta_r - \tfrac{1}{3}\pi\right)\right)i_{as} + \left(L_{ls} + \overline{L}_m - L_{\Delta m}\cos 2\left(\theta_r - \tfrac{2}{3}\pi\right)\right)i_{bs}$$

$$+ \left(-\tfrac{1}{2}\overline{L}_m - L_{\Delta m}\cos 2\theta_r\right)i_{cs} + \psi_m \sin\left(\theta_r - \tfrac{2}{3}\pi\right), \tag{6.41}$$

$$\psi_{cs} = \left(-\tfrac{1}{2}\overline{L}_m - L_{\Delta m}\cos 2\left(\theta_r + \tfrac{1}{3}\pi\right)\right)i_{as} + \left(-\tfrac{1}{2}\overline{L}_m - L_{\Delta m}\cos 2\theta_r\right)i_{bs}$$

$$+ \left(L_{ls} + \overline{L}_m - L_{\Delta m}\cos 2\left(\theta_r + \tfrac{2}{3}\pi\right)\right)i_{cs} + \psi_m \sin\left(\theta_r + \tfrac{2}{3}\pi\right)$$

From Equation 6.41, one has

$$\psi_{abcs} = \mathbf{L}_s \mathbf{i}_{abcs} + \psi_m$$

$$= \begin{bmatrix} L_{ls} + \bar{L}_m - L_{\Delta m}\cos 2\theta_r & -\frac{1}{2}\bar{L}_m - L_{\Delta m}\cos 2\left(\theta_r - \frac{1}{3}\pi\right) & -\frac{1}{2}\bar{L}_m - L_{\Delta m}\cos 2\left(\theta_r + \frac{1}{3}\pi\right) \\ -\frac{1}{2}\bar{L}_m - L_{\Delta m}\cos 2\left(\theta_r - \frac{1}{3}\pi\right) & L_{ls} + \bar{L}_m - L_{\Delta m}\cos 2\left(\theta_r - \frac{2}{3}\pi\right) & -\frac{1}{2}\bar{L}_m - L_{\Delta m}\cos 2\theta_r \\ -\frac{1}{2}\bar{L}_m - L_{\Delta m}\cos 2\left(\theta_r + \frac{1}{3}\pi\right) & -\frac{1}{2}\bar{L}_m - L_{\Delta m}\cos 2\theta_r & L_{ls} + \bar{L}_m - L_{\Delta m}\cos 2\left(\theta_r + \frac{2}{3}\pi\right) \end{bmatrix}\begin{bmatrix} i_{as} \\ i_{bs} \\ i_{cs} \end{bmatrix}$$

$$+ \psi_m \begin{bmatrix} \sin\theta_r \\ \sin\left(\theta_r - \frac{2}{3}\pi\right) \\ \sin\left(\theta_r + \frac{2}{3}\pi\right) \end{bmatrix}$$

The inductance matrix $\mathbf{L}_s$ is given by

$$\mathbf{L}_s = \begin{bmatrix} L_{ls} + \bar{L}_m - L_{\Delta m}\cos 2\theta_r & -\frac{1}{2}\bar{L}_m - L_{\Delta m}\cos 2\left(\theta_r - \frac{1}{3}\pi\right) & -\frac{1}{2}\bar{L}_m - L_{\Delta m}\cos 2\left(\theta_r + \frac{1}{3}\pi\right) \\ -\frac{1}{2}\bar{L}_m - L_{\Delta m}\cos 2\left(\theta_r - \frac{1}{3}\pi\right) & L_{ls} + \bar{L}_m - L_{\Delta m}\cos 2\left(\theta_r - \frac{2}{3}\pi\right) & -\frac{1}{2}\bar{L}_m - L_{\Delta m}\cos 2\theta_r \\ -\frac{1}{2}\bar{L}_m - L_{\Delta m}\cos 2\left(\theta_r + \frac{1}{3}\pi\right) & -\frac{1}{2}\bar{L}_m - L_{\Delta m}\cos 2\theta_r & L_{ls} + \bar{L}_m - L_{\Delta m}\cos 2\left(\theta_r + \frac{2}{3}\pi\right) \end{bmatrix}$$

It was shown that $\bar{L}_m$ and $L_{\Delta m}$ are

$$\bar{L}_m = \frac{1}{3}\left(\frac{N_s^2}{\Re_{mq}} + \frac{N_s^2}{\Re_{md}}\right) \quad \text{and} \quad L_{\Delta m} = \frac{1}{3}\left(\frac{N_s^2}{\Re_{md}} - \frac{N_s^2}{\Re_{mq}}\right)$$

Permanent-magnet synchronous microtransducers are round-rotor electrical machines. (The magnetic paths in the quadrature and direct magnetic axes are identical, and $\Re_{mq} = \Re_{md}$.) Thus,

$$\bar{L}_m = \frac{2N_s^2}{3\Re_{mq}} = \frac{2N_s^2}{3\Re_{md}} \quad \text{and} \quad L_{\Delta m} = 0$$

Therefore, the inductance matrix is

$$\mathbf{L}_s = \begin{bmatrix} L_{ls} + \bar{L}_m & -\frac{1}{2}\bar{L}_m & -\frac{1}{2}\bar{L}_m \\ -\frac{1}{2}\bar{L}_m & L_{ls} + \bar{L}_m & -\frac{1}{2}\bar{L}_m \\ -\frac{1}{2}\bar{L}_m & -\frac{1}{2}\bar{L}_m & L_{ls} + \bar{L}_m \end{bmatrix}$$

From Equation 6.41 the expressions for the flux linkages are

$$\psi_{as} = (L_{ls} + \bar{L}_m)i_{as} - \frac{1}{2}\bar{L}_m i_{bs} - \frac{1}{2}\bar{L}_m i_{cs} + \psi_m \sin\theta_r$$

$$\psi_{bs} = -\frac{1}{2}\bar{L}_m i_{as} + (L_{ls} + \bar{L}_m)i_{bs} - \frac{1}{2}\bar{L}_m i_{cs} + \psi_m \sin\left(\theta_r - \frac{2}{3}\pi\right) \qquad (6.42)$$

$$\psi_{cs} = -\frac{1}{2}\bar{L}_m i_{as} - \frac{1}{2}\bar{L}_m i_{bs} + (L_{ls} + \bar{L}_m)i_{cs} + \psi_m \sin\left(\theta_r + \frac{2}{3}\pi\right)$$

or in matrix form,

$$\mathbf{\psi}_{abcs} = \mathbf{L}_s \mathbf{i}_{abcs} + \mathbf{\psi}_m = \begin{bmatrix} L_{ls} + \overline{L}_m & -\frac{1}{2}\overline{L}_m & -\frac{1}{2}\overline{L}_m \\ -\frac{1}{2}\overline{L}_m & L_{ls} + \overline{L}_m & -\frac{1}{2}\overline{L}_m \\ -\frac{1}{2}\overline{L}_m & -\frac{1}{2}\overline{L}_m & L_{ls} + \overline{L}_m \end{bmatrix} \begin{bmatrix} i_{as} \\ i_{bs} \\ i_{cs} \end{bmatrix} + \psi_m \begin{bmatrix} \sin\theta_r \\ \sin\left(\theta_r - \frac{2}{3}\pi\right) \\ \sin\left(\theta_r + \frac{2}{3}\pi\right) \end{bmatrix}$$

Using Equation 6.40 and Equation 6.42, we have

$$\mathbf{u}_{abcs} = \mathbf{r}_s \mathbf{i}_{abcs} + \frac{d\mathbf{\psi}_{abcs}}{dt} = \mathbf{r}_s \mathbf{i}_{abcs} + \mathbf{L}_s \frac{d\mathbf{i}_{abcs}}{dt} + \frac{d\mathbf{\psi}_m}{dt}$$

where

$$\frac{d\mathbf{\psi}_m}{dt} = \psi_m \begin{bmatrix} \omega_r \cos\theta_r \\ \omega_r \cos\left(\theta_r - \frac{2}{3}\pi\right) \\ \omega_r \cos\left(\theta_r + \frac{2}{3}\pi\right) \end{bmatrix}$$

Cauchy's form can be found by making use of $\mathbf{L}_s^{-1}$. In particular,

$$\frac{d\mathbf{i}_{abcs}}{dt} = -\mathbf{L}_s^{-1}\mathbf{r}_s \mathbf{i}_{abcs} - \mathbf{L}_s^{-1}\frac{d\mathbf{\psi}_m}{dt} + \mathbf{L}_s^{-1}\mathbf{u}_{abcs}$$

The state-space stator circuitry dynamics in Cauchy's form is given as

$$\begin{bmatrix} \dfrac{di_{as}}{dt} \\ \dfrac{di_{bs}}{dt} \\ \dfrac{di_{cs}}{dt} \end{bmatrix} = \begin{bmatrix} -\dfrac{r_s(2L_{ss} - \overline{L}_m)}{2L_{ss}^2 - L_{ss}\overline{L}_m - \overline{L}_m^2} & -\dfrac{r_s\overline{L}_m}{2L_{ss}^2 - L_{ss}\overline{L}_m - \overline{L}_m^2} & -\dfrac{r_s\overline{L}_m}{2L_{ss}^2 - L_{ss}\overline{L}_m - \overline{L}_m^2} \\ -\dfrac{r_s\overline{L}_m}{2L_{ss}^2 - L_{ss}\overline{L}_m - \overline{L}_m^2} & -\dfrac{r_s(2L_{ss} - \overline{L}_m)}{2L_{ss}^2 - L_{ss}\overline{L}_m - \overline{L}_m^2} & -\dfrac{r_s\overline{L}_m}{2L_{ss}^2 - L_{ss}\overline{L}_m - \overline{L}_m^2} \\ -\dfrac{r_s\overline{L}_m}{2L_{ss}^2 - L_{ss}\overline{L}_m - \overline{L}_m^2} & -\dfrac{r_s\overline{L}_m}{2L_{ss}^2 - L_{ss}\overline{L}_m - \overline{L}_m^2} & -\dfrac{r_s(2L_{ss} - \overline{L}_m)}{2L_{ss}^2 - L_{ss}\overline{L}_m - \overline{L}_m^2} \end{bmatrix} \begin{bmatrix} i_{as} \\ i_{bs} \\ i_{cs} \end{bmatrix}$$

$$+ \begin{bmatrix} -\dfrac{\psi_m(2L_{ss} - \overline{L}_m)}{2L_{ss}^2 - L_{ss}\overline{L}_m - \overline{L}_m^2} & -\dfrac{\psi_m\overline{L}_m}{2L_{ss}^2 - L_{ss}\overline{L}_m - \overline{L}_m^2} & -\dfrac{\psi_m\overline{L}_m}{2L_{ss}^2 - L_{ss}\overline{L}_m - \overline{L}_m^2} \\ -\dfrac{\psi_m\overline{L}_m}{2L_{ss}^2 - L_{ss}\overline{L}_m - \overline{L}_m^2} & -\dfrac{\psi_m(2L_{ss} - \overline{L}_m)}{2L_{ss}^2 - L_{ss}\overline{L}_m - \overline{L}_m^2} & -\dfrac{\psi_m\overline{L}_m}{2L_{ss}^2 - L_{ss}\overline{L}_m - \overline{L}_m^2} \\ -\dfrac{\psi_m\overline{L}_m}{2L_{ss}^2 - L_{ss}\overline{L}_m - \overline{L}_m^2} & -\dfrac{\psi_m\overline{L}_m}{2L_{ss}^2 - L_{ss}\overline{L}_m - \overline{L}_m^2} & -\dfrac{\psi_m(2L_{ss} - \overline{L}_m)}{2L_{ss}^2 - L_{ss}\overline{L}_m - \overline{L}_m^2} \end{bmatrix} \begin{bmatrix} \omega_r \cos\theta_r \\ \omega_r \cos\left(\theta_r - \frac{2}{3}\pi\right) \\ \omega_r \cos\left(\theta_r + \frac{2}{3}\pi\right) \end{bmatrix}$$

$$+ \begin{bmatrix} \dfrac{2L_{ss} - \overline{L}_m}{2L_{ss}^2 - L_{ss}\overline{L}_m - \overline{L}_m^2} & \dfrac{\overline{L}_m}{2L_{ss}^2 - L_{ss}\overline{L}_m - \overline{L}_m^2} & \dfrac{\overline{L}_m}{2L_{ss}^2 - L_{ss}\overline{L}_m - \overline{L}_m^2} \\ \dfrac{\overline{L}_m}{2L_{ss}^2 - L_{ss}\overline{L}_m - \overline{L}_m^2} & \dfrac{2L_{ss} - \overline{L}_m}{2L_{ss}^2 - L_{ss}\overline{L}_m - \overline{L}_m^2} & \dfrac{\overline{L}_m}{2L_{ss}^2 - L_{ss}\overline{L}_m - \overline{L}_m^2} \\ \dfrac{\overline{L}_m}{2L_{ss}^2 - L_{ss}\overline{L}_m - \overline{L}_m^2} & \dfrac{\overline{L}_m}{2L_{ss}^2 - L_{ss}\overline{L}_m - \overline{L}_m^2} & \dfrac{2L_{ss} - \overline{L}_m}{2L_{ss}^2 - L_{ss}\overline{L}_m - \overline{L}_m^2} \end{bmatrix} \begin{bmatrix} u_{as} \\ u_{bs} \\ u_{cs} \end{bmatrix}$$

Here $L_{ss} = L_{ls} + \overline{L}_m$.

In expanded form, we have the following nonlinear differential equations that allow the designer to model the circuitry transient behavior:

$$\frac{di_{as}}{dt} = -\frac{r_s(2L_{ss}-\overline{L}_m)}{2L_{ss}^2 - L_{ss}\overline{L}_m - \overline{L}_m^2} i_{as} - \frac{r_s\overline{L}_m}{2L_{ss}^2 - L_{ss}\overline{L}_m - \overline{L}_m^2} i_{bs} - \frac{r_s\overline{L}_m}{2L_{ss}^2 - L_{ss}\overline{L}_m - \overline{L}_m^2} i_{cs}$$

$$-\frac{\psi_m(2L_{ss}-\overline{L}_m)}{2L_{ss}^2 - L_{ss}\overline{L}_m - \overline{L}_m^2}\omega_r\cos\theta_r - \frac{\psi_m\overline{L}_m}{2L_{ss}^2 - L_{ss}\overline{L}_m - \overline{L}_m^2}\omega_r\cos\left(\theta_r - \tfrac{2}{3}\pi\right)$$

$$-\frac{\psi_m\overline{L}_m}{2L_{ss}^2 - L_{ss}\overline{L}_m - \overline{L}_m^2}\omega_r\cos\left(\theta_r + \tfrac{2}{3}\pi\right) + \frac{2L_{ss}-\overline{L}_m}{2L_{ss}^2 - L_{ss}\overline{L}_m - \overline{L}_m^2} u_{as} + \frac{\overline{L}_m}{2L_{ss}^2 - L_{ss}\overline{L}_m - \overline{L}_m^2} u_{bs}$$

$$+\frac{\overline{L}_m}{2L_{ss}^2 - L_{ss}\overline{L}_m - \overline{L}_m^2} u_{cs},$$

$$\frac{di_{bs}}{dt} = -\frac{r_s\overline{L}_m}{2L_{ss}^2 - L_{ss}\overline{L}_m - \overline{L}_m^2} i_{as} - \frac{r_s(2L_{ss}-\overline{L}_m)}{2L_{ss}^2 - L_{ss}\overline{L}_m - \overline{L}_m^2} i_{bs} - \frac{r_s\overline{L}_m}{2L_{ss}^2 - L_{ss}\overline{L}_m - \overline{L}_m^2} i_{cs}$$

$$-\frac{\psi_m\overline{L}_m}{2L_{ss}^2 - L_{ss}\overline{L}_m - \overline{L}_m^2}\omega_r\cos\theta_r - \frac{\psi_m(2L_{ss}-\overline{L}_m)}{2L_{ss}^2 - L_{ss}\overline{L}_m - \overline{L}_m^2}\omega_r\cos\left(\theta_r - \tfrac{2}{3}\pi\right)$$

$$-\frac{\psi_m\overline{L}_m}{2L_{ss}^2 - L_{ss}\overline{L}_m - \overline{L}_m^2}\omega_r\cos\left(\theta_r + \tfrac{2}{3}\pi\right) + \frac{\overline{L}_m}{2L_{ss}^2 - L_{ss}\overline{L}_m - \overline{L}_m^2} u_{as}$$

$$+\frac{2L_{ss}-\overline{L}_m}{2L_{ss}^2 - L_{ss}\overline{L}_m - \overline{L}_m^2} u_{bs} + \frac{\overline{L}_m}{2L_{ss}^2 - L_{ss}\overline{L}_m - \overline{L}_m^2} u_{cs},$$

$$\frac{di_{cs}}{dt} = -\frac{r_s\overline{L}_m}{2L_{ss}^2 - L_{ss}\overline{L}_m - \overline{L}_m^2} i_{as} - \frac{r_s\overline{L}_m}{2L_{ss}^2 - L_{ss}\overline{L}_m - \overline{L}_m^2} i_{bs} - \frac{r_s(2L_{ss}-\overline{L}_m)}{2L_{ss}^2 - L_{ss}\overline{L}_m - \overline{L}_m^2} i_{cs}$$

$$-\frac{\psi_m\overline{L}_m}{2L_{ss}^2 - L_{ss}\overline{L}_m - \overline{L}_m^2}\omega_r\cos\theta_r - \frac{\psi_m\overline{L}_m}{2L_{ss}^2 - L_{ss}\overline{L}_m - \overline{L}_m^2}\omega_r\cos\left(\theta_r - \tfrac{2}{3}\pi\right)$$

$$-\frac{\psi_m(2L_{ss}-\overline{L}_m)}{2L_{ss}^2 - L_{ss}\overline{L}_m - \overline{L}_m^2}\omega_r\cos\left(\theta_r + \tfrac{2}{3}\pi\right) + \frac{\overline{L}_m}{2L_{ss}^2 - L_{ss}\overline{L}_m - \overline{L}_m^2} u_{as}$$

$$+\frac{\overline{L}_m}{2L_{ss}^2 - L_{ss}\overline{L}_m - \overline{L}_m^2} u_{bs} + \frac{2L_{ss}-\overline{L}_m}{2L_{ss}^2 - L_{ss}\overline{L}_m - \overline{L}_m^2} u_{cs}.$$

(6.43)

Having derived the differential equations to model the circuitry dynamics, the transient behavior of the rotor (mechanical system) must be incorporated. One cannot solve Equation 6.43 when the electrical angular velocity ω_r and angular displacement θ_r are used as the state variables.

Making use of Newton's second law, $T_e - B_m\omega_{rm} - T_L = J\, d^2\theta_{rm}/dt^2$, we have a set of two differential equations:

$$\frac{d\omega_{rm}}{dt} = \frac{1}{J}(T_e - B_m\omega_{rm} - T_L)$$

$$\frac{d\theta_{rm}}{dt} = \omega_{rm}$$

The expression for the electromagnetic torque developed must be found using the coenergy

$$W_c = \frac{1}{2}[i_{as} \quad i_{bs} \quad i_{cs}]\mathbf{L}_s \begin{bmatrix} i_{as} \\ i_{bs} \\ i_{cs} \end{bmatrix} + [i_{as} \quad i_{bs} \quad i_{cs}] \begin{bmatrix} \psi_m \sin\theta_r \\ \psi_m \sin\left(\theta_r - \frac{2}{3}\pi\right) \\ \psi_m \sin\left(\theta_r + \frac{2}{3}\pi\right) \end{bmatrix} + W_{PM}$$

Here W_{PM} is the energy stored in the permanent magnet.

For round-rotor synchronous microtransducers, we have

$$\mathbf{L}_s = \begin{bmatrix} L_{ls} + \overline{L}_m & -\frac{1}{2}\overline{L}_m & -\frac{1}{2}\overline{L}_m \\ -\frac{1}{2}\overline{L}_m & L_{ls} + \overline{L}_m & -\frac{1}{2}\overline{L}_m \\ -\frac{1}{2}\overline{L}_m & -\frac{1}{2}\overline{L}_m & L_{ls} + \overline{L}_m \end{bmatrix}$$

The inductance matrix $\mathbf{L}_s$ and W_{PM} are not functions of θ_r. One obtains the following formula to calculate the electromagnetic torque for three-phase P-pole permanent-magnet synchronous micromotors:

$$T_e = \frac{P}{2}\frac{\partial W_c}{\partial \theta_r} = \frac{P\psi_m}{2}\left(i_{as}\cos\theta_r + i_{bs}\cos\left(\theta_r - \frac{2}{3}\pi\right) + i_{cs}\cos\left(\theta_r + \frac{2}{3}\pi\right)\right)$$

Therefore, we have

$$\frac{d\omega_{rm}}{dt} = \frac{P\psi_m}{2J}\left(i_{as}\cos\theta_r + i_{bs}\cos\left(\theta_r - \frac{2}{3}\pi\right) + i_{cs}\cos\left(\theta_r + \frac{2}{3}\pi\right)\right) - \frac{B_m}{J}\omega_{rm} - \frac{1}{J}T_L$$

$$\frac{d\theta_{rm}}{dt} = \omega_{rm}$$

Using the electrical angular velocity ω_r and displacement θ_r, related to the mech-anical angular velocity and displacement as $\omega_{rm} = (2/P)\omega_r$ and $\theta_{rm} = (2/P)\theta_r$, the following differential equations to model the torsional-mechanical transient dynamics finally result:

$$\frac{d\omega_r}{dt} = \frac{P^2\psi_m}{4J}\left(i_{as}\cos\theta_r + i_{bs}\cos\left(\theta_r - \frac{2}{3}\pi\right) + i_{cs}\cos\left(\theta_r + \frac{2}{3}\pi\right)\right) - \frac{B_m}{J}\omega_r - \frac{P}{2J}T_L$$

$$(6.44)$$

$$\frac{d\theta_r}{dt} = \omega_r$$

From Equation 6.43 and Equation 6.44, one obtains a nonlinear mathematical model of permanent-magnet synchronous micromotors in Cauchy's form as given by a system of five highly nonlinear differential equations:

$$\frac{di_{as}}{dt} = -\frac{r_s(2L_{ss} - \bar{L}_m)}{2L_{ss}^2 - L_{ss}\bar{L}_m - \bar{L}_m^2} i_{as} - \frac{r_s\bar{L}_m}{2L_{ss}^2 - L_{ss}\bar{L}_m - \bar{L}_m^2} i_{bs} - \frac{r_s\bar{L}_m}{2L_{ss}^2 - L_{ss}\bar{L}_m - \bar{L}_m^2} i_{cs}$$

$$-\frac{\psi_m(2L_{ss} - \bar{L}_m)}{2L_{ss}^2 - L_{ss}\bar{L}_m - \bar{L}_m^2} \omega_r \cos\theta_r - \frac{\psi_m\bar{L}_m}{2L_{ss}^2 - L_{ss}\bar{L}_m - \bar{L}_m^2} \omega_r \cos\left(\theta_r - \tfrac{2}{3}\pi\right)$$

$$-\frac{\psi_m\bar{L}_m}{2L_{ss}^2 - L_{ss}\bar{L}_m - \bar{L}_m^2} \omega_r \cos\left(\theta_r + \tfrac{2}{3}\pi\right) + \frac{2L_{ss} - \bar{L}_m}{2L_{ss}^2 - L_{ss}\bar{L}_m - \bar{L}_m^2} u_{as}$$

$$+\frac{\bar{L}_m}{2L_{ss}^2 - L_{ss}\bar{L}_m - \bar{L}_m^2} u_{bs} + \frac{\bar{L}_m}{2L_{ss}^2 - L_{ss}\bar{L}_m - \bar{L}_m^2} u_{cs}$$

$$\frac{di_{bs}}{dt} = -\frac{r_s\bar{L}_m}{2L_{ss}^2 - L_{ss}\bar{L}_m - \bar{L}_m^2} i_{as} - \frac{r_s(2L_{ss} - \bar{L}_m)}{2L_{ss}^2 - L_{ss}\bar{L}_m - \bar{L}_m^2} i_{bs} - \frac{r_s\bar{L}_m}{2L_{ss}^2 - L_{ss}\bar{L}_m - \bar{L}_m^2} i_{cs}$$

$$-\frac{\psi_m\bar{L}_m}{2L_{ss}^2 - L_{ss}\bar{L}_m - \bar{L}_m^2} \omega_r \cos\theta_r - \frac{\psi_m(2L_{ss} - \bar{L}_m)}{2L_{ss}^2 - L_{ss}\bar{L}_m - \bar{L}_m^2} \omega_r \cos\left(\theta_r - \tfrac{2}{3}\pi\right)$$

$$-\frac{\psi_m\bar{L}_m}{2L_{ss}^2 - L_{ss}\bar{L}_m - \bar{L}_m^2} \omega_r \cos\left(\theta_r + \tfrac{2}{3}\pi\right) + \frac{\bar{L}_m}{2L_{ss}^2 - L_{ss}\bar{L}_m - \bar{L}_m^2} u_{as}$$

$$+\frac{2L_{ss} - \bar{L}_m}{2L_{ss}^2 - L_{ss}\bar{L}_m - \bar{L}_m^2} u_{bs} + \frac{\bar{L}_m}{2L_{ss}^2 - L_{ss}\bar{L}_m - \bar{L}_m^2} u_{cs}$$

$$\frac{di_{cs}}{dt} = -\frac{r_s\bar{L}_m}{2L_{ss}^2 - L_{ss}\bar{L}_m - \bar{L}_m^2} i_{as} - \frac{r_s\bar{L}_m}{2L_{ss}^2 - L_{ss}\bar{L}_m - \bar{L}_m^2} i_{bs} - \frac{r_s(2L_{ss} - \bar{L}_m)}{2L_{ss}^2 - L_{ss}\bar{L}_m - \bar{L}_m^2} i_{cs}$$

$$-\frac{\psi_m\bar{L}_m}{2L_{ss}^2 - L_{ss}\bar{L}_m - \bar{L}_m^2} \omega_r \cos\theta_r - \frac{\psi_m\bar{L}_m}{2L_{ss}^2 - L_{ss}\bar{L}_m - \bar{L}_m^2} \omega_r \cos\left(\theta_r - \tfrac{2}{3}\pi\right)$$

$$-\frac{\psi_m(2L_{ss} - \bar{L}_m)}{2L_{ss}^2 - L_{ss}\bar{L}_m - \bar{L}_m^2} \omega_r \cos\left(\theta_r + \tfrac{2}{3}\pi\right) + \frac{\bar{L}_m}{2L_{ss}^2 - L_{ss}\bar{L}_m - \bar{L}_m^2} u_{as}$$

$$+\frac{\bar{L}_m}{2L_{ss}^2 - L_{ss}\bar{L}_m - \bar{L}_m^2} u_{bs} + \frac{2L_{ss} - \bar{L}_m}{2L_{ss}^2 - L_{ss}\bar{L}_m - \bar{L}_m^2} u_{cs}$$

$$\frac{d\omega_r}{dt} = \frac{P^2\psi_m}{4J}\left(i_{as}\cos\theta_r + i_{bs}\cos\left(\theta_r - \tfrac{2}{3}\pi\right) + i_{cs}\cos\left(\theta_r + \tfrac{2}{3}\pi\right)\right) - \frac{B_m}{J}\omega_r - \frac{P}{2J}T_L$$

(6.45)

$$\frac{d\theta_r}{dt} = \omega_r$$

The state-space form of the lumped-parameter mathematical model can be derived. In particular, from Equation 6.45, we have

$$
\begin{bmatrix}
\dfrac{di_{as}}{dt} \\[2ex]
\dfrac{di_{bs}}{dt} \\[2ex]
\dfrac{di_{cs}}{dt} \\[2ex]
\dfrac{d\omega_r}{dt} \\[2ex]
\dfrac{d\theta_r}{dt}
\end{bmatrix}
=
\begin{bmatrix}
-\dfrac{r_s(2L_{ss}-\bar{L}_m)}{2L_{ss}^2-L_{ss}\bar{L}_m-\bar{L}_m^2} & -\dfrac{r_s\bar{L}_m}{2L_{ss}^2-L_{ss}\bar{L}_m-\bar{L}_m^2} & -\dfrac{r_s\bar{L}_m}{2L_{ss}^2-L_{ss}\bar{L}_m-\bar{L}_m^2} & 0 & 0 \\[2ex]
-\dfrac{r_s\bar{L}_m}{2L_{ss}^2-L_{ss}\bar{L}_m-\bar{L}_m^2} & -\dfrac{r_s(2L_{ss}-\bar{L}_m)}{2L_{ss}^2-L_{ss}\bar{L}_m-\bar{L}_m^2} & -\dfrac{r_s\bar{L}_m}{2L_{ss}^2-L_{ss}\bar{L}_m-\bar{L}_m^2} & 0 & 0 \\[2ex]
-\dfrac{r_s\bar{L}_m}{2L_{ss}^2-L_{ss}\bar{L}_m-\bar{L}_m^2} & -\dfrac{r_s\bar{L}_m}{2L_{ss}^2-L_{ss}\bar{L}_m-\bar{L}_m^2} & -\dfrac{r_s(2L_{ss}-\bar{L}_m)}{2L_{ss}^2-L_{ss}\bar{L}_m-\bar{L}_m^2} & 0 & 0 \\[2ex]
0 & 0 & 0 & -\dfrac{B_m}{J} & 0 \\[2ex]
0 & 0 & 0 & 1 & 0
\end{bmatrix}
\begin{bmatrix}
i_{as} \\[2ex] i_{bs} \\[2ex] i_{cs} \\[2ex] \omega_r \\[2ex] \theta_r
\end{bmatrix}
$$

$$
+
\begin{bmatrix}
-\dfrac{\psi_m(2L_{ss}-\bar{L}_m)}{2L_{ss}^2-L_{ss}\bar{L}_m-\bar{L}_m^2}\,\omega_r & -\dfrac{\psi_m\bar{L}_m}{2L_{ss}^2-L_{ss}\bar{L}_m-\bar{L}_m^2}\,\omega_r & -\dfrac{\psi_m\bar{L}_m}{2L_{ss}^2-L_{ss}\bar{L}_m-\bar{L}_m^2}\,\omega_r \\[3ex]
-\dfrac{\psi_m\bar{L}_m}{2L_{ss}^2-L_{ss}\bar{L}_m-\bar{L}_m^2}\,\omega_r & -\dfrac{\psi_m(2L_{ss}-\bar{L}_m)}{2L_{ss}^2-L_{ss}\bar{L}_m-\bar{L}_m^2}\,\omega_r & -\dfrac{\psi_m\bar{L}_m}{2L_{ss}^2-L_{ss}\bar{L}_m-\bar{L}_m^2}\,\omega_r \\[3ex]
-\dfrac{\psi_m\bar{L}_m}{2L_{ss}^2-L_{ss}\bar{L}_m-\bar{L}_m^2}\,\omega_r & -\dfrac{\psi_m\bar{L}_m}{2L_{ss}^2-L_{ss}\bar{L}_m-\bar{L}_m^2}\,\omega_r & -\dfrac{\psi_m(2L_{ss}-\bar{L}_m)}{2L_{ss}^2-L_{ss}\bar{L}_m-\bar{L}_m^2}\,\omega_r \\[3ex]
\dfrac{P^2\psi_m}{4J}\,i_{as} & \dfrac{P^2\psi_m}{4J}\,i_{bs} & \dfrac{P^2\psi_m}{4J}\,i_{cs} \\[3ex]
0 & 0 & 0
\end{bmatrix}
\begin{bmatrix}
\cos\theta_r \\[3ex]
\cos\!\left(\theta_r-\tfrac{2}{3}\pi\right) \\[3ex]
\cos\!\left(\theta_r+\tfrac{2}{3}\pi\right)
\end{bmatrix}
$$

$$
+
\begin{bmatrix}
\dfrac{2L_{ss}-\bar{L}_m}{2L_{ss}^2-L_{ss}\bar{L}_m-\bar{L}_m^2} & \dfrac{\bar{L}_m}{2L_{ss}^2-L_{ss}\bar{L}_m-\bar{L}_m^2} & \dfrac{\bar{L}_m}{2L_{ss}^2-L_{ss}\bar{L}_m-\bar{L}_m^2} \\[3ex]
\dfrac{\bar{L}_m}{2L_{ss}^2-L_{ss}\bar{L}_m-\bar{L}_m^2} & \dfrac{2L_{ss}-\bar{L}_m}{2L_{ss}^2-L_{ss}\bar{L}_m-\bar{L}_m^2} & \dfrac{\bar{L}_m}{2L_{ss}^2-L_{ss}\bar{L}_m-\bar{L}_m^2} \\[3ex]
\dfrac{\bar{L}_m}{2L_{ss}^2-L_{ss}\bar{L}_m-\bar{L}_m^2} & \dfrac{\bar{L}_m}{2L_{ss}^2-L_{ss}\bar{L}_m-\bar{L}_m^2} & \dfrac{2L_{ss}-\bar{L}_m}{2L_{ss}^2-L_{ss}\bar{L}_m-\bar{L}_m^2} \\[3ex]
0 & 0 & 0 \\[2ex]
0 & 0 & 0
\end{bmatrix}
\begin{bmatrix}
u_{as} \\[2ex] u_{bs} \\[2ex] u_{cs}
\end{bmatrix}
-
\begin{bmatrix}
0 \\[2ex] 0 \\[2ex] 0 \\[2ex] \dfrac{P}{2J} \\[2ex] 0
\end{bmatrix}
T_L
$$

To control the angular velocity, one regulates the currents fed or voltages applied to the stator *abc* windings. Neglecting the viscous friction coefficient, the analysis of Newton's second law, $T_e - T_L = J(d\omega_{rm}/dt)$, indicates that

The angular velocity ω_{rm} increases if $T_e > T_L$.

The angular velocity ω_{rm} decreases if $T_e < T_L$.

The angular velocity ω_{rm} is constant ($\omega_{rm} = const$) if $T_e = T_L$.

That is, to regulate the electromagnetic torque, which was found as

$$T_e = \frac{P\psi_m}{2}\left(i_{as}\cos\theta_r + i_{bs}\cos\left(\theta_r - \tfrac{2}{3}\pi\right) + i_{cs}\cos\left(\theta_r + \tfrac{2}{3}\pi\right)\right),$$

phase currents must be changed.

If, using ICs, the *abc* windings are fed by a balanced three-phase current set

$$i_{as}(t) = \sqrt{2}i_M\cos(\omega_r t) = \sqrt{2}i_M\cos(\omega_e t) = \sqrt{2}i_M\cos\theta_r$$

$$i_{bs}(t) = \sqrt{2}i_M\cos\left(\omega_r t - \tfrac{2}{3}\pi\right) = \sqrt{2}i_M\cos\left(\omega_e t - \tfrac{2}{3}\pi\right) = \sqrt{2}i_M\cos\left(\theta_r - \tfrac{2}{3}\pi\right)$$

$$i_{cs}(t) = \sqrt{2}i_M\cos\left(\omega_r t + \tfrac{2}{3}\pi\right) = \sqrt{2}i_M\cos\left(\omega_e t + \tfrac{2}{3}\pi\right) = \sqrt{2}i_M\cos\left(\theta_r + \tfrac{2}{3}\pi\right)$$

taking note of the trigonometric identity

$$\cos^2\theta_r + \cos^2\left(\theta_r - \tfrac{2}{3}\pi\right) + \cos^2\left(\theta_r + \tfrac{2}{3}\pi\right) = \tfrac{3}{2},$$

one obtains

$$T_e = \frac{P\psi_m}{2}\sqrt{2}i_M\left(\cos^2\theta_r + \cos^2\left(\theta_r - \tfrac{2}{3}\pi\right) + \cos^2\left(\theta_r + \tfrac{2}{3}\pi\right)\right) = \frac{3P\psi_m}{2\sqrt{2}}i_M$$

One concludes that to regulate the angular velocity, i_M must be changed. Furthermore, the phase currents $i_{as}(t)$, $i_{bs}(t)$, and $i_{cs}(t)$, which are shifted by $\tfrac{2}{3}\pi$, are the functions of the electrical angular displacement θ_r (measured using the Hall-effect microsensors).

If the voltage-fed ICs are used, one changes the magnitude of voltages $u_{as}(t)$, $u_{bs}(t)$, and $u_{cs}(t)$. The angular displacement θ_r must be measured (or estimated) in order to generate phase voltages.

In particular, the *abc* voltages needed to be supplied are

$$u_{as}(t) = \sqrt{2}u_M\cos(\theta_r + \varphi_u), \quad u_{bs}(t) = \sqrt{2}u_M\cos\left(\theta_r - \tfrac{2}{3}\pi + \varphi_u\right) \quad \text{and}$$
$$u_{cs}(t) = \sqrt{2}u_M\cos\left(\theta_r + \tfrac{2}{3}\pi + \varphi_u\right).$$

Neglecting the circuitry transients (assuming that inductances are negligibly small), we have

$$u_{as}(t) = \sqrt{2}u_M\cos\theta_r, \quad u_{bs}(t) = \sqrt{2}u_M\cos\left(\theta_r - \tfrac{2}{3}\pi\right) \quad \text{and}$$
$$u_{cs}(t) = \sqrt{2}u_M\cos\left(\theta_r + \tfrac{2}{3}\pi\right).$$

Using a set of nonlinear differential equations (Equation 6.45), the block diagram is developed and documented in Figure 6.67. Here

$$T_s = \frac{r_s(2L_{ss} - \overline{L}_m)}{2L_{ss}^2 - L_{ss}\overline{L}_m - \overline{L}_m^2}$$

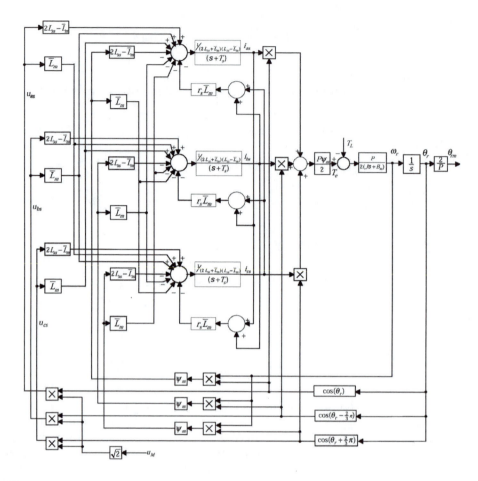

FIGURE 6.67

Block diagram of radial topology three-phase permanent-magnet synchronous micromotors controlled by applying three-phase balanced voltage set $u_{as}(t) = \sqrt{2}u_M \cos\theta_r$, $u_{bs}(t) = \sqrt{2}u_M \cos(\theta_r - \frac{2}{3}\pi)$, and $u_{cs}(t) = \sqrt{2}u_M \cos(\theta_r + \frac{2}{3}\pi)$.

Example 6.31   THE LAGRANGE EQUATIONS of MOTION and DYNAMICS of PERMANENT-MAGNET SYNCHRONOUS MICROMOTORS

Having derived the mathematical model for three-phase permanent-magnet synchronous micromotors using Kirchhoff's voltage law (to model the circuitry dynamics), Newtonian mechanics (to model the torsional-mechanical dynamics), and the coenergy concept (to find the electromagnetic torque), develop the mathematical model for three-phase permanent-magnet synchronous motors utilizing Lagrange's concept.

SOLUTION

The generalized coordinates are the electric charges in the *abc* stator windings $q_1 = i_{as}/s$, $\dot{q}_1 = i_{as}$, $q_2 = i_{bs}/s$, $\dot{q}_2 = i_{bs}$, $q_3 = i_{cs}/s$, $\dot{q}_3 = i_{cs}$, and the angular displacement $q_4 = \theta_r$, $\dot{q}_4 = \omega_r$.

The generalized forces are the applied voltages to the *abc* windings $Q_1 = u_{as}$, $Q_2 = u_{bs}$, $Q_3 = u_{cs}$ and the load torque $Q_4 = -T_L$.

The resulting Lagrange equations are

$$\frac{d}{dt}\left(\frac{\partial \Gamma}{\partial \dot{q}_1}\right) - \frac{\partial \Gamma}{\partial q_1} + \frac{\partial D}{\partial \dot{q}_1} + \frac{\partial \Pi}{\partial q_1} = Q_1$$

$$\frac{d}{dt}\left(\frac{\partial \Gamma}{\partial \dot{q}_2}\right) - \frac{\partial \Gamma}{\partial q_2} + \frac{\partial D}{\partial \dot{q}_2} + \frac{\partial \Pi}{\partial q_2} = Q_2$$

$$\frac{d}{dt}\left(\frac{\partial \Gamma}{\partial \dot{q}_3}\right) - \frac{\partial \Gamma}{\partial q_3} + \frac{\partial D}{\partial \dot{q}_3} + \frac{\partial \Pi}{\partial q_3} = Q_3$$

$$\frac{d}{dt}\left(\frac{\partial \Gamma}{\partial \dot{q}_4}\right) - \frac{\partial \Gamma}{\partial q_4} + \frac{\partial D}{\partial \dot{q}_4} + \frac{\partial \Pi}{\partial q_4} = Q_4$$

The total kinetic energy includes kinetic energies of electrical and mechanical systems. In particular,

$$\Gamma = \Gamma_E + \Gamma_M = \tfrac{1}{2}L_{asas}\dot{q}_1^2 + \tfrac{1}{2}(L_{asbs} + L_{bsas})\dot{q}_1\dot{q}_2 + \tfrac{1}{2}(L_{ascs} + L_{csas})\dot{q}_1\dot{q}_3 + \tfrac{1}{2}L_{bsbs}\dot{q}_2^2 + \tfrac{1}{2}(L_{bscs} + L_{csbs})\dot{q}_2\dot{q}_3$$

$$+ \tfrac{1}{2}L_{cscs}\dot{q}_3^2 + \psi_m\dot{q}_1\sin q_4 + \psi_m\dot{q}_2\sin\left(q_4 - \tfrac{2}{3}\pi\right) + \psi_m\dot{q}_3\sin\left(q_4 + \tfrac{2}{3}\pi\right) + \tfrac{1}{2}J\dot{q}_4^2$$

Then we have

$$\frac{\partial \Gamma}{\partial q_1} = 0, \quad \frac{\partial \Gamma}{\partial \dot{q}_1} = L_{asas}\dot{q}_1 + \tfrac{1}{2}(L_{asbs} + L_{bsas})\dot{q}_2 + \tfrac{1}{2}(L_{ascs} + L_{csas})\dot{q}_3 + \psi_m\sin q_4$$

$$\frac{\partial \Gamma}{\partial q_2} = 0, \quad \frac{\partial \Gamma}{\partial \dot{q}_2} = \tfrac{1}{2}(L_{asbs} + L_{bsas})\dot{q}_1 + L_{bsbs}\dot{q}_2 + \tfrac{1}{2}(L_{bscs} + L_{csbs})\dot{q}_3 + \psi_m\sin\left(q_4 - \tfrac{2}{3}\pi\right)$$

$$\frac{\partial \Gamma}{\partial q_3} = 0, \quad \frac{\partial \Gamma}{\partial \dot{q}_3} = \tfrac{1}{2}(L_{ascs} + L_{csas})\dot{q}_1 + \tfrac{1}{2}(L_{bscs} + L_{csbs})\dot{q}_2 + L_{cscs}\dot{q}_3 + \psi_m\sin\left(q_4 + \tfrac{2}{3}\pi\right)$$

$$\frac{\partial \Gamma}{\partial q_4} = \psi_m\dot{q}_1\cos q_4 + \psi_m\dot{q}_2\cos\left(q_4 - \tfrac{2}{3}\pi\right) + \psi_m\dot{q}_3\cos\left(q_4 + \tfrac{2}{3}\pi\right), \quad \frac{\partial \Gamma}{\partial \dot{q}_4} = J\dot{q}_4$$

The total potential energy is $\Pi = 0$.

The total dissipated energy is found as a sum of the heat energy dissipated by the electrical system and the heat energy dissipated by the mechanical system. That is,

$$D = \tfrac{1}{2}\left(r_s\dot{q}_1^2 + r_s\dot{q}_2^2 + r_s\dot{q}_3^2 + B_m\dot{q}_4^2\right)$$

The differentiation of D with respect to the generalized coordinates gives

$$\frac{\partial D}{\partial \dot{q}_1} = r_s\dot{q}_1, \quad \frac{\partial D}{\partial \dot{q}_2} = r_s\dot{q}_2, \quad \frac{\partial D}{\partial \dot{q}_3} = r_s\dot{q}_3, \quad \text{and} \quad \frac{\partial D}{\partial \dot{q}_4} = B_m\dot{q}_4$$

Taking note of $\dot{q}_1 = i_{as}$, $\dot{q}_2 = i_{bs}$, $\dot{q}_3 = i_{cs}$, and $\dot{q}_4 = \omega_r$, the Lagrange equations lead us to four differential equations:

$$L_{asas}\frac{di_{as}}{dt} + \tfrac{1}{2}(L_{asbs} + L_{bsas})\frac{di_{bs}}{dt} + \tfrac{1}{2}(L_{ascs} + L_{csas})\frac{di_{cs}}{dt} + \psi_m\omega_r\cos\theta_r + r_s i_{as} = u_{as}$$

$$\tfrac{1}{2}(L_{asbs} + L_{bsas})\frac{di_{as}}{dt} + L_{bsbs}\frac{di_{bs}}{dt} + \tfrac{1}{2}(L_{bscs} + L_{csbs})\frac{di_{cs}}{dt} + \psi_m\omega_r\cos\left(\theta_r - \tfrac{2}{3}\pi\right) + r_s i_{bs} = u_{bs}$$

$$\tfrac{1}{2}(L_{ascs} + L_{csas})\frac{di_{as}}{dt} + \tfrac{1}{2}(L_{bscs} + L_{csbs})\frac{di_{bs}}{dt} + L_{cscs}\frac{di_{cs}}{dt} + \psi_m\omega_r\cos\left(\theta_r + \tfrac{2}{3}\pi\right) + r_s i_{cs} = u_{cs}$$

$$J\frac{d^2\theta_r}{dt^2} - \psi_m i_{as}\cos\theta_r - \psi_m i_{bs}\cos\left(\theta_r - \tfrac{2}{3}\pi\right) - \psi_m i_{cs}\cos\left(\theta_r + \tfrac{2}{3}\pi\right) + B_m\frac{d\theta_r}{dt} = -T_L$$

For round-rotor permanent-magnet synchronous microtransducers, one obtains

$$(L_{ls} + \bar{L}_m)\frac{di_{as}}{dt} - \tfrac{1}{2}\bar{L}_m\frac{di_{bs}}{dt} - \tfrac{1}{2}\bar{L}_m\frac{di_{cs}}{dt} + \psi_m\omega_r\cos\theta_r + r_s i_{as} = u_{as}$$

$$-\tfrac{1}{2}\bar{L}_m\frac{di_{as}}{dt} + (L_{ls} + \bar{L}_m)\frac{di_{bs}}{dt} - \tfrac{1}{2}\bar{L}_m\frac{di_{cs}}{dt} + \psi_m\omega_r\cos\left(\theta_r - \tfrac{2}{3}\pi\right) + r_s i_{bs} = u_{bs}$$

$$-\tfrac{1}{2}\bar{L}_m\frac{di_{as}}{dt} - \tfrac{1}{2}\bar{L}_m\frac{di_{bs}}{dt} + (L_{ls} + \bar{L}_m)\frac{di_{cs}}{dt} + \psi_m\omega_r\cos\left(\theta_r + \tfrac{2}{3}\pi\right) + r_s i_{cs} = u_{cs}$$

$$J\frac{d\omega_r}{dt} + B_m\omega_r - \psi_m\left[i_{as}\cos\theta_r + i_{bs}\cos\left(\theta_r - \tfrac{2}{3}\pi\right) + i_{cs}\cos\left(\theta_r + \tfrac{2}{3}\pi\right)\right] = -T_L$$

$$\frac{d\theta_r}{dt} = \omega_r$$

From the fourth differential equation one finds the electromagnetic torque as

$$T_e = \psi_m\left[i_{as}\cos\theta_r + i_{bs}\cos\left(\theta_r - \tfrac{2}{3}\pi\right) + i_{cs}\cos\left(\theta_r + \tfrac{2}{3}\pi\right)\right]$$

Differential equations in Cauchy's form result, as given in the state-space form by Equation 6.45 for *P*-pole permanent-magnet synchronous micromotors. It was demonstrated that, applying Lagrange's concept, a complete mathematical model for permanent-magnet synchronous microtransducers can be straightforwardly developed.

Example 6.32   THREE-PHASE PERMANENT-MAGNET SYNCHRONOUS MICROGENERATORS

Develop the mathematical model for a three-phase permanent-magnet synchronous microgenerator.

SOLUTION

For permanent-magnet synchronous microgenerators, as shown in Figure 6.65, the mathematical model can be developed using Kirchhoff's second law,

$$
\mathbf{u}_{abcs} = -\mathbf{r}_s \mathbf{i}_{abcs} + \frac{d\mathbf{\psi}_{abcs}}{dt}, \quad
\begin{bmatrix} u_{as} \\ u_{bs} \\ u_{cs} \end{bmatrix}
= -
\begin{bmatrix} r_s & 0 & 0 \\ 0 & r_s & 0 \\ 0 & 0 & r_s \end{bmatrix}
\begin{bmatrix} i_{as} \\ i_{bs} \\ i_{cs} \end{bmatrix}
+
\begin{bmatrix} \dfrac{d\psi_{as}}{dt} \\[2mm] \dfrac{d\psi_{bs}}{dt} \\[2mm] \dfrac{d\psi_{cs}}{dt} \end{bmatrix}
$$

$$
\mathbf{\psi}_{abcs} = -\mathbf{L}_s \mathbf{i}_{abcs} + \mathbf{\psi}_m
$$

$$
= - \begin{bmatrix}
L_{ls} + L_m - L_{\Delta m}\cos 2\theta_r & -\tfrac{1}{2}\overline{L}_m - L_{\Delta m}\cos 2\left(\theta_r - \tfrac{1}{3}\pi\right) & -\tfrac{1}{2}\overline{L}_m - L_{\Delta m}\cos 2\left(\theta_r + \tfrac{1}{3}\pi\right) \\
-\tfrac{1}{2}\overline{L}_m - L_{\Delta m}\cos 2\left(\theta_r - \tfrac{1}{3}\pi\right) & L_{ls} + L_m - L_{\Delta m}\cos 2\left(\theta_r - \tfrac{2}{3}\pi\right) & -\tfrac{1}{2}\overline{L}_m - L_{\Delta m}\cos 2\theta_r \\
-\tfrac{1}{2}\overline{L}_m - L_{\Delta m}\cos 2\left(\theta_r + \tfrac{1}{3}\pi\right) & -\tfrac{1}{2}\overline{L}_m - L_{\Delta m}\cos 2\theta_r & L_{ls} + L_m - L_{\Delta m}\cos 2\left(\theta_r + \tfrac{2}{3}\pi\right)
\end{bmatrix}
$$

$$
\times \begin{bmatrix} i_{as} \\ i_{bs} \\ i_{cs} \end{bmatrix} + \psi_m \begin{bmatrix} \sin\theta_r \\ \sin\left(\theta_r - \tfrac{2}{3}\pi\right) \\ \sin\left(\theta_r + \tfrac{2}{3}\pi\right) \end{bmatrix} \Bigg|
$$

and Newton's second law of motion, $-T_e - B_m\omega_{rm} + T_{pm} = J\,d^2\theta_{rm}/dt^2$, which gives

$$
\frac{d\omega_{rm}}{dt} = \frac{1}{J}\left(-T_e - B_m\omega_{rm} + T_{pm}\right)
$$

$$
\frac{d\theta_{rm}}{dt} = \omega_{rm}
$$

Here T_{pm} is the electromagnetic torque of the prime mover that rotates the microgenerator. The application of the results presented for the permanent-magnet synchronous micro-motors results in the following set of differential equations:

$$
\frac{di_{as}}{dt} = -\frac{r_s(2L_{ss} - \overline{L}_m)}{2L_{ss}^2 - L_{ss}\overline{L}_m - \overline{L}_m^2}\,i_{as} - \frac{r_s\overline{L}_m}{2L_{ss}^2 - L_{ss}\overline{L}_m - \overline{L}_m^2}\,i_{bs} - \frac{r_s\overline{L}_m}{2L_{ss}^2 - L_{ss}\overline{L}_m - \overline{L}_m^2}\,i_{cs}
$$

$$
+ \frac{\psi_m(2L_{ss} - \overline{L}_m)}{2L_{ss}^2 - L_{ss}\overline{L}_m - \overline{L}_m^2}\,\omega_r\cos\theta_r + \frac{\psi_m\overline{L}_m}{2L_{ss}^2 - L_{ss}\overline{L}_m - \overline{L}_m^2}\,\omega_r\cos\left(\theta_r - \tfrac{2}{3}\pi\right)
$$

$$
+ \frac{\psi_m\overline{L}_m}{2L_{ss}^2 - L_{ss}\overline{L}_m - \overline{L}_m^2}\,\omega_r\cos\left(\theta_r + \tfrac{2}{3}\pi\right) - \frac{2L_{ss} - \overline{L}_m}{2L_{ss}^2 - L_{ss}\overline{L}_m - \overline{L}_m^2}\,u_{as}
$$

$$
- \frac{\overline{L}_m}{2L_{ss}^2 - L_{ss}\overline{L}_m - \overline{L}_m^2}\,u_{bs} - \frac{\overline{L}_m}{2L_{ss}^2 - L_{ss}\overline{L}_m - \overline{L}_m^2}\,u_{cs}
$$

$$\frac{di_{bs}}{dt} = -\frac{r_s \overline{L}_m}{2L_{ss}^2 - L_{ss}\overline{L}_m - \overline{L}_m^2} i_{as} - \frac{r_s(2L_{ss} - \overline{L}_m)}{2L_{ss}^2 - L_{ss}\overline{L}_m - \overline{L}_m^2} i_{bs} - \frac{r_s \overline{L}_m}{2L_{ss}^2 - L_{ss}\overline{L}_m - \overline{L}_m^2} i_{cs}$$

$$+\frac{\psi_m \overline{L}_m}{2L_{ss}^2 - L_{ss}\overline{L}_m - \overline{L}_m^2} \omega_r \cos\theta_r + \frac{\psi_m(2L_{ss} - \overline{L}_m)}{2L_{ss}^2 - L_{ss}\overline{L}_m - \overline{L}_m^2} \omega_r \cos\left(\theta_r - \frac{2}{3}\pi\right)$$

$$+\frac{\psi_m \overline{L}_m}{2L_{ss}^2 - L_{ss}\overline{L}_m - \overline{L}_m^2} \omega_r \cos\left(\theta_r + \frac{2}{3}\pi\right) - \frac{\overline{L}_m}{2L_{ss}^2 - L_{ss}\overline{L}_m - \overline{L}_m^2} u_{as}$$

$$-\frac{2L_{ss} - \overline{L}_m}{2L_{ss}^2 - L_{ss}\overline{L}_m - \overline{L}_m^2} u_{bs} - \frac{\overline{L}_m}{2L_{ss}^2 - L_{ss}\overline{L}_m - \overline{L}_m^2} u_{cs}$$

$$\frac{di_{cs}}{dt} = -\frac{r_s \overline{L}_m}{2L_{ss}^2 - L_{ss}\overline{L}_m - \overline{L}_m^2} i_{as} - \frac{r_s \overline{L}_m}{2L_{ss}^2 - L_{ss}\overline{L}_m - \overline{L}_m^2} i_{bs} - \frac{r_s(2L_{ss} - \overline{L}_m)}{2L_{ss}^2 - L_{ss}\overline{L}_m - \overline{L}_m^2} i_{cs}$$

$$+\frac{\psi_m \overline{L}_m}{2L_{ss}^2 - L_{ss}\overline{L}_m - \overline{L}_m^2} \omega_r \cos\theta_r + \frac{\psi_m \overline{L}_m}{2L_{ss}^2 - L_{ss}\overline{L}_m - \overline{L}_m^2} \omega_r \cos\left(\theta_r - \frac{2}{3}\pi\right)$$

$$\text{(6.46)}$$

$$+\frac{\psi_m(2L_{ss} - \overline{L}_m)}{2L_{ss}^2 - L_{ss}\overline{L}_m - \overline{L}_m^2} \omega_r \cos\left(\theta_r + \frac{2}{3}\pi\right) - \frac{\overline{L}_m}{2L_{ss}^2 - L_{ss}\overline{L}_m - \overline{L}_m^2} u_{as}$$

$$-\frac{\overline{L}_m}{2L_{ss}^2 - L_{ss}\overline{L}_m - \overline{L}_m^2} u_{bs} - \frac{2L_{ss} - \overline{L}_m}{2L_{ss}^2 - L_{ss}\overline{L}_m - \overline{L}_m^2} u_{cs}$$

$$\frac{d\omega_r}{dt} = -\frac{P^2 \psi_m}{4J}\left(i_{as}\cos\theta_r + i_{bs}\cos\left(\theta_r - \frac{2}{3}\pi\right) + i_{cs}\cos\left(\theta_r + \frac{2}{3}\pi\right)\right) - \frac{B_m}{J}\omega_r + \frac{P}{2J}T_{pm}$$

$$\frac{d\theta_r}{dt} = \omega_r$$

In the state-space form, from Equation 6.46, we have the following mathematical model of three-phase permanent-magnet synchronous microgenerators:

$$\begin{bmatrix} \dfrac{di_{as}}{dt} \\[3mm] \dfrac{di_{bs}}{dt} \\[3mm] \dfrac{di_{cs}}{dt} \\[3mm] \dfrac{d\omega_r}{dt} \\[3mm] \dfrac{d\theta_r}{dt} \end{bmatrix} = \begin{bmatrix} -\dfrac{r_s(2L_{ss} - \overline{L}_m)}{2L_{ss}^2 - L_{ss}\overline{L}_m - \overline{L}_m^2} & -\dfrac{r_s \overline{L}_m}{2L_{ss}^2 - L_{ss}\overline{L}_m - \overline{L}_m^2} & -\dfrac{r_s \overline{L}_m}{2L_{ss}^2 - L_{ss}\overline{L}_m - \overline{L}_m^2} & 0 & 0 \\[3mm] -\dfrac{r_s \overline{L}_m}{2L_{ss}^2 - L_{ss}\overline{L}_m - \overline{L}_m^2} & -\dfrac{r_s(2L_{ss} - \overline{L}_m)}{2L_{ss}^2 - L_{ss}\overline{L}_m - \overline{L}_m^2} & -\dfrac{r_s \overline{L}_m}{2L_{ss}^2 - L_{ss}\overline{L}_m - \overline{L}_m^2} & 0 & 0 \\[3mm] -\dfrac{r_s \overline{L}_m}{2L_{ss}^2 - L_{ss}\overline{L}_m - \overline{L}_m^2} & -\dfrac{r_s \overline{L}_m}{2L_{ss}^2 - L_{ss}\overline{L}_m - \overline{L}_m^2} & -\dfrac{r_s(2L_{ss} - \overline{L}_m)}{2L_{ss}^2 - L_{ss}\overline{L}_m - \overline{L}_m^2} & 0 & 0 \\[3mm] 0 & 0 & 0 & -\dfrac{B_m}{J} & 0 \\[3mm] 0 & 0 & 0 & 1 & 0 \end{bmatrix}\begin{bmatrix} i_{as} \\[3mm] i_{bs} \\[3mm] i_{cs} \\[3mm] \omega_r \\[3mm] \theta_r \end{bmatrix}$$

$$
+\begin{bmatrix}
\dfrac{\psi_m(2L_{ss}-\bar{L}_m)}{2L_{ss}^2 - L_{ss}\bar{L}_m - \bar{L}_m^2}\omega_r & \dfrac{\psi_m\bar{L}_m}{2L_{ss}^2 - L_{ss}\bar{L}_m - \bar{L}_m^2}\omega_r & \dfrac{\psi_m\bar{L}_m}{2L_{ss}^2 - L_{ss}\bar{L}_m - \bar{L}_m^2}\omega_r \\[2ex]
\dfrac{\psi_m\bar{L}_m}{2L_{ss}^2 - L_{ss}\bar{L}_m - \bar{L}_m^2}\omega_r & \dfrac{\psi_m(2L_{ss}-\bar{L}_m)}{2L_{ss}^2 - L_{ss}\bar{L}_m - \bar{L}_m^2}\omega_r & \dfrac{\psi_m\bar{L}_m}{2L_{ss}^2 - L_{ss}\bar{L}_m - \bar{L}_m^2}\omega_r \\[2ex]
\dfrac{\psi_m\bar{L}_m}{2L_{ss}^2 - L_{ss}\bar{L}_m - \bar{L}_m^2}\omega_r & \dfrac{\psi_m\bar{L}_m}{2L_{ss}^2 - L_{ss}\bar{L}_m - \bar{L}_m^2}\omega_r & \dfrac{\psi_m(2L_{ss}-\bar{L}_m)}{2L_{ss}^2 - L_{ss}\bar{L}_m - \bar{L}_m^2}\omega_r \\[2ex]
-\dfrac{P^2\psi_m}{4J}i_{as} & -\dfrac{P^2\psi_m}{4J}i_{bs} & -\dfrac{P^2\psi_m}{4J}i_{cs} \\[2ex]
0 & 0 & 0
\end{bmatrix}
\begin{bmatrix}
\cos\theta_r \\[1ex]
\cos\!\left(\theta_r - \tfrac{2}{3}\pi\right) \\[1ex]
\cos\!\left(\theta_r + \tfrac{2}{3}\pi\right)
\end{bmatrix}
$$

$$
-\begin{bmatrix}
\dfrac{2L_{ss}-\bar{L}_m}{2L_{ss}^2 - L_{ss}\bar{L}_m - \bar{L}_m^2} & \dfrac{\bar{L}_m}{2L_{ss}^2 - L_{ss}\bar{L}_m - \bar{L}_m^2} & \dfrac{\bar{L}_m}{2L_{ss}^2 - L_{ss}\bar{L}_m - \bar{L}_m^2} \\[2ex]
\dfrac{\bar{L}_m}{2L_{ss}^2 - L_{ss}\bar{L}_m - \bar{L}_m^2} & \dfrac{2L_{ss}-\bar{L}_m}{2L_{ss}^2 - L_{ss}\bar{L}_m - \bar{L}_m^2} & \dfrac{\bar{L}_m}{2L_{ss}^2 - L_{ss}\bar{L}_m - \bar{L}_m^2} \\[2ex]
\dfrac{\bar{L}_m}{2L_{ss}^2 - L_{ss}\bar{L}_m - \bar{L}_m^2} & \dfrac{\bar{L}_m}{2L_{ss}^2 - L_{ss}\bar{L}_m - \bar{L}_m^2} & \dfrac{2L_{ss}-\bar{L}_m}{2L_{ss}^2 - L_{ss}\bar{L}_m - \bar{L}_m^2} \\[2ex]
0 & 0 & 0 \\[1ex]
0 & 0 & 0
\end{bmatrix}
\begin{bmatrix}
u_{as} \\ u_{bs} \\ u_{cs}
\end{bmatrix}
+
\begin{bmatrix}
0 \\ 0 \\ 0 \\ \dfrac{P}{2J} \\ 0
\end{bmatrix} T_{pm}
$$

One concludes that the lumped-parameter nonlinear mathematical model of permanent-magnet synchronous microgenerators is derived to be used in analysis, modeling, and control.

6.8.4.3 Mathematical Models of Permanent-Magnet Synchronous Micromachines in the Arbitrary, Rotor, and Synchronous Reference Frames

By fixing the reference frame with the rotor and making use of the direct Park transformations

$$
\mathbf{u}_{qd0s} = \mathbf{K}_s \mathbf{u}_{abcs}, \quad \mathbf{i}_{qd0s} = \mathbf{K}_s \mathbf{i}_{abcs}, \quad \boldsymbol{\psi}_{qd0s} = \mathbf{K}_s \boldsymbol{\psi}_{abcs},
$$

$$
\mathbf{K}_s = \frac{2}{3}
\begin{bmatrix}
\cos\theta & \cos\!\left(\theta - \tfrac{2}{3}\pi\right) & \cos\!\left(\theta + \tfrac{2}{3}\pi\right) \\[1ex]
\sin\theta & \sin\!\left(\theta - \tfrac{2}{3}\pi\right) & \sin\!\left(\theta + \tfrac{2}{3}\pi\right) \\[1ex]
\tfrac{1}{2} & \tfrac{1}{2} & \tfrac{1}{2}
\end{bmatrix}
$$

Equation 6.40, $\mathbf{u}_{abcs} = \mathbf{r}_s \mathbf{i}_{abcs} + d\boldsymbol{\psi}_{abcs}/dt$, is rewritten in the $qd0$ variables as

$$
\mathbf{K}_s^{-1}\mathbf{u}_{qd0s} = \mathbf{r}_s \mathbf{K}_s^{-1}\mathbf{i}_{qd0s} + \frac{d\left(\mathbf{K}_s^{-1}\boldsymbol{\psi}_{qd0s}\right)}{dt}, \quad
\mathbf{K}_s^{-1} =
\begin{bmatrix}
\cos\theta & \sin\theta & 1 \\[1ex]
\cos\!\left(\theta - \tfrac{2}{3}\pi\right) & \sin\!\left(\theta - \tfrac{2}{3}\pi\right) & 1 \\[1ex]
\cos\!\left(\theta + \tfrac{2}{3}\pi\right) & \sin\!\left(\theta + \tfrac{2}{3}\pi\right) & 1
\end{bmatrix}
$$

Multiplying left and right sides by $\mathbf{K}_s$, one obtains

$$\mathbf{K}_s \mathbf{K}_s^{-1} \mathbf{u}_{qdos} = \mathbf{K}_s \mathbf{r}_s \mathbf{K}_s^{-1} \mathbf{i}_{qd0s} + \mathbf{K}_s \frac{d\mathbf{K}_s^{-1}}{dt} \boldsymbol{\psi}_{qd0s} + \mathbf{K}_s \mathbf{K}_s^{-1} \frac{d\boldsymbol{\psi}_{qd0s}}{dt}$$

The matrix $\mathbf{r}_s$ is diagonal, and thus $\mathbf{K}_s \mathbf{r}_s \mathbf{K}_s^{-1} = \mathbf{r}_s$. From

$$\frac{d\mathbf{K}_s^{-1}}{dt} = \omega \begin{bmatrix} -\sin\theta & \cos\theta & 0 \\ -\sin\left(\theta - \frac{2}{3}\pi\right) & \cos\left(\theta - \frac{2}{3}\pi\right) & 0 \\ -\sin\left(\theta + \frac{2}{3}\pi\right) & \cos\left(\theta + \frac{2}{3}\pi\right) & 0 \end{bmatrix}$$

we have

$$\mathbf{K}_s \frac{d\mathbf{K}_s^{-1}}{dt} = \omega \begin{bmatrix} 0 & 1 & 0 \\ -1 & 0 & 0 \\ 0 & 0 & 0 \end{bmatrix}$$

Hence, Equation 6.40 is rewritten in the $qd0$ variables as

$$\mathbf{u}_{qd0s} = \mathbf{r}_s \mathbf{i}_{qd0s} + \omega \begin{bmatrix} \psi_{ds} \\ -\psi_{qs} \\ 0 \end{bmatrix} + \frac{d\boldsymbol{\psi}_{qd0s}}{dt} \tag{6.47}$$

Using the Park transformation, the quadrature-, direct-, and zero-axis components of stator flux linkages are found as

$$\boldsymbol{\psi}_{qd0s} = \mathbf{K}_s \boldsymbol{\psi}_{abcs}$$

where

$$\boldsymbol{\psi}_{abcs} = \mathbf{L}_s \mathbf{i}_{abcs} + \boldsymbol{\psi}_m = \begin{bmatrix} L_{ls} + \overline{L}_m & -\frac{1}{2}\overline{L}_m & -\frac{1}{2}\overline{L}_m \\ -\frac{1}{2}\overline{L}_m & L_{ls} + \overline{L}_m & -\frac{1}{2}\overline{L}_m \\ -\frac{1}{2}\overline{L}_m & -\frac{1}{2}\overline{L}_m & L_{ls} + \overline{L}_m \end{bmatrix} \begin{bmatrix} i_{as} \\ i_{bs} \\ i_{cs} \end{bmatrix} + \psi_m \begin{bmatrix} \sin\theta_r \\ \sin\left(\theta_r - \frac{2}{3}\pi\right) \\ \sin\left(\theta_r + \frac{2}{3}\pi\right) \end{bmatrix}$$

Hence,

$$\boldsymbol{\psi}_{qd0s} = \mathbf{K}_s \mathbf{L}_s \mathbf{K}_s^{-1} \mathbf{i}_{qd0s} + \mathbf{K}_s \boldsymbol{\psi}_m \tag{6.48}$$

where

$$\mathbf{K}_s \mathbf{L}_s \mathbf{K}_s^{-1} = \begin{bmatrix} L_{ls} + \frac{3}{2}\overline{L}_m & 0 & 0 \\ 0 & L_{ls} + \frac{3}{2}\overline{L}_m & 0 \\ 0 & 0 & L_{ls} \end{bmatrix}$$

$$\mathbf{K}_s \boldsymbol{\psi}_m = \tfrac{2}{3}\begin{bmatrix} \cos\theta & \cos(\theta - \tfrac{2}{3}\pi) & \cos(\theta + \tfrac{2}{3}\pi) \\ \sin\theta & \sin(\theta - \tfrac{2}{3}\pi) & \sin(\theta + \tfrac{2}{3}\pi) \\ \tfrac{1}{2} & \tfrac{1}{2} & \tfrac{1}{2} \end{bmatrix} \boldsymbol{\psi}_m \begin{bmatrix} \sin\theta_r \\ \sin(\theta_r - \tfrac{2}{3}\pi) \\ \sin(\theta_r + \tfrac{2}{3}\pi) \end{bmatrix} = \boldsymbol{\psi}_m \begin{bmatrix} -\sin(\theta - \theta_r) \\ \cos(\theta - \theta_r) \\ 0 \end{bmatrix}$$

From Equation 6.48 we obtain the following expression:

$$\boldsymbol{\psi}_{qd0s} = \begin{bmatrix} L_{ls} + \tfrac{3}{2}\overline{L}_m & 0 & 0 \\ 0 & L_{ls} + \tfrac{3}{2}\overline{L}_m & 0 \\ 0 & 0 & L_{ls} \end{bmatrix} \mathbf{i}_{qd0s} + \boldsymbol{\psi}_m \begin{bmatrix} -\sin(\theta - \theta_r) \\ \cos(\theta - \theta_r) \\ 0 \end{bmatrix}$$

Using Equation 6.47, one finds

$$\mathbf{u}_{qd0s} = \mathbf{r}_s \mathbf{i}_{qd0s} + \omega \begin{bmatrix} \psi_{ds} \\ -\psi_{qs} \\ 0 \end{bmatrix} + \begin{bmatrix} L_{ls} + \tfrac{3}{2}\overline{L}_m & 0 & 0 \\ 0 & L_{ls} + \tfrac{3}{2}\overline{L}_m & 0 \\ 0 & 0 & L_{ls} \end{bmatrix} \frac{d\mathbf{i}_{qd0s}}{dt} + \boldsymbol{\psi}_m \frac{d\begin{bmatrix} -\sin(\theta - \theta_r) \\ \cos(\theta - \theta_r) \\ 0 \end{bmatrix}}{dt}$$

The differential equation in matrix form that models the permanent-magnet circuitry dynamics in the *arbitrary* reference frame is

$$\mathbf{u}_{qd0s} = \mathbf{r}_s \mathbf{i}_{qd0s} + \omega \begin{bmatrix} \psi_{ds} \\ -\psi_{qs} \\ 0 \end{bmatrix} + \begin{bmatrix} L_{ls} + \tfrac{3}{2}\overline{L}_m & 0 & 0 \\ 0 & L_{ls} + \tfrac{3}{2}\overline{L}_m & 0 \\ 0 & 0 & L_{ls} \end{bmatrix} \frac{d\mathbf{i}_{qd0s}}{dt} + \boldsymbol{\psi}_m \frac{d\begin{bmatrix} -\sin(\theta - \theta_r) \\ \cos(\theta - \theta_r) \\ 0 \end{bmatrix}}{dt}$$

In the rotor reference frame, the electrical angular velocity is equal to the synchronous angular velocity. We assign the angular velocity of the reference frame to be $\omega = \omega_r = \omega_e$. Then, taking note of $\theta = \theta_r$, we have the Park transformation matrix

$$\mathbf{K}_s^r = \tfrac{2}{3}\begin{bmatrix} \cos\theta_r & \cos(\theta_r - \tfrac{2}{3}\pi) & \cos(\theta_r + \tfrac{2}{3}\pi) \\ \sin\theta_r & \sin(\theta_r - \tfrac{2}{3}\pi) & \sin(\theta_r + \tfrac{2}{3}\pi) \\ \tfrac{1}{2} & \tfrac{1}{2} & \tfrac{1}{2} \end{bmatrix}$$

One finds

$$\mathbf{K}_s^r \boldsymbol{\psi}_m = \tfrac{2}{3}\begin{bmatrix} \cos\theta_r & \cos(\theta_r - \tfrac{2}{3}\pi) & \cos(\theta_r + \tfrac{2}{3}\pi) \\ \sin\theta_r & \sin(\theta_r - \tfrac{2}{3}\pi) & \sin(\theta_r + \tfrac{2}{3}\pi) \\ \tfrac{1}{2} & \tfrac{1}{2} & \tfrac{1}{2} \end{bmatrix} \boldsymbol{\psi}_m \begin{bmatrix} \sin\theta_r \\ \sin(\theta_r - \tfrac{2}{3}\pi) \\ \sin(\theta_r + \tfrac{2}{3}\pi) \end{bmatrix} = \boldsymbol{\psi}_m \begin{bmatrix} 0 \\ 1 \\ 0 \end{bmatrix}$$

From Equation 6.48, we have

$$\boldsymbol{\psi}^r_{qd0s} = \begin{bmatrix} L_{ls} + \frac{3}{2}\overline{L}_m & 0 & 0 \\ 0 & L_{ls} + \frac{3}{2}\overline{L}_m & 0 \\ 0 & 0 & L_{ls} \end{bmatrix} \mathbf{i}^r_{qd0s} + \begin{bmatrix} 0 \\ \psi_m \\ 0 \end{bmatrix}$$

In expanded form, the quadrature-, direct-, and zero-flux linkages are found to be

$$\psi^r_{qs} = \left(L_{ls} + \tfrac{3}{2}\overline{L}_m\right)i^r_{qs}, \quad \psi^r_{ds} = \left(L_{ls} + \tfrac{3}{2}\overline{L}_m\right)i^r_{ds} + \psi_m \quad \text{and} \quad \psi^r_{0s} = L_{ls}i^r_{0s}$$

In the rotor reference frame using Equation 6.47, one finds

$$\frac{di^r_{qs}}{dt} = -\frac{r_s}{L_{ls} + \frac{3}{2}\overline{L}_m}i^r_{qs} - \frac{\psi_m}{L_{ls} + \frac{3}{2}\overline{L}_m}\omega_r - i^r_{ds}\omega_r + \frac{1}{L_{ls} + \frac{3}{2}\overline{L}_m}u^r_{qs}$$

$$\frac{di^r_{qs}}{dt} = -\frac{r_s}{L_{ls} + \frac{3}{2}\overline{L}_m}i^r_{ds} + i^r_{qs}\omega_r + \frac{1}{L_{ls} + \frac{3}{2}\overline{L}_m}u^r_s \qquad (6.49)$$

$$\frac{di^r_{0s}}{dt} = -\frac{r_s}{L_{ls}}i^r_{0s} + \frac{1}{L_{ls}}u^r_{0s}$$

The electromagnetic torque

$$T_e = \frac{P\psi_m}{2}\left[i_{as}\cos\theta_r + i_{bs}\cos\left(\theta_r - \tfrac{2}{3}\pi\right) + i_{cs}\cos\left(\theta_r + \tfrac{2}{3}\pi\right)\right]$$

should be found in terms of the quadrature, direct, and zero currents.
   Using the Park transformation

$$\begin{bmatrix} i_{as} \\ i_{bs} \\ i_{cs} \end{bmatrix} = \begin{bmatrix} \cos\theta_r & \sin\theta_r & 1 \\ \cos\left(\theta_r - \tfrac{2}{3}\pi\right) & \sin\left(\theta_r - \tfrac{2}{3}\pi\right) & 1 \\ \cos\left(\theta_r + \tfrac{2}{3}\pi\right) & \sin\left(\theta_r + \tfrac{2}{3}\pi\right) & 1 \end{bmatrix}\begin{bmatrix} i^r_{qs} \\ i^r_{ds} \\ i^r_{0s} \end{bmatrix}$$

and substituting

$$i_{as} = \cos\theta_r i^r_{qs} + \sin\theta_r i^r_{ds} + i^r_{0s}, \quad i_{bs} = \cos\left(\theta_r - \tfrac{2}{3}\pi\right)i^r_{qs} + \sin\left(\theta_r - \tfrac{2}{3}\pi\right)i^r_{ds} + i^r_{0s} \quad \text{and}$$
$$i_{cs} = \cos\left(\theta_r + \tfrac{2}{3}\pi\right)i^r_{qs} + \sin\left(\theta_r + \tfrac{2}{3}\pi\right)i^r_{ds} + i^r_{0s}$$

in the expression for T_e, one finds

$$T_e = \frac{3P\psi_m}{4}i^r_{qs}$$

For *P*-pole permanent-magnet synchronous micromotors, the torsional-mechanical dynamics are

$$\frac{d\omega_r}{dt} = \frac{3P^2\psi_m}{8J} i^r_{qs} - \frac{B_m}{J}\omega_r - \frac{P}{2J}T_L$$

$$\frac{d\theta_r}{dt} = \omega_r$$

(6.50)

Augmenting Equation 6.49 and Equation 6.50, we have the lumped-parameter mathematical model of three-phase permanent-magnet synchronous micromotors in the rotor reference frame:

$$\frac{di^r_{qs}}{dt} = -\frac{r_s}{L_{ls}+\frac{3}{2}\overline{L}_m} i^r_{qs} - \frac{\psi_m}{L_{ls}+\frac{3}{2}\overline{L}_m}\omega_r - i^r_{ds}\omega_r + \frac{1}{L_{ls}+\frac{3}{2}\overline{L}_m}u^r_{qs}$$

$$\frac{di^r_{ds}}{dt} = -\frac{r_s}{L_{ls}+\frac{3}{2}\overline{L}_m} i^r_{ds} + i^r_{qs}\omega_r + \frac{1}{L_{ls}+\frac{3}{2}\overline{L}_m}u^r_{ds}$$

$$\frac{di^r_{0s}}{dt} = -\frac{r_s}{L_{ls}} i^r_{0s} + \frac{1}{L_{ls}}u^r_{0s}$$

(6.51)

$$\frac{d\omega_r}{dt} = \frac{3P^2\psi_m}{8J} i^r_{qs} - \frac{B_m}{J}\omega_r - \frac{P}{2J}T_L$$

$$\frac{d\theta_r}{dt} = \omega_r$$

In the state-space form, the mathematical model of permanent-magnet synchronous micromotors in the rotor reference frame is given by

$$
\begin{bmatrix}
\dfrac{di^r_{qs}}{dt} \\[2ex]
\dfrac{di^r_{ds}}{dt} \\[2ex]
\dfrac{di^r_{0s}}{dt} \\[2ex]
\dfrac{d\omega_r}{dt} \\[2ex]
\dfrac{d\theta_r}{dt}
\end{bmatrix}
=
\begin{bmatrix}
-\dfrac{r_s}{L_{ls}+\frac{3}{2}\overline{L}_m} & 0 & 0 & -\dfrac{\psi_m}{L_{ls}+\frac{3}{2}\overline{L}_m} & 0 \\[3ex]
0 & -\dfrac{r_s}{L_{ls}+\frac{3}{2}\overline{L}_m} & 0 & 0 & 0 \\[3ex]
0 & 0 & -\dfrac{r_s}{L_{ls}} & 0 & 0 \\[3ex]
\dfrac{3P^2\psi_m}{8J} & 0 & 0 & -\dfrac{B_m}{J} & 0 \\[3ex]
0 & 0 & 0 & 1 & 0
\end{bmatrix}
\begin{bmatrix}
i^r_{qs} \\[2ex]
i^r_{ds} \\[2ex]
i^r_{0s} \\[2ex]
\omega_r \\[2ex]
\theta_r
\end{bmatrix}
$$

$$
+ \begin{bmatrix} -i_{ds}^r \omega_r \\[4pt] i_{qs}^r \omega_r \\[4pt] 0 \\[4pt] 0 \\[4pt] 0 \end{bmatrix} + \begin{bmatrix} \dfrac{1}{L_{ls} + \frac{3}{2}\bar{L}_m} & 0 & 0 \\[10pt] 0 & \dfrac{1}{L_{ls} + \frac{3}{2}\bar{L}_m} & 0 \\[10pt] 0 & 0 & \dfrac{1}{L_{ls}} \\[10pt] 0 & 0 & 0 \\[6pt] 0 & 0 & 0 \end{bmatrix} \begin{bmatrix} u_{qs}^r \\[4pt] u_{ds}^r \\[4pt] u_{0s}^r \end{bmatrix} - \begin{bmatrix} 0 \\[4pt] 0 \\[4pt] 0 \\[4pt] \dfrac{P}{2J} \\[4pt] 0 \end{bmatrix} T_L
$$

A balanced three-phase current set, to be fed to the stator windings, is

$$
i_{as}(t) = \sqrt{2}\,i_M \cos\theta_r, \quad i_{bs}(t) = \sqrt{2}\,i_M \cos\!\left(\theta_r - \tfrac{2}{3}\pi\right), \quad \text{and} \quad i_{cs}(t) = \sqrt{2}\,i_M \cos\!\left(\theta_r - \tfrac{2}{3}\pi\right)
$$

Using the direct Park transformation

$$
\begin{bmatrix} i_{qs}^r \\[4pt] i_{ds}^r \\[4pt] i_{0s}^r \end{bmatrix} = \frac{2}{3}\begin{bmatrix} \cos\theta_r & \cos\!\left(\theta_r - \tfrac{2}{3}\pi\right) & \cos\!\left(\theta_r + \tfrac{2}{3}\pi\right) \\[6pt] \sin\theta_r & \sin\!\left(\theta_r - \tfrac{2}{3}\pi\right) & \sin\!\left(\theta_r + \tfrac{2}{3}\pi\right) \\[6pt] \tfrac{1}{2} & \tfrac{1}{2} & \tfrac{1}{2} \end{bmatrix}\begin{bmatrix} i_{as} \\[4pt] i_{bs} \\[4pt] i_{cs} \end{bmatrix}
$$

one obtains the quadrature, direct, and zero currents to regulate the angular velocity of permanent-magnet synchronous micromotors and guarantee the balanced operating conditions. We have

$$
\begin{bmatrix} i_{qs}^r \\[4pt] i_{ds}^r \\[4pt] i_{0s}^r \end{bmatrix} = \frac{2}{3}\begin{bmatrix} \cos\theta_r & \cos\!\left(\theta_r - \tfrac{2}{3}\pi\right) & \cos\!\left(\theta_r + \tfrac{2}{3}\pi\right) \\[6pt] \sin\theta_r & \sin\!\left(\theta_r - \tfrac{2}{3}\pi\right) & \sin\!\left(\theta_r + \tfrac{2}{3}\pi\right) \\[6pt] \tfrac{1}{2} & \tfrac{1}{2} & \tfrac{1}{2} \end{bmatrix}\begin{bmatrix} \sqrt{2}\,i_M \cos\theta_r \\[4pt] \sqrt{2}\,i_M \cos\!\left(\theta_r - \tfrac{2}{3}\pi\right) \\[4pt] \sqrt{2}\,i_M \cos\!\left(\theta_r + \tfrac{2}{3}\pi\right) \end{bmatrix}
$$

Hence, one obtains

$$
i_{qs}^r(t) = \sqrt{2}\,i_M, \quad i_{ds}^r(t) = 0, \quad i_{0s}^r(t) = 0
$$

Due to the self-inductances, the *abc* voltages should be supplied with advanced phase shifting. One supplies the following phase voltages:

$$
u_{as}(t) = \sqrt{2}\,u_M \cos\!\left(\theta_r + \varphi_u\right), \quad u_{bs}(t) = \sqrt{2}\,u_M \cos\!\left(\theta_r - \tfrac{2}{3}\pi + \varphi_u\right), \quad \text{and}
$$

$$
u_{cs}(t) = \sqrt{2}\,u_M \cos\!\left(\theta_r + \tfrac{2}{3}\pi + \varphi_u\right)
$$

Taking note of the direct Park transformation

$$
\begin{bmatrix} u_{qs}^r \\ u_{ds}^r \\ u_{0s}^r \end{bmatrix} = \frac{2}{3} \begin{bmatrix} \cos\theta_r & \cos\left(\theta_r - \frac{2}{3}\pi\right) & \cos\left(\theta_r + \frac{2}{3}\pi\right) \\ \sin\theta_r & \sin\left(\theta_r - \frac{2}{3}\pi\right) & \sin\left(\theta_r + \frac{2}{3}\pi\right) \\ \frac{1}{2} & \frac{1}{2} & \frac{1}{2} \end{bmatrix} \begin{bmatrix} u_{as} \\ u_{bs} \\ u_{cs} \end{bmatrix}
$$

one finds

$$
\begin{bmatrix} u_{qs}^r \\ u_{ds}^r \\ u_{0s}^r \end{bmatrix} = \frac{2}{3} \begin{bmatrix} \cos\theta_r & \cos\left(\theta_r - \frac{2}{3}\pi\right) & \cos\left(\theta_r + \frac{2}{3}\pi\right) \\ \sin\theta_r & \sin\left(\theta_r - \frac{2}{3}\pi\right) & \sin\left(\theta_r + \frac{2}{3}\pi\right) \\ \frac{1}{2} & \frac{1}{2} & \frac{1}{2} \end{bmatrix} \begin{bmatrix} \sqrt{2}u_M \cos\left(\theta_r + \varphi_u\right) \\ \sqrt{2}u_M \cos\left(\theta_r - \frac{2}{3}\pi + \varphi_u\right) \\ \sqrt{2}u_M \cos\left(\theta_r + \frac{2}{3}\pi + \varphi_u\right) \end{bmatrix}
$$

Using the trigonometric identities, we have

$$
u_{qs}^r(t) = \sqrt{2}u_M \cos\varphi_u, \quad u_{ds}^r(t) = -\sqrt{2}u_M \sin\varphi_u, \quad u_{0s}^r(t) = 0
$$

Due to small inductances, $\varphi_u \approx 0$, and the following voltages must be applied:

$$
u_{qs}^r(t) = \sqrt{2}u_M, \quad u_{ds}^r(t) = 0 \quad u_{0s}^r(t) = 0
$$

To visualize the results, an *s*-domain block diagram in the *qd*0 variables is developed using Equation 6.51. (See Figure 6.68.)

Analyzing permanent-magnet synchronous microtransducers in the synchronous reference frame, one specifies the angular velocity of the reference frame to be $\omega = \omega_e$. Hence, $\theta = \theta_e$, and the Park transformation matrix is given as

$$
\mathbf{K}_s^e = \frac{2}{3} \begin{bmatrix} \cos\theta_e & \cos\left(\theta_e - \frac{2}{3}\pi\right) & \cos\left(\theta_e + \frac{2}{3}\pi\right) \\ \sin\theta_e & \sin\left(\theta_e - \frac{2}{3}\pi\right) & \sin\left(\theta_e + \frac{2}{3}\pi\right) \\ \frac{1}{2} & \frac{1}{2} & \frac{1}{2} \end{bmatrix}
$$

Substituting $\omega_r = \omega_e$ in Equation 6.51 we have the following system of differential equations that model the permanent-magnet micromotor dynamics in the synchronous reference frame:

$$
\frac{di_{qs}^e}{dt} = -\frac{r_s}{L_{ls} + \frac{3}{2}\overline{L}_m} i_{qs}^e - \frac{\psi_m}{L_{ls} + \frac{3}{2}\overline{L}_m} \omega_r - i_{ds}^e \omega_r + \frac{1}{L_{ls} + \frac{3}{2}\overline{L}_m} u_{qs}^e
$$

$$
\frac{di_{ds}^e}{dt} = -\frac{r_s}{L_{ls} + \frac{3}{2}\overline{L}_m} i_{ds}^e + i_{qs}^e \omega_r + \frac{1}{L_{ls} + \frac{3}{2}\overline{L}_m} u_{ds}^e
$$

$$
\frac{di_{0s}^e}{dt} = -\frac{r_s}{L_{ls}} i_{0s}^e + \frac{1}{L_{ls}} u_{0s}^e
$$

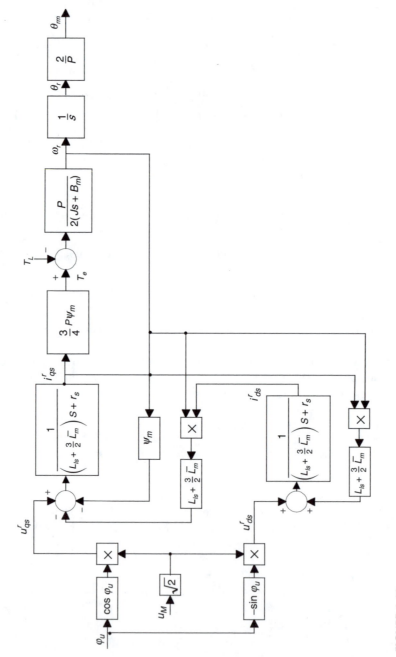

FIGURE 6.68
s-domain block diagram of permanent-magnet synchronous micromotors in the rotor reference frame.

$$\frac{d\omega_r}{dt} = \frac{3P^2\psi_m}{8J} i_{qs}^e - \frac{B_m}{J}\omega_r - \frac{P}{2J}T_L$$

$$\frac{d\theta_r}{dt} = \omega_r$$

The quadrature, direct, and zero currents that must be fed to guarantee the balanced operation are $i_{qs}^e(t) = \sqrt{2}i_M$, $i_{ds}^e(t) = 0$, and $i_{0s}^e(t) = 0$.

To control the angular velocity (in the drive application) of permanent-magnet synchronous micromotors or the displacement (in servo-system application), one supplies the phase voltages to the *abc* stator windings as a function of the angular displacement (measured by the Hall-effect sensors). Thus, ICs must be used, and the permanent-magnet synchronous micromotors can be derived by the Motorola MC33035 ICs.

6.8.4.4 Simulation and Analysis of Permanent-Magnet Synchronous Micromotors in SIMULINK

In this section we will simulate and analyze radial topology three-phase synchronous micromotor using SIMULINK. The studied permanent-magnet synchronous micromotors are described by five nonlinear differential equations. (See Equation 6.45.)

The following phase voltages are applied to guarantee the balance motor operation:

$$u_{as}(t) = \sqrt{2}u_M \cos\theta_r, \quad u_{bs}(t) = \sqrt{2}u_M \cos\left(\theta_r - \frac{2}{3}\pi\right), \quad \text{and}$$

$$u_{cs}(t) = \sqrt{2}u_M \cos\left(\theta_r + \frac{2}{3}\pi\right)$$

These differential equations integrated with the balanced voltage set (to properly supply voltages, the rotor angular displacement must be measured or observed) can be straightforwardly simulated in SIMULINK. The SIMULINK block diagram (mdl model) to simulate permanent-magnet synchronous motors with the symbolic parameters (to be numerically assigned to perform numerical simulations) is documented in Figure 6.69.

Having developed the SIMULINK model, the micromachine parameters must be downloaded. Let the micromotor parameters be assigned as $u_M = 5$ V, $r_s = 200$ ohm, $L_{ss} = 0.002$ H, $L_{ls} = 0.00015$ H, $\bar{L}_m = 0.0012$ H, $\psi_m = 0.1$ V-sec/rad [N-m/A], $B_m = 0.000000004$ N-m-sec/rad, and $J = 0.0000000005$ kg-m^2.

The micromotor dynamics are studied as the microactuator accelerates as the rated voltage (with the magnitude $\sqrt{2}5$ V) is supplied to the stator windings, that is, $u_{as}(t) = \sqrt{2}5\cos\theta_r$, $u_{bs}(t) = \sqrt{2}5\cos(\theta_r - \frac{2}{3}\pi)$, and $u_{cs}(t) = \sqrt{2}5\cos(\theta_r + \frac{2}{3}\pi)$.

The motor parameters are downloaded using the following statement typed in the Command window:

```
% Parameters of the permanent-magnet synchronous micromotor
P=2; um=10; rs=100; Lss=0.002; Lls=0.00015; fm=0.1;
Bm=0.000000005; J=0.0000000001;
Lmb=2*(Lss-Lls)/3;
```

The motor accelerates from stall, and Figure 6.70 illustrates the evolution of the angular velocity. The motor reaches the steady-state angular velocity (710 rad/sec) at 0.0002 sec. The current dynamics are also reported in Figure 6.70.

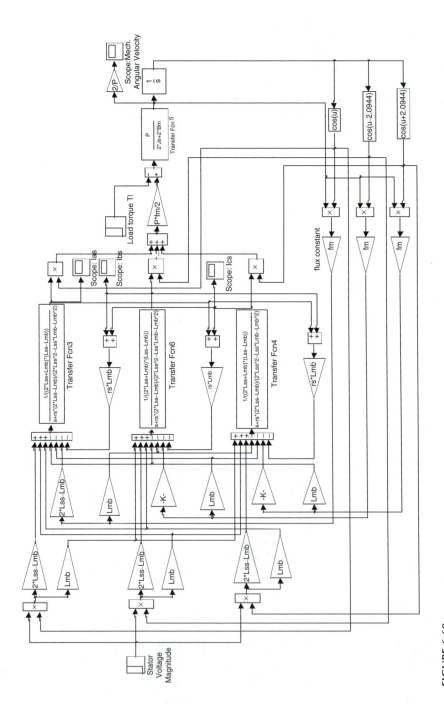

FIGURE 6.69
SIMULINK block diagram to simulate permanent-magnet synchronous motors (`c6_8_2.md1`).

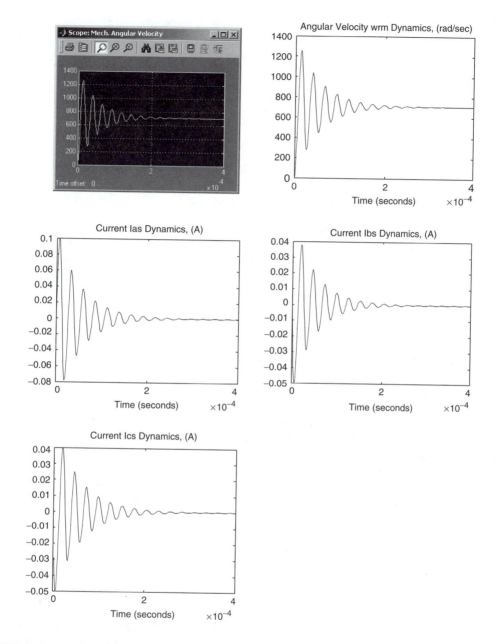

FIGURE 6.70

Transient dynamics of the permanent-magnet synchronous motor variables.

It should be emphasized that all state variables (phase currents, angular velocity, and angular displacement) were plotted using the plot function. In particular, the following statements are used to plot the transient dynamics data stored in the MATLAB memory:

```
% Plots of the transient dynamics of the permanent-magnet
% synchronous micromotor
plot(Ias(:,1),Ias(:,2)); xlabel('Time (seconds)');
title('Current Ias Dynamics, [A]');
```

```
plot(Ibs(:,1),Ibs(:,2)); xlabel('Time (seconds)');
title('Current Ibs Dynamics, [A]');
plot(Ics(:,1),Ics(:,2)); xlabel('Time (seconds)');
title('Current Ics Dynamics, [A]');
plot(wrm(:,1),wrm(:,2)); xlabel('Time (seconds)');
title('Angular Velocity wrm Dynamics, [rad/sec]');
```

6.9 Permanent-Magnet Stepper Micromotors

In MEMS and microscale devices, permanent-magnet stepper micromotors can be used. Translational and rotational stepper micromotors (which are synchronous micromachines) have been designed, fabricated, and tested. These micromotors (microactuators) develop high electromagnetic torque, while the mechanical angular velocity is relatively low. Therefore, permanent-magnet stepper micromotors can be utilized in the direct-drive microservos. This direct connection of micromotors without matching mechanical coupling allows one to achieve a remarkable level of efficiency, reliability, and performance.

Stepper micromotors must be controlled to ensure stability, precision tracking, desired steady-state and dynamic performance, disturbance rejection, and zero steady-state error. To approach the analysis, simulation, and control, complete nonlinear mathematical models of stepper micromotors must be found. The operating principles in control of stepper micromotors are well known. In particular, by energizing the stator windings in the proper sequence, the rotor rotates in the counterclockwise or clockwise direction due to the electromagnetic torque developed. In particular, the rotor displaces by full or half steps. Hence, when energizing windings, one achieves the angular increment equal to a full or half step. The angular velocity is regulated by changing the frequency of the phase currents fed or voltages supplied to the phase windings, as was shown for permanent-magnet synchronous micromotors. Due to the possibilities to operate stepper motors in the open-loop modes properly energizing the windings, the stepper motors were among the first electric machines to be fabricated and tested in the early 1990s (Technical University of Berlin, Kiev Polytechnic Institute, and University of Wisconsin, Madison). (See Figure 6.71.)

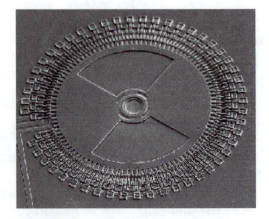

FIGURE 6.71
Micromachined stepper motor.

6.9.1 Mathematical Model in the Machine Variables

For two-phase permanent-magnet stepper micromotors, we have

$$u_{as} = r_s i_{as} + \frac{d\psi_{as}}{dt}$$
$$u_{bs} = r_s i_{bs} + \frac{d\psi_{bs}}{dt} \qquad (6.52)$$

where the flux linkages are given as

$$\psi_{as} = L_{asas} i_{as} + L_{asbs} i_{bs} + \psi_{asm}$$
$$\psi_{bs} = L_{bsas} i_{as} + L_{bsbs} i_{bs} + \psi_{bsm} \qquad (6.53)$$

The electrical angular velocity and displacement are found using the number of rotor teeth RT, $\omega_r = RT\omega_{rm}$, and $\theta_r = RT\theta_{rm}$. The flux linkages are the functions of the number of the rotor teeth and displacement,

$$\psi_{asm} = \psi_m \cos(RT\theta_{rm})$$

$$\psi_{bsm} = \psi_m \sin(RT\theta_{rm}) \qquad (6.54)$$

The self-inductance of the stator windings is

$$L_{ss} = L_{asas} = L_{bsbs} = L_{ls} + \bar{L}_m \qquad (6.55)$$

The stator windings are displaced by 90 electrical degrees. Hence, the mutual inductances between the stator windings are zero, $L_{asbs} = L_{bsas} = 0$.

From Equation 6.53, Equation 6.54, and Equation 6.55, we have

$$\psi_{as} = L_{ss} i_{as} + \psi_m \cos(RT\theta_{rm})$$
$$\psi_{bs} = L_{ss} i_{bs} + \psi_m \sin(RT\theta_{rm}) \qquad (6.56)$$

Taking note of Equation 6.52 and Equation 6.56, one has

$$u_{as} = r_s i_{as} + \frac{d(L_{ss} i_{as} + \psi_m \cos(RT\theta_{rm}))}{dt} = r_s i_{as} + L_{ss}\frac{di_{as}}{dt} - RT\psi_m \omega_{rm} \sin(RT\theta_{rm})$$

$$u_{bs} = r_s i_{bs} + \frac{d(L_{ss} i_{bs} + \psi_m \sin(RT\theta_{rm}))}{dt} = r_s i_{bs} + L_{ss}\frac{di_{bs}}{dt} + RT\psi_m \omega_{rm} \cos(RT\theta_{rm})$$

Therefore,

$$\frac{di_{as}}{dt} = -\frac{r_s}{L_{ss}} i_{as} + \frac{RT\psi_m}{L_{ss}} \omega_{rm} \sin(RT\theta_{rm}) + \frac{1}{L_{ss}} u_{as},$$

$$\frac{di_{bs}}{dt} = -\frac{r_s}{L_{ss}} i_{bs} - \frac{RT\psi_m}{L_{ss}} \omega_{rm} \cos(RT\theta_{rm}) + \frac{1}{L_{ss}} u_{bs} \qquad (6.57)$$

Using Newton's second law we have

$$\frac{d\omega_{rm}}{dt} = \frac{1}{J}(T_e - B_m\omega_{rm} - T_L)$$

$$\frac{d\theta_{rm}}{dt} = \omega_{rm}$$

The expression for the electromagnetic torque developed by permanent-magnet stepper micromotors must be found. Taking note of

$$W_c = \tfrac{1}{2}\left(L_{ss}i_{as}^2 + L_{ss}i_{bs}^2\right) + \psi_m i_{as}\cos(RT\theta_{rm}) + \psi_m i_{bs}\sin(RT\theta_{rm}) + W_{PM}$$

one finds the electromagnetic torque:

$$T_e = \frac{\partial W_c}{\partial \theta_{rm}} = -RT\psi_m[i_{as}\sin(RT\theta_{rm}) - i_{bs}\cos(RT\theta_{rm})]$$

Hence, the transient evolution of the rotor angular velocity ω_{rm} and displacement θ_{rm} is modeled by the following differential equations:

$$\frac{d\omega_{rm}}{dt} = -\frac{RT\psi_m}{J}[i_{as}\sin(RT\theta_{rm}) - i_{bs}\cos(RT\theta_{rm})] - \frac{B_m}{J}\omega_{rm} - \frac{1}{J}T_L$$

$$\frac{d\theta_{rm}}{dt} = \omega_{rm}$$

$$(6.58)$$

Augmenting Equation 6.57 and Equation 6.58, one has

$$\frac{di_{as}}{dt} = -\frac{r_s}{L_{ss}}i_{as} + \frac{RT\psi_m}{L_{ss}}\omega_{rm}\sin(RT\theta_{rm}) + \frac{1}{L_{ss}}u_{as}$$

$$\frac{di_{bs}}{dt} = -\frac{r_s}{L_{ss}}i_{bs} - \frac{RT\psi_m}{L_{ss}}\omega_{rm}\cos(RT\theta_{rm}) + \frac{1}{L_{ss}}u_{bs}$$

$$(6.59)$$

$$\frac{d\omega_{rm}}{dt} = -\frac{RT\psi_m}{J}[i_{as}\sin(RT\theta_{rm}) - i_{bs}\cos(RT\theta_{rm})] - \frac{B_m}{J}\omega_{rm} - \frac{1}{J}T_L$$

$$\frac{d\theta_{rm}}{dt} = \omega_{rm}$$

These four nonlinear differential equations are rewritten in the state-space form as

$$
\begin{bmatrix} \dfrac{di_{as}}{dt} \\[2mm] \dfrac{di_{bs}}{dt} \\[2mm] \dfrac{d\omega_{rm}}{dt} \\[2mm] \dfrac{d\theta_{rm}}{dt} \end{bmatrix} =
\begin{bmatrix} -\dfrac{r_s}{L_{ss}} & 0 & 0 & 0 \\[2mm] 0 & -\dfrac{r_s}{L_{ss}} & 0 & 0 \\[2mm] 0 & 0 & -\dfrac{B_m}{J} & 0 \\[2mm] 0 & 0 & 1 & 0 \end{bmatrix}
\begin{bmatrix} i_{as} \\[1mm] i_{bs} \\[1mm] \omega_{rm} \\[1mm] \theta_{rm} \end{bmatrix}
$$

$$
+ \begin{bmatrix} \dfrac{RT\psi_m}{L_{ss}}\omega_{rm}\sin(RT\theta_{rm}) \\[3mm] -\dfrac{RT\psi_m}{L_{ss}}\omega_{rm}\cos(RT\theta_{rm}) \\[3mm] -\dfrac{RT\psi_m}{J}[i_{as}\sin(RT\theta_{rm})-i_{bs}\cos(RT\theta_{rm})] \\[3mm] 0 \end{bmatrix}
+ \begin{bmatrix} \dfrac{1}{L_{ss}} & 0 \\[2mm] 0 & \dfrac{1}{L_{ss}} \\[2mm] 0 & 0 \\[2mm] 0 & 0 \end{bmatrix}
\begin{bmatrix} u_{as} \\[1mm] u_{bs} \end{bmatrix}
- \begin{bmatrix} 0 \\[1mm] 0 \\[1mm] \dfrac{1}{J} \\[1mm] 0 \end{bmatrix} T_L
$$

From Equation 6.59, an s-domain block diagram is developed and illustrated in Figure 6.72.

The analysis of the torque equation

$$
T_e = -RT\psi_m[i_{as}\sin(RT\theta_{rm})-i_{bs}\cos(RT\theta_{rm})]
$$

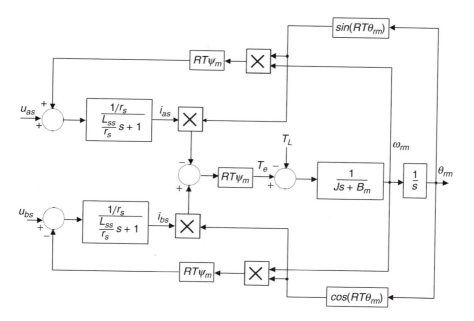

FIGURE 6.72
Block diagram of permanent-magnet stepper micromotors.

guides one to the conclusion that the expressions for a balanced two-phase current sinusoidal set is

$$i_{as} = -\sqrt{2}i_M \sin(RT\theta_{rm})$$
$$i_{bs} = \sqrt{2}i_M \cos(RT\theta_{rm})$$
(6.60)

because the electromagnetic torque is a function of the current magnitude i_M, and

$$T_e = \sqrt{2}RT\psi_m i_M$$

Using ICs, the phase currents that must be fed (Equation 6.60) are the functions of the rotor angular displacement. Assuming that the inductances are negligibly small, the following phase voltages must be supplied:

$$u_{as} = -\sqrt{2}u_M \sin(RT\theta_{rm})$$
$$u_{bs} = \sqrt{2}u_M \cos(RT\theta_{rm})$$
(6.61)

An s-domain block diagram of permanent-magnet stepper micromotors that is controlled by changing the phase voltages using ICs, as given by Equation 6.61, is shown in Figure 6.73.

6.9.2 Mathematical Models of Permanent-Magnet Stepper Micromotors in the Rotor and Synchronous Reference Frames

It was shown that using the machine variables, Kirchhoff's voltage law results in two nonlinear differential equations:

$$u_{as} = r_s i_{as} + L_{ss}\frac{di_{as}}{dt} - RT\psi_m\omega_{rm}\sin(RT\theta_{rm}) \quad \text{and} \quad u_{bs} = r_s i_{bs} + L_{ss}\frac{di_{bs}}{dt} + RT\psi_m\omega_{rm}\cos(RT\theta_{rm})$$

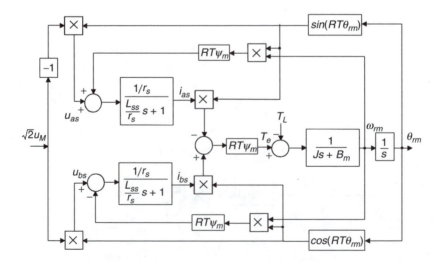

FIGURE 6.73
s-domain diagram of permanent-magnet stepper micromotors, $u_{as} = -\sqrt{2}u_M \sin(RT\theta_{rm})$ and $u_{bs} = \sqrt{2}u_M \cos(RT\theta_{rm})$.

Applying the direct Park formation, which in the rotor reference frame is given as

$$\begin{bmatrix} u_{qs}^r \\ u_{ds}^r \end{bmatrix} = \begin{bmatrix} -\sin(RT\theta_{rm}) & \cos(RT\theta_{rm}) \\ \cos(RT\theta_{rm}) & \sin(RT\theta_{rm}) \end{bmatrix} \begin{bmatrix} u_{as} \\ u_{bs} \end{bmatrix},$$

$$\begin{bmatrix} i_{qs}^r \\ i_{ds}^r \end{bmatrix} = \begin{bmatrix} -\sin(RT\theta_{rm}) & \cos(RT\theta_{rm}) \\ \cos(RT\theta_{rm}) & \sin(RT\theta_{rm}) \end{bmatrix} \begin{bmatrix} i_{as} \\ i_{bs} \end{bmatrix}$$

the following differential equations in the qd quantities are found:

$$u_{qs}^r = r_s i_{qs}^r + L_{ss} \frac{di_{qs}^r}{dt} + RT\psi_m \omega_{rm} + RTL_{ss} i_{ds}^r \omega_{rm}$$

$$u_{ds}^r = r_s i_{ds}^r + L_{ss} \frac{di_{ds}^r}{dt} - RTL_{ss} i_{qs}^r \omega_{rm}$$

Hence, the resulting nonlinear circuitry dynamics are shown by

$$\begin{aligned} \frac{di_{qs}^r}{dt} &= -\frac{r_s}{L_{ss}} i_{qs}^r - \frac{RT\psi_m}{L_{ss}} \omega_{rm} - RT i_{ds}^r \omega_{rm} + \frac{1}{L_{ss}} u_{qs}^r \\ \frac{di_{ds}^r}{dt} &= -\frac{r_s}{L_{ss}} i_{ds}^r + RT i_{qs}^r \omega_{rm} + \frac{1}{L_{ss}} u_{ds}^r \end{aligned} \tag{6.62}$$

From

$$T_e = -RT\psi_m [i_{as} \sin(RT\theta_{rm}) - i_{bs} \cos(RT\theta_{rm})]$$

using the inverse Park transformation

$$\begin{bmatrix} i_{as} \\ i_{bs} \end{bmatrix} = \begin{bmatrix} -\sin(RT\theta_{rm}) & \cos(RT\theta_{rm}) \\ \cos(RT\theta_{rm}) & \sin(RT\theta_{rm}) \end{bmatrix} \begin{bmatrix} i_{qs}^r \\ i_{ds}^r \end{bmatrix}$$

we have

$$T_e = RT\psi_m i_{qs}^r$$

From Newton's second law of motions, one has

$$\begin{aligned} \frac{d\omega_{rm}}{dt} &= \frac{RT\psi_m}{J} i_{qs}^r - \frac{B_m}{J} \omega_{rm} - \frac{1}{J} T_L \\ \frac{d\theta_{rm}}{dt} &= \omega_{rm} \end{aligned} \tag{6.63}$$

Augmenting Equation 6.62 and Equation 6.63, the following lumped-parameter mathe-matical model of permanent-magnet synchronous micromotors in the rotor reference frame results:

$$\frac{di_{qs}^r}{dt} = -\frac{r_s}{L_{ss}}i_{qs}^r - \frac{RT\psi_m}{L_{ss}}\omega_{rm} - RTi_{ds}^r\omega_{rm} + \frac{1}{L_{ss}}u_{qs}^r$$

$$\frac{di_{ds}^r}{dt} = -\frac{r_s}{L_{ss}}i_{ds}^r + RTi_{qs}^r\omega_{rm} + \frac{1}{L_{ss}}u_{ds}^r$$

$$\frac{d\omega_{rm}}{dt} = \frac{RT\psi_m}{J}i_{qs}^r - \frac{B_m}{J}\omega_{rm} - \frac{1}{J}T_L \tag{6.64}$$

$$\frac{d\theta_{rm}}{dt} = \omega_{rm}$$

Four nonlinear differential equations that describe the circuitry and torsional-mechan-ical dynamics are derived. These nonlinear differential equations can be used for anal-ysis, simulation, control, and performance analysis of permanent-magnet stepper micromotors.

In the state-space form, we have

$$\begin{bmatrix} \dfrac{di_{qs}^r}{dt} \\[2mm] \dfrac{di_{ds}^r}{dt} \\[2mm] \dfrac{d\omega_{rm}}{dt} \\[2mm] \dfrac{d\theta_{rm}}{dt} \end{bmatrix} = \begin{bmatrix} -\dfrac{r_s}{L_{ss}} & 0 & -\dfrac{RT\psi_m}{L_{ss}} & 0 \\[2mm] 0 & -\dfrac{r_s}{L_{ss}} & 0 & 0 \\[2mm] \dfrac{RT\psi_m}{J} & 0 & -\dfrac{B_m}{J} & 0 \\[2mm] 0 & 0 & 1 & 0 \end{bmatrix} \begin{bmatrix} i_{qs}^r \\[2mm] i_{ds}^r \\[2mm] \omega_{rm} \\[2mm] \omega_{rm} \end{bmatrix}$$

$$+ \begin{bmatrix} -RTi_{ds}^r\omega_{rm} \\[2mm] RTi_{qs}^r\omega_{rm} \\[2mm] 0 \\[2mm] 0 \end{bmatrix} + \begin{bmatrix} \dfrac{1}{L_{ss}} & 0 \\[2mm] 0 & \dfrac{1}{L_{ss}} \\[2mm] 0 & 0 \\[2mm] 0 & 0 \end{bmatrix} \begin{bmatrix} u_{qs}^r \\[2mm] u_{ds}^r \end{bmatrix} - \begin{bmatrix} 0 \\[2mm] 0 \\[2mm] \dfrac{1}{J} \\[2mm] 0 \end{bmatrix} T_L$$

It is evident that these nonlinear differential equations cannot be linearized. Straightfor-ward analytical and numerical analysis can be performed using the developed lumped-parameter mathematical models.

The phase currents and voltages supplied to the *ab* windings must be fed using the rotor angular displacement. We have the two-phase current

$$i_{as} = -\sqrt{2}i_M \sin(RT\theta_{rm}), \quad i_{bs} = \sqrt{2}i_M \cos(RT\theta_{rm})$$

and two-phase voltage

$$u_{as} = -\sqrt{2}u_M \sin(RT\theta_{rm}), \quad u_{bs} = \sqrt{2}u_M \cos(RT\theta_{rm})$$

balanced sets.

From the Park transformation, as given by,

$$\begin{bmatrix} i_{qs}^r \\ i_{ds}^r \end{bmatrix} = \begin{bmatrix} -\sin(RT\theta_{rm}) & \cos(RT\theta_{rm}) \\ \cos(RT\theta_{rm}) & \sin(RT\theta_{rm}) \end{bmatrix} \begin{bmatrix} i_{as} \\ i_{bs} \end{bmatrix}$$

one obtains

$$i_{qs}^r = -i_{as} \sin(RT\theta_{rm}) + i_{bs} \cos(RT\theta_{rm})$$

$$i_{ds}^r = i_{as} \cos(RT\theta_{rm}) + i_{bs} \sin(RT\theta_{rm})$$

Therefore,

$$i_{qs}^r = \sqrt{2}i_M \sin^2(RT\theta_{rm}) + \sqrt{2}i_M \cos^2(RT\theta_{rm}) = \sqrt{2}i_M$$

and

$$i_{ds}^r = -\sqrt{2}i_M \sin(RT\theta_{rm})\cos(RT\theta_{rm}) + \sqrt{2}i_M \sin(RT\theta_{rm})\cos(RT\theta_{rm}) = 0$$

Thus, $i_{qs}^r = \sqrt{2}i_M$ and $i_{ds}^r = 0$.

Similarly, for the quadrature and direct voltages, from the following relationship derived using the Park transformation matrix

$$\begin{bmatrix} u_{qs}^r \\ u_{ds}^r \end{bmatrix} = \begin{bmatrix} -\sin(RT\theta_{rm}) & \cos(RT\theta_{rm}) \\ \cos(RT\theta_{rm}) & \sin(RT\theta_{rm}) \end{bmatrix} \begin{bmatrix} u_{as} \\ u_{bs} \end{bmatrix}$$

one has the expressions for the quadrature and direct voltages to guarantee the balance operation. The multiplication and simplification gives $u_{qs}^r = \sqrt{2}u_M$ and $u_{ds}^r = 0$.

If the advanced shifting is used, we obtain

$$u_{qs}^r = \sqrt{2}u_M \cos\varphi_u$$
$$u_{ds}^r = -\sqrt{2}u_M \sin\varphi_u$$

(6.65)

Using Equation 6.64, the block diagram of permanent-magnet stepper micromotors, modeled in the rotor reference frame and controlled by changing the quadrature and direct voltages, is developed. The resulting block diagram is illustrated in Figure 6.74.

Synchronous micromotors rotate with the synchronous angular velocity, $\omega_r = \omega_e$. From Equation 6.64, the resulting model of permanent-magnet stepper micromotors in the

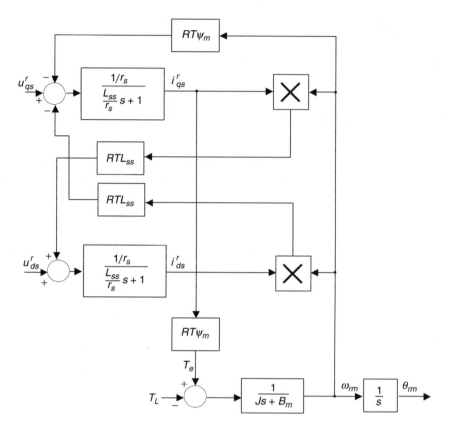

FIGURE 6.74
Block diagram of permanent-magnet stepper micromotors modeled in the rotor reference frame.

synchronous reference frame is

$$\frac{di^e_{qs}}{dt} = -\frac{r_s}{L_{ss}} i^e_{qs} - \frac{RT\psi_m}{L_{ss}} \omega_{rm} - RTi^e_{ds}\omega_{rm} + \frac{1}{L_{ss}} u^e_{qs}$$

$$\frac{di^e_{ds}}{dt} = -\frac{r_s}{L_{ss}} i^e_{ds} + RTi^e_{qs}\omega_{rm} + \frac{1}{L_{ss}} u^e_{ds}$$

$$\frac{d\omega_{rm}}{dt} = \frac{RT\psi_m}{J} i^e_{qs} - \frac{B_m}{J} \omega_{rm} - \frac{1}{J} T_L$$

$$\frac{d\theta_{rm}}{dt} = \omega_{rm}$$

To control the angular velocity, the ICs energize the *as* and *bs* windings (the so-called step-by-step open-loop operation). As an example, the Motorola monolithic MC3479 ICs driver-controller can be used.

6.10 Piezotransducers

It is known that the generation of a potential difference across the opposite faces of certain nonconducting crystals as a result of the applied mechanical stress between these faces is called the piezoelectric effect. The piezoelectric effect, exhibited by quartz, tourmaline, and Rochelle salt, is quite modest. However, polycrystalline ferroelectric ceramic materials (for example, $BaTiO_3$, and lead zirconate titanate) exhibit strong piezoelectric effect. Electroactive, magnetoactive, and electrooptoactive materials change their geometry (shape) in response to electromagnetic stimuli. The application of these materials allows one to achieve induced-strain actuation and strain sensing without commonly used conventional kinematics arrangements (bearing, gears, sleeves, etc.) that were examined for translational and rotational MEMS in this chapter. This may result in direct and efficient energy conversion (electromagnetic to magnetic energy and vice versa). However, it will be illustrated that some limitations are imposed. For example, from the actuation viewpoint, the strain is 0.1% of the actuator length, that is, a 1000-μm actuator will result in 1 μm displacement, which may not be sufficient. In the sensing application, piezoelectric strain sensors are used, and well-defined signals with high signal-to-noise ratio result, simplifying the signal conditioning, processing, and amplification ICs. Therefore, these sensors can be used for vibroacoustic sensing and attenuation, high-accuracy pointing, active flow control, etc. Among active materials that have been under development for many years and have been widely commercialized, the following should be emphasized: piezoelectric ceramics (PZT), electrostrictive ceramics (PMN), piezoelectric polymers (PVDF), and magnetostrictive compounds (Terfenol-D). Ferroelectric ceramics (which become piezoelectric when poled) comprises randomly oriented crystals or grains, each having one or a few domains. With the dipoles randomly oriented, the material is isotropic and does not exhibit the piezoelectric effect. Applying a strong DC electric field (poling is performed using electrodes creating the specified field for the specific material that is usually greater than 2000 V/m), the dipoles will tend to align themselves parallel to the field, and the material will have a permanent (or remnant) polarization becoming piezoelectric. After poling, the material has a remnant polarization P_r and remanent stress S_r. For example, commonly used lead zirconate titanate (PZT) crystallites are centrosymmetric cubic (isotropic) before poling, and, after poling, they exhibit tetragonal symmetry (anisotropic structure) below the Curie temperature. (Above the Curie temperature, crystals lose their piezoelectric properties.) The Curie temperature is the temperature at which the crystal structure changes from a nonsymmetric to a symmetric form. The charge separation between the positive and negative ions results in electric dipole behavior (dipole groups with parallel orientation are called the Weiss domains). The Weiss domains are randomly oriented in the raw piezoelectric materials before the poling. (When the field is applied, the material expands along the axis of the electric field and contracts perpendicular to that axis.) The electric dipoles align and stay (hopefully) in the alignment. The material has a remanent polarization.

Piezoelectricity was discovered by Pierre and Jacques Curie in 1880 based on their study of the symmetry of crystalline matter of quartz, topaz, tourmaline, and Rochelle salt (sodium potassium tartrate). In particular, they measured the surface crystal charges subject to the mechanical stress. However, discovering the direct piezoelectric effect (electric field due to the applied stress), they did not study the converse piezoelectric effect (stress in response to the applied electric field). This phenomenon was mathematically proven using thermodynamic principles by Lippmann in 1881, and the Curies experimentally confirmed the converse effect, thus illustrating a complete reversibility of electroelastomechanical deformations in piezoelectric crystals. The reversible exchange of electrical and mechanical energy and the application of thermodynamics in analysis of complex phenomena and relationships among mechanical, thermal, and electrical variables were made. In addition, the classification of piezoelectric crystals on the basis of asymmetric crystal structure was made. Twenty crystal classes (in which piezoelectric effects occur) and 18 piezoelectric coefficients were published by Voigt in 1910. Based on these fundamental results, the application of the piezoelectric phenomena started. In 1916, to detect submarines, Paul Langevin designed an underwater ultrasonic detector using piezoelectric quartz sandwiched between two steel plates. (The resonant frequency of the detector was 50 kHz.) The success in sonar developments stimulated intensive activities in resonating and nonresonating piezoelectric-based devices. However, the time for advanced actuators and sensors, microphones, accelerometers, ultrasonic transducers, filters, and other devices had not come yet.

During World War II, scientists from the United States, U.S.S.R., Japan, and Germany discovered ferroelectric ceramics (prepared by sintering metallic oxide powders) with improved piezoelectric characteristics and with dielectric constants hundreds of times higher than common crystals. In particular, barium titanate piezoceramics were devised (this led to the PZT), advanced complex perovskite crystal structures were devised with the corresponding studies of their effect on the electromechanical characteristics, and doping with metallic impurities was performed in order to achieve desired properties (dielectric constant, stiffness, piezoelectric coupling coefficients, poling, robustness, etc.). For example, some common perovskite (Russian mineralogist Count Lev Perovski, 1792–1856) are $(Zr, Yi)O_3$ (piezoelectric transducers), $Pb(Mg, Nb)O_3$ (electrostrictive actuators), $(Ca, La)MnO_3$ (ferromagnets), $LiNbO_3$ (optical switch and optical steering), etc. These advances significantly contributed to establishing entirely new techniques and perspectives in piezoelectric device development through controlling (determining) the material properties and characteristics to specific applications such as sonars, smart piezotransducers (actuators-sensors), filter devices, etc. Many commercial piezoelectric ceramics are typically made of simple perovskites and solid-solution perovskite alloys. Typical examples of simple perovskites are $BaTiO_3$ and $PbTiO_3$, while the solid-solution perovskite alloys are $Pb(Zr, Ti)O_3$, $(Pb, La)(Zr, Ti)O_3$ (lead lanthanum zirconate titanate oxide), and ternary ceramics (for example, BaO-TiO_2-$RE2O_3$, where RE is a rare earth metal).

The piezoelectric transducers are widely used in optic devices (image and beam stabilization, scanning microscopy, autofocusing, interferometry, alignment, switching, mirror scanners and petitioners, active optics, tuning, vibroacoustic attenuation and control), disk drives (actuation, testing, calibration, alignment, vibration attenuation), microelectronics (nanoactuation and measurements, wafer and mask positioning and alignment, lithography and etching), precision electromechanics (vibroacoustic attenuation and control, high-accuracy actuation and sensing, nano- and microrobotics and mechatronics, assembling, nanometrology), medicine (microdrives and actuators, sensors, micromanipulators and robots, mircostimulus devices), etc. Piezoelectric materials (actuators and sensors) are typically integrated in the controlled structures as the patched and embedded composite structures, and these piezotransducers are controlled using ICs.

Piezoelectric materials are used to convert electrical energy into mechanical energy and vice versa. For nano- and micropositioning, precise (high-accuracy) and fast motion results due to the electric field (voltage) applied to the piezoelectric. (The resulting strain in the order of 1/1000, that is, 0.1%, means that the 10 mm long actuator has a displacement of 10 micrometers.) The piezoelectric actuator is illustrated in Figure 6.75. The advantages of the piezoelectric actuators are repeatable high-accuracy (nanometers) fast positioning achieved without moving parts (which lead to robustness, durability, efficiency), high force and torque (high force and torque density), high efficiency, integrity,

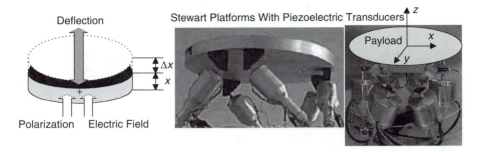

FIGURE 6.75
Piezoelectric actuator and smart piezotransducers application in the Stewart platforms.

affordability, controllability, etc. The piezoelectric actuators can generate the force and acceleration in the order of thousands of Newtons and g's, directly converting electrical energy into mechanical energy. Piezotransducers are widely used in active vibroacoustic and noise-control devices, vibration- and noise-cancellation applications, geometry control, accurate pointing (including multidegree-of-freedom systems such as the Stewart platforms illustrated in Figure 6.75), etc. The Stewart and inverted Stewart platforms are kinematically designed platforms to attain control in six degrees of freedom (x, y, z, pitch, roll, and yaw). Therefore, Stewart platforms found applications in aerospace, manufacturing, medicine, robotics, etc. In general Stewart platforms are based on triangle-mounted frames actuated by high-performance actuators. Usually two actuators are attached at each corner of the platform to guarantee the desired actuation with six-degrees-of-freedom manipulation features.

The relationships between the applied voltage and the resulting forces and displacement, and vice versa, depend on the piezoelectromechanical properties of the ceramic, size and shape, direction of the electrical and mechanical excitation, temperature, etc. Piezoceramics are usually studied in the Cartesian coordinate system (three-dimensional orthogonal set of axes) using x, y, and z axes. The polarization direction is established during manufacturing by applying a DC voltage (electric field) to a pair of electrodes. These poling electrodes are then removed and replaced by electrodes deposited on a second pair of faces. Piezoelectric materials are anisotropic, and their electromechanical properties differ for electrical and mechanical excitation along different directions. Therefore, for the systematic tabulation of electromechanical properties, the directions are standardized and specified. In particular, one defines the axes by numerals: 1 corresponds to the x axis, 2 corresponds to the y axis, and 3 corresponds to the z axis.

Piezoelectric ceramics are isotropic and become piezoelectric after poling. (Once polarized, piezoceramics are anisotropic.) The direction of the poling electric field (applied DC voltage) is distinguished in three directions. The poling electric field can be applied in such ways that the ceramic will exhibit piezoelectric responses in various directions or a combination of directions. The poling process permanently changes the properties of the ceramics.

6.10.1 Piezoactuators: Steady-State Models and Characteristics

Simplifying the analysis, we consider the transverse effect in the rectangular piezoelectric film. (See Figure 6.76.) The first (1) and second (2) axes define a plane that is parallel to the film surface, while the third (3) axis is perpendicular to the surface. (Third axis points in the opposite direction to the electric field applied to pole the piezoelectric films.)

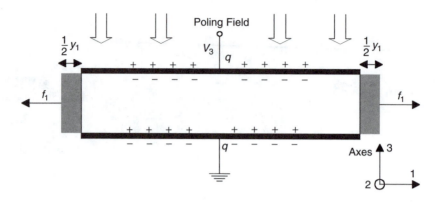

FIGURE 6.76
Piezoelectric thin film.

The film length, width, and thickness are within the first, second, and third axes, respectively. A voltage is measured (or applied) across the upper and lower electrodes, which have free charges $+q$ and $-q$. A tensile force f_1 and an elongation displacement y_1 are measured along the first axis. It must be emphasized that if piezoelectric films are used as actuators, a voltage V_3 is applied across the film, and the induced polarization in the piezoelectric generates a tensile force f_1 and a displacement y_1. If piezoelectric films are used as sensors, a tensile force f_1 is applied, stretching the film by y_1, and the induced polarization in the piezoelectric causes an increase of free charges generating a voltage V_3.

Thus, steady-state piezoelectric behavior can be modeled as

$$y_1 = s^E f_1 + d V_3, \quad q = d f_1 + c^T V_3 \text{ (Force and voltage are the left-side variables.)}$$
$$f_1 = \tfrac{1}{s^E} y_1 - e V_3, \quad q = e y_1 + c^S V_3, \text{ (Displacement and voltage are the left-side variables.)}$$

where s^E is the piezoelectric film compliance at the constant electric field; d is the piezoelectric film charge to force ratio; c^T is the piezoelectric film capacitance at the constant stress; e is the piezoelectric film charge to displacement ratio; and c^S is the piezoelectric film capacitance at the constant strain.

It is important to perform high-fidelity modeling and analysis of piezotransducers. In general piezoelectricity and piezoelectromechanical properties are described by the constitutive equations that define how the vectors of the stress ($\mathbf{T} \in \mathbb{R}^{6\times1}$, N/m²), strain ($\mathbf{S} \in \mathbb{R}^{6\times1}$), electric charge-density displacement ($\mathbf{D} \in \mathbb{R}^{3\times1}$, C/m²), and electric field ($\mathbf{E} \in \mathbb{R}^{3\times1}$, N/C) are related.

In the conventional form, there are 21 independent elastic constants, 18 independent piezoelectric constants, and six independent dielectric constants. Four forms of the steady-state piezoelectric constitutive equations are given in Table 6.3.

TABLE 6.3

Steady-State Piezoelectric Constitutive Equations

Strain-Charge Form	Stress-Charge Form	Strain-Voltage Form	Stress-Voltage Form
$\mathbf{S} = \mathbf{s}_E \cdot \mathbf{T} + \mathbf{d}^T \cdot \mathbf{E}$	$\mathbf{T} = \mathbf{c}_E \cdot \mathbf{S} - \mathbf{e}^T \cdot \mathbf{E}$	$\mathbf{S} = \mathbf{s}_D \cdot \mathbf{T} + \mathbf{g}^T \cdot \mathbf{D}$	$\mathbf{T} = \mathbf{c}_D \cdot \mathbf{S} - \mathbf{d}^T \cdot \mathbf{D}$
$\mathbf{D} = \mathbf{d} \cdot \mathbf{T} + \varepsilon_T \cdot \mathbf{E}$	$\mathbf{D} = \mathbf{e} \cdot \mathbf{S} - \varepsilon_S \cdot \mathbf{E}$	$\mathbf{E} = -\mathbf{g} \cdot \mathbf{T} + \varepsilon_T^{-1} \cdot \mathbf{D}$	$\mathbf{E} = -\mathbf{q} \cdot \mathbf{S} + \varepsilon_S^{-1} \cdot \mathbf{D}$

TABLE 6.4

Matrix Transformations for Converting Piezoelectric Constitutive Equations

Strain-Charge to Stress-Charge	Strain-Charge to Strain-Voltage	Stress-Charge to Stress-Voltage	Strain-Voltage to Stress-Voltage
$\mathbf{c}_E = \mathbf{s}_E^{-1}$	$\mathbf{s}_D = \mathbf{s}_E - \mathbf{d}^T \cdot \varepsilon_T^{-1} \cdot \mathbf{d}$	$\mathbf{c}_D = \mathbf{c}_E + \mathbf{e}^T \cdot \varepsilon_S^{-1} \cdot \mathbf{e}$	$\mathbf{c}_D = \mathbf{s}_D^{-1}$, $\mathbf{q} = \mathbf{g} \cdot \mathbf{s}_D^{-1}$
$\mathbf{e} = \mathbf{d} \cdot \mathbf{s}_E^{-1}$	$\mathbf{g} = \varepsilon_T^{-1} \cdot \mathbf{d}$	$\mathbf{q} = \varepsilon_S^{-1} \cdot \mathbf{e}$	$\varepsilon_S^{-1} = \varepsilon_T^{-1} + \mathbf{g} \cdot \mathbf{s}_D^{-1} \cdot \mathbf{g}^T$
$\varepsilon_S = \varepsilon_T - \mathbf{d} \cdot \mathbf{s}_E^{-1} \cdot \mathbf{d}^T$			

Here, $\mathbf{s} \in \mathbb{R}^{6 \times 6}$ is the matrix of compliance coefficients [m²/N]; $\mathbf{d} \in \mathbb{R}^{3 \times 6}$ is the matrix of the piezoelectric coupling coefficients [C/N]; $\mathbf{c} \in \mathbb{R}^{6 \times 6}$ is the matrix of stiffness coefficients [N/m²]; $\mathbf{e} \in \mathbb{R}^{3 \times 6}$ is the matrix of the piezoelectric coupling coefficients [C/m²]; $\mathbf{g} \in \mathbb{R}^{3 \times 6}$ is the matrix of the piezoelectric coupling coefficients [m²/C]; $\varepsilon \in \mathbb{R}^{3 \times 6}$ is the matrix of electric permittivity [F/m]; and $\mathbf{q} \in \mathbb{R}^{3 \times 6}$ is the matrix of the piezoelectric coupling coefficients [N/C].

The matrix transformations for converting piezoelectric constitutive relationships from one form into another form are given in Table 6.4.

The subscripts describe the conditions under which the piezoelectric material property data was measured. For example, the subscript E on the compliance matrix $\mathbf{s}_E$ means that the compliance data was measured under the constant (zero) electric field. The subscript S on the permittivity matrix ε_s means that the permittivity data was measured under the constant (zero) strain.

Other commonly used pairs of the constitutive equations are given in the tensor form as

$$S_{ij} = s_{ijkl}^E T_{kl} + d_{kij} E_k \quad \text{(strain-charge form)}$$
$$D_i = d_{ikl} T_{kl} + \varepsilon_{ik}^T E_k, \quad \text{or} \quad T_{ij} = c_{ijkl}^E S_{kl} - e_{kij} E_k \quad \text{(stress-charge form)}$$
$$D_i = e_{ikl} S_{kl} + \varepsilon_{ij}^S E_k, \quad \text{or} \quad S_{ij} = s_{ijkl}^D T_{kl} + g_{kij} D_k$$
$$E_i = -g_{ikl} T_{kl} + \beta_{ik}^T D_k, \quad \text{or} \quad T_{ij} = c_{ijkl}^D S_{kl} - h_{kij} D_k$$
$$E_i = -h_{ikl} S_{kl} + \beta_{ik}^S D_k$$

where S_{ij} is the mechanical strain; T_{kl} is the mechanical stress; E_k is the electric field; D_j is the electrical displacement; s_{ijkl}^E is the mechanical compliance of the material measured at the zero electric field; d_{kij} is the piezoelectric coupling between the electrical and mechanical variables; ε_{jk}^T is the dielectric permittivity measured at the zero mechanical stress; c_{ijkl}^E is the stiffness of the material measured at the zero electric field; e_{ikl} and h_{ikl} are the piezoelectric constants; $i, j, k = 1, 2, 3$; and $p, q, r = 1, 2, 3, 4, 5, 6$. (See the ANSI/IEEE Standard 176-1987 and references therein[1].)

Integrating the temperature and thermal expansion, the electric and mechanical interaction of linear piezoelectric materials is described by the constitutive relations that utilize mechanical and electrical variables. In particular, we have[1]

$$S_{ij} = s_{ijkl}^E T_{kl} + d_{ijk} E_k + \delta_{ij} \alpha_i^E \Theta \tag{6.66}$$

and

$$D_j = d_{jkl} T_{kl} + \varepsilon_{jk}^T E_k + d_{j\Theta} \Theta \tag{6.67}$$

where Θ is the temperature variable; α_i^E is the coefficient of thermal expansion under constant electric field; $d_{j\Theta}$ is the electric displacement temperature coefficient; and δ_{ij} is the Kroneker delta ($\delta_{ij} = 1$ if $i = j$, and $\delta_{ij} = 0$ otherwise).

It is evident that the stress and strain variables are second-order tensors, while the electric field and the electric displacement are first-order tensors. The thermal effects only influence the diagonal terms, and, hence, coefficients α_i and d_j have single subscripts.

The superscripts T, S, D, and E mean that the quantities are measured at zero stress ($T = 0$), zero strain ($S = 0$), zero electric displacement ($D = 0$), and zero electric field ($E = 0$). The zero electric displacement condition corresponds to open circuit (zero current across the electrodes), while the zero electric field corresponds to closed circuit (zero voltage across the electrodes).

Equation 6.66 and Equation 6.67 are the actuation equations. Equation 6.66 gives strain due to: (1) stress, (2) electric field, and (3) temperature. For the reference, the so-called induced strain actuation is $S_{ij} = d_{ijk} E_k$. Equation 6.67 gives the electric displacement (charge per unit area) required to accommodate the simultaneous influence of the: (1) stress, (2) electric field, and (3) temperature.

Equation 6.66 can be rewritten in the matrix form using the Voigt notations by arranging the stress and strain tensors as six-component vectors. (The first three components represent direct stress and strain, while the last three components represent shear stress and strain.) In particular,

$$\begin{Bmatrix} S_{11} \\ S_{22} \\ S_{33} \\ S_{23} \\ S_{31} \\ S_{12} \end{Bmatrix} \rightarrow \begin{Bmatrix} \varepsilon_{xx} \\ \varepsilon_{yy} \\ \varepsilon_{zz} \\ \varepsilon_{yz} \\ \varepsilon_{zx} \\ \varepsilon_{xy} \end{Bmatrix} \equiv \begin{Bmatrix} \varepsilon_1 \\ \varepsilon_2 \\ \varepsilon_3 \\ \varepsilon_4 \\ \varepsilon_5 \\ \varepsilon_6 \end{Bmatrix} \quad \text{and} \quad \begin{Bmatrix} T_{11} \\ T_{22} \\ T_{33} \\ T_{23} \\ T_{31} \\ T_{12} \end{Bmatrix} \rightarrow \begin{Bmatrix} \sigma_{xx} \\ \sigma_{yy} \\ \sigma_{zz} \\ \sigma_{yz} \\ \sigma_{zx} \\ \sigma_{xy} \end{Bmatrix} \equiv \begin{Bmatrix} \sigma_1 \\ \sigma_2 \\ \sigma_3 \\ \sigma_4 \\ \sigma_5 \\ \sigma_6 \end{Bmatrix}$$

Therefore, Equation 6.66 in the matrix form is

$$\begin{Bmatrix} \varepsilon_1 \\ \varepsilon_2 \\ \varepsilon_3 \\ \varepsilon_4 \\ \varepsilon_5 \\ \varepsilon_6 \end{Bmatrix} = \begin{bmatrix} S_{11} & S_{12} & S_{13} & & & \\ S_{21} & S_{22} & S_{23} & & & \\ S_{31} & S_{32} & S_{33} & & & \\ & & & S_{44} & & \\ & & & & S_{55} & \\ & & & & & S_{66} \end{bmatrix} \begin{Bmatrix} \sigma_1 \\ \sigma_2 \\ \sigma_3 \\ \sigma_4 \\ \sigma_5 \\ \sigma_6 \end{Bmatrix} + \begin{bmatrix} d_{11} & d_{21} & d_{31} \\ d_{12} & d_{22} & d_{32} \\ d_{13} & d_{23} & d_{33} \\ d_{14} & d_{24} & d_{34} \\ d_{15} & d_{25} & d_{35} \\ d_{16} & d_{26} & d_{36} \end{bmatrix} \begin{Bmatrix} E_1 \\ E_2 \\ E_3 \end{Bmatrix} + \begin{bmatrix} \alpha_1 \\ \alpha_2 \\ \alpha_3 \\ \\ \\ \end{bmatrix} \Delta\Theta \qquad (6.68)$$

or

$$\{\varepsilon\} = [s]\{\sigma\} + [d]\{E\} + [\alpha]\Delta\Theta$$

where $\Delta\Theta$ is the temperature difference.

Many of the piezoelectric coefficients d_{ij} ($i = 1, \ldots, 6; j = 1, 2, 3$) are negligibly small because the piezoelectric materials respond preferentially along certain directions. Therefore, d_{13}, d_{23}, d_{33}, d_{15}, and d_{25} are usually not zero coefficients. Using the symmetry of the compliance matrix and the above mentioned d_{ij} coefficients, we can simplify Equation 6.68. The values of the piezoelectric coupling coefficients vary significantly. These non-zero d_{13}, d_{23}, d_{33}, d_{15}, and d_{25} vary through crystal cut orientation.

In analysis of piezoelectric actuators and sensors, we use the electric field E_i and electric displacement D_i as the variables. These variables are used due to the fact that E_i and D_j directly correspond to the measured voltage and current. However, the electromagnetic theory uses the polarization P_j instead of the electric displacement D_j. The polarization, electric displacement, and electric field intensity are related using the dielectric permittivity ε as $D_i = \varepsilon E_i + P_i$. It also known that $D_i = \varepsilon_{ij} E_j$, where ε_{ij} is the effective dielectric permittivity of the material. Thus, we have $P_i = (\varepsilon_{ij} - \varepsilon) E_j = \kappa_{ij} E_j$. Equation 6.66 and Equation 6.67 are rewritten as

$$S_{ij} = s^E_{ijkl} T_{kl} + d_{ijk} E_k + \delta_{ij} \alpha^E_i \Theta \quad \text{and} \quad P_j = d_{jkl} T_{kl} + \kappa^T_{jk} E_k + p_{j\Theta} \Theta$$

where $p_{j\Theta}$ is the coefficient of pyroelectric polarization.

The physics of polarization are reported. Polarization is a phenomenon observed in dielectrics, and it gives the separation of positive and negative electric charges at different ends of the dielectric material as an external electric field is applied. The piezoelectricity is the property of a material to display electric charges on its surface under the application of an external mechanical stress. Thus, a piezoelectric material changes its polarization under stress. Piezoelectricity is related to permanent polarization, and the change in permanent polarization produces a mechanical deformation (strain). It is interesting to note that the prefix *piezo-* came from the Greek word for *force*.

Having examined the actuation equations, the sensing equations are reported. In particular, we have

$$S_{ij} = s^D_{ijkl} T_{kl} + g_{ijk} D_k + \delta_{ij} \alpha^D_i \Theta \tag{6.69}$$

and

$$E_j = g_{jkl} T_{kl} + \beta^T_{jk} D_k + e_{j\Theta} \Theta \tag{6.70}$$

where $e_{j\Theta}$ are the pyroelectric voltage coefficients and represent how much electric field is induced per unit temperature change; and g_{ijk} are the piezoelectric voltage constants (which represent how much electric field is induced per unit stress).

Equation 6.69 and Equation 6.70 are important sensing equations. For example, the tensor relationship (Equation 6.70) gives the electric field (voltage per unit thickness) generated by "squeezing" the piezoelectric material due to the piezoelectric effect.

In general the linear piezoelectricity theory is based on the first law of thermodynamics,

$$\dot{U} = T_{ij} \dot{S}_{ij} + E_i \dot{D}_i$$

where U is the stored energy density for the piezoelectric continuum.

Defining the electric enthalpy density as

$$H = U - E_i D_i$$

we have

$$\dot{H} = T_{ij}\dot{S}_{ij} - D_i\dot{E}_i, \quad T_{ij} = \frac{\partial H}{\partial S_{ij}}, \quad \text{and} \quad D_i = -\frac{\partial H}{\partial E_i}$$

Thus, we have

$$U = \tfrac{1}{2}c_{ijkl}^E S_{ij}S_{kl} + \tfrac{1}{2}\varepsilon_{ij}^S E_i E_j \quad \text{and} \quad H = \tfrac{1}{2}c_{ijkl}^E S_{ij}S_{kl} - e_{kij}E_k S_{ij} - \tfrac{1}{2}\varepsilon_{ij}^S E_i E_j$$

The catalog data provided by the piezoelectric ceramics manufacturers is given using the reported notations. The double subscript of piezoelectric coefficients relate electrical and mechanical quantities. In particular, the first subscript specifies the direction of the electrical field associated with the voltage applied or the charge produced. The second subscript specifies the direction of the mechanical stress or strain. The piezoceramic material constants are given using superscripts that specify either a mechanical or electrical boundary condition. The superscripts are T (constant or zero stress), E (constant or zero electric field), D (constant or zero charge-density displacement), and S (constant or zero strain). The piezoelectric constants relating to the mechanical strain produced by an applied electric field are called the strain constants or d coefficients. We have

$$d = \frac{\text{mechanical strain developed}}{\text{electric field applied}}$$

The superscripts describe external factors (electrical and mechanical conditions) that affect the piezoelectric (piezoelectromechanical) characteristics. The subscripts describe the relationship of the piezoelectric (piezoelectromechanical) characteristics with respect to the poling axis. The subscripts define the axes of a component in terms of orthogonal axes. (1 corresponds to the x axis, 2 to the y axis, and 3 to the z axis.) The first subscript gives the direction of the action, and the second subscript gives the direction of the response. For example, for the piezoelectric coupling constant d, the first subscript refers to the direction of the electric field and the second refers to the direction of the strain. For the converse piezoelectric constant g, the first subscript refers to the stress and the second subscript refers to the voltage (electric field).

The mechanical stiffness properties of the piezoelectric ceramics are described using Young's modulus, which is the ratio of stress (force per unit area) to strain (change in length per unit length), and

$$Y = \frac{\text{stress}}{\text{strain}}$$

Mechanical stress produces the electric response (voltage) that opposes the resultant strain. Therefore, the effective Young's modulus with short-circuited electrodes is lower than with the open-circuited electrodes. It should be emphasized that the stiffness is different in the third direction from that in the first and second directions. These effects must be distinguished, described, and characterized. For example, Y_{33}^E and Y_{33}^D are the ratios of stress to strain (Young's modulus) in the third direction at constant electric field E (electrodes short-circuited) and if the electrodes are open-circuited.

Due to the wide range of operating conditions, the piezoelectric materials must be properly chosen, studying dielectric, ferroelectric, electromechanical, thermal, and other characteristics.

The piezoelectric coefficients can be measured in accordance with the ANSI/IEEE 176-1987 standard. In engineering applications, using the Impedance Analyzer, one finds the effective electromechanical coupling coefficient k_e as

$$k_e = \sqrt{\frac{f_{max}^2 - f_{min}^2}{f_{max}^2}}$$

where f_{max} and f_{min} are the maximum and minimum impedance frequencies.

The piezoelectric coefficients for the transverse (k_{31}, d_{31}, and g_{31}) and longitudinal (k_{33}, d_{33}, and g_{33}) modes of operation are found by measuring the resonance properties of the thickness-poled and length-poled ceramics. In particular, these constants are found as

$$k_{31} = \sqrt{\frac{\frac{1}{2}\pi \frac{f_{min}}{f_{max}} \tan\left(\frac{1}{2}\pi \frac{f_{max}-f_{min}}{f_{min}}\right)}{1 + \frac{1}{2}\pi \frac{f_{min}}{f_{max}} \tan\left(\frac{1}{2}\pi \frac{f_{max}-f_{min}}{f_{min}}\right)}} \, , \quad s_{11}^E = \frac{1}{4\rho f_{min}^2 l^2}$$

$$d_{31} = k_{31}\sqrt{\varepsilon_0 k_3 s_{11}^E} \, , \quad g_{31} = \frac{d_{31}}{\varepsilon_0 k_3}$$

$$k_{33} = \sqrt{\frac{\pi}{2}\frac{f_m}{f_n} \tan\left(\frac{\pi}{2}\frac{f_n-f_m}{f_n}\right)} \, , \quad s_{33}^D = \frac{1}{4\rho f_{max}^2 l^2} \, , \quad s_{33}^E = \frac{s_{33}^D}{1-k_{33}^2}$$

$$d_{33} = k_{33}\sqrt{\varepsilon_0 k_3 s_{33}^E} \, , \quad g_{33} = \frac{d_{33}}{\varepsilon_0 k_3}$$

Here, k_{31} and k_{33} are the transverse and extensional coupling coefficients; s_{11} and s_{33} are the elastic compliance constants [m²/N] (superscripts D and E denote the measurements at the constant electric displacement and electric field); d_{31} and d_{33} are the transverse and extensional strain constants [m/V]; g_{31} and g_{33} are the transverse and extensional voltage constants [V-m/N]; k_3 is the dielectric constant; l is the specimen length [m]; ε_0 is the permittivity of free space, $\varepsilon_0 = 8.85 \times 10^{-12}$ F/m; and ρ is the density [kg/m³].

The dielectric, ferroelectric, and piezoelectric properties of piezoelectric ceramics (PZT-4 and PZT-5 with the density ρ = 7.5 kg/cm³) for T = 25°C assuming the dimensional requirements—the diameter of the circular piezoelectric material is at least 10 times greater than the thickness, which is in the micrometer range, and the length of the rectangular piezoelectric material is at least three times greater than the width—are given below in Table 6.5.

Almost all piezoelectric materials exhibit lower dielectric constant values at low temperature, and as the temperature increases, the dielectric constant increases. For example, for the PZT-4 and PZT-5A, the dielectric constant increases steadily as a function of temperature up to the Curie temperature, and the temperature range is from –150 to 250°C. It must be emphasized that in general, the dielectric constants are nonlinear functions of the temperature. For example, for the PZT-5H ceramics, the maximum dielectric constant is observed at 185°C. The dissipation factors depend on the temperature and the measurement frequency. In particular, for PZT-4, tan δ = 0.05 (measured from 1 to 100 kHz) over the temperature range from –150 to 180°C, and for PZT-5A, the tan δ varies from 0.015 to 0.025

TABLE 6.5

Dielectric (at 1000 Hz), Ferroelectric, and Piezoelectric Properties of the Piezoelectric Ceramics

	k_3	E_c (kV/cm)	P_R (μC/cm^2)	P_{sat} (μC/cm^2)	k_e	d_{33} (m/V)	g_{33} (V-m/N)	k_{33}	d_{31} (m/V)	G_{31} (V-m/N)	k_{31}
PZT4	1400	14	31	40	0.5	22.5×10^{-12}	0.009	0.35	-85×10^{-12}	-0.008	0.22
PZT5	3400	5.8	13	20	0.5	590×10^{-12}	0.0013	0.6	-270×10^{-12}	-0.009	0.37

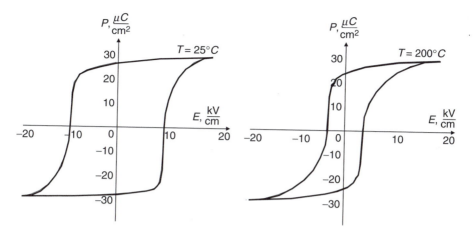

FIGURE 6.77
Ferroelectric polarization versus electric field characteristics (*P-E* curves) of the PZT-5A piezoceramics and 25°C and 200°C.

(measured at 100 Hz to 10 kHz) over the entire temperature range (from −150 to 250°C). For PZT-5H, the tan δ is a nonlinear function of the temperature, and the tan δ varies from 0.01 to 0.05 (at 100 Hz to 10 kHz) with the peak value 0.05 at 165°C (10 kHz). It must be emphasized that for PZT-5H, the tan δ varies from 0.035 to 0.17 (at 100 kHz) with the peak value 0.17 at 185°C. The resistivity of the PZT-4 and PZT-5 also varies as a function of temperature. For example, though relatively small (up to 20%) variations observed for the temperature range from −150 to 100°C for the PZT-4, the resistivity decreases in the high temperature region. The polarization is significantly influenced by the temperature. Usually the maximum remanent polarization occurs in the temperature range from 0 to 50°C. The polarization versus electric field characteristics, given as the *P-E* curves, of the PZT-5A at temperatures 25°C and 200°C are illustrated in Figure 6.77.

The discussion and data provided indicate that thorough consideration must be carried out, selecting the piezoelectric actuators for given operating conditions. To illustrate another trade-off, it should be emphasized that the power consumption must also be studied. In particular, the power *P*, consumed by the piezoelectric actuators at the sinusoidal operations is proportional to the capacitance *C*, operating frequency *f*, and the squired peak-to-peak operating voltage V_{pp}. In particular,

$$P = \pi C f V_{pp}^2$$

In micro- and nanoelectromechanical systems, devices, and structures, PZT thin films are widely used. The characteristics of sol-gel thin films depend on the orientation of $PbZr_{1-x}Ti_x$ substrates. For example, sol-gel PZT films on Pt(111)/Ti/SiO$_2$/Si are 100-oriented. However, 111-oriented PZT thin films can be made on the same substrate when the films are pyrolyzed at 350°C. The PZT ($PbZr_{65}Ti_{35}$, $PbZr_{53}Ti_{47}$, and $PbZr_{35}Ti_{65}$) thin films, made on Pt(111)/SiO$_2$/Si and Pt(200)/SiO$_2$/Si substrates, exhibit the hysteresis. (It must be emphasized that all piezoelectric ceramics exhibit the hysteresis effect.) The ferroelectric polarization versus voltage (electric field) characteristics (*P-V* or *P-E* curves) are illustrated in Figure 6.78.

The saturation magnetization (*B-H* curves) and the force-displacement curves are similar to the *P-E* curves. Thus, the hysteresis is the important effect to be thoroughly studied.

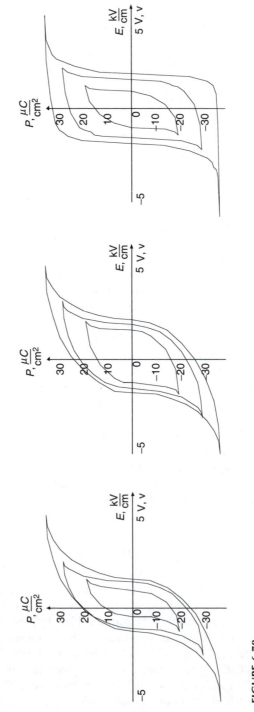

FIGURE 6.78
Ferroelectric polarization versus applied voltage (electric field) characteristics (*P-V* or *P-E* curves) for the sol-gel PZT (PbZr$_{65}$Ti$_{35}$, PbZr$_{53}$Ti$_{47}$, and PbZr$_{35}$Ti$_{65}$) thin films.

In addition to the nonlinear effects, parameter variations, and other phenomena and effects observed in piezotransducers, it is extremely important to integrate the piezotransducer dynamics. In fact, the steady-state analysis does not allow one to fully examine the system performance and make a conclusion based on requirements and specifications imposed. This is particularly important in micro- and nanopositioning applications (vibration and noise attenuation and control, actuators and drives, etc.).

6.10.2 Mathematical Models of Piezoactuators: Dynamics and Nonlinear Equations of Motion

The mathematical models of piezotransducers are given in the form of differential equations that allow the designer to attain steady-state and dynamic analysis.

Consider a circular plate piezoactuator (polarized ceramics of classes C_3, C_{3v}, C_6, or C_{6v}, circular surface normal to the three- or sixfold axis) assuming that the actuator surfaces are completely covered with electrodes. (See Figure 6.79.)

The following partial differential equation models the dynamic motion (radial displacement x_r) of the circular piezoactuator[1]

$$\frac{\partial^2 x_r}{\partial r^2} + \frac{1}{r}\frac{\partial x_r}{\partial r} - \frac{x_r}{r^2} = \rho \frac{\left(s_{11}^E\right)^2 - \left(s_{12}^E\right)^2}{s_{11}^E}\frac{\partial^2 x_r}{\partial t^2}$$

with the constitutive equations

$$T_{rr} = \frac{s_{11}^E}{\left(s_{11}^E\right)^2 - \left(s_{12}^E\right)^2}\frac{\partial x_r}{\partial r} - \frac{s_{12}^E}{\left(s_{11}^E\right)^2 - \left(s_{12}^E\right)^2}\frac{x_r}{r} + \frac{1}{t_x}\frac{d_{31}}{s_{11}^E + s_{12}^E}V_3$$

$$T_{rr}\big|_{r=R} = 0$$

$$D_3 = \frac{d_{31}}{s_{11}^E + s_{12}^E}\frac{1}{r}\frac{\partial(rx_r)}{\partial r} + \frac{1}{t_x}\left(\frac{2d_{31}^2}{s_{11}^E + s_{12}^E} - \varepsilon_{33}^T\right)V_3$$

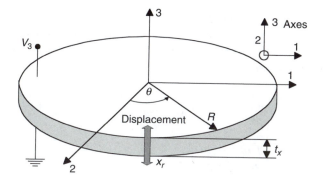

FIGURE 6.79
Circular plate piezoactuator.

These equations can be simulated in the high-performance environments (e.g., MATLAB and MATEMATICA), and, making use of the piezoceramic parameters, the coefficients of the equations are straightforwardly found.

In addition to high-fidelity modeling, which is based on the solution of the partial differential equations, lumped-parameter models of piezotransducers provide manageable and practical results. It must be emphasized that the hysteresis phenomena cannot be neglected and thus must be integrated in mathematical modeling, simulation, analysis, and control.

The differential equations of motion of piezoelectric actuators are represented as

$$m\frac{d^2x}{dt^2} + k_x\frac{dx}{dt} + k_1x + k_2x^2 + k_3x^3 = k_vV - k_zz$$

and the hysteresis equation is

$$\frac{dz}{dt} = k_{z1}\frac{dV}{dt} - k_{z2}\left|\frac{dV}{dt}\right|z - k_{z3}\frac{dV}{dt}|z|$$

Here, x is the actuator displacement; V is the applied voltage; z is the hysteresis variable; and k_l are the time-varying coefficients.

Other contemporary equations of motion can also be derived and used to attain accurate analysis and design. For example, nonlinear dynamics with the hysteresis effect can be modeled as

$$m\frac{d^2x}{dt^2} + k_x\frac{dx}{dt} + k_1x + k_2x^2 + k_3x^3 = k_vV - k_ff$$

$$\frac{df}{dt} = \frac{dx}{dt} - \frac{1}{k_{fx}}\left|\frac{dx}{dt}\right|f$$

where k_f and k_{fx} are the coefficients that form and shape the hysteresis curve.

It must be emphasized that using the available high-performance software environments, the documented differential equations can be easily solved. The challenges result in the design of control algorithms. In particular, the hysteresis effect degrades the piezoactuator's performance. Therefore, nonlinear phenomena must be integrated in analysis, control, and optimization.

In the x-y-z plane, if three actuators are used, one has sets of nonlinear diffe-rential equations. For example, for the identical PZT-4 piezoactuators with diameter 2 mm and thickness 0.01 mm, we have the following set of nonlinear differential equations (using lumped-parameter model of piezoactuators):

$$\frac{d^2x}{dt^2} + k_x\frac{dx}{dt} + k_1x + k_2x^2 + k_3x^3 = k_vV_x$$

$$\frac{d^2y}{dt^2} + k_y\frac{dy}{dt} + k_1y + k_2y^2 + k_3y^3 = k_vV_y$$

$$\frac{d^2z}{dt^2} + k_y\frac{dz}{dt} + k_1z + k_2z^2 + k_3z^3 = k_vV_z$$

TABLE 6.6

Properties of Terfenol-D

μ_{r33}^T	μ_{r33}^s	d_{33}	k_{33}	Elastic Modulus c_{33}^H (N/mm²)	Elastic Modulus c_{33}^B (N/mm²)	Comprehensive Strength T_t (N/mm²)	Tensile Strength T_p (N/mm²)
9.4	4.5	1.5×10^{-8}	0.75	28×10^3	52×10^3	710	29

where V_x, V_y, and V_z are the applied voltages to the x-, y-, and z-axis piezoactuators to displace actuators.

The coefficients of the differential equations are

$$k_x = k_y = 1000, \quad k_1 = 1 \times 10^7, \quad k_2 = -1 \times 10^6, \quad k_3 = 5 \times 10^4, \quad \text{and} \quad k_v = 1.5 \times 10^5$$

The simulation, using the MATLAB environment, can be straightforwardly performed to study the dynamic and steady-state responses of the piezotransducers using the piezo-electromechanical characteristics reported, as well as the given differential equations.

The giant magnetostrictive materials, fabricated using rare-earth alloys, have enhanced electromechanical characteristics, compared with PZT. Table 6.6 documents some Terfenol-D ($Tb_{0.3}Dy_{0.7}Fe_2$) characteristics for $T = 25°C$. It must be emphasized that Terfenol-D (which has a Curie temperature of 380°C and density $\rho = 9.3 \times 10^3$ kg/m³) is frequently used to fabricate high-performance magnetostrictive transducers.

Having discussed the basic fundamentals, the fabrication should be emphasized and covered. Electroactive ceramics display strong piezoelectric and electrostrictive response. Electroactive ceramics consist of polycrystalline structures of ferroelectric perovskites with strong piezoelectric and electrostrictive properties. The electroactive ceramics are synthetic compounds that can be fabricated with the desired characteristics and phenomena to meet specified requirements. Fabrication of ferroelectric ceramics has been performed as sequential steps: (1) synthesis of the ferroelectric perovskite powders; (2) sintering and compaction of the perovskite powders into ferroelectric ceramics, and (3) electric poling of the ferroelectric ceramics. Consider fabrication of the PZT compound $Pb(Zr_xTi_{1-x})O_3$. The synthesis of the ferroelectric perovskite compound utilizes the oxide mixing process. A stoichiometric mixture of lead oxide (PbO), zirconium dioxide (ZrO_2), and titanium dioxide (TiO_2) powders is calcined in an oven at a temperature from 800 to 1000°C (below the melting point of all constituents) for 1–2 hours. The calcination process causes a solid-state reaction that results in aggregates of the PZT perovskite. These PZT perovskite aggregates are crushed into smaller sizes and milled into a fine powder. During Step 2, PZT perovskite powder is mixed with a binder mix and laid in thin sheets. These preforms are sintered and compacted. A ferroelectric ceramic is obtained through sintering fine particles of ferroelectric PZT perovskite to attain sufficient mechanical strength in the final product without distorting significantly from the initial shape. During sintering, diffusion of the constituent atoms takes place at the interface between particles. This diffusion reconstructs the crystal bonding at the interface and provides binding. The diffusion rate is controlled by changing the temperature. The sintering of PZT perovskite powder is performed in a furnace at 1200°C. The third step is electric poling. During cooling, the PZT ceramics undergo phase transformation from a paraelectric state to a ferroelectric state. (This transformation takes place as the material cools to below the Curie temperature.) The resulting ferroelectric ceramic has a polycrystalline structure with randomly oriented ferroelectric domains. Due to the random orientation of the electric domains, individual polarizations cancel each other, and the net polarization of the virgin ferroelectric ceramic is zero. This random orientation should be transformed into a preferred orientation (net polarization) using the poling process. Poling of piezoceramics is performed at elevated temperatures at high electric DC field 1000 to 4000 V/mm in order to align the crystalline domains. This alignment is permanent as the piezoceramic is cooled (permanent polarization). During poling, the orientation of the piezoelectric domains also results in mechanical deformation.

A number of novel fabrication methods and processes have been developed to enhance characteristics of ferroelectric ceramics. For example, doping improves the solid-reaction process. Coprecipitation and alkoxide hydrolysis (sol-gel method) are wet chemical methods that can produce better perovskite powder precursors. Thin-film fabrications are applied to obtain very thin films of ferroelectric perovskites. Thin-film fabrication methods are classified as liquid-phase epitaxial methods (which promote the growth of the perovskite layer on the face of a substrate such that the perovskite layer has the same crystal orientation as the substrate) and gas-phase methods (electron beam sputtering, RF sputtering, and chemical vapor deposition). Single-crystal fabrication has been used for the growth of quartz (SiO_2) crystals through hydrothermal synthesis, as well as $LiNbO_3$ and $LiTaO_3$ perovskites by the Czochralski method.

6.11 Modeling of Electromagnetic Radiating Energy Microdevices

The electromagnetic power is generated and radiated by antennae. Time-varying current radiates electromagnetic waves (radiated electromagnetic fields). Radiation pattern, beam width, directivity, and other major characteristics can be studied using Maxwell's equations [5–9, 13, 14, 17]. We use the vectors of the electric field intensity *E*, electric flux density *D*, magnetic field intensity *H*, and magnetic flux density *B*. The constitutive equations are

$$\mathbf{D} = \varepsilon\mathbf{E} \quad \text{and} \quad \mathbf{B} = \mu\mathbf{H}$$

where ε is the permittivity, and μ is the permeability.

It was shown in Section 2.2 that in the static (time-invariant) fields, electric and magnetic field vectors form separate and independent pairs. That is, *E* and *D* are not related to *H* and *B*, and vice versa. However, for time-varying electric and magnetic fields, we have the following fundamental electromagnetic equations:

$$\nabla \times \mathbf{E}(x,y,z,t) = -\mu\frac{\partial \mathbf{H}(x,y,z,t)}{\partial t}$$

$$\nabla \times \mathbf{H}(x,y,z,t) = \sigma\mathbf{E}(x,y,z,t) + \varepsilon\frac{\partial \mathbf{E}(x,y,z,t)}{\partial t} + \mathbf{J}(x,y,z,t)$$

$$\nabla \cdot \mathbf{E}(x,y,z,t) = \frac{\rho_v(x,y,z,t)}{\varepsilon}$$

$$\nabla \cdot \mathbf{H}(x,y,z,t) = 0$$

where *J* is the current density, and using the conductivity σ, we have $\mathbf{J} = \sigma\mathbf{E}$; and ρ_v is the volume charge density.

The total current density is the sum of the source current $\mathbf{J}_S$ and the conduction current density $\sigma\mathbf{E}$ (due to the field created by the source $\mathbf{J}_s$). Thus,

$$\mathbf{J}_\Sigma = \mathbf{J}_S + \sigma\mathbf{E}$$

The equation of conservation of charge (continuity equation) is

$$\oint_s \mathbf{J}\cdot d\mathbf{s} = -\frac{d}{dt}\int_v \rho_v dv$$

and in the point form one obtains

$$\nabla\cdot\mathbf{J}(x,y,z,t) = -\frac{\partial\rho_v(x,y,z,t)}{\partial t}$$

Therefore, the net outflow of current from a closed surface results in decrease of the charge enclosed by the surface.

Electromagnetic waves transfer the electromagnetic power. That is, the energy is delivered by means of electromagnetic waves. Using

$$\nabla\times\mathbf{E} = -\mu\frac{\partial\mathbf{H}}{\partial t} \quad\text{and}\quad \nabla\times\mathbf{H} = \varepsilon\frac{\partial\mathbf{E}}{\partial t} + \mathbf{J}$$

we have

$$\nabla\cdot(\mathbf{E}\times\mathbf{H}) = \mathbf{H}\cdot(\nabla\times\mathbf{E}) - \mathbf{E}\cdot(\nabla\times\mathbf{H}) = -\mathbf{H}\cdot\mu\frac{\partial\mathbf{H}}{\partial t} - \mathbf{E}\cdot\left(\varepsilon\frac{\partial\mathbf{E}}{\partial t} + \mathbf{J}\right)$$

In a medium where the constituent parameters are constant (time invariant) we have the so-called point-function relationship:

$$\nabla\cdot(\mathbf{E}\times\mathbf{H}) = -\frac{\partial}{\partial t}\left(\tfrac{1}{2}\varepsilon E^2 + \tfrac{1}{2}\mu H^2\right) - \sigma E^2$$

In integral form one obtains

$$\oint_s (\mathbf{E}\times\mathbf{H})\cdot d\mathbf{s} = \underbrace{-\frac{\partial}{\partial t}\int_v\left(\tfrac{1}{2}\varepsilon E^2 + \tfrac{1}{2}\mu H^2\right)dv}_{\substack{\text{time–rate of change of energy stored in}\\ \text{the electric field E and magnetic field H}}} - \underbrace{\int_v \sigma E^2 dv}_{\substack{\text{ohmic power dissipated}\\ \text{in the presence of E}}}$$

The right side of the derived equation gives the rate of decrease of the electric and magnetic energies stored minus the ohmic power dissipated as heat in the volume v. The pointing vector, which is a power density vector, represents the power flows per unit area, and

$$\mathbf{P} = \mathbf{E}\times\mathbf{H}$$

Furthermore,

$$\oint_s (\mathbf{E} \times \mathbf{H}) \cdot d\mathbf{s} = \oint_s \mathbf{P} \cdot d\mathbf{s} = \frac{\partial}{\partial t} \int_v (w_E + w_H) dv + \int_v \rho_\sigma dv$$

$$\underbrace{\phantom{\oint_s (\mathbf{E} \times \mathbf{H}) \cdot d\mathbf{s}}}_{\text{power leaving the enclosed volume}}$$

where $w_E = \frac{1}{2}\varepsilon E^2$ and $w_H = \frac{1}{2}\mu H^2$ are the electric and magnetic energy densities; and $\rho_\sigma = \sigma E^2 = (1/\sigma)J^2$ is the ohmic power density.

The important conclusion is that the total power transferred into a closed surface s at any instant equals the sum of the rate of increase of the stored electric and magnetic energies and the ohmic power dissipated within the enclosed volume v.

If the source charge density $\rho_v(x,y,z,t)$ and the source current density $\mathbf{J}(x,y,z,t)$ vary sinusoidally, the electromagnetic fields also vary sinusoidally. Hence, we have to deal with the so-called time-harmonic electromagnetic fields. The sinusoidal time-varying electromagnetic fields will be studied. Hence, the phasor analysis is applied. For example,

$$\mathbf{E}(\mathbf{r}) = E_x(\mathbf{r})\mathbf{a}_x + E_y(\mathbf{r})\mathbf{a}_y + E_z(\mathbf{r})\mathbf{a}_z$$

The electric field intensity components are the complex functions, and

$$E_x(\mathbf{r}) = E_{x\,\text{Re}} + jE_{x\,\text{Im}}, \quad E_y(\mathbf{r}) = E_{y\,\text{Re}} + jE_{y\,\text{Im}}, \quad E_z(\mathbf{r}) = E_{z\,\text{Re}} + jE_{z\,\text{Im}}$$

For the real electromagnetic field, we have

$$E_x(\mathbf{r},t) = E_{x\,\text{Re}}(\mathbf{r})\cos \omega t - E_{x\,\text{Im}}(\mathbf{r})\sin \omega t$$

One obtains the time-harmonic electromagnetic field equations. In particular,

Faraday's law $\nabla \times \mathbf{E} = -j\omega\mu\mathbf{H}$

Generalized (by Maxwell's) Amphere's law $\nabla \times \mathbf{H} = \sigma\mathbf{E} + j\omega\varepsilon\mathbf{E} + \mathbf{J} = j\omega(\frac{\sigma}{j\omega} + \varepsilon)\mathbf{E} + \mathbf{J}$

Gauss's law $\nabla \cdot \mathbf{E} = \dfrac{\rho_v}{\frac{\sigma}{j\omega} + \varepsilon}$

Continuity of magnetic flux $\nabla \cdot \mathbf{H} = 0$

Continuity law $\nabla \cdot \mathbf{J} = -j\omega\rho_v$ (6.71)

where $(\sigma/j\omega + \varepsilon)$ is the complex permittivity. However, for simplicity we will use ε, keeping in mind that the expression for the complex permittivity $\sigma/j\omega + \varepsilon$ must be applied.

The electric field intensity E, electric flux density D, magnetic field intensity H, magnetic flux density B, and current density J are complex-valued functions of spatial coordinates.

From Equation 6.71, taking the curl of $\nabla \times \mathbf{E} = -j\omega\mu\mathbf{H}$, which is rewritten as $\nabla \times \mathbf{E} = -j\omega\mathbf{B}$, and using $\nabla \times \mathbf{H} = j\omega\mathbf{D} + \mathbf{J}$, one obtains

$$\nabla \times \nabla \times \mathbf{E} = \omega^2 \mu\varepsilon\mathbf{E} = k_v^2\mathbf{E} = -j\omega\mu\mathbf{J}$$

where k_v is the wave constant, $k_v = \omega\sqrt{\mu\varepsilon}$, and in free space $k_{v0} = \omega\sqrt{\mu_0\varepsilon_0} = \omega/c$ because the speed of light is $c = 1/\sqrt{\mu_0\varepsilon_0}$, $c = 3 \times 10^8$ m/sec.

The wavelength is found as

$$\lambda_v = \frac{2\pi}{k_v} = \frac{2\pi}{\omega\sqrt{\mu\varepsilon}}$$

and in free space

$$\lambda_{v0} = \frac{2\pi}{k_{v0}} = \frac{2\pi c}{\omega}$$

Using the magnetic vector potential A, we have $\mathbf{B} = \nabla \times \mathbf{A}$.

Hence, $\nabla \times (\mathbf{E} + j\omega\mathbf{A}) = 0$, and thus $\mathbf{E} + j\omega\mathbf{A} = -\nabla\Lambda$, where Λ is the scalar potential. To guarantee that $\nabla \times \mathbf{H} = j\omega\mathbf{D} + \mathbf{J}$ holds, it is required that

$$\nabla \times \mu\mathbf{H} = \nabla \times \nabla \times \mathbf{A} = \nabla\nabla \cdot \mathbf{A} - \nabla^2\mathbf{A} = j\omega\mu\varepsilon\mathbf{E} + \mu\mathbf{J}$$

Therefore, one finally finds the equation that must be solved:

$$\nabla^2\mathbf{A} + k_v^2\mathbf{A} = \nabla(\nabla \cdot \mathbf{A} + j\omega\mu\varepsilon\Lambda) - \mu\mathbf{J}$$

Taking note of the Lorentz condition, $\nabla \cdot \mathbf{A} = -j\omega\mu\varepsilon\Lambda$, one obtains $\nabla^2\mathbf{A} + k_v^2\mathbf{A} = -\mu\mathbf{J}$. Thus, the equation for Λ is found. In particular,

$$\nabla^2\Lambda + k_v^2\Lambda = -\frac{\rho_v}{\varepsilon}$$

The equation for the magnetic vector potential is found solving the following inhomogeneous Helmholtz equation: $\nabla^2\mathbf{A} + k_v^2\mathbf{A} = -\mu\mathbf{J}$. The expression for the electromagnetic field intensity, in terms of the vector potential, is

$$\mathbf{E} = -j\omega\mathbf{A} + \frac{\nabla\nabla \cdot \mathbf{A}}{j\omega\mu\varepsilon}$$

To derive $\mathbf{E}$, one must have $\mathbf{A}$. The Laplacian for $\mathbf{A}$ in different coordinate systems can be found. For example, we have

$$\nabla^2 A_x + k_v^2 A_x = -\mu J_x$$

$$\nabla^2 A_y + k_v^2 A_y = -\mu J_y$$

$$\nabla^2 A_z + k_v^2 A_z = -\mu J_z$$

It was shown that the magnetic vector potential and the scalar potential obey the time-dependent inhomogeneous wave equation

$$\left(\nabla^2 - k\frac{\partial^2}{\partial t^2}\right)\Omega(\mathbf{r}, t) = -F(\mathbf{r}, t)$$

The solution of this equation is found using Green's function as

$$\Omega(\mathbf{r},t) = -\int\int\int\int F(\mathbf{r}',t')G(\mathbf{r}-\mathbf{r}';t-t')dt'd\tau'$$

where

$$G(\mathbf{r}-\mathbf{r}';t-t') = -\frac{1}{4\pi|\mathbf{r}-\mathbf{r}'|}\delta(t-t'-k|\mathbf{r}-\mathbf{r}'|)$$

The so-called retarded solution is

$$\Omega(\mathbf{r},t) = -\int\int\int \frac{F(\mathbf{r}',t'-k|\mathbf{r}-\mathbf{r}'|)}{|\mathbf{r}-\mathbf{r}'|}d\tau'$$

For sinusoidal electromagnetic fields, we apply the Fourier analysis to obtain

$$\Omega(\mathbf{r}) = -\frac{1}{4\pi}\int\int\int \frac{e^{-jk_v|\mathbf{r}-\mathbf{r}'|}}{|\mathbf{r}-\mathbf{r}'|}F(\mathbf{r}')d\tau'$$

Thus, we have the expressions for the phasor retarded potentials:

$$\mathbf{A}(\mathbf{r}) = \frac{\mu}{4\pi}\int_v \frac{e^{-jk_v|\mathbf{r}-\mathbf{r}'|}}{|\mathbf{r}-\mathbf{r}'|}\mathbf{J}(\mathbf{r}')dv$$

$$\mathbf{\Lambda}(\mathbf{r}) = \frac{1}{4\pi\varepsilon}\int_v \frac{e^{-jk_v|\mathbf{r}-\mathbf{r}'|}}{|\mathbf{r}-\mathbf{r}'|}\rho(\mathbf{r}')dv$$

Example 6.33

Consider a short (*dl*) thin filament of current located in the origin. (See Figure 6.80.) Derive the expressions for magnetic vector potential and electromagnetic field intensities.

SOLUTION

The magnetic vector potential has only a *z* component, and thus, from $\nabla^2\mathbf{A}+k_v^2\mathbf{A}=-\mu\mathbf{J}$, we have

$$\nabla^2 A_z + k_v^2 A_z = -\mu J_z = -\mu\frac{i}{ds}$$

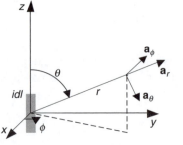

FIGURE 6.80
Current filament in the spherical coordinate system.

where ds is the cross-sectional area of the filament.

Taking note of the spherical symmetry, we conclude that the magnetic vector potential A_z is not a function of the polar and azimuth angles θ and ϕ. In particular, the following equation results:

$$\frac{1}{r^2}\frac{\partial}{\partial r}r^2\frac{\partial A_z}{\partial r}+k_v^2A_z=0$$

It is well known that the solution of equation $d^2\psi/d\psi^2+k_v^2\psi=0$ has two components. In particular, $e^{jk_v r}$ (outward propagation) and $e^{-jk_v r}$ (inward propagation). The inward propagation is not a part of solution for the filament located in the origin. Thus, we have $\psi(t,r)=ae^{j\omega t-jk_v r}$ (outward propagating spherical wave).

In free space, we have $\psi(t,r)=ae^{j\omega(t-r/c)}$.

Substituting $A_z=\psi/r$, one obtains

$$A_z(r)=\frac{a}{r}e^{-\frac{j\omega r}{c}}$$

To find the constant a, we use the volume integral

$$\int_v \nabla^2 A_z dv=\oint_s \nabla A_z\cdot\mathbf{a}_r r_d^2\sin\theta d\theta d\phi=-\frac{\omega^2}{c^2}\int_v A_z dv-\mu_0\int_v J_z dv$$

where the differential spherical volume is $dv=r_d^2\sin\theta d\theta d\phi dr$; and r_d is the differential radius.

Making use of

$$\nabla A_z\cdot\mathbf{a}_r=\frac{\partial A_z}{\partial r}=-\left(1+j\frac{\omega}{c}r\right)\frac{a}{r^2}e^{-j\frac{\omega}{c}r}$$

we have

$$\lim_{r_d\to 0}\int_0^{2\pi}\int_0^{\pi}-\left(1+j\frac{\omega}{c}r_d\right)ae^{-j\frac{\omega}{c}r_d}\sin\theta d\theta d\phi=-4\pi a=-\mu_0 idl$$

$$a=\frac{\mu_0 idl}{4\pi}$$

Thus, the following expression results

$$A_z(r)=\frac{\mu_0 idl}{4\pi r}e^{-\frac{j\omega r}{c}}$$

Therefore, the final equation for the magnetic vector potential (outward propagating spherical wave) is

$$\mathbf{A}(r)=\frac{\mu_0 idl}{4\pi r}e^{-\frac{j\omega r}{c}}\mathbf{a}_z$$

From $\mathbf{a}_z = \mathbf{a}_r \cos\theta - \mathbf{a}_\theta \sin\theta$, we have

$$\mathbf{A}(r) = \frac{\mu_0 idl}{4\pi r} e^{-\frac{j\omega r}{c}} (\mathbf{a}_r \cos\theta - \mathbf{a}_\theta \sin\theta)$$

The magnetic and electric field intensities are found using

$$\mathbf{B} = \nabla \times \mathbf{A} \quad \text{and} \quad \mathbf{E} = -j\omega\mathbf{A} + \frac{\nabla\nabla\cdot\mathbf{A}}{j\omega\mu\varepsilon}$$

Then one finds

$$\mathbf{H}(r) = \frac{1}{\mu_0}\nabla \times \mathbf{A}(r) = \frac{idl\sin\theta}{4\pi r}\left(\frac{j\omega}{cr} + \frac{1}{r^2}\right) e^{-\frac{j\omega r}{c}} \mathbf{a}_\phi$$

$$\mathbf{E}(r) = \frac{j\sqrt{\frac{\mu_0}{\varepsilon_0}}cidl}{4\pi\omega}\cos\theta\left(\frac{j\omega}{cr^2} + \frac{1}{r^3}\right)e^{-\frac{j\omega r}{c}}\mathbf{a}_r - \frac{j\sqrt{\frac{\mu_0}{\varepsilon_0}}cidl}{4\pi\omega}\sin\theta\left(-\frac{\omega^2}{c^2 r} + \frac{j\omega}{cr^2} + \frac{1}{r^3}\right)e^{-\frac{j\omega r}{c}}\mathbf{a}_\theta$$

The intrinsic impedance is given as

$$Z_0 = \sqrt{\frac{\mu_0}{\varepsilon_0}}, \quad \text{and} \quad Y_0 = \frac{1}{Z_0} = \sqrt{\frac{\varepsilon_0}{\mu_0}}$$

Near-field and far-field electromagnetic radiation fields can be found, simplifying the expressions for $\mathbf{H}(r)$ and $\mathbf{E}(r)$.

For near-field, we have

$$\mathbf{H}(r) = \frac{1}{\mu_0}\nabla \times \mathbf{A}(r) = j\frac{idl\sin\theta\omega}{4\pi cr^2}e^{-\frac{j\omega r}{c}}\mathbf{a}_\phi$$

and

$$\mathbf{E}(r) = j\frac{\sqrt{\frac{\mu_0}{\varepsilon_0}}cidl\omega}{4\pi c^2 r}\sin\theta\mathbf{a}_\theta$$

The complex Pointing vector can be found as

$$\tfrac{1}{2}\mathbf{E}(r) \times \mathbf{H}^*(r)$$

The following expression for the complex power flowing out of a sphere of radius r results:

$$\tfrac{1}{2}\oint_s (\mathbf{E}(r) \times \mathbf{H}^*(r)) \cdot d\mathbf{s} = \frac{\omega^2 \mu_0 \sqrt{\mu_0 \varepsilon_0} i^2 dl^2}{12\pi} = \frac{\omega\mu_0 k_v i^2 dl^2}{12\pi}$$

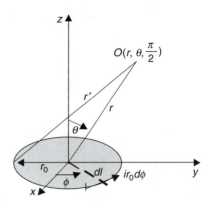

FIGURE 6.81
Current loop in the *x-y* plane.

The real quality is found, and the power is dissipated in the sense that it travels away from source and cannot be recovered.

Example 6.34

Derive the expressions for the magnetic vector potential and electromagnetic field intensities for a magnetic dipole (small current loop), which is shown in Figure 6.81.

SOLUTION

The magnetic dipole moment is

$$\mathbf{M} = \pi r_0^2 i \mathbf{a}_z = M \mathbf{a}_z$$

For the short current filament, it was derived in Example 6.33 that

$$\mathbf{A}(r) = \frac{\mu_0 i dl}{4\pi r} e^{-\frac{j\omega r}{c}} \mathbf{a}_z$$

In contrast, we have

$$\mathbf{A} = \frac{\mu_0 i}{4\pi} \oint_l \frac{1}{r'} dl$$

The distance between the source element *dl* and point $O(r, \theta, \frac{\pi}{2})$ is denoted as r'. It should be emphasized that the current filament lies in the *x-y* plane, and

$$dl = \mathbf{a}_\phi r_0 d\phi = (-\mathbf{a}_x \sin\phi + \mathbf{a}_y \cos\phi) r_0 d\phi$$

Thus, due to symmetry

$$\mathbf{A} = \mathbf{a}_\phi \frac{\mu_0 i r_0}{2\pi} \int_{-\pi/2}^{\pi/2} \frac{\sin\phi}{r'} d\phi$$

where using the trigonometric identities one finds

$$r'^2 = r^2 + r_0^2 - 2rr_0 \sin\theta \sin\phi$$

Assuming that $r^2 \gg r_0^2$, we have

$$\frac{1}{r'} \approx \frac{1}{r}\left(1 + \frac{r_0}{r}\sin\theta\sin\phi\right).$$

Therefore,

$$\mathbf{A} = \mathbf{a}_\phi \frac{\mu_0 i r_0}{2\pi} \int\limits_{-\pi/2}^{\pi/2} \frac{\sin\phi}{r'}d\phi = \mathbf{a}_\phi \frac{\mu_0 i r_0}{2\pi r} \int\limits_{-\pi/2}^{\pi/2} \left(1 + \frac{r_0}{r}\sin\theta\sin\phi\right)\sin\phi\,d\phi = \mathbf{a}_\phi \frac{\mu_0 i r_0^2}{4r^2}\sin\theta$$

Having obtained the explicit expression for the vector potential, the magnetic flux density is found. In particular,

$$\mathbf{B} = \nabla \times \mathbf{A} = \nabla \times \mathbf{a}_\phi \frac{\mu_0 i r_0^2}{4r^2}\sin\theta = \frac{\mu_0 i r_0^2}{4r^3}(2\mathbf{a}_r\cos\theta + \mathbf{a}_\theta\sin\theta)$$

Taking note of the expression for the magnetic dipole moment $\mathbf{M} = \pi r_0^2 i \mathbf{a}_z$, one has

$$\mathbf{A} = \mathbf{a}_\phi \frac{\mu_0 i r_0^2}{4r^2}\sin\theta = \frac{\mu_0}{4\pi r^2}\mathbf{M} \times \mathbf{a}_r$$

It was shown that using

$$\mathbf{A} = \frac{\mu_0 i}{4\pi}\oint\limits_l \frac{1}{r'}dl$$

the desired results are obtained.

Let us apply

$$\mathbf{A} = \frac{\mu_0 i}{4\pi}\oint\limits_l \frac{e^{-j\frac{\omega}{c}r'}}{r'}dl$$

From

$$e^{-j\frac{\omega}{c}r'} \approx \left[1 - j\frac{\omega}{c}(r'-r)\right]e^{-j\frac{\omega}{c}r}$$

we have

$$\mathbf{A} = \frac{\mu_0 i}{4\pi}\oint\limits_l \frac{[1 - j\frac{\omega}{c}(r'-r)]e^{-j\frac{\omega}{c}r}}{r'}dl = \mathbf{a}_\phi \frac{\mu_0 M}{4\pi r^2}\left(1 + j\frac{\omega}{c}r\right)e^{-j\frac{\omega}{c}r}\sin\theta$$

Therefore, one finds

$$E_\phi = j \frac{\mu_0 \omega^3 M}{4c^2 \pi} \left(\frac{1}{j\frac{\omega}{c} r} - \frac{1}{\frac{\omega^2}{c^2} r^2} \right) e^{-j\frac{\omega}{c} r} \sin\theta$$

$$H_r = j \frac{2\mu_0 \omega^3 M}{4c^2 \sqrt{\frac{\mu_0}{\varepsilon_0}} \pi} \left(\frac{1}{\frac{\omega^2}{c^2} r^2} + \frac{1}{j\frac{\omega^3}{c^3} r^3} \right) e^{-j\frac{\omega}{c} r} \cos\theta$$

$$H_\theta = -j \frac{\mu_0 \omega^3 M}{4c^2 \sqrt{\frac{\mu_0}{\varepsilon_0}} \pi} \left(\frac{1}{j\frac{\omega}{c} r} - \frac{1}{\frac{\omega^2}{c^2} r^2} - \frac{1}{j\frac{\omega^3}{c^3} r^3} \right) e^{-j\frac{\omega}{c} r} \sin\theta$$

The electromagnetic fields in near- and far-fields can be straightforwardly derived, and thus the corresponding approximations for E_ϕ, H_r, and H_θ can be obtained.

Let the current density distribution in the volume be given as $J(r_0)$, and for far-field from Figure 6.82 one has $r \approx r' - r_0$.

The formula to calculate far-field magnetic vector potential is

$$\mathbf{A}(\mathbf{r}) = \frac{\mu}{4\pi r} e^{-jk_v r} \int_v \mathbf{J}(\mathbf{r}_0) e^{-jk_v \mathbf{r}_0} dv$$

and the electric and magnetic field intensities are found using

$$\mathbf{E} = -j\omega \mathbf{A} + \frac{\nabla \nabla \cdot \mathbf{A}}{j\omega \mu \varepsilon}$$

and

$$\mathbf{B} = \Delta \times \mathbf{A}.$$

We have

$$\mathbf{E}(\mathbf{r}) = \frac{jk_v Z_v}{4\pi r} e^{-jk_v r} \int_v [\mathbf{a}_r \cdot \mathbf{J}(\mathbf{r}_0) \mathbf{a}_r - \mathbf{J}(\mathbf{r}_0)] e^{-jk_v \mathbf{r}_0} dv$$

$$\mathbf{H}(\mathbf{r}) = Y_v \mathbf{a}_r \times \mathbf{E}(\mathbf{r})$$

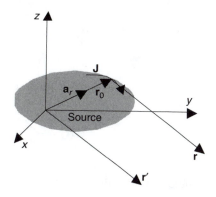

FIGURE 6.82
Radiation from volume current distribution.

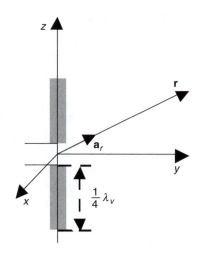

FIGURE 6.83
Half-wave dipole antenna.

Example 6.35

Consider a half-wave dipole antenna fed from a two-wire transmission line, as shown in Figure 6.83. The antenna is one-quarter wavelength; that is, $-\frac{1}{4}\lambda_v \le z \le \frac{1}{4}\lambda_v$. The current distribution is $i(z) = i_0 \cos k_v z$. Obtain the equations for electromagnetic field intensities and radiated power.

SOLUTION

The wavelength is given as

$$\lambda_v = \frac{2\pi}{k_v} = \frac{2\pi}{\omega\sqrt{\mu\varepsilon}}$$

Thus, in free space we have

$$\lambda_{v0} = \frac{2\pi}{k_{v0}} = \frac{2\pi c}{\omega}$$

It was emphasized that $k_{v0} = \omega\sqrt{\mu_0\varepsilon_0}$.

Making use of

$$\mathbf{E}(\mathbf{r}) = \frac{jk_v Z_v}{4\pi r} e^{-jk_v r} \int_v \left[\mathbf{a}_r \cdot \mathbf{J}(\mathbf{r}_0)\mathbf{a}_r - \mathbf{J}(\mathbf{r}_0) \right] e^{-jk_v \mathbf{r}_0} dv$$

we have the following line integral:

$$\mathbf{E}(\mathbf{r}) = \frac{jk_v Z_v}{4\pi r} e^{-jk_v r} \oint_l \left[(\mathbf{a}_r \cdot \mathbf{a}_l)\mathbf{a}_r - \mathbf{a}_l \right] i(l) e^{-jk_v \mathbf{r}_0} dl$$

where $\mathbf{a}_l$ is the unit vector in the current direction.

Then

$$\mathbf{E}(\mathbf{r}) = \frac{jk_v Z_v i_0}{4\pi r} e^{-jk_v r} \int_{-\frac{1}{4}\lambda_v}^{\frac{1}{4}\lambda_v} (\mathbf{a}_r \cos\theta - \mathbf{a}_z) \cos k_v z e^{-jk_v z \cos\theta} dz = \frac{jZ_v i_0}{2\pi r \sin\theta} \frac{\cos(\frac{1}{2}\pi\cos\theta)}{} e^{-jk_v r} \mathbf{a}_\theta$$

Having found the magnetic field intensity as

$$\mathbf{H}(\mathbf{r}) = Y_v \mathbf{a}_r \times \mathbf{E}(\mathbf{r}) = H_\phi \mathbf{a}_\phi = \frac{ji_0 \cos(\frac{1}{2}\pi\cos\theta)}{2\pi r \sin\theta} e^{-jk_v r} \mathbf{a}_\phi$$

the power flux per unit area is

$$\frac{1}{2}\text{Re}\left(\mathbf{E}(\mathbf{r}) \times \mathbf{H}(\mathbf{r})^* \cdot \mathbf{a}_r\right) = \frac{1}{2}E_\phi H_\phi^* = \frac{|i_0|^2 Z_0 \cos^2(\frac{1}{2}\pi\cos\theta)}{8\pi^2 r^2 \sin^2\theta}$$

Integrating the derived expression over the surface

$$\frac{|i_0|^2 Z_0}{8\pi^2} \int_0^{2\pi}\int_0^\pi \frac{\cos^2(\frac{1}{2}\pi\cos\theta)}{\sin^2\theta} \sin\theta \, d\theta \, d\phi$$

the total radiated power is found to be $36.6|i_0|^2$.

If the current density distribution is known, the radiation field can be found. Using Maxwell's equations, taking note of the electric and magnetic vector potentials $\mathbf{A}_E$ and $\mathbf{A}_H$, we have the following equations:

$$\left(\nabla^2 + k_v^2\right)\mathbf{A}_E = -\mu\mathbf{J}_E, \quad \left(\nabla^2 + k_v^2\right)\mathbf{A}_H = -\varepsilon\mathbf{J}_H$$

$$\mathbf{E} = -j\omega\mathbf{A}_E + \frac{1}{j\omega\mu\varepsilon}\nabla\nabla\cdot\mathbf{A}_E - \frac{1}{\varepsilon}\nabla\times\mathbf{A}_H$$

$$\mathbf{H} = -j\omega\mathbf{A}_H + \frac{1}{j\omega\mu\varepsilon}\nabla\nabla\cdot\mathbf{A}_H - \frac{1}{\mu}\nabla\times\mathbf{A}_E$$

The solutions are

$$\mathbf{A}_E(\mathbf{r}) = \frac{\mu}{4\pi r} e^{-jk_v r} \int_v e^{jk_v \mathbf{r}} \mathbf{J}_E(\mathbf{r}) d\mathbf{r}$$

$$\mathbf{A}_H(\mathbf{r}) = \frac{\varepsilon}{4\pi r} e^{-jk_v r} \int_v e^{jk_v \mathbf{r}} \mathbf{J}_H(\mathbf{r}) d\mathbf{r}$$

Example 6.36

Consider the slot (one-half wavelength long slot is dual to the half-wave dipole antenna studied in Example 6.35), which is exited from the coaxial line. (See Figure 6.84.)

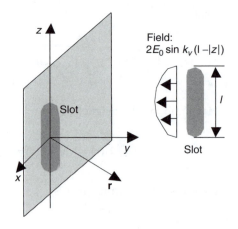

FIGURE 6.84
Slot antenna.

The electric field intensity in the z-direction is $E = E_0 \sin k_v(l - |z|)$. Derive the expressions for the magnetic vector potential and electromagnetic field intensities.

SOLUTION

Using the magnetic current density $\mathbf{J}_H$, from

$$\int_s \nabla \times \mathbf{E} \cdot d\mathbf{s} = \oint_l \mathbf{E} \cdot dl = -\int_s j\omega\mathbf{B} \cdot d\mathbf{s} - \int_s \mathbf{J}_H \cdot d\mathbf{s}$$

the boundary conditions for the magnetic current sheet are found as

$$\mathbf{a}_n \times \mathbf{E}_1 - \mathbf{a}_n \times \mathbf{E}_2 = -\mathbf{J}_H$$

The slot antenna is exited by the magnetic current with strength $2E_0 \sin k_v(l - |z|)$ in the z-axis.

For half-wave slot we have $i_H = i_0 \sin k_v(l - |z|)$, and

$$\left(\nabla^2 + k_v^2\right)\mathbf{A}_H = -\varepsilon\mathbf{J}_H$$

$$\mathbf{H} = -j\omega\mathbf{A}_H + \frac{\nabla\nabla \cdot \mathbf{A}_H}{j\omega\mu\varepsilon} = j\frac{i_0 Y_0 \cos(\frac{1}{2}\pi\cos\theta)}{2\pi r \sin\theta} e^{-jk_v r}\mathbf{a}_\theta$$

$$\mathbf{E} = -\frac{\nabla \times \mathbf{A}_H}{\varepsilon} = -j\frac{i_0 \cos(\frac{1}{2}\pi\cos\theta)}{2\pi r \sin\theta} e^{-jk_v r}\mathbf{a}_\phi$$

$$\mathbf{A}_H(\mathbf{r}) = \frac{\varepsilon}{16\pi} e^{-jk_v r}\int_s e^{jk_v \mathbf{r}}\mathbf{J}(\mathbf{r})ds$$

The boundary condition

$$\mathbf{a}_n \times \mathbf{E} = -\tfrac{1}{2}\mathbf{J}_H = \mathbf{a}_n \times \mathbf{a}_x E_0 \sin k_v\left(l - |z|\right)$$

is satisfied by the radiated electromagnetic field.

The radiation pattern of the slot antenna is the same as for the dipole antenna.

6.12 Thermodynamics, Thermoanalysis, and Heat Equation

Thermodynamics is the study of the behavior and patterns of energy changes. (*Thermo* refers to energy and *dynamics* means behavior changes.) Many books have been written in thermodynamics, and this section does not intend to rewrite any existing outstanding books. Thus, only introductory issues are covered here. Discussing exchanges that can occur between the system and environment, let us mention energy exchange (heat, work, friction, radiation, etc.) and matter exchange (movement of molecules across the boundary of the system and environment). The systems can be defined as isolated systems (no exchange of matter or energy), closed systems (no exchange of matter but exchange of energy), and open systems (exchange of both matter and energy). One should also note distinct thermodynamic processes that can be reversible and irreversible. In general thermodynamics describes the behavior of systems containing a large number of particles. The analysis covered in this section is of particular importance to MEMS and NEMS, where thermodynamic phenomena are very important and processes must be coherently described. Thermodynamics is a science that comprises the study of energy transformations and relationships among the various physical qualities or properties that are affected by these transformations. Micro- and nanosystems can be examined using temperature, sizing features, materials, number and the type of particles, uniformity, environment, etc. It is impossible to study MEMS and NEMS using individual particles (atoms), and from a modeling viewpoint it is valuable to examine mechanical, electromechanical, electronic, optical, chemical, biological, and other systems from a generic viewpoint.

The first law of thermodynamics states that when a closed system is altered adiabatically, the work is the same for all possible paths that connect two given equilibrium states. That is, in the closed adiabatic system, the work depends on the initial and end states. This concept can be extended to the open systems (many MEMS and NEMS are open systems) where the transfer exits the boundaries. The first law of thermodynamics is a conservation law for energy transformations. Regardless of the type of energy involved (mechanical, thermal, electrical, elastic, etc.) the change in the energy of a system is equal to the difference between energy input and energy output.

The concept of equilibrium is very important in the analysis of properties of matter and the conservation of energy. The second law of thermodynamics states that any system having specific constraints and bounds can reach from any initial state a stable equilibrium state with no net effect on the environment. This law is a directional and limiting law because process tends to proceed in one direction, attaining a final equilibrium state, and may not be reversed. From this standpoint, the state of systems can be described by the total energy, entropy, and other parameters.

The change in energy can be caused by heat, work, or particles, and the mathematical expression is

$$dE = dH + dW + \eta dN$$

where E is the total energy; H is the heat; W is the work; η is the energy added to a system when adding one particle without adding either heat or work; and N is the number of particles.

Fermi energy (E_F) is the energy associated with a particle, which is in thermal equilibrium with the system of interest and does not consist of heat or work energies. The Fermi function $f(E)$,

$$f(E) = \frac{1}{1 + e^{\frac{E - E_F}{k_B T}}}$$

applies to all particles with half-integer spin. (These particles are called Fermions and obey the Pauli exclusion principle. That is, no two Fermions in a given system can have the exact same set of quantum numbers.) Here k_B is the Boltzmann constant. Considering an energy level at energy E, which is in thermal equilibrium with a large system characterized by a temperature T and Fermi energy E_F, $f(E)$ gives the probability that an electron occupies such energy level.

It is known that the heat propagates (flows) in the direction of decreasing temperature, and the rate of propagation is proportional to the gradient of the temperature. Using the thermal conductivity of the media k_t and the temperature $T(t, x, y, z)$, one has the following equation to calculate the velocity of the heat flow:

$$\vec{\mathbf{v}}_h = -k_t \nabla T(t, x, y, z) \tag{6.72}$$

Consider the region $\mathbf{R}$ and let s be the boundary surface. Using the divergence theorem, from Equation 6.72 one obtains the partial differential equation (heat equation), which is expressed as

$$\frac{\partial T(t, x, y, z)}{\partial t} = k^2 \nabla^2 T(t, x, y, z) \tag{6.73}$$

where k is the thermal diffusivity of the media, $k = k_t / k_h k_d$; and k_h and k_d are the specific heat and density constants.

Solving Equation 6.73, which is subject to initial and boundary conditions, one finds the temperature of the homogeneous media. In the Cartesian coordinate system, one has

$$\nabla^2 T(t, x, y, z) = \frac{\partial^2 T(t, x, y, z)}{\partial x^2} + \frac{\partial^2 T(t, x, y, z)}{\partial y^2} + \frac{\partial^2 T(t, x, y, z)}{\partial z^2}$$

Using the Laplacian of T in the cylindrical and spherical coordinate systems, one can reformulate the thermoanalysis problem using different coordinates in order to straight-forwardly solve the problem.

If the heat flow is steady (time-invariant), then $\partial T(t, x, y, z)/\partial t = 0$.

Hence, the three-dimensional heat equation (Equation 6.73) becomes Laplace's equation, as given by $k^2 \nabla^2 T(t, x, y, z) = 0$.

The two-dimensional heat equation is

$$\frac{\partial T(t, x, y)}{\partial t} = k^2 \nabla^2 T(t, x, y) = k^2 \left(\frac{\partial^2 T(t, x, y)}{\partial x^2} + \frac{\partial^2 T(t, x, y)}{\partial y^2} \right)$$

If $\partial T(t, x, y)/\partial t = 0$, one has

$$k^2 \nabla^2 T(t, x, y) = k^2 \left(\frac{\partial^2 T(t, x, y)}{\partial x^2} + \frac{\partial^2 T(t, x, y)}{\partial y^2} \right) = 0$$

Using initial and boundary conditions, this partial differential equation can be solved using Fourier series, Fourier integrals, or Fourier transforms.

The so-called one-dimensional heat equation is

$$\frac{\partial T(t, x)}{\partial t} = k^2 \frac{\partial^2 T(t, x)}{\partial x^2}$$

with initial and boundary conditions $T(t_0, x) = T_t(x)$, $T(t, x_0) = T_0$, and $T(t, x_f) = T_f$.

A large number of analytical and numerical methods are available to solve the heat equation.

The analytic solution of $T(t, x_0) = 0$ and $T(t, x_f) = 0$ is given as

$$T(t, x) = \sum_{i=1}^{\infty} B_i \sin \frac{i \pi x}{x_f} e^{-i^2 \frac{k^2 \pi^2}{x_f^2} t}$$

$$B_i = \frac{2}{x_f} \int_{x_0}^{x_f} T_t(x) \sin \frac{i \pi x}{x_f} dx$$

Assuming that $T_t(x)$ is piecewise continuous in $x \in [x_0 \ x_f]$ and has one-sided derivatives at all interior points, one finds the coefficients of the Fourier sine series B_i.

Example 6.37

Consider the copper bar with length 0.1 mm. The thermal conductivity, specific heat, and density constants are $k_t = 1$, $k_h = 0.09$, and $k_d = 9$. The initial and boundary conditions are $T(0, x) = T_t(x) = 0.2 \sin(\pi x / 0.001)$, $T(t, 0) = 0$, and $T(t, 0.001) = 0$. Find the temperature in the bar as a function of the position and time.

SOLUTION

From the general solution

$$T(t, x) = \sum_{i=1}^{\infty} B_i \sin \frac{i \pi x}{x_f} e^{-i^2 \frac{k^2 \pi^2}{x_f^2} t}$$

using the initial condition given, we have

$$T(0, x) = \sum_{i=1}^{\infty} B_i \sin \frac{i \pi x}{x_f} = 0.2 \sin \frac{\pi x}{0.001}$$

Thus, $B_1 = 0.2$ and all other B_i coefficients are zero. Hence, the solution for the temperature is found to be

$$T(t, x) = \sum_{i=1}^{\infty} B_i \sin \frac{i \pi x}{x_f} e^{-i^2 \frac{k^2 \pi^2}{x_f^2} t} = B_1 \sin \frac{\pi x}{x_f} e^{-\frac{k^2 \pi^2}{x_f^2} t} = 0.2 \sin \frac{\pi x}{0.001} e^{-1.5 \times 10^7 t}$$

It is evident that the temperature is a nonlinear function of the position x and time t.

Homework Problems

1. A spherical electrostatic actuator, as documented in Figure 6.85, is designed using spherical conducting shells separated by the flexible material. (For example, parylene, teflon, and polyethylene have relative permittivity ε_r from 2 to 3).

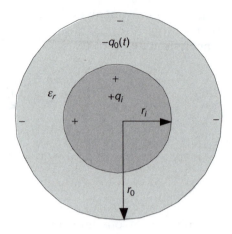

FIGURE 6.85
Magnetic levitation systems.

The inner shell has total charge $+q_i$ and diameter r_i. The charge of the outer shell $q_o(t)$ is seminegative and time varying. The diameter of the outer shell is denoted as r_o.

a. Derive the expression for the capacitance and calculate the numerical value for capacitance if $r_i = 50 \ \mu m$, $r_i = 100 \ \mu m$, $q_i = 1 \ \mu C$, and $q_o(t) = 10(\sin(t) - 1)$.

b. Derive the expression for the electrostatic force and calculate the force between the inner and outer shells.

c. For a flexible material (parylene, teflon, or polyethylene), find the resulting displacements. (You should use the expression for the restoring force of the flexible medium chosen.)

d. Assign the dimensions of the microactuator, and perform the design. Choose the materials that can be used to fabricate this device emphasizing high performance, packaging, and affordability. (You can hypothetically assume that you are able to utilize any fabrication technologies, processes, and materials, but please keep in mind that the size, thermal expansion, permittivity, and other factors should be accounted for.) Develop the possible technology (processes, materials, and chemicals) to fabricate the studied microactuator.

2. Assume that the magnetic field is uniform, and the top and low magnets or current loop vibrate such that the magnetic field relative to the current loop varies as $B(t) = \sin(t) + \sin(2t) + \cos(5t) + \sin(t)*\cos^2(t)$. That is, permanent magnet(s) can be stationary or movable, and conduction current loop can be movable and stationary. For example, they can be attached to any vibrating structures. The variations of the magnetic field are shown in Figure 6.86. The area of the conducting current loop in the magnetic field is $100 \ \mu m^2$. The total circuit resistance is 10 ohm.

a. Find the induced (generated) emf and the current in the circuit. Assume that the number of turns is 1 and 100. How does the emf change if $N = 1$ and $N = 100$?

Note that the Partial Differential Toolbox in the MATLAB environment can be used to plot the resulting variables, perform differentiation, and complete any calculations. In particular, the following MATLAB statement is helpful:

```
t=sym('t','positive');
B=sin(t)+sin(2*t)+cos(5*t)+sin(t).* cos(t)^2;
ezplot(B); dBdt=diff(B)
```

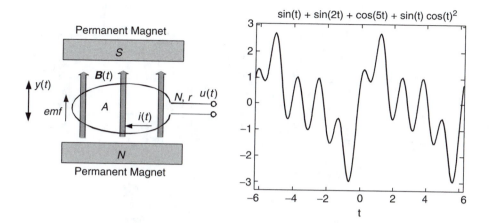

FIGURE 6.86
Microgeneration system and assumed variations of the magnetic field.

The resulting plot for the magnetic field and derivative for the magnetic field density are found. Specifically, in the MATLAB command window we have the result

```
dBdt =

cos(t)=2*cos(2*t)-5*sin(5*t)=cos(t)^3-2*sin(t)^2*cos(t)
```

b. Assign the dimensions of the microdevice and perform the design. Choose the materials that can be used to fabricate this device emphasizing high performance, packaging, and affordability. (You can hypothetically assume that you are able to utilize any fabrication technologies, processes, and materials, but please keep in mind that the size, thermal expansion, magnetic properties, and other factors should be accounted for.) Develop the possible technology (processes, materials, and chemicals) to fabricate the microgeneration device.

3. One-, two- and three-dimensional magnetic levitation systems have been foreseen to be used in control of underwater and flight vehicles (so-called moving mass concept). Microscale magnetic levitation systems can be utilized in micropropulsion systems. For a magnetic levitation system with a ball, as illustrated in Figure 6.87,

a. Derive a mathematical model. (You can apply Newton's and Kirchhoff's laws as well as find the expression for the electromagnetic force developed by the microelectromagnets using the coenergy.) Note that the reluctance in the magnetic system varies. You may use literature resources or approximate the reluctance (inductance) using the analytic expression, for example, $L(x) = k_1/(k_2 + x)$, where k_1 and k_2 are constants. You can also develop the magnetic circuit analog in order to derive the mathematical model. However, mathematical models can be found using electromagnetic laws, and thus you may decide not to use the magnetic circuit analogs.

b. Assign magnetic levitation system dimensions and derive the parameters that are used in the mathematical model you derived. Depending on the dimensions you assign, distinct levitation system designs will result. For example, say the length is 1000 μm. Assuming that the diameter of copper wire is 20 μm,

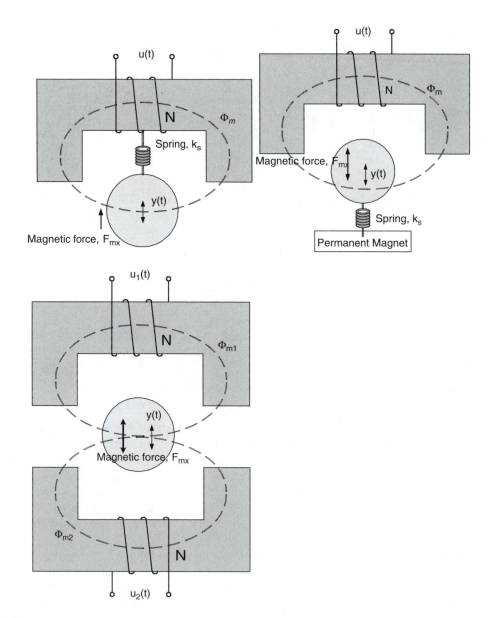

FIGURE 6.87
Magnetic levitation systems.

one layer winding can include 25–30 turns, but you may have multilayered winding. (Note that the standard AWG 39 copper wire is 0.09 mm in diameter with the current carry density of 0.02 A.) The geometry (shape) and diameter of the moving mass (ball) and its media density will result in the value for mass m, etc. You should use the parameters for media such as resistivity (define winding resistance), permeability (stationary member and moving mass), etc.

c. Using any software environment, develop the files and perform numerical simulations of the studied magnetic levitation systems. Analyze the results. For example, simulate it in MATLAB, analyze the dynamics, and make conclusions.

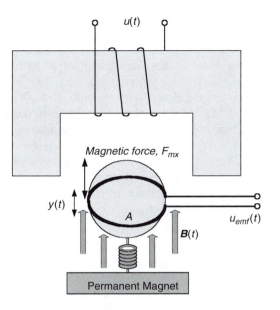

FIGURE 6.88
Magnetic levitation systems with the conducting current loop.

  d. Develop the possible technology (processes, materials, and chemicals) to fabricate the magnetic levitation system.

  e. Examine what will happen if the suspended mass (coated with an insulator) is surrounded by N turns conducting current loop as illustrated in Figure 6.88. Assume that the current loop has a resistance r and area A.

 4. Consider a microsolenoid as documented in Figure 6.89. This solenoid integrates movable (plunger) and stationary (fixed) members made from high-permeability ferromagnetic materials. The windings are wound in a helical pattern.

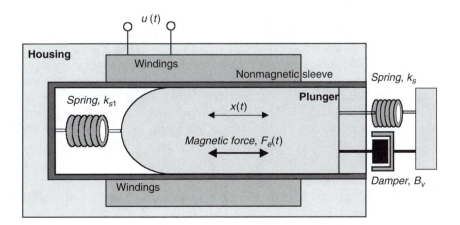

FIGURE 6.89
Schematic of a microsolenoid.

The performance of solenoids strongly depends on the electromagnetic system, mechanical geometry, magnetic permeability, winging resistivity, inductance, friction, etc. When the voltage is supplied to the winding, current flows in the winding, and the electromagnetic force is developed causing the plunger to move. When the applied voltage becomes zero, the plunger may resume its original position due to the springs (assuming that there are no residual magnetism and static friction). Different materials for the sleeve and plunger coating should be chosen to attain minimum friction and minimize wear. Glass-filled nylon and brass for the guide and copper plating or other low friction coatings for the plunger and lubrication are possible solutions. In many cases you must trade off a variety of mechanical, electrical, thermal, acoustical, and other physical properties in order to attain the desired overall microsolenoid characteristics.

a. Assign the dimensions of microsolenoid (say 500 μm length and 100 μm outer diameter) and make the estimates for all other dimensions. Choose the materials that can be used to fabricate your microsolenoid, emphasizing high performance, packaging, and affordability. (You can hypothetically assume that you are able to utilize any fabrication technologies, processes, and materials, but please keep in mind that the size, thermal expansion, permeability, electromagnetic loads, and other factors should be accounted for.) Develop the possible technology (processes, materials, and chemicals) to fabricate the proposed microsolenoid.

b. Calculate the inductance and winding resistance as well as other parameters of interest that will be used in modeling, simulation, and analysis, in particular, friction coefficient, relative permeabilities of movable and stationary members, resistivity of materials, spring constants, etc.

c. Derive a mathematical model. (You can apply Newton's and Kirchhoff's laws as well as find the expression for the electromagnetic force developed using coenergy.) Note that the reluctance in the magnetic system varies as a function of $x(t)$. You can also develop the magnetic circuit analog in order to derive the mathematical model. However, mathematical models can be found using electromagnetic laws, and thus you may decide not to use the magnetic circuit analogs.

d. Using any software environment, develop the files and perform numerical simulations of the studied microsolenoid. Analyze the results. For example, simulate it in MATLAB or ANSYS, analyze the dynamics, and make conclusions.

5. Consider the microelectromechanical motion device as illustrated in Figure 6.90. (Let the outer diameter be 500 μm). Note that in general you have N turns. (Figure 6.90 illustrates only one turn). Use $N = 5$. Design the fabrication processes to make the microdevice using surface or buck micromachining technologies. (Document the needed masks and describe fabrication processes involved using 100 μm Si wafers.) Please note that permanent magnets can be glued to the structure and may (or may not) be fabricated using MEMS fabrication technologies. Consider both. You can develop masks using Mentor Graphics or other software. (Document your masks and drawings.)

    a. Choose the materials to design and fabricate a high-performance microelectromechanical motion device. The critical data (from design and fabrication) must be reported.

    b. Classify the studied microdevice using the MEMS classifier.

    c. For the proposed permanent-magnet micromotor (carrying the high-performance design as the major goal), perform the electromagnetic (energy, **B**, ψ_m, etc.),

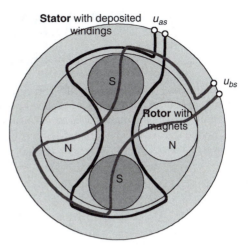

FIGURE 6.90
Schematic of a two-phase permanent-magnet synchronous micromotor.

      mechanical and vibroacoustic analysis (for example thermal analysis, vibration, etc.) relevant to the design.

d. For the given volume, perform the energy estimates using the scaling laws. That is, use the energy densities $\rho_{We} = \frac{1}{2}\varepsilon E^2$ and $\rho_{Wm} = \frac{1}{2}(B^2/\mu) = \frac{1}{2}\mu H^2$ for electrostatic and magnetic (assigned) motion microdevices. Using energy-based analysis, estimate the electromagnetic torque. Make the conclusions regarding electrostatic versus magnetic motion devices.

References

1. An American National Standard, *IEEE Standard on Piezoelectricity*, ANSI/IEEE Standard 176-1987, IEEE Inc., 1987.
2. Berg, H.C., The rotary motor of bacterial flagella. *J. Annual Rev. Biochem.*, 72, 19–54, 2003.
3. Drexler, E.K., *Nanosystems: Molecular Machinery, Manufacturing, and Computations*, Wiley-Interscience, New York, 1992.
4. Collin, R.E., *Antennas and Radiowave Propagation*, McGraw-Hill, New York, 1985.
5. Giurgiutiu, V. and Lyshevski, S.E., *Micromechatronics: Modeling, Analysis, and Desing With MATLAB®*, CRC Press, Boca Raton, FL, 2004.
6. Hayt, W.H., *Engineering Electromagnetics*, McGraw-Hill, New York, 1989.
7. Ikeda, T., *Fundamentals of Piezoelectricity*, Oxford University Press, 1996.
8. Janocha, H., *Adaptronics and Smart Structures*, Springer Verlag, 1999.
9. Krause, J.D. and Fleisch, D.A., *Electromagnetics With Applications*, McGraw-Hill, New York, 1999.
10. Krause, P.C. and Wasynczuk, O., *Electromechanical Motion Devices*, McGraw-Hill, New York, 1989.
11. Kuo, B.C., *Automatic Control Systems*, Prentice Hall, Englewood Cliffs, NJ, 1995.
12. Lines, M.E. and Glass, A.M., *Principles and Applications of Ferroelectrics and Related Materials*, Clarendon Press, Oxford, 2001.
13. Lyshevski, S.E., *NEMS and NEMS: Systems, Devices, and Structures*, CRC Press, Boca Raton, FL, 2002.
14. Lyshevski, S.E., *Micro- and Nanoelectromechanical Systems, Fundamentals of Micro- and Nanoengineering*, CRC Press, Boca Raton, FL, 2000.
15. Lyshevski, S.E., *Engineering and Scientific Computations Using MATLAB®*, Wiley, Hoboken, NJ, 2003.

16. MATLAB 6.5 Release 13, CD-ROM, MathWorks, Inc., 2002.

17. Paul, C.R., Whites, K.W., and Nasar, S.A., *Introduction to Electromagnetic Fields*, McGraw-Hill, New York, 1998.

18. Seeman, N.C., DNA engineering and its application to nanotechnology, *Nanotechnology, 17,* 437–443, 1999.

19. Krygowski, T.W., Rodgers, M.S., Sniegowski, J.J., Miller, S.M., and Jakubczak, J., A low-voltage actuator fabricated using a five-level polysilicon surface micromachining technology, in *Technical Digest - International Electron Devices Meeting,* pp. 697–700, 1999.

20. Yasseen, A.A., Mitchell, J.N., Klemic, J.F., Smith, D.A., and Mehregany, M., Rotary electrostatic micromotor 1_8 optical switch, *IEEE Jour. on Selected Topics in Quan. Electron.,* 5, 1, 26–32, 1999.

7

Quantum Mechanics and Its Applications

7.1 Atomic Structures and Quantum Mechanics

7.1.1 Introduction

Utilizing fundamental results in nanosystems, applied research, engineering, and technology have undergone major developments in recent years. High-performance nanostructures and microdevices have been fabricated, tested, and implemented (accelerometers, microphones, actuators, sensors, molecular wires, transistors, etc.). Nanoengineering and nanoscience study nanosystems, as well as their subsystems, devices, and structures that are made from atoms, molecules, or molecular complexes. Students and engineers have obtained the necessary background in physics classes. Complex phenomena, effects, properties, and performance of materials (media) are examined and understood through the analysis of the atomic structure, and the electron is considered a fundamental particle.

The atomic structures were studied by Planck, Rutherford, and Einstein (in the 1900s), Heisenberg and Dirac (in the 1920s), Schrödinger, Bohr, and many other scientists. In 1910 Louis de Broglie examined Einstein's theory of relativity and the photoelectric effect studying the wave behavior of light. The theory of quantum electrodynamics studies the interaction of electrons and photons. In the 1940s, the major breakthrough appears in augmentation of the electron dynamics with electromagnetic fields. One can control molecules and molecular complexes (nanostructures) using electromagnetic fields. Micro- and nanoscale devices have been fabricated, and some problems in structural design and optimization have been addressed and solved. However, these nano- and microscale devices must be controlled, and one faces an extremely challenging problem to design nanosystems integrating control, optimization, self-organization, decision making, diagnostics, self-repairing, signal processing, communication, and other functional features. All media are made of atoms. The medium properties and observed phenomena depend on the atomic structure. Recalling Rutherford's structure of the atomic nuclei, we can view atomic models omitting some details because three subatomic particles (proton, neutron, and electron) have major impact. The nucleus of the atom bears the major mass. It is an extremely dense region, which contains positively charged protons and neutral neutrons. It occupies a small amount of the atomic volume compared with the cloud of negatively charged electrons attracted to the positively charged nucleus by the force that exists between the particles of opposite electric charge.

For the atom of the element, the number of protons is the same, but the number of neutrons may vary. Atoms of a given element, which differ in number of neutrons (and consequently in mass), are called isotopes. For example, carbon always has six protons, but it may have six neutrons as well. In this case, it is called "carbon-12" (^{12}C). The simplified two-dimensional representation of the carbon atom electron configuration is

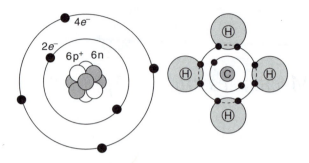

FIGURE 7.1

Simplified two-dimensional representation of electron configuration for carbon atom ($_6$C): Six electrons (e^-, black dots), orbiting the nucleus, occupying two shells. Electrons in each energy level (filling electron shell) represented as dots on concentric rings. The first shell is closest to the nucleus, and the electrons in this shell have the lowest energy. Four electrons on the second shell have higher energy. An electron can change its shell by absorbing or losing an amount of energy equal to the difference in potential energy between the original and final shell. Six protons (p+, dashed) and six neutrons (n, white) are in centrally located nucleus. Methane is formed from one carbon and four hydrogen atoms using covalent bonds.

given in Fig. 7.1. The orbitals do not define electrons pathway because the electron orbitals are three-dimensional with the varying probability of being found in the given volume.

There exists the endless diversity of organic carbon-based molecules, and the versatility of carbon in molecular architectures is fascinating. Valence is the number of bonds an atom can form, and the number of bonds is equal to the number of electrons required to complete the valence (outermost) electron shell. For $_6$C, the electron configuration is $1s^2\, 2s^2\, 2p^2$. Atoms with incomplete valence shells interact with certain other atoms to complete valence shells by sharing electrons. A covalent bond is the sharing of a pair of valence electrons by two atoms. Hydrogen has one valence electron in the first shell, but the shell capacity is two electrons. Therefore, for example, four hydrogen atoms satisfy the valence of one carbon atom forming methane (atoms share valence electrons), and the molecular formula is

(see Fig. 7.1). With six electrons in its second electron shell, oxygen ($_8$O) with an electron configuration of $1s^2\, 2s^2\, 2p^4$ needs two more electrons to complete this valence shell. Two oxygen atoms form a molecule by sharing two pairs of valence electrons, and double covalent bond is formed, i.e., O=O. Each atom sharing electrons has a bond capacity (number of covalent bonds that must be formed for the atom to have a full complement of valence electrons). The bond capacity is called the atom's valence. Carbon $_6$C has four valence electrons. Therefore, its bond capacity (or valence) is four.

Hence, the valences for hydrogen, oxygen, hydrogen, and carbon are 1, 2, 3, and 4, respectively.

An atom has no net charge due to the equal number of positively charged protons in the nucleus and negatively charged electrons around it. For example, carbon has six protons and six electrons. If electrons are lost or gained by the neutral atom due to the chemical reaction, a charged particle called an ion is formed.

When one deals with subatomic particles, the dual nature of matter places a fundamental limitation on how accurate we can describe location, momentum, and other quantities. Austrian physicist Erwin Schrödinger in 1926 derived an equation that describes the wave particle natures. This fundamental equation led to the new area in physics, called quantum mechanics, which enables us to deal with subatomic particles. The complete solution to Schrödinger's equation gives a set of wave functions and set of corresponding energies. These wave functions are related to the orbitals. A collection of orbitals with the same principal quantum number, which describes the orbit, is called the electron shell. Each shell is divided into the number of subshells with the equal principal quantum number. Each subshell consists of a number of orbitals. Each shell may contain only two electrons of the opposite spin (Pauli exclusion principle). When the electron is in the lowest-energy orbital, the atom is in its ground state. When the electron enters the higher orbital, the atom is in an excited state. To move the electron to the excited-state orbital, a photon of the appropriate energy should be absorbed as the external energy.

When the size of the orbital increases and the electron spends more time farther from the nucleus, it possesses more energy and is less tightly bound to the nucleus. It was reported that the most outer shell is the valence shell. The electrons that occupy it are referred to as valence electrons. Inner-shell electrons are called the core electrons. The valence electrons contribute to the bond formation between atoms when molecules are formed. For ion formation, the electrons are removed from the electrically neutral atom and the positively charged cation is formed. They possess the highest ionization energies (the energy that measures the ease of removing the electron from the atom) and occupy the energetically weakest orbital since it is the most remote orbital from the nucleus. The valence electrons removed from the valence shell become free electrons, transferring the energy from one atom to another.

The electric conductivity of a medium is predetermined by the density of free electrons, and good conductors have a free electron density in the range of 1×10^{23} free electrons per cubic centimeter. In contrast, the free electron density of good insulators is in the range of 10 free electrons per cubic centimeter. The free electron density of semiconductors is in the range from $1 \times 10^{7}/cm^3$ to $1 \times 10^{15}/cm^3$ (for example, the free electron concentrations in silicon at 25°C and 100°C are $2 \times 10^{10}/cm^3$ and $2 \times 10^{12}/cm^3$, respectively). The free electron density is determined by the energy gap between valence and conduction (free) electrons. That is, the properties of the media (conductors, semiconductors, and insulators) are determined by the atomic structure.

Molecules (or atoms) can form different structures and complexes. There are many challenging problems needed to be solved such as mathematical modeling, analysis, simulation, design, and optimization, not to mention fabricating these molecular structures emphasizing testing, characterization, implementation, and deployment. To build nano-systems and nanodevices, advanced fabrication technologies and processes must be developed and applied. To fabricate nanosystems at the molecular level, the fundamental problems in atomic-scale positional assembly, self-controlled assembly and self-replication (systems are able to build copies of themselves, e.g., crystal growth or complex DNA-RNA-protein processes that accurately copy tens of millions of atoms), and intelligent automata problems must be solved. Organic, inorganic, and hybrid nanosystems and devices are distinct. Fighter aircraft and eagle, submarine and whale, microrobot and ant, microaircraft and dragonfly, and nanopropulsor and *E. coli* are enormously different, even though all can fly and sail. Limited progress has been accomplished even in high-fidelity modeling, heterogeneous simulation, and data-intensive analysis at nanoscale due to formidable challenges. This chapter covers basic quantum theory as applied to nanosystems with the ultimate goal to coherently understand the basic phenomena in nanoscale and utilize these phenomena in order to design novel nanosystems.

7.1.2 Some Basic Fundamentals

The quantitative explanation, analysis, and simulation of natural phenomena can be approached using comprehensive mathematical models that model and describe essential features. Newton's laws, the Lagrange equations of motion, and the Hamilton concept allow one to model electromechanical phenomena in microsystems, and the Maxwell equations can be applied to model electromagnetic phenomena. In the 1920s, new theoretical developments and concepts were made to develop the *quantum mechanics* to model and analyze atomic particles. In general, atomic-scale systems do not obey the classical laws of physics and mechanics. In 1900, Max Planck discovered the effect of quantization of energy. He found that the radiated (emitted) energy is

$$E = nh\upsilon$$

where n is the nonnegative integer, $n = 0, 1, 2, \ldots$; h is the Planck constant, $h = 6.62606876 \times 10^{-34}$ J $\cdot$ sec ($h = 4.13566727 \times 10^{-15}$ eV $\cdot$ sec); υ is the frequency of radiation, $\upsilon = c/\lambda$, c is the speed of light, $c = 299{,}792{,}458$ m/sec, and λ is the wavelength, $\lambda = c/\upsilon$.

The modified Planck constant $\hbar = h/2\pi = 1.054571596 \times 10^{-34}$ J $\cdot$ sec ($\hbar = 6.58211889 \times 10^{-16}$ eV $\cdot$ sec) is usually used. The following discrete energy values result: $E_0 = 0$, $E_1 = h\upsilon$, $E_2 = 2h\upsilon$, $E_3 = 3h\upsilon$, etc.

The observation of discrete energy spectra suggests that each particle has the energy $h\upsilon$ (the radiation results due to N particles), and the particle with the energy $h\upsilon$ is called a *photon*. Figure 7.2 demonstrates the photon emission when an electron jumps from the state (orbital) n_1 to the state n_2. The radiated energy is $h\upsilon$. The *photon* has the momentum $p = h\upsilon/c = h/\lambda$.

Soon, Einstein demonstrated the discrete nature of light. In 1913, the Danish physicist Niels Bohr developed the model of the hydrogen atom using the planetary system analog (see Fig. 7.2). It is clear that if the electron has planetary-type orbits, it can be excited to an outer orbit and can "fall" to the inner orbits. Therefore, to develop the model, Bohr postulated that the electron has a certain stable circular orbit (that is, the orbiting electron does not produce the radiation because otherwise the electron would lose the energy and change the path); the electron changes higher or lower energy orbitals by receiving or radiating a discrete amount of energy; etc.

To attain uniform circular motion, from Newton's law, the electrostatic (Coulomb) force must be equal to the radial force. The radius R_n is found using the kinetic energy. The attractive Coulomb force is

$$F_C = \frac{1}{4\pi\varepsilon_0} \frac{q_{electron} q_{nucleus}}{R_n^2}$$

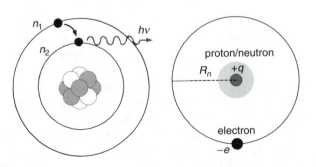

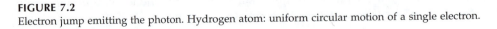

FIGURE 7.2
Electron jump emitting the photon. Hydrogen atom: uniform circular motion of a single electron.

For a single-electron hydrogen atom, taking into the account the centripetal acceleration $\frac{v^2}{R_n}$, we have

$$F_C = \frac{1}{4\pi\varepsilon_0}\frac{e^2}{R_n^2} = \frac{mv^2}{R_n}$$

Using the kinetic energy

$$\Gamma = \tfrac{1}{2}mv^2 = \tfrac{1}{2}m\left(\frac{n\hbar}{mR_n}\right)^2 = \frac{e^2}{8\pi\varepsilon_0 R_n}$$

one obtains the expression for the radius of the nth orbital R_n. In particular,

$$R_n = \frac{4\pi\varepsilon_0\hbar^2}{me^2}n^2 = r_0 n^2$$

where the Bohr radius is

$$r_0 = \frac{4\pi\varepsilon_0\hbar^2}{me^2} = 5.291772082\times 10^{-11} \text{ m}$$

That is, the radius of the hydrogen atom is approximately 0.0529 nm.

By making use the potential energy $\Pi = -e^2/4\pi\varepsilon_0 R_n$, the total energy of the system is

$$E = \Gamma + \Pi = -\frac{e^2}{8\pi\varepsilon_0 R_n}$$

Using the index for energy levels n, the total energy of the electron in the nth orbit is found to be

$$E_n = \Gamma_n + \Pi_n = -\frac{me^4}{32\pi^2\varepsilon_0^2\hbar^2 n^2}$$

It should be emphasized that the application of quantum mechanics leads to the same expression for the quantized energy, i.e.,

$$E_n = -\frac{me^4}{32\pi^2\varepsilon_0^2\hbar^2 n^2}$$

That is, the energy depends on the quantum number n. Fig. 7.3 illustrates the quantized energy for different n. The MATLAB statement to perform calculations and plot the results is given below (electron mass is $9.10938188\times 10^{-31}$ kg and charge is $1.602176462\times 10^{-19}$ C):

```
>> n = 1:1:5; m = 9.10938188e - 31; e = 1.602176462e - 19;
eps = 8.854187817e - 12; h = 1.054571596e - 34;
En = -(m*e^4)./(32*pi*pi*eps*eps*h*h*n.*n);
stem(n,En); title('Quantized Energy, [J]'); xlabel('n');
ylabel('En');
```

Using the conversion 1 eV = $1.602176462\times 10^{-19}$ J, the second plot provides the quantized energy E_n in the eV units that are commonly used.

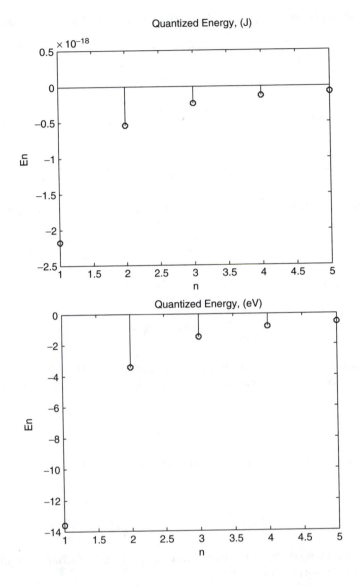

FIGURE 7.3
Quantized energy E_n for different n.

Denoting the nucleus charge as Ze, the derived equations can be modified. For example, the energy is expressed as

$$E_n = -\frac{m(Ze^2)^2}{32\pi^2\varepsilon_0^2\hbar^2 n^2} = -13.6\frac{Z^2}{n^2}\text{ eV}$$

One finds the energy difference between the orbitals as

$$\Delta E = E_{n1} - E_{n2} = \frac{me^4}{32\pi^2\varepsilon_0^2\hbar^2}\left(\frac{1}{n_2^2} - \frac{1}{n_1^2}\right)$$

The evaluation of the $me^4/32\pi^2\varepsilon_0^2\hbar^2$ term gives a 13.6 eV difference in the energy levels.

Using the difference between the energy levels, from $\Delta E = E_{n1} - E_{n2} = h\upsilon$ one finds

$$\upsilon = \frac{me^4}{64\pi^3\varepsilon_0^2\hbar^3}\left(\frac{1}{n_2^2} - \frac{1}{n_1^2}\right)$$

The wavelength of the emitted radiation is

$$\lambda = \frac{c}{\upsilon} = \frac{64\pi^3\varepsilon_0^2\hbar^3 c}{me^4}\left(\frac{n_1^2 n_2^2}{n_1^2 - n_2^2}\right) = \frac{1}{R_R}\left(\frac{n_1^2 n_2^2}{n_1^2 - n_2^2}\right)$$

where R_R is the Rydberg constant, $R_R = 1.097373 \times 10^7$ m^{-1}.

Example 7.1

Let the electron jump from $n_1 = 3$ to $n_2 = 2$ and from $n_1 = 6$ to $n_2 = 5$. One can calculate the corresponding wavelengths. We have

$$\lambda = \frac{1}{1.097373 \times 10^7}\left(\frac{3^2 2^2}{3^2 - 2^2}\right) = 656 \text{ nm};$$

$$\lambda = \frac{1}{1.097373 \times 10^7}\left(\frac{6^2 5^2}{6^2 - 5^2}\right) = 7455.8 \text{ nm}$$

The dynamics of particles can be described by different equations of motion using Newtonian, Lagrangian, and quantum mechanics. A number of quantities and variables are used, for example, mass m, position r, velocity v, charge q, etc. The kinetic energy of the translational motion is $\Gamma = \frac{1}{2}mv^2$. In addition, particles can experience wave dynamics. Waves are described using wavelength λ, angular frequency ω, amplitude A, velocity, etc. For waves, the parameters are related to each other, and for light traveling at speed c, we have $\lambda\upsilon = c$, and the energy of a wave is $E \propto A^2$. Classical theory has been widely used. However, quantum mechanics should be used for many problems. For example, let us consider the body radiation. When objects are hot, they emit radiation. Radiation emitted depends on the temperature (hot metal rod can be red or blue). An ideal emitter is called a blackbody for which the emitted radiation is in thermal equilibrium with the object. The radiation output from a blackbody can be measured and described by the energy distribution ρ, which is a measure of how much energy is emitted at a specific wavelength. The energy distribution of blackbodies depends on wavelength and temperature. Classical theory models precisely for long wavelengths the dependence of the energy on wavelength, but fails to model for short wavelengths. In particular, instead of predicting that the distribution will drop to zero, classical Rayleigh-Jeans theory predicts the "ultraviolet catastrophe" that the distribution will become infinite, and the energy density is $\rho = 8\pi kT/\lambda^4$. Using the MATLAB statement

```
kT = 1; L = 0.1:0.01:0.5; R = 8*pi*kT./(L.*L.*L.*L);
plot(L,R); title('Energy Density Plot');
xlabel ('L'); ylabel('R');
```

the resulting calculations and plot are illustrated in Fig. 7.4.

Max Planck predicted that the energy is proportional to the frequency, i.e., $E = nh\upsilon$. This equation predicts that light can have energy only at certain discrete values, i.e., light is quantized. This formulation predicts that the energy distribution will decrease at small wavelengths. Blackbody radiation is associated with a temperature and arises from motion of the waves of the blackbody. In order for something to vibrate with a frequency υ, it

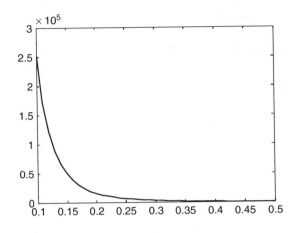

FIGURE 7.4
Energy distribution due to the classical Rayleigh-Jeans theory.

needs an energy $h\upsilon$. As υ increases, there is not enough thermal energy kT available to vibrate. The energy distribution is given as

$$\rho = \frac{8\pi hc}{\lambda^5} \left(\frac{1}{e^{hc/\lambda kT} - 1} \right)$$

For long wavelengths,

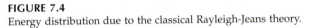

$$\frac{hc}{\lambda kT} \ll 1, \quad \text{and} \quad e^{\frac{hc}{\lambda kT}} - 1 \approx \frac{hc}{\lambda kT}$$

Thus, one obtains

$$\rho \approx \frac{8\pi hc}{\lambda^5} \frac{\lambda kT}{hc} = \frac{8\pi kT}{\lambda^4}$$

Hence, for long wavelengths, the equation

$$\rho = \frac{8\pi hc}{\lambda^5} \left(\frac{1}{e^{hc/\lambda kT} - 1} \right)$$

results in the classical Rayleigh-Jeans theory. In general, classical theory agrees with the quantum theory to a certain degree.

   When objects are heated, they emit light that is characteristic for each element. The emitted light is discrete. In classical physics, there is no explanation for this phenomenon. However, in quantum mechanics, discrete emission naturally follows from electrons revolving around the nucleus in the orbitals with light being emitted when an electron changes orbitals from higher to lower energy orbitals.

7.1.3   Quantum Theory: Basic Principles

Bohr's model was expanded and generalized by Heisenberg and Schrödinger using the *wave mechanics*. The characteristics of particles and waves are augmented, replacing the trajectory consideration by the waves using continuous, finite, and single-valued wave function

$\Psi(x, y, z, t)$ (Cartesian coordinate system), $\Psi(r, \phi, z, t)$ (cylindrical coordinate system), and $\Psi(r, \theta, \phi, t)$ (spherical coordinate system). The wave function gives the dependence of the wave amplitude on space coordinates and time.

The Schrödinger equation can be derived utilizing the classical mechanics paradigm. Using classical mechanics, for a particle of mass m with energy E moving in the Cartesian coordinate system, one has

$$\underset{\text{total energy}}{E(x, y, z, t)} = \underset{\text{kinetic energy}}{\Gamma(x, y, z, t)} + \underset{\text{potential energy}}{\Pi(x, y, z, t)}$$

$$= \frac{p^2(x, y, z, t)}{2m} + \Pi(x, y, z, t) = \underset{\text{Hamiltonian}}{H(x, y, z, t)}$$

Thus, we have

$$p^2(x, y, z, t) = 2m[E(x, y, z, t) - \Pi(x, y, z, t)]$$

Using the formula for the wavelength (Broglie's equation for the particle wavelength) $\lambda = h/p = h/mv$, one finds

$$\frac{1}{\lambda^2} = \left(\frac{p}{h}\right)^2 = \frac{2m}{h^2}[E(x, y, z, t) - \Pi(x, y, z, t)]$$

This expression is substituted into the *Helmholtz* equation

$$\nabla^2\Psi + \frac{4\pi^2}{\lambda^2}\Psi = 0$$

which gives the evolution of the wave function.

We obtain the Schrödinger equation as

$$E(x, y, z, t)\Psi(x, y, z, t) = -\frac{\hbar^2}{2m}\nabla^2\Psi(x, y, z, t) + \Pi(x, y, z, t)\Psi(x, y, z, t)$$

or

$$E(x, y, z, t)\Psi(x, y, z, t) = -\frac{\hbar^2}{2m}\left(\frac{\partial^2\Psi(x, y, z, t)}{\partial x^2} + \frac{\partial^2\Psi(x, y, z, t)}{\partial y^2} + \frac{\partial^2\Psi(x, y, z, t)}{\partial z^2}\right)$$

$$+ \Pi(x, y, z, t)\Psi(x, y, z, t)$$

Our goal is to examine the equations of motion to study particle dynamics. The physical properties of nanosystems (charge transport, radiation, polarization, position, velocity, acceleration, etc.) are functions of time. Using classical mechanics, the evolution (motion or dynamics) of particles is given using physical variables called the *canonically coupled variables*. Quantum mechanics offers a different framework. In general, when a system is "in motion" due to different $\left\{\begin{smallmatrix}\text{"force"}\\\text{"potential"}\end{smallmatrix}\right\}$ the dynamic and steady-state behavior of the system is found by solving $\left\{\begin{smallmatrix}\text{Newton / Lagrange / Hamilton}\\\text{Schrödinger}\end{smallmatrix}\right\}$ equations. The $\left\{\begin{smallmatrix}\text{variable}\\\text{wave function}\end{smallmatrix}\right\}$ is continuous across the boundary $\left\{\begin{smallmatrix}\text{"force"}\\\text{"potential"}\end{smallmatrix}\right\}$ as finite.

Quantum mechanics describes particles (for example, electrons) using wave functions. There is a direct correspondence between the physical quantities of particles, e.g., energy E and momentum $\mathbf{p}$, and frequency ω and the wave vector $\mathbf{k}$. We recall the de Broglie relation, i.e., $\mathbf{p} = \hbar\mathbf{k}$ and $E = \hbar\omega$.

Example 7.2

Derive the Schrödinger equation using the conservation of energy concept and de Broglie's equation.

Making use the conservation of energy concept, we have $E = \Gamma + \Pi$.

The kinetic energy is $\Gamma = \frac{1}{2}mv^2 = \frac{1}{2}(p^2/m)$. By taking note of the de Broglie equation $p = \hbar k = \hbar (2\pi/\lambda)$, we have $\Gamma = \frac{1}{2}(\hbar^2 k^2/m)$. Here, p is the momentum and k is the wave number. Using the analogs from classical mechanics and electromagnetism (string, membrane, Maxwell's equation, etc.), one defines the free-particle de Broglie wave as $\Psi(t, x) = A\sin(kx - \omega t)$. For $t = 0$, the differentiation of $\Psi(t, x) = A\sin kx$ gives

$$\frac{d^2\Psi}{dx^2} = -k^2\Psi = -\frac{2m}{\hbar^2}\Gamma\Psi = -\frac{2m}{\hbar^2}(E-\Pi)\Psi$$

Hence, one obtains the Schrödinger equation

$$-\frac{\hbar^2}{2m}\frac{d^2\Psi}{dx^2} + \Pi\Psi = E\Psi$$

In 1926, Erwine Schrödinger derived the equation

$$-(\hbar^2/2m)\,\nabla^2\Psi + \Pi\Psi = E\Psi$$

which is related to the Hamiltonian function $H = -(\hbar^2/2m)\,\nabla + \Pi$. Thus, $H\Psi = E\Psi$.

The Schrödinger equation determines possible wave functions and energies of the system given by the Hamiltonian H. For the Cartesian coordinate system, we have

$$\nabla^2\Psi(x, y, z, t) = \frac{\partial^2\Psi(x, y, z, t)}{\partial x^2} + \frac{\partial^2\Psi(x, y, z, t)}{\partial y^2} + \frac{\partial^2\Psi(x, y, z, t)}{\partial z^2}$$

or

$$\nabla^2\Psi(\mathbf{r}, t) = \frac{\partial^2}{\partial x^2}\Psi(\mathbf{r}, t) + \frac{\partial^2}{\partial y^2}\Psi(\mathbf{r}, t) + \frac{\partial^2}{\partial z^2}\Psi(\mathbf{r}, t)$$

For the cylindrical system, we have

$$\nabla^2\Psi(r, \phi, z, t) = \frac{1}{r}\frac{\partial}{\partial r}\left(r\frac{\partial\Psi(r, \phi, z, t)}{\partial r}\right) + \frac{1}{r^2}\frac{\partial^2\Psi(r, \phi, z, t)}{\partial\phi^2} + \frac{\partial^2\Psi(r, \phi, z, t)}{\partial z^2}$$

For the spherical system, we have

$$\nabla^2\Psi(r, \theta, \phi, t) = \frac{1}{r^2}\frac{\partial}{\partial r}\left(r^2\frac{\partial\Psi(r, \theta, \phi, t)}{\partial r}\right) + \frac{1}{r^2\sin\theta}\frac{\partial}{\partial\theta}\left(\sin\theta\frac{\partial\Psi(r, \theta, \phi, t)}{\partial\theta}\right)$$

$$+ \frac{1}{r^2\sin^2\theta}\frac{\partial^2\Psi(r, \theta, \phi, t)}{\partial\phi^2}$$

All quantitative and qualitative characteristics of nanoscale particles, devices, and systems are defined by the *wave function* $\Psi(\mathbf{r}, t)$. The wave function is not a physical variable, but it contains quantitative and qualitative information about the phenomena, effects, and properties that the system exhibits. A wave function is complex and may be expressed using the complex exponential function as

$$\Psi(\mathbf{r}, t) = \Psi_0 e^{i(\omega t - \mathbf{k} \cdot \mathbf{r})}$$

where ω is the angular frequency, and $\mathbf{k}$ is the wave (propagation) vector.

A wave function gives the probability of finding a particle in space. If $\Psi(\mathbf{r}, t)$ is a wave function of a particle, then the probability that this particle is in a differential volume dV is $\Psi\Psi^* dV$, where Ψ^* is the complex conjugate of Ψ. The probability leads to the *normalization condition*

$$\int_V \Psi^*(\mathbf{r}) \Psi(\mathbf{r}) \, dV = \int_V |\Psi(\mathbf{r})|^2 \, dV = 1.$$

Here, V is the space volume where the particle can be encountered.

The Schrödinger partial differential equation must be solved using the boundary conditions, and the wave function is normalized using the probability density $\int |\Psi(\mathbf{r})|^2 dV = 1$ or $\int_V \Psi^*(\mathbf{r}) \Psi(\mathbf{r}) \, dV = 1$. This normalization condition can be justified and the physical meaning can be explained. A particle is never at a specific location (volume V) but only has a probability of being there. The probability of finding a particle at a specific position is given by the square of the wave function $\Psi^*(\mathbf{r})\Psi(\mathbf{r})dV$. In general, the probability of finding a particle within a volume V is $\int_V \Psi^*(\mathbf{r})\Psi(\mathbf{r})dV$.

Example 7.3

The finite, single-valued, smooth, and continuous wave functions may be imaginary; however, the probability is always real. In fact, let

$$\Psi(x) = A(x) + iB(x) \quad \text{and} \quad \Psi^*(x) = A(x) - iB(x)$$

Then,

$$\Psi^*(x)\Psi(x) = (A(x) - iB(x))(A(x) + iB(x)) = A^2(x) + B^2(x) \geq 0$$

The probability of being anywhere is space must be 1. Therefore, $\int_{-\infty}^{\infty} \Psi^*(\mathbf{r}) \Psi(\mathbf{r}) \, dV = 1$, where in the Cartesian coordinate system $dV = dxdydz$.

Every physical observable p has an associated operator, and the average expectation value of the observable is given by

$$\langle p \rangle = \int \Psi^*(\mathbf{r}) \, \hat{p} \, \Psi(\mathbf{r}) \, dV$$

The Schrödinger equation is an eigenfunction, and each of the solutions are orthogonal (solutions are unique and not related to each other).

Example 7.4   CLASSICAL PHYSICS AND WAVE EQUATIONS

In classical physics, there is a number of wave equations. For example, the following equation is widely used:

$$\frac{\partial^2 u(t, x)}{\partial x^2} = \frac{1}{v^2} \frac{\partial^2 u(t, x)}{\partial t^2}$$

where t is the time, $u(t, x)$ is the displacement, and v is the velocity of propagation.

This equation can be solved by separating the variables. One has $u(t, x) = w(x)e^{2\pi i v t}$. The solution of the differential equation is found. The Schrödinger equation belongs to the class of the wave equations and must be solved using the boundary and normalization conditions. Compared with classical mechanics, the normalization condition is used.

Schrödinger derived the time-dependent equation that satisfies the de Broglie principle and describes the atomic-scale motion to study the dynamics of atoms. The Schrödinger equation defines the energy of atoms as electrons move around. For example, solving the Schrödinger equation for N-electron atoms, one derives the position of the electron relative to the nucleus and calculates how the wave function varies as a function of the radial distance from the nucleus. For the hydrogen atom, the 1s, 2s, and 2p wave functions are

$$\Psi_{1s}(r) = \frac{1}{\sqrt{\pi r_0^3}} e^{-\frac{r}{r_0}} = \frac{1}{\sqrt{\pi}} r_0^{-1.5} e^{-\frac{r}{r_0}}, \quad \Psi_{2s}(r) = \frac{1}{\sqrt{8}} r_0^{-1.5} \left(2 - \frac{r}{r_0}\right) e^{-\frac{r}{2r_0}}, \quad \text{and}$$

$$\Psi_{2p}(r) = \frac{r}{\sqrt{24 r_0}} r_0^{-1.5} e^{-\frac{r}{2r_0}}$$

where r is the radial distance from the nucleus, and r_0 is the Bohr radius, which is defined as the most probable distance of the electron from the nucleus when it is in the 1s orbital ($r_0 = 0.0529$ nm).

Example 7.5

The ground-state wave function for the hydrogen atom is

$$\Psi_{1s}(r) = \frac{1}{\sqrt{\pi r_0^3}} e^{-\frac{r}{r_0}}$$

Let us verify the normalization condition, e.g., $\int |\Psi(\mathbf{r})|^2 dV = 1$. We have

$$\int |\Psi(\mathbf{r})|^2 dV = \int |\Psi_{1s}(\mathbf{r})|^2 dV = \int_0^\infty \frac{1}{\pi r_0^3} e^{-2\frac{r}{r_0}} 4\pi r^2 dr = \frac{4}{r_0^3} \int_0^\infty r^2 e^{-2\frac{r}{r_0}} dr$$

From

$$\int r^2 e^{-2\frac{r}{r_0}} dr = \left(-\frac{1}{2} r_0 r^2 - \frac{1}{2} r_0^2 r - \frac{1}{4} r_0^3\right) e^{-2\frac{r}{r_0}}$$

the evaluation with the lower and upper limits results in

$$\int |\Psi_{1s}(\mathbf{r})|^2 dV = \frac{4}{r_0^3} \int_0^\infty r^2 e^{-2\frac{r}{r_0}} dr = \frac{4}{r_0^3} \frac{1}{4} r_0^3 = 1$$

If one would like to find the probability p that the electron is at a distance less than r_0 from the nucleus, the upper limit is changed to the specified r_1. Here, $r_1 < r_0$. In particular, one performs the integration

$$p = \int |\Psi_{1s}(\mathbf{r})|^2 dV = \int_0^{r_1} \frac{1}{\pi r_0^3} e^{-2\frac{r}{r_0}} 4\pi r^2 dr$$

and for the specified r_1, the probability p results.

The Schrödinger equation can be solved using a number of methods. It is very difficult to analytically solve the Schrödinger equation except for simple atoms. One can apply various methods, e.g., the Hartree self-consistent field theory (each electron moves in a spherically symmetrical field, and the Schrödinger equation can be solved for each particle independently, and then the solution is found by averaging), the Hartree-Fock approximation (electron spin and other motions of the electron are integrated in the self-consistent field theory), density functional theory, etc.

Example 7.6

Examine the one- and three-dimensional problem and demonstrate the correlation between classical and quantum mechanics. Briefly examine the possible solution.

SOLUTION

In general, the problem is formulated quite broadly, and there are many possible solutions. The classical theory was modified to provide a mathematical formalism describing the quantum effects by Schrödinger. The energy of a particle is given by the sum of the kinetic and potential energies, e.g.,

$$E = \frac{1}{2} mv^2 + \Pi(x) = \frac{p^2}{2m} + \Pi(x)$$

In quantum mechanics, this expression is modified using the conventional variables as well as introducing the "operators," e.g.,

$$t \to t, \quad x \to x, \quad \Pi \to \Pi, \quad p_x \to -i\hbar \frac{\partial}{\partial x} \quad \text{and} \quad E \to i\hbar \frac{\partial}{\partial t}$$

where i is the complex number, $i = \sqrt{-1}$.

The classical expression for a one-dimensional case then becomes

$$E = \frac{p^2}{2m} + \Pi(x)$$

and

$$i\hbar \frac{\partial}{\partial t} \Psi(t, x) = -\frac{\hbar^2 \partial^2}{2m\partial x^2} \Psi(t, x) + \Pi(x) \Psi(t, x)$$

For a three-dimensional case, we have

$$i\hbar \frac{\partial}{\partial t} \Psi(t, \mathbf{r}) = -\frac{\hbar^2}{2m} \nabla^2 \Psi(t, \mathbf{r}) + \Pi(\mathbf{r}) \Psi(t, \mathbf{r})$$

where $\mathbf{r}$ is the three-dimensional position vector.

The resulting time-varying equation is very complex, and usually time-invariant equations are studied. A general form for the wave function can be given as

$$\Psi(t, \mathbf{r}) = \Psi(\mathbf{r}) e^{-i\omega t} = \Psi(\mathbf{r}) e^{-i\frac{E}{\hbar}t}, \quad \omega = \frac{E}{\hbar}$$

Then we have

$$ i\hbar \frac{\partial \left(\Psi(\mathbf{r})\, e^{-i\frac{E}{\hbar}t} \right)}{\partial t} = -\frac{\hbar^2}{2m} \nabla^2 \left(\Psi(\mathbf{r})\, e^{-i\frac{E}{\hbar}t} \right) + \Pi(\mathbf{r})\Psi(\mathbf{r})\, e^{-i\frac{E}{\hbar}t} $$

Hence,

$$ e^{-i\frac{E}{\hbar}t}\left[i\hbar \frac{(-iE)}{\hbar}\, \Psi(\mathbf{r}) \right] = e^{-i\frac{E}{\hbar}t}\left[-\frac{\hbar^2}{2m} \nabla^2\, \Psi(\mathbf{r}) + \Pi(\mathbf{r})\, \Psi(\mathbf{r}) \right] $$

and

$$ E\, \Psi(\mathbf{r}) = -\frac{\hbar^2}{2m} \nabla^2\, \Psi(\mathbf{r}) + \Pi(\mathbf{r})\, \Psi(\mathbf{r}) $$

7.1.4 Harmonic Oscillator: Newtonian Mechanics, Schrödinger Equation, and Quantum Theory

When a mass m is attached to a stationary object by a spring with a spring constant k, the mass experiences harmonic motion. This motion can be exp-ressed by using Newton's and Hooke's laws. In particular, the force F on the mass for any displacement of the spring from its equilibrium x is given as $F = -kx$, and Newton's law gives the differential equation $m(d^2x/dt^2) = ma = F$.

Thus, considering only the spring force, one has the following second-order differential equation:

$$ m\frac{d^2x}{dt^2} = -kx $$

The solution of this equation, which gives the harmonic motion, is found to be

$$ x(t) = A\sin(\omega t) $$

Making use of $m(d^2x/dt^2) = -kx$, as well as the general solution of this differential equation in the form $x(t) = A\sin(\omega t)$, we have

$$ \frac{d^2x}{dt^2} = \frac{d}{dt}\left(\frac{d}{dt} A\sin(\omega t) \right) = \frac{d}{dt}(A\omega\cos(\omega t)) = -A\omega^2 \sin(\omega t) = -\omega^2 x $$

Therefore, for $m\,(d^2x/dt^2) = -kx$, the resulting equation is

$$ m(-\omega^2 x) = -kx $$

and

$$ \omega = \sqrt{\frac{k}{m}} $$

The object (mass) oscillates with a frequency ω.

The harmonic oscillator can be studied using quantum theory by applying the Schrödinger equation. To solve the Schrödinger equation for this problem, we should use the potential energy Π, which is $\Pi = \frac{1}{2}kx^2$. This well-known expression can be derived as follows:

$$ \Pi = -\int F dx = -\int (-kx)dx = k\int x dx = \tfrac{1}{2}kx^2 $$

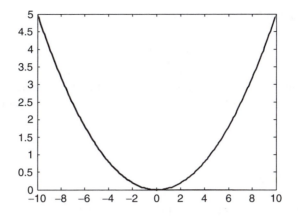

FIGURE 7.5
Potential energy if $k = 0.1$.

The resulting plot, numerically calculated using the following MATLAB statement

```
>> k = 0.1; x = -10:0.1:10; P = k*x.*x/2; plot(x,P);
title('Potential Energy Plot');
xlabel('Displacement, x');
ylabel('Potential Energy')
```

is illustrated in Fig. 7.5.

Using the potential energy for the harmonic oscillator, the Schrödinger equation becomes

$$-\frac{\hbar^2}{2m}\frac{d^2}{dx^2}\Psi(x) + \tfrac{1}{2}kx^2\Psi(x) = E\Psi(x)$$

This equation is solved letting $y = x/a$. We have

$$y = \frac{x}{a}\frac{d}{dx} = \frac{d}{dy}\frac{dy}{dx} = \frac{1}{a}\frac{d}{dy} \quad \text{and} \quad \frac{d^2}{dx^2} = \frac{d}{dx}\frac{d}{dx} = \frac{1}{a^2}\frac{d^2}{dy^2}$$

Thus, one has

$$-\frac{\hbar^2}{2m}\frac{1}{a^2}\frac{d^2}{dy^2}\Psi(y) + \tfrac{1}{2}k(ay)^2\Psi(y) = E\Psi(y)$$

Multiplying this equation by $-(2m/\hbar^2)a^2$, we obtain

$$\frac{d^2}{dy^2}\Psi(y) - \frac{2ma^2}{\hbar^2}\tfrac{1}{2}k(ay)^2\Psi(y) = -\frac{2ma^2}{\hbar^2}E\Psi(y)$$

The resulting equation is

$$\frac{d^2}{dy^2}\Psi(y) + \frac{2ma^2}{\hbar^2}E\Psi(y) - \frac{ma^4k}{\hbar^2}y^2\Psi(y) = 0$$

Letting

$$a = \left(\frac{\hbar^2}{mk}\right)^{\frac{1}{4}} \quad \text{and} \quad b = \frac{2ma^2}{\hbar^2}E = \frac{2mE}{\hbar^2}\sqrt{\frac{\hbar^2}{mk}} = \frac{2E}{\hbar}\sqrt{\frac{m}{k}} = \frac{2E}{\hbar\omega}$$

we obtain Hermite's equation:

$$\frac{d^2}{dy^2}\Psi(y)+(b-y^2)\Psi(y)=0$$

The general solution of this equation is known and can be derived by examining the limits. In particular, if $y \to \infty$, then $y^2 \gg b$, and $b - y^2 \approx -y^2$. The equation becomes

$$\frac{d^2}{dy^2}\Psi(y)-y^2\Psi(y)=0$$

The solution for this equation is

$$\Psi(y)=e^{-\frac{1}{2}y^2}, \quad \text{and} \quad \frac{d}{dy}\Psi(y)=e^{-\frac{1}{2}y^2}(-y)$$

Furthermore,

$$\frac{d^2}{dy^2}\Psi(y)=\frac{d}{dt}\left(e^{-\frac{1}{2}y^2}(-y)\right)=e^{-\frac{1}{2}y^2}+ye^{-\frac{1}{2}y^2}(-y)\approx y^2e^{-\frac{1}{2}y^2}=y^2\Psi(y)$$

The general form of the solution is

$$\Psi_z(y)=N_zM_z(y)e^{-\frac{1}{2}y^2}, \quad z=0, 1, 2, \ldots$$

where N_z are the normalization constants that are different for each term,

$$N_z=\frac{1}{\sqrt{a\sqrt{\pi}\,2^z z!}}$$

$M_z(y)$ is Hermite's function, and $M_z(y)$ are polynomials with the first four terms as given by $z = 0$, $M_0(y) = 1$, $z = 1$, $M_1(y) = 2y$, $z = 2$, $M_2(y) = 4y^2 - 2$, and $z = 3$, $M_3(y) = 8y^3 - 12y$.

The ground-state wave function and its probability distribution can be calculated (due to the exponential drop-off, both functions remain nonzero for all values of x). The solutions have symmetry about the origin. MATLAB can be straightforwardly used, and for $z = 2$ we can plot the wave function as illustrated in Fig. 7.6 using the following statement:

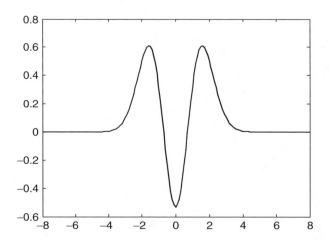

FIGURES 7.6
Wave function $\Psi_z(y) = N_zM_z(y)e^{-\frac{1}{2}y^2}$ for $z = 2$.

```
a = 1; z = 2; y = -8:0.1:8;
N = 1/(sqrt(a*sqrt(pi)*(2^(z))*z));
M = 4*y.*y-2;
p = N*M.*exp(-y.*y/2); plot(y,p);
title('Wavefunction Plot'); xlabel('y'); ylabel('p');
```

We used $y = x/a$. It is in our interest to check the solutions for all values of x by substituting the different solutions into the Schrödinger equation. For example, using the ground-state solution and its derivatives, we have

$$\Psi_0(x) = N_0 e^{-\frac{1}{2a^2}x^2}$$

$$\frac{d}{dx}\Psi_0(x) = N_0 \frac{d}{dx}\left(e^{-\frac{1}{2a^2}x^2}\right) = N_0 e^{-\frac{1}{2a^2}x^2}\left(-\frac{x}{a^2}\right)$$

$$\frac{d^2}{dx^2}\Psi_0(x) = -N_0 \frac{d}{dx}\left(\frac{x}{a^2}e^{-\frac{1}{2a^2}x^2}\right) = N_0 e^{-\frac{1}{2a^2}x^2}\left(\frac{x^2}{a^4} - \frac{1}{a^2}\right) = \Psi_0(x)\left(\frac{x^2}{a^4} - \frac{1}{a^2}\right)$$

Substituting the derived expressions in the Schrödinger equation results in

$$-\frac{\hbar^2}{2m}\Psi_0(x)\left(\frac{x^2}{a^4} - \frac{1}{a^2}\right) + \frac{1}{2}kx^2\Psi_0(x) = E_0\Psi_0(x)$$

and

$$\Psi_0(x)\left[x^2\left(-\frac{\hbar^2}{2ma^4} + \frac{1}{2}k\right) + \left(\frac{\hbar^2}{2ma^2} - E_0\right)\right] = 0$$

Due to the fact that

$$-\frac{\hbar^2}{2ma^4} + \frac{1}{2}k = -\frac{\hbar^2}{2m}\frac{mk}{\hbar^2} + \frac{1}{2}k = 0$$

the term on the right-hand side must be zero. This gives

$$E_0 = \frac{\hbar^2}{2ma^2} = \frac{\hbar^2}{2m}\sqrt{\frac{mk}{\hbar^2}} = \frac{1}{2}\hbar\sqrt{\frac{k}{m}} = \frac{1}{2}\hbar\omega$$

This is the ground-state energy. Substitution of the zth wave function results in the energy

$$E_z = \left(z + \frac{1}{2}\right)\hbar\omega$$

The mass attached to the spring vibrates back and forth and is restricted to some region. The maximum displacement of the mass from the equilibrium position x_e is where the total energy is all potential energy. Therefore, we have

$$E = \frac{1}{2}kx_e^2 \quad \text{and} \quad x_e = \sqrt{2\frac{E}{k}}$$

The probability p of finding the mass outside the "allowed" region is

$$p = \int_{x_e}^{\infty} \Psi^*(x)\Psi(x)dx = \int_{y_e}^{\infty} \Psi^*(y)\Psi(y)a\,dy$$

Recalling that

$$a = \left(\frac{\hbar^2}{mk} \right)^{\frac{1}{4}}$$

the value of y_e is given as

$$y_e = \frac{x_e}{a} = \sqrt{2\frac{E}{k} \left(\frac{mk}{\hbar^2} \right)^{\frac{1}{4}}} = \sqrt{2\frac{(z+\frac{1}{2})\hbar\omega}{k} \left(\frac{mk}{\hbar^2} \right)^{\frac{1}{4}}}$$

Making use of $\omega = \sqrt{\frac{k}{m}}$, one finds

$$y_e = \sqrt{2\frac{(z+\frac{1}{2})\hbar\omega}{k} \left(\frac{mk}{\hbar^2} \right)^{\frac{1}{4}}} = \sqrt{2z+1}$$

For the ground state, $z = 0$, and substitution of the wave function

$$\Psi_0(y) = \sqrt{\frac{1}{a\sqrt{\pi}}} e^{-\frac{1}{2}y^2}$$

gives

$$p = \int_{y_e}^{\infty} \Psi^*(y)\Psi(y) a \, dy = \int_{1}^{\infty} \frac{1}{a\sqrt{\pi}} e^{-y^2} a \, dy = \frac{1}{\sqrt{\pi}} \int_{1}^{\infty} e^{-y^2} dy$$

Taking note of the probability integral

$$\mathrm{erf} X = 1 - \frac{2}{\sqrt{\pi}} \int_{X}^{\infty} e^{-y^2} dy$$

the values can be derived from the table of the error function available in any calculus textbook. The resulting plot, calculated and plotted using the MATLAB script

```
X = -3:0.1:3; Y = erf(X); plot(X,Y);
title('Error Function'); xlabel('X'); ylabel('Y');
```

is illustrated in Fig. 7.7. For example, let $X = 1$. The error function is 0.8427 and can be found using MATLAB as

```
>> X = 1; Y(1) = erf(X); disp(Y(1))
0.8427
```

The probability is found to be $p = 0.079$. Thus, there is a 7.9 % probability of finding the mass past the turning point.

Unlike the particle in a box, the energy levels for the harmonic oscillator are evenly spaced as

$$\Delta E = \left(z+1+\frac{1}{2} \right)\hbar\omega - \left(z+\frac{1}{2} \right)\hbar\omega = \hbar\omega$$

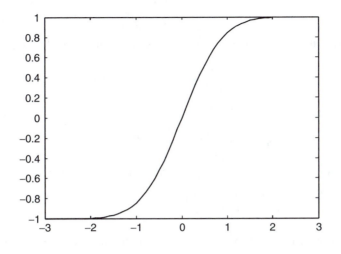

FIGURE 7.7
Plot of the error function.

Example 7.7

Consider a proton ($m = 1.67 \times 10^{-27}$ kg) held to a spring with the constant $k = 100$ N/m. Find the angular frequency of vibration, energy, and wavelength.

We have

$$\omega = \sqrt{\frac{k}{m}} = \sqrt{\frac{100}{1.67 \times 10^{-27}}} = 2.447 \times 10^{14} \text{ } \tfrac{\text{rad}}{\text{sec}}$$

The energy is found to be $E = \hbar\omega = 1.055 \times 10^{-34} \times 2.447 \times 10^{-14} = 2.58 \times 10^{-20}$ J.

The wavelength is $\lambda = c/v = 3.4 \times 10^{-6}m$.

In general, transitions of vibrational states are always associated with infrared light and involve energies that are smaller than the energies associated with the electronic levels.

Example 7.8

Illustrate the application of the Schrödinger equation studying the particle in a box (one-dimensional). Let the particle move from $x = 0$ to $x = x_f$ (see Fig. 7.8). Assume that the walls are rigid and the collisions are perfectly elastic (infinite square well potential). Find wave functions and energy levels.

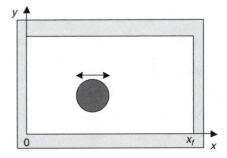

FIGURE 7.8
Particle in the box: particle moves in one dimension, i.e., confined in $0 \leq x \leq x_f$.

SOLUTION

We assume that the particle moves in the x direction (translational motion). We have,

$$-\frac{\hbar^2}{2m}\frac{\partial^2 \Psi(x)}{\partial x^2} + \Pi(x)\Psi(x) = E(x)\Psi(x)$$

The Hamiltonian function is given as

$$H(x,p) = \frac{p^2(x)}{2m} + \Pi(x) = -\frac{\hbar^2}{2m}\frac{d^2}{dx^2} + \Pi(x)$$

The particle moves from $x = 0$ to $x = x_f$, and the potential energy is

$$\Pi(x) = \begin{cases} 0, 0 \leq x \leq x_f \\ \infty, x < 0 \text{ and } x > x_f \end{cases}$$

Thus, the motion of the particle is bounded in the infinite square well potentials, and

$$\Psi(x) = \begin{cases} \text{continuous if } 0 \leq x \leq x_f \\ 0 \text{ if } x < 0 \text{ and } x > x_f \end{cases}$$

If $0 \leq x \leq x_f$, the potential energy is zero, and we have

$$-\frac{\hbar^2}{2m}\frac{d^2\Psi(x)}{dx^2} = E\Psi(x), 0 \leq x \leq x_f$$

The general solution of the resulting second-order differential equation

$$\frac{d^2\Psi(x)}{dx^2} + k^2\Psi(x) = 0, \quad k = \sqrt{\frac{2mE}{\hbar^2}}$$

is $\Psi(x) = ae^{ikx} + be^{-ikx} = a(\cos kx + i\sin kx) + b(\cos kx - i\sin kx)$

$$= c\sin kx + d\cos kx.$$

The wave number k is given as $k = 2\pi/\lambda$.

The solution can be easily verified by substituting the solution into the left side of the differential equation

$$-\frac{\hbar^2}{2m}\frac{d^2\Psi(x)}{dx^2} = E\Psi(x)$$

and we have

$$E\Psi(x) = E\Psi(x)$$

The kinetic energy of the particle is $p^2/2m$, where $p = kh$.

It is obvious that one must use the boundary conditions.

We have $\Psi(x)\big|_{x=0} = \Psi(0) = 0$, and from $\Psi(0) = c\sin kx + d\cos kx = 0$, we conclude that $d = 0$.

From $\Psi(x)\big|_{x=x_f} = \Psi(x_f) = 0$ using $c\sin kx_f = 0$, one must find the constant c and the expression for kx_f.

Assuming that $c \neq 0$, from $c\sin kx_f = 0$, we have

$$kx_f = n\pi, \quad n = 1, 2, 3, \ldots$$

where n is a positive or negative integer (if $n = 0$, the wave function vanishes everywhere, and thus, $n \neq 0$).

One uses the values of k such that the wave function is equal to zero at $x = x_f$, i.e., $\Psi(x)|_{x=x_f} = \Psi(x_f) = 0$. Thus, $kx_f = n\pi$. The wavelength is found to be

$$\lambda = \frac{2\pi}{k} = \frac{2x_f}{n}, \quad n = 1, 2, 3, \ldots$$

From $k = \sqrt{2mE/\hbar^2}$ and making use of $kx_f = n\pi$, we have the expression for the energy (discrete values of the energy which allow for solution of the Schrödinger equation) as

$$E_n = \frac{\hbar^2\pi^2}{2mx_f^2}n^2, \quad n = 1, 2, 3, \ldots$$

where the integer n designates the allowed energy level (n is called the quantum number).

Thus, we derived the possible energy levels for a particle in a box. Each energy level has its quantum number n and the corresponding wave function. For example, if $n = 1$ and $n = 2$, we have $E_0 = \hbar^2\pi^2/2mx_f^2$ (the lowest possible energy, called the ground state, and denoted as E_0) and $E_{n=2} = 2\hbar^2\pi^2/mx_f^2$.

Thus, we have illustrated that the energy of the particle is quantized.

The expression for the wave function is found to be

$$\Psi_n(x) = c\sin kx + d\cos kx = c\sin\left(\frac{n\pi}{x_f}x\right)$$

This is not the final solution because the constant c is not found yet. The normalization condition (the probability of finding the particle along the x-axis) is 100 %, i.e.,

$$\int_{-\infty}^{+\infty} \Psi_n^2(x)dx = 1$$

However, the wave function is equal to zero except when $0 \leq x \leq x_f$.

Using the probability density, we normalize the wave function as

$$\int_0^{x_f} \Psi_n^2(x)dx = c^2\int_0^{x_f} \sin^2\left(\frac{n\pi}{x_f}x\right)dx = c^2\frac{x_f}{n\pi}\int_0^{n\pi} \sin^2 g\,dg = 1, \; g = \frac{n\pi}{xf}x,$$

$$c^2\frac{x_f}{n\pi}\frac{n\pi}{2} = c^2\frac{x_f}{2} = 1,$$

Hence, $c = \sqrt{2/x_f}$, and one obtains

$$\Psi_n(x) = \sqrt{\frac{2}{x_f}}\sin\frac{n\pi}{x_f}x, \quad 0 \leq x \leq x_f, \quad n = 1, 2, 3, \ldots$$

For $n = 1$ and $n = 2$, we have

$$\Psi_1(x) = \sqrt{\frac{2}{x_f}}\sin\frac{\pi}{x_f}x \quad \text{and} \quad \Psi_2(x) = \sqrt{\frac{2}{x_f}}\sin\frac{2\pi}{x_f}x$$

Using the formula for the probability density $\rho = \Psi^T \Psi$, one has

$$\rho_n(x) = \frac{2}{x_f} \sin^2 \frac{n\pi}{x_f} x$$

The probability of finding a particle between $x = 0$ and $x = l$ is given by

$$p = \int_0^l \Psi^*(x)\Psi(x)dx = \frac{2}{x_f} \int_0^l \sin^2\left(\frac{n\pi}{x_f}x\right)dx$$

$$= \frac{1}{x_f}\left[x - \frac{x_f}{2n\pi}\sin\left(\frac{2n\pi}{x_f}x\right)\right]\Bigg|_0^l = \frac{1}{x_f}\left[l - \frac{x_f}{2n\pi}\sin\left(\frac{2n\pi}{x_f}l\right)\right]$$

The derived probability can be analyzed. If $l = x_f$, we have $p = 1$, and $p = 0.5$ when $l = x_f/2$.

One of the special properties of eigenfunctions is that the solutions are orthogonal: (Although the solutions are related to each other, their overlap is zero, that is, $\int_V \Psi^*(x)\Psi(x)dV = 0$).

We only determine the expectation (average value) of a parameter. For example, the average momentum is

$$\bar{p} = \int_0^{x_f} \Psi^*(x)\frac{\hbar}{i}\frac{\partial(\Psi(x))}{\partial x}dx = \frac{2}{x_f}\int_0^{x_f}\sin\left(\frac{n\pi}{x_f}x\right)\frac{\hbar}{i}\frac{\partial\left[\sin\left(\frac{n\pi}{x_f}x\right)\right]}{\partial x}dx$$

$$= \frac{2\hbar n\pi}{ix_f^2}\int_0^{x_f}\sin\left(\frac{n\pi}{x_f}x\right)\cos\left(\frac{n\pi}{x_f}x\right)dx = \frac{i\hbar}{x_f}\left[\sin^2\left(\frac{n\pi}{x_f}x\right)\right]\Bigg|_0^{x_f} = 0$$

In fact, the average momentum is zero because the particle will travel to the right and to the left with equal probability, resulting in zero net momentum. In contrast, the average square of the momentum is not zero, e.g.,

$$\bar{p}^2 = \int_0^{x_f} \Psi^*(x)\left(\frac{\hbar}{i}\frac{\partial}{\partial x}\right)^2 \Psi(x)dx = \frac{h^2 n^2}{4x_f^2} \neq 0$$

The expectation value of position is

$$\bar{x} = \int_0^{x_f} \Psi^*(x)x\Psi(x)dx = \frac{2}{x_f}\int_0^{x_f}\sin\left(\frac{n\pi}{x_f}x\right)x\sin\left(\frac{n\pi}{x_f}x\right)dx = \frac{2}{x_f}\int_0^{x_f}\sin^2\left(\frac{n\pi}{x_f}x\right)xdx = \tfrac{1}{2}x_f$$

We conclude that the particle has an average position at $\tfrac{1}{2}x_f$, e.g., the particle is most likely to be in the middle of the box.

Figure 7.9 documents the energy levels, wave functions, and probability densities Ψ^2 for the distinct states (wave functions and probability densities are regularized to 1 by dividing by the maximum value). The ground state corresponds to $n = 1$, and the data for $n = 5$ is reported by assigning $x_f = 1 \times 10^{-12}$ and $m = 1 \times 10^{-29}$. The equations

$$E_n = \frac{\hbar^2\pi^2}{2mx_f^2}n^2 \quad \text{and} \quad \Psi_n(x) = \sqrt{\frac{2}{x_f}}\sin\frac{n\pi}{x_f}x$$

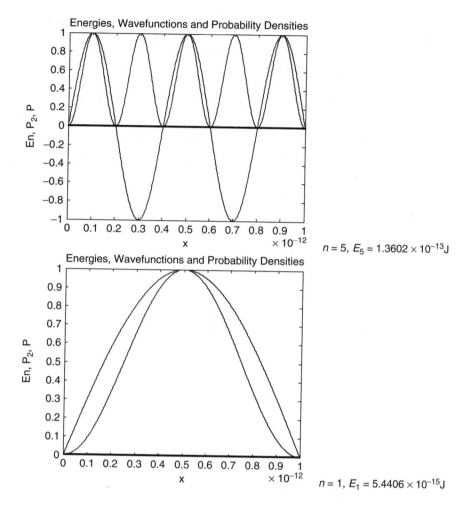

$n = 5$, $E_5 = 1.3602 \times 10^{-13}$J

$n = 1$, $E_1 = 5.4406 \times 10^{-15}$J

FIGURE 7.9
Energy levels, wave functions, and probability densities (normalized) for $n = 1$ and $n = 5$.

are solved in MATLAB using the following statement:

```
xf = 1e - 12; m = 1e - 29; x = 0:1e - 15:xf; n = 5;
h = 1.055e - 34; En = h*h*pi*pi*n*n/(2*m*xf*xf);
P = sqrt(0.5*xf)*sin(n*pi*x/xf);
plot(x,En,x,P/max(P),x, P.*P/max(P)^2);
title('Energies, Wavefunctions and Probability Densities');
xlabel('x'); ylabel('En, P, P^2');
```

For $n = 1$ and $n = 5$, the values for energy levels E_n are $E_1 = 5.4406 \times 10^{-15}$ J and $E_5 = 1.3602 \times 10^{-13}$ J, respectively. It must be emphasized that the states for which $n > 1$ are called the excited states. For $n = 1$, the probability density is maximum at $x_f/2$, and the probability density falls to zero at the edges (in contrast, classical mechanics gives equal probability at any position in the box). For $n = 5$, the probability density is maximum at $x_f/10$, $3x_f/10$, $5x_f/10$, $7x_f/10$, and $9x_f/10$ with zero probability density at 0, $x_f/5$, $2x_f/5$, $3x_f/5$, $4x_f/5$, and x_f. The particle travels and can be found (with maximum probabilities) at $x_f/10$, $3x_f/10$, $5x_f/10$, $7x_f/10$, and $9x_f/10$ and cannot be found at positions 0, $x_f/5$, $2x_f/5$, $3x_f/5$, $4x_f/5$, and x_f. How can the particle travel from $x_f/10$ to $3x_f/10$ and not be found at $x_f/5$? Quantum physics requires us to examine phenomena in terms of waves (not particle position).

Example 7.9

Assume that the particle moves in the x and y directions (two-dimensional case) (see Fig. 7.10). In particular, consider a particle confined in a rectangular area where the potential is zero and the potential is infinite outside the box. Set up the problem to find energy levels and wave functions.

SOLUTION

The Schrödinger equation becomes

$$-\frac{\hbar^2}{2m}\left(\frac{\partial^2\Psi(x,y)}{\partial x^2}+\frac{\partial^2\Psi(x,y)}{\partial y^2}\right)+\Pi(x,y)\Psi(x,y)=E(x,y)\Psi(x,y)$$

This equation can be solved by using the separation-of-variables technique. In particular, we let

$$\Psi(x,y)=X(x)Y(y)$$

where

$$X(x)=c_x\sin k_x x+d_x\cos k_x x \quad \text{and} \quad Y(y)=c_y\sin k_y y+d_y\cos k_y y$$

Substitution of this relationship into the Schrödinger equation results in

$$-\frac{\hbar^2}{2m}\left(Y(y)\frac{d^2X(x)}{dx^2}+X(x)\frac{d^2Y(y)}{dy^2}\right)=E(x,y)X(x)Y(y)$$

The division by

$$-\frac{2m}{\hbar^2}\frac{1}{X(x)Y(y)}$$

gives

$$\frac{1}{X(x)}\frac{d^2X(x)}{dx^2}+\frac{1}{Y(y)}\frac{d^2Y(y)}{dy^2}=-\frac{2m}{\hbar^2}E(x,y)$$

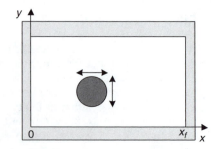

FIGURE 7.10
Particle in the box: particle moves in two dimensions.

By making an assumption about the energy, we have a set of two equations:

$$\frac{1}{X(x)}\frac{d^2X(x)}{dx^2} = -\frac{2m}{\hbar^2}E_x(x), \quad \text{e.g.,} \quad \frac{d^2X(x)}{dx^2} = -\frac{2m}{\hbar^2}E_x(x)X(x)$$

and

$$\frac{1}{Y(y)}\frac{d^2Y(y)}{dy^2} = -\frac{2m}{\hbar^2}E_y(y), \quad \text{e.g.,} \quad \frac{d^2Y(y)}{dy^2} = -\frac{2m}{\hbar^2}E_y(y)Y(y)$$

The derived equations are similar to the one-dimensional case considered in Example 7.8. The solution is

$$\Psi(x, y) = X(x)Y(y) = \sqrt{\frac{2}{x_f}}\sin\left(\frac{n_x \pi x}{x_f}\right)\sqrt{\frac{2}{y_f}}\sin\left(\frac{n_y \pi y}{y_f}\right)$$

$$E(x, y) = E_x(x) + E_y(y) = \frac{\hbar^2 \pi^2}{8mx_f^2}\left(n_x^2 + n_y^2\right)$$

Compared with the one-dimensional problem, two states now can have equal energies (states are degenerate) if $x_f = y_f$.

The normalized condition is

$$\int_{-\infty}^{+\infty}\int_{-\infty}^{+\infty}\Psi_n^2(x, y)dxdy = 1.$$

Example 7.10

An electron is trapped in a one-dimensional region (infinite square well) with dimensions $x = 0$ to $x_f = 1 \times 10^{-10}$ m. Find the external energy needed to excite the electron from the ground state to the first and second exited states ($n = 2$ and $n = 3$). What is the probability of finding the electron in the region $(0.09 \times 10^{-10}$ to $0.11 \times 10^{-10})$ m in the ground state? For the first exited state, what is the probability of finding the electron in the region 0 to 0.25×10^{-10} m?

SOLUTION

Using the results derived in Example 7.8, one has

$$E_n = \frac{\hbar^2 \pi^2}{2mx_f^2}n^2, \quad n = 1, 2, 3, \ldots$$

Therefore, we obtain $E_0 = \hbar^2\pi^2/2mx_f^2$.

Thus,

$$E_0 = \frac{\hbar^2\pi^2}{2mx_f^2} = \frac{(1.055\times10^{-34})^2 3.14^2}{2(9.1\times10^{-31})(1\times10^{-10})^2} = 6\times10^{-18} \text{ J} = 37 \text{ eV}$$

For the fist and second excited states, the differences are 4 and 9 times due to the term n^2 in the equation $E_n = (\hbar^2\pi^2/2mx_f^2)n^2$. Correspondingly, the external energies needed to excite the electron from the ground state to the first ($n = 2$) and second ($n = 3$) excited states are found to be

$$E_{ext2} = 4E_0 - E_0 = 3E_0 = 111 \text{ eV}$$
$$E_{ext3} = 9E_0 - E_0 = 8E_0 = 296 \text{ eV}$$

Using the results reported in Example 7.8 to find the probability of finding a particle between $x = x_1$ and $x = x_2$, we have for $n = 1$

$$p = \int_{x_1}^{x_2} \Psi^2(x)dx = \frac{2}{x_f} \int_{x_1}^{x_2} \sin^2\left(\frac{\pi}{x_f}x\right)dx = \frac{1}{x_f}\left[x - \frac{x_f}{2\pi}\sin\left(\frac{2\pi}{x_f}x\right)\right]\Bigg|_{x_1}^{x_2} = 0.0038$$

Finally, in the first excited state ($n = 1$), the probability of finding the electron is

$$p = \int_{x_1}^{x_2} \Psi^2(x)dx = \frac{2}{x_f} \int_{x_1}^{x_2} \sin^2\left(\frac{2\pi}{x_f}x\right)dx = \frac{1}{x_f}\left[x - \frac{x_f}{4\pi}\sin\left(\frac{4\pi}{x_f}x\right)\right]\Bigg|_{x_1}^{x_2} = 0.25$$

The plane-wave solution of the Schrödinger equation is $\Psi(x,t) = \Psi_0 e^{\pm i(\omega t - kx)}$. For a particle with a momentum $p_x = k\hbar$ and energy $E = \hbar\omega = k^2\hbar^2/2m$ moving in the x direction toward $+x$ to the right, we have $\Psi(x,t) = \Psi_0 e^{-i\omega t + ikx}$. If the particle moves toward $-x$, one obtains $\Psi(x,t) = \Psi_0 e^{i\omega t - ikx}$.

Let the potential of a particle vary in the x direction, while the potential in the y and z directions is constant. The solution of the Schrödinger equation

$$\left\{\frac{d^2}{dx^2} + \frac{2m}{\hbar^2}[E - \Pi(x)]\right\}\Psi(x) = 0$$

can be easily derived. For example, assuming that $\Pi(x) = \Pi_0 = \text{const}$, the solution of the Schrödinger equation is a plane wave

$$\Psi(x,t) = \Psi_0 e^{\pm i(\omega t - kx)}$$

where $k = (1/\hbar)\sqrt{2m(E - \Pi_0)}$.

If one studies the time-independent energy of the particle, the solution is $\Psi(x,t) = \Psi_0 e^{\pm ikx}$. The minus sign is used for the particle moving to the left, while the plus sign corresponds to the case where the particle moves to the right. For the constant potential, similar solutions are obtained for the movement in the y and z directions. If those movements are independent, the three-dimensional Schrödinger equation will have the wave function as

$$\Psi(x,y,z) = \Psi(x)\Psi(y)\Psi(z) = \Psi_0 e^{ik_x x} e^{ik_y y} e^{ik_z z} = \Psi_0 e^{i\mathbf{k}\cdot\mathbf{r}}$$

The relationship between energy E and the wave vector $\mathbf{k}$ for a constant potential is

$$E = \Pi_0 + \frac{\hbar^2 k^2}{2m}$$

Example 7.11

Consider a particle that moves from the region of negative values of coordinate x to the region of positive values of x. At $x = 0$, the particle encounters a potential barrier of height Π_0 with width L (see Fig. 7.11). Using classical and quantum mechanics, find the probabilities that the particle passes the barrier and is reflected by the barrier.

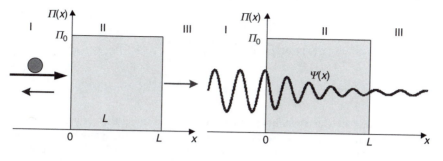

FIGURE 7.11

Particle and the rectangular potential barrier (arrows show the directions of the particle before and after the interaction with the barrier). The possible wave function for a particle tunneling through the potential-energy barrier wave functions joint continuously at the boundaries.

SOLUTION

We examine the probability that the barrier reflects the particle and the probability that the particle passes the barrier, i.e., p_R and p_T. It is obvious that $p = p_R + p_T = 1$.

In general, the reflection and transmission coefficients are denoted as R and T.

Using classical mechanics, one examines the relation between the particle energy E and the barrier height $\Pi(x)$. For $E > \Pi(x)$, the result is $R = 0$ and $T = 1$. For $E < \Pi(x)$, using classical mechanics, one finds that $R = 1$ and $T = 0$.

In quantum mechanics, the results are different. If the barrier potential is finite, there is always a finite probability for transmission and reflection of the quantum particle by the barrier. Outside of the barrier (areas I and III), $\Pi(x) = 0$. Therefore, the particle is free, and its wave function is $\Psi(x) = \Psi_0 e^{\pm ikx}$. In particular, we have

$$\Psi(x) = \Psi_{F0} e^{ikx} + \Psi_{R0} e^{-ikx}, \quad x \le 0$$

$$\Psi(x) = \Psi_{T0} e^{ikx}, \quad x \ge L$$

where Ψ_{F0} is the amplitude for the particle falling on the barrier, and Ψ_{R0} and Ψ_{T0} are the amplitudes for the particle reflected and transmitted.

The reflection and the transmission coefficients are related to Ψ_{F0}, Ψ_{R0}, and Ψ_{T0} as

$$R = \left| \Psi_{R0} / \Psi_{F0} \right|^2 \quad \text{and} \quad T = \left| \Psi_{T0} / \Psi_{F0} \right|^2$$

For the rectangular barrier, if $E < \Pi_0$,

$$T = \frac{16E(\Pi_0 - E)}{\Pi_0^2} e^{-\frac{2L}{\hbar}\sqrt{2m(\Pi_0 - E)}}$$

This finite transmission (classical mechanics gives zero) is called the *tunneling effect*. The tunneling probability depends exponentially on the width of the barrier L and energy difference $(\Pi_0 - E)$. It is clear that if the barrier is very thin (L is small), the transmission coefficient can be large.

Let the solutions are

$$\Psi(x) = ae^{ik_1x} + be^{-ik_1x}, \quad k_1 = \sqrt{\frac{2mE}{\hbar^2}}, \quad x < 0$$

$$\Psi(x) = ce^{ik_2x} + de^{-ik_2x}, \quad k_2 = \sqrt{\frac{2m(E - \Pi_0)}{\hbar^2}}, \quad x \geq 0$$

where a, b, c, and d are the constants that can be obtained using the continuity, boundary, and normalization conditions. Here, $|a|^2$ gives the intensity of the wave incident from the left, $|b|^2$ is proportional to the intensity of the reflected wave, $|c|^2$ is the intensity of the transmitted wave, $|d|^2$ and gives the intensity of the particle moving to the left in the region $x > 0$. Let $d = 0$. For the multiparticle case, the transmission and reflection are found as

$$T = \frac{|c|^2 \hbar k_2}{|a|^2 \hbar k_1} = \frac{|c|^2 k_2}{|a|^2 k_1} \quad \text{and} \quad R = \frac{|b|^2 \hbar k_1}{|a|^2 \hbar k_1} = \frac{|b|^2}{|a|^2}$$

because the number of particles incident from the left per second is $|a|^2 \hbar k_1$, the number of particles transmitted to the right per second is $|c|^2 \hbar k_2$, and the number of particles reflected to the left per second is $|b|^2 \hbar k_1$.

Using classical mechanics and taking note of the varying potential when $E > \Pi(x)$ and $E < \Pi(x)$, the transmission coefficient is given as

$$T = e^{-2\int_{x_i}^{x_j} \frac{\sqrt{2m[\Pi(x)-E]}}{\hbar} dx}, \quad \Pi(x) > E$$

$$T = 0, \quad \Pi(x) < E$$

where x_i and x_j are the turning points when $\Pi(x) > E$ (at x_i and x_j, classical physics predicts changes from reflection to transmission or from transmission to reflection depending on the relationship between $\Pi(x)$ and E). The tunneling phenomena and probability for a microscopic particle does not depend on a particular barrier width, but it depends on the effective field E and potential $\Pi(x)$.

Among the examples of tunneling, the proton transfer reaction can be mentioned:

Example 7.12

Consider a *quantum well* as illustrated in Figure 7.12. This quantum well represents a finite potential well. Outside of the well, the potential is Π_0, and in the studied three regions (I, II, and III) the potential is given as

$$\begin{cases} \Pi(x) = 0, & -L \leq x \leq L \\ \Pi(x) = \Pi_0, & |x| > L \end{cases}$$

This example to some extent is similar to Example 7.11, and the particle moves in these regions. At $x = L$, the particle encounters a barrier with the potential. Find the wave

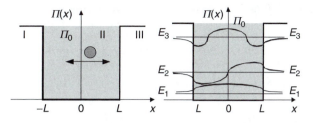

FIGURE 7.12
Particle in *quantum well* with a squire-well finite potential and energy-level diagram with the wave functions for three bound states.

functions and derive the expression for energy. Examine the energy for different potential and width L.

SOLUTION

In Newtonian mechanics, a particle trapped (localized) in a potential well can vibrate with periodic motion if $E < \Pi_0$. We apply the quantum mechanics theory.

The symmetry of the potential

$$\begin{cases} \Pi(x) = 0, & -L \leq x \leq L \\ \Pi(x) = \Pi_0, & |x| > L \end{cases}$$

leads to the symmetry of the Hamiltonian. Correspondingly, the wave function will satisfy $\Psi(x) = \Psi(-x)$ for even, and $\Psi(x) = -\Psi(-x)$ for odd. Hence, we may solve the problem for $0 \leq x < \infty$ and then derive the solution in $-\infty < x < \infty$.

In the Schrödinger equation

$$\left\{ \frac{d^2}{dx^2} + \frac{2m}{\hbar^2} [E - \Pi(x)] \right\} \Psi(x) = 0$$

we denote

$$\gamma^2 = \frac{2mE}{\hbar^2} \quad \text{and} \quad \beta^2 = \frac{2m}{\hbar^2}(\Pi_0 - E) = \frac{2m\Pi_0}{\hbar^2} - \gamma^2$$

Therefore, the Schrödinger equation in two regions (II and III) is

$$\left(\frac{d^2}{dx^2} + \gamma^2 \right) \Psi_{II} = 0, \quad 0 \leq x \leq L \text{ (inside the well)}$$

$$\left(\frac{d^2}{dx^2} - \beta^2 \right) \Psi_{III} = 0, \quad x \geq L \text{ (outside the well)}$$

A finite solution for $x \to \infty$ (in region III) is found as $\Psi_{III} = a_3 e^{-\beta x}$, $\beta > 0$.

In region II, we have two solutions for even and odd, i.e., $\Psi_{II} = b_2 \cos\gamma x$ and $\Psi_{II} = c_2 \sin\gamma x$.

The constants a_3, b_2, and c_2 are determined using the boundary and normalization conditions. For the even case, using $x = L$, we have

$$b_2 \cos\gamma L = a_3 e^{-\beta L} \quad \text{and} \quad \gamma b_2 \sin\gamma L = \beta a_3 e^{-\beta L}$$

Dividing $\gamma b_2 \sin \gamma L = \beta a_3 e^{-\beta L}$ by $b_2 \cos \gamma L = a_3 e^{-\beta L}$, we obtain

$$\gamma \tan(\gamma L) = \beta = \sqrt{\frac{2m\Pi_0}{\hbar^2} - \gamma^2} = \sqrt{\frac{2m}{\hbar^2}(\Pi_0 - E)}$$

For the odd case, using $c_2 \sin \gamma L = a_3 e^{-\beta L}$ and $\gamma c_2 \cos \gamma L = -\beta a_3 e^{-\beta L}$, the division gives

$$\gamma \operatorname{ctn}(\gamma L) = -\beta = -\sqrt{\frac{2m\Pi_0}{\hbar^2} - \gamma^2} = -\sqrt{\frac{2m}{\hbar^2}(\Pi_0 - E)}$$

We obtain two transcendental equations that result in positive values of the wave number a that correspond to the energy levels of the particle. In fact, using a and b, we have the circle equation, i.e.,

$$\gamma^2 + \beta^2 = \frac{2m\Pi_0}{\hbar^2}$$

Let us multiply the left and right side by L^2. We obtain the constant

$$r^2 = \frac{2m\Pi_0 L^2}{\hbar^2}$$

and

$$(\gamma^2 + \beta^2)L^2 = \frac{2m\Pi_0}{\hbar^2}L^2$$

For distinct values of m, L, Π_0, and E, one can draw the circle, and the unknown constants a_3, b_2, and c_2 result. Therefore, these transcendental equations can be numerically solved.

Figure 7.12 shows an energy-level diagram with the wave functions for three bound states.

By making use of $\gamma^2 = 2mE/\hbar^2$, the quantized energy is found to be

$$E_n = \frac{\hbar^2 \gamma_n^2}{2m}, \quad n = 1, 2, 3, \dots$$

In the case of a quantum well of infinite depth ($\Pi_0 \to \infty$), one has $\gamma_n = n\pi/2L$, and

$$E_n = \frac{\hbar^2 \pi^2}{2m(2L)^2} n^2, \quad n = 1, 2, 3, \dots$$

When the width L of the well increases, the energy decreases. When $L \to \infty$, the particle becomes free and has a continuous energy as given by $E = \Pi_0 + (\hbar^2 k^2/2m)$.

Example 7.13

Examine the Schrödinger equation, Hamiltonian, and Lorenz equations in the electromagnetic field. Use the potential instead of the wave function.

SOLUTION

It was shown that $H\Psi = E\Psi$, where $H = -(\hbar^2/2m)\nabla + \Pi$. Using the CGS (centimeter/gram/second) units, when the electromagnetic field is quantized, the potential can be used instead of the wave function. In particular, using the momentum operator due to electron orbital angular momentum **L**, the classical Hamiltonian for electrons in an electromagnetic field is

$$H = \frac{1}{2m}\left(\mathbf{p} + \frac{e}{c}\mathbf{A}\right)^2 - e\phi$$

The angular momentum of a particle is $\mathbf{L} = \mathbf{r} \times \mathbf{p}$, where $\mathbf{r} = (x, y, z)$ and $\boldsymbol{p} = (p_x, p_y, p_x)$.

From the Hamilton equations

$$\dot{q} = \frac{\partial H}{\partial p}, \quad \dot{p} = -\frac{\partial H}{\partial q}$$

by making use of

$$\frac{d\mathbf{r}}{dt} = \frac{1}{m}\left(\mathbf{p} + \frac{e}{c}\mathbf{A}\right), \quad \mathbf{p} = m\mathbf{v} - \frac{e}{c}\mathbf{A} \quad \text{and} \quad \dot{\mathbf{p}} = -\frac{e}{mc}\left(\mathbf{p} + \frac{e}{c}\mathbf{A}\right) \cdot \frac{\partial \mathbf{A}}{\partial \mathbf{x}} + e\frac{\partial \phi}{\partial \mathbf{x}}$$

one finds the Lorentz force equation to be

$$\mathbf{F} = -\frac{e}{c}\mathbf{v} \times \mathbf{B} - e\mathbf{E}$$

This equation gives the force due to motion in a magnetic field and the force due to electric field.

It is important to emphasize that the following equation results

$$-\frac{\hbar^2}{2m}\nabla^2\Psi + \frac{e}{2mc}\mathbf{B} \cdot \mathbf{L}\Psi + \frac{e^2}{8mc^2}(r^2B^2 - (\mathbf{r} \cdot \mathbf{B})^2)\Psi = (E + e\phi)\Psi$$

to study the quantized Hamilton equation, where the dominant term due to magnetic field is

$$\frac{e}{2mc}\mathbf{B} \cdot \mathbf{L} = -\boldsymbol{\mu} \cdot \mathbf{B}$$

where μ is the magnetic momentum due to the electron orbital angular momentum (the so-called Zeeman effect), and $\mu = -(e/2mc)\mathbf{L}$.

7.1.5 Schrödinger Equation for the Hydrogen Atom

We consider the hydrogen atom and apply the Schrödinger equation. For the hydrogen atom, the potential is due to the electrostatic interaction between the negatively charged electron and the positively charged nucleus. The potential energy is found using the distance between the electron and nucleus (r) and the electron charge (e). In particular,

$$\Pi(r) = -\frac{e^2}{4\pi\varepsilon_0 r}$$

Using the Schrödinger equation and making use of the potential energy, in the spherical polar coordinates, we have

$$-\frac{\hbar^2}{2m}\nabla^2(r, \theta, \phi)\Psi(r, \theta, \phi) - \frac{e^2}{4\pi\varepsilon_0 r}\Psi(r, \theta, \phi) = E\Psi(r, \theta, \phi)$$

To solve the equation, we will use the separation-of-variables approach. This will lead to three separate equations, each with only one of the three variables. We obtain three eigenequations and three quantum numbers (n, l, and m_l). These quantum numbers have mathematical meaning (indexes for wave functions) and geometric interpretation. The potential energy has only a radial dependence r. Therefore, we first consider the radial component separately and define and angular component as

$$\Lambda^2(\theta,\phi) = \frac{1}{\sin^2\theta}\frac{\partial^2}{\partial\phi^2} + \frac{1}{\sin\theta}\frac{\partial}{\partial\theta}\left(\sin\theta\frac{\partial}{\partial\theta}\right)$$

Using this definition, we can write the gradient squared as

$$\nabla^2(r,\theta,\phi)\,\Psi(r,\theta,\phi) = \frac{1}{r}\frac{\partial^2(r\Psi(r,\theta,\phi))}{\partial r^2} + \frac{1}{r^2}\Lambda^2(\theta,\phi)\Psi(r,\theta,\phi)$$

This results in the following Schrödinger equation:

$$-\frac{\hbar^2}{2m}\left\{\frac{1}{r}\frac{\partial^2(r\Psi(r,\theta,\phi))}{\partial r^2} + \frac{1}{r^2}\Lambda^2(\theta,\phi)\Psi(r,\theta,\phi)\right\} + \frac{-e^2}{4\pi\varepsilon_0 r}\Psi(r,\theta,\phi) = E\,\Psi(r,\theta,\phi)$$

This equation should be solved. The separation of variables technique can be used. In particular, one can express $\Psi(r,\theta,\phi) = R(r)\Theta(\theta)\Phi(\phi)$ using single-variable functions. Three differential equations (each for a single variable) result. In particular, the wave function can be found as $\Psi_{n,l,m_l}(r,\theta,\phi) = R_{n,l}(r)\Theta_{l,m1}(\theta)\Phi_{m1}(\phi)$, where n, l, and m_1 are the indices. However, the angular components may not be separated. Correspondingly, we apply

$$\Psi(r,\theta,\phi) = R(r)Y(\theta,\phi)$$

The Schrödinger equation becomes

$$-\frac{\hbar^2}{2m}\left\{Y(\theta,\phi)\frac{1}{r}\frac{\partial^2(rR(r))}{\partial r^2} + R(r)\frac{1}{r^2}\Lambda^2(\theta,\phi)Y(\theta,\phi)\right\} - \frac{e^2}{4\pi\varepsilon_0 r}R(r)Y(\theta,\phi) = E\,R(r)Y(\theta,\phi)$$

This equation is multiplied by $r^2/[R(r)\,Y(\theta,\phi)]$. We obtain

$$-\frac{\hbar^2}{2m}\left\{\frac{r}{R(r)}\frac{\partial^2(rR(r))}{\partial r^2} + \frac{1}{Y(\theta,\phi)^2}\Lambda^2(\theta,\phi)Y(\theta,\phi)\right\} - \frac{e^2 r}{4\pi\varepsilon_0} = E\,r^2$$

Grouping the terms with the same variables gives

$$\left(-\frac{\hbar^2}{2m}\frac{r}{R(r)}\frac{\partial^2}{\partial r^2} - \frac{e^2 r}{4\pi\varepsilon_0} - E\,r^2\right) - \left(\frac{\hbar^2}{2m}\frac{1}{Y(\theta,\phi)}\Lambda^2 Y(\theta,\phi)\right) = 0$$

The term in the first bracket depends on the variable r. The second term depends on θ and ϕ. We define the separation constant as

$$\frac{\Lambda^2 Y(\theta,\phi)}{Y(\theta,\phi)} = -l(l+1)$$

The radial equation becomes

$$-\frac{\hbar^2}{2m}\frac{r}{R(r)}\frac{d^2(r\,R(r))}{d r^2} - \frac{e^2 r}{4\pi\varepsilon_0} - r^2 E + \frac{\hbar^2}{2m}l(l+1) = 0$$

The angular part of the equation is given as

$$\Lambda^2(\theta, \phi)Y(\theta, \phi) = -l(l+1)Y(\theta, \phi)$$

Thus, using Λ one obtains

$$\frac{1}{\sin^2\theta}\frac{\partial^2}{\partial\phi^2}Y(\theta, \phi) + \frac{1}{\sin\theta}\frac{\partial}{\partial\phi}\left(\sin\theta\frac{\partial}{\partial\theta}\right)Y(\theta, \phi) = -l(l+1)\,Y(\theta, \phi)$$

The separation of angular variables gives

$$Y(\theta, \phi) = \Theta(\theta)\,\Phi(\phi)$$

and multiplying the Schrödinger equation by $\frac{\sin^2\theta}{\Theta(\theta)\,\Phi(\phi)}$ results in

$$\frac{1}{\Phi(\phi)}\frac{\partial^2\Phi(\phi)}{\partial\phi^2} + \frac{\sin\theta}{\Theta(\theta)}\frac{\partial\left(\sin\theta\dfrac{\partial\Theta(\theta)}{\partial\theta}\right)}{\partial\theta} + l(l+1)\sin^2\theta = 0$$

Thus, the following two equations are obtained:

$$\frac{1}{\Phi(\phi)}\frac{d^2\Phi(\phi)}{d\phi^2} = -m_l^2$$

$$\sin\theta\frac{d\left(\sin\theta\dfrac{d\Theta(\theta)}{d\theta}\right)}{d\theta} + \left(l(l+1)\sin^2\theta - m_l^2\right)\Theta(\theta) = 0$$

The solution to the first equation for ϕ is $\Phi(\phi) = A\,e^{im_l\phi}$. One can verify this solution by substitution of the solution in the differential equation. The normalization constant is determined using the condition that integral over every value of ϕ should yield a value of 1. That is,

$$1 = \int_0^{2\pi}\Phi^*(\phi)\Phi(\phi)d\phi = A^2\int_0^{2\pi}e^{-im_l\phi}e^{im_l\phi}d\phi = A^2\int_0^{2\pi}d\phi = 2\pi A^2$$

Therefore, the normalized constant is $A = 1/\sqrt{2\pi}$.

The values of the constant m_l are fixed by the physical constraint that if the object is fully rotated by 2π, then the solution must be the same. This gives

$$\Phi(\phi + 2\pi) = \Phi(\phi) \quad \text{and} \quad Ae^{im_l\phi} = Ae^{im_l(\phi+2\pi)} = Ae^{im_l\phi}e^{im_l 2\pi}$$

To satisfy these, we need

$$e^{im_l 2\pi} = 1$$

with $m_l = 0, \pm 1, \pm 2, \pm 3, \ldots$

The second equation for $\Theta(\theta)$ is called the Legendre equation. This equation has a series solutions of the form denoted as $P_l(\cos\Theta)$ for which there are restrictions on the separation constants

$$\sin\theta\frac{d\left(\sin\theta\dfrac{d\Theta(\theta)}{d\theta}\right)}{d\theta} + \left(l(l+1)\sin^2\theta - m_l^2\right)\Theta(\theta) = 0, \quad l = 0, 1, 2, \ldots \quad \text{and}$$

$$m_l = -l, -l+1, -l+2, \ldots$$

The combined angular terms are called spherical harmonics with solutions given as

$$Y(\Theta, \Phi) = \frac{1}{\sqrt{4\pi}} \quad \text{(for } l = 0 \text{ and } m_l = 0\text{)},$$

$$Y(\Theta, \Phi) = -\sqrt{\frac{3}{4\pi}} \cos\theta \quad \text{(for } l = 1 \text{ and } m_l = 0\text{)}$$

$$Y(\Theta, \Phi) = -\sqrt{\frac{3}{\pi}} \sin\theta\, e^{i\phi} \quad \text{(for } l = 1 \text{ and } m_l = 1\text{)},$$

$$Y(\Theta, \Phi) = \sqrt{\frac{15}{16\pi}} (3\cos^2\theta - 1) \quad \text{(for } l = 2 \text{ and } m_l = 1\text{)}, \text{ etc.}$$

It is shown that the solutions are polynomials of trigonometric functions.

Let us verify the results. By substituting $\Theta(\theta) = A$, where A is a constant, the equation is

$$0 + \left[l(l+1)\sin^2\theta - m_l^2 \right] A = 0$$

This is met if $l = 0$ and $m_l = 0$.

Another possible solution is $\Theta(\theta) = A_1 \cos\theta$. Substitution gives

$$\sin\theta\left(-2A_1 \sin\theta \cos\theta\right)\theta + \left(l(l+1)\sin^2\theta - m_l^2\right)\Theta(\theta) = 0$$

This result corresponds to $l = 1$ and $m_l = 0$.

The radial equation was found to be

$$-\frac{\hbar^2}{2m} \frac{r}{R(r)} \frac{d^2(rR(r))}{dr^2} - \frac{e^2 r}{4\pi\varepsilon_0} - r^2 E + \frac{\hbar^2}{2m} l(l+1) = 0$$

Multiplying this equation by $R(r)/r$, we have

$$-\frac{\hbar^2}{2m} \frac{d^2(rR(r))}{dr^2} - \frac{e^2 R(r)}{4\pi\varepsilon_0} + \frac{\hbar^2}{2m} l(l+1) \frac{R(r)}{r} = rR(r)E$$

This is the Laguerre equation. The eigenfunctions and solutions can be found.

Let $\Sigma(r) = R(r)r$. We have

$$-\frac{\hbar^2}{2m} \frac{d^2}{dr^2} \Sigma(r) + \left(\frac{\hbar^2 l(l+1)}{2mr^2} - \frac{e^2}{4\pi\varepsilon_0 r} \right)\Sigma(r) = E\Sigma(r)$$

Consider the case when $l = 0$. Multiplying the equation by $-2m/\hbar^2$, one obtains

$$\frac{d^2\Sigma(r)}{dr^2} + \frac{2m}{\hbar^2}\left(\frac{e^2}{4\pi\varepsilon_0 r} + E \right)\Sigma(r) = 0$$

The solution is $\Sigma(r) = re^{-Br}$, and we have

$$\frac{d}{dr}\Sigma(r) = -rBe^{-Br} + e^{-Br},$$

$$\frac{d^2\Sigma(r)}{dr^2} = -B(-Bre^{-Br} + e^{-Br}) - Be^{-Br} = B^2 re^{-Br} - 2Be^{-Br}$$

Thus,

$$B^2 r e^{-Br} - 2Be^{-Br} + \frac{2m}{\hbar^2}\left(\frac{e^2}{4\pi\varepsilon_0 r} + E\right)re^{-Br} = 0$$

and

$$re^{-Br}\left(B^2 + \frac{2mE}{\hbar^2}\right) + e^{-Br}\left(-2B + \frac{e^2}{4\pi\varepsilon_0}\frac{2m}{\hbar^2}\right) = 0$$

Each of these two terms must be zero, and the second term gives the expression for the constant B, i.e.,

$$B = \frac{m}{\hbar^2}\frac{e^2}{4\pi\varepsilon_0}$$

Therefore,

$$R(r) = e^{-\frac{m}{\hbar^2}\frac{e^2}{4\pi\varepsilon_0}r} \quad \text{and} \quad E = -\frac{me^4}{32\pi^2\varepsilon_0\hbar^2}$$

It must be emphasized that the solution is found for a specific set of quantum numbers ($n = 1$, $l = 0$, and $m_l = 0$). In general, the solutions for the Laguerre equation $L_w^k(x)$ can be expressed in the following form:

$$L_w^k(x) = \frac{e^x x^{-k}}{w!}\frac{d(e^{-x} x^{w+k})}{dx^w}$$

where w is the index with integral values starting with 0, and k is the index greater than −1.

We conclude that for the hydrogen atom, the solutions are given by the product of the three terms, and the wave function is

$$\Psi(r, \theta, \phi)_{n,l,m_l} = R(r)_{n,l} Y_l^{m_l}(\theta, \phi)$$

with

$$E_n = -\frac{me^4}{32\pi^2\varepsilon_0^2\hbar^2}\frac{1}{n^2} = -\frac{hcR_R}{n^2}$$

where R_R is the Rydberg constant, $R_R = me^4/8\varepsilon_0^2 h^3 c$, $R_R = 1.097373 \times 10^7$ m^{-1}.

The solutions can be written in terms of the individual solutions that we found using three quantum numbers. In particular,

$$\Psi_{n,l,m_l}(r, \theta, \phi) = A_{n,l,m_l}\, P_l^{m_l}(\cos\theta)\, e^{im_l\phi}\, R_{n,l}(\rho)\, e^{-\rho/2}$$

where $\rho = 2N_p r/nr_0$, $r_0 = 4\pi\varepsilon_0\hbar^2/me^2 = 0.0529$ nm, and N_p is the number of protons ($N_p = 1$ for hydrogen atom).

The quantum numbers are the principal quantum number ($n = 1, 2, 3, \ldots$), the angular momentum quantum number ($l = 0, 1, 2, \ldots, n - 1$), and the magnetic quantum number ($m_l = l, l - 1, l - 2, \ldots, -l$). The states (orbitals) are dependent only on the quantum number n. Thus, the orbitals are degenerate in energy (n^2-fold degenerate). There are two electrons per orbital (one spin up and one spin down).

The angular momentum of a particle is

$$\mathbf{L} = \mathbf{r} \times \mathbf{p}$$

where $\mathbf{r} = (x, y, z)$ and $\mathbf{p} = (p_x, p_y, p_z)$.

For example, considering the z component, we have

$$L_z = xp_y - yp_x = x \frac{\hbar}{i} \frac{\partial}{\partial y} - y \frac{\hbar}{i} \frac{\partial}{\partial x} = \frac{\hbar}{i} \frac{\partial}{\partial \phi}$$

The variable ϕ is the solution of the Schrödinger equation, and we have

$$L_z \Psi(r, \theta, \phi) = A_{n,l,m_l} P(\theta) R(\rho) e^{-\rho/2} \left(\frac{\hbar}{i} \frac{\partial}{\partial \phi} \right) e^{im_l \phi} = m_l \hbar \Psi(r, \theta, \phi)$$

and the angular momentum is quantized by 1, and the projection of the angular momentum along the z direction is quantized by m_l.

The solutions to the Schrödinger equation provide all properties for molecules. The solutions to the Schrödinger equation consist of the product of a radial term and an angular term, i.e.,

$$\psi_{n,l,m_l} = R_{n,l} Y_{l,m_l}$$

where solutions are uniquely defined by the three indices

$$n = 1, 2, 3, \ldots; \quad l = 0, 1, 2, \ldots; \quad \text{and} \quad m_l = l, l - 1, l, -2, -l$$

with n viewed as the principal quantum number due to quantization of energy as $E_n = -hcR_h/n^2$, $R_h = m_e e^4 / 8 \varepsilon_0^2 h^3 c$, l is the angular momentum quantum number due to quantization of the angular momentum with the magnitude $\sqrt{l(l+1)}\,\hbar$, and m_l is due to quantization of the z-component of angular momentum $m_l \hbar$.

Now the properties of s-orbitals and p-orbitals can be examined using the wave functions. For the s-orbital, $l = 0$. Thus, the angular component is a constant and the dependence is radial. For the $1s$-orbital, the wave function is

$$\Psi_{100} = \sqrt{\frac{1}{\pi r_0}} \, e^{-\frac{r}{r_0}}$$

The higher orbitals have the same general appearance, as there is an exponential dependence multiplied by a polynomial term. For example, for the $2s$-orbital, we have

$$\Psi_{200} = \frac{1}{2\sqrt{2}} \sqrt{\frac{1}{4\pi r_0^3}} \left(2 - \tfrac{1}{2} \rho \right) e^{-\frac{\rho}{4}}$$

The probability of finding an electron at a certain point is space is $\Psi(\mathbf{r})^* \Psi(\mathbf{r}) dV$. One may find the probability of finding the electron at a certain radius (not at a specific point) using the following formula: $\int \Psi^* \Psi (4\pi r^2) dr$. The most probable radial position of the electron is the peak position of this term that can be obtained by setting the derivative to zero, i.e.,

$$\frac{d(4\pi r^2 \Psi^* \Psi)}{dr} = 0$$

For the ground state, we have

$$\frac{d\left(4\pi r^2 \frac{1}{\pi r_0^3} e^{-\frac{2r}{r_0}}\right)}{dr} = \frac{4}{r_0^3} \frac{d\left(r^2 e^{-\frac{2r}{r_0}}\right)}{dr} = 0$$

and

$$\frac{8}{r_0^3} e^{-\frac{2r}{r_0}} r\left(1 - \frac{r}{r_0}\right) = 0$$

resulting in $r = r_0$.

The probability of finding the electron within this radius is derived as

$$\int_0^{a_0} \Psi^* \Psi \, dV = \int_0^{a_0} \frac{1}{\pi r_0^3} e^{-\frac{2r}{r_0}} (4\pi r^2) \, dr = \frac{4}{r_0^3} \int_0^{a_0} r^2 e^{-\frac{2r}{r_0}} \, dr$$

Letting $y = r/r_0$, we have

$$(4/r_0^3)\int_0^1 (r_0 y)^2 e^{-2y} (r_0 dy) = 4\int_0^1 x^2 e^{-2y} dy$$

Using the product rule for integration $\int u \, dv = uv - \int v \, du$, we set $u = y^2$, $du = 2y \, dy$, $v = -\frac{1}{2} e^{-2y}$, and $dv = e^{-2y} dy$. We have

$$4\int_0^1 y^2 e^{-2y} dy = -4y^2 \frac{1}{2} e^{-2y} - 4\int_0^1 \frac{1}{2} e^{-2y} 2y \, dy = -2ye^{-2y} - 4\int_0^1 e^{-2y} y \, dy$$

and the second substitution is $u = y$, $du = dy$, $v = -\frac{1}{2} e^{-2y}$, and $dv = e^{-2y} \, dy$. These result in

$$\int_0^1 e^{-2y} y \, dy = \frac{1}{2} ye^{-2y} - \int_0^1 \frac{1}{2} e^{-2y} dy = \frac{1}{2} ye^{-2y} - \frac{1}{2} e^{-2y}\left(-\frac{1}{2}\right)$$

One obtains

$$4\int_0^1 y^2 e^{-2y} dy = -e^{-2y}(2y^2 + 2y + 1)\Big|_0^1 = 0.32$$

Thus, the probability of finding the electron with the radius a_0 is 32%.

One can average the radius, and the following integral should be used:

$$\int_0^\infty \Psi^* r \Psi dV = \int_0^\infty r\left(\frac{1}{\pi r_0}\right) e^{-\frac{2r}{r_0}} 4\pi r^2 dr = \frac{4}{r_0} \int_0^\infty r^3 e^{-\frac{2r}{r_0}} dr$$

Making use of $y = r/r_0$ and using the product rule, we have

$$\int_0^\infty \Psi^* r \Psi dV = \frac{4}{r_0} \int_0^\infty r^3 e^{-\frac{2r}{r_0}} dr = 4r_0 \int_0^\infty y^3 e^{-2y} dy = 4r_0 e^{-2y}\left(-\frac{1}{2} y^3 - \frac{3}{4} y^2 - \frac{3}{4} y - \frac{3}{8}\right)\Big|_0^\infty = \frac{3}{2} r_0$$

For p-orbitals, $l = 1$, and the electron has a nonzero angular momentum $\sqrt{2}\,\hbar$. This results in nonradial symmetry for the distribution as for the s-orbitals. While the use and applications of the Schrödinger equation lead to the coherent high-fidelity analysis, one can utilize this meaningful concept for only a limited class of problems due to formidable complexity.

7.1.6   Some Applications of the Schrödinger Equation and Quantum Mechanics

7.1.6.1   Mathematical Modeling of Atoms with Many Electrons

As was emphasized, the solutions for the Schrödinger equation present serious difficulties if there is more than one electron. The Schrödinger equation can be solved by assumptions, approximations, and simplifications incorporating the interactions between the electrons. These approaches are based on the assumption that the wave functions are already known or need only minor adjustments.

7.1.6.2   Empirical Constants Concept

One concept is to modify the constants used in the expressions to include the existing interactions between electrons and nuclei. This can be accomplished by modifying the nuclear charge Z. The presence of the additional negative electrons effectively decreases the positive charge of the nucleus. For the hydrogen atom, the energy of an electron is given by the equation

$$E_n = -\frac{Z^2 \, m_e \, e^4}{32\pi^2 \, \varepsilon_0^2 \, \hbar^2} \frac{1}{n^2}$$

The energy can be measured because the atomic spectrum has a series of lines (energy required to completely remove the electron is called the ionization energy). The analysis of the spectral lines shows that the energy is lower than expected and the energy is proportional to Z^2. One obtains

$$Z_{eff}^{atom} = \sqrt{\frac{E^{atom}}{E^H}}$$

For example, for helium, substitution of the measured value of 2372 kJ/mol compared to the value of 1312 kJ/mol for hydrogen gives $Z_{eff}^{atom} = 1.34$. One must compare equivalent orbitals. For lithium, an ionization energy is 513 kJ/mol, and comparison to the 1312 kJ/mol value for hydrogen yields the wrong results $Z_{eff}^{atom} = 0.62$ (this is much lower than expected because two electrons will produce the effective charge in the range of 2). The problem is that the ionization energy for lithium is for the 2s-orbital and not the 1s-orbital. For lithium, the ionization energy for the 1s-orbital is 7298 kJ/mol resulting in $Z_{eff}^{atom} = 2.4$. Using the 2s value for hydrogen of 1.27 leads to $Z_{eff}^{atom} = 1.27$. Therefore, the correct comparisons give the effective charges in the expected ranges of values. Though this concept can be used for some media, this approach, in general, is limited and cannot be effectively applied to even simple nanostructures.

7.1.6.3   Hartree-Fock Modeling Method: Self-Consistent Field Theory

Another approach is to calculate how the wave functions should be changed in response to the interactions between electrons. The potential for i electrons in the system is derived as

$$\Pi = -\sum_i \frac{Ze^2}{4\pi\varepsilon_0 r_i} + \frac{1}{2}\sum_{ij} \frac{e^2}{4\pi\varepsilon_0 r_{ij}} = \Pi_0 + \Pi_1$$

In this equation, Π_0 denotes the potential between the nucleus and each electron, while Π_1 represents the interactions between electrons. The assumption is that $\Pi_0 \gg \Pi_1$. If $\Pi_0 \gg \Pi_1$, the wave functions can be given as

$$\Psi(\mathbf{r}) = \Psi_0(\mathbf{r}) + \Psi_1(\mathbf{r}) \quad \text{and} \quad E(\mathbf{r}) = E_0(\mathbf{r}) + E_1(\mathbf{r})$$

where the second term is much smaller than the first term. The Schrödinger equation becomes

$$-\frac{\hbar^2}{2m} \nabla^2 (\Psi_0(\mathbf{r}) + \Psi_1(\mathbf{r})) + (\Pi_0(\mathbf{r}) + \Pi_1(\mathbf{r}))(\Psi_0(\mathbf{r}) + \Psi_1(\mathbf{r})) = (E_0(\mathbf{r}) + E_1(\mathbf{r}))(\Psi_0(\mathbf{r}) + \Psi_1(\mathbf{r}))$$

This equation can be simplified because Ψ_0 is the solution to the zero terms, and

$$\Psi_1(\mathbf{r}) = \sum a \, \Psi_0(\mathbf{r})$$

Though the method can be viewed as a well-defined approach, for even simple nanostructures, formidable difficulties and challenges arise because extremely complex (from the quantum theory viewpoint) interactive molecules ultimately cannot be examined even if $\Pi_0 \gg \Pi_1$ is guaranteed.

7.2 Molecular and Nanostructure Dynamics

Conventional, miniscale, and microscale electromechanical systems can be modeled, simulated, and analyzed using electromagnetic and circuitry theories, classical mechanics, and thermodynamics. Other fundamental concepts are applied as well. The mathematical models of microelectromechanical systems (nonlinear ordinary and partial differential equations explicitly describe the spectrum of electromagnetic and electromechanical phenomena, processes, and effects) are not ambiguous, and numerical algorithms to solve the equations have been derived. Computationally efficient software and environments to support heterogeneous simulation and data-intensive analysis are available. Nanoscale structures, devices, and systems, in general, cannot be studied using the conventional concepts.

The ability to find equations (mathematical models) that adequately describe nanosystem properties, phenomena, and effects is a key problem in modeling, analysis, synthesis, optimization, control, fabrication, manufacturing, and implementation of nanosystems, nanodevices, and nanostructures. In this section, using classical and quantum mechanics, functional density concept, and electromagnetic theory, we utilize distinct concepts to model, analyze, and simulate nanosystems. The reported developments support the existing paradigms in modeling, analysis, and design of nanosystems. The documented fundamental results allow the designer to solve a broad spectrum of problems compared with currently applied methods. The reported theoretical and applied results are verified and demonstrated. Studying novel phenomena in nanosystems using quantum methods is critical to overcome current obstacles to complete understanding of processes and phenomena in nanoscale. In particular, the long-standing goal is the development of fundamental and experimental tools to design nanosystems.

The fundamental and applied research in molecular nanotechnology and nanosystems is concentrated on design, modeling, simulation, and fabrication of molecular-scale structures and devices. The design, modeling, and simulation of nano- and microsystems and their components can be attacked using advanced theoretical developments and simulation concepts. Comprehensive analysis must be performed before the designer engages in costly fabrication (nanoscale structures and devices, molecular machines, and subsystems can be fabricated with atomic precision) because through modeling and simulation, evaluation and prototyping can be performed, thereby facilitating significant advantages and manageable perspectives to attain the desired objectives. With advanced computer-aided design tools, nanostructures, nanodevices, and nanosystems can be designed, analyzed, and evaluated.

Classical quantum mechanics does not allow the designer to perform analytical and numerical analysis even for simple nanostructures that consist of a couple of molecules. Steady-state three-dimensional modeling and simulation are also restricted to simple nanostructures. Our goal is to introduce fundamental concepts that model phenomena at nanoscale with emphasis on their further applications in nanodevices, nanosubsystems, and nanosystems. The objective is the development of theoretical fundamentals to perform 3D+ (three-dimensional dynamics in the time domain) modeling and simulation.

The atomic-level dynamics can be studied using the wave function. The Schrödinger equation for complex systems must be solved. However, this problem cannot be solved even for simple nanostructures. In previous studies,[1-5] the density functional theory was developed, and the charge density is used rather than the electron wave functions. In particular, the N-electron problem is formulated as N one-electron equations where each electron interacts with all other electrons via an effective exchange-correlation potential. These interactions are augmented using the charge density. Plane wave sets and total energy pseudo-potential methods can be used to solve the Kohn-Sham one-electron equations.[1-3] The Feynman theory can be applied to calculate the forces solving the molecular dynamics problem.[4,6]

7.2.1 Schrödinger Equation and Wave Function Theory

For two point charges, Coulomb's law is given as

$$\mathbf{F} = \frac{q_1 q_2}{4\pi\varepsilon d^2}\mathbf{a}_r = \frac{q_1 q_2}{4\pi\varepsilon}\frac{(\mathbf{r}-\mathbf{r}')}{|\mathbf{r}-\mathbf{r}'|^3}$$

and in the Cartesian coordinate systems one has

$$\mathbf{F} = \frac{q_1 q_2}{4\pi\varepsilon d^2}\mathbf{a}_r = \frac{q_1 q_2}{4\pi\varepsilon d^2}\frac{(x-x')\mathbf{a}_x + (y-y')\mathbf{a}_y + (z-z')\mathbf{a}_z}{\sqrt{(x-x')^2 + (y-y')^2 + (z-z')^2}}$$

In the case of charge distribution, using the volume charge density ρ_v, the net force exerted on q_1 by the entire volume charge distribution is the vector sum of the contribution from all differential elements of charge within this distribution. In particular,

$$\mathbf{F} = \frac{q_1}{4\pi\varepsilon}\int_v \rho_v \frac{(\mathbf{r}-\mathbf{r}')}{|\mathbf{r}-\mathbf{r}'|^3}dv$$

(see Fig. 7.13).

In the electrostatic field, the potential energy stored in a region of continuous charge distribution is found as

$$\Pi_V = \frac{1}{2}\int_v \mathbf{D}\cdot\mathbf{E}dv = \frac{1}{2}\int_v \varepsilon\mathbf{E}^2 dv = \frac{1}{2}\int_v \rho_v(\mathbf{r})V(\mathbf{r})dv$$

where $V(\mathbf{r})$ is the potential, and v is the volume containing ρ_v.

The charge distribution can be given in terms of volume, surface, and line charges. In particular, we have

$$V(\mathbf{r}) = \int_v \frac{\rho_v(\mathbf{r}')}{4\pi\varepsilon|\mathbf{r}-\mathbf{r}'|}dv', \quad V(\mathbf{r}) = \int_s \frac{\rho_s(\mathbf{r}')}{4\pi\varepsilon|\mathbf{r}-\mathbf{r}'|}ds' \quad \text{and} \quad V(\mathbf{r}) = \int_l \frac{\rho_l(\mathbf{r}')}{4\pi\varepsilon|\mathbf{r}-\mathbf{r}'|}dl'$$

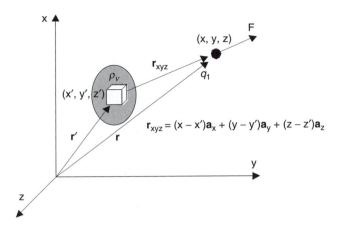

FIGURE 7.13
Coulomb's law.

It should be emphasized that that the electric field intensity is found as

$$\mathbf{E}(\mathbf{r}) = \int_v \frac{\rho_v(\mathbf{r}')}{4\pi\varepsilon} \frac{(\mathbf{r}-\mathbf{r}')}{|\mathbf{r}-\mathbf{r}'|^3} dv'$$

Thus, the energy of an electric field or a charge distribution is stored in the field.
  The energy, stored in the steady magnetic field is

$$\Pi_M = \frac{1}{2}\int_v \mathbf{B}\cdot\mathbf{H} dv$$

7.2.1.1 *Mathematical Models: Energy-Based Quantum and Classical Mechanics*

To perform the comprehensive modeling and analysis of nanostructures in the time domain, there is a critical need to develop and apply advanced theories using fundamental physical laws. Classical and quantum mechanics are widely used, and this section illustrates that the Schrödinger equation can be found using Hamilton's concept (it is well known that the Euler-Lagrange equations, given in terms of the generalized coordinates and forces, can be straightforwardly derived applying the variational principle). Quantum mechanics gives the system evolution in the form of the Schrödinger equations.
  Newton's second law $\sum \vec{F}(t,\mathbf{r}) = m\vec{a}$ in terms of the linear momentum $\vec{p} = m\vec{v}$ is given by

$$\sum \vec{F} = \frac{d\vec{p}}{dt} = \frac{d(m\vec{v})}{dt}$$

Using the potential energy $\Pi(\vec{r})$, for the conservative mechanical system we have

$$\sum \vec{F}(\vec{r}) = -\nabla\Pi(\vec{r})$$

Hence, $m(d^2\vec{r}/dt^2) + \nabla\Pi(\vec{r}) = 0$.
  For the system of N particles, the equations of motion are

$$m_N \frac{d^2\vec{r}_N}{dt^2} + \nabla\Pi(\vec{r}_N) = 0, \quad m_i \frac{d^2(\vec{x}_i, \vec{y}_i, \vec{z}_i)}{dt^2} + \frac{\partial\Pi(\vec{x}_i, \vec{y}_i, \vec{z}_i)}{\partial(\vec{x}_i, \vec{y}_i, \vec{z}_i)} = 0, \quad i = 1, \dots, N$$

The total kinetic energy of the particle is $\Gamma = \frac{1}{2}mv^2$. For N particles, one has

$$\Gamma\left(\frac{d\vec{x}_i}{dt}, \frac{d\vec{y}}{dt}_i, \frac{d\vec{z}_i}{dt}\right) = \frac{1}{2}\sum_{i=1}^{N}m_i\left(\frac{d\vec{x}_i}{dt}, \frac{d\vec{y}}{dt}_i, \frac{d\vec{z}_i}{dt}\right)$$

Using the generalized coordinates $(q_1, \ldots, q_n)$ and generalized velocities $(dq_1/dt, \ldots, dq_n/dt)$, one finds the total kinetic $\Gamma(q_1, \ldots, q_n, dq_1/dt, \ldots, dq_n/dt)$ and potential $\Pi(q_1, \ldots, q_n)$ e-nergies. Thus, Newton's second law of motion can be given as

$$\frac{d}{dt}\left(\frac{\partial\Gamma}{\partial\dot{q}_i}\right) + \frac{\partial\Pi}{\partial q_i} = 0$$

That is, the generalized coordinates q_i are used to model multibody systems, and

$$(q_1, \ldots, q_n) = (\vec{x}_1, \vec{y}_1, \vec{z}_1, \ldots, \vec{x}_N, \vec{y}_N, \vec{z}_N)$$

The obtained results are connected to the Lagrange equations of motion. Using the total kinetic $\Gamma(t, q_1, \ldots, q_n, dq_1/dt, \ldots, dq_n/dt)$, dissipation $D(t, q_1, \ldots, q_n, dq_1/dt, \ldots, dq_n/dt)$, and potential $\Pi(t, q_1, \ldots, q_n)$ energies, the Lagrange equations of motion are

$$\frac{d}{dt}\left(\frac{\partial\Gamma}{\partial\dot{q}_i}\right) - \frac{\partial\Gamma}{\partial q_i} + \frac{\partial D}{\partial\dot{q}_i} + \frac{\partial\Pi}{\partial q_i} = Q_i$$

Here, q_i and Q_i are the generalized coordinates and the generalized forces (applied forces and disturbances).

The Hamilton concept allows one to model the system dynamics, and the differential equations are found using the generalized momenta p_i, $p_i = \frac{\partial L}{\partial\dot{q}_i}$. The Lagrangian function $L(t, q_1, \ldots, q_n, dq_1/dt, \ldots, dq_n/dt)$ for the conservative systems is the difference between the total kinetic and potential energies. We have

$$L\left(t, q_1, \ldots, q_n, \frac{dq_1}{dt}, \ldots, \frac{dq_n}{dt}\right) = \Gamma\left(t, q_1, \ldots, q_n, \frac{dq_1}{dt}, \ldots, \frac{dq_n}{dt}\right) - \Pi(t, q_1, \ldots, q_n)$$

Thus, $L(t, q_1, \ldots, q_n, dq_1/dt, \ldots, dq_n/dt)$ is the function of $2n$ independent variables, and

$$dL = \sum_{i=1}^{n}\left(\frac{\partial L}{\partial q_i}dq_i + \frac{\partial L}{\partial\dot{q}_i}d\dot{q}_i\right) = \sum_{i=1}^{n}(\dot{p}_idq_i + p_id\dot{q}_i)$$

Define the Hamiltonian function as

$$H(t, q_1, \ldots, q_n, p_1, \ldots, p_n) = -L\left(t, q_1, \ldots, q_n, \frac{dq_1}{dt}, \ldots, \frac{dq_n}{dt}\right) + \sum_{i=1}^{n}p_i\dot{q}_i$$

$$dH = \sum_{i=1}^{n}(-\dot{p}_idq_i + \dot{q}_idp_i)$$

where

$$\sum_{i=1}^{n}p_i\dot{q}_i = \sum_{i=1}^{n}\frac{\partial L}{\partial\dot{q}_i}\dot{q}_i = \sum_{i=1}^{n}\frac{\partial\Gamma}{\partial\dot{q}_i}\dot{q}_i = 2\Gamma$$

Thus,

$$H\left(t, q_1, \ldots, q_n, \frac{dq_1}{dt}, \ldots, \frac{dq_n}{dt}\right) = \Gamma\left(t, q_1, \ldots, q_n, \frac{dq_1}{dt}, \ldots, \frac{dq_n}{dt}\right) + \Pi(t, q_1, \ldots, q_n)$$

or

$$H(t, q_1, \ldots, q_n, p_1, \ldots, p_n) = \Gamma(t, q_1, \ldots, q_n, p_1, \ldots, p_n) + \Pi(t, q_1, \ldots, q_n)$$

One concludes that the Hamiltonian function, which is equal to the total energy, is expressed as a function of the generalized coordinates and generalized momenta. The equations of motion are governed by the following equations:

$$\dot{p}_i = -\frac{\partial H}{\partial q_i}, \quad \dot{q}_i = \frac{\partial H}{\partial p_i}$$

which are the Hamiltonian equations of motion.

The Hamiltonian function

$$H = \underbrace{-\frac{\hbar^2}{2m}\nabla^2}_{\text{one-electron kinetic energy}} + \underbrace{\Pi}_{\text{potential energy}}$$

can be used to derive the one-electron Schrödinger equation. To describe the behavior of electrons in a media, one must use the N-dimensional Schrödinger equation to obtain the N-electron wave function $\Psi(t, \mathbf{r}_1, \mathbf{r}_2, \ldots, \mathbf{r}_{N-1}, \mathbf{r}_N)$.

The Hamiltonian for an isolated N-electron atomic system is

$$H = -\frac{\hbar^2}{2m}\sum_{i=1}^{N}\nabla_i^2 - \frac{\hbar^2}{2M}\nabla^2 - \sum_{i=1}^{N}\frac{1}{4\pi\varepsilon}\frac{e_i q}{|\mathbf{r}_i - \mathbf{r}_n'|} + \sum_{i\neq j}^{N}\frac{1}{4\pi\varepsilon}\frac{e^2}{|\mathbf{r}_i - \mathbf{r}_j'|}$$

where q is the potential due to nucleus, and $e = 1.6 \times 10^{-19}$ C.

For an isolated N-electron, Z-nucleus molecular system, the Hamiltonian function (Hamiltonian operator) is found to be

$$H = -\frac{\hbar^2}{2m}\sum_{i=1}^{N}\nabla_i^2 - \sum_{k=1}^{Z}\frac{\hbar^2}{2m_k}\nabla_k^2 - \sum_{i=1}^{N}\sum_{k=1}^{Z}\frac{1}{4\pi\varepsilon}\frac{e_i q_k}{|\mathbf{r}_i - \mathbf{r}_k'|} + \sum_{i\neq j}^{N}\frac{1}{4\pi\varepsilon}\frac{e^2}{|\mathbf{r}_i - \mathbf{r}_j'|} + \sum_{k\neq m}^{Z}\frac{1}{4\pi\varepsilon}\frac{q_k q_m}{|\mathbf{r}_k - \mathbf{r}_m'|}$$

where q_k are the potentials due to nuclei.

The first and second terms of the Hamiltonian function

$$-\frac{\hbar^2}{2m}\sum_{i=1}^{N}\nabla_i^2 \quad \text{and} \quad -\sum_{k=1}^{Z}\frac{\hbar^2}{2m_k}\nabla_k^2$$

are the multibody kinetic energy operators.

The term

$$-\sum_{i=1}^{N}\sum_{k=1}^{Z}\frac{1}{4\pi\varepsilon}\frac{e_i q_k}{|\mathbf{r}_i - \mathbf{r}_k'|}$$

maps the interaction of the electrons with the nuclei at $\mathbf{R}$ (the electron-nucleus attraction energy operator).

In the Hamiltonian, the fourth term

$$\sum_{i \neq j}^{N} \frac{1}{4\pi\varepsilon} \frac{e^2}{|\mathbf{r}_i - \mathbf{r}_j'|}$$

gives the interactions of electrons with each other (the electron-electron repulsion energy operator).

The term

$$\sum_{k \neq m}^{Z} \frac{1}{4\pi\varepsilon} \frac{q_k q_m}{|\mathbf{r}_k - \mathbf{r}_m'|}$$

describes the interaction of the Z nuclei at $\mathbf{R}$ (the nucleus-nucleus repulsion energy operator).

For an isolated N-electron, Z-nucleus atomic or molecular system in the Born-Oppenheimer nonrelativistic approximation, we have $H\Psi = E\Psi$. The Schrödinger equation is

$$\left[-\frac{\hbar^2}{2m} \sum_{i=1}^{N} \nabla_i^2 - \sum_{k=1}^{Z} \frac{\hbar^2}{2m_k} \nabla_k^2 - \sum_{i=1}^{N}\sum_{k=1}^{Z} \frac{1}{4\pi\varepsilon} \frac{e_i q_k}{|\mathbf{r}_i - \mathbf{r}_k'|} + \sum_{i \neq j}^{N} \frac{1}{4\pi\varepsilon} \frac{e^2}{|\mathbf{r}_i - \mathbf{r}_j'|} + \sum_{k \neq m}^{Z} \frac{1}{4\pi\varepsilon} \frac{q_k q_m}{|\mathbf{r}_k - \mathbf{r}_m'|} \right]$$

$$\times \Psi(t, \mathbf{r}_1, \mathbf{r}_2, \ldots, \mathbf{r}_{N-1}, \mathbf{r}_N) = E(t, \mathbf{r}_1, \mathbf{r}_2, \ldots, \mathbf{r}_{N-1}, \mathbf{r}_N)\Psi(t, \mathbf{r}_1, \mathbf{r}_2, \ldots, \mathbf{r}_{N-1}, \mathbf{r}_N) \qquad (7.1)$$

The total energy $E(t, \mathbf{r}_1, \mathbf{r}_2, \ldots, \mathbf{r}_{N-1}, \mathbf{r}_N)$ must be found using the nucleus-nucleus Coulomb repulsion energy as well as the electron energy.

It is very difficult, or impossible, to solve analytically or numerically the nonlinear partial differential Equation 7.1. Taking into account only the Coulomb force (electrons and nuclei are assumed to interact due to the Coulomb force only), the Hartree approximation is applied. In particular, the N-electron wave function $\Psi(t, \mathbf{r}_1, \mathbf{r}_2, \ldots, \mathbf{r}_{N-1}, \mathbf{r}_N)$ is expressed as a product of N one-electron wave functions as

$$\Psi(t, \mathbf{r}_1, \mathbf{r}_2, \ldots, \mathbf{r}_{N-1}, \mathbf{r}_N) = \psi_1(t, \mathbf{r}_1)\psi_2(t, \mathbf{r}_2) \ldots \psi_{N-1}(t, \mathbf{r}_{N-1})\psi_N(t, \mathbf{r}_N)$$

The one-electron Schrödinger equation for jth electron is

$$\left(-\frac{\hbar^2}{2m} \nabla_j^2 + \Pi(t, \mathbf{r}) \right) \psi_j(t, \mathbf{r}) = E_j(t, \mathbf{r})\psi_j(t, \mathbf{r}) \qquad (7.2)$$

In Equation 7.2, the first term $-(\hbar^2/2m)\nabla_j^2$ is the one-electron kinetic energy, and $\Pi(t, \mathbf{r}_j)$ is the total potential energy. The potential energy includes the potential that the jth electron feels from the nucleus (considering the ion, the repulsive potential in the case of anion, or the attractive potential in the case of cation). It is obvious that the jth electron feels the repulsion (repulsive forces) from other electrons. Assume that the negative electrons' charge density $\rho(\mathbf{r})$ is smoothly distributed in $\mathbf{R}$. Hence, the potential energy due to interaction (repulsion) of an electron in $\mathbf{R}$ is

$$\Pi_{Ej}(t, \mathbf{r}) = \int_{\mathbf{R}} \frac{e\rho(\mathbf{r}')}{4\pi\varepsilon|\mathbf{r} - \mathbf{r}'|} d\mathbf{r}'$$

We made some assumptions, and the results derived contradict with some fundamental principles. The Pauli exclusion principle requires that the multisystem wave function be antisymmetric under the interchange of electrons. For two electrons, we have

$$\Psi(t, \mathbf{r}_1, \mathbf{r}_2, \ldots, \mathbf{r}_j, \ldots, \mathbf{r}_{j+i}, \mathbf{r}_{N-1}, \mathbf{r}_N) = -\Psi(t, \mathbf{r}_1, \mathbf{r}_2, \ldots, \mathbf{r}_{j+i}, \mathbf{r}_j, \ldots, \mathbf{r}_{N-1}, \mathbf{r}_N)$$

This principle is not satisfied, and the generalization is needed to integrate the asymmetry phenomenon using the asymmetric coefficien ±1. The Hartree-Fock equation is

$$\left[-\frac{\hbar^2}{2m}\nabla_j^2 + \Pi(t, \mathbf{r})\right]\psi_j(t, \mathbf{r}) - \sum_i \int_R \frac{\psi_i^*(t, \mathbf{r}')\psi_j(t, \mathbf{r}')\psi_i(t, \mathbf{r})\psi_j^*(t, \mathbf{r})}{|\mathbf{r}-\mathbf{r}'|}d\mathbf{r}' = E_j(t, \mathbf{r})\psi_j(t, \mathbf{r}) \quad (7.3)$$

The so-called Hartree-Fock nonlinear partial differential Equation 7.3, which is difficult to solve, is the approximation because the multibody electron interactions should be considered in general. Thus, the explicit equation for the total energy must be used. This phenomenon can be integrated using the charge density function.

7.2.2 Density Functional Theory

There is a critical need to develop computationally efficient and accurate procedures to perform quantum modeling of nanoscale structures. This section reports the related results and gives the formulation of the modeling problem to avoid the complexity associated with many-electron wave functions that result if the classical quantum mechanics formulation is used. The complexity of the Schrödinger equation is enormous even for very simple molecules. For example, the carbon atom has six electrons. Can one visualize six-dimensional space? Furthermore, the simplest carbon nanotube molecule has six carbon atoms. That is, one has 36 electrons, and a 36-dimensional problem results. The difficulties associated with the solution of the Schrödinger equation drastically limit the applicability of conventional quantum mechanics. The analysis of properties, processes, phenomena, and effects in even the simplest nanostructures cannot be studied and comprehended. The problems can be solved applying the Hohenberg-Kohn density functional theory.

The statistical consideration, proposed by Thomas and Fermi in 1927, gives the distribution of electrons in atoms. The following assumptions were used: electrons are distributed uniformly, and there is an effective potential field that is determined by the nuclei charge and the distribution of electrons. Considering electrons distributed in a three-dimensional box, the energy analysis can be performed. Summing all energy levels, one finds the energy. Thus, one can relate the total kinetic energy and the electron charge density. The statistical consideration can be used in order to approximate the distribution of electrons in an atom. The relation between the total kinetic energy of N electrons E and the electron density was derived using the local density approximation concept.

The Thomas-Fermi kinetic energy functional is

$$\Gamma_F(\rho_e(\mathbf{r})) = 2.87 \int_R \rho_e^{5/3}(\mathbf{r})d\mathbf{r}$$

and the exchange energy is found to be

$$E_F(\rho_e(\mathbf{r})) = 0.739 \int_R \rho_e^{4/3}(\mathbf{r})d\mathbf{r}$$

For homogeneous atomic systems, applying the electron charge density $\rho_e(\mathbf{r})$, Thomas and Fermi derived the following energy functional:

$$E_F(\rho_e(\mathbf{r})) = 2.87 \int_R \rho_e^{5/3}(\mathbf{r})d\mathbf{r} - q\int_R \frac{\rho_e(\mathbf{r})}{r}d\mathbf{r} + \int_R\int_R \frac{1}{4\pi\varepsilon}\frac{\rho_e(\mathbf{r})\rho_e(\mathbf{r}')}{|\mathbf{r}-\mathbf{r}'|}d\mathbf{r}d\mathbf{r}'$$

considering electrostatic electron-nucleus attraction and electron-electron repulsion.

Following this idea, instead of the many-electron wave functions, the charge density for N-electron systems can be used. Only knowledge of the charge density is needed to perform analysis of molecular dynamics. The charge density is the function that describes the number of electrons per unit volume (function of three spatial variables x, y, and z in the Cartesian coordinate system). Quantum mechanics and quantum modeling must be applied to understand and analyze nanostructures and nanodevices because they operate under the quantum effects.

The total energy of N-electron system under the external field is defined in terms of the three-dimensional charge density $\rho(\mathbf{r})$. The complexity is significantly decreased because the problem of modeling N-electron, Z-nucleus systems becomes equivalent to the solution of the equation for one electron. The total energy is given as

$$E(t,\rho(\mathbf{r})) = \underbrace{\Gamma_1(t,\rho(\mathbf{r})) + \Gamma_2(t,\rho(\mathbf{r}))}_{\text{kinetic energy}} + \underbrace{\int_{\mathbf{R}} \frac{e\rho(\mathbf{r}')}{4\pi\varepsilon|\mathbf{r}-\mathbf{r}'|}d\mathbf{r}'}_{\text{potential energy}} \qquad (7.4)$$

where $\Gamma_1(t,\rho(\mathbf{r}))$ and $\Gamma_2(t,\rho(\mathbf{r}))$ are the interacting (exchange) and noninteracting kinetic energies of a single electron in an N-electron, Z-nucleus system,

$$\Gamma_1(t,\rho(\mathbf{r})) = \int_{\mathbf{R}} \gamma(t,\rho(\mathbf{r}))\rho(\mathbf{r})d\mathbf{r} \quad \text{and} \quad \Gamma_2(t,\rho(\mathbf{r})) = -(\hbar^2/2m)\sum_{j=1}^{N}\int_{\mathbf{R}} \Psi_j^*(t,\mathbf{r})\nabla_j^2\Psi(t,r)dr$$

and $\gamma(t,\rho(\mathbf{r}))$ is the parameterization function.

It should be emphasized that the Kohn-Sham electronic orbitals are subject to the following orthogonal condition: $\int_{\mathbf{R}} \Psi_i^*(t,\mathbf{r})\Psi_j(t,\mathbf{r})d\mathbf{r} = \delta_{ij}$.

The state of a substance (media) depends largely on the balance between the kinetic energies of the particles and the interparticle energies of attraction.

The expression for the total potential energy is easily justified.

The term

$$\int_{\mathbf{R}} \frac{e\rho(\mathbf{r}')}{4\pi\varepsilon|\mathbf{r}-\mathbf{r}'|}d\mathbf{r}'$$

represents the Coulomb interaction in $\mathbf{R}$, and the total potential energy is a function of the charge density $\rho(\mathbf{r})$.

The total kinetic energy (interactions of electrons and nuclei, and electrons) is integrated into the equation for the total energy. The total energy, as given by Equation 7.4, is stationary with respect to variations in the charge density. The charge density is found by taking note of the Schrödinger equation. The first-order Fock-Dirac electron charge density matrix is

$$\rho_e(\mathbf{r}) = \sum_{j=1}^{N} \Psi_j^*(t,\mathbf{r})\Psi_j(t,\mathbf{r}) \qquad (7.5)$$

The three-dimensional electron charge density is a function in three variables (x, y, and z in the Cartesian coordinate system). Integrating the electron charge density $\rho_e(\mathbf{r})$, one obtains the charge of the total number of electrons N. Thus,

$$\int_{\mathbf{R}} \rho_e(\mathbf{r})\,d\mathbf{r} = Ne$$

Hence, $\rho_e(\mathbf{r})$ satisfies the following properties:

$$\rho_e(\mathbf{r}) < 0, \quad \int_{\mathbf{R}} \rho_e(\mathbf{r})\,d\mathbf{r} = Ne, \quad \int_{\mathbf{R}} \left|\sqrt{\nabla\rho_e(\mathbf{r})}\right|^2 d\mathbf{r} < \infty \quad \text{and} \quad \int_{\mathbf{R}} \nabla^2\rho_e(\mathbf{r})\,d\mathbf{r} = \infty$$

For the nuclei charge density, we have $\rho_n(\mathbf{r}) > 0$ and

$$\int_R \rho_n(\mathbf{r})d\mathbf{r} = \sum_{k=1}^{Z} q_k \,.$$

There exist an infinite number of antisymmetric wave functions that give the same $\rho(\mathbf{r})$. The minimum-energy concept (energy-functional minimum principle) is applied. The total energy is a function of $\rho(\mathbf{r})$, and the ground state Ψ must minimize the expectation value $\langle E(\rho) \rangle$.

The searching density functional $F(\rho)$, which searches all Ψ in the N-electron Hilbert space H_Ψ to find $\rho(r)$ and guarantee the minimum to the energy expectation value, is expressed as

$$F(\rho) \leq \min_{\substack{\Psi \to \rho \\ \Psi \in H_\Psi}} \langle \Psi | E(\rho) | \Psi \rangle$$

where H_ψ is any subset of the N-electron Hilbert space.

Using the variational principle, we have

$$\frac{\Delta E(\rho)}{\Delta f(\rho)} = \int_R \frac{\Delta E(\rho)}{\Delta \rho(\mathbf{r}')} \frac{\Delta \rho(\mathbf{r}')}{\Delta f(\mathbf{r})} d\mathbf{r}' = 0$$

where $f(\rho)$ is the nonnegative function. Thus,

$$\left. \frac{\Delta E(\rho)}{\Delta f(\rho)} \right|_N = \text{const}$$

The solution to the system of equations in Equation 7.2 is found using the charge density as given by Equation 7.5. Hence, to perform the analysis, one studies the molecular dynamics. Distinct phenomena at the molecular level can be examined. Substituting the expression for the total kinetic and potential energies in Equation 7.4, where the charge density is given by Equation 7.5, the total energy $E(t, \rho(\mathbf{r}))$ results. It should be emphasized that the external energy is supplied to control nanodevices, and one has

$$E_\Sigma(t, \mathbf{r}) = E_{external}(t, \mathbf{r}) + E(t, \rho(\mathbf{r}))$$

Then, for example, studying mechanical motions, the force at position $\mathbf{r}_r$ is

$$\mathbf{F}_r(t, \mathbf{r}) = -\frac{dE_\Sigma(t, \mathbf{r})}{d\mathbf{r}_r} = -\frac{\partial E_\Sigma(t, \mathbf{r})}{\partial \mathbf{r}_r} - \sum_j \frac{\partial E(t, \mathbf{r})}{\partial \psi_j(t, \mathbf{r})} \frac{\partial \psi_j(t, \mathbf{r})}{\partial \mathbf{r}_r} - \sum_j \frac{\partial E(t, \mathbf{r})}{\partial \psi_j^*(t, \mathbf{r})} \frac{\partial \psi_j^*(t, \mathbf{r})}{\partial \mathbf{r}_r} \quad (7.6)$$

Taking note of

$$\sum_j \frac{\partial E(t, \mathbf{r})}{\partial \psi_j(t, \mathbf{r})} \frac{\partial \psi_j(t, \mathbf{r})}{\partial \mathbf{r}_r} + \sum_j \frac{\partial E(t, \mathbf{r})}{\partial \psi_j^*(t, \mathbf{r})} \frac{\partial \psi_j^*(t, \mathbf{r})}{\partial \mathbf{r}_r} = 0$$

the expression for the force is found from Equation 7.6. In particular, one finds

$$\mathbf{F}_r(t, \mathbf{r}) = -\frac{\partial E_{external}(t, \mathbf{r})}{\partial \mathbf{r}_r} - \int_R \rho(t, \mathbf{r}) \frac{\partial [\Pi_r(t, \mathbf{r}) + \Gamma_r(t, \mathbf{r})]}{\partial \mathbf{r}_r} d\mathbf{r} - \int_R \frac{\partial E_\Sigma(t, \mathbf{r})}{\partial \rho(t, \mathbf{r})} \frac{\partial \rho(t, \mathbf{r})}{\partial \mathbf{r}_r} d\mathbf{r}$$

As the wave functions converge (the conditions of the Feynman theorem are satisfied), we have

$$\int_R \frac{\partial E(t,\mathbf{r})}{\partial \rho(t,\mathbf{r})} \frac{\partial \rho(t,\mathbf{r})}{\partial \mathbf{r}_r} d\mathbf{r} = 0$$

Utilizing the functional density theory, one can deduce the expression for the wave functions, find the charge density, examine the phenomena, and study processes at nanoscale.

7.2.3 Nanostructures and Molecular Dynamics

Atomistic modeling can be performed using the force field method. The effective interatomic potential for a system of N particles is found as the sum of the second-, third-, fourth-, and higher-order terms as

$$\Pi(\mathbf{r}_1,\ldots,\mathbf{r}_N) = \sum_{i,j=1}^{N} \Pi^{(2)}(\mathbf{r}_{ij}) + \sum_{i,j,k=1}^{N} \Pi^{(3)}(\mathbf{r}_i,\mathbf{r}_j,\mathbf{r}_k) + \sum_{i,j,k,l=1}^{N} \Pi^{(4)}(\mathbf{r}_i,\mathbf{r}_j,\mathbf{r}_k,\mathbf{r}_l) + \cdots$$

Usually, the interatomic effective pair potential $\sum_{i,j=1}^{N}\Pi^{(2)}(\mathbf{r}_{ij})$, which depends on the interatomic distance r_{ij} between the nuclei i and j, dominates. For example, the three-body interconnection terms cannot be omitted only if the angle-dependent potentials are considered. Using the effective ionic charges Q_i and Q_j, we have

$$\Pi^{(2)} = \underbrace{\frac{Q_i Q_j}{4\pi\varepsilon r_{ij}}}_{electrostatic} + \underbrace{\phi(\mathbf{r}_{ij})}_{short-range}$$

where $\phi(\mathbf{r}_{ij})$ is the short-range interaction energy due to the repulsion between electron charge clouds, Van der Waals attraction, bond bending, and stretching phenomena.

For ionic and partially ionic media we have

$$\phi(r_{ij}) = k_{1ij}e^{-k_{2ij}r_{ij}} - k_{3ij}r_{ij}^{-6} + k_{4ij}r_{ij}^{-12}$$

where $k_{1ij} = \sqrt{k_{1i}k_{1j}}$, $k_{2ij} = \sqrt{k_{2i}k_{2j}}$, $k_{3ij} = \sqrt{k_{3i}k_{3j}}$, and $k_{4ij} = \sqrt{k_{4i}k_{4j}}$; k_i are the bond energy constants (for example, for Si we have $Q = 2.4$, $k_3 = 0.00069$, and $k_4 = 104$, for Al one has $Q = 1.4$, $k_3 = 1690$, and $k_4 = 278$, and for Na^+ we have $Q = 1$, $k_3 = 0.00046$, and $k_4 = 67423$).

Another commonly used approximation is $\phi(r_{ij}) = k_{5ij}(r_{ij} - r_{Eij})$, where r_{ij} is the bond length, $r_{ij} = |\mathbf{r}_j - \mathbf{r}_i|$, and r_{Eij} is the equilibrium bond distance.

Performing the summations in the studied $\mathbf{R}$, one finds the potential energy, and the force results. The position (displacement) is represented by the vector $\mathbf{r}$, which in the Cartesian coordinate system has the components x, y, and z. Taking note of the expression for the potential energy $\Pi(\vec{\mathbf{r}}) = \Pi(\mathbf{r}_1,\ldots,\mathbf{r}_N)$, one has

$$\sum \vec{F}(\vec{\mathbf{r}}) = -\nabla\Pi(\vec{\mathbf{r}})$$

From Newton's second law for the system of N particles, we have the following equation of motion:

$$m_N \frac{d^2\vec{\mathbf{r}}_N}{dt^2} + \nabla\Pi(\vec{\mathbf{r}}_N) = 0$$

or

$$m_i \frac{d^2(\vec{x}_i, \vec{y}_i, \vec{z}_i)}{dt^2} + \frac{\partial \Pi(\vec{x}_i, \vec{y}_i, \vec{z}_i)}{\partial(\vec{x}_i, \vec{y}_i, \vec{z}_i)} = 0, \quad i = 1, \ldots, N$$

To perform molecular modeling, one applies the energy-based methods. It was shown that electrons can be considered explicitly. However, it can be assumed that electrons will obey the optimum distribution once the positions of the nuclei in **R** are known. This assumption is based on the Born-Oppenheimer approximation of the Schrödinger equation. This approximation is satisfied because nuclei mass is much greater than electron mass, and thus, nuclei motions (vibrations and rotations) are slow compared with the electrons' mo-tions. Therefore, nuclei motions can be studied separately from electrons' dynamics. Molecules can be studied as Z-body systems of elementary masses (nuclei) with springs (bonds between nuclei). The molecule potential energy (potential energy equation) is found using the number of nuclei and bond types (bending, stretching, lengths, geometry, angles, and other parameters), van der Waals radius, parameters of media, etc. The molecule potential energy surface is

$$E_T = E_{bs} - E_b - E_{sb} - E_{ts} - E_W - E_{dd}$$

Here, the energy due to bond stretching is found using the equation similar to Hook's law. In particular,

$$E_{bs} = k_{bs1}(l - l_0) + k_{bs3}(l - l_0)^3$$

where k_{bs1} and k_{bs3} are the constants, and l and l_0 are the actual and natural bond length (displacement).

The equations for energies due to bond angle bending E_b, stretch-bend interactions E_{sb}, torsion strain E_{ts}, van der Waals interactions E_W, and dipole-dipole interactions E_{dd} are well known and can be readily applied.

Example 7.14

Consider two separated hydrogen atoms (*A* and *B*) with electrons as documented in Fig. 7.14. If these atoms are infinitely separated, there can be no interaction between them, and the wave function of the system can be written as $\Psi(\mathbf{r}_1, \mathbf{r}_2, \mathbf{R}) = \Psi_{1s,A}(\mathbf{r}_1)\Psi_{1s,B}(\mathbf{r}_2)$. However, if the atoms interact, the potential energy is given as

$$\Pi(\mathbf{r}_A, \mathbf{r}_B, \mathbf{r}_1, \mathbf{r}_2, \mathbf{r}) = \frac{e^2}{4\pi\varepsilon}\left[-\left(\frac{Z_A}{|\mathbf{r}_1 - \mathbf{r}_A|} + \frac{Z_A}{|\mathbf{r}_2 - \mathbf{r}_A|} + \frac{Z_B}{|\mathbf{r}_1 - \mathbf{r}_B|} + \frac{Z_B}{|\mathbf{r}_2 - \mathbf{r}_B|}\right) + \frac{1}{|\mathbf{r}_1 - \mathbf{r}_2|} + \frac{Z_A Z_B}{|\mathbf{r}_A - \mathbf{r}_B|}\right].$$

The terms describe the electron-nucleus attraction, electron-electron repulsion, and nucleus-nucleus repulsion.

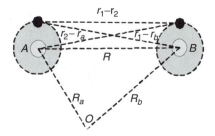

FIGURE 7.14
Two hydrogen atoms with electrons.

7.2.4   Electromagnetic Fields and Their Quantization

The mathematical models for energy conversion (energy storage, transport, and dissipation) and electromagnetic field (propagation, radiation, and other major characteristics) in media are found using Maxwell's equations. The vectors of electric field intensity *E*, electric flux density D, magnetic field intensity H, and magnetic flux density B are used as the cornerstone variables. The finite element analysis concept cannot be viewed as a meaningful paradigm because it provides one with the steady-state solution, which does not describe the important electromagnetic phenomena and effects even for conventional systems. The time-independent and frequency-domain Maxwell's and Schrödinger equations also have serious limitations in nano- and microscale system analysis. Therefore, complete mathematical models must be developed without simplifications and assumptions to understand, analyze, and comprehend a wide spectrum of phenomena and effects.

Maxwell's equations can be quantized. This concept provides a meaningful means for interpreting, understanding, predicting, and analyzing complex time-dependent behavior at nanoscale without invoking all the intricacy and difficulty of quantum electrodynamics, electromagnetism, and mechanics.

The Lorentz force on the charge *q* moving at the velocity *v* is found as

$$\mathbf{F}(t, \mathbf{r}) = q\mathbf{E}(t, \mathbf{r}) - q\mathbf{v} \times \mathbf{B}(t, \mathbf{r})$$

Using the electromagnetic potential *A*, we have

$$\mathbf{F} = q\left(-\frac{\partial \mathbf{A}}{\partial t} + \nabla V + \mathbf{v} \times (\nabla \times \mathbf{A})\right)$$

$$\mathbf{v} \times (\nabla \times \mathbf{A}) = \nabla(\mathbf{v} \cdot \mathbf{A}) - \frac{d\mathbf{A}}{dt} + \frac{\partial \mathbf{A}}{\partial t}$$

where $V(t, \mathbf{r})$ is the scalar electrostatic potential function (potential difference).

Four Maxwell's equations in the time domain are

$$\nabla \times \mathbf{E}(t, \mathbf{r}) = -\mu\frac{\partial \mathbf{H}(t, \mathbf{r})}{\partial t} - \mu\frac{\partial \mathbf{M}(t, \mathbf{r})}{\partial t}$$

$$\nabla \times \mathbf{H}(t, \mathbf{r}) = \mathbf{J}(t, \mathbf{r}) + \varepsilon\frac{\partial \mathbf{E}(t, \mathbf{r})}{\partial t} + \frac{\partial \mathbf{P}(t, \mathbf{r})}{\partial t}$$

$$\nabla \cdot \mathbf{E}(t, \mathbf{r}) = \frac{\rho_v(t, \mathbf{r})}{\varepsilon} - \frac{\nabla \mathbf{P}(t, \mathbf{r})}{\varepsilon}$$

$$\nabla \cdot \mathbf{H}(t, \mathbf{r}) = 0$$

where *J* is the current density, and using the conductivity σ, we have $\mathbf{J} = \sigma\mathbf{E}$; ε is the permittivity; μ is the permeability; and ρ_v is the volume charge density.

Using the electric *P* and magnetic *M* polarizations (dipole moment per unit volume) of the medium, one obtains the constitutive equations as

$$\mathbf{D}(t, \mathbf{r}) = \varepsilon\mathbf{E}(t, \mathbf{r}) - \mathbf{P}(t, \mathbf{r}) \quad \text{and} \quad \mathbf{B}(t, \mathbf{r}) = \mu\mathbf{H}(t, \mathbf{r}) + \mu\mathbf{M}(t, \mathbf{r})$$

The electromagnetic waves transfer the electromagnetic power, and we have

$$\int_v \nabla \cdot (\mathbf{E} \times \mathbf{H}) dv = \oint_s (\mathbf{E} \times \mathbf{H}) \cdot d\mathbf{s} = -\int_v \frac{\partial}{\partial t} \left(\tfrac{1}{2} \varepsilon \mathbf{E} \cdot \mathbf{E} + \tfrac{1}{2} \mu \mathbf{H} \cdot \mathbf{H} \right) dv$$

total power flowing into volume bounded by s — rate of change of the electromagnetic stored energy in electromagnetic fields

$$- \int_v \mathbf{E} \cdot \mathbf{J} dv - \int_v \mathbf{E} \cdot \frac{\partial \mathbf{P}}{\partial t} dv - \int_v \mu \mathbf{H} \cdot \frac{\partial \mathbf{M}}{\partial t} dv$$

power expended by the field on moving charges — power expended by the field on electric dipoles

The Poynting vector $\mathbf{E} \times \mathbf{H}$, which is a power density vector, represents the power flows per unit area. Furthermore, the electromagnetic momentum is found as $\mathbf{M} = \frac{1}{c^2} \int_v \mathbf{E} \times \mathbf{H} dv$.

The electromagnetic field can be studied using the magnetic vector potential A. In particular,

$$\mathbf{B}(t,\mathbf{r}) = \nabla \times \mathbf{A}(t,\mathbf{r}), \quad \mathbf{E}(t,\mathbf{r}) = -\frac{\partial \mathbf{A}(t,\mathbf{r})}{\partial t} - \nabla V(t,\mathbf{r})$$

Making use of the *Coulomb gauge* $\nabla \cdot \mathbf{A}(t,\mathbf{r}) = 0$, for the free electromagnetic field (A is determined by the transverse current density), we have

$$\nabla^2 \mathbf{A}(t,\mathbf{r}) - \frac{1}{c^2} \frac{\partial^2 \mathbf{A}(t,\mathbf{r})}{\partial t^2} = 0$$

where c is the speed of light, $c = 1/\sqrt{\mu_0 \varepsilon_0}$, $c = 3 \times 10^8 \ \frac{m}{sec}$.

The solution of the partial differential equation is

$$\mathbf{A}(t,\mathbf{r}) = \frac{1}{2\sqrt{\varepsilon}} \sum_s a_s(t) \mathbf{A}_s(\mathbf{r})$$

and using the separation of variables technique we have

$$\nabla^2 \mathbf{A}_s(\mathbf{r}) + \frac{\omega_s}{c^2} \mathbf{A}_s(\mathbf{r}) = 0, \quad \frac{d^2 a_s(t)}{dt^2} + \omega_s a_s(t) = 0$$

where ω_s is the separation constant that determines the eigenfunctions.

The stored electromagnetic energy $\langle W(t) \rangle = \frac{1}{2v} \int_v (\varepsilon \mathbf{E} \cdot \mathbf{E} + \mu \mathbf{H} \cdot \mathbf{H}) dv$ is given by

$$\langle W(t) \rangle = \frac{1}{4v} \int_v (\omega_s \omega_{s'} \mathbf{A}_s \cdot \mathbf{A}_{s'}^* + c^2 \nabla \times \mathbf{A}_s \cdot \nabla \times \mathbf{A}_{s'}^*) a_s(t) a_{s'}^*(t) dv$$

$$= \frac{1}{4v} \sum_{s,s'} (\omega_s \omega_{s'} + \omega_{s'}^2) a_s(t) a_{s'}^*(t) \int_v \mathbf{A}_s \cdot \mathbf{A}_{s'}^* dv = \frac{1}{2} \sum_s \omega_s^2 a_s(t) a_s^*(t)$$

The Hamiltonian is $H = \frac{1}{2v} \int_v (\varepsilon \mathbf{E} \cdot \mathbf{E} + \mu \mathbf{H} \cdot \mathbf{H}) dv$.

Let us apply quantum mechanics to examine very important features. The Hamiltonian function is found using the kinetic and potential energies Γ and Π. For a particle of mass m with energy E moving in the Cartesian coordinate system one has

$$\underset{\text{total energy}}{E(x,y,z,t)} = \underset{\text{kinetic energy}}{\Gamma(x,y,z,t)} + \underset{\text{potential energy}}{\Pi(x,y,z,t)} = \frac{p^2(x,y,z,t)}{2m} + \Pi(x,y,z,t) = \underset{\text{Hamiltonian}}{H(x,y,z,t)}$$

Thus, $p^2(x,y,z,t) = 2m[E(x,y,z,t) - \Pi(x,y,z,t)]$.

Using the formula for the wavelength (de Broglie's equation) $\lambda = h/p = h/mv$, one finds

$$\frac{1}{\lambda^2} = \left(\frac{p}{h}\right)^2 = \frac{2m}{h^2}[E(x, y, z, t) - \Pi(x, y, z, t)]$$

This expression is substituted in the Helmholtz equation $\nabla^2 \Psi + (4\pi^2/\lambda^2)\Psi = 0$, which gives the evolution of the wave function.

We obtain the Schrödinger equation as

$$E(x, y, z, t)\Psi(x, y, z, t) = -\frac{\hbar^2}{2m}\nabla^2\Psi(x, y, z, t) + \Pi(x, y, z, t)\Psi(x, y, z, t)$$

or

$$E(x, y, z, t)\Psi(x, y, z, t) = -\frac{\hbar^2}{2m}\left(\frac{\partial^2\Psi(x, y, z, t)}{\partial x^2} + \frac{\partial^2\Psi(x, y, z, t)}{\partial y^2} + \frac{\partial^2\Psi(x, y, z, t)}{\partial z^2}\right)$$
$$+ \Pi(x, y, z, t)\Psi(x, y, z, t)$$

Here, the modified Planck constant is $\hbar = h/2\pi = 1.055 \times 10^{-34}$ J $\cdot$ sec.

The Schrödinger equation

$$-\frac{\hbar^2}{2m}\nabla^2\Psi + \Pi\Psi = E\Psi$$

is related to the Hamiltonian $H = -(\hbar^2/2m)\nabla + \Pi$, and

$$H\Psi = E\Psi$$

The Schrödinger partial differential equation must be solved, and the wave function is normalized using the probability density $\int |\Psi|^2\, dV = 1$.

Let us illustrate the application of the Schrödinger equation.

Example 7.15

Consider a one-degree-of-freedom harmonic oscillator 7, 8. We have,

$$-\frac{\hbar^2}{2m}\frac{d^2\Psi(q)}{dq^2} + \Pi(q)\Psi(q) = E(q)\Psi(q)$$

and the Hamiltonian function is

$$H(q, p) = \frac{p^2(q)}{2m} + \Pi(q) = \frac{1}{2m}\left(-\hbar^2\frac{d^2}{dq^2} + m^2\omega^2 q^2\right) = -\frac{\hbar^2}{2m}\frac{d^2}{dq^2} + \frac{m\omega^2 q^2}{2}$$

The Hamilton equations of motion relating dq/dt to p, and dp/dt to q, are

$$\frac{dq}{dt} = \frac{\partial H}{\partial p} = \frac{p}{m}$$

$$\frac{dp}{dt} = -\frac{\partial H}{\partial q} = -m\omega^2 q$$

Therefore, one has the following second-order homogeneous equation of motion:

$$\ddot{q} = -\omega^2 q^2$$

Making use of the initial conditions, the solution is found to be

$$q(t) = q_0 \cos \omega t + \frac{p_0}{m\omega} \sin \omega t$$

$$p(t) = -q_0 m\omega \sin \omega t + p_0 \cos \omega t$$

The quantization is performed by considering p and q equivalent to the momentum and coordinate operators.

Applying the variables

$$a(t) = \sqrt{\frac{m}{2\omega}} \left(\omega q + i \frac{p}{m} \right) \quad \text{and} \quad a^+(t) = \sqrt{\frac{m}{2\omega}} \left(\omega q - i \frac{p}{m} \right)$$

one has

$$q = \frac{1}{\sqrt{2m\omega}} (a + a^+) \quad \text{and} \quad p = -i\sqrt{\tfrac{1}{2} m\omega} (a - a^+)$$

Thus, the Hamiltonian is given by

$$H = \omega a^+ a$$

Therefore Hamilton equations

$$\frac{dq}{dt} = \frac{\partial H}{\partial p} = \frac{p}{m}$$

$$\frac{dp}{dt} = -\frac{\partial H}{\partial q} = -m\omega^2 q$$

are written in the decoupled form as

$$i\frac{da}{dt} = \frac{\partial H}{\partial a^+} = \omega a \quad \text{and} \quad i\frac{da^+}{dt} = -\frac{\partial H}{\partial a} = -\omega a^+$$

The q and p, as well as a and a^+, are used as the variables. We apply the Hermitian operators q and p that satisfy the commutative relations

$$[\mathbf{q}, \mathbf{q}] = 0, \quad [\mathbf{p}, \mathbf{p}] = 0 \quad \text{and} \quad [\mathbf{q}, \mathbf{p}] = i\hbar\delta$$

The Schrödinger representation of the energy eigenvector $\Psi_n(q) = \langle q | E_n \rangle$ satisfies the following equations:

$$\langle q | H | E_n \rangle = E_n \langle q | E_n \rangle, \quad \left\langle q \left| \frac{1}{2m} (\mathbf{p}^2 + m^2\omega^2 \mathbf{q}^2) \right| E_n \right\rangle = E_n \langle q | E_n \rangle$$

and

$$\left(-\frac{\hbar^2}{2m} \frac{d^2}{dq^2} + \frac{m\omega^2 q^2}{2} \right) \Psi_n(q) = E_n \Psi_n(q)$$

The solution is

$$\Psi_n(q) = \sqrt{\frac{1}{2^n n!} \sqrt{\frac{\omega}{\pi\hbar}}} H_n \left(\sqrt{\frac{m\omega}{\hbar}} q \right) e^{-\frac{m\omega}{2\hbar} q^2}$$

where

$$H_n\left(\sqrt{\frac{m\omega}{\hbar}}q\right)$$

is the Hermite polynomial, and the energy eigenvalues that correspond to the given eigenstates are $E_n = \hbar\omega_n = \hbar\omega(n+\frac{1}{2})$.

The eigenfunctions can be generated using the following procedure. Using non-Hermitian operators

$$\mathbf{a} = \sqrt{\frac{m}{2\hbar\omega}}\left(\omega\mathbf{q} + i\frac{\mathbf{p}}{m}\right) \quad \text{and} \quad \mathbf{a}^+ = \sqrt{\frac{m}{2\hbar\omega}}\left(\omega\mathbf{q} - i\frac{\mathbf{p}}{m}\right)$$

we have

$$\mathbf{q} = \sqrt{\frac{\hbar}{2m\omega}}(\mathbf{a} + \mathbf{a}^+) \quad \text{and} \quad \mathbf{p} = -i\sqrt{\frac{m\hbar\omega}{2}}(\mathbf{a} - \mathbf{a}^+)$$

The commutation equation is $[\mathbf{a}, \mathbf{a}^+] = 1$, and one obtains

$$H = \frac{\hbar\omega}{2}(\mathbf{aa}^+ + \mathbf{a}^+\mathbf{a}) = \hbar\omega\left(\mathbf{aa}^+ + \frac{1}{2}\right)$$

The Heisenberg equations of motion are

$$\frac{d\mathbf{a}}{dt} = \frac{1}{i\hbar}[\mathbf{a}, H] = -i\omega\mathbf{a} \quad \text{and} \quad \frac{d\mathbf{a}^+}{dt} = \frac{1}{i\hbar}[\mathbf{a}^+, H] = i\omega\mathbf{a}^+$$

with solutions

$$\mathbf{a} = \mathbf{a}_s e^{-i\omega t} \quad \text{and} \quad \mathbf{a}^+ = \mathbf{a}_s^+ e^{i\omega t}$$

and

$$\mathbf{a}|0\rangle = 0, \quad \mathbf{a}|n\rangle = \sqrt{n}\,|n-1\rangle \quad \text{and} \quad \mathbf{a}^+|n\rangle = \sqrt{n+1}\,|n+1\rangle$$

Using the *state vector generating rule* $|n\rangle = \frac{\mathbf{a}^{+n}}{\sqrt{n!}}|0\rangle$, one has the eigenfuction generator

$$\langle q|n\rangle = \frac{1}{\sqrt{n!}}\left(\sqrt{\frac{m}{2\hbar\omega}}\right)^n\left(\omega q - \frac{\hbar}{m}\frac{d}{dq}\right)^n\langle q|0\rangle$$

for the equation

$$\Psi_n(q) = \sqrt{\frac{1}{2^n n!}}\sqrt{\frac{\omega}{\pi\hbar}}\,H_n\left(\sqrt{\frac{m\omega}{\hbar}}q\right)e^{-\frac{m\omega}{2\hbar}q^2}$$

Comparing the equation for the stored electromagnetic energy $\langle W(t)\rangle = \frac{1}{2}\sum_s \omega_s^2 a_s(t)a_s^*(t)$ and the Hamiltonian $H = \omega a^+ a$, we have

$$\sqrt{\frac{\omega_s}{2}}a_s \Rightarrow a = \sqrt{\frac{m}{2\omega}}\left(\omega q + i\frac{p}{m}\right) \quad \text{and} \quad \sqrt{\frac{\omega_s}{2}}a_s^* \Rightarrow a^+ = \sqrt{\frac{m}{2\omega}}\left(\omega q - i\frac{p}{m}\right)$$

Therefore, to perform the canonical quantization, the electromagnetic field variables are expressed as the field operators using

$$\sqrt{\frac{\omega_s}{2}}a_s \Rightarrow \sqrt{\frac{m}{2\omega}}\left(\omega\mathbf{q}+i\frac{\mathbf{p}}{m}\right) = \sqrt{\hbar}\mathbf{a}$$

and

$$\sqrt{\frac{\omega_s}{2}}a_s^* \Rightarrow \sqrt{\frac{m}{2\omega}}\left(\omega\mathbf{q}-i\frac{\mathbf{p}}{m}\right) = \sqrt{\hbar}\mathbf{a}^+$$

The following equations finally result:

$$H = \sum_s \frac{\hbar\omega_s}{2}[\mathbf{a}_s(t)\mathbf{a}_s^+(t)+\mathbf{a}_s^+(t)\mathbf{a}_s(t)]$$

$$\mathbf{E}(t,\mathbf{r}) = \sum_s \sqrt{\frac{\hbar\omega_s}{2\varepsilon}}[\mathbf{a}_s(t)\mathbf{A}_s(\mathbf{r})-\mathbf{a}_s^+(t)\mathbf{A}_{s'}^*(\mathbf{r})]$$

$$\mathbf{H}(t,\mathbf{r}) = c\sum_s \sqrt{\frac{\hbar}{2\mu\omega_s}}[\mathbf{a}_s(t)\nabla\times\mathbf{A}_s(\mathbf{r})+\mathbf{a}_s^+(t)\nabla\times\mathbf{A}_{s'}^*(\mathbf{r})]$$

$$\mathbf{A}(t,\mathbf{r}) = \sum_s \sqrt{\frac{\hbar}{2\varepsilon\omega_s}}[\mathbf{a}_s(t)\mathbf{\acute{A}}_s(\mathbf{r})+\mathbf{a}_s^+(t)\mathbf{\acute{A}}_{s'}^*(\mathbf{r})]$$

The derived expressions can be straightforwardly applied. For example, for a single mode field we have

$$\mathbf{E}(t,\mathbf{r}) = i\mathbf{e}\sqrt{\frac{\hbar\omega}{2\varepsilon v}}(\mathbf{a}(t)e^{i\mathbf{k}\cdot\mathbf{r}-i\omega t}-\mathbf{a}^+(t)e^{-i\mathbf{k}\cdot\mathbf{r}-i\omega t})$$

$$\mathbf{H}(t,\mathbf{r}) = i\sqrt{\frac{\hbar\omega}{2\mu v}}\mathbf{k}\times\mathbf{e}(\mathbf{a}(t)e^{i\mathbf{k}\cdot\mathbf{r}-i\omega t}-\mathbf{a}^+(t)e^{-i\mathbf{k}\cdot\mathbf{r}-i\omega t})$$

$$\langle n|\mathbf{E}|n\rangle = 0, \quad \langle n|\mathbf{H}|n\rangle = 0$$

$$\Delta\mathbf{E} = \sqrt{\frac{\hbar\omega}{\varepsilon v}\left(n+\tfrac{1}{2}\right)}, \quad \Delta\mathbf{H} = \sqrt{\frac{\hbar\omega}{\mu v}\left(n+\tfrac{1}{2}\right)}$$

$$\Delta\mathbf{E}\Delta\mathbf{H} = \sqrt{\frac{1}{\varepsilon\mu}}\frac{\hbar\omega}{v}\left(n+\tfrac{1}{2}\right)$$

where $E_p = \sqrt{\hbar\omega/2\varepsilon v}$ is the electric field per photon.

The complete Hamiltonian of a coupled system is $H_\Sigma = H + H_{ex}$, where H_{ex} is the interaction Hamiltonian. For example, the photoelectric interaction Hamiltonian is found as

$$H_{exp} = -e\sum_n \mathbf{r}_n\cdot\mathbf{E}(t,\mathbf{R})$$

where r_n are the relative spatial coordinates of the electrons bound to a nucleus located at R.

Taking note of quantum electrodynamics, electromagnetism, and mechanics, as well as the reported results in quantization and relationship between Maxwell's and Schrödinger equations, the documented paradigm is illustrated in Fig. 7.15.

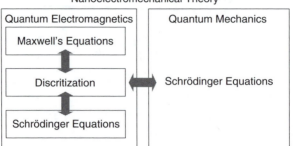

FIGURE 7.15
Interactive quantum electrodynamics, electromagnetism, and mechanics.

7.3   Quantum Mechanics and Energy Bands

The analytic solution to model and examine particle dynamics and energetics in a periodic field (quantum wells) can be derived. The wave functions can be obtained using different methods. We study a one-dimensional problem known as the Kronig-Penney model. In this model, the complex self-consistent potential problem is formulated utilizing our solutions for the rectangular potential and quantum well as reported in Section 7.1. We consider periodic quantum wells of width a separated by barriers with potential Π_0 and width b (barriers can be represented as the Dirac delta functions) (see Fig. 7.16).

Consider the electron tunneling process through rectangular barriers. The Schrödinger equation for two regions (a and b) is given as

$$\left(\frac{d^2}{dx^2} + \frac{2mE}{\hbar^2} \right) \Psi_a = 0, \quad 0 \le x \le a$$

and

$$\left(\frac{d^2}{dx^2} - \frac{2m}{\hbar^2}(\Pi_0 - E) \right) \Psi_b = 0, \quad -b \le x \le 0$$

In these equations, we define

$$\gamma^2 = \frac{2mE}{\hbar^2} \quad \text{and} \quad \beta^2 = \frac{2m}{\hbar^2}(\Pi_0 - E) = \frac{2m\Pi_0}{\hbar^2} - \gamma^2$$

The solution is found using the Bloch function $\psi_k(x) = u_k(x)e^{ikx}$.

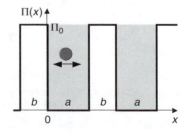

FIGURE 7.16
Periodic quantum wells of width a separated by barriers with potential Π_0 and width b.

For an infinite lattice, the Bloch amplitude is periodic with a period $(a + b)$. Thus, $u_k (x + a + b) = u_k (x)$.

For $0 \leq x \leq a$, we have

$$(d^2 u_a / dx^2) + 2ik(du_a / dx) + (\gamma^2 - k^2)u_a = 0$$

with the solution

$$u_a(x) = Ae^{i(\gamma - k)x} + Be^{-i(\gamma + k)x}$$

For $-b \leq x \leq 0$, one obtains

$$(d^2 u_b / dx^2) + 2ik(du_b / dx) - (\beta^2 + k^2)u_b = 0.$$

The solution is given as

$$u_b(x) = Ce^{i(\beta - k)x} + De^{-i(\beta + k)x}$$

Having found $u_a(x)$ and $u_b(x)$, the wave functions $\psi_k(x) = u_k(x)e^{ikx}$ result.

The unknown coefficients A, B, C, and D are found using the boundary conditions

$$u_a(0) = u_b(0), \quad u_a(a) = u_b(-b), \quad \left. \frac{du_a}{dx} \right|_{x=0} = \left. \frac{du_b}{dx} \right|_{x=0} \quad \text{and} \quad \left. \frac{du_a}{dx} \right|_{x=a} = \left. \frac{du_b}{dx} \right|_{x=-b}$$

Taking note of these boundary conditions, one finds a set of the following equations:

$$A - B = C - D$$

$$Ae^{i(\gamma - k)a} - Be^{-i(\gamma + k)a} = Ce^{-(\beta - ik)b} - De^{(\beta + ik)b}$$

$$Ai(\gamma - k) - Bi(\gamma + k) = C(\beta - ik) - D(\beta + ik)$$

$$Ai(\gamma - k)e^{i(\gamma - k)a} - Bi(\gamma + k)e^{-i(\gamma + k)a} = C(\beta - ik)e^{-(\beta - ik)b} - D(\beta + ik)e^{(\beta + ik)b}$$

In general, these equations can be solved numerically (for example, using MATLAB). Correspondingly, one finds $\psi_a(x) = u_a(x)e^{ikx}$ and $\psi_b(x) = u_b(x)e^{ikx}$.

The analytic solution can be found with some assumptions. We consider Dirac delta function potentials (the barrier field exists at 0, a, $2a$, . . .). Assuming that $E < \Pi_0$, β is real. We estimate the binding energy of an electron (having energy E) to the atomic core in a medium (for example, crystal) as

$$E_B = \lim_{b \to 0, \beta \to \infty} \tfrac{1}{2} ab\beta^2$$

Hence, we can use limits $b \to 0$ and $\beta \to \infty$ (here, $\beta^2 b = \text{const}$, but $\beta b \to 0$). A set of four equations (defined using the boundary conditions) is reduced to the following two equations:

$$A\left(1 - e^{i(\gamma - k)a}\right) + B\left(1 - e^{-i(\gamma + k)a}\right) = 0$$

$$A\left[\frac{2E_B}{a} - i(\gamma - k)\left(1 - e^{i(\gamma - k)a}\right)\right] + B\left[\frac{2E_B}{a} + i(\gamma + k)\left(1 - e^{-i(\gamma + k)a}\right)\right] = 0$$

From these equations, we obtain

$$E_B \frac{\sin \gamma a}{\gamma a} + \cos \gamma a = \cos ka$$

This expression should be examined in detail. The condition that the Bloch functions (which quantitatively and qualitatively describe the electron) exist is equivalent to the condition that this equation has a solution. Assigning $E_B = 4$ and $E_B = 0.04$, the MATLAB statement to calculate the variations of the left side of the equation $E_B (\sin \gamma a / \gamma a) + \cos \gamma a = \cos ka$ (e.g., changes of $E_B(\sin \gamma a / \gamma a) + \cos \gamma a$) by varying γa is

```
Eb = 4; ga = 0:0.01:25*pi; COSka = Eb*sinc(ga) + cos(ga);
ONE = ones(size(ga));
plot(ga,COSka,ga,ONE,ga, - ONE);
```

The resulting plot is documented in Fig. 7.17.

The solution $E_B(\sin \gamma a / \gamma a) - \cos \gamma a$, related to the waves described by the Bloch functions, is real. Therefore, the wave number k is real. Imaginary values of k correspond to exponentially vanishing localized states. To obtain real values of k, the right side of equation $E_B(\sin \gamma a / \gamma a) - \cos \gamma a = \cos ka$ has to have the values from -1 to $+1$ (depending on E_B, these limits can be exceeded [see Figure 7.17]). However, constraint $[-1, +1]$ imposes limits on the allowed values of γa, but not on E_B. One can obtain the "allowed" values of γa when the condition $[-1, +1]$ is met, as shown in Fig. 7.17.

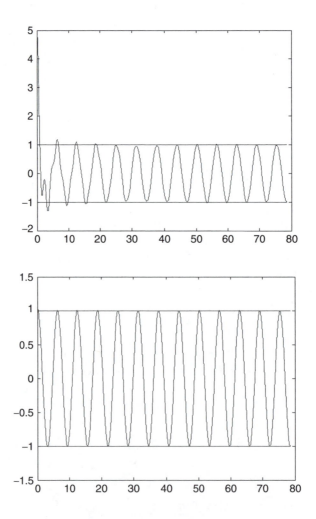

FIGURE 7.17
Variations $E_B \frac{\sin \gamma a}{\gamma a} + \cos \gamma a$ as a function of γa for $E_B = 4$ and $E_B = 0.04$.

The equation $E = \hbar^2\gamma^2/2m$ means that the electron moving in the periodic potential can have the energies lying in the allowed bands, with forbidden energy gaps in between them. An electron with the energy from the range of the forbidden energy has to be localized on the media surface (this electron is not described by the Bloch function). The allowed energy bandwidth decreases, and the gap width increases with the increase of E_B, i.e., with the increase in the electron binding energy. For $E_B \to \infty$ (when the electron is localized in an infinite quantum well), each energy band is reduced to a discrete energy level. In this case, there exist only the solutions for $\gamma = \pm n\pi/a$, $n = 1, 2, 3, \ldots$. Taking into account $E = \hbar^2\gamma^2/2m$, one obtains for this case the following allowed energies:

$$E_n = \frac{\hbar^2\pi^2}{2ma^2}n^2$$

These discrete energy levels describe the allowed energies in an infinite quantum well. For $E_B = 0$ (free unbound electron), we obtain $\gamma = k$, which gives the expression $E = \hbar^2k^2/2m$ (energy of a free particle).

Thus, in the case of deep wells surrounded by wide barriers, the electrons stay inside the wells and fill the allowed energy levels. The media with such a potential and energy levels will be an insulator because the electrons cannot move freely. When the height or width of the barrier decreases, the electrons begin to tunnel to the neighboring wells. The electrons interact, and this interaction leads to splitting of the energy levels into energy bands. We may have the strong interaction for the wide bands and narrow width between the gaps. When the barriers disappear, energy bands merge into one energy band of a free electron in a medium.

The value E_B characterizes the binding energy of electrons to the media atoms. The difference in E_B is due to different chemical bonds. Therefore, media have different energy gaps.

The function $\cos ka$ is periodic. Therefore, the dispersion relation must be periodic with the same period. Thus,

$$E(k) = E\left(k \pm n\frac{\pi}{a}\right), \quad n = 0, \pm 1, \pm 2, \pm 3 \ldots$$

Different values of k are contained in the first Brillouin zone, and

$$-n\frac{\pi}{a} \leq k \leq +n\frac{\pi}{a}, \quad n = 1, 2, 3 \ldots.$$

The electron energy levels are determined by the equation

$$\left[-\frac{\hbar^2}{2m}\nabla^2 + \Pi(\mathbf{r})\right]\Psi(\mathbf{r}) = E\Psi(\mathbf{r})$$

The media (for example, crystal) potential $\Pi(\mathbf{r})$ has a symmetry of the crystal structure—for example, the translation symmetry of the Bravais lattice as well as other symmetries including the rotational symmetry. The solution of the Schrödinger equation is sought in the form of the Bloch functions

$$\psi_{n\mathbf{k}} = u_{n\mathbf{k}}(\mathbf{r})e^{\mathrm{i}\mathbf{k}\mathbf{r}}$$

where n is the band number, and $\mathbf{k}$ is the wave vector (can be limited to the first Brillouin zone, i.e., solid).

For these Bloch functions, the energies are defined as

$$E_{n,\mathbf{k}} = E_n(\mathbf{k})$$

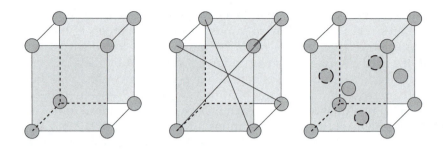

FIGURE 7.18
The cubic, body-centered cubic, and face-centered cubic lattices.

In the three-dimensional case, $E_n(\mathbf{k})$ is a function of three variables k_x, k_y, and k_z. As the first Brillouin zone (solid) is a construction in the space of the reciprocal lattice, the shape of this solid is determined by the Bravais lattice of the crystal. The Brillouin zone of a crystal can be the body-centered cube, face-centered cube and other lattices. The majority of semiconductors have the face-centered crystallographic lattice.

The simple cubic structure has an atom located at each corner of the unit cell, the body-centered cubic lattice has an additional atom at the center of the cube, and the face-centered cubic unit cell has atoms at the eight corners and centered on the six faces. These lattices are reported in Fig. 7.18. The basic lattice structure for many semiconductors is the diamond lattice (including Si and Ge). In many compound semiconductors, atoms are arranged in a basic diamond structure but are different on alternating sites. This is called a zincblende lattice and is typical of III-V compounds. The diamond lattice can be thought of as a face-centered cubic structure with an extra atom placed at $\frac{1}{4}\mathbf{a}+\frac{1}{4}\mathbf{b}+\frac{1}{4}\mathbf{c}$ from each of the face-centered cubic atoms.

The symmetry of the Brillouin zone simplifies the calculation of the energy bands. For example, the $E(\mathbf{k})$ can be expressed using the symmetric directions or isoenergetic surfaces where $E(\mathbf{k})$ can be constant. The electron energy bands can be labeled by the symbols of the points or the directions of the Brillouin zone with the band number.

The bands and energies are determined by the crystal symmetry. A particular feature of the three-dimensional bands is the degeneracy in some zone points or along directions of high symmetry (this feature is not observed in the one-dimensional case). Due to the fact that there exist the maximal and minimal values of $E(\mathbf{k})$, at those points the gradient of $E(\mathbf{k})$ is zero, e.g., $\nabla_{\mathbf{k}} E(\mathbf{k}) = 0$. It is also important that relation $E(-\mathbf{k}) = E(\mathbf{k})$ (surface $E(\mathbf{k})$ = const has the reflection symmetry with respect to $\mathbf{k} = 0$) is valid even if the bands are degenerated. Thus, the isoenergetic surfaces have both full symmetry of the crystal lattice and inversion symmetry. If a crystal of the cubic symmetry has an extreme $E(\mathbf{k})$ in a "general" point of the Brillouin zone $\mathbf{k}_0$, the extreme appears in the remaining 48 symmetric points of the zone. If the extreme exists in a point lying on a line of high symmetry, then the number of equivalent zone points will reduce. For example, if the point lies on the (100) direction, the total number of the equivalent points is equal to six. In this case, the isoenergetic surfaces are six equivalent ellipsoids of revolution lying on three mutually perpendicular space directions. If, in addition, the extreme point lies at the first Brillouin zone border, then only halves of the ellipsoids are lying within the first Brillouin zone. This means that the first zone contains effectively three ellipsoids. If an extreme point $\mathbf{k}_0$ lies on the (111) direction, then the total number of the extreme points is eight. In this case, the isoenergetic surfaces are eight ellipsoids of revolution directed along the main diagonals of the cube, i.e., along the (111) directions. If the extreme points lie at the zone border, and only the halves of the ellipsoids are inside the zone, then the first Brillouin zone will contain effectively four ellipsoids.

Assume that $E(\mathbf{k})$ has an extreme at $\mathbf{k} = \mathbf{k}_0$. Then, we have

$$E(\mathbf{k}) = E(\mathbf{k}_0) + \tfrac{1}{2} \sum_{ij} \left(\frac{\partial^2 E(\mathbf{k})}{\partial k_i \partial k_j} \right)_{\mathbf{k}=\mathbf{k}_0} (k_i - k_{0i})(k_j - k_{0j}), \quad i, j = 1, 2, 3$$

The second term in this equation is a tensor of the second rank that can be defined as the tensor of the inverse of the effective mass, i.e.,

$$\frac{1}{\hbar^2} \left(\frac{\partial^2 E(\mathbf{k})}{\partial k_i \partial k_j} \right)_{\mathbf{k}_0} = \frac{1}{m_{ij}}$$

Using the principal axes, one finds

$$\frac{1}{\hbar^2} \left(\frac{\partial^2 E(\mathbf{k})}{\partial k_z^2} \right)_{\mathbf{k}_0} = \frac{1}{m_z}, \quad z = 1, 2, 3$$

where z is the index that denotes three directions of the principal axes. Thus, by making use of the principal axes, we have the following equation

$$E(\mathbf{k}) = E(\mathbf{k}_0) + \frac{1}{2} \sum_z \frac{\hbar^2}{m_z} (k_z - k_{0z})^2$$

Taking note of the *pseudo-momentum*, as given by $\mathbf{p} = \hbar(\mathbf{k} - \mathbf{k}_0)$, using the effective mass approximation, one obtains

$$E(\mathbf{k}) - E(\mathbf{k}_0) = \sum_z \frac{p_s^2}{2m_z}$$

If the values of wave vector $\mathbf{k}$ are limited to the first Brillouin zone, the pseudo-momentum has the properties of the momentum defined in classical mechanics. Because the electron energy is a periodic function of $\mathbf{k}$, for an arbitrary vector of the reciprocal lattice $\mathbf{K}$ we have $E(\mathbf{k} + \mathbf{K}) = E(\mathbf{k})$. This expression for the energy has the form of the classical kinetic energy with an anisotropic mass having components m_z along the principal axes. These components have the dimension of mass, and they represent the effective mass tensor in the diagonal form. The effective mass is called the density of states effective mass. If all of the mass tensor components are equal, the scalar effective mass m_{ef} can be applied, and, in particular, $E(\mathbf{k}) - E(\mathbf{k}_0) = \mathbf{p}^2 / 2m_{ef}$.

Example 7.16

Consider an electron of energy E moving in the crystal that should overcome the periodic potential Π_0 in order to leave the crystal and become free in conductor. Examine the energy assuming that k is complex.

SOLUTION

If the energy of the electron is $E < \Pi_0$, then the work function represents an infinite quantum barrier for the electron, and the wave function rapidly decreases in the barrier. To study the behavior of an electron, we solve the Schrödinger equation for $E < \Pi_0$, i.e.,

$$-\frac{\hbar^2}{2m} \frac{d^2 \Psi}{dx^2} + \Pi(x)\Psi = E\Psi$$

For $x < 0$, the solutions of the Schrödinger equation

$$\Psi_1 = A e^{\frac{\sqrt{2m(\Pi_0 - E)}}{\hbar} x}$$

decays to zero at $x \to -\infty$.

For $x > 0$, for the periodic potential, the solution is sought in the form of the Bloch functions. In particular,

$$\Psi_2 = u_k(x) e^{ikx}$$

The solution exists, and k is real assuming that the function does not vanish. However, let k be complex as assigned, i.e., $k = k_1 + ik_2$. Thus, the wave function is expressed as $\Psi_2 = u_k(x) e^{ik_1 x} e^{-k_2 x}$ where $k_2 > 0$.

Using the boundary and continuity conditions for the wave function, we have

$$\Psi_1(0) = \Psi_2(0) \quad \text{and} \quad \left. \frac{\partial \Psi_1}{\partial x} \right|_{x=0} = \left. \frac{\partial \Psi_2}{\partial x} \right|_{x=0}$$

Thus, one finds

$$A = u_k(0) \quad \text{and} \quad A \frac{\sqrt{2m(\Pi_0 - E)}}{\hbar} = u_k'(0) + u_k(0)(ik_1) + u_k(0)(-k_2)$$

Solution gives the energy as

$$E = \Pi_0 - \frac{\hbar^2}{2m} \left(\frac{u_k'(0)}{u_k(0)} + ik_1 - k_2 \right)^2$$

Assuming that k is complex, complex energy is found. The obtained energy is real, but it falls in a forbidden region for the band electrons (surface levels or Tamm levels). The wave functions of the surface levels, therefore, describe localized states at the crystal surface. In a three-dimensional crystal, the Tamm surface states form a surface band. In this band, electrons can move freely contributing to the electric conductance. These surface effects can usually be neglected in classical electromagnetism. However, the existence of the surface states is important for the nanocontacts. The surface states can exist not only in the forbidden gap, but also in the valence or a conduction band.

Utilizing the results reported, the band structures of semiconductors can be discussed. Such semiconductors as silicon and germanium have the structure of diamond, while gallium arsenide has the structure of sphalerite. As the Bravais lattices of these crystals are the same, their electronic bands are similar as the reader is aware from semiconductor and solid-state physics courses. Let us provide useful details using quantum mechanics.

There are two types of current carriers in semiconductors: electrons and holes. The electrons that filled the valence band can receive sufficient energy (usually thermal) to cross the relatively small forbidden band (typically less than 1.1 eV) into the conduction band. The vacancies left by these electrons represent unfilled energy states in the valence band, and the vacancy is called a hole. Thus, in semiconductors the highest band filled completely with electrons is called the *valence band*. This valence band is a band of states associated with the highest filled maximum of the band. The lowest empty band is called the *conduction band*. The conduction band is a band of states associated with the lowest empty minimum of the band. The distance between the valence band maximum and the conduction band minimum is called the *forbidden energy gap* or the *band gap*, and the band gap energy is denoted as E_{gap}. If the valence band maximum and the conduction band minimum lie at the same point of the Brillouin zone, then the band gap is called the *direct band gap*. Otherwise, the band gap is called the *indirect band gap*. The band structure provides very important information that is needed to study nanoelectronic devices. Let us discuss the role of photons and phonons in the electron excitations from the valence to the conduction band. The reader may be familiar with phonons. Phonons are quasiparticles associated with the thermal vibrations of a crystal while photons are associated with the vibrations of the electromagnetic field. Phonons are quanta of the thermal energy of the crystal, while phonons are represented by the angular frequency ω and wave vector **q**. Thus, phonons have energy $\hbar\omega$ and momentum $\hbar\mathbf{q}$. To excite an electron from the valence to conduction band, one has to supply it with energy $E \geq E_{gap}$. This E is the energy $E = \hbar\omega$ of a photon (and phonon) having the frequency ω. During the energy transfer, the annihilation of the exciting particle and momentum transfer (conservation) can be examined. In the case of a photon with the wave vector **q**, the magnitude of the momentum is $p = \hbar q = \hbar\omega/c = E/c$. For a phonon, one has $p = E/u_s$, where u_s is the sound velocity (from 1 to 10 km/sec for different media). Taking into account that u_s is much less than c, for the same amount of transferred energy, the transfer of the momentum by the photon is small compared to the phonon. Hence, for Si and Ge, the electrons can be transferred from the valence to conduction band by the thermal excitations only, and not by the optical ones. These indirect gaps can be called the *thermal gaps*. As the optical excitations of electrons in semiconductors having the indirect gaps are forbidden, these semiconductors (in the bulk crystal form) cannot be used for the optoelectronic applications. In contrast, GaAs has a direct gap, and GaAs is used for the opto-electronic applications. The average energy transferred by a phonon is on the order of kT. Since the energy gap of Si is considerably larger than in Ge, fewer electrons will be thermally transferred by the energy gap of Si. Therefore, the electrical properties of silicon will be much less dependent on temperature. This means that the electronic devices made using silicon will be thermally robust compared with Ge. With the increase in the energy bandwidth, the forbidden gap between two neighbor bands decreases. Therefore, the effective mass in the conduction band decreases with the gap between the valence and the conduction band. The smallest electron effective masses are observed in the narrow-gap semiconductors (the electron effective mass can be less than 1% of the free electron mass).

Silicon-based microelectronics is reaching its physical and technological limits. Quantum mechanics allows one to discover and deploy novel electronic nanodevices uniquely utilizing basic physics, advanced materials, and nanotechnology. Though it is unlikely that silicon CMOS technology will be completely replaced by nanoelectronics, the CMOS technology will play a complementary role for low-end electronics in future nanoelectronics. Having emphasized silicon, it should be emphasized that diamond (i.e., carbon-based) technology has been already explored to complement Si CMOS-based devices. Diamond is one of the best possible materials with the potential of meeting a number of semiconductor device requirements. From the power, energetics, thermal, mechanical, and other considerations, diamond is an ideal material for ultra-high-speed electronics. The large energy gap of diamond (5.5 eV) results in low leakage or standby current (several orders of magnitude better than silicon). Considering the dynamic power (switching power consumption), the large band gap is undesirable. However, two major factors are taken into account considering switching speed and power. The use of Johnson's and Keys's criteria shows that diamond is better than Si by a factor of 10 or higher for higher-speed switching due to the higher carrier saturation velocity. It is possible to obtain sufficient current in the sub-threshold region of operation of a diamond device to drive smaller devices at these higher speeds and thus avoid the problem associated with the high threshold voltage (due to the larger band gap).

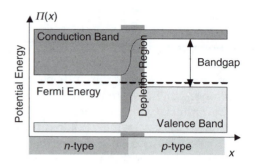

FIGURE 7.19
A *pn*-junction with conduction and valence bands.

Two-dimensional electronic devices (diodes, transistors, switches, etc.) have been designed by utilizing potential barriers to control the flow of electrons. In transistors, one utilizes *pn*-junctions (*npn* bipolar transistor) and electrostatic depletion regions (field effect transistors). The *pn*-junction is the basic functional device that can be fabricated from any semiconductor material. It consists of a single silicon crystal, half of which is *n*-doped, while other half is *p*-doped. At the junction between the *n*- and *p*-type semiconductor materials, a potential barrier is formed. This barrier gradually increases in height from the *n*-type silicon to the *p*-type silicon as documented in Fig. 7.19. The device illustrated functions as a diode.

An *npn* bipolar transistor has two *pn*-junctions, integrated back-to-back. The emitter is an *n*-type, the base is a *p*-type, and the collector is an *n*-type semiconductor. As the voltage across the base region is increased (with respect to the emitter), the height of the barrier on the emitter side is decreased. As it decreases, more and more electrons entering the emitter are then able to cross the base and exit through the collector. The use of some form of barrier to control current flow from one end of the device to the other is the basis of all transistors. However, the minimum width of the depletion region forming the potential barrier is in the range of 20 nm. This, as well as other fundamental limitations, restricts the minimum dimension of devices. One may utilize totally different phenomena and effects observed at the nanoscale to design electronic nanodevices. For example, a *heterojunction* between two different semiconductors with different band gaps can be formed and utilized. Heterojunctions can be used to build *tunnel barriers* and *quantum wells*. As was emphasized, a tunnel barrier is a wide-band-gap material surrounded by narrow-band-gap material. If the layer is very thin (compared with the electron wavelength), electrons have the possibility of tunneling through it. A quantum well is the opposite (narrow-band-gap material surrounded by wide-band-gap material). If the well is made narrow enough, the energy levels confined within it are well spaced out, instead of smearing together into wide energy bands. These topologies are reported in Fig. 7.20. Nanotechnology offers the opportunity to implement novel concepts that are based on quantum mechanics. For example, nanotechnology-based nano-transistors can be designed using the tunnel barrier with the base as the barrier, and the emitter and collector are on other sides.

The electrical properties of semiconductors depend on many factors, including purity and crystal structure. Impurity atoms can donate electrons to the conduction band (in this band, the electrons are free electric carriers contributing to the electric current). The impurities donating electrons to a conduction band are called *donors*. There are also impurities that accept electrons from the valence band creating holes in that band. These impurities are called *acceptors*. Consider the $_{51}$Sb (antimony) atom in the crystal lattice of Si. The Sb atom has $5s^2\,5p^3$ configuration, while for $_{14}$Si, we have $3s^2\,3p^2$. The extra Sb valence electron will not contribute to the covalent bonds of the crystal, but it can move freely in the electric field of the impurity atom. Due to the weak bonds, small thermal energy can cause this electron to become a conduction electron. These acceptor

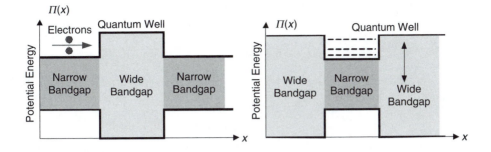

FIGURE 7.20
Tunnel barrier (electrons incident to the barrier have a finite probability of passing through it) and quantum well (consists of discrete energy levels).

electrons can be examined utilizing the effective mass studying the electron as a particle with an effective mass m_{ef}. The potential is given as $\Pi(\mathbf{r}) = -e^2/4\pi\varepsilon r$. The energy is found to be $E_n = -13.6(m_{ef}/\varepsilon_r m)(1/n^2)\mathrm{eV}$. Here, $n = 1, 2, 3, \ldots$ is the main quantum number; ε_r is the relative dielectric constant. For some semiconductors,

$$\frac{m_{ef}}{\varepsilon_r m} \approx \frac{1}{1000}$$

We conclude that the mass of a conduction electron is different from the mass of a free electron due to the periodic potential of the nuclei. For the free electron, we have $p = mv$, $E = \frac{1}{2}mv^2 = p^2/2m$, and $p = \hbar k$, where the wave number is $k = 2\pi/\lambda$. For the electron moving through the lattice, the group velocity (electron velocity) is given by

$$v_g = \frac{\partial\omega}{\partial k} = \frac{\partial(E/\hbar)}{\partial k}$$

That is, the electron is accelerated by the applied external field. The acceleration is

$$a = \frac{dv_g}{dt} = \frac{1}{\hbar}\frac{d}{dt}\left(\frac{dE}{dk}\right)$$

and the force acting on electron is $F = m(dv/dt)$. Therefore,

$$a = \frac{dv_g}{dt} = \frac{1}{\hbar}\frac{dk}{dt}\left(\frac{d^2E}{dk^2}\right) = \frac{1}{\hbar^2}F\frac{d^2E}{dk^2}$$

By making use of the relationship $F = \frac{dp}{dt} = \hbar\frac{dk}{dt}$, one finds

$$F = \hbar^2\left(\frac{d^2E}{dk^2}\right)^{-1}\frac{dv_g}{dt}$$

The comparison

$$F = m\frac{dv}{dt} \quad \text{and} \quad F = \hbar^2\left(\frac{d^2E}{dk^2}\right)^{-1}\frac{dv_g}{dt}$$

leads to the conclusion that the term

$$\hbar^2 \left(\frac{d^2E}{dk^2} \right)^{-1}$$

can be considered as an equivalent to the mass of the free electron. Correspondingly, the effective mass of the electron is

$$m_{ef} = \hbar^2 \left(\frac{d^2E}{dk^2} \right)^{-1}$$

For silicon, the electron effective mass can be $m_{ef,n} = 1.1 m_e$, while the hole effective mass is $m_{ef,p} = 0.56 m_e$, where m_e is the rest mass of the electron, $m_e = 9.11 \times 10^{-31}$ kg.

Consider germanium ($_{32}$Ge) and arsenic ($_{33}$As). Arsenic is in group V and has five valence electrons, while germanium has four valence electrons. When one of the arsenic electrons is removed, the remaining electron structure is essentially identical to that of germanium. However, the atomic nucleus of $_{33}$As has a charge $+ 33e$ versus $+ 32e$ for $_{32}$Ge. An arsenic atom can take the place of a germanium atom as shown in Fig. 7.21. In fact, four of its five valence electrons form the covalent bonds. The fifth valence electron is loosely bound and does not participate in the covalent bonds. For $n = 4$, the bonding energy is very small (0.01 eV, not 0.85 eV that one can anticipate making calculations $13.6(1/n^2) = 13.6\frac{1}{4^2} = 0.85$ eV) due to the probability distribution and many-atomic system. The energy level of this fifth electron corresponds to the energy level lying below E_F, and this level is called a donor level. Therefore, the impurity atom is called a donor. All elements in group V (N, P, As, Sb, and Bi) can serve as donors. In contrast, the elements (B, Al, Ga, In, Tl) of group III have three valence electrons. For example, gallium ($_{31}$Ga) forms covalent bonds with germanium $_{32}$Ge but has only three outer electrons. It can seize the electron from a neighboring germanium atom to complete the required covalent bonds (see Figure 7.21). The hole is formed. The hole acts as a positive charge that can move through the crystal. The gallium atom is called the acceptor. In the semiconductor that is doped with acceptors, the conductivity is due to positive-charge (hole) motion. These semiconductors are called *p*-type semiconductors with *p*-type impurities.

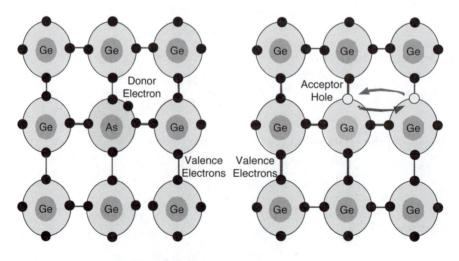

FIGURE 7.21
Donor (*n*-type) and acceptor (*p*-type) impurities.

In the absence of any disturbing potential, the valence band states are given by the dispersion law

$$E(\mathbf{k}) = -E_{gap} - \frac{\hbar^2 \mathbf{k}^2}{2|m_{ef}|}$$

where E_{gap} is the energy gap between the top valence band and bottom conduction band, i.e., $E_{gap} > 0$.

Substituting $\mathbf{k} \to -i\nabla_k$, we have

$$-\frac{\hbar^2}{2|m_{ef}|} \Delta \psi_n(\mathbf{r}) - \Pi(\mathbf{r})\psi_n(\mathbf{r}) = -(E_{gap} + E_n)\psi_n(\mathbf{r})$$

The negative sign of the energy is due to the fact that we consider the electrons in the valence band, and the zero energy is assumed at the bottom of conduction band.

The semiconductor with the electrical conduction dominated by donors is called the *n*-type semiconductor (electrons from the donor levels are thermally activated to the conduction band levels). The semiconductor with the electrical conduction dominated by acceptors is called the *p*-type semiconductor (electrons of the valence band are thermally activated to the transitions from the band to the acceptor levels, leaving holes in the valence band). Free electrons or free holes produced by this mechanism can conduct the electric current. If in semiconducting materials donors coexist with acceptors, then their energy levels compensate each other. If the donor concentration is N_d and the acceptor concentration is N_a, the effective concentration of the impurities is $|N_d - N_a|$. Depending on which of the concentrations is larger, the corresponding *n*- and *p*-type semiconductors result.

If the periodic potential of a crystal is periodic, the electron momentum $\hbar\mathbf{k}$ in such a crystal would not be disturbed by the internal electric fields. On the other hand, in the presence on an external electric field $\mathbf{E}$, the momentum would change as described by equation

$$\frac{d}{dt}\hbar\mathbf{k} = -e\mathbf{E}$$

In crystals, the potential is not periodic, and the electron is affected by a perturbed potential. This perturbed potential can be viewed as a superposition of the ideal periodic potential and localized additional potential. This additional potential scatters the electron. The scatterings change the electron trajectory, and the scattering potentials can be associated with impurities or crystal thermal vibrations. These scatterings cause electron transitions from the occupied to empty states in the Brillouin zone.

Using the classical solid-state theory, the density of state in the conduction band represents the number of energy states that could be occupied by electrons. The density of state is expressed as

$$N(E) = 4\pi \left(\frac{2m_{ef,n}}{h^2}\right)^{\frac{3}{2}} \sqrt{E} = \frac{\sqrt{2}}{\pi^2} \frac{m_{ef,n}^{3/2}}{\hbar^3} \sqrt{E}$$

The number of electrons in the conduction band (electron concentration) is

$$N = \int_0^\infty N(E)f(E)dE$$

where

$$f(E) = \frac{1}{e^{\frac{E-E_F}{k_B T}} - 1}$$

is the Fermi-Dirac distribution function (probability function that gives the probability that an available energy state at E will be occupied by an electron at absolute temperature T), $f(E) = 0.5$ if $E = E_F$, and for $(E - E_F) > 3k_BT$, $e^{E-E_F/k_BT} \gg 1$; k_B is the Boltzmann constant, $k_B = 1.38 \times 10^{-23}$ J/K; E_F is the Fermi energy level; and T is the temperature.

For $E = 0 \rightarrow E_c$, one obtains

$$N = 2\left(\frac{2\pi m_{ef,n}k_BT}{h^2}\right)^{\frac{3}{2}} e^{-\frac{E_c-E_F}{k_BT}} = N_c e^{-\frac{E_c-E_F}{k_BT}}$$

where N_c is the effective density of states in the conduction band,

$$N_c = 2\left(\frac{2\pi m_{ef,n}k_BT}{h^2}\right)^{\frac{3}{2}}$$

The hole density in the valence band is

$$P = 2\left(\frac{2\pi m_{ef,p}k_BT}{h^2}\right)^{\frac{3}{2}} e^{-\frac{E_F-E_v}{k_BT}} = N_v e^{-\frac{E_F-E_v}{k_BT}}$$

where N_v is the effective density of states in the valence band,

$$N_v = 2\left(\frac{2\pi m_{ef,p}k_BT}{h^2}\right)^{\frac{3}{2}}$$

In intrinsic semiconductors, $N = P$, and $N_c e^{-\frac{E_c-E_F}{k_BT}} = N_v e^{-\frac{E_F-E_v}{k_BT}}$. Hence,

$$E_F = \frac{1}{2}(E_c + E_v) + \frac{3}{4}k_BT\ln\left(\frac{m_{ef,p}}{m_{ef,n}}\right) \quad \text{and} \quad NP = N_cN_v e^{-\frac{E_{gap}}{k_BT}} = N_i^2$$

where $N_i = \sqrt{N_cN_v}\,e^{-\frac{E_{gap}}{2k_BT}}$ (in Si, $N_i = 1.45 \times 10^{10}$ cm^{-3}).

The electron mobility $\mu_n = q\tau_c/m_{ef,n}$ depends on the mean free time τ_c (average time between collisions is 1×10^{-12} sec) and the effective mass. Taking note that the momentum applied to an electron during the free motion between collisions is equal to the momentum gained by the electron $-qE\tau_c = m_{ef,n}v_n$, one finds the expression for the drift velocity as $v_n = -(q\tau_c/m_{ef,n})E = -\mu_nE$. The mobility describes how strongly the motion of an electron is influenced by an applied electric field. For holes, $v_p = \mu_pE$, where $\mu_p = q\tau_c/m_{ef,p}$.

The electron current density is found as

$$J_n = \sum_{i=0}^{N}(-qv_i) = -qNv_n = qN\mu_nE$$

while for hole current density,

$$J_p = qP\mu_pE$$

The total current is

$$J = J_n + J_p = q(N\mu_n + P\mu_p)\,E = \sigma E$$

where σ is the conductivity, $\sigma = q(N\mu_n + P\mu_p)$.

The resistivity is given as

$$\rho = \frac{1}{\sigma} = \frac{1}{q(N\mu_n + P\mu_p)}$$

The reported results are well known from classical solid-state physics. However, in nanoelectronic devices, quantum theory must be applied. For example, the probability function is defined using the wave function. Correspondingly, one is not able to apply classical microelectronics concepts to many nanoelectronics problems.

Nano- and microscale systems' performance directly depends on their design, topologies, fabrication, materials, processes, and other qualitative and quantitative characteristics. The coherent level of knowledge regarding the studied systems can be obtained through mathematical modeling with subsequent tasks in simulation, analysis, optimization, etc. High-fidelity modeling allows one to examine nanosystems as well as possibly devise novel phenomena and new operational principles. These guarantee synthesis of superior nanosystems with enhanced functionality through complementary design and optimization. We conclude that to design high-performance nanosystems, fundamental, applied, and experimental research must be performed to devise new nanosystems and to further develop existing nanosystems. Several fundamental electromagnetic and mechanical laws that have been utilized are quantum mechanics, Maxwell's equations, Hamiltonian concept, etc. These paradigms were covered and demonstrated in this chapter. High-fidelity mathematical models were developed, and these models can be used to solve a variety of problems in data-intensive analysis, heterogeneous simulation, optimization, and control of nanosystems.

Homework Problems

1. What does the equation $E = nh\nu$ represent?
2. Using the moment of inertia and angular momentum, compare the energy and the frequency ν of electrons using classical and quantum mechanics. Illustrate the difference for distinct orbitals and generalize the results.
3. What is the potential energy of the harmonic oscillator?
4. In the potential-free region, $\Pi(x) = 0$. Derive the one-dimensional Schrödinger equation

$$-\frac{\hbar^2}{2m}\frac{\partial^2 \Psi(t,x)}{\partial x^2} = i\hbar \frac{\partial \Psi(t,x)}{\partial t}$$

   Use the wave function as

$$\Psi(t,x) = \frac{1}{\sqrt{2\pi\hbar}} \int \phi(p) e^{i\frac{(px-Et)}{\hbar}} dp$$

5. Write the expression using Kronecker's delta function that illustrates the orthogonality of wave functions.
6. Explain when and why the *normalization condition*

$$\int_V \Psi^*(\mathbf{r})\Psi(\mathbf{r})dV = \int_V |\Psi(\mathbf{r})|^2 dV = 1 \text{ is used.}$$

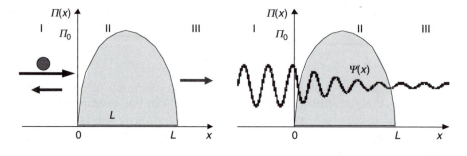

FIGURE 7.22
Particle and the potential barrier. The possible wave function for a particle tunneling through the potential energy barrier (wave functions joint continuously at the boundaries) are reported.

7. Examine and compare the energy levels for a particle in a box and harmonic oscillator.

8. Consider a particle whose normalized wave function is

$$\Psi(t,x) = \begin{cases} 2a\sqrt{a}\,xe^{-ax} & \text{for } x \geq 0 \\ 0 & \text{for } x < 0 \end{cases}$$

Derive Δx and Δp. Using the results, prove that $\Delta p \Delta x \propto \hbar$.
Hint: $\Delta x = \sqrt{\langle x^2 \rangle - \langle x \rangle^2}$ and $\Delta p = \sqrt{\langle p^2 \rangle - \langle p \rangle^2}$.

9. Consider a particle that moves from the region of negative values of coordinate x to the region of positive values of x. The particle encounters a potential barrier $\Pi(x)$ with width L (see Fig. 7.22). Using quantum mechanics, find the probabilities that the particle passes the barrier and is reflected by the barrier. Find the wave functions and derive the expression for energy. *Hint:* Use the sinusoidal function to describe $\Pi(x)$.

10. Using the analogies to semiconductors, explain the possible donor (*n*-type) and acceptor (*p*-type) impurities for the carbon-based nanoelectronics. Justify your reasoning based on the organic carbon complexes (DNA and amino acids).

References

1. Hohenberg, P. and Kohn, W, Inhomogeneous electron gas, *Phys. Rev.*, 136, B864–B871, 1964.
2. Kohn, W. and Driezler, R.M., Time-dependent density-functional theory: conceptual and practical aspects, *Phys. Rev. Lett.*, 56, 1993–1995, 1986.
3. Kohn, W. and Sham, L.J., Self-consistent equations including exchange and correlation effects, *Phys. Rev.*, 140, A1133–A1138, 1965.
4. Lyshevski, S.E., *MEMS and NEMS: Systems, Devices, and Structures*, CRC Press, Boca Raton, FL, 2002.
5. Parr, R.G. and Yang, W., *Density-Functional Theory of Atoms and Molecules*, Oxford University Press, New York, NY, 1989.
6. Davidson, E.R., *Reduced Density Matrices in Quantum Chemistry*, Academic Press, New York, NY, 1976.
7. Yariv, A., *Quantum Electronics*, John Wiley and Sons, New York, 1989.

8. Jones, V.R., *Optical Physics and Quantum Electronics*, lecture notes, Harvard University, 2000.

9. Ellenbogen, J.C. and Love, J.C., *Architectures for Molecular Electronic Computers*, MP 98W0000183, MITRE Corporation, 1999.

10. Tian, W.T., Datta, S., Hong, S., Reifenberger, R., Henderson, J.I., and Kubiak, C.P., Conductance spectra of molecular wires, *Int. J. Chem. Phys.*, 109, 2874–2882, 1998.

11. Colbert, D.T. and Smalley, R.E., Past, present and future of fullerene nanotubes: Buckytubes, in *Perspectives of Fullerene Nanotechnology*, Osawa, E., Ed., Kluwer Academic Publisher, London, 2002, pp. 3–10.

8

Molecular and Carbon Nanoelectronics

8.1   Past, Current, and Future of Electronics with Prospects for 2020 and Beyond

Even within the most pioneering expected developments, it is anticipated that the CMOS transistors may be scaled approximately to 100 nm in the total size by 2010 if the 18-nm gate-length projection will be met. Currently, 65-nm nanolithography technology is emerging and applied to fabricate high-yield high-performance planar (two-dimensional) multilayered ICs with hundreds of millions of transistors on a single 1-cm^2 die (2004 Pentium quarter-size processor integrate more than 300 million transistors). However, further rather revolutionary (not evolutionary) developments are needed, and novel advances are emerging. Future generations of nano-ICs will be built using nanotechnology-based fabricated electronics. It should be emphasized that the nanoscale ICs can be designed and fabricated as nanotechnology-enhanced microelectronics, which is not nanoelectronics. Devising, integration, and implementation of new affordable high-yield electronic nanodevices are critical to meet the currently projected (foreseen) needs. For microelectronics, Moore's laws are commonly used. The power- and size-centered version of this law for high-yield room-temperature mass-produced electronics is reported in Figure 8.1. It is envisioned by the microelectronics industry that after 2010 the rates will be changed due to formidable fundamental and technological problems. For example, it is highly unlikely that microelectronic devices can operate with a power dissipation of one electron per device and that feature size less than 0.01 μm can be achieved by 2020. Revolutionary changes are needed. It is foreseen that nanotechnology will enter microelectronics by 2010, first complementing microelectronics, and then as nanotechnology matures, microelectronics will complement nanoelectronics. In particular, microelectronics will be applied to the low-end devices and subsystems. The reported data and foreseen trends can be viewed as controversial, and, indeed, it is a subject for adjustments. However, the major tendencies are obvious, and most likely, the general trends documented cannot be seriously argued and disputed. In addition to the foreseen departure from the first Moore's law, there is a necessity to depart from the so-called second Moore's law that foresees that the cost of a microelectronics fabrication facility will be hundreds of billions by 2020. High-yield affordable nanoelectronics will ensure significant decrease in cost, and these revolutionary trends can be anticipated by 2010 as microelectronics approach the limit.[1-6] Furthermore, novel nanoscale technologies will likely ensure the relatively steady infrastructure cost (due to the variety of possible paradigms, solutions and processes, adaptability and robustness of the nanoelectronics-oriented nanotechnology, etc.) with a significant decrease from the microelectronics-based fabrication-oriented projections. It is likely that nanoelectronics will utilize low-cost microelectronics and will mature as a hybrid technology, gradually growing and progressing to implementation of nano-ICs in processors, memories, and

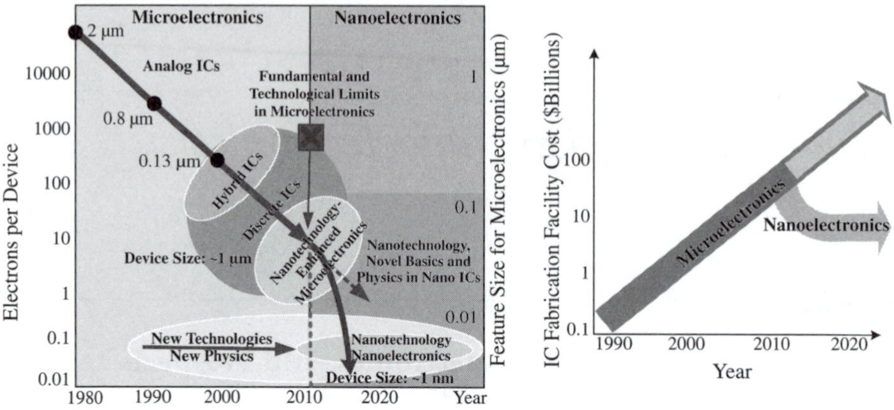

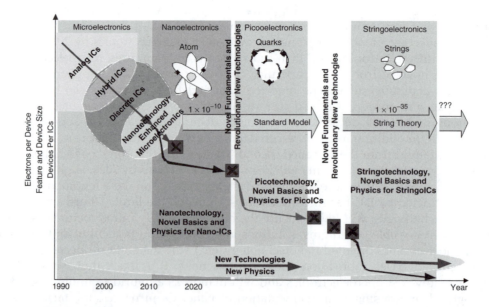

FIGURE 8.1 (See color insert following page 332.)
Projected revolutionary developments in nanoelectronics and nano-ICs compared with microelectronics (Moore's first and second laws for microelectronics are documented with the expected revolutionary changes that will emerge after 2010, drastically affecting the electronics industry and society).

other high-performance units and systems employing microelectronics for low-end units (power electronics, etc.).

Atoms (approximately 10^{-10} m in diameter) are composed of subatomic particles: protons, neutrons, and electrons. Protons and neutrons form the nucleus with the diameter in the range of 10^{-15} m. Novel physics and technologies will result in further revolutionary advances in electronics. Figure 8.2 reports the progress that may be achieved. One should realize the difference between the atom and string. The diameter of the hydrogen atom is 1.058×10^{-10} m, while the size of the string is estimated to be 1.6161×10^{-35} m. Thus, there is 6.56×10^{24} difference, and the string is smaller than the hydrogen atom by

FIGURE 8.2 (See color insert.)
Projected nano-, pico-, . . . , and stringoelectronics and beyond.

6,560,000,000,000,000,000,000,000 times. The author believes that electronics will not stop at the nanoscale level, and, in the future, electronics will progress to pico- (1×10^{-12}), femto- (1×10^{-15}), atto- (1×10^{-18}), . . . , and stringo- (1×10^{-35}) electronics (see Figure 8.2). The stringoelectronics will likely progress to even smaller dimensions, and there is no end to the progress that society will experience. When will the pico- or stringoelectronics emerge? It may take hundreds or thousands of years. As Leucippus and Democritus envisioned atoms (in 400 B.C.), it took almost 2300 years to utilize this discovery, culminating in Mendeleyev's periodic table of the elements reported in 1868. It may not take other 2300 years to progress to the Planck constant dimension from 1900 when this constant was introduced (in 1900 Max Planck discovered the effect of quantization of energy, and he found that the radiated energy is $E = nh\nu$). However, sooner or later, pico-, femto-, . . . , and stringoelectronics will be as mature and advanced as microelectronics is today. It took only 50 years to accomplish tremendous progress in microelectronics. It is likely that it will take 50 years for nanoelectronics to mature. However, the author's predictions regarding further progress may be quite speculative and exploratory. In 2004, the microelectronics industry's revenue was more than 150 billion dollars. The yearly revenue will double or triple by the end of this decade, and revolutionary changes have been pursued by the leading high-technology companies.

As one departs from microelectronics, conventional semiconductor devices, and CMOS technology, it is very difficult to accurately predict basic physics of future electronic nanodevices as well as potential fabrication technologies even within a 10-year horizon. However, it is easy to envision self-assembled functional 3D nano-ICs made from multi-terminal electronic nanodevices. For example, doped carbon molecules or functional carbon-based complexes including DNA molecules lead to a density of billions of devices in a 1-mm³ volume, see Figure 8.3. In fact, a DNA strand a few nanometers long (3.4 nm per helix turn) and 2 nm in diameter can be attached to metal electrodes. Two DNA strands are held together by hydrogen bonds between the bases that are paired in the interior of the double helix. Base pairs are 0.34 nm apart, and there are 10 pairs per helix turn. DNA strands exhibit semiconductor characteristics with a wide band gap to transfer electrons from the valence band to a conduction band. Hence, applying a sufficient voltage, one controls the electron transfer in DNA strands utilizing DNA as electronic nanodevices. Metalic and insulating DNA have also been found. The dimension (length) of the fabricated and characterized electronic nanodevices, such as fullerenes and DNA, is at least 100 times less than newly designed microelectronic devices including nano-FETs with 10-nm gate length. Thus, the application of fullerenes leads to a volume reduction of up to 1,000,000 times, not to mention revolutionary enhanced functionality due to multi-terminal and three-dimensional features.

FIGURE 8.3 (See color insert.)
Three-dimensional nano-ICs made of functional carbon-based complexes, doped carbon complexes in $2 \times 2 \times 2$-nm cube, and double-stranded DNA.

In Chapters 1 and 2, nanotechnology was defined as a combination of research, engineering, and technological developments at the atomic, molecular, or macromolecular levels with the following goals:

- To provide a coherent fundamental understanding of interactive phenomena and effects in materials and complexes at the nanoscale
- To devise, design, and implement new organic, inorganic, and hybrid nanoscale structures, devices, and systems that exhibit novel properties due to their basic physics or fabrication technologies, and utilize these novel phenomena and effects
- To provide a coherent fundamental understanding of controlling matter at nanoscale, and develop paradigms to coherently control and manipulate at the molecular level with atomic precision to fabricate advanced and novel complex devices and systems

Emphasizing nanotechnology-based nanoelectronics as one of the core nanotechnology areas, one can state that nanoelectronics is a fundamental/applied/experimental research that envisions and strives to devise (discover) new enhanced-functionality high-performance atomic- and molecular-scale structures, devices, and systems (discovering, understanding, and utilizing novel phenomena and effects) and fabricate them to develop novel nanotechnologies and processes for supercomplex high-yield high-performance nano-ICs that will revolutionize electronics and information technologies.

Microelectronics and its derivatives, including nanotechnology-enhanced microelectronics, cannot be viewed as nanoelectronics. The nanoelectronic devices studied in this chapter are not submicron microelectronic devices that can possibly reach nanoscale dimensions by scaling down their size. If the total size of the FET will be downscaled to 100 nm, it does not mean that this erroneously called "nano-FET," which is rather a submicron FET, is a nanoelectronic and nanotechnology-based device. There are fundamental theoretical and technological differences for nanoelectronic devices versus microelectronic (possibly nanoscale in size) devices, nanoelectronics versus microelectronics, and nano-ICs versus ICs. These enormous differences are due to distinct basic physics, novel phenomena and effects observed at the nanoscale, application and unique utilization of these novel phenomena, distinct fabrication technologies, different architectures, novel enabling topologies, etc. For example, DNA and carbon-based molecular electronic nanodevices are greatly different compared with silicon technology devices. These microelectronic devices can be fabricated implementing nanotechnology-enhanced materials, processes, and techniques, for example, using carbon nanotubes, fullerenes, and nanolithography. Microelectronics can be nanotechnology-enhanced, and one can define it as nanotechnology-enhanced microelectronics. Hence, microelectronic nanoscale-size devices cannot be classified as nanotechnology-based electronic nanodevices.

Different molecular wires, nanodevices, and simple nano-ICs have been devised and tested. For example, molecular wires can consist of molecules with organic, inorganic, or hybrid interconnects. Molecular wires and interconnects interface electronic nanodevices ensuring functionality of nano-ICs. Extremely high current density can be achieved. For example, the current densities in carbon nanotubes, 1,4-dithiol benzene (molecular wire), and copper are 10^{11}, 10^{12}, and 10^6 electron/sec $\cdot$ nm^2, respectively. According to this data, the maximum current density in carbon nanotubes is 100,000 higher than in copper. Significant progress has been made. For example, carbon nanotubes can be filled with other media such as organic and inorganic materials.

8.2 Fundamentals

Nanostructures and devices should be examined studying underlying atomic structures. Consider covalent bonds. These bonds occur from sharing electrons between two atoms. Covalent bonds represent the interactions of two nonmetallic elements, or metallic and nonmetallic elements. Let us study the electron density around the nuclei of two atoms. If the electron cloud's overlap region passes through on the line joining two nuclei, the bond is called a σ bond (see Figure 8.4). The overlap may occur between orbitals perpendicularly oriented to the internuclear axis. The resulting covalent bond produces overlap above and below the internuclear axis. Such a bond is called a π bond. There is no probability of finding the electron on the internuclear axis in a π bond, and the overlap in it is less than in the σ bond. Therefore, π bonds are generally weaker than σ bonds.

The following example illustrates the application of classical electromagnetic theory to study molecular systems and structures.

Example 8.1

Examine the electric field inside and outside a spherical cloud of electrons in the carbon nanostructure assuming a uniform charge density $\rho = -\rho_0$ for $0 \le r < R$ and $\rho = 0$ for $r \ge R$.

SOLUTION

The solution will be found applying the Poisson and Laplace equations for the electric potential, which is defined as $\mathbf{E} = -\nabla V$. In linear isotropic media, $\nabla \cdot \varepsilon \mathbf{E} = \rho$. In general, ε is a function of position, but in the homogeneous media, ε is constant. Thus, we obtain

$$\nabla^2 V = -\frac{\rho}{\varepsilon}$$

where the Laplacian operator stands for the divergence of the gradient.

The equation $\nabla^2 V = -\rho/\varepsilon$ is called Poisson's equation, which can be solved in different coordinate systems. In the Cartesian system,

$$\nabla^2 V(x,y,z,t) = \frac{\partial^2 V(x,y,z,t)}{\partial x^2} + \frac{\partial^2 V(x,y,z,t)}{\partial y^2} + \frac{\partial^2 V(x,y,z,t)}{\partial z^2} = -\frac{\rho}{\varepsilon}$$

In the cylindrical system,

$$\nabla^2 V(r,\phi,z,t) = \frac{1}{r}\frac{\partial}{\partial r}\left(r\frac{\partial V(r,\phi,z,t)}{\partial r}\right) + \frac{1}{r^2}\frac{\partial^2 V(r,\phi,z,t)}{\partial \phi^2} + \frac{\partial^2 V(r,\phi,z,t)}{\partial z^2}$$

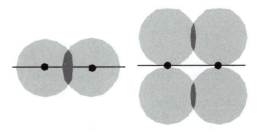

FIGURE 8.4
σ and π covalent bonds.

In the spherical system,

$$\nabla^2 V(r,\theta,\phi,t) = \frac{1}{r^2}\frac{\partial}{\partial r}\left(r^2\frac{\partial V(r,\theta,\phi,t)}{\partial r}\right) + \frac{1}{r^2\sin\theta}\frac{\partial}{\partial\theta}\left(\sin\theta\frac{\partial V(r,\theta,\phi,t)}{\partial\theta}\right)$$

$$+\frac{1}{r^2\sin^2\theta}\frac{\partial^2 V(r,\theta,\phi,t)}{\partial\phi^2}$$

One must use the boundary conditions to solve Poisson's equation.

Assuming that $\rho = 0$, the Poisson equation becomes the Laplace equation, i.e., $\nabla^2 V = 0$.

The results reported are applied to the problem under consideration.

Inside the electron cloud, we must solve the Poisson equation, and in the spherical coordinate system, we have

$$\nabla^2 V(r) = \frac{1}{r^2}\frac{d}{dr}\left(r^2\frac{dV(r)}{dr}\right) = \frac{\rho_0}{\varepsilon_0} \quad \text{or} \quad \frac{d}{dr}\left(r^2\frac{dV(r)}{dr}\right) = r^2\frac{\rho_0}{\varepsilon_0}$$

Integration results in the following expression:

$$\frac{dV(r)}{dr} = \frac{\rho_0}{3\varepsilon_0}r + \frac{c_1}{r^2}$$

The electric field intensity is found to be

$$\mathbf{E} = -\nabla V = -\mathbf{a}_r\frac{dV(r)}{dr} = -\mathbf{a}_r\left(\frac{\rho_0}{3\varepsilon_0}r + \frac{c_1}{r^2}\right)$$

We need to find the integration constant c_1. The electric field cannot be infinite at $r = 0$. Thus, $c_1 = 0$. Therefore, one finds

$$\mathbf{E} = -\mathbf{a}_r\frac{\rho_0}{3\varepsilon_0}r \quad \text{for} \quad 0 \le r < R$$

Outside the cloud, $\rho = 0$ for $r \ge R$. Correspondingly, one must solve the Laplace equation

$$\nabla^2 V = 0 \quad \text{or} \quad \frac{1}{r^2}\frac{d}{dr}\left(r^2\frac{dV(r)}{dr}\right) = 0$$

Integration gives $dV(r)/dr = c_2/r^2$. Thus,

$$\mathbf{E} = -\nabla V = -\mathbf{a}_r\frac{dV(r)}{dr} = -\mathbf{a}_r\frac{c_2}{r^2}$$

The integration constant c_2 is found using the boundary condition at $r = R$. In particular, from $c_2/R^2 = (\rho_0/3\varepsilon_0)R$, one has $c_2 = (\rho_0/3\varepsilon_0)R^3$. Thus, we obtain

$$\mathbf{E} = -\mathbf{a}_r\frac{\rho_0}{3\varepsilon_0}\frac{R^3}{r^2} \quad \text{for} \quad r \ge R$$

Metallic solids (conductors, for example, copper, silver, and iron) consist of metal atoms. These metallic solids usually have hexagonal, cubic, or body-centered cubic close-packed structures (see Figure 8.5). Each atom has 8 or 12 adjacent atoms. The bonding is due to valence electrons that are delocalized throughout the entire solid. The mobility of electrons is examined to study the conductivity properties.

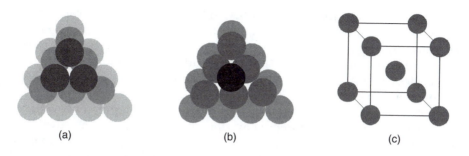

FIGURE 8.5

Close packing of metal atoms: (a) cubic packing; (b) hexagonal packing; (c) body-centered cubic packing.

More than two electrons can fit in an orbital. Furthermore, these two electrons must have two opposite spin states (spin-up and spin-down). Therefore, the spins are said to be paired. Two opposite directions in which the electron spins (up $+\frac{1}{2}$ and down $-\frac{1}{2}$) produce oppositely directed magnetic fields. For an atom with two electrons, the spins may be either parallel (S = 1) or opposite, and thus cancel (S = 0). Because of spin pairing, most molecules have no net magnetic field, and these molecules are called *diamagnetic*. In the absence of the external magnetic field, the net magnetic field produced by the magnetic fields of the orbiting electrons and the magnetic fields produced by the electron spins is zero. The external magnetic field will produce no torque on the *diamagnetic* atom as well as no realignment of the dipole fields. Accurate quantitative analysis can be performed using quantum theory. Using the simplest atomic model, we assume that a positive nucleus is surrounded by electrons that orbit it in various circular orbits (an electron on the orbit can be studied as a current loop, and the direction of current is opposite to the direction of the electron rotation). The torque tends to align the magnetic field, produced by the orbiting electron, with the external magnetic field. The electron can have a spin magnetic moment of $\pm\, 9 \times 10^{-24}$ A $\cdot$ m^2. The plus and minus signs indicate that there are two possible electron alignments—in particular, aiding or opposing the external magnetic field. The atom has many electrons, and only the spins of those electrons in shells that are not completely filled contribute to the atom's magnetic moment. The nuclear spin contributes only slightly to the atom moment. The magnetic properties of the media (diamagnetic, paramagnetic, superparamagnetic, ferromagnetic, antiferromagnetic, ferrimagnetic) result due to the combination of the listed atom's moments.

Let us briefly emphasize the *paramagnetic* materials. The atom can have a small magnetic moment, but the random orientation of the atoms results in a net torque of zero. Thus, the media do not show the magnetic effect in the absence of the external magnetic field. As the external magnetic field is applied, due to the atom moments, the atoms will align with the external field. If the atom has a large dipole moment (due to electron spin moments), the material is called ferromagnetic. In antiferromagnetic materials, the net magnetic moment is zero, and thus the antiferromagnetic media are only slightly affected by the external magnetic field.

Single bonds are usually σ bonds. Double bonds, which are much stronger, consist of one σ bond and one π bond, and the triple bond (the strongest one) consists of one σ bond and two π bonds. In the case of carbon nanotubes, the strong interaction among the carbon atoms is guaranteed by the strength of the C—C single bond, which holds carbon atoms together in the honeycomb- like hexagon unit (open-ended nanotube).

Carbon has played an important role in the evolution of life on Earth. A single cell consists of 70 to 95% water, and the rest consists mainly of carbon-based complexes and compounds. Biosystems are made mainly from carbon, oxygen, hydrogen, and nitrogen.

We reported in previous chapters that DNA, amino acids, proteins, carbohydrates, and other molecules are composed of carbon atoms bonded to one another as well as to atoms of other elements. In particular, hydrogen (H), oxygen (O), nitrogen (N), sulfur (S), and phosphorous (P) are common composites of the carbon-based molecules and complexes. These molecules can be simple or complex (thousands of atoms). Carbon chains form the skeletons of organic molecules. For example, straight, branched, closed, and other various carbon skeletons result. These skeletons possess bonding sites for atoms of other elements. Isomer molecules (formed from carbon) have the same formula but have distinctive structures and properties. Functional groups consist of specific groups of atoms that covalently bond to carbon skeletons and provide the molecule with distinct physical and chemical properties. In this chapter, we will focus our attention on carbon-based electronics.

Carbon nanotubes, fullerenes, and other carbon structures have a valence band generated by π and σ valence electrons. The π electrons are mobile and highly polarizable and define electronic and electromechanical properties, while σ electrons are usually localized and contribute to mechanical properties.

8.3   Carbon Nanotubes

8.3.1   Analysis of Carbon Nanotubes

Carbon fibers (nanometers in diameter) have been known for many years. In 1889, U.S. patent 405,480 was awarded to T. V. Hughes and C. R. Chambers for technology to grow carbon filaments from "swamp gas" (methane) using metallic crucibles that catalyze the reaction. In 1960, R. Bacon discovered submicrometer-diameter scroll morphology graphite whiskers (concentric scroll tubes and rolled-up graphite layer sheets) that were grown in a DC carbon arc under high pressure of an inert gas. He published his findings in *J. Applied Physics*, vol. 31, 1960.[7] Multiwall carbon nanotubes, observed by Iijima in 1991 in the soot of carbon arc, were made utilizing DC carbon arc process.[7-9] These nanotubes are found to be molecular structures that consist of graphite cylinders closed at either end with caps containing pentagonal rings, as well as concentric fullerenes. Carbon nanotubes are produced by vaporizing carbon graphite with an electric arc under an inert atmosphere. The carbon molecules organize a perfect network of hexagonal graphite rolled up onto itself to form a hollow tube. In 1993, Iijima demonstrated that single-wall carbon nanotubes (*buckytubes*) formed in carbon arc with transition metal catalysts result in a small number of buckytubes. These strong and flexible buckytubes, which exhibit remarkable properties, can be single- or multi-walled. The standard arc-evaporation method produces mainly multi-walled nanotubes, and the single-wall uniform carbon nanotubes were synthesized in the late 1990s. One can fill nanotubes with some atoms, molecules, and even biological materials. The carbon nanotubes can be conducting or insulating depending on their structure. Recently, cost-effective chemical-vapor-deposition-based processes were developed to produce carbon nanotubes. Among the feasible processes, the HiPco process (Rice University, Carbon Nanotechnologies, Inc.) can be emphasized. A cool high-pressure stream of metastable CO feedstock gas (CO has the strongest diatomic bonds, 11 eV) and a thermodynamically stable transition metal carbonyl catalyst precursor, e.g., $Fe(CO)_3$, are used. This CO-catalyst media is quickly (1 μsec) heated to 1000°C by mixing it with CO preheated to 1200°C. The thermodynamically stable $Fe(CO)_3$ at high temperature (1000°C) becomes unstable. As $Fe(CO)_3$ dissociates, it collides with other iron carbonyl fragments and nucleates tiny Fe nanoparticles. CO has a catalytic path of producing CO_2, leaving C atoms as tiny clusters that form self-assembled aromatic carbon sheets. At high temperature

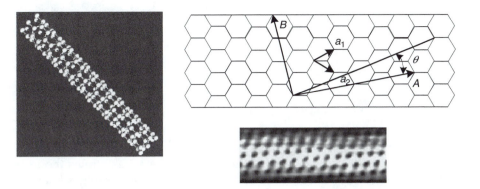

FIGURE 8.6
Single-wall carbon nanotube.

and pressure, these carbon sheets curve to minimize the edge energy resulting in carbon nanotubes and a small number of spheroidal fullerenes (C_{xx}).

Carbon nanotubes can be visualized as sheets of graphite rolled into hollow (but closed-ended) cylinders. The vector *AB* defines the rolling (see Figure 8.6). This *AB* vector is expressed as a linear combination of the lattice unit vectors a_1 and a_2, as well as indices n and m. In particular,

$$AB = na_1 + ma_2$$

The pair of integer indices (n, m), $0 \le |m| \le n$, defines the diameter and chirality of single-wall carbon nanotubes. Specifically, the chiral vector is given as $C_h = na_1 + ma_2 \equiv (n, m)$.

For example, $m = 0$ for the "zigzag" nanotubes ($\theta = 0°$), $n = m$ for the "armchair" nanotubes ($\theta = 30°$), and for "chiral" nanotubes one has (n, m) and $0° < |\theta| < 30°$. The chiral angle θ (or simply chirality) is found using the inner product of C_h and a_1, i.e.,

$$\cos\theta = \frac{C_h \cdot \alpha_1}{|C_h||\alpha_1|} = \frac{2n + m}{\sqrt{n^2 + nm + m^2}}$$

A single-wall carbon nanotube and an unrolled honeycomb lattice of a nanotube that consists of carbon molecules (two-dimensional graphite sheet) are illustrated in Figure 8.6.

The diameter of the carbon nanotube d_{CNT} is found using the circumference length L (which is a function of C_h) as

$$d_{CNT} = \frac{L}{\pi} = \frac{|C_h|}{\pi} = \frac{\sqrt{C_h \cdot C_h}}{\pi} = \frac{a\sqrt{n^2 + nm + m^2}}{\pi}$$

where a is the lattice constant (for honeycomb lattice, $a = 2.49$ Å).

This equation can be straightforwardly applied. For example, the armchair nanotube $C_h = na_1 + ma_2 \equiv (5, 5)$ closed by the hemisphere C_{60} fullerene has a diameter of 6.88 Å.

Vectors a_1 and a_2 are not orthogonal. In particular, $a_1 \cdot a_1 = a_2 \cdot a_2 = a^2$ and $a_1 \cdot a_2 = a^2/2$.

The application of these nanotubes, which are a few carbon atoms in diameter, provides the possibility to fabricate devices on an atomic and molecular scale. The diameter of a nanotube is 100,000 times less than the diameter of a sewing needle. Carbon nanotubes, which are much stronger than steel wire (in the comparison, it is assumed that the diameter is the same), are the perfect conductor (better than silver) and have thermal conductivity better than diamond. Carbon nanotubes exhibit high Young's modulus and tensile strength. In carbon nanotubes, carbon atoms bond together forming the pattern. Single-wall carbon

FIGURE 8.7
Single carbon nanotube ring with six atoms.

nanotubes are manufactured using chemical vapor deposition, laser vaporization, arc technology, vapor growth, as well as other methods. Figure 8.7 illustrates the carbon ring with six atoms. When such a sheet rolls itself into a tube so that its edges join seamlessly together, a nanotube is formed.

Carbon nanotubes can be widely used in MEMS and NEMS, as well as in nanoelectronics. Two slightly displaced (twisted) nanotube molecules, joined end-to-end, act as the diode. Molecular-scale transistors can be manufactured using different alignments. There are strong relationships between the nanotube electromagnetic properties and its diameter and chirality. In fact, the electronic and electromagnetic properties of the carbon nanotubes depend on the molecule's twist, and Figure 8.8 illustrates the possible configurations. If the graphite sheet forming the single-wall carbon nanotube is rolled up perfectly (all its hexagons line up along the molecule's axis), the nanotube is a perfect conductor. If the graphite sheet rolls up at a twisted angle, the nanotube exhibits semiconductor properties. The carbon nanotubes, which are stronger than steel wire, can be added to the media to make the conductive composite materials as well as to change its mechanical properties.

Example 8.2

Consider the semiconducting graphite (carbon) sheet assuming the following one-dimensional approximation for the charge distribution:

$$\rho = \rho_0 \, \text{sech} \, \frac{x}{L} \tanh \frac{x}{L}$$

Find the total charge, electric field intensity, and potential field.

SOLUTION

For the carbon layer sheet with the length 100 nm and $\rho_0 = 2 \times 10^{-10}$, the charge distribution is reported in Figure 8.9 as calculated and plotted using the following MATLAB statement:

```
x =-50e - 9:1e - 9:50e - 9;  L = 10e - 9;  rho0 = 2e - 10;
rho = rho0*sech(x/L).*tanh(x/L);

plot(x/L,rho);
```

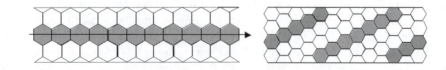

FIGURE 8.8
Carbon nanotubes.

```
xlabel('Distance x/L'),ylabel('Charge Distribution');
title('Charge Distribution in the Graphite Sheet');
```

Figure 8.9 also documents the electric field intensity. The expression for E_x will be found as

$$E_x = -\frac{\rho_0}{\varepsilon} L \operatorname{sech} \frac{x}{L}$$

The total positive charge is found to be

$$Q = A \int_0^{2L} \rho_0 \operatorname{sech} \frac{x}{L} \tanh \frac{x}{L} dx = 2A\rho_0 L$$

where A is the junction cross-sectional area.

The Poisson equation $\nabla^2 V = -\rho/\varepsilon$ must be solved to find the electric field intensity and potential. In particular, we have

$$\frac{d^2 V}{dx^2} = -\frac{\rho_0}{\varepsilon} \operatorname{sech} \frac{x}{L} \tanh \frac{x}{L}$$

Integration results in the following expression:

$$\frac{dV}{dx} = \frac{\rho_0}{\varepsilon} L \operatorname{sech} \frac{x}{L} + c_1$$

Hence, the one-dimensional electric field intensity is found to be

$$E_x = -\frac{\rho_0}{\varepsilon} L \operatorname{sech} \frac{x}{L} - c_1$$

We use the boundary conditions to solve Poisson's equation. In particular, the integration constant c_1 must be obtained. The electric field must tend to zero as x tends to infinity. Therefore, $c_1 = 0$. Hence,

$$E_x = -\frac{\rho_0}{\varepsilon} L \operatorname{sech} \frac{x}{L}$$

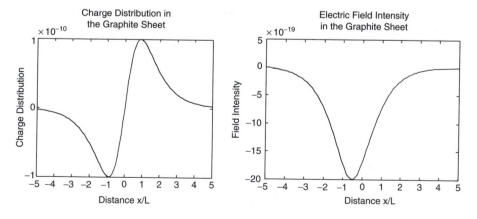

FIGURE 8.9
Charge distribution and electric field intensity in the semiconducting graphite sheet.

This electric field intensity, reported in Figure 8.9, is calculated and plotted using the following MATLAB statement:

```
x = -50e - 9:1e - 9:50e - 9; L = 10e - 9; rho0 = 2e - 10; eps = 1;
E = -rho0*L*sech(x/L)/eps;
plot(x/L,E);
xlabel('Distance x/L'),ylabel('Field Intensity');
title('Electric Field Intensity in the Graphite Sheet');
```

The potential is found by integrating

$$\frac{dV}{dx} = \frac{\rho_0}{\varepsilon} L \operatorname{sech}\frac{x}{L} + c_1, \quad c_1 = 0$$

In particular, we have

$$V = 2\frac{\rho_0}{\varepsilon} L^2 \tan^{-1} e^{\frac{x}{L}} + c_2$$

The constant c_2 is found by setting the potential at any point on the carbon nanotube. Let $V = 0$ at $x = 0$. Then, from

$$V_{x=0} = 2\frac{\rho_0}{\varepsilon} L^2 \tfrac{1}{4}\pi + c_2 = 0$$

one finds c_2.

Thus, we have

$$V = 2\frac{\rho_0}{\varepsilon} L^2 \left(\tan^{-1} e^{\frac{x}{L}} - \tfrac{1}{4}\pi \right)$$

One can derive the basic electromagnetic properties in the carbon nanotubes using the electromagnetic theory. Assume that a single-wall metallic carbon nanotube forms an ideal cylindrical surface. Then, the expression for the charge density is given as

$$\frac{\partial J}{\partial t} = -\frac{\rho_{se} e^2}{m} \nabla V, \quad \frac{\partial \rho_p}{\partial t} + \nabla J = 0$$

where ρ_{se} is the surface electron density of the graphite, $\rho_{se} = 16/3\sqrt{3}b^2$; e and m are the electron charge and mass of the electron; V is the potential on the surface of the carbon nanotube including induced potential of all charges on the surface; and ρ_p is the variable charge density due to the plasmon mode,

$$\rho_{p\,hl} = \frac{1}{4\pi} \frac{V_{hl}}{r K_l(kr) M_l(kr)}$$

where $K_l(kr)$ and $M_l(kr)$ are the Bessel functions of imaginary argument and of order l taken at the carbon nanotube radius r, and h and l are the coaxial momentum and the circumference quantum number of the mode, respectively.

Examining carbon nanotubes as structural composites, the energy-based analysis is straightforward. In general, one applies the following equation: $\partial E/\partial \mathbf{r} = 0$. The total energy is given as

$$E(\mathbf{r}, V) = T(\mathbf{r}) - E_{ES}(\mathbf{r}, V, C) - W(\mathbf{r}, \varsigma, \alpha)$$

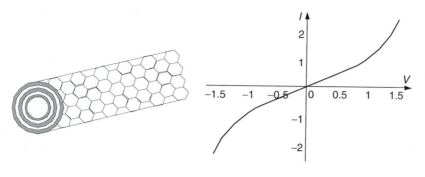

FIGURE 8.10
Multiwall carbon nanotube.

where $T(\mathbf{r})$ is the elastic strain energy, $T(x) = \frac{1}{2}k(x_0 - x)^2$; E_{ES} is the electrostatic energy, $E_{ES}(x, C, V) = \frac{1}{2}C(x)V^2$; W is the van der Waals energy, $W(x, \varsigma, \alpha) = \varsigma x^{-\alpha}$; and ς and α are the media coefficient and dispersion force coefficient (α is usually in the range from 5 to 7).

All three energies are nonlinear functions containing not only variables but also the parameters (elasticity, radius, permittivity, etc.). One may study carbon nanotubes and fullerenes as ideal conducting cylinders (cylinders for multiwall nanotubes) and spheres. The capacitance of the cylinder of length l and sphere (with the inner and outer diameters r_i and r_o) are known to be

$$C = 2\pi\varepsilon \frac{l}{\ln(r_o/r_i)} \quad \text{and} \quad C = 4\pi\varepsilon \frac{r_o r_i}{r_o - r_i}$$

The vapor-grown carbon nanotubes with N layers are illustrated in Figure 8.10, and the industrially manufactured nanotubes have nanometer diameter and micrometer length.

The BN-covered aligned B–C–N semiconducting multiwall nanotubes have been synthesized utilizing the aligned CN_z ($z < 0.1$) nanotube mats with boron oxide (B_2O_3) and nitrogen (N_2) environment with a CuO promoter at temperatures from 2000 to 2100 K. The resulting inner B–C–N layers have shown semiconducting characteristics. The overall composition is found to be 1 carbon, 0.46 boron, and 0.22 nitrogen with some variations. The *I-V* characteristic for the B–C–N multiwall carbon nanotube is documented in Figure 8.10. It should be emphasized that the resistivity of the B–C–N carbon nanotubes is 1×10^8 ohm (and higher), while pure carbon nanotube resistivity is usually 1×10^5 ohm or less.

The carbon nanotube energy is found to be[10]

$$E(N, r) = \underbrace{\frac{9N}{8r^2} E_c}_{\text{curvature energy}} + \underbrace{\frac{4\pi\varsigma r}{\sqrt{3}} E_b}_{\text{dangling energy}}$$

where $E_c = 0.9$ eV; E_b is the dangling bond energy, $E_b = 2.36$ eV; ς is the dangling bond coefficient ($\varsigma = 1$ for the zigzag tubes, and $\varsigma = 2/\sqrt{3}$ for the armchair tubes); N is the number of atoms; and r is the dimensionless carbon nanotube radius. The total perimeter $4\pi r$ was reduced to reduce the number of dangling bonds.

Closed and open carbon nanotubes can be examined using energy-based analysis. Specifically, carbon nanotubes and other carbon nanoclusters are rolled and closed to minimize the energy. We examine the transformation of a planar fragment of a graphite monolayer into a spherical closed shell. The carbon spherical-shell closed cluster has the lowest energy; however, curving of a plane fragment into a segment has an energy barrier. We assume that the carbon bonds are the same for all atoms and the curvature is constant. The rolling energy for carbon clusters (with a curved surface to form closed spherical shells) is given using three terms: total energy of 12-pentagonal rings, correction-curved energy due to curved bonds, and the energy of the nanotube spherical cluster related to the experimentally derived energy of C_{60}. In particular,

$$E_t = E_{t0} + \frac{16\pi}{\sqrt{3}} E_c - \frac{N_c}{N} E_c$$

where E_{t0} is the total energy of 12-pentagonal rings (electrons at the pentagon have the energy, which is different from electrons on a graphite-like hexagon) accounting for non-equivalent bonds, $E_{t0} = 17.7$ eV; and N_c is the characteristic number (to fit the energy of the nanotube spherical cluster to the experimentally derived energy of C_{60}), $N_c = 1161$.

Carbon nanotubes can be also used as nanoelectromechanical devices.[11] Among possible applications, the following can be emphasized: cantilever beams, carbon springs, deformable suspended structures, mechanical switches, etc. For example, using carbon nanotubes, one can design electromechanical nanoswitches as schematically illustrated in Figure 8.11. It should be mentioned that these electromechanical carbon nanotube–based devices are extremely difficult to fabricate, and their possible application is limited. However, the suspended aligned carbon nanotubes and flexible carbon fabrics can be relatively straightforwardly deposited, and these devices have found some applications.

Carbon nanotubes are 10 times stronger and 5 times lighter than steel (for the same size). Carbon nanotubes have conductivity greater than silver, and they transmit heat better than diamond. The current density of carbon nanotubes and copper are 10^{11} and 10^6 electron/sec·nm^2. The current technologies and processes allow one to fill carbon nanotubes with inorganic and organic materials. It is evident that carbon-based nanoelectronics may be implemented.[6] For electronics applications, it is important to fabricate nanodevices that exhibit the desired physical properties, as well as form the desired pattern and specified topologies, make robust interface and low-resistance contact with electrodes, etc. Different paradigms in modeling, analysis, and simulations of electronic nanodevices are reported in this book and elsewhere.[12–14] Challenges in the synthesis, assembling, processing, and fabrication of carbon-based nanoelectronics (nanotubes, fullerenes, and other carbon molecules and complexes) exist. For example, for carbon nanotubes, different catalysts have been utilized—Fe, Mo, Co, Ni, Al, Y, etc. Chemical vapor deposition, integrated with e-beam lithography, provides meaningful processes to fabricate carbon nanotube–based nanoelectronic devices.

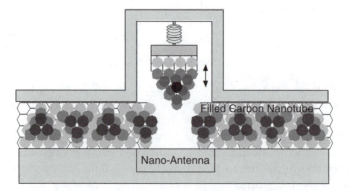

FIGURE 8.11

Schematic diagram of the possible application of carbon nanotubes as a nanoswitch.

8.3.2 Classification of Carbon Nanotubes

This section covers classification of carbon nanotubes. Though the classification can be performed through the direct characterization of carbon nanotubes measuring *I-V* characteristics, this concept has significant limitations in practice. One high-throughput, indirect, accurate, and efficient method used to characterize carbon nanotubes is based on Raman spectroscopy. In nanoelectronics, it is important to fabricate carbon nanotubes that exhibit the desired physical properties, form the desired pattern, make low-resistance contact with electrodes, etc.

Carbon nanotubes exhibit metallic, semimetallic, and semiconducting properties depending on chirality and diameter. The energy band gap analysis shows that armchair (n, n) carbon nanotubes with $n = m$ are metallic. In general, if $(n - m) = 3k$, one concludes that these carbon nanotubes are metallic. Here, k is the integer. The resistance is usually in the range of hundreds of kilohms at room temperature. The (n, m) nanotubes are semiconducting if $(m - n) \neq 3k$, and the energy gap is found to be

$$E_{gap} \propto \frac{1}{d_{CNT}}$$

where d_{CNT} is the nanotube diameter. Thus, the energy band gap E_{gap} is inversely proportional to the diameter of the nanotube. More specifically, E_{gap} can be estimated using the following formula:

$$E_{gap} = 2\frac{E_{cc}a_{cc}}{d_{CNT}}$$

where E_{cc} is the carbon-to-carbon tight-binding overlap energy, and a_{cc} is the nearest-neighbor carbon-to-carbon distance, $a_{cc} = 0.142$ nm. The curvature of nanotubes leads to nonparallel π-orbitals interconnecting with σ-orbitals causing semiconducting properties due to the band gap openings. The band gap usually is in the range from 1 to 10 meV. Finally, the (m, n) single-wall carbon nanotubes are semimetallic with zero band gap if $(m - n) = 3k$.

Raman spectroscopy and Raman spectra allow one to obtain important qualitative and quantitative data. In particular, the G-, D-, and G′-bands allow one to examine the diameter and chirality (chiral angle) of carbon nanotubes, thereby allowing one to determine whether a carbon nanotube is metallic, semimetallic, or semiconducting. When the incident or scattered photons in the Raman process are in resonance with an electronic transition between the valence and conducting bands at the specific energy states, the Raman signal becomes large due to strong coupling between electrons and photons in the carbon nanotubes under the resonant conditions. The frequency of the Raman spectra is usually from 0 to 3000 cm^{-1}.

The Raman spectra for a carbon nanotube is shown in Figure 8.12, and the cubic spline interpolation is reported as well (it should be emphasized that one cannot blindly apply different interpolation techniques because incorrect interpolation results, e.g., intensity is higher than 100).

The experimental data set contains measurements on several variables and characteristics. In particular, the measured data includes D-, G-, and G′-band intensities as well as D-, G-, and G′-band frequencies. For SWCNTs, using the Raman spectra, the problem is to determine to which group those nanotubes belong and what properties they can exhibit. For 300 metallic, semimetallic, and semiconducting carbon nanotubes, the three-dimensional plots for the bands magnitudes and frequencies (D-, G-, and G′-bands) are reported in Figure 8.13. This data is used for the classification of carbon nanotubes.

We apply the discriminant analysis (linear and quadratic discriminant methods) to classify carbon nanotubes. By making use of only the G′-band, 24% of carbon nanotubes were misclassified. Utilizing all three bands, 7% of carbon nanotubes are misclassified.

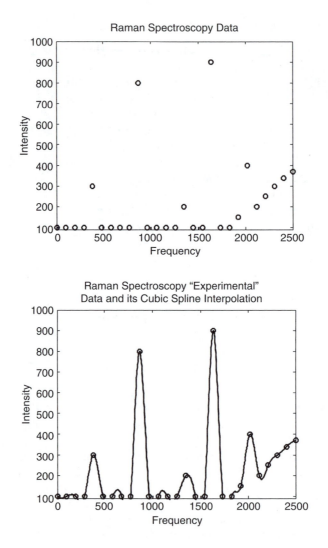

FIGURE 8.12
Raman spectra for a carbon nanotube.

However, discriminant analysis can be effectively used assuming that the measurements have a multivariate normal distribution. If this assumption is not guaranteed, nonparametric classification methods (for example, decision tree) should be applied. Decision tree is a set of linguistic rules. For example, "if the gap intensity is less or more than specified using lower and upper bounds (B_l and B_u), classify the CNT as xxx." Decision trees do not require any assumptions about the distribution of the measurements (in general, measurements can be categorical, continuous, discrete, or hybrid). In fact, cluttered decision trees use well-posed lower and upper bound rules. To determine the properties of carbon nanotubes, one starts at the top node and applies the rule. If the point satisfies the rule, we take the left path, and if not, we go to the right path. Ultimately, one reaches a terminal node that assigns the specified properties (metallic, semimetallic, semiconducting, and undefined).

We obtain the complex decision tree. It is usually possible to find a simpler tree. The following sequential and well-defined procedure to derive a tree that classifies one particular

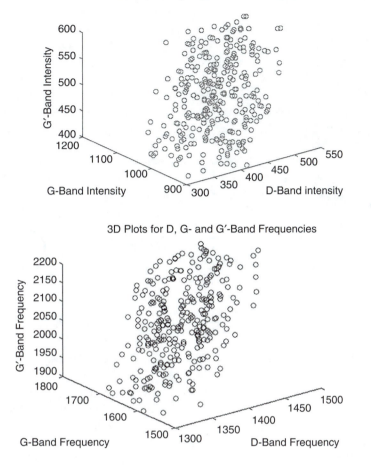

FIGURE 8.13
Three-dimensional plots for bands intensities and frequencies used to classify 300 carbon nanotubes as metallic, semimetallic, and semiconducting.

data set is applied. Derive the error rate and use this tree to classify new data. Estimate the error by classifying the second data set directly (full trees and subsets of trees can be used). Let us assume the worst case, i.e., there is no second data set. The classification can be performed in this situation as well by doing the cross-validation of a single set. In particular, using a subset ($x\%$) of the data, build a tree using the other ($100 - x$)%. Then, one uses the decision tree to classify the removed $x\%$. One can repeat this by removing each of X subsets one at a time. For each subset, we may find that smaller trees give smaller errors than the full tree. The "resubstitution error" is computed, i.e., one finds the proportion of original observations that were misclassified by various subsets of the original tree. The next step is to use the cross-validation to estimate the true error for trees of various sizes. Usually, the analysis shows that the resubstitution error is overly optimistic. It decreases as the tree size grows, but the cross-validation results show that beyond a certain point (error cost), increasing the tree size increases the error rate. We select the decision tree with the smallest cross-validation error, and a simple tree may be as descriptive as a complex tree. The standard error of the minimum is calculated and used. The computed cutoff value is equal to the minimum cost plus standard error. The optimal

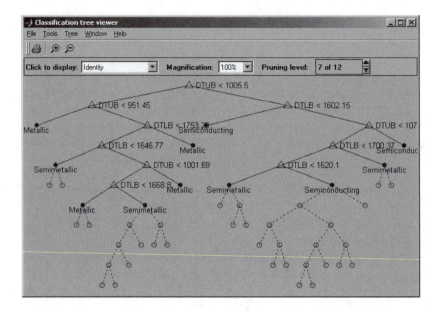

FIGURE 8.14
Application of decision tree for classification of carbon nanotubes.

classification level is calculated. The proposed decision tree classification guarantees 3.7% misclassification.

We conclude that by making use of distinct classification methods, including the decision trees, the carbon nanotubes were classified. The application of the reported decision tree method with cross-validation is demonstrated in Figure 8.14 and Figure 8.15. It must be emphasized that using a single band, it is also possible to classify carbon nanotubes using decision trees; however, 19% of carbon nanotubes were misclassified.

The classified carbon nanotubes are reported in Figure 8.16.

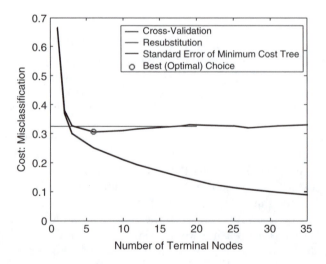

FIGURE 8.15
Cross-validation in classification of carbon nanotubes.

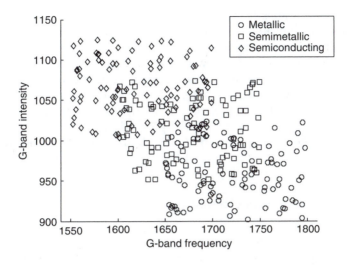

FIGURE 8.16 (See color insert.)
Metallic, semimetallic, and semiconducting carbon nanotubes.

8.4 Carbon-Based Nanoelectronics and Three-Dimensional Nano-ICs

This section covers and discusses carbon-based nanoelectronics, in particular, (1) fullerene-centered nanoelectronics and (2) biocentered nanoelectronics. The diameter of fullerene molecules is comparable to that of the double helix of DNA. We will cover and examine novel phenomena and how these effects can be uniquely utilized.

Wide spectra of complex unsolved fundamental, applied, experimental, and technological problems remain. Many critical problems and issues have not even been addressed. This chapter introduces the reader to carbon-based nanoelectronics, nanodevices, interconnects, and nanofabrication, as well as possible topologies and architectures. It is difficult to fabricate carbon nanotubes with uniform and controllable electronic properties, and small changes result in totally distinct properties of carbon nanotubes, e.g., they become semiconducting, metallic, or semimetallic. Other solutions are needed, and fullerene-centered nanoelectronics is a possible candidate for nano-ICs. Different methods for fabrication of functional fullerenes and fullerene complexes have been developed. Correspondingly, novel nanoelectronic devices can be devised and made using fullerenes as building blocks. The encapsulation of distinct media in fullerenes leads to defining and controlling of electronic properties of endohedral fullerenes.

The carbon-based electronic nanodevices, fabricated utilizing nanotechnology, have been considered as one of the most feasible technologies for nano-ICs. We already briefly covered carbon nanotubes. Let us focus our attention on a family of carbon-based electronics that utilizes fullerenes C_{xx} and their derivatives. We first examine fullerene-centered nanoelectronics. Carbon bonding plays an important role in organic molecules. A carbon atom contains six electrons. The lowest energy level is occupied by two electrons with oppositely pairing electron spins. Four remaining electrons on $2s$, $2p_x$, $2p_y$, and $2p_z$ orbitals participate in bonding. These $2s$ and $2p$ orbitals are hybridized to form three types of chemical bonds of carbon with other atoms. This sp^n hybridization leads to (1) diagonal sp^1 hybridization with the resulting bond angle of 180°; (2) triagonal sp^2 hybridization with the resulting 120° bond angle; and (3) tetrahedral sp^3 hybridization with the resulting bond angle of 109°28'. Other bond angles have been found. In particular, the cubane C_8H_8 was synthesized with bond angles from 108° to 110°. There exists a great variety of complex organic carbon-based molecules. We concentrate our attention on fullerenes and their derivatives. C_{60} (with 12 pentagon faces) named fullerene after the architect Buckminster Fuller, who designed a geodesic dome with the same fundamental symmetry. This C_{60} is the roundest and one of the symmetrically largest known molecules. It consists of 60 carbon atoms arranged in a series of interlocking hexagons and pentagons, forming a structure that looks similar to a soccer ball. C_{60} forms truncated icosahedron (consisting of 12 pentagons and 20 hexagons with hollow spheroid [see Figure 8.17]) and was reported in 1985 by professors Sir Harry Kroto (U.K.), Richard E. Smalley, and Robert F. Curl, Jr. They were jointly awarded the 1996 Nobel Prize in chemistry.

8.4.1 Fullerene-Centered Nanoelectronics

Fullerenes (C_{60}, C_{70}, C_{76}, C_{80}, C_{84}, C_{102}, etc.) can be used as high-performance multi-terminal electronic nanodevices that exhibit novel phenomena and lead to new functionality with revolutionary enhanced features. In addition to unique electronic properties, robust mechanical structures are formed. Fullerenes have linkage groups and therefore can be interconnected in the cubic and other structures forming complex functional topologies uniquely suited for nano-ICs. Fullerenes can ensure the desirable characteristics (conductivity, resistivity, *I-V* characteristics, stiffness, strength, robustness, permeability, permittivity, thermal expansion, etc.), affordability, manufacturability, etc. It is evident that many fullerene characteristics are predetermined by bonds (weak and strong). The charge distribution, density of state, field, *I-V* characteristics, resistivity, permeability, and permittivity are of interest to us. The fullerene assemblies exhibit unique electronic properties.[15–20] For example, fullerenes can exhibit semiconductor properties. Fullerenes can contain a great variety of noncarbon atoms inside carbon cages as well as on the carbon cages. Carbon nanotubes can be filled with fullerenes. For example, if a single-wall carbon nanotubes, filled with gadolinium fullerenes ($Gd@C_{80}$), these endohedral fullerenes are self-lined inside carbon nanotubes with 1-nm displacement (after evaporation process) changing electronic properties at room temperature.[8] Metal atoms (for example, alkali, rare earth metals, and others) were known to change the electronic structure of fullerenes, significantly affecting their electronic properties. Figure 8.17 illustrates C_{60}, endohedral C_{60} with one metal atom ($M@C_{60}$), functional fullerene complex, 1.4-nm-diameter single-wall carbon nanotube with fullerenes, and fullerenes on silicon. Numerous stable symmetric and asymmetric metallofullerene complexes have been synthesized and characterized.

Fullerenes can be utilized in nanoelectronics with far-reaching application to 3D nano-ICs. Despite the large amount of research in fullerenes and their derivatives, the electronic structure, electronic characteristics, bond structures, and other characteristics have not been fully examined. In particular, fullerenes behave as a molecular crystal in which the

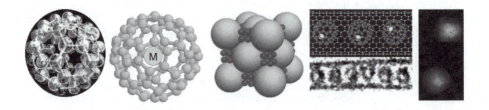

FIGURE 8.17 (See color insert.)
Empty fullerene C_{60}, endohedral fullerene $M@C_{60}$, functional fullerene complex, single-wall carbon nanotube with fullerenes, and endohedral fullerenes on Si.

molecular properties (vibration states and electronic excitations) are weakly perturbed by the crystal symmetry. Fullerenes behave as a semiconductor with a wide band gap, and they can exhibit metallic and superconducting characteristics. Fullerenes differ from other allotropic forms of carbon (graphite and diamond) due to the hybridization of the atomic s- and p-orbitals. Carbon has $1s^2\,2s^2\,2p^2$ electronic configuration. In diamond, there is the sp^3 hybridization with four identical σ-bonds with 109.5° bond angle. In graphite, the sp^2 hybridization results in three strong planar σ-bonds with a 120° angle. The out-of-plane p-orbitals result in π-bonds and weak interplanar bonding. In C_{60}, the curvature of the surface at carbon atoms is achieved by forming bonds with intermediate hybridization. In particular, three σ-bonds lead to $sp^{2.278}$ hybridization, and the π-bonds result in $s^{0.09}p$ hybridization. The effect of the curvature on the hybridization is often neglected, and the σ-bonds are considered to be $2s$ orbitals hybridized with the two in-plane $2p_x$ and $2p_y$ orbitals. The remaining $2p_z$ orbitals are mixed to form molecular π-bonds. Due to the strong σ-overlap between neighboring carbon atoms, the σ-states depart from the Fermi level. As a result, the low-energy excitations (several eV) involve predominantly the π-states, and C_{60} fullerenes can be modeled and examined using 60π electrons. Assuming spherical geometry, using the radius r, the kinetic energy is given as

$$\Gamma_n = \frac{\hbar n(n+1)}{2m_{ef}r^2}$$

The number of states with energy $\Gamma \leq \Gamma_n$ is $2(n+1)^2$. Considering 60π electrons, the states up to $n = 4$ would be filled by 50 electrons, and the remaining 10 electrons would be distributed within the 22-fold degenerate $n = 5$ shell. Thus, the idealized sphere model does not represent the closed-shell configuration. This model is refined using the so-called linear combination of atomic orbitals (orbitals are decomposed using subshells). The molecular orbitals for C_{60} have been examined.[19] The gap between the highest occupied molecular orbital (HOMO) and lowest unoccupied molecular orbital (LUMO) is found to be approximately 2.3 eV.

Fullerenes that are semiconducting at room temperature have been fabricated by thermal sublimation. The resistivity can be significantly decreased by using alkali metals. In M_xC_{xx}, as x in M_x increases, the effective resistivity decreases, reaching a minimum for the metallic stoichiometry ($x = 3$), corresponding to a half-filled conduction band. By increasing the M_x concentration, the resistivity increases until an insulator results at $x = 6$. The metallic ($x = 3$) state is superconducting with a superconducting transition temperature —for example, 30 K for Rb as M. These useful properties have found limited application because the alkali-doped media is unstable in air. This problem can be overcome by encapsulating the metal atoms inside the fullerene's carbon cage. The endohedral $M_x@C_{xx}$ fullerene cage structure is utilized as shown in Figure 8.17. Thus, it is possible to synthesize robust endohedral fullerenes with alkali metals. Endohedral fullerenes can be fabricated with one, two, or three metal atoms encapsulated and confined within the fullerene cages. Though encapsulation, isolation, and purification remain a challenge, promising results have been achieved for $Li_x@C_{xx}$, $Na_x@C_{xx}$, $K_x@C_{xx}$, $Rb_x@C_{xx}$, $La_x@C_{xx}$, and other complexes. The conductivity is maximized by utilizing metals that transfer three electrons to the fullerene, thus half-filling the lowest unoccupied molecular orbital of C_{60}. The possibility to engineer endohedral fullerene complexes with controlled, predicted, and desired electronic properties is a significant step toward feasibility of nano-ICs utilizing carbon-based nanoelectronics that can be fullerene-centered.

One concludes that different classes of fullerenes exist. In this section we cover (1) heterofullerenes (doped fullerenes) and (2) endohedral fullerenes (can be metallo- and nonmetallofullerenes). Table 8.1 documents major elements (see shaded area) from the periodic table of the elements that have been integrated in distinct fullerene molecules.

TABLE 8.1

Elements (Shaded) That Have Been Utilized in Heterofullerenes and Endohedral Fullerenes

IA	IIA	III	IV	V	VI	VII	VIII			IB	IIB	III	IV	V	VI	VII	VIIIA
H																	He
Li	Be											B	C	N	O	F	Ne
Na	Mg											Al	Si	P	S	Cl	Ar
K	Ca	Sc	Ti	V	Cr	Mn	Fe	Co	Ni	Cu	Zn	Ga	Ge	As	Se	Br	Kr
Rb	Sr	Y	Zr	Nb	Mo	Tc	Ru	Rh	Pd	Ag	Cd	In	Sn	Sb	Te	I	Xe
Cs	Ba	La–Lu	Hf	Ta	W	Re	Os	Ir	Pt	Au	Hg	Tl	Pb	Bi	Po	At	Rn
Lanthanides		La	Ce	Pr	Nd	Pm	Sm	Eu	Gd	Tb	Dy	Ho	Er	Tm	Yb	Lu	

8.4.1.1 Heterofullerenes

It is possible to substitute carbon with the heteroatoms (alkali, alkaline earth metals, transition metals, etc.) in the exterior surface of fullerenes. For example, silicon can be doped, and three-dimensional Si_xC_{xx} complexes result. Silicon-doped fullerenes have been fabricated utilizing laser ablation processes (using Si/C composite rods), direct growth, photofragmentation, chemical vapor deposition, etc. Considering SiC_{xx} (xx = 58–61) and Si_2C_{xx} (xx = 58, 60, . . .), it is evident that complex bridge-, edge-, and other Si_xC_{xx} topological configurations result. This allows one to synthesize complex three-dimensional nano-IC functional topologies. In addition to the silicon-doped fullerenes, boron/nitrogen-doped heterofullerenes $C_{60-2x}(BN)_x$ are another important inroad for fullerene-centered nanoelectronics. Replacing one carbon on the fullerene cage with nitrogen leads to $C_{59}N$, and one electron is added to the π-system of C_{60}. The modified anions of C_{60} exhibit distinct properties. The $C_{59}N$ heterofullerene is a radical with an unpaired electron localized on the carbon atom C_1 closest to the nitrogen. In this doped fullerene, C_1 becomes fourfold coordinated as $C_{59}HN$ or $(C_{59}N)_2$. Among other heterofullerenes, $C_{48}N_{12}$ and $C_{48}B_{12}$ are very promising for nanoelectronics. Those and other doped fullerenes can be uniquely utilized.

Many doped carbon molecules are kinetically stable, such as n- and p-type impurity-doped bulk silicon. The B=N bond is from 1.42 to 1.45 Å, i.e., almost the same as the C–C bond distance in fullerenes, which varies from 1.4 to 1.47 Å. For example, the C–C, C–N, C–B, and B–N distances in $C_{58}BN$ are 1.45, 1.43, 1.4, and 1.41 Å. To use $C_{60-2x}(BN)_x$ as nanoelectronic devices, the band gap, HOMO and LUMO energy levels, and atomic charges (in fullerenes on B and N atoms) are extremely important. The data is reported in Table 8.2.[21] It should also be emphasized that the data may not be fully conclusive because different methods lead to distinct results, and, correspondingly, sources provide distinct data (for example, the band gap for C_{60} was reported to be from 2.3 to 2.5 eV versus 1.6 eV as documented in Table 8.2). This is due to the fact that distinct theories, methods, and techniques are used, and experiments provide one with indecisive accuracy.

One-, two-, and three-dimensional fullerene complexes are reported in Figure 8.18. The neighboring fullerene cages are linked by the four-membered rings from carbon atoms at a fusion between two hexagons of a molecule. Each C_{60} in the polymeric chain is connected to two neighbors and belongs to symmetry point group D_{2h}. In the tetragonal and hexagonal layers, each individual molecule having four or six neighbors belongs to the symmetry point group D_{2h} and D_{3d}. Three-dimensional cubic complex can be made attaching C_{60} molecules in a perpendicular direction to the tetragonal layer.

Band structures, stability, and electron density distribution for fullerene orbitals can be examined. Fabricated linear chain, tetragonal and hexagonal two-dimensional complexes, and three-dimensional fullerenes with cubic lattice can be studied using an empirical tight-binding method. The energy gap varies, and the smallest gap is found to be for the

TABLE 8.2
Band Gap, HOMO and LUMO Energy Levels, and Charges for C_{60} and Heterofullerenes

Fullerene	Bandgap	HOMO	LUMO	Atomic Charge
C_{60}	1.6	−11.5	−9.9	0
$C_{59}N$	0.19	−10.1	−9.9	−0.14
$C_{59}B$	0.28	−11.42	−11.14	0.7
$C_{58}N_2$	0.27	−10.18	−9.91	−1.2
$C_{58}B_2$	0.38	−11.4	−11.02	0.6
$C_{58}BN$	1.4	−11.32	−9.92	0.78 (B), −0.13 (N)
$C_{56}(BN)_2$	1.26	−11.26	−10	0.85 (B), −0,7 (N)
$C_{54}(BN)_3$	1.45	−11.37	−9.92	+1 (B), −0.87 (N)

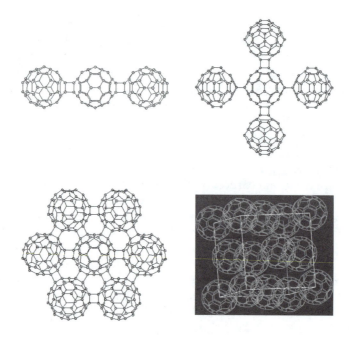

FIGURE 8.18
One-, two-, and three-dimensional tetragonal and hexagonal fullerene C_{60} complexes.

hexagonal structures. Molecules of C_{60} are linked by strong covalent bonds. The crystal orbitals (characterized by the electron density localization on an individual carbon cage) are separated in the electronic structure. The polymerized C_{60} forms are less stable than icosahedral fullerene molecules. At ambient conditions, solid C_{60} is a molecular crystal, in which the molecules are disordered (orientationally) while their centers of mass form a face-centered cubic lattice. At higher temperature and pressure, the fullerite C_{60} is polymerized, forming various crystal structures depending on the synthetic conditions. The treatment of the fullerite for pressures below 8 GPa and temperatures below 750 K results in the formation of linear C_{60} chains along the 110 direction of the original face-centered cubic lattice leading to an orthorhombic phase. Higher temperature results in additional intermolecular bonding in either the 111 or 100 plane and the synthesis of a rhombohedral or tetragonal phase. These phases have a two-dimensional structure with the interlayer distance of 9.8 Å (see Figure 8.18). Above 8 GPa pressure, rigid three-dimensional polymeric fullerene complexes form.

Studies by infrared, nuclear magnetic resonance, and Raman spectroscopy showed the lowering in symmetry of the fullerene molecules and the presence of sp^3 intermolecular bonds in the polymerized C_{60} complexes. Fullerenes are linked with each other by $2+2$ cycloaddition mechanism forming a four-membered ring between the neighbors as documented in Figure 8.18. There are strong covalent bonds between cages. As a result of these bonds, there is a reduction in the number of π-electrons in C_{60}, changing the electronic properties of the multifullerene complexes. Analysis of the electronic structure of fullerenes using quantum mechanics is very complex and important in order to examine phenomena and assess characteristics. From tight-binding calculation, a linear C_{60} chain has been found to exhibit semiconducting properties with 1.15 eV band gap and predominant σ-type intermolecular bonding. Assuming weak van der Waals interlayer interactions in the rhombohedral and tetragonal complexes, a theoretical prediction on stability and conductivity is made carrying out the tight-binding calculation for a single layer. Different numbers of intermolecular covalent bonds in the tetragonal and rhombohedral structures lead to 1.2 and 1 eV band gaps. Making use of the electronic structure for three-dimensional rhombohedral C_{60} complex, the local-density approximation method gives the 0.35 eV semiconductor band gap.

The fullerene complexes considered for nano-ICs are one-dimensional, two-dimensional tetragonal, two-dimensional hexagonal, and three-dimensional cubic. The total energy of the fullerene complex can be expressed as

$$E_{tot} = E_{bs} + E_{rep}$$

where E_{bs} is the sum of electronic eigenvalues for all occupied electronic states, and E_{rep} is the short-range repulsive energy.

The electronic eigenvalues are derived solving the empirical tight-binding Hamiltonian. The hopping potentials $V_{pp\pi}$ and $V_{sp\sigma}$ describe the interaction of the π-electrons as well as s- and p-states of hybridization in fullerenes. Using the theoretical and experimental spectra for C_{60}, the following values are obtained: $E_s = -3.65$ eV, $E_p = 3.65$ eV, $V_{ss\sigma} = -3.63$ eV, $V_{sp\sigma} = 4.50$ eV, $V_{pp\sigma} = 5.38$ eV, and $V_{pp\pi} = -3.04$ eV. The off-diagonal potentials $V_{ss\sigma}$, $V_{sp\sigma}$, $V_{pp\sigma}$, and $V_{pp\pi}$ are derived with interatomic separation r as a function $s(r)$. The repulsive energy is

$$E_{rep} = \sum_i f \left(\sum_j \varphi(r_{ij}) \right)$$

where $\varphi(r_{ij})$ is the potential between atoms i and j, and f is the fourth-order polynomial. The potential $\varphi(r)$ and scaling function $s(r)$ are

$$\phi(r) = \phi_0 \frac{d_0^m}{r^m} e^{m\left(-(r/d_c)^{m_c} + (d_0/d_c)^{m_c}\right)} \quad \text{and} \quad s(r) = \frac{r_0^n}{r^n} e^{n\left(-(r/r_c)^{n_c} + (r_0/r_c)^{n_c}\right)}$$

The change of $V_{sp\sigma}$ and $V_{pp\pi}$ require the reparameterization of these functions. The parameters are found by fitting the dependence of cohesive energy versus nearest-neighbor interatomic separation for graphite and diamond. The coefficients are $n = 2$, $n_c = 6.5$, $r_c = 2.18$ Å, $r_0 = 1.536$ Å, $\varphi_0 = 4.165$ eV, $m = 4.2$, $m_c = 8.185$, $d_c = 2.1$ Å, $d_0 = 1.64$ Å, $c_0 = -2.59$, $c_1 = 0.572$, $c_2 = -1.789 \times 10^{-3}$, $c_3 = 2.353 \times 10^{-5}$, and $c_4 = -1.242 \times 10^{-7}$. For these parameters, the bond lengths in diamond and graphite at equilibrium are 1.542 and 1.446 Å (experiments give 1.546 and 1.42 Å).

Consider the linear C_{60} chain reported in Figure 8.18. Molecular orbitals of the icosahedral C_{60} may be divided into radial ρ-orbitals (directed toward the center of the carbon cage) and tangential τ-orbitals aligned with the local normal of the cage surface. For C_{60}, the HOMO is a fivefold degenerated orbital with h_u symmetry. In contrast, the LUMO is a threefold orbital with t_{1u} symmetry. These molecular orbitals are ρ-orbitals. Distortion of the fullerenes during polymerization changes their electronic structure. A deviation from the icosahedral structure causes the splitting of the degenerated molecular orbitals and, hence, the energetic broadening of ρ- and τ-orbitals. The splitting depends on the molecular distortion. The covalent bonding of the carbon cages in the polymeric chain results in the dispersion of energy levels. Electronic density from the ρ-orbital is mainly localized on 12 hexagons of carbon cage, while on the atoms providing the intermolecular bonding it is practically equal to zero. According to the band structure calculation, the linear C_{60} chain is a semiconductor. A direct band gap is 2.2 eV. The polymer chain formation changes the icosahedral C_{60} HOMO-LUMO gap to 0.92 eV. The highest occupied and lowest unoccupied bands of the C_{60} chain are not degenerate. The 0.13 eV width of the lowest unoccupied band indicates a small value of the transfer integral of ρ-electrons along the polymer chain. As the lowest unoccupied band is separated from higher states by 0.13 eV, the electron doping of the one-dimensional C_{60} polymer will result in a single-band conducting system. The analysis of the electron density distribution predicts the reactivity of C_{60} complexes.

Considering three-dimensional cubic fullerene complexes, the carbon cage is symmetric and has the greatest number of neighbors. The band structure calculation shows that the cubic complex has a direct energy gap of 2.235 eV. The increase of neighboring carbons to six widens the energy gap by 0.05 eV compared with the tetragonal structure. The formation of covalent bonds along the additional direction enhances the overlapping of ρ-electrons, and, thus, their energy broadening. Correspondingly, the energy dispersion of the low unoccupied band increases to 0.956 eV, and the unoccupied states form one continuous conduction band. In two-dimensional hexagonal complexes, the symmetry of C_{60} is strongly distorted around the molecular equator due to the formation of 12 covalent bonds. The complex has an indirect 0.58-eV gap. The low unoccupied and highest occupied bands are twofold-degenerated in the center of the Brillouin zone. The calculated HOMO-LUMO gap for icosahedral C_{60}, linear chain, tetragonal layer, hexagonal layer, and cubic fullerene complexes are 3.1, 2.2, 1.85, 0.58, and 2.235 eV, respectively.

8.4.1.2   *Endohedral Fullerenes*

Different fullerenes have been fabricated including endohedral fullerenes, e.g., $M_x@C_{xx}$. Here, the symbol @ means that x atoms (other than carbon) are encaged (encapsulated) inside the fullerene's carbon cage, which has xx carbon atoms. The diameters of C_{60} and C_{120} are approximately 7.1 and 10 Å (0.71 and 1 nm), measured from the nucleus of one carbon to the nucleus of the carbon on the opposite side. Fullerenes' carbon cages have cavities. Taking into account that the van der Waals diameter of the carbon atoms is 3.4 Å, for C_{60}, we obtain a cavity with a 3.7-Å diameter. The complexes can be formed as packed fullerenes, each of them connected to neighboring fullerenes. For example, C_{60} can be connected to six neighbors by 2+2 cycloaddition of double bonds. An equilibrium inter-molecular distance for C_{60} is 9.2 Å. Rhombohedral C_{60} solid has higher energy than free C_{60} molecules by 2.1 eV/molecule, and the barrier for dissociation to free C_{60} molecules is approximately 1.6 eV/molecule. The lowest energy conformation of this new phase of solid C_{60} is a semiconductor, but defects or engineered changes in the intermolecular bonding pattern lead to semimetal properties.

The endohedral metallofullerenes have been mainly centered on utilizing Ca, Sr, Ba, Sc, Y, and La as M in $M_x@C_{xx}$. In addition, the lanthanide (Ce, Pr, Nd, Gd, Tb, Dy, Ho, Er, Tm, and Lu) metallofullerenes, denoted as $R_x@C_{xx}$ (usually $R_3@C_{80}$ and $R_3@C_{82}$) have been fabricated and characterized. The experimental results show that the charge states for the lanthanum ($_{57}$La $1s^2\,2s^2\,2p^6\,3s^2\,3p^6\,3d^{10}\,4s^2\,4p^6\,4d^{10}\,5s^2\,5p^6\,5p^1$) and yttrium metallofullerenes are $La^{3+}@C_{82}^{3-}$ and $Y^{3+}@C_{82}^{3-}$. Hence, lanthanum is strongly ionized, and three outer electrons from metals are transferred to the fullerene cage. That is, we have a positively charged core metal within a negatively charged carbon cage that can be viewed as an analog to semiconductor spherical heterostructures. It should be emphasized that other endohedral fullerenes are reported, for example, $La@C_{82}$ and $Y@C_{82}$ as shown in Figure 8.19. The C–La distance is 2.16 Å (with 0.035-Å double-bond electron elongation). There exist a great number of mixed dimetallofullerenes ($YLa@C_{80}$) as well as mono-, di-, tri-, and tetrafullerenes. For example, the scandium metallofullerenes are $Sc@C_{82}$, $Sc_2@C_{82}$, $Sc_3@C_{82}$, and $Sc_4@C_{82}$, etc.

Consider $La_2@C_{72}$ that exhibits a dipole moment and has distinct electronic configurations confirmed by the chromatography analysis $La_2^{3+}@C_{72}^{6-}$ and $La_2^{2+}@C_{72}^{4-}$. The endohedral lanthanum metallofullerenes are fabricated by the direct-current discharge (25 V and 100 A and higher). Core rods are packed with a mixture of graphite (graphite rods with a diameter of a couple of millimeters) and either pure La, La_2O_3, lanthanum carbide, or other composite rod under helium (He) flow at high pressure (usually from 25 kPa). The graphite rods can be filled with La_2O_3 with a metal concentration doping of 0.5 to 5%. The resulting fullerene-rich soot deposits on the cooled chamber walls. Mono-, di-, and tri-$La_x@C_{xx}$ can be fabricated. The control is accomplished by the metal concentration (1% for $La@C_{xx}$, greater than 2% for $La_2@C_{xx}$, greater than 4% for $La_3@C_{xx}$), temperature, pressure, etc. In general, one obtains $La_2@C_{xx}$ metallofullerenes, e.g., $La_2@C_{70}$, $La_2@C_{72}$, ..., $La_2@C_{82}$, ..., $La_2@C_{140}$ and higher orders of carbon cages. The high-yield metallofullerenes are extracted from the electric-arc-generated soot by dissolving them in toluene CS_2 or 1,2-dichlorobenzene. Other metallofullerenes that were made are $Ca@C_{60}$ ($Ca^{2+}@C_{60}^{2-}$), $U@C_{60}$, etc.

The electronic configurations of metallofullerenes are distinct. For example, the possible electronic configurations for $Sc_x@C_{xx}$ are $Sc^+@C_{82}^-$, $Sc_2^{2+}@C_{84}^{4-}$, and $Sc_3^{3+}@C_{82}^{3-}$. Let us examine the dimensions of the $Sc_3@C_{82}$ metallofullerene. The Sc^+–Sc^+ distance in the equilateral triangular Sc^{3+} cluster is 2.3 Å (this distance is shorter compared with 3 Å for a neutral Sc_3). The Sc–Sc distance 2.98 Å in $Sc_3^{2+}@C_{82}^{2-}$ is found using the Huckel calculations. The scandium cluster is encapsulated in the C_{82} carbon cage with an average inner fullerene diameter (cavity diameter) of 5.3 Å. The outer $Sc_3@C_{82}$ metallofullerene diameter is 8.28 Å. The distances between the center of the Sc_3 trimer and carbon atoms and the Sc–C distances are 3.4 Å and 2.52 Å. The scandium metallofullerenes are fabricated by the direct-current discharge (500 A and higher) of the Sc/C or other Sc/C-composite rods under He flow at high pressure (usually from 5kPa and higher). The resulting shoot is extracted in carbon disulfide, and the fullerenes are separated utilizing liquid chromatography (the purity reaches 99.99%).

Other functional fullerenes are emerging. For example, $ErSc_2N@C_{80}$, $Er_2ScN@C_{80}$, and $Er_3N@C_{80}$, as well as other lanthanide derivatives such as $R_3N@C_{80}$, $R_{2(1)}R_{1(2)}N@C_{80}$, and $R_{1(2)}R_{2(2)}N@C_{80}$. In particular, $R_3N@C_{80}$ endohedral metallofullerene is reported in Figure 8.19. Fullerenes react with the carbon derivatives resulting in functional fullerenes that can be used as multiterminal electronic nanodevices. For example, $Sc_3N@C_{80}$ reacts with the diene precursor 6,7-dimethoxyisochroman-three-one forming $Sc_3N@C_{80}$– $C_{10}H_{12}O_2$ complex. Figure 8.19 documents a fullerene with the attached benzoyloxybenzoyl (diphenyl) phosphine-chloro-carbonyl-iridium group. In $R_x@C_{xx}$ fullerenes, the bonds have different lengths and fullerenes can be slightly distorted. In the di-scandium fullerene $Sc_2@C_{84}$, the

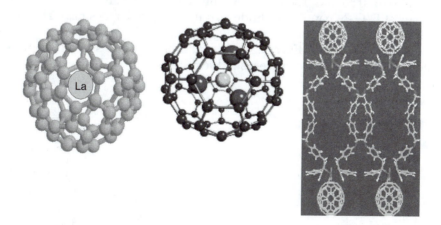

FIGURE 8.19

$La@C_{82}$ fullerene, $R_3N@C_{80}$ endohedral metallofullerene, and endohedral fullerene with benzoyloxybenzoyl (diphenyl) phosphine-chloro-carbonyl-iridium group.

Sc–C distances are 2.4 Å. For $Sc_3N@C_{80}$ metallofullerene, the C–C bond distances are 1.4 Å (6:5 C–C bond distance is 1.45 Å, while the 6:6 C–C bond distance is 1.42 Å), the C–Sc distances are 2.25 Å, and the Sc–N distances are 2 Å (for bulk ScN, the Sc–N distance is 2.25 Å). The energy band gap and the transition barrier for $Sc_3N@C_{80}$ are found to be 2 eV and 0.5 eV, respectively.[16]

We consider $Sc_3N@C_{80}$. For the ground state one has $_6C$ $1s^2$ $2s^2$ $2p^2$, $_7N$ $1s^2$ $2s^2$ $2p^3$, and $_{21}Sc$ $1s^2$ $2s^2$ $2p^6$ $3s^2$ $3p^6$ $3d^1$ $4s^2$. The Sc_3N complex can donate six electrons to an electron acceptor, and the icosahedral symmetric (I_h) C_{80} is an acceptor. The $Sc_3N–C_{80}$ bonds significantly alter the $I_h–C_{80}$ symmetry. The HOMO of the $I_h–C_{80}$ is fourfold degenerated but it is occupied by two electrons. The degenerated LUMO C_{80} can take up to six electrons from the Sc_3N. The charge and electron configuration of the nitrogen in the isolated Sc_3N complex are $q = -0.8$ and $s^{1.7}p^{4.1}$, whereas the charge and electron configuration of the nitrogen in the $Sc_3N@C_{80}$ fullerene are $q = -0.9$ and $s^{1.7}p^{4.2}$. In the isolated Sc_3N complex, taking note of the scandium electron configuration, $s^1p^{0.2}d^{1.5}$, one concludes that 0.3 electrons per scandium atom are transferred to nitrogen. In the endohedral $Sc_3N@C_{80}$ fullerene, the electronic configuration of the nitrogen remains almost the same (from $s^{1.7}p^{4.1}$ to $s^{1.7}p^{4.2}$), but the electronic configuration of the scandium becomes $s^{0.2}p^{0.2}d^{1.4}$ (from $s^1p^{0.2}d^{1.5}$ for nonencapsulated scandium in the isolated Sc_3N complex) with $q = +1.2$. Hence, each scandium atom donates one s electron to the carbon cage. In addition to this ionic interaction (a significant charge transfer from the Sc_3N complex to the C_{80} LUMO with the HOMO-LUMO band gap 2 eV), there are the $p_x–d_x$ exchanges. The endohedral isolated $Sc_3N@C_{80}$ exhibits a remarkable stability.

The electronic properties of endohedral fullerenes have been studied using cyclic voltammetry. The electronic characteristics of endohedral metallofullerenes differ greatly compared with the empty fullerenes. For example, $La@C_{82}$ is a strong electron acceptor and electron donor. In particular, according to [19], at least five electrons can be transferred to the C_{82} cage while maintaining 3+ charge state of the encaged La atom, i.e., $(La^{3+}@C_{82}^{3-})^{5-}$. The schematic energy level diagram for $La@C_{82}$ is reported in Figure 8.20.[19]

Having introduced different metallofullerenes, it is important to emphasize that p- and n-type dopants for crystalline diamond (fabricated through chemical vapor deposition) have been studied. The carbon can be n-doped by nitrogen and p-doped by boron (forming a shallow acceptor with 0.35 eV above the highest valence band). Nitrogen leads to a 1.7 eV gap below the conduction band, while phosphorous and sulfur induce shallow acceptor level with 0.56 and 0.38 eV below the conduction band. As previously emphasized, there exist B–C–N nanotubes. Endohedral $N@C_{60}$ fullerenes have also been made, though relatively low yield has been received to date. The importance of these media (nitrogen, phosphorus, and sulfur) is the fact that they are parts of DNA, proteins, and other carbon-based molecular complexes in biosystems.

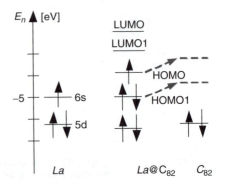

FIGURE 8.20
Energy level diagram for $La@C_{82}$.

Paramagnetic atoms (^{4}N, ^{4}P, ^{3}O, and ^{3}S) can be encapsulated in carbon cages—for example, (N, P, O, or S)@C_{60}. The encapsulated N and P do not donate electrons to the carbon cage, whereas La can donate three unpaired electrons. Assuming the I_h symmetry, the distances between N (P) and C_{60} centroid for N@C_{60} and P@C_{60} fullerenes are 3.56 Å. For the N...C_6H_6 and C_6H_6...N...C_6H_6 benzene complexes, the distance is 3.3 Å, while for the P...C_6H_6 and C_6H_6...P...C_6H_6 complexes, the distance is 3.8 Å. The N@C_{60} and P@C_{60} fullerenes are stable due to high energies: –7.4 and –9.4 kcal/mol, respectively. In contrast, the N...C_6H_6, C_6H_6...N...C_6H_6, P...C_6H_6, and C_6H_6...P...C_6H_6 complexes have energies of –0.73, –1.4, –0.9, and –1.74 kcal/mol. It is also interesting that the spin density in the N@C_{60} fullerene for carbon atoms is –0.00006, and for N, the spin density is 0.004.

The use of the endohedral fullerene topologies requires development of novel fabrication technologies. Conventional CMOS technologies are not suitable to fabricate fullerenes, and typical microelectronics patterning and deposition must be advanced. The integration of the shadow masking with scanning probe methods allows one to deposit structures locally using pin apertures within the proximity of the cantilever tip. Predefined excursions of the sample may lead to the fabrication of moderately complex structures with the nanometer pattern scale. The material composition of the deposited line can be varied allowing for the formation of junctions within a single layer. Two-dimensional (planar) structures can be fabricated utilizing enhanced CMOS technology. Samples can be placed underneath a cantilever by piezoactuators using a preprogrammed sequence of two-dimensional (planar) lateral excursions. This technique is still based on the nanotechnology-enhanced CMOS technology utilizing nanolithography with the capability to produce 20-nm-wide lines with 20-nm spacing. This technique can be used to develop nanotechnology-enhanced microelectronics approaching 10-nm patterning and using fullerenes as nanodevices. However, novel technologies must be developed to ensure three-dimensional capabilities thereby fully utilizing the opportunities offered by multi-terminal three-dimensional electronic nanodevices. In general, interconnect technologies can be considered to be one of the primary obstacles to standard sub-micron processes due to reliability problems, performance degradation, high cost, etc. However, nanotechnology, novel topologies, and new nano-IC architectures will overcome these limitations. Novel processes have emerged to synthesize complex and functional fullerene topologies. It is shown that fabrication and technological issues are promised to be solved by encapsulating media atoms in fullerenes (or other carbon cages), polymerizing fullerenes in rigid structures, interconnecting fullerenes attaining functionality, etc. Thus, the fullerenes paradigm provides a viable pathway for nanoelectronics. Typical subtractive processing result in dangling bonds that can be passivated by various impurities. The fullerene processing and novel additive nanofabrication techniques allow one to produce interconnects, junctions, and spacings with no contamination between functional multiterminal fullerene complexes. Studies for functional fullerenes provide a far-reaching implication for nanoelectronics and nano-ICs. These devices have not been coherently examined and characterized. From the fundamental viewpoint, the reported carbon-based molecular electronics is very complex, and even computational methods are not developed. The proposed concept, which is based on endohedral and doped fullerenes, promises to design complex three-dimensional nano-ICs with superior computing, logic, signal processing, and memory features. Utilizing 1-nm fullerenes as multiterminal nanodevices, there are 1×10^{18} interconnected functional fullerenes packed into a 1-mm^3 volume. Hence, 1,000,000,000,000,000,000 novel enhanced-functionality nanodevices are packed into 1 mm^3, guaranteeing tremendous increase of device density in future three-dimensional nano-IC architectures. A wide spectrum of very important issues that have not been addressed in this book arises, e.g., nanoarchitectronics, design, aggregation, optimization, analysis, etc. It is virtually impossible to even address these problems in one book. Most important, this book is not aimed to solve far-reaching problems in nanoarchitectronics that emerge as a vital nanoelectronics paradigm.

Nano- and microsystem performance directly depends on (1) fundamental principles and basic physics (e.g., synthesis and discovery with the corresponding modeling, analysis, simulation, design, optimization, architectures, logic/topological design, CAD, and other tasks); (2) fabrication technologies (e.g., processes, techniques, and materials); and (3) fabrication infrastructure. Synthesis, structural/dynamic optimization, analysis, design, classification, and other tasks will allow the designer to devise new operational principles (that uniquely utilize novel phenomena) guaranteeing synthesis of nano- and microsystems with superior performance and enhanced functionality. To design these nano- and microsystems, fundamental, applied, and experimental research must be performed. Fundamental laws (quantum mechanics, Maxwell's equations, electromechanics, etc.) will be likely complemented by new discoveries that will ensure further progress to the unknown nanoscale world and beyond.

8.4.2   Biocentered Nanoelectronics

Having covered fullerene-centered nanoelectronics, let us turn our attention to biocentered nanoelectronics. DNA, amino acids, proteins, carbohydrates, and other molecules are composed of carbon atoms bonded to one another as well as to hydrogen, oxygen, nitrogen, sulfur, and phosphorus. These H, O, N, S, and P atoms are common composites of the carbon-based biomolecules and complexes. DNA molecules consist of two polynucleotide chains (strands) that spiral around, forming a double helix. These polynucleotide chains are held together by hydrogen bonds between the paired bases. DNA is a linear double-stranded polymer of four nitrogeneous bases: adenine (A), thymine (T), guanine (G), and cytosine (C). The base-pairing rule (A always pairs with T as A=T, while G always pairs with C as G≡C) specifies that two strands of the double helix are complementary (see Figure 8.21).

Proteins are large biomolecules that are chemically similar because they are made of amino acids linked together in long chains. Amino acids are organic molecules, and cells build their proteins from 20 amino acids. These 20 amino acids are specified by codon sequences. For amino acids, their molecular formulas are as follows: Ala (alanine) $C_3H_7NO_2$, Arg (arginine) $C_6H_{14}N_4O_2$, Asn (asparagine) $C_4H_8N_2O_3$, Asp (aspartic) $C_4H_7NO_4$, Cys (cysteine) $C_3H_7NO_2S$, Gln (glutamine) $C_5H_9NO_4$, Glu (glutamic) $C_5H_{10}N_2O_3$, Gly (glycine) $C_2H_5NO_2$, His (histidine) $C_6H_9N_3O_2$, Ile (isoleucine) $C_6H_{13}NO_2$, Lys (lysine) $C_6H_{13}NO_2$, Leu (leucine) $C_6H_{14}N_2O_2$, Met (methionine) $C_5H_{11}NO_2S$, Phe (phenylalanine) $C_9H_{11}NO_2$,

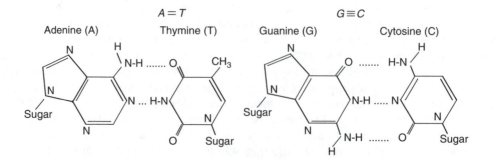

FIGURE 8.21
Paired nitrogeneous bases and their structures.

Pro (praline) $C_5H_9NO_2$, Ser (serine) $C_3H_7NO_3$, Thr (threonine) $C_4H_9NO_3$, Trp (tryptophane) $C_{11}H_{12}N_2O_2$, Tyr (tyrosine) $C_9H_{11}NO_3$, and Val (valine) $C_5H_{11}NO_2$.

It is evident that each amino acid has a hydrogen atom, a carboxyl group, and an amino acid group bonded to the alpha (α) carbon. It is important to note that all 20 amino acids that make up proteins differ only in what they are attached to by the fourth bond to the α carbon. Physical and chemical properties of the side chain define the unique characteristics of amino acids.

Let us discuss the role of photons and phonons in the electron excitations from the valence to the conduction band. Phonons are quasiparticles associated with the thermal vibrations of a crystal, while photons are associated with the vibrations of the electromagnetic field. Phonons are quanta of the thermal energy of the crystal and are represented by the angular frequency ω and wave vector $\mathbf{q}$. A phonon has energy $\hbar\omega$ and momentum $\hbar\mathbf{q}$. To excite an electron from the valence to conduction band, one has to supply it with energy $E \geq E_{gap}$. This E is the energy $E = \hbar\omega$ of a photon (and phonon) having the frequency ω. During the energy transfer, the annihilation of the exciting particle and momentum transfer (conservation) are studied. In the case of a photon with the wave vector $\mathbf{q}$, the magnitude of the momentum is $p = \hbar q = \hbar\omega/c = E/c$. For a phonon, one has $p = E/u_s$, where u_s is the sound velocity (from 1 to 10 km/sec for different media). Taking into account that u_s is much less than c for the same amount of transferred energy, the transfer of the momentum by the photon is small compared with the phonon. Hence, for Si and Ge, the electrons can be transferred from the valence to conduction band by the thermal excitations only, and not by the optical ones. These indirect gaps can be called the *thermal gaps*. As the optical excitations of electrons in semiconductors having the indirect gaps are forbidden, these semiconductors (in the bulk crystal form) cannot be used for the optoelectronic applications. In contrast, GaAs has a direct gap, and GaAs is used for the optoelectronic applications. The average energy transferred by a phonon is on the order of kT. The energy gap of Si is considerably larger than in Ge, and, thus, fewer electrons will be thermally transferred by the energy gap of Si. Therefore, the electrical properties of silicon are less dependent on temperature. This means that the electronic devices made from silicon will be thermally robust compared with Ge. With the increase in the energy bandwidth, the forbidden gap between two neighboring bands decreases. Therefore, the effective mass in the conduction band decreases with the gap between the valence and the conduction band. The smallest electron effective mass is observed in the narrow-gap semiconductors (the electron effective mass can be less than 1% of the free electron mass).

Charge transport is observed in DNA molecules. It has been shown that the overlap of the π-orbitals along the stacked base pairs leads to the charge transfer over a distance of tens of nanometers. A hopping mechanism was developed to explain the charge transfer in DNA, and the G≡C pair is considered a hole donor due to its low ionization potential compared with the A=T base pair. Electron transport depends largely on the distance between the base pair and their sequence. It also has been suggested that a hole can be created in the A=T pair by thermal activation from a G≡C base pair when the charge tunneling is unlikely due to the long distance. Therefore, long-range charge hopping can be supported. In addition to hopping, polaron motion has also been examined as a possible mechanism to explain the charge transfer. In fact, the charge coupling in DNA can lead to a polaron that can be considered a Brownian particle that collides with the low-energy excitations in the thermal environment. The dynamic properties of a polaron, induced by the carrier interaction with acoustic and optical photons, are very complex. The electron and hole transfer in DNA decreases exponentially as $e^{-\beta x}$, where β is a constant that varies from 0.2 to 1.6 $Å^{-1}$ (for proteins, β varies from 1.4 to 1.6 $Å^{-1}$). The variations of β are due to the hopping transport (series of nearest-neighbor charge transfer steps with photon assistance between isoenergetic bases). Another mechanism that has been studied for many years is the proton tunneling in the isolated base pairs. It is evident that different theories exist that are supported by the experimental results. Different *I-V* characteristics for distinct DNA that consist of different base sequences and DNA sticky end–electrode contacts are shown in Figure 8.22. These results document semiconducting and semimetallic behaviors.[22] The conductivity of DNA for the phonon-assisted charge transfer can be estimated as

$$\sigma = \frac{Ne^2 x_{b-b}^2}{T\tau_e}$$

where N is the effective concentration of electrons, x_{b-b} is the hopping length (base-to-base distance), T is the thermodynamic temperature, and τ_e is the electronic time (from $\tau_e = m_{ef} x_{b-b}^2$ one concludes that τ_e is within the femtosecond range). For semiconduction and metallic DNA, the conductivities were found to be in the range from 0.1 to $1 \times 10^4 \, \Omega^{-1} \text{cm}^{-1}$. Consider DNA nucleotides and base pairs. In guanine, one of its electrons is more weakly bound than any other electrons in other DNA nucleotides. The hole migration over 60 base pairs in purified DNA has been experimentally confirmed, and theory suggests it could travel through many base pairs. Figure 8.22 documents how the "hole" migrates along the DNA molecule, probably by hopping, from guanine to guanine.

DNA can be examined as a one-dimensional linear chain of base pairs. In the ground state, every base pair contains bound electrons only. Then, the charge is carried through DNA by excited electrons (or holes), which can jump between the base pairs. Assume that (1) the excited electron states at the base pairs are separated by an energy spacing larger

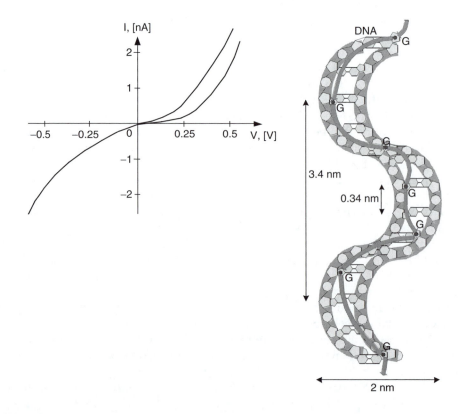

FIGURE 8.22
I-V characteristics for the DNA strands and hole propagation in DNA.

than the temperature (not thermally excited); (2) thermal motions of the DNA base pairs are elastic vibrations with a characteristic frequency ω_b; (3) efficient charge transfer between neighboring base pairs takes place only as "controlled" event; and (4) the Coulomb interactions between the electrons and holes can be neglected for describing the hopping transport. The characteristic frequency ω_b is estimated to be from 10^{11} to 10^{12} 1/sec. The characteristic electronic hopping time τ is used, and the experiments show that $\omega_b\tau$ is from 10^2 to 10^3. The above reasoning leads to a one-dimensional hopping Hamiltonian for the electrons in a noninteracting electron system

$$H = \hbar \sum_i \left(\varsigma_i a_i^+ a_{i+1} + \varsigma_i^* a_{i+1}^+ a_i \right)$$

where a_i and a_i^+ are electronic annihilation and creation operators at the ith base pair, and ς_i are the hopping amplitudes (time-dependent quantities).

The equations for the Heisenberg operators a_i are $\partial a_i/\partial t = -i(\varsigma_i a_{i+1} + \varsigma_{i-1}^* a_{i-1})$.

Different ς_i possess independent statistics because ς_i are related to independent thermal base-pair fluctuations. The hopping matrix element ς_i can be examined as $\langle \varsigma_i \rangle$ which describes the coherent charge carrier motion in a completely rigid lattice and a fluctuating part. This one-dimensional theoretical model is based on hopping Hamiltonians and can be used to examine the charge transport in DNA and other complexes including proteins, RNA, etc.

Consider the tunneling in the G≡C base pairs. Using the Pauli matrices σ^E and σ^T, for the hydrogen bonds we have $H_H = \frac{1}{2}(T\sigma^T - E\sigma^E)$, where T is the constant energy element of the tunneling matrix, and E is the energy bias between two localized protons.

When energy ratio T/E is small, the proton is localized on one side of the hydrogen bond. If the ratio T/E is large, the proton is delocalized. The typical G≡C base pair is in the low energy state, while its tautometric form G^t≡C^t is in the excited state. When $T \ll E$, the probability of having the G^t≡C^t form is small. However, the radical cation in G$^+$≡C results in the energy comparable to the energy level of G^t≡C^t. The proton state becomes delocalized in the hydrogen bonds. Figure 8.23 provides the interpretation of the proton transfer along ≡C base pairs.

The equation for polarons can be derived using quantum mechanics, and the Hamiltonian is

$$H = \sum_{i=1}^{M} \left[E_i a_i^+ a_i - \left(t_0 - c(x_{i+1} - x_i) \right) \left(a_i^+ a_{i+1} + a_{i+1}^+ a_i \right) + \tfrac{1}{2} K(x_{i+1} - x_i)^2 + \tfrac{1}{2} m_s \dot{x}_i^2 \right]$$

where E_i is the on-site energy, t_0 is the overlap integral of π orbitals between neighboring sites, c is the electron-photon coupling map, K is the elastic constant, and m_s is the site mass.

Having emphasized the one-dimensional model, our goal is to cover high-fidelity mathematical models that will allow one to examine complex DNA. Figure 8.24 documents the double-stranded DNA configuration. In this representation, we simplify extremely complex DNA topological bonds and consider the π–electrons with the highest occupied molecular orbital states. One may consider complex π–π interaction between the nitrogeneous base pairs as documented in Figure 8.24.

The Hamiltonian for the proposed DNA schematics is given as

$$H = H_A + H_B + H_C + H_{AB} + H_{BC}$$

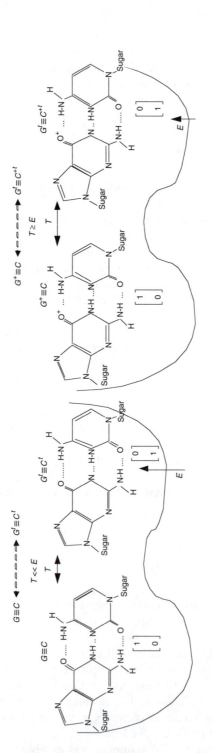

FIGURE 8.23
Proton transfer in DNA with G≡C base pairs with distinct energies.

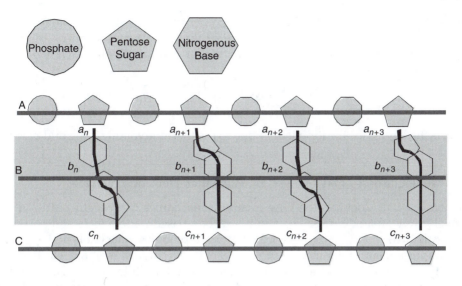

FIGURE 8.24
DNA Nucleotides (monomers of nucleic acids composed of three smaller molecular building blocks: a nitrogen base purine or pyrimidine, pentose sugar, and phosphate group).

where

$$H_A = \sum_{n=1}^{N} (A_{n,n}|A{:}n\rangle\langle A{:}n| - A_{n+1,n}|A{:}n+1\rangle\langle A{:}n| + \delta)$$

$$H_B = \sum_{n=1}^{N} (B_{n,n}|B{:}n\rangle\langle B{:}n| - B_{n+1,n}|B{:}n+1\rangle\langle B{:}n| + \delta)$$

$$H_C = \sum_{n=1}^{N} (C_{n,n}|C{:}n\rangle\langle C{:}n| - C_{n+1,n}|C{:}n+1\rangle\langle C{:}n| + \delta)$$

$$H_{AB} + H_{BC} = -\sum_{n=1}^{N} (a_n|A{:}n\rangle\langle B{:}n| + c_n|C{:}n\rangle\langle B{:}n| + \delta)$$

Here, $A_{n,n}$, $B_{n,n}$, and $C_{n,n}$ are the on-site energies at site n; $|A{:}n\rangle$, $|B{:}n\rangle$, and $|C{:}n\rangle$ are the orthonormalized sets; $|A{:}n+1\rangle$, $|B{:}n+1\rangle$, and $|C{:}n+1\rangle$ are the hopping integrals between n and $(n+1)$ sites; and a_n, b_n, and c_n are the hopping integrals from the corresponding chains (for example, a_n from A to chain C) at site n.

The Schrödinger equation $H|\Psi\rangle = E|\Psi\rangle$ for the examined DNA topology is given as

$$A_{n,n-1}\psi_{n-1}^A + A_{n,n}\psi_n^A + A_{n+1,n}\psi_{n+1}^A + a_n\psi_n^B = E\psi_n^A$$

$$B_{n,n-1}\psi_{n-1}^B + B_{n,n}\psi_n^B + B_{n+1,n}\psi_{n+1}^B + a_n\psi_n^A + c_n\psi_n^C = E\psi_n^B$$

$$C_{n,n-1}\psi_{n-1}^C + C_{n,n}\psi_n^C + C_{n+1,n}\psi_{n+1}^C + c_n\psi_n^B = E\psi_n^C$$

where $\psi_n^B = \langle B{:}n|\psi\rangle$

The energy dependence of the Lyapunov exponent should be found. In particular, the Lyapunov exponent is related to the density of state $\rho(E)$ as $\sum_{i=1}^{3} v_i(E) \approx \int \ln|E - E'|\rho(E')dE'$.

The possible singularity of the Lyapunov expo-nents is related to the singularity of the density of states. The electronic conductance of the DNA strands can be related to the Lyapunov conductance (equivalent to the Landauer conductance), which is defined as

$$v_c = \sum_{i=1}^{3} \left(\cosh\left(\frac{2m}{\varsigma_i}\right) - 1 \right)^{-1} \approx e^{-\frac{2m}{\varsigma_3}}$$

where ζ_3 is the largest localization length due to the thermodynamic limit, and $\varsigma_3 = 1/v_3(E)$.

8.4.3   Nanoelectronics and Analysis of Molecular Electronic Devices

In Chapter 7 it was documented how quantum mechanics can be utilized to model and analyze electronic nanodevices. The studied carbon-based nanodevices can be aggregated in large-scale two- and three-dimensional complex topologies that enable high-performance robust computing, fault tolerance, highest level of hierarchy, fast memory, etc. For example, the density of nano-ICs designed and manufactured using fullerenes or DNA will exceed by thousands of times the density of ICs developed using most advanced CMOS-based technologies even if nanolithography is applied. In addition to the device density, other characteristics can be improved. The electronic properties of the carbon-based nanodevices depend on many factors. These carbon-based electronic nanodevices exhibit metallic (conductor), semimetallic, and semiconducting properties. Carbon-based nanoelectronics can be uniquely utilized in super-high-density, high-speed, robust, immune, low-power and compatible logic gates, switches, diodes, latches, converters and other devices. These logic gates, switches, and converters have a direct application to all nanocomputer units, wireless communication, and networks. Common qualitative symmetric and asymmetric current–voltage (*I-V*) and conductance–voltage (*G-V*) characteristics for carbon-based nanodevices are reported in Figure 8.25.

We covered some major chemical and physical properties of molecular assemblies and devices essential from electronic standpoints in order to coherently examine them. Efforts to fabricate these carbon-based molecular devices, which are based on molecular components, have been driven by the fundamental interest to coherently utilize unique physics at the atomic and molecular level focusing on the foreseen technological expectations and short- and medium-term needs. The question that may be asked is, why molecules and molecular devices? There are many reasons, and, keeping in mind previously discussed advantages, some other important benefits are mentioned in the following paragraphs.

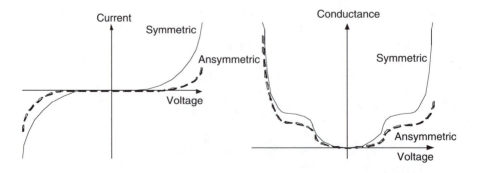

FIGURE 8.25
Symmetric and asymmetric current–voltage and conductance–voltage (d*I*/d*V*) characteristics.

Molecules and functional molecular complexes can be chemically synthesized with a wide range of desired physical properties that are observed at the nanoscale.

Molecules have a tendency to self-assemble and interface (bond), resulting in complex nanoscale topologies and architectures that have unique controlled properties. This self-assembly is due to molecular (natural) forces, and molecular structures can be synthesized (interfaced) on the desired substrate surfaces.

Synthetic chemistry allows one to build molecular materials from the bottom up (from its atomic constituents) and to prepare many copies of the same molecule and complexes in parallel. The flexibility of molecular chemistry may be used to prepare molecules with a wide range of required electronic, optical, electromagnetic, and thermomechanical properties. The molecules may be functionalized to direct their assembly on the metallic, semimetallic, semiconducting, and insulating surfaces. The electronic coupling between molecules and metal electrodes (interconnect) can be controlled.

A wide range of fundamental methods, applied research, and experiments have been carried out to understand (and utilize) unique electronic, optical, and chemoelectromechanical properties of individual molecules, molecular assemblies, and clusters. These methods and techniques range from basics (quantum mechanics) to instrumentation (scanning tunneling microscopy and electronic measurements), from computing to nanotechnology processes and materials, etc. The flexibility in the synthesis and assembly of molecules, functionality and physical properties that can be designed and achieved, and nanodevices make nanoelectronics an exciting prospect.

Molecular nanodevices can be analyzed and examined using quantum mechanics (covered in Chapter 7) as well as other methods that also utilize quantum mechanics—for example, the energy concepts that are based on examining Fermi energy (E_F), energy level broadening (E_B), and charge density.[12–14] The difference between the Fermi energy and the highest/lowest occupied molecular orbitals is usually the most important one to examine the *I-V* characteristics.

The application of functional carbon-based nanodevices (as any other atoms, molecules, and clusters made from any media) requires interfacing and networking. A lot of issues are associated with logic design, topology aggregation, optimization, and other problems in two- and three-dimensional nano-ICs. These issues are not covered in this book because the major emphasis is on device-level nanoelectronics. Figure 8.26 illustrates two- and multiterminal electronic nanodevices that comprise of the carbon molecule(s) with atomic bonding interconnects to sources and drains. The interatomic and molecular junctions are significantly different compared with bulk junctions that form contacts to source and drain.

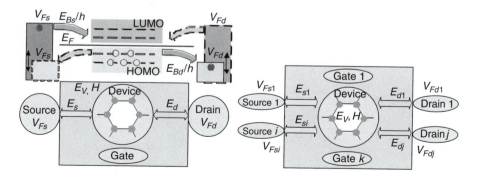

FIGURE 8.26 (See color insert.)
One-dimensional schematic representation of two- and multichannel electronic nanodevices with Hamiltonian H, $H = H_0 + V_{SC} = H_0 + f_\rho(\rho)$.

The atom-to-atom interconnects and gate of electronic nanodevices are illustrated in Figure 8.26. The concept that will be reported may not directly fit into dynamic analysis of super-large-scale nano-IC topologies due to different atomic interconnects, very complex electron transport, lack of accurate information, etc. Even the Fermi energy calculations using distinct methods lead to different values significantly influencing the results and assessments. Novel methods are needed to accomplish high-fidelity modeling tasks with data-intensive analysis features as a coproduct of heterogeneous modeling.

Consider the application of the Green function paradigm, or the so-called nonequilibrium Green function formalism,[12-14] for the electronic nanodevice as reported in Figure 8.26. The source and drain potentials (electrochemical potentials) are denoted as V_{Fs} and V_{Fd}, while E_s and E_d are the self-energy functions of the source and drain. Potential V_{Fs} and V_{Fd} vary, and there is no electron transport if $V_{Fs} = V_{Fd}$. The HOMO and LUMO orbitals and the Fermi level are documented. Depending on the HOMO and LUMO levels, as well as E_F, the electron transport takes place through the particular orbitals, and the electron transfer rates are E_{Bs}/h and E_{Bd}/h (see Figure 8.26).

The Green function is a function of energies, and, in general,

$$G(E) = (E - E_V + iE_B)^{-1}$$

where E_V is the single-energy potential that depends on the charge density $\rho(E)$ or number of electrons N, $E_V = E_{V0} + V_{SC}$; and V_{SC} is the self-consistent potential to be determined by solving the Poisson equation using the charge density, $V_{SC} = f_\rho(\rho)$ or $V_{SC} = V(N - N_0)$.

For two-terminal nanodevices with the source and drain broadening energies E_{Bs} and E_{Bd}, we have

$$G(E) = \left[E - E_V + \tfrac{1}{2} i (E_{Bs} + E_{Bd}) \right]^{-1}$$

The broadening can be estimated using the time τ_i required for an electron that occupies the specific orbital on the source or drain to escape into the electron transfer path (connect). In particular, $E_{Bi} = \hbar/\tau_i$. Thus, the electron transfer rate is E_{Bi}/h. Taking into account the spin (making use of the coefficient 2), the number of electrons and current are estimated as functions of the broadening energies as

$$N = 2\frac{E_{Bs}}{E_{Bs} + E_{Bd}} \quad \text{and} \quad I = \frac{eNE_{Bd}}{h} = \frac{2eE_{Bs}E_{Bd}}{h(E_{Bs} + E_{Bd})}$$

However, more general results are obtained using the electrochemical potentials V_{Fs} and V_{Fd} as well as E_F.[12-14] In particular,

$$N = 2\frac{E_{Bs} f(E_V, V_{Fs}) + E_{Bd} f(E_V, V_{Fd})}{E_{Bs} + E_{Bd}}$$

and

$$I = \frac{eNE_{Bd}}{h} = \frac{2eE_{Bs}E_{Bd}[f(E_V, V_{Fs}) - f(E_V, V_{Fd})]}{h(E_{Bs} + E_{Bd})}$$

where $f(E_V, V_{Fs})$ and $f(E_V, V_{Fd})$ are the Fermi-Dirac functions,

$$f(E_V, V_{Fs}) = \left(1 + e^{\frac{E_V - V_{Fs}}{kT}} \right)^{-1} \quad \text{and} \quad f(E_V, V_{Fd}) = \left(1 + e^{\frac{E_V - V_{Fd}}{kT}} \right)^{-1}$$

In general, the density matrix will be used. For the simplified consideration, using the number of electrons at the equilibrium N_0

$$N_0 = 2f(E_{V0}, E_F) = 2\left(1 + e^{\frac{E_{V0} - E_F}{kT}}\right)^{-1},$$

one has the following relationships for the self-consistent potential, number of electrons, and density of states:

$$V_{SC} = V(N - N_0)$$

$$N = 2\int_{-\infty}^{\infty} D(E)\frac{E_{Bs}f(E_V, V_{Fs}) + E_{Bd}f(E_V, V_{Fd})}{E_{Bs} + E_{Bd}} dE$$

$$D(E) = \frac{E_{Bs} + E_{Bd}}{2\pi\left((E - E_V)^2 + \frac{1}{4}(E_{Bs} + E_{Bd})^2\right)}$$

The current in a molecular nanodevice with broadening can be found as

$$I = \frac{2e}{h}\int_{-\infty}^{\infty} D(E)\frac{E_{Bs}E_{Bd}[f(E_V, V_{Fs}) - f(E_V, V_{Fd})]}{E_{Bs} + E_{Bd}} dE$$

In general, the total number of electrons N and N_0 are found using the trace operator as

$$N = \text{tr}(\rho S) \quad \text{and} \quad N_0 = \text{tr}(\rho_{\text{equilibrium}} S)$$

where S is the identity matrix for orthogonal basis functions.

To perform qualitative and quantitative analysis, the Green function is utilized. The Green function can be examined as the wave function at $\mathbf{r}$ (position in the multidimensional domain) resulting from an excitation applied at $\mathbf{r}_E$. Due to the fact that we examine the retarded Green function that represents the response of electronic nanodevices at the impulse excitation, we have

$$(E - H)G(\mathbf{r}, \mathbf{r}_E) = \delta(\mathbf{r} - \mathbf{r}_E)$$

where H is the Hamiltonian of the infinite system.

It is evident that the reported concept uses the boundary conditions on the interface. In general, the boundary conditions must be satisfied for both transport equation as well as for Poisson's equation (equations for the gate and interfaces). To examine the finite system (study primarily the electron transport problem), the molecular Hamiltonian of the isolated systems H_0 (isolated system energy) and the complex self-energy functions E_i are used instead of E_V and broadening energies. In the matrix notations, using the overlap identity matrix S, the Green function is given as

$$[G(E)] = (E[S] - [H] - [V_{SC}] - \sum_i [E_i])^{-1}$$

where S is the identity overlap matrix for orthogonal basis functions.

This generic concept is applied to the studied multiterminal electronic nanodevices. We have

$$[E_i] = [S_i][G_i][S_i^*]$$

where S_i is the geometry-dependent coupling (contact) matrix between the surface of the source/drain and the device. Therefore, the imaginary non-Hermitian (unlike Hamiltonian) self-energy functions of the source and drain (first and second electron reservoirs) E_s and E_d are given as

$$[E_s] = [S_s][G_s][S_s^*] \quad \text{and} \quad [E_d] = [S_d][G_d][S_d^*]$$

where G_s and G_d are the surface Green functions for the source and drain that are found by applying the recursive methods, and S_s and S_d are the geometry-dependent coupling (contact) matrices between the surface of the source and drain, and the device.

Taking note of the Green function, the density of state $D(E)$ is given as

$$D(E) = -\frac{1}{\pi} \text{Im}\{G(E)\}$$

It should be emphasized that the spectral function $A(E)$ is the anti-Hermitian term of the Green function, i.e.,

$$A(E) = i[G(E) - G^*(E)] = -2\,\text{Im}[G(E)], \quad \text{and} \quad D(E) = (1/2\pi)\text{tr}[A(E)S].$$

Using the broadening energy functions E_{Bs} and E_{Bd}, we obtain the source and drain spectral functions as

$$[A_s(E)] = [G(E)][E_{Bs}(E)][G^*(E)] \quad \text{and} \quad [A_d(E)] = [G(E)][E_{Bd}(E)][G^*(E)].$$

The nonequilibrium density matrix for multiterminal electronic nanodevices is

$$[\rho(E)] = \frac{1}{2\pi} \int_{-\infty}^{\infty} \sum_{i,j=s,d} f(E_V, V_{Fi.j})[A_{i,j}(E)]dE$$

$$= \frac{1}{2\pi} \int_{-\infty}^{\infty} \sum_{i,j=s,d} f(E_V, V_{Fi.j})[G(E)][E_{Bi,j}(E)][G^*(E)]dE$$

where V_{Fs} and V_{Fd} are the source and drain potential related to the Fermi levels.

For the two-terminal nanodevice given in Figure 8.26, we have

$$\rho(E) = \frac{1}{2\pi} \int_{-\infty}^{\infty} [f(E_V, V_{Fs})G(E)E_{Bs}G^*(E) + f(E_V, V_{Fd})G(E)E_{Bd}G^*(E)]dE$$

Utilizing the transmission matrix $T(E) = \text{tr}[E_{Bs}G(E)E_{Bd}G^*(E)]$ and taking note of the Green function and broadening, the current density between two junctions (contacts) is found to be

$$I = \frac{2e}{h} \int_{-\infty}^{+\infty} \text{tr}[E_{Bs}G(E)E_{Bd}G^*(E)][f(E_V, V_{Fs}) - f(E_V, V_{Fd})]dE$$

As a molecule of the electronic nanodevice is coupled to the contact, there is the energy transfer between the molecule and contact. The contact–molecule– device–molecule–contact system attains equilibrium with the Fermi level (depending on materials, the Fermi level can reach –10 eV, and for gold it is –5.1 eV).

The current–voltage characteristics can be obtained by self-consistently solving the coupled transport–capacitive Poisson's equation.[12–14] As was emphasized, the reported concept may have a limited application to the specialized tasks such as logic or topological designs of

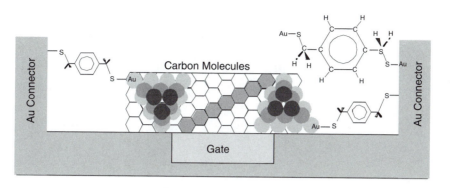

FIGURE 8.27
Nanoswitch with carbon nanotube and molecular wire (1, 4-dithiol benzene) for interconnect.

super-large-scale nano-ICs where broadening is not a major issue. For these tasks, mathematical models can be simplified to describe the dominant nonlinear input-output effect. Our main efforts have been concentrated on functionality of multiterminal electronic nanodevices utilizing unique phenomena such as electron transport, electron switching, charge transport, etc. Despite some limitations, the Green function paradigm can be used to examine electronic nanodevices— for example, carbon molecules as well as carbon nanotube (that can be filled with media) complexes with interconnect that can be used as nanoswitches (see Figure 8.27).

For the nanodevice documented in Figure 8.27, the current I is a function of the applied voltage V, and Landauer's formula is

$$I = \frac{2e}{h} \int_{-\infty}^{\infty} T(E, V) \left(\frac{1}{1 + e^{\frac{E - V_{p1}}{kT}}} - \frac{1}{1 + e^{\frac{E - V_{p2}}{kT}}} \right) dE$$

where V_{p1} and V_{p2} are the electrochemical potentials that can be expressed as $V_{p1} = E_F + \frac{1}{2}eV$ and $V_{p2} = E_F - \frac{1}{2}eV$; E_F is the equilibrium Fermi energy of the source; and $T(E,V)$ is the transmission function obtained using the molecular energy levels and coupling.

Using previously reported results,[12–14] we have

$$I = \frac{2e}{h} \int_{V_{p1}}^{V_{p2}} T(E, V) \frac{1}{4kT} \operatorname{sech}^2\left(\frac{E}{2kT} \right) dE$$

where $kT = 26$ meV.

The molecular wire conductance is found to be

$$G = \frac{\partial I}{\partial V} \approx \frac{e^2}{h} [T(V_{p1}) + T(V_{p2})]$$

Example 8.3

Consider the endohedral $N@C_{60}$ fullerene as an electronic nanodevice (see Figure 8.28). The interatomic C—C and C—N bond lengths are 1.43 and 1.4 Å. The energy band gap is 0.3 eV. The neighboring carbon atoms are interconnected to other fullerenes to form the source and drain with the varying potentials V_{Fs} and V_{Fd} that can vary from 5 to –5 V. Analytically or numerically derive the wave functions. Calculate the *I-V* and *G-V*

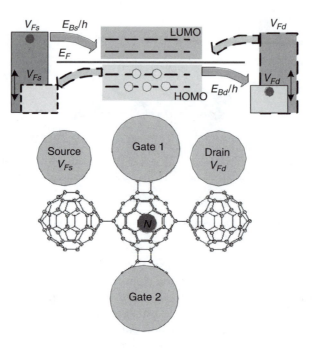

FIGURE 8.28
Endohedral $N@C_{60}$ fullerene as an electronic nanodevice.

characteristics. Draw conclusions for the possible application and control of the studied fullerene as an electronic nanodevice. Quantum mechanics should be used.

SOLUTION

We first examine the wave function representing the $N@C_{60}$ as a one-dimensional chain of three atoms, C–N–C. Considering an electron traveling through this chain, the potential energy is assumed to be a square barrier located at the N atom with a width 140 pm equal to the diameter of the N atom. Its maximum energy is equal to the band gap of the molecule, i.e., 0.3 eV. There is a length equal to the diameter of the C atom of zero potential both before and after the potential energy barrier. The separation of the nuclei of C and N is 140 pm. The unbound valence shell radii of C and N are both approximately 75 pm. These radii are reduced to 70 pm to coincide with the dimensions of the $N@C_{60}$ molecule. The linear axis begins with the first C atom and continues to the last C atom—a total of 420 pm.

The wave function is determined by numerically solving the Schrödinger equation using the MATLAB differential equation solver ode45. The 1-D potential energy described above is used in conjunction with the electron's incident (kinetic) energy E. The determined wave function's magnitude is normalized such that the integral of its magnitude over all time is equal to 1. The probability of the incident electron traversing the potential barrier is calculated as the integral of the wave function's magnitude in the region of the C atom on the opposite side of the potential barrier. Given the normalized wave function $\Psi(x, E)$ determined for an incident energy E, the probability of transmission is

$$T(E) = \int\limits_{280 \text{ pm}}^{420 \text{ pm}} |\Psi(x, E)|^2 dE$$

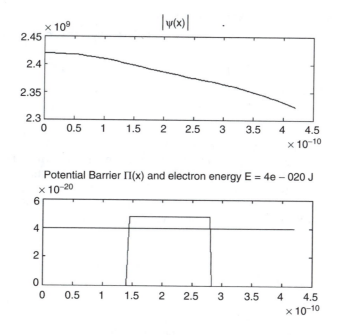

FIGURE 8.29
Plots of wave function and potential barrier.

Plots for the wave function magnitude $|\Psi|^2$ and potential barrier are given in Figure 8.29. The results demonstrate that the wave function has a very low frequency and that the probability of transmission is significant. The MATLAB file used and the results displayed in the Command Window are reported below.

```
function tp = calc_trans_prob(E)
m = 9.1e - 31; % mass of electron [kg]
meff = 0.06*m; % effective mass [kg]
h_bar = 1.0546e - 34; % J s
kappa_sqrd = 2*meff/(h_bar*h_bar);
[x, psi] = ode45(@schrod, [0, 420e - 12], [1 0], [ ], kappa_sqrd, E, @pot_energy);
psi = psi(:,1);
ii = num_int(x, psi .* psi, -1, 1);
a = sqrt(1/ii);
psi = a * psi;
tp = num_int(x, psi .* psi, 280e - 12, 420e - 12);
subplot(2,1,1); plot(x, psi .* psi);
title(sprintf('|\\psi(x)|'));
subplot(2,1,2);
p = x;
for i = 1:length(x)
   p(i) = pot_energy(x(i));
end
```

```
plot(x, p, x, E*ones(size(x)));
title(sprintf('Potential Barrier \\Pi(x) and electron energy E = %g J', E));
>> E=4e-20
E = 4.0000e-020
>> calc_trans_prob(E)
ans = 0.3294
```

To calculate the current characteristics, the Landauer formula is used. Landauer derived a formula to determine the current through an arbitrary molecule surrounded by electron source and sink contacts (see IBM, *J. Res. Dev.*, 1, 223, 1957). His function is based on the probability of an electron tunneling through the molecule given a variety of incident energies E. The method accounts for a potential voltage applied across the molecule and uses the Fermi energy levels of the contacts. The Landauer equation is given as

$$I(V) = \frac{2e}{h} \int_{-\infty}^{\infty} T(E) \left(\frac{1}{1 + e^{\frac{E - V_{p1}}{kT}}} - \frac{1}{1 + e^{\frac{E - V_{p2}}{kT}}} \right) dE$$

where $T(E)$ is the probability that an electron can tunnel through the molecule given an energy E; V_{p1} and V_{p2} are the biased potential energies of the contacts; k is the Boltzmann constant; and T is the Kelvin temperature.

The functions $V_{p1}(V)$ and $V_{p2}(V)$ are based on the Fermi level of the contacts. Given the Fermi levels E_F at the source and the drain contacts, these potentials are

$$V_{p1}(V) = E_{Fs} + \tfrac{1}{2}eV \quad \text{and} \quad V_{p2}(V) = E_{Fd} + \tfrac{1}{2}eV$$

The accuracy of the results is directly related to the correctness of $T(E)$ and the potentials. Figure 8.30 documents the *I-V* characteristic. Two MATLAB files developed to perform these calculations and plotting are reported below.

```
function curr = calc_current(v)
e = 1.602176462e-19; % elementary charge [J]
h = 6.62606876e-34; % Planck constant [J s]
[E, c] = ode45(@diff_current, [-10*e 10*e], 0, [], v);
curr = 2*e/h * c(length(c));
```

and

```
function [v,curr] = plot_current
v = linspace(-5, 5, 100);
curr = v;
for i = 1 : length(v)
    curr(i) = calc_current(v(i));
end;
plot(v, curr);
```

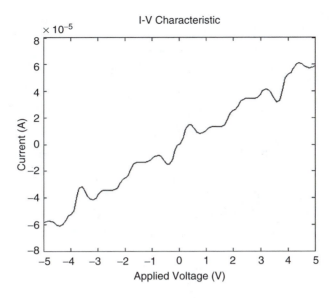

FIGURE 8.30
I-V characteristic in *N@C*$_{60}$.

```
title('I-V Characteristic');
xlabel('Applied Voltage [V]'); ylabel('Current [A]');
```

Figure 8.31 reports the *G-V* characteristic for the endohedral *N@C*$_{60}$ fullerene.

The numerical results reported are in agreement with the experimental results. However, in general, three-dimensional modeling and analysis are needed, and high-fidelity models must be derived.

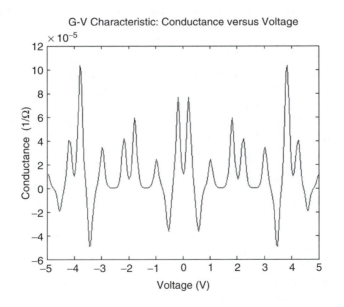

FIGURE 8.31
G-V characteristic for *N@C*$_{60}$.

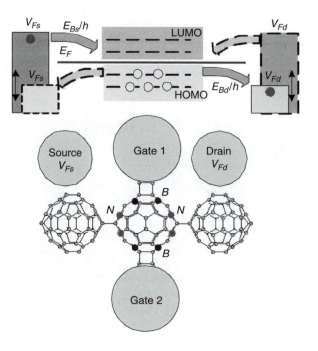

FIGURE 8.32
$C_{44}(BN)_8$ heterofullerene as electronic nanodevice.

Other functional fullerenes can be examined. Among the prospective heterofullerenes, $C_{44}(BN)_8$ can be utilized as an electronic nanodevice for nano-IC as shown in Figure 8.32. The interatomic C–C, C–N, and C–B bond lengths are 1.45, 1.42, and 1.41 Å. The HOMO and LUMO energy levels are –11.3 and –9.9 eV. The neighboring carbon atoms to N are interconnected to other fullerenes to form the source and drain with the varying potentials V_{Fs} and V_{Fd}. The wave functions can be analytically and numerically derived, and the *I-V* and *G-V* characteristics result.

Chapter 3 covered neurons that perform extremely important functions—in particular, receiving, processing (computing), and sending information. Recently, nanoelectronics has been extensively researched, and novel quantum electronic nanodevices have been devised and demonstrated. The majority of quantum nanodevices operate at low temperatures to avoid thermal decoherence. However, quantum entanglement (Schrödinger introduced the term *entanglement* to describe the connection between quantum systems) and Einstein–Podolsky–Rosen effects occur at room temperature. Many effects and phenomena at the nanoscale have not been comprehended, and mechanisms for generating quantum states and avoiding thermal decoherence are unknown. The brain is viewed as an adapting hierarchical system composed of hierarchical layers with bottom-up and top-down feedback. There are complex interactions among layers. Many researchers have assumed that the bottom level is the neuron's membrane protein, implying that ion channels and receptors are the fundamental basic units in analysis of information processing and representation of the nervous system. However, the standard doctrine of neuro-science that neurons with synapses can be viewed as fundamental units can be regarded as an attempt to simplify the understanding and present the brain as a well-defined comprehended organization that is analogous to the existing computing architectures. These issues and various sound and unfounded assumptions were examined and discussed by other researchers.[23–26]

The established view is that the neuronal network circuits involved in cognition and perception through the neural–neural interaction utilize axonal–dendritic chemical synaptic transmission (see Chapter 3). The idea is that the terminal axon of the neuron N_1 releases a neurotransmitter, which then binds to a postsynaptic receptor on a dendritic spine on the neuron N_2. This changes the neuron N_2 local dendritic membrane potential, which interacts with those on neighboring spines and dendrites on neuron N_2 and can reach the threshold to trigger axonal depolarizations to the N_2 axon. These move along the axon and result in the release of neurotransmitter vesicles from the N_2 presynaptic axon terminal into another synapse. The neurotransmitter then binds to postsynaptic dendritic spine receptors on the neuron N_3. Hence, the information and communication pathway results. The basic idea is that in each neuron, multi-input and analog processing in dendrites reaches a threshold for axonal firings, and this discrete event (output) can be viewed as the fundamental unit (biobit) of information. Electrochemical synaptic network patterns, presumed to correlate with a particular overall state, arise and are changed dynamically by the strengths of the chemical synapses. The synaptic strengths change the network dynamics and are responsible for cognitive effects such as learning and adaptation by selecting particular neural network patterns.

However, there are other possibilities for information processing mechanisms in neurons, including dendritic–dendritic processing (processing among dendrites in the same neuron), interconnected electronic gap junction processing (cytoplasm flows through the 4-nm gap, and the cells are synchronously coupled electrically via gap junctions), etc. These gap junctions may be important for macroscopic broadening of quantum states among neurons and glia because electron tunneling can occur up to a distance of 5 nm, and 4-nm separation may enable tunneling through the gaps leading to the quantum processing or computing.[26] The major information analyzing neuronal behavior within the brain has been derived from electrophysiological recordings. Coherent measurement equipment to detect quantum-based processing does not exist, and the high-frequency signals referred to as noise are averaged or filtered. However, these signals can be essential in computing, information processing, communication, and other biosystem functions.

Synapses and membrane proteins, considered in conventional approaches as the bottom unit, are short-lived and are synthesized in the cell body and delivered to the synapse through unknown events. Adaptation, reconfiguration, as well as active and passive control mechanisms in the nucleus have been debated for many years. How are intracellular information processing, learning, computing, decision making, and control accomplished? Furthermore, single-cell organisms (bacteria) exhibit intelligence, e.g., avoid obstacles, find food, form complex swarm patterns, perform sensing, control, locomotion, etc. Single cells do not have nervous systems and there are no synapses. This example also leads to the need to further examine information processing, control, computing, and other mechanisms.

In terms of real-time dynamical regulation of cellular activity, the conformational states of proteins are one of the most important factors. This may include an ion channel opening or closing, receptor shape changing upon binding of neurotransmitter, enzyme catalyzing reactions, messenger-based protein shape changing to signal to facilitate movement or transport, etc. The protein functionality strongly depends on shape, or conformation. Proteins are synthesized as linear chains of amino acids that fold into three-dimensional conformations. The precise folding depends on attractive and repellent forces among various amino acid side groups. Intermediate protein conformations are not the final one. Predicting three-dimensional folded proteins based on modeling and simulation has proven to be very difficult. The main force in protein folding occurs as uncharged nonpolar amino acid groups join together, repelled by solvent water. These hydrophobic groups attract each other by dipole couplings resulting in van der Waals forces. Intra-protein hydrophobic pockets compose side groups of nonpolar polarizable amino acids such as leucine, isoleucine, phenylalanine, tryptophan, tyrosine, and valine. Volumes of these hydrophobic pockets (0.4 nm^3) are in the range of 1% of the total single-protein volume. Van der Waals forces in hydrophobic pockets establish protein shape during folding, and these forces also control dynamic conformational changes. Proteins in a living state are dynamic systems. The conformational transitions in which proteins vary and function occur in nanoseconds, i.e., from 1×10^{-9} sec to 1×10^{-11} sec. Proteins are not robust, and their conformation is a balance of three-dimensional forces. The ionic forces, hydrogen bonds, and interactive dipole forces are examined. Dipole-to-dipole interactions (van der Waals forces) include permanent dipole–permanent dipole, permanent dipole–induced dipole, and induced dipole–induced dipole (the weakest but most purely nonpolar and known as London's dispersion forces). Though London's forces are much weaker than hydrogen bonds, these forces are numerous. London's forces in hydrophobic pockets can govern protein conformational states. Atoms and molecules that are electrically neutral and spherically symmetrical may have instantaneous electric dipoles due to asymmetry in their electron distribution. The electric field from each fluctuating dipole couples to others in electron clouds of nonpolar amino acid side groups. Due to inherent uncertainty in electron localization, London's forces may result in quantum effects. These quantum dipole oscillations within hydrophobic pockets were proposed by other researchers[27,28] to regulate protein conformation and engage in macroscopic coherence. One paper[29] reports functional protein vibrations that depend on quantum effects centered in two hydrophobic phenylalanine residues. Another article[30] suggests that quantum coherent states exist in the protein ferritin. Protein (tubulin) can flip (switch between two conformational states) at 1×10^{-9} sec between two states due to van der Waals interactions in a hydrophobic pocket. The localization of paired electrons (London's force) within a hydrophobic pocket is documented in Figure 8.33.

Purified DNA molecules conduct charge transport. In guanine, one of its electrons is more weakly bound than any other electrons in other DNA nucleotides. The hole migration over 60 base pairs in purified DNA has been experimentally confirmed, and theory suggests that hole could travel as far as 100 base pairs. Figure 8.22 documents how the hole may migrate along the DNA molecule, probably by the hopping effect. In particular, the hole propagates from guanine to guanine bases.

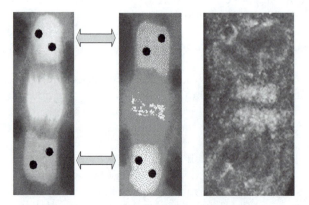

FIGURE 8.33
Localization of paired electrons (London's force) within hydrophobic pockets.

Though significant research has been performed, and meaningful results have been reported examining information processing, computing, evolutionary learning, massive parallelisms, memories, networking, communication, and control and adaptation of the central nervous system and brain, there are many unsolved problems [26–29]. For example, how are super-high-performance and super-fast robust computing, processing, information propagation, and networking accomplished, and how is robustness guaranteed? The existing hypotheses fail to meet simple entropy and energetics analyses from viewpoints of computing, information processing, memory, adaptation, robustness, etc. The assumptions that information processing and propagation in axons occur at very low speed (hyperpolarization and repolarization/depolarization of the axon segments) are not well supported. The following hypothesis can be made: In the central nervous system, computing, information processing, networking, memory, control, and other tasks are performed by three-dimensional reconfigurable nanobioelectronics based on superfast adaptive reversible/irreversible computing, memories interfacing in neuronal proteins, and ballistic information propagating in the *nanobiofabric information channels*. The reported neuroscience-related results clearly document the existence of high-performance robust 3D nanobiocircuitry with biomolecules. Though many problems are unsolved, the documented fundamentals support 3D nano-ICs and the multi-terminal electronic nanodevices paradigm covered in this chapter.

Homework Problems

1. Explain the impact of nanotechnology on nanoelectronics.

2. Explain the difference between nanotechnology, nanoscience, and nanoengineering from the nanoelectronics viewpoint.

3. Assume that by 2015, the total length of nanotechnology-enhanced CMOS-based FET will be downscaled to 100 nm. Explain if this FET should be classified as a micro- or nanoelectronic device. Justify your reasoning.

4. In your judgment, what is the most promising scientific, engineering, and technological solution for future nanoelectronics?

5. What is the difference between organic and inorganic nanoelectronics? Provide examples.

6. What is the distance between the carbon atoms in fullerenes?

7. What is the length of the DNA strand that has 20 base pairs?

8. Can doped $C_{59}B$ and $C_{59}N$ fullerenes be applied in nanoelectronics? How and in what devices can they be utilized (provide the analogies to solid-state electronics and semiconducting devices). Comparing with the classical semiconductors (silicon), provide the analogies for $C_{59}B$ and $C_{59}N$.

9. Consider the simplified one-dimensional electron transport between two opposite carbon atoms in the functional endohedral fullerene. For distinct energy band gaps and fullerene dimensions reported in Figure 8.34, using quantum theory, examine the electron transport, energy, and wave functions. Identify the areas with the carbon atoms. Why are distinct energy levels and contacts documented? How can these contacts be implemented? What bounds can serve as contacts between fullerenes? Formulate and set the problem to derive and calculate the *I-V* characteristics. How can this nanodevice be controlled?

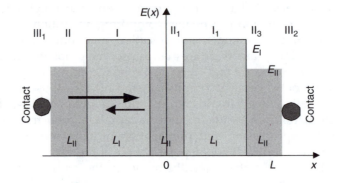

FIGURE 8.34
Simplified one-dimensional energy diagram in a functional fullerene.

References

1. Appenzeller, J., Martel, R., Solomon, P., Chan, K., Avouris, P., Knoch, J., Benedict, J., Tanner, M., Thomas, S., Wang, L.L., and Del Alamo, J.A., A 10 nm MOSFET concept, *Microelectron. Eng.*, 56, 213–219, 2001.
2. Bohr, M.T., Nanotechnology goals and challenges for electronic applications, *IEEE Trans. Nanotechnol.*, 1, 56–62, 2002.
3. Derycke, V., Martel, R., Appenzeller, J., and Avouris, P., Carbon nanotube inter- and intramolecular logic gates, *Nano Lett.*, 2001.
4. Ellenbogen, J.C. and Love, J.C., Architectures for molecular electronic computers: Logic structures and an adder designed from molecular electronic diodes, *Proc. IEEE*, 88, 386–426, 2000.
5. Kamins, T.I., Williams, R.S., Basile, D.P., Hesjedal, T., and Harris, J.S., Ti-catalyzed Si nanowires by chemical vapor deposition: Microscopy and growth mechanism, *J. Appl. Phys.*, 89, 1008–1016, 2001.
6. Lyshevski, S.E., *Nanocomputers and Nanoarchitectronics*, *Handbook of Nanoscience, Engineering and Technology*, Goddard, W., Brenner, D., Lyshevski, S., and Iafrate, G., Eds., CRC Press, Boca Raton, FL, 2002, pp. 6.1–6.39.
7. Colbert, D.T. and Smalley, R.E., Past, present and future of fullerene nanotubes: Buckytubes, in *Perspectives of Fullerene Nanotechnology*, Osawa, E., Ed., Kluwer Academic Publisher, London, 2002, pp. 3–10.
8. Hirahara, K., Bandow, S., Suenaga, K., Kato, H., Okazaki, T., Shinahara, H., and Iijima, S., Electron diffraction study of one-dimensional crystals of fullerenes, *Phys. Rev. B*, 64, 115420–115424, 2001.
9. Saito, R., Dresselhaus, G., and Dresselhaus, M.S., *Physical Properties of Carbon Nanotubes*, Imperial College Press, London, 1999.
10. Dequesnes, M., Rotkin, S.V., and Aluru, N.R., Calculation of pull-in voltages for carbon-nanotube-based nanoelectromechanical systems, *Nanotechnology*, 13, 120–131, 2002.
11. Lyshevski, S.E., *MEMS and NEMS: Systems, Devices, and Structures*, CRC Press, Boca Raton, FL, 2002.
12. Ren, Z., Venugopal, R., Goasguen, S., Datta, S., and Lundstrom, M.S., NanoMOS 2.5: A two-dimensional simulator for quantum transport on double-gated MOSFETs, *IEEE Trans. Electron Devices*, 50, 1914–1925, 2003.
13. Rahman, A., Guo, J., Datta, S., and Lundstrom, M.S., Theory of ballistic nanotransistors, *IEEE Trans. Electron Devices*, 50, 1853–1864, 2003.
14. Tian, W.T., Datta, S., Hong, S., Reifenberger, R., Henderson, J.I., and Kubiak, C.P., Conductance spectra of molecular wires, *Int. J. Chem. Phys.*, 109, 2874–2882, 1998.

15. Andreoni, W., Computational approach to the physical chemistry of fullerenes and their derivatives, *Annu. Rev. Phys. Chem.* , 49, 405–439, 1998.
16. Krause, M., Kuzmany, H., Georgi, P., Dunsch, L., Vietze, K., and Seifert, G., Structure and stability of endohedral fullerene $Sc_3N@C_{80}$: A Raman, infrared, and theoretical analysis, *J. Chem. Phys.* , 115, 2001.
17. Lee, H.M., Olmstead, M.M., Iezzi, E., Duchamp, J.C., Dorn, H.C., and Balch, A.L., Crystallographic characterization and structural analysis of the first organic functionalization product of the endohedral fullerene $Sc_3N@C_{80}$, *J. Amer. Chem. Soc.* , 124, 3494–3495, 2002.
18. Forro, L. and Mihaly, L., Electronic properties of doped fullerenes, *Rep. Prog. Phys.* , 64, 649–699, 2001.
19. Shinohara, H., Endohedral metallofullerenes, *Rep. Prog. Phys.* , 63, 843–892, 2000.
20. Stevenson, S., Burbank, P., Harich, K., Sun, Z., Dorn, H.C., van Loosdrecht, P.H.M., deVries, M.S., Salem, J.R., Kiang, C.H., Johnson, R.D., and Bethune, D.S., $La_2@C_{72}$: Metal-mediated stabilization of a carbon cage, *J. Phys. Chem. A* , 102, 2833–2837, 1998.
21. Yee, K.A., Yi, H., Lee, S., Kang, S.K., Song, J.S., and Seong, S., The electronic structure and stability of the heterofullerene $C_{(60-2x)}(BN)_x$, *Bull. Korean Chem. Soc.* , 24, 494–498, 2003.
22. Rakitin, A., Aich, P., Papadopoulos, C., Kobzar, Yu., Vedeneev, A.S., Lee, J.S., and Xu, J.M., Metallic conduction through engineered DNA: DNA nanoelectronic building blocks, *Phys. Rev. Lett.* , 86, 3670–3673, 2001.
23. Beck, F. and Eccles, J.C., Quantum aspects of brain activity and the role of consciousness, *Proc. Nat. Acad. Sci. U.S.A.* , 89, 11357–11361, 1992.
24. Lyshevski, S.E. and Renz, T., Microtubules and neuronal nanobioelectronics, *Proc. IEEE Conf. Nanotechnol.*, Munich, Germany, 2004.
25. Wallace, R., Quantum computation in the neural membrane, in *Toward a Science of Consciousness* , Hameroff, S.R., Kaszniak, A., and Scott, A.C., Eds., MIT Press, Cambridge, MA, 1996, pp. 419–424.
26. Hameroff, S.R. and Penrose, R., Orchestrated reduction of quantum coherence in brain microtubules: A model for consciousness, in *Toward a Science of Consciousness* , Hameroff, S.R., Kaszniak, A., and Scott, A.C., Eds., MIT Press, Cambridge, MA, 1996, pp. 507–540.
27. Frohlich, H., Long-range coherence and energy storage in biological systems, *Int. J. Quantum Chem.*, 2, 641–649, 1968.
28. Roitberg, A., Gerber, R.B., Elber, R., and Ratner, M.A., Anharmonic wave functions of proteins: quantum self-consistent field calculations of BPTI, *Science*, 268, 1319–1322, 1995.
29. Tejada, J., Garg, A., Gider, S., Awschalom, D.D., DiVincenzo, D.P., and Loss, D., Does macroscopic quantum coherence occur in ferritin?, *Science*, 272, 424–426, 1996.

9

Control of MEMS and NEMS

9.1 Continuous-Time and Discrete-Time MEMS

High-fidelity mathematical modeling, heterogeneous simulation, data-intensive analysis, robust control, and optimization have become important problems that need to be solved for MEMS and NEMS. These problems are very complex due to the fact that the majority of microsystems belong to the class of large-scale nonlinear systems that include a variety of complex subsystems (actuators, sensors, ICs, etc.). It was illustrated that the behavior of MEMS and NEMS is modeled and examined using nonlinear differential equations. The analysis of these systems has been performed by numerically and analytically solving these equations. Newtonian and Lagrangian mechanics, as well as Kirchhoff's laws, energy conservation principles, and Maxwell's equations were used to develop mathematical models of micro- and some nanosystems. Model development efforts are driven by the final goal, which is to derive accurate mathematical description of systems with the ultimate objective to design closed-loop systems to satisfy the desired performance characteristics as measured against a spectrum of imposed specifications and requirements. Analog and digital controllers can be designed and implemented for a large class of dynamic systems.[1-11] The application of control theory to micro- and nanosystems is covered in publications.[7,8] The specifications imposed on closed-loop microsystems are given in the performance domain. For example, the criteria under consideration can be as follows:

- Stability with the desired stability margins in the full operating envelope
- Robustness to parameter variations, structural, kinematical, and environmental changes
- Tracking accuracy that is assessed using dynamic and steady-state accuracies
- Disturbance and noise attenuation
- Transient response specifications (settling and peak times, maximum overshoot, etc.)

Correspondingly, the system's performance is measured and assessed against multiple criteria (stability, robustness, transient behavior, accuracy, disturbance attenuation, dynamic behavior, etc.). The specifications are dictated by the requirements imposed on systems in the full operating envelope. Mathematically, the performance characteristics are assigned and assessed using performance criteria in the form of performance functionals (for continuous-time systems) and performance indexes (for discrete-time systems). For example, in the behavioral domain, the designer can examine and optimize the input-output transient dynamics. Denoting the reference (command) and output variables as $r(t)$ and $y(t)$, the tracking error $e(t) = y(t) - r(t)$ is minimized and dynamics is optimized

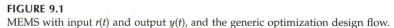
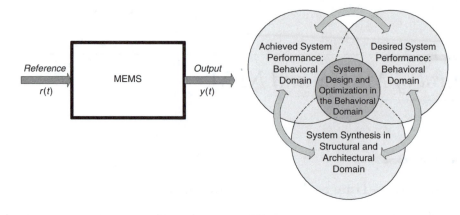

FIGURE 9.1
MEMS with input *r*(*t*) and output *y*(*t*), and the generic optimization design flow.

(controlled) using different performance criteria emphasized. To illustrate the input-output dynamic system behavior, MEMS with the reference *r*(*t*) and output *y*(*t*) is illustrated in Figure 9.1. In this chapter, MEMS and NEMS dynamics will be examined and optimized in the behavioral domain, examining and optimizing the steady-state responses and dynamic transient dynamics. Using the design flow, as reported in Figure 9.1, the optimization will be performed utilizing the synthesized systems by making use of the mathematical models.

In Chapter 2, the microsystem architecture and dynamics were reported applying the system topology as shown in Figure 2.19. We modify this conceptual diagram as illustrated in Figure 9.2.

How can one design controllers for MEMS and NEMS in order to optimize the system performance? In general, the functional block diagrams of controlled (closed-loop) microsystems can be very complex, For example, the possible closed-loop system configuration is reported in Figure 9.3. Robust control laws must be designed, examined, verified, and implemented. We will apply and demonstrate different methods to design closed-loop microsystems. To implement these analog and digital controllers, different ICs or nano-ICs should be designed and used. These computing, controlling, processing, and signal conditioning ICs are the components of MEMS as were emphasized in this book.

To optimize microsystem dynamics, one can minimize the tracking error and settling time. Transient responses can be optimized using different performance functionals and

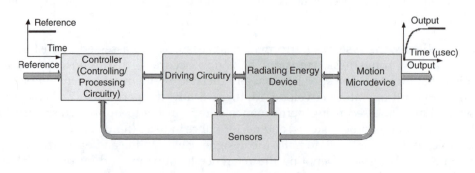

FIGURE 9.2
Functional block diagram of the closed-loop MEMS with the input-output dynamics.

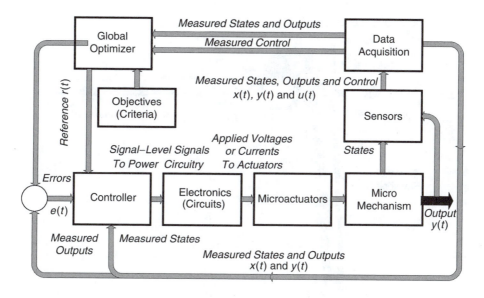

FIGURE 9.3
High-level functional block diagram of the closed-loop MEMS.

performance indexes. Thus, the performance specifications and requirements are examined. For example, utilizing the settling time t and the tracking error $e(t) = y(t) - r(t)$ as variables to design the performance criteria, one can set

$$J = \min_{t,e} \int_0^{\infty} t |e| \, dt$$

The functionals that incorporate only the tracking error are

$$J = \min_{e} \int_0^{\infty} |e| \, dt \quad \text{or} \quad J = \min_{e} \int_0^{\infty} e^2 \, dt$$

A great number of specifications are imposed. The system output dynamics is usually prioritized making use of the constraints imposed on the system variables. Correspondingly, the output transient dynamics and the evolution envelopes are examined. Consider the output transient response as illustrated in Figure 9.4 for the step-type reference input $r(t) = \text{const}$. It is obvious that the microsystem dynamics is stable because output converges to the steady-state value $y_{\text{steady state}}$, i.e., $\lim_{t \to \infty} y(t) = y_{\text{steady-state}}$, and the tracking error is zero. The output can be studied within the specified evolution envelopes. Two evolution envelopes (I and II) are illustrated. The system dynamics is within the evolution envelope II. However, the desired system performance within evolution envelope I is not guaranteed. To attain the optimal MEMS performance, controllers are used to improve transient behavior, ensuring the system dynamics within the specified (desired) evolution envelopes. It must be emphasized that due to system limits (rated torque, force, power, and other variables), the designer may not be able to achieve the desired performance criteria while guaranteeing the maximum *achievable* performance. A system redesign may be needed. This evolution envelope can be assigned in terms of the settling time, overshoot, etc.

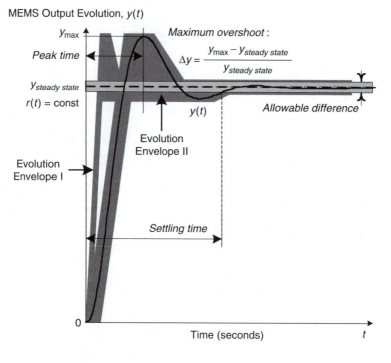

MEMS Output Evolution, $y(t)$

Maximum overshoot :

$$\Delta y = \frac{y_{\max} - y_{steady\ state}}{y_{steady\ state}}$$

Allowable difference

FIGURE 9.4
MEMS output transient response and evolution envelopes, $r(t)$ = const.

The microsystem's performance in the behavioral domain is examined using the well-defined criteria of settling time, overshoot, and accuracy as reported in Figure 9.4. The settling time is the time needed for the system output $y(t)$ to reach and stay within the steady-state value $y_{steady\ state}$. The absolute allowable difference between $y(t)$ and $y_{steady\ state}$ is used to find the settling time (this difference varies; for example, in high-accuracy pointing systems, the required accuracy can be microradians). The settling time is the minimum time after which the system response remains within the desired accuracy taking into account the steady-state value $y_{steady\ state}$ and the command. The maximum overshoot is the difference between the maximum peak value of the system output $y(t)$ and the steady-state value $y_{steady\ state}$ divided by $y_{steady\ state}$, i.e.,

$$\Delta y = \frac{y_{\max} - y_{steady\ state}}{y_{steady\ state}} \times 100\%$$

The rise time is the time needed for the system output $y(t)$ to increase from 10 to 90% of the steady-state value $y_{steady\ state}$ if $r(t)$ = const. The delay time is the time needed for the system output $y(t)$ to reach 50% of $y_{steady\ state}$ if the reference is the step. The peak time is the time required for the system output $y(t)$ to reach the first peak of the overshoot.

In microsystems, due to nonlinearities, bounds and time-varying parameter variations, some performance criteria are stringent. In particular, the most important criteria are stability and robustness to parameter variations in the full operating envelope. The MEMS transient evolution envelope must be examined within the operating envelope. This chapter reports different viable methods to solve challenging problems in the design of closed-loop microsystems. We will focus our efforts on the dynamic optimization assuming that the steady-state performance (efficiency, reliability, robustness, vibroacoustics, kinematics, etc.) was designed and optimized performing synthesis and optimization tasks. However, efficiency, vibroacoustics, and other system performance characteristics can be further improved by designing control laws.

We will report the basic methods used to solve the motion control problem in designing analog and digital proportional-integral-derivative (PID) controllers as well as state-space control laws. The PID control laws use the tracking error $e(t)$, while the state-space controllers use both the tracking error $e(t)$ and the state variables $x(t)$. Proportional-integral-derivative and state-space control algorithms are widely applied to guarantee stability, attain tracking and disturbance attenuation, ensure robustness, guarantee accuracy, etc. Linear microsystems can be described (modeled) in the s and z domains using transfer functions $G_{sys}(s)$ and $G_{sys}(z)$ (note that there is a very limited class of microsystems and microdevices that can be modeled by linear differential equations). These transfer functions are used to synthesize PID-type controllers.

For MEMS modeled in the state-space form, we use n states $x \in \mathbb{R}^n$ and m controls $u \in \mathbb{R}^m$. The transient dynamics of linear MEMS is described by a set of n linear first-order differential equations

$$\frac{dx_1}{dt} = a_{11}x_1 + a_{12}x_2 + \cdots + a_{1n-1}x_{n-1} + a_{1n}x_n + b_{11}u_1 + b_{12}u_2 + \cdots + b_{1m-1}u_{m-1}$$
$$+ b_{1m}u_m, \ x_1(t_0) = x_{10}$$

$$\frac{dx_2}{dt} = a_{21}x_1 + a_{22}x_2 + \cdots + a_{2n-1}x_{n-1} + a_{2n}x_n + b_{21}u_1 + b_{22}u_2 + \cdots + b_{2m-1}u_{m-1}$$
$$+ b_{2m}u_m, \ x_2(t_0) = x_{20}$$

$$\vdots$$

$$\frac{dx_{n-1}}{dt} = a_{n-11}x_1 + a_{n-12}x_2 + \cdots + a_{n-1n-1}x_{n-1} + a_{n-1n}x_n + b_{n-11}u_1 + b_{n-12}u_2$$
$$+ \cdots + b_{n-1m-1}u_{m-1} + b_{m-1m}u_m, \ x_{n-1}(t_0) = x_{n-10}$$

$$\frac{dx_n}{dt} = a_{n1}x_1 + a_{n2}x_2 + \cdots + a_{nn-1}x_{n-1} + a_{nn}x_n + b_{n1}u_1 + b_{n2}u_2 + \cdots$$
$$+ b_{nm-1}u_{m-1} + b_{nm}u_m, \ x_n(t_0) = x_{n0}$$

and the resulting matrix form is given as

$$\frac{dx}{dt} = \begin{bmatrix} \dfrac{dx_1}{dt} \\ \dfrac{dx_2}{dt} \\ \vdots \\ \dfrac{dx_{n-1}}{dt} \\ \dfrac{dx_n}{dt} \end{bmatrix} = \begin{bmatrix} a_{11} & a_{12} & \cdots & a_{1n-1} & a_{1n} \\ a_{21} & a_{22} & \cdots & a_{2n-1} & a_{2n} \\ \vdots & \vdots & \ddots & \vdots & \vdots \\ a_{n-11} & a_{n-12} & \cdots & a_{n-1n-1} & a_{n-1n} \\ a_{n1} & a_{n2} & \cdots & a_{nn-1} & a_{nn} \end{bmatrix} \begin{bmatrix} x_1 \\ x_2 \\ \vdots \\ x_{n-1} \\ x_n \end{bmatrix}$$

$$+ \begin{bmatrix} b_{11} & b_{12} & \cdots & b_{1m-1} & b_{1m} \\ b_{21} & b_{22} & \cdots & b_{2m-1} & b_{2m} \\ \vdots & \vdots & \ddots & \vdots & \vdots \\ b_{n-11} & b_{n-12} & \cdots & b_{n-1m-1} & b_{n-1m} \\ b_{n1} & b_{n2} & \cdots & b_{nm-1} & b_{nm} \end{bmatrix} \begin{bmatrix} u_1 \\ u_2 \\ \vdots \\ u_{m-1} \\ u_m \end{bmatrix} = Ax + Bu, \quad x\,(t_0) = x_0$$

Assuming that matrices $A \in \mathbb{R}^{n \times n}$ and $B \in \mathbb{R}^{n \times m}$ are constant-coefficients (microsystem parameters are constant), we have the characteristic equation as

$$sI - A = 0 \quad \text{or} \quad a_n s^n + a_{n-1} s^{n-1} + \cdots + a_1 s + a_0 = 0$$

Here $I \in \mathbb{R}^{n \times n}$ is the identity matrix.

Solving the characteristic equation, one finds the eigenvalues (characteristic roots). The system is stable if real parts of all eigenvalues are negative. It must be emphasized that the stability analysis using the eigenvalues is valid only for linear dynamic systems.

The transfer function $G\,(s) = Y\,(s)/U\,(s)$ can be found using the state-space equations. Consider the linear time-invariant MEMS as described by $dx/dt = Ax + Bu$, $y = Hx$. Here, $y = Hx$ is the output equation. Taking the Laplace transform of the state-space $dx/dt = Ax + Bu$ and output $y = Hx$ equations, we obtain

$$sX(s) - x(t_0) = AX(s) + BU(s), \quad Y(s) = HX\,(s)$$

Assuming that the initial conditions are zero, we have $X(s) = (sI - A)^{-1} BU(s)$. Hence, $Y(s) = HX(s) = H(sI - A)^{-1} BU(s)$. Therefore, the transfer function is found as

$$G(s) = \frac{Y(s)}{U(s)} = H(sI - A)^{-1} B$$

Assuming that the initial conditions are zero, we apply the Laplace transform to both sides of the n-order differential equation

$$\sum_{i=0}^{n} a_i \frac{d^i y(t)}{dt^i} = \sum_{i=0}^{m} b_i \frac{d^i u(t)}{dt^i}$$

Taking note of

$$\left(\sum_{i=0}^{n} a_i s^i \right) Y(s) = \left(\sum_{i=0}^{m} b_i s^i \right) U(s)$$

one concludes that the transfer function, as the ratio of numerator and denominator, is given by

$$G(s) = \frac{Y(s)}{U(s)} = \frac{b_m s^m + b_{m-1} s^{m-1} + \cdots + b_1 s + b_0}{a_n s^n + a_{n-1} s^{n-1} + \cdots + a_1 s + a_0}$$

By setting the denominator polynomial of the transfer function to zero, one obtains the characteristic equation. The stability of linear time-invariant systems is guaranteed if all characteristic eigenvalues, obtained by solving the characteristic equation $a_n s^n + a_{n-1} s^{n-1} + \cdots + a_1 s + a_0 = 0$, have negative real parts.

In general, the MEMS dynamics are described by nonlinear differential equations in the state-space form

$$\dot{x}(t) = F(x,r,d) + B(x)u, \quad y = H(x), \quad u_{min} \le u \le u_{max}, \quad x(t_0) = x_0$$

where $x \in X \subset \mathbb{R}^n$ is the state vector (displacement, position, velocity, current, voltage, etc.); $u \in U \subset \mathbb{R}^m$ is the bounded control vector (supplied voltage); $r \in R \subset \mathbb{R}^b$ and $y \in Y \subset \mathbb{R}^b$ are the measured reference and output vectors; $d \in D \subset \mathbb{R}^s$ is the disturbance vector (load and friction torques, etc.); $F(\cdot): \mathbb{R}^n \times \mathbb{R}^b \times \mathbb{R}^s \to \mathbb{R}^c$ and $B(\cdot): \mathbb{R}^n \to \mathbb{R}^{n \times m}$ are nonlinear maps; and $H(\cdot): \mathbb{R}^n \to \mathbb{R}^b$ is the nonlinear map defined in the neighborhood of the origin, $H(0) = 0$. The output equation is given as $y = H(x)$. The control bounds are represented as $u_{min} \le u \le u_{max}$.

Consider n-degree-of-freedom discrete microsystems modeled using difference equations. Using n-dimensional state, m-dimensional control, and b-dimensional output vectors, the MEMS state, control, and output variables are

$$x_k = \begin{bmatrix} x_{k1} \\ x_{k2} \\ \vdots \\ x_{k\,n-1} \\ x_{kn} \end{bmatrix}, \quad u_k = \begin{bmatrix} u_{k1} \\ u_{k2} \\ \vdots \\ u_{k\,n-1} \\ u_{kn} \end{bmatrix} \quad \text{and} \quad y_k = \begin{bmatrix} y_{k1} \\ y_{k2} \\ \vdots \\ y_{k\,b-1} \\ y_{kb} \end{bmatrix}$$

In matrix form, the state-space equations are given as

$$x_{k+1} = \begin{bmatrix} x_{k+1,1} \\ x_{k+1,2} \\ \vdots \\ x_{k+1,n-1} \\ x_{k+1,n} \end{bmatrix} = \begin{bmatrix} a_{k11} & a_{k12} & \cdots & a_{k1\,n-1} & a_{k1n} \\ a_{k21} & a_{k22} & \cdots & a_{k2\,n-1} & a_{k2n} \\ \vdots & \vdots & \ddots & \vdots & \vdots \\ a_{kn-11} & a_{kn-12} & \cdots & a_{kn-1\,n-1} & a_{kn-1n} \\ a_{kn1} & a_{kn2} & \cdots & a_{kn\,n-1} & a_{knn} \end{bmatrix} \begin{bmatrix} x_{k1} \\ x_{k2} \\ \vdots \\ x_{k\,n-1} \\ x_{kn} \end{bmatrix}$$

$$+ \begin{bmatrix} b_{k11} & b_{k12} & \cdots & b_{k1\,m-1} & b_{k1m} \\ b_{k21} & b_{k22} & \cdots & b_{k2\,m-1} & b_{k2m} \\ \vdots & \vdots & \ddots & \vdots & \vdots \\ b_{kn-11} & b_{kn-12} & \cdots & b_{kn-1\,m-1} & b_{kn-1m} \\ b_{kn1} & b_{kn2} & \cdots & b_{kn\,m-1} & b_{knm} \end{bmatrix} \begin{bmatrix} u_{k1} \\ u_{k2} \\ \vdots \\ u_{k\,m-1} \\ u_{km} \end{bmatrix} = A_k x_k + B_k u_k, \quad x_{k=k_0} = x_{k0}$$

Here, $A_k \in \mathbb{R}^{n \times n}$ and $B_k \in \mathbb{R}^{n \times m}$ are the matrices of coefficients.

The output equation that integrates the microsystem's measured output and state variables is

$$y_k = H_k x_k$$

where $H_k \in \mathbb{R}^{b \times b}$ is the matrix of the constant coefficients.

The n-order linear difference equation is written as

$$\sum_{i=0}^{n} a_i y_{n-i} = \sum_{i=0}^{m} b_i u_{n-i}, \quad n \geq m$$

Assuming that the coefficients are time-invariant (constant-coefficient), using the z-transform and assuming that the initial conditions are zero, one has

$$\left(\sum_{i=0}^{n} a_i z^i \right) Y(z) = \left(\sum_{i=0}^{m} b_i z^i \right) U(z)$$

Therefore, the transfer function is

$$G(z) = \frac{Y(z)}{U(z)} = \frac{b_m z^m + b_{m-1} z^{m-1} + \cdots + b_1 z + b_0}{a_n z^n + a_{n-1} z^{n-1} + \cdots + a_1 z + a_0}$$

Nonlinear MEMS can be modeled using nonlinear difference equations. In particular, we have

$$x_{n+1} = F(x_n, r_n, d_n) + B(x_n) u_n, \quad u_{n\min} \leq u_n \leq u_{n\max}$$

This chapter covers analog and digital control of MEMS emphasizing the application of PID control, linear quadratic regulator, Hamilton-Jacobi, and other concepts for linear and nonlinear nano- and microscale systems.

9.2 Analog Control of MEMS Using Transfer Functions

In this section, we cover the topics in analog control of MEMS utilizing PID controllers. It was emphasized that all MEMS are nonlinear dynamic systems, e.g., the transient dynamics is modeled by nonlinear differential equations. However, as a starting point, linear systems will be considered first. The state-space methods that are based on the LQR and Hamilton-Jacobi concepts will be covered.

9.2.1 Analog PID Controllers

The majority of MEMS evolve in continuous-time domain. Therefore, these systems are continuous-time and modeled using differential equations. The simplest control algorithms available are the PID-type controllers. The linear analog PID control law is

$$u(t) = \underbrace{k_p e(t)}_{\text{proportional term}} + \underbrace{k_i \frac{e(t)}{s}}_{\text{integral term}} + \underbrace{k_d s e(t)}_{\text{derivative term}} = \underbrace{k_p e(t) + k_i \int e(t) dt + k_d \frac{de(t)}{dt}, s = \frac{d}{dt}}_{\text{Linear control with three terms}} \quad (9.1)$$

where $e(t)$ is the error between the reference signal and the system output, and k_p, k_i, and k_d are the proportional, integral, and derivative feedback gains.

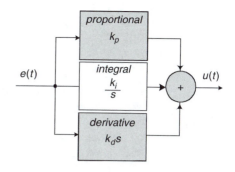

FIGURE 9.5
Analog PID controller with linear control surface.

The block diagram of the analog PID controller (see Equation 9.1) with linear control surface is shown in Figure 9.5.

The variety of controllers can be obtained utilizing Equation 9.1. Setting k_d equal to zero, the proportional-integral (PI) controller results:

$$u(t) = k_p e(t) + k_i \frac{e(t)}{s} = k_p e(t) + k_i \int e(t) dt$$

Assigning the integral feedback coefficient k_i to be zero, we have the proportional-derivative (PD) controller as given by

$$u(t) = k_p e(t) + k_d s e(t) = k_p e(t) + k_d \frac{de(t)}{dt}$$

If k_i and k_i are set to be zero, the proportional (P) control law results as $u(t) = k_p e(t)$.

Using the Laplace transform, from Equation 9.1, we have

$$U(s) = \left(k_p + \frac{k_i}{s} + k_d s \right) E(s)$$

One finds the transfer function of the analog PID controllers as

$$G_{PID}(s) = \frac{U(s)}{E(s)} = \frac{k_d s^2 + k_p s + k_i}{s} = \frac{k_i \left(T_c^2 s^2 + 2 T_c \xi_c s + 1 \right)}{s}$$

Different PID-type analog control algorithms can be designed and implemented (e.g., proportional, proportional-integral, linear, nonlinear, unbounded, and constrained). The closed-loop MEMS with PID-type controller in the time- and s-domains with the negative output feedback loop are represented in Figure 9.6.

If the system output $y(t)$ converges to the reference signal $r(t)$ as time approaches infinity (the steady-state value of the output is equal to the reference input), one concludes that tracking of the reference input is accomplished, and the error vector $e(t) = r(t) - y(t)$ is zero. That is, tracking is achi-eved if $e(t) = r(t) - y(t) = 0$ as $t \to \infty$, or $\lim e(t) = 0$.

In the time domain, the tracking error is given as $e(t) = r(t) - y(t)$. Then, the Laplace transform of the error signal is $E(s) = R(s) - Y(s)$. For the closed-loop system, as given in Figure 9.6, the Laplace transform of the output $y(t)$ is obtained as

$$Y(s) = G_{sys}(s) U(s) = G_{sys}(s) G_{PID}(s) E(s) = G_{sys}(s) G_{PID}(s) [R(s) - Y(s)]$$

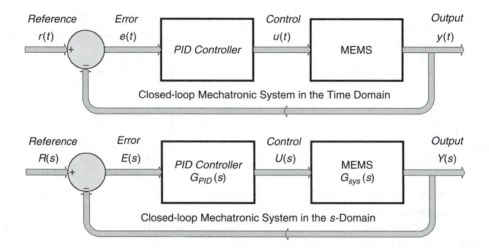

FIGURE 9.6
Time- and s-domain diagrams of closed-loop MEMS with PID controllers.

Hence, the following transfer function of the closed-loop MEMS systems with PID control law results:

$$G(s) = \frac{Y(s)}{R(s)} = \frac{G_{sys}(s)G_{PID}(s)}{1 + G_{sys}(s)G_{PID}(s)}$$

In the frequency domain, one obtains

$$G(j\omega) = \frac{Y(j\omega)}{R(j\omega)} = \frac{G_{sys}(j\omega)G_{PID}(j\omega)}{1 + G_{sys}(j\omega)G_{PID}(j\omega)}$$

The characteristic equations of the closed-loop MEMS can be easily found, and the stability can be guaranteed by adjusting the proportional, integral, and derivative feedback gains. In particular, k_p, k_i, and k_d feedback gains affect the closed-loop system performance. Using the closed-loop MEMS configuration, as reported in Figure 9.6, the closed-loop transfer function is

$$G(s) = \frac{Y(s)}{R(s)} = \frac{G_{sys}(s)G_{controller}(s)}{1 + G_{sys}(s)G_{controller}(s)}$$

where $G_{controller}(s)$ is the transfer function of the controller.
Augmenting the controller and MEMS dynamics, one finds

$$G_\Sigma(s) = G_{sys}(s)G_{controller}(s), \quad \text{and} \quad G(s) = \frac{Y(s)}{R(s)} = \frac{G_\Sigma(s)}{1 + G_\Sigma(s)}$$

The number of the eigenvalues at the origin defines the error. Using the constant factor k, eigenvalues at the origin, real and complex-conjugate eigenvalues and zeros, one can write

$$G_{\Sigma}(s) = \frac{k(T_{n1}s+1)(T_{n2}s+1)\cdots\left(T_{n,l-1}^2 s^2 + 2\xi_{n,l-1}T_{n,l-1}s+1\right)\left(T_{n,l}^2 s^2 + 2\xi_{n,l}T_{n,l}s+1\right)}{s^M(T_{d1}s+1)(T_{d2}s+1)\cdots\left(T_{d,p-1}^2 s^2 + 2\xi_{d,p-1}T_{d,p-1}s+1\right)\left(T_{d,p}^2 s^2 + 2\xi_{d,p}T_{d,p}s+1\right)}$$

where T_i and ξ_i are the time constants and damping coefficients, and M is the order of the eigenvalues at the origin.

The controller transfer function can be found by assigning the location of the characteristic eigenvalues and zeros. For example, if the PID controller (see Equation 9.1) is used, the feedback gains k_p, k_i, and k_d can be derived to attain the specific location of the principal characteristic eigenvalues because other eigenvalues can be located far out left in the complex plane (see Figure 9.7). Many textbooks on feedback control, which are listed in the references, cover the controller design using the root-locus methods. However, this method can be applied only for linear systems.

A *generic* PID control law with the nonlinear proportional control surface is given by

$$u(t) = \underbrace{\sum_{j=1}^{2N_p-1} k_{p(2j-1)}e^{2j-1}(t)}_{proportional} + \underbrace{\sum_{j=1}^{2N_i-1} k_{i(2j-1)}\frac{e(t)}{s^{2j-1}}}_{integral} + \underbrace{\sum_{j=1}^{2N_d-1} k_{d(2j-1)}\frac{d^{2j-1}e(t)}{dt^{2j-1}}}_{derivative} \qquad (9.2)$$

where N_p, N_i, and N_d are the positive integers, and $k_{p(2j-1)}$, $k_{i(2j-1)}$, and $k_{d(2j-1)}$ are the proportional, integral, and derivative feedback coefficients.

In Equation 9.2, N_p, N_i, and N_d integers are assigned by the designer. Setting $N_p = 1$, $N_i = 1$, and $N_d = 1$, we have the PID controller given by Equation 9.1. Letting, $N_p = 2$, $N_i = 2$ and $N_d = 1$, from Equation 9.2, we have the PP³P⁵II³I⁵D controller as given by

$$u(t) = k_{p1}e(t) + k_{p3}e^3(t) + k_{p5}e^5(t) + k_{i1}\frac{e(t)}{s} + k_{i3}\frac{e(t)}{s^3} + k_{i5}\frac{e(t)}{s^5} + k_{d1}\frac{de(t)}{dt}$$

It is evident that the control is a nonlinear function of the tracking error.

Nonlinear control laws with nonlinear control surfaces (nonlinear error maps) are used to improve MEMS performance. There exists a variety of equations to describe nonlinear

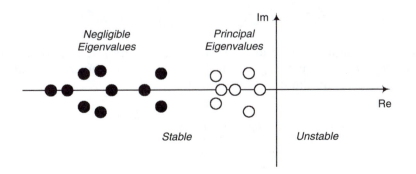

FIGURE 9.7
Eigenvalues in the complex plane.

PID controllers. In particular, nonlinear PID-type controllers can be described using the following equation:

$$u(t) = \sum_{j=0}^{\varsigma} k_{pj} e^{\frac{2j+1}{2\beta+1}}(t) + \sum_{j=0}^{\sigma} k_{ij} \frac{1}{s} e^{\frac{2j+1}{2\mu+1}}(t) + \sum_{j=0}^{\eta} k_{dj} se(t)^{\frac{2j+1}{2\gamma+1}} \quad (9.3)$$

$$\underbrace{\qquad\qquad}_{\text{proportional}} \qquad \underbrace{\qquad\qquad}_{\text{integral}} \qquad \underbrace{\qquad\qquad}_{\text{derivative}}$$

where $\varsigma, \beta, \sigma, \mu, \eta$, and γ are the nonnegative integers assigned by the designer.

In Equation 9.3, the nonlinear proportional, integral, and derivative feedback terms are given in terms of the error vector. Nonnegative integers $\varsigma, \beta, \sigma, \mu, \eta$, and γ are assigned by the designer to attain the specified criteria in stability and tracking, accuracy and disturbance attenuation, etc. Setting $\varsigma = \beta = \sigma = \mu = \eta = \gamma = 0$, the linear PID controller (Equation 9.1) results. If the designer assigns nonnegative integers to be $\varsigma = \beta = 1, \sigma = \mu = 2$ and $\eta = \gamma = 3$, the following nonlinear control law results:

$$u(t) = k_{p0}e^{\frac{1}{3}} + k_{p1}e + k_{i0}\frac{1}{s}e^{\frac{1}{5}} + k_{i1}\frac{1}{s}e^{\frac{3}{5}} + k_{i2}\frac{1}{s}e + k_{d0}se^{\frac{1}{7}} + k_{d1}se^{\frac{3}{7}} + k_{d2}se^{\frac{5}{7}} + k_{d3}se$$

That is, the tracking error is in the corresponding power. It must be emphasized that one cannot calculate the eigenvalues because the closed-loop system with PID-type nonlinear controller (Equation 9.3) contains nonlinear terms.

In MEMS, the controls, states, and outputs are bounded. For example, motion microdevices are controlled by using ICs that change the applied voltage. This voltage, applied to the windings, is bounded. Furthermore, the angular velocity of microactuators is constrained. In addition, the bounds on the current, force, torque, and power are imposed. Mechanical limits are imposed on the maximum angular and linear velocities of actuators. The duty cycle and IC variables (voltages and currents) are bounded. These rated (maximum allowed) voltages and currents, angular and linear velocities, and displacements are specified as a part of the MEMS design flow. For example, taking into account that the applied voltages supplied to the microactuator windings cannot exceed the rated voltage, and the currents in windings cannot be greater than the maximum allowed peak and continuous rated currents, the designer examines the closed-loop MEMS performance by analyzing the data. We conclude that due to the limits imposed, the allowed control is bounded, and the system variables must be within the maximum admissible (rated) set. The closed-loop MEMS system with saturated control law is shown in Figure 9.8.

The bounded control, as a function of the error vector, can be found as

$$u(t) = \text{sat}_{u_{\min}}^{u_{\max}}\left(k_p e(t) + k_i \int e(t)dt + k_d \frac{de(t)}{dt} \right), \quad u_{\min} \le u \le u_{\max} \quad (9.4)$$

Thus, the control signal u is bounded between the minimum and maximum values $u_{\min} \le u \le u_{\max}$, $u_{\min} < 0$, and $u_{\max} > 0$. In the linear region, the control varies between the maximum $u_{\max}$ and minimum $u_{\min}$ values, and

$$u(t) = k_p e(t) + k_i \int e(t)dt + k_d \frac{de(t)}{dt}$$

If

$$k_p e(t) + k_i \int e(t)dt + k_d \frac{de(t)}{dt} > u_{\max}$$

$$u(t) = \text{sat}_{u_{\min}}^{u_{\max}} \left(\sum_{j=0}^{\varsigma} k_{pj} e^{\frac{2j+1}{2\beta+1}} + \sum_{j=0}^{\sigma} k_{ij} \frac{1}{s} e^{\frac{2j+1}{2\mu+1}} + \sum_{j=0}^{\eta} k_{dj} se^{\frac{2j+1}{2\gamma+1}} \right)$$

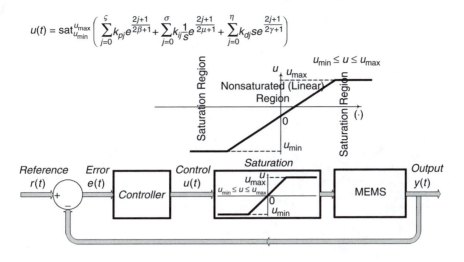

FIGURE 9.8
Closed-loop MEMS with the bounded control, $u_{\min} \le u \le u_{\max}$.

the control is bounded, and $u(t) = u_{\max}$. For

$$k_p e(t) + k_i \int e(t)dt + k_d \frac{de(t)}{dt} < u_{\min}$$

we have $u(t) = u_{\min}$. Thus, the control is bounded.

Other PID-type nonlinear and bounded controllers are

$$u(t) = \text{sat}_{u_{\min}}^{u_{\max}} \left(\sum_{j=1}^{2N_p - 1} k_{p(2j-1)} e^{2j-1}(t) + \sum_{j=1}^{2N_i - 1} k_{i(2j-1)} \frac{e(t)}{s^{2j-1}} + \sum_{j=1}^{2N_d - 1} k_{d(2j-1)} \frac{d^{2j-1} e(t)}{dt^{2j-1}} \right)$$

$$u(t) = \text{sat}_{u_{\min}}^{u_{\max}} \left(\sum_{j=0}^{\varsigma} k_{pj} e^{\frac{2j+1}{2\beta+1}} + \sum_{j=0}^{\sigma} k_{ij} \frac{1}{s} e^{\frac{2j+1}{2\mu+1}} + \sum_{j=0}^{\eta} k_{dj} se(t)^{\frac{2j+1}{2\gamma+1}} \right)$$

where $u_{\min} \le u \le u_{\max}$.

9.2.2 Control of a Microsystem with Permanent-Magnet DC Micromotor Using PID Controller

Consider a MEMS that integrates a permanent-magnet DC micromotor that actuates rotational microstage (see Figure 9.9). This geared micromotor (with planetary micro-gearhead fabricated utilizing MEMS technology) is directly attached to the microstage. Our goal is to design the control law in order to attain the desired performance (fast displacement of the stage, no tracking error, etc.). The stage angular displacement is a function of the rotor displacement, and taking into account the microgear ratio k_{gear}, one obtains the output equation as $y(t) = k_{gear} \theta_r(t)$. To change the angular velocity and displacement, one regulates the voltage applied to the armature winding u_a. The analog PID

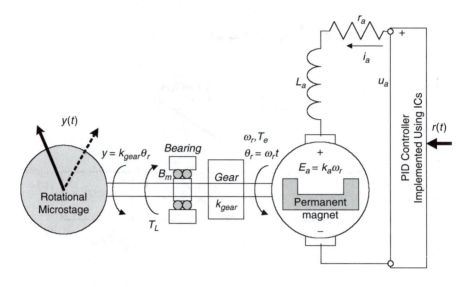

FIGURE 9.9
Schematic diagram of a microsystem with DC micromotor.

control law should be designed, and the feedback coefficients must be found. The rated armature voltage for the micromotor is $\pm u_{max}$ V. The rated (maximum) current is i_{max}, and the maximum angular velocity is ω_{max}. The bounds and parameters must be assigned. For a 2-mm-diameter actuator, we have u_{max} = 30 V, i_{max} = 0.015 A, ω_{max} = 200 rad/sec, r_a = 200 ohm, L_a= 0.002 H, k_a = 0.2 V · sec/rad, k_a = 0.2 N · m/A, J = 0.00000002 kg · m², and B_m = 0.00000005 N · m · sec/rad. The reduction microgear ratio is 100 : 1.

The mathematical model of permanent-magnet DC micromotors was developed in Chapter 6. For the studied MEMS, we have the following differential equations of motion:

$$\frac{di_a}{dt} = -\frac{r_a}{L_a}i_a - \frac{k_a}{L_a}\omega_r + \frac{1}{L_a}u_a$$

$$\frac{d\omega_r}{dt} = \frac{1}{J}(T_e - T_{viscous} - T_L) = \frac{1}{J}(k_a i_a - B_m\omega_r - T_L)$$

$$\frac{d\omega_r}{dt} = \omega_r$$

Using the Laplace operator $s = d/dt$, one obtains the following equations in the s-domain:

$$\left(s+\frac{r_a}{L_a}\right)i_a(s) = -\frac{k_a}{L_a}\omega_r(s) + \frac{1}{L_a}u_a(s), \quad \left(s+\frac{B_m}{J}\right)\omega_r(s) = \frac{1}{J}k_a i_a(s) - \frac{1}{J}T_L(s),$$

$$s\theta_r(s) = \omega_r(s)$$

These equations allow us to design the s-domain block diagram. In particular, making use of the output equation, $y(t) = k_{gear}\theta_r(t)$, $y(s) = k_{gear}\theta_r(s)$, one obtains the block diagram of the open-loop MEMS as documented in Figure 9.10.

MEMS with permanent-magnet DC micromotor
Permanent-magnet DC micromotor

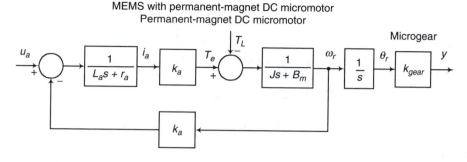

FIGURE 9.10
Block diagram of the open-loop MEMS.

The transfer functions for open-loop MEMS is

$$G_{sys}(s) = \frac{Y(s)}{U(s)} = \frac{k_{gear}k_a}{s\left(L_aJs^2 + (r_aJ + L_aB_m)s + r_aB_m + k_a^2\right)}$$

Using analog linear PID controller

$$u_a(t) = k_p e(t) + k_i \frac{e(t)}{s} + k_d se(t)$$

we have

$$G_{PID}(s) = \frac{U(s)}{E(s)} = \frac{k_d s^2 + k_p s + k_i}{s}$$

The closed-loop block diagram is documented in Figure 9.11.
    The closed-loop transfer function is found to be

$$G(s) = \frac{Y(s)}{R(s)} = \frac{G_{sys}(s)G_{PID}(s)}{1 + G_{sys}(s)G_{PID}(s)} = \frac{k_{gear}k_a(k_d s^2 + k_p s + k_i)}{s^2(L_aJs^2 + (r_aJ + L_aB_m)s + r_aB_m + k_a^2) + k_{gear}k_a(k_d s^2 + k_p s + k_i)}$$

$$= \frac{\dfrac{k_d}{k_i}s^2 + \dfrac{k_p}{k_i}s + 1}{\dfrac{L_aJ}{k_{gear}k_a k_i}s^4 + \dfrac{(r_aJ + L_aB_m)}{k_{gear}k_a k_i}s^3 + \dfrac{\left(r_aB_m + k_a^2 + k_{gear}k_a k_d\right)}{k_{gear}k_a k_i}s^2 + \dfrac{k_p}{k_i}s + 1}.$$

We use the following micromotor parameters: r_a = 200 ohm, L_a= 0.002 H, k_a = 0.2 V · sec/rad, k_a = 0.2 N · m/A, J = 0.00000002 kg · m², and B_m= 0.00000005 N · m · sec/rad. The rated armature voltage for the micromotor studied is ±30 V. The reduction gear ratio is 100 : 1. The numerical values of the numerator and denominator coefficients in the MEMS transfer function

$$G_{sys}(s) = \frac{Y(s)}{U(s)} = \frac{k_{gear}k_a}{s\left(L_aJs^2 + (r_aJ + L_aB_m)s + r_aB_m + k_a^2\right)}$$

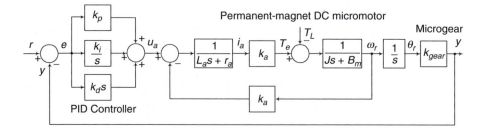

FIGURE 9.11
Block diagram of the closed-loop MEMS with PID controller.

are found using the following MATLAB script:

```
% MEMS parameters
ra=200; La=0.002; ka=0.2; J=0.00000002; Bm=0.00000005;
kgear=0.01;
% Numerator and denominator of the open-loop MEMS
% transfer function
format short e
num_s=[ka*kgear]; den_s=[La*J ra*J+La*Bm ra*Bm+ka^2 0];
num_s, den_s
```

Using the following numerical results obtained in the Command Window

```
num_s =
    2.0000e-003
den_s =
    4.0000e-011   4.0001e-006   4.0010e-002   0
```

We have the open-loop MEMS transfer function as

$$G_{sys}(s) = \frac{Y(s)}{U(s)} = \frac{2 \times 10^{-3}}{s(4 \times 10^{-11}s^2 + 4 \times 10^{-6}s + 4 \times 10^{-2})}$$

The open-loop system is unstable because the eigenvalues are found to be positive. In particular,

```
>> Eigenvalues=roots(den_s)
Eigenvalues =
             0
   -8.8729e+004
   -1.1273e+004
```

To stabilize the studied MEMS and attain the desired dynamic performance, control algorithms must be designed. The characteristic equation of the closed-loop transfer function with PID controller (Equation 9.1) is

$$\frac{L_a J}{k_{gear} k_a k_i} s^4 + \frac{(r_a J + L_a B_m)}{k_{gear} k_a k_i} s^3 + \frac{\left(r_a B_m + k_a^2 + k_{gear} k_a k_d\right)}{k_{gear} k_a k_i} s^2 + \frac{k_p}{k_i} s + 1 = 0$$

It is obvious that the proportional k_p, integral k_i, and derivative k_d feedback coefficients of the controller

$$u_a(t) = k_p e(t) + k_i \frac{e(t)}{s} + k_d s e(t)$$

influence the location of the eigenvalues. Let $k_p = 25000$, $k_i = 250$, and $k_d = 25$. Then, the PID controller is

$$u_a(t) = 25000 e(t) + 250 \int e(t) dt + 25 \frac{de(t)}{dt}$$

The particular location of the characteristic eigenvalues of the closed-loop system results. To derive the eigenvalues values, the following MATLAB script is developed and used:

```
% MEMS parameters
ra=200; La=0.002; ka=0.2; J=0.00000002; Bm=0.00000005;
kgear=0.01;
%   Feedback coefficients
kp=25000; ki=250; kd=25;
% Denominator of the closed-loop MEMS transfer function
den_c=[(La*J)/(kgear*ka*ki) (ra*J+La*Bm)/...
(kgear*ka*ki) (ra*Bm+ka^2+kgear*ka*kd)/(kgear*ka*ki) kp/ki 1];
% Eigenvalues of the closed-loop MEMS
Eigenvalues_Closed_Loop=roots(den_c)
```

The numerator and denominator of the closed-loop system are found. The resulting eigenvalues of the closed-loop system

```
Eigenvalues_Closed_Loop =
  -6.6393e+004
  -3.3039e+004
  -5.6983e+002
  -1.0000e-002
```

allow us to conclude that the closed-loop system is stable because the real parts are negative. Since there are no imaginary parts, it is evident that there is no overshoot (overshoot is usually an undesirable phenomenon in the high-performance positioning systems). The closest eigenvalue is –0.01, and the settling time can be evaluated to be less than 0.04 sec (four times the maximum time constant). The transient dynamics is our particular interest to make the final conclusions regarding the closed-loop MEMS performance.

The following MATLAB script allows the user to simulate the closed-loop MEMS:

```
% MEMS parameters
ra=200; La=0.002; ka=0.2; J=0.00000002; Bm=0.00000005;
kgear=0.01;
%   Feedback coefficients
kp=25000; ki=250; kd=25;
ref=1.5; % reference (command) input is 1.5 rad
```

```
% Numerator and denominator of the closed-loop transfer function
num_c=[kd/ki kp/ki 1];
den_c=[(La*J)/(kgear*ka*ki) (ra*J+La*Bm)/...
(kgear*ka*ki) (ra*Bm+ka^2+kgear*ka*kd)/(kgear*ka*ki) kp/ki 1];
t=0:0.0001:0.025;
u=ref*ones(size(t));
y=lsim(num_c,den_c,u,t);
plot(t,y,'-',y,u,':');
title('MEMS Output and Reference Evolutions');
xlabel('Time (seconds)');
ylabel('Output y(t) and Reference r(t)');
axis([0 0.025,0 2])   % axis limits
```

The closed-loop MEMS output (angular displacement) and reference $r(t)$ is illustrated in Figure 9.12 if $r(t) = 1.5$ rad.

The feedback gains significantly influence the stability and dynamics. Let us reduce the proportional gain k_p and increase the integral feedback k_i keeping the derivative feedback coefficients as before. In particular, let $k_p = 2500$, $k_i = 25000$, and $k_d = 25$. We calculate the eigenvalues and simulate the MEMS performance using the following MATLAB script:

```
% MEMS parameters
ra=200; La=0.002; ka=0.2; J=0.00000002; Bm=0.00000005;
kgear=0.01;
%  Feedback coefficients
kp=2500; ki=25000; kd=25;
```

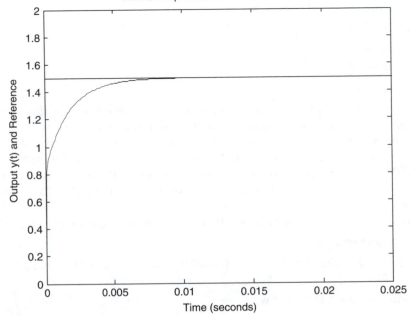

FIGURE 9.12
Dynamics of the closed-loop MEMS with analog PID controller.

```
ref=1.5; % reference (command) input is 1.5 rad
% Numerator and denominator of the closed-loop transfer function
num_c=[kd/ki kp/ki 1];
den_c=[(La*J)/(kgear*ka*ki) (ra*J+La*Bm)/...
(kgear*ka*ki) (ra*Bm+ka^2+kgear*ka*kd)/(kgear*ka*ki) kp/ki 1];
t=0:0.0001:0.25;
u=ref*ones(size(t));
y=lsim(num_c,den_c,u,t);
plot(t,y,'-',y,u,':');
title('MEMS Output and Reference Evolutions');
xlabel('Time (seconds)');
ylabel('Output y(t) and Reference r(t)');
axis([0 0.25,0 2])   % axis limits
% Denominator of the closed-loop MEMS transfer function
den_c = [(La*J)/(kgear*ka*ki) (ra*J+La*Bm)/...
(kgear*ka*ki) (ra*Bm+ka^2+kgear*ka*kd)/(kgear*ka*ki) kp/ki 1];
% Eigenvalues of the closed-loop MEMS
Eigenvalues_Closed_Loop=roots(den_c)
```

The characteristic eigenvalues are

```
Eigenvalues_Closed_Loop =
 -6.5869e+004
 -3.4078e+004
 -4.2586e+001
 -1.3076e+001
```

The time constant is significantly increased, and, hence, the settling time is increased. The MEMS dynamics if $r(t)$ = 1.5 rad is documented in Figure 9.13.

It was illustrated that one must properly design the PID controller feedback gains. Different approaches can be used to simulate MEMS in the MATLAB environment. The application of SIMULINK is illustrated. The resulting diagram to perform simulations is documented in Figure 9.14. In this diagram, the PID Controller building block is used. The system parameters and feedback gain coefficients must be downloaded in the Command Window. In particular,

```
% MEMS parameters
ra=200; La=0.002; ka=0.2; J=0.00000002; Bm=0.00000005;
kgear=0.01;
%  Feedback coefficients
kp=25000; ki=250; kd=25;
```

Running the mdl-file (clicking the *Simulation* button), the simulation results become available. To plot the state variables $x_1(t)$ and $x_2(t)$ as well as the output angular displacement $y(t) = k_{gear}\theta_r(t)$ and applied voltage, one types in the Command window

```
>> plot(x1(:,1),x1(:,2));
title('Armature Current, [A]'); xlabel('Time, [sec]');
>> plot(x2(:,1),x2(:,2));
title('Angular Velocity, [rad/sec]');xlabel('Time, [sec]');
```

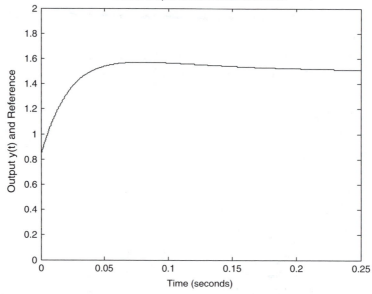

FIGURE 9.13
Closed-loop MEMS, $k_p = 2500$, $k_i = 25,000$, and $k_d = 25$.

```
>> plot(y(:,1),y(:,2));
title('Output: Angular Displacement, [rad]'); xlabel('Time, [sec]');
>> plot(ua(:,1),ua(:,2));
title('Applied Armature Voltage [V]');
xlabel('Time, [sec]');
```

Here, $x_1(t)$ denotes the armature current $i_a(t)$, and $x_2(t)$ corresponds to the angular velocity $\omega_r(t)$. The transient dynamics of the microsystem variables (armature current x_1 and angular velocity x_2) and the output (angular displacement y) are documented in Figures 9.15.

The analysis of the transient dynamics indicates that the settling time is 0.01 sec and there is no overshoot. Though the closed-loop MEMS is stable, and one may tentatively conclude that the desired performance is reached, many things are incorrect. The peak value for the current is 140 A (rated current is 150 mA), and the peak value for the angular velocity is 80,000 rad/sec (the rated angular velocity is 150 rad/sec). The reason is simple: the armature voltage that was supposed to be supplied to the windings reached 4000 V. This is completely unacceptable. The questions are, What is wrong with the design? Where were the mistakes made? What must be changed?

The answer is that the control bound and state constraints must be integrated in the modeling, simulations, and design. The rated armature voltage is ±30 V, i.e., $-30 \leq u_a \leq 30$ V.

Correspondingly, the bounded PID controller is

$$u_a(t) = \text{sat}_{-30}^{+30}\left(25000e(t) + 250\int e(t)dt + 25\frac{de(t)}{dt}\right)$$

Integrating these bounds, the simulation is performed in SIMULINK. The SIMULINK model is built utilizing the `Saturation` block as illustrated in Figure 9.16.

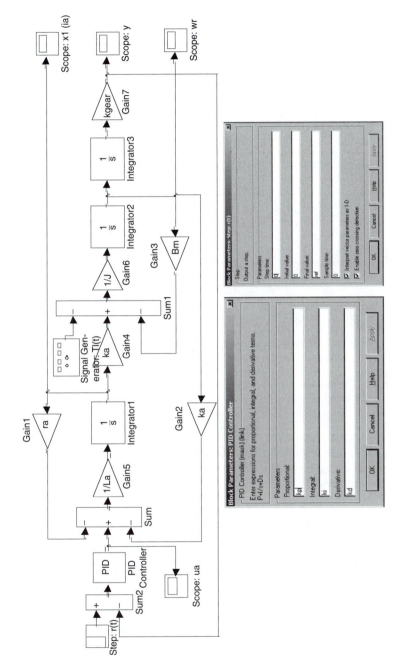

FIGURE 9.14

SIMULINK diagram to model MEMS built using the Step, Sum, PID Controller, Game, Integration, Signal Generator, and Scope building blocks from the SIMULINK block library (ch9_1.mdl).

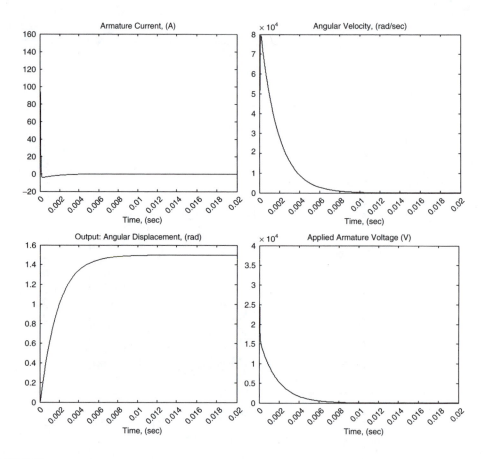

FIGURE 9.15
Dynamics of the closed-loop MEMS with PID controller, $k_p = 25{,}000$, $k_i = 250$, and $k_d = 25$.

For the reference angular displacement $r(t) = 1.5$ rad, the resulting states and output transient responses are documented in Figure 9.17 if $T_L = 0\,\text{N} \cdot \text{m}$. The evolution of the state variables, the output, and the bounded voltage are illustrated in Figure 9.17. The control bounds $-30 \leq u_a \leq 30$ are downloaded as

```
>> Umax=30;
```

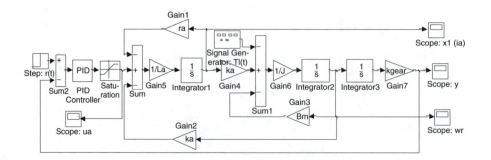

FIGURE 9.16
SIMULINK diagram of the closed-loop MEMS with saturation (`ch9_2.mdl`).

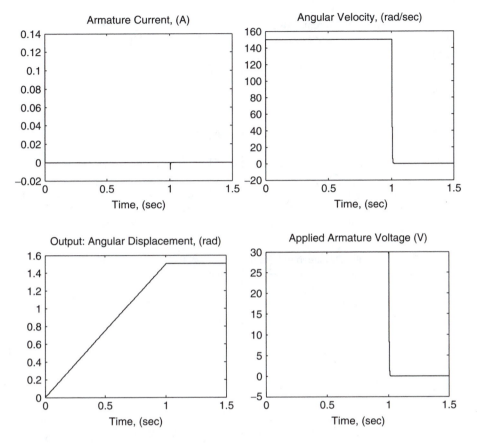

FIGURE 9.17

Dynamics of the closed-loop MEMS with bounded PID controller ($k_p = 25,000$, $k_i = 250$, and $k_d = 25$) if $r(t) = 1.5$ rad.

The comparison of the simulation results, reported in Figure 9.15 and Figure 9.17, provides one with the evidence that the imposed constraints significantly increase the settling time. However, the bounds imposed have been guaranteed. For the reference angular displacement $r(t) = 0.015$ rad, the resulting states and output transient responses in the studied bounded MEMS ($-30 \leq u_a \leq 30$ V) are reported in Figure 9.18.

That is, the reference (command) input significantly influences the settling time and MEMS evolution. For $r(t) = 0.015$ rad, the settling time is 0.015 sec with no overshoot. One concludes that system nonlinearities must be integrated and used to attain accurate data-intensive analysis. Figure 9.19 demonstrates some build-in nonlinear blocks that are available from the SIMULINK block library. These blocks must be integrated as one incorporates the friction, backlash, dead zone, and other nonlinear phenomena in the system studied.

9.3 The Hamilton-Jacobi Theory and Optimal Control of MEMS and NEMS

We examined and demonstrated the application of the PID-type controllers to attain the desirable features of closed-loop microsystems. This section reports different methods to design control algorithms minimizing performance functionals, which depend on the

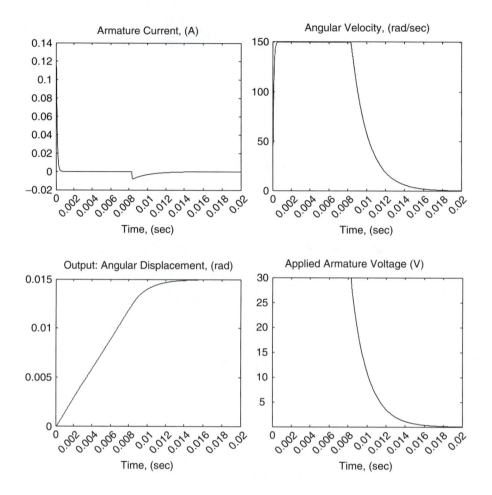

FIGURE 9.18
Dynamics of the closed-loop MEMS with bounded PID controller (k_p = 25,000, k_i = 250, and k_d = 25) if $r(t)$ = 0.015 rad.

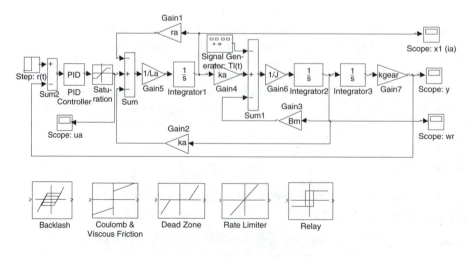

FIGURE 9.19
SIMULINK diagram of the closed-loop MEMS with nonlinearities (`ch9_3.mdl`).

system variables (states x, outputs y, tracking errors e, and control u). These variables measure MEMS and NEMS performance. In general, the performance functional can be maximized. A general problem formulation in design of control laws using the Hamilton-Jacobi concept is to find the *admissible* (bounded or unbounded) time-invariant or time-varying control law as a nonlinear function of error $e(t)$ and state $x(t)$ vectors

$$u = \phi(t, e, x) \tag{9.5}$$

minimizing the performance functional

$$J(x(\cdot), y(\cdot), e(\cdot), u(\cdot)) = \int_{t_0}^{t_f} W_{xyeu}(x, y, e, u)dt \tag{9.6}$$

subject to the microsystem dynamics.

In the functional (Equation 9.6), $W_{xyeu}(\cdot):\mathbb{R}^n \times \mathbb{R}^b \times \mathbb{R}^b \times \mathbb{R}^m \to \mathbb{R}_{\geq 0}$ is the positive-definite and continuously differentiable integrand function synthesized and assigned by the designer, and t_0 and t_f are the initial and final time.

It was illustrated that MEMS and NEMS are described by linear and nonlinear differential equations. Furthermore, the control $u(t)$ (for example, applied voltage to the actuator windings, fed currents to the windings, or regulated signal to the comparator to control the converter duty ratio) is bounded. The performance functionals given by Equation 9.6, which play a very important role, have not been examined yet. Therefore, illustrative examples and explanation are needed. One may minimize the squired tracking error and control; that is, we obtain the following performance functional

$$J(e(\cdot), u(\cdot)) = \int_{t_0}^{t_f} (e^2 + u^2)dt$$

to be minimized. The quadratic performance integrand $W_{eu}(e, u) = e^2 + u^2$ is used. The performance functional must be positive definite, and the performance integrand is positive definite. For example, $W_{eu}(e, u) = |e| + |u|$, $W_{eu}(e, u) = |e|e^2 + |u|u^2$, or $W_{eu}(e, u) = |e|e^4 + |u|u^6$.

One can design and minimize a great variety of functionals when attempting to achieve better performance. For example, nonquadratic functionals

$$J(e(\cdot), u(\cdot)) = \int_{t_0}^{t_f} (e^8 + u^6)dt$$

or

$$J(e(\cdot), u(\cdot)) = \int_{t_0}^{t_f} \left(|e| + e^2 u^4 + |u|u^{\frac{4}{3}} + u^4\right)dt$$

may be utilized. The question is, Why are the quadratic functionals with the squired quantity commonly used? The reason is that the application of quadratic performance functionals guarantee the simplicity because design problems involving squired terms are straightforwardly analytical and numerically solvable. In general, the application of nonquadratic integrands results in mathematical complexity. However, the system performance is improved as nonlinear performance functionals are used to synthesize control laws.

Once the optimization problem is formulated, and the performance functional (Equation 9.6) is chosen by synthesizing the integrand functions, the design involves the straightforward minimization or maximization problem using the Hamilton-Jacobi concept, dynamic programming, maximum principle, variational calculus, nonlinear programming, or other paradigms. The ultimate objective is to design control algorithms (see Equation 9.5) to attain optimal dynamic responses.

The microsystem dynamics is modeled by nonlinear differential equations

$$\dot{x}(t) = F(x) + B(x)u \,,\, x(t_0) = x_0 \tag{9.7}$$

To find an optimal control, the necessary conditions for optimality must be studied.

The Hamiltonian function is

$$H\left(x,u,\frac{\partial V}{\partial x}\right) = W_{xu}(x,u) + \left(\frac{\partial V}{\partial x}\right)^T (F(x) + B(x)u)$$

where $V(\cdot):\mathbb{R}^n \to \mathbb{R}_{\geq 0}$ is the smooth and bounded return function, $V(0) = 0$.

The controller (Equation 9.5) is found using the first-order necessary condition for optimality

(n1)
$$\frac{\partial H\left(x,u,\dfrac{\partial V}{\partial x}\right)}{\partial u} = 0$$

The second-order necessary condition for optimality that needs to be satisfied is

(n2)
$$\frac{\partial^2 H\left(x,u,\dfrac{\partial V}{\partial x}\right)}{\partial u \times \partial u^T} > 0$$

Example 9.1

Consider the micromass assuming that the applied force is a control variable u, and the velocity is the state variable x. That is, from $\dot{x}(t) = (1/m)\sum F$, taking into account the viscous friction, the MEMS dynamics may be modeled as

$$\dot{x}(t) = \frac{1}{m}\sum F = \frac{1}{m}(F_e - B_m x)$$

or

$$\dot{x}(t) = ax + bu,$$

where $a = -B_m/m$ and $b = 1/m$.

This first-order differential equation models the input-output dynamics of the MEMS under consideration. Using the quadratic functional

$$J(x(\cdot), u(\cdot)) = \tfrac{1}{2} \int_{t_0}^{t_f} (qx^2 + gu^2) dt, \quad q \geq 0, g > 0$$

find a control law. Here, the weighting coefficients are denoted as q and g.

SOLUTION

The Hamiltonian function is

$$H\left(x, u, \frac{\partial V}{\partial x}\right) = \underbrace{\tfrac{1}{2}(qx^2 + gu^2)}_{\substack{\text{Performance Functional} \\ J(x,u)=\int_{t_0}^{t_f}(qx^2+gu^2)dt}} + \frac{\partial V}{\partial x} \underbrace{\frac{dx}{dt}}_{\substack{\text{System Dynamics} \\ \dot{x}(t)=ax+bu}} = \tfrac{1}{2}(qx^2 + gu^2) + \frac{\partial V}{\partial x}(ax + bu)$$

The Hamiltonian function is minimized using the first-order necessary condition for optimality (n1). Specifically, making use of

$$\frac{\partial H\left(x, u, \dfrac{\partial V}{\partial x}\right)}{\partial u} = 0$$

we obtain

$$gu + \frac{\partial V}{\partial x} b = 0$$

From this equation, the controller is

$$u = -\frac{b}{g} \frac{\partial V}{\partial x} = -g^{-1} b \frac{\partial V}{\partial x}$$

Let the return function $V(x)$ be given in the quadratic form, i.e., $V(x) = \tfrac{1}{2} kx^2$.

Then, the controller is

$$u = -g^{-1} bkx$$

The unknown coefficient k is found by solving the Riccati equation as will be illustrated later.

Making use of the system dynamics $\dot{x}(t) = ax + bu$ and the controller $u = -g^{-1} bkx$, the closed-loop system is expressed in the form

$$\dot{x}(t) = (a - g^{-1} b^2 k)x$$

The closed-loop system is stable if $a - g^{-1} b^2 k < 0$.

This condition for stability will be guaranteed.

9.3.1 Stabilization Problem for Linear MEMS and NEMS

The linear control law was synthesized in Example 9.1 for the first-order linear dynamic system. Here, we will generalize the results. Consider a linear time-invariant MEMS and NEMS described by the linear differential equations

$$\dot{x}(t) = Ax + Bu , x(t_0) = x_0 \tag{9.8}$$

where $A \in \mathbb{R}^{n \times n}$ and $B \in \mathbb{R}^{n \times m}$ are the constant-coefficient matrices.

Making use of the quadratic integrands, the quadratic performance functional is found to be

$$J(x(\cdot), u(\cdot)) = \tfrac{1}{2} \int_{t_0}^{t_f} (x^T Q x + u^T G u) dt, Q \geq 0, \quad G > 0 \tag{9.9}$$

where $Q \in \mathbb{R}^{n \times n}$ is the positive semidefinite constant-coefficient weighting matrix, and $G \in \mathbb{R}^{m \times m}$ is the positive-definite constant-coefficient weighting matrix.

From Equation 9.8 and Equation 9.9, the Hamiltonian function is found to be

$$H\left(x, u, \frac{\partial V}{\partial x}\right) = \tfrac{1}{2}(x^T Q x + u^T G u) + \left(\frac{\partial V}{\partial x}\right)^T (Ax + Bu) \tag{9.10}$$

The Hamilton-Jacobi functional equation is

$$-\frac{\partial V}{\partial t} = \min_{u} \left[\tfrac{1}{2}(x^T Q x + u^T G u) + \left(\frac{\partial V}{\partial x}\right)^T (Ax + Bu) \right]$$

The derivative of the Hamiltonian function $H(x, u, \partial V/\partial x)$ exists, and control function $u(\cdot):[t_0, t_f) \rightarrow \mathbb{R}^m$ is found by using the first-order necessary condition

$$\frac{\partial H\left(x, u, \frac{\partial V}{\partial x}\right)}{\partial u} = 0$$

From

$$\frac{\partial H\left(x, u, \frac{\partial V}{\partial x}\right)}{\partial u} = u^T G + \left(\frac{\partial V}{\partial x}\right)^T B$$

one finds the following optimal controller:

$$u = -G^{-1}B^T \frac{\partial V}{\partial x} \tag{9.11}$$

This optimal controller is found by minimizing the quadratic performance functional as given by Equation 9.9.

The second-order necessary condition for optimality (n2)

$$\frac{\partial^2 H\left(x, u, \frac{\partial V}{\partial x}\right)}{\partial u \times \partial u^T} > 0$$

is guaranteed. In fact, the weighting matrix G is positive definite, and, thus,

$$\frac{\partial^2 H\left(x, u, \frac{\partial V}{\partial x}\right)}{\partial u \times \partial u^T} = G > 0$$

Substituting the controller (Equation 9.11) into Equation 9.10, we obtain the partial differential equation

$$-\frac{\partial V}{\partial t} = \frac{1}{2}\left(x^T Q x + \left(\frac{\partial V}{\partial x}\right)^T BG^{-1}B^T \frac{\partial V}{\partial x}\right) + \left(\frac{\partial V}{\partial x}\right)^T Ax - \left(\frac{\partial V}{\partial x}\right)^T BG^{-1}B^T \frac{\partial V}{\partial x}$$

$$= \frac{1}{2}x^T Q x + \left(\frac{\partial V}{\partial x}\right)^T Ax - \frac{1}{2}\left(\frac{\partial V}{\partial x}\right)^T BG^{-1}B^T \frac{\partial V}{\partial x} \tag{9.12}$$

This equation must be solved. Equation 9.12 is satisfied by the quadratic return function

$$V(x) = \frac{1}{2}x^T K(t)x \tag{9.13}$$

where $K \in \mathbb{R}^{n \times n}$ is the symmetric matrix.
    The unknown matrix

$$K = \begin{bmatrix} k_{11} & k_{12} & \cdots & k_{1n-1} & k_{1n} \\ k_{21} & k_{22} & \cdots & k_{2n-1} & k_{2n} \\ \vdots & \vdots & \ddots & \vdots & \vdots \\ k_{n-11} & k_{n-12} & \cdots & k_{n-1n-1} & k_{n-1n} \\ k_{n1} & k_{n2} & \cdots & k_{nn-1} & k_{nn} \end{bmatrix}, \quad k_{ij} = k_{ji}$$

must be positive definite because positive semidefinite and positive-definite constant-coefficient weighting matrices Q and G have been used in the performance functional (Equation 9.9). The positive definiteness of the quadratic return function $V(x)$ can be verified using the Sylvester criterion.
    Taking note of Equation 9.13 and using the matrix identity $x^T KAx = \frac{1}{2}x^T(A^T K + KA)x$, from Equation 9.12, one has

$$-\frac{\partial\left(\frac{1}{2}x^T Kx\right)}{\partial t} = \frac{1}{2}x^T Q x + \frac{1}{2}x^T A^T Kx + \frac{1}{2}x^T KAx - \frac{1}{2}x^T KBG^{-1}B^T Kx \tag{9.14}$$

Using the boundary conditions

$$V(t_f, x) = \frac{1}{2} x^T K(t_f) x = \frac{1}{2} x^T K_f x \qquad (9.15)$$

the following nonlinear differential equation (Riccati equation) must be solved to find the unknown symmetric matrix K:

$$-\dot{K} = Q + A^T K + K^T A - K^T B G^{-1} B^T K, \ K(t_f) = K_f \qquad (9.16)$$

From Equation 9.11 and Equation 9.13, the controller is

$$u = -G^{-1} B^T K x \qquad (9.17)$$

The feedback gain matrix is found to be

$$K_F = -G^{-1} B^T K$$

Augmenting Equation 9.8 and Equation 9.17, we have the closed-loop microsystem as given by

$$\dot{x}(t) = Ax + Bu = Ax - BG^{-1}B^T Kx = (A - BG^{-1}B^T K)x = (A - BK_F)x \qquad (9.18)$$

The eigenvalues of the matrix $(A - BG^{-1}B^T K) = (A - BK_F) \in \mathbb{R}^{n \times n}$ have negative real parts. If $t_f = \infty$ in the performance functional (Equation 9.9), matrix K is found as

$$0 = -Q - A^T K - K^T A + K^T B G^{-1} B^T K$$

The Riccati equation solver lqr is available in MATLAB. In particular,

```
>> help lqr
  LQR  Linear-quadratic regulator design for continuous-time systems.
  [K,S,E] = LQR(A,B,Q,R,N)  calculates the optimal gain matrix K
  such that the state-feedback law  u = -Kx  minimizes the cost function
     J = Integral {x'Qx + u'Ru + 2*x'Nu} dt
  subject to the state dynamics  x = Ax + Bu.
  The matrix N is set to zero when omitted.  Also returned are the
  Riccati equation solution S and the closed-loop eigenvalues E:
                    -1
  SA + A'S - (SB+N)R  B'S+N') + Q = 0 ,     E = EIG(A-B*K).
  See also  LQRY, DLQR, LQGREG, CARE, and REG.
```

That is, using the lqr function, one finds the feedback gain matrix K_F, matrix K, and eigenvalues of the closed-loop system.

Example 9.2

Consider the system studied in Example 9.1 assuming that $m = 1$. The viscous friction can be neglected. That is, the motion microdevice is model as

$$dx / dt = u.$$

Find the control law minimizing the performance functional

$$J(x(\cdot), u(\cdot)) = \frac{1}{2} \int_0^{t_f} (x^2 + u^2) dt$$

Apply MATLAB to solve the problem for the functional

$$J(x(\cdot), u(\cdot)) = \frac{1}{2} \int_0^{\infty} (x^2 + u^2) dt$$

Examine the system stability.

SOLUTION

From $dx / dt = u$, one finds that in the model $\dot{x}(t) = Ax + Bu$, the matrices A, B, Q, and G are

$$A = [0], \quad B = [1], \quad Q = [1], \quad \text{and } G = [1]$$

The matrix K is unknown. From Equation 9.16 we have

$$-\dot{K}(t) = 1 - K^2(t), \quad K(t_f) = 0$$

The analytic solution of this nonlinear differential equation is

$$K(t) = \frac{1 - e^{-2(t_f - t)}}{1 + e^{-2(t_f - t)}}$$

Hence, an optimal control law that guarantees the minimum of the quadratic functional

$$J(x(\cdot), u(\cdot)) = \frac{1}{2} \int_0^{t_f} (x^2 + u^2) dt$$

with respect to the system dynamics $dx / dt = u$ is found using Equation 9.17 as

$$u(t) = -\frac{1 - e^{-2(t_f - t)}}{1 + e^{-2(t_f - t)}} x$$

Using the lqr function in MATLAB, one finds the feedback, return function coefficients and eigenvalues if $t_f = \infty$. That is, minimizing functional

$$J(x(\cdot), u(\cdot)) = \frac{1}{2} \int_0^{\infty} (x^2 + u^2) dt$$

we have the following numerical results in the controller design, feedback gain, and eigenvalue:

```
>> [K_feedback,K,Eigenvalues] = lqr(0,1,1,1,0)
K_feedback =
      1
```

```
K =

     1
Eigenvalues =

     -1.0000
```

Since the eigenvalue has a negative real part, the closed-loop system is stable.

One also concludes that these numerical results correspond to the analytic results documented.

Example 9.3

For the second-order microsystem (assuming that the states are the velocity and displacement, and there is no viscous friction)

$$\frac{dx_1}{dt} = x_2, \quad \frac{dx_2}{dt} = u$$

find a control law minimizing the quadratic functional

$$J(x(\cdot), u(\cdot)) = \frac{1}{2} \int_0^\infty \left(x_1^2 + q_{22} x_2^2 + g u^2 \right) dt, \quad q_{22} \geq 0, g > 0$$

Letting $q_{22} = 10$ and $g = 100$, apply MATLAB to design the controller. Express the closed-loop system using differential equations.

SOLUTION

Using the state-space notations, we express the system evolution matrix differential equation $\dot{x}(t) = Ax + Bu$ as

$$\begin{bmatrix} \dot{x}_1 \\ \dot{x}_2 \end{bmatrix} = \begin{bmatrix} 0 & 1 \\ 0 & 0 \end{bmatrix} \begin{bmatrix} x_1 \\ x_2 \end{bmatrix} + \begin{bmatrix} 0 \\ 1 \end{bmatrix} u, \quad A = \begin{bmatrix} 0 & 1 \\ 0 & 0 \end{bmatrix}, \quad B = \begin{bmatrix} 0 \\ 1 \end{bmatrix}$$

The performance functional is

$$J(x(\cdot), u(\cdot)) = \frac{1}{2} \int_0^\infty \left(\begin{bmatrix} x_1 & x_2 \end{bmatrix} \begin{bmatrix} 1 & 0 \\ 0 & q_{22} \end{bmatrix} \begin{bmatrix} x_1 \\ x_2 \end{bmatrix} + g u^2 \right) dt, \quad Q = \begin{bmatrix} 1 & 0 \\ 0 & q_{22} \end{bmatrix},$$

$$G = g, \quad q_{22} \geq 0, \quad g > 0$$

Using the quadratic return function (Equation 9.15)

$$V(x) = \frac{1}{2} k_{11} x_1^2 + k_{12} x_1 x_2 + \frac{1}{2} k_{22} x_2^2 = \frac{1}{2} \begin{bmatrix} x_1 & x_2 \end{bmatrix} \begin{bmatrix} k_{11} & k_{12} \\ k_{21} & k_{22} \end{bmatrix} \begin{bmatrix} x_1 \\ x_2 \end{bmatrix}, \quad k_{12} = k_{21}$$

the controller (Equation 9.17) is

$$u = -G^{-1} B^T Kx = -g^{-1} \begin{bmatrix} 0 & 1 \end{bmatrix} \begin{bmatrix} k_{11} & k_{12} \\ k_{21} & k_{22} \end{bmatrix} \begin{bmatrix} x_1 \\ x_2 \end{bmatrix} = -\frac{1}{g} (k_{21} x_1 + k_{22} x_2)$$

The unknown matrix

$$K = \begin{bmatrix} k_{11} & k_{12} \\ k_{21} & k_{22} \end{bmatrix}$$

is found by solving the Riccati equation (Equation 9.16).

For the example studied, this matrix Riccati equation is

$$-Q - A^T K - K^T A + K^T B G^{-1} B^T K$$

$$= -\begin{bmatrix} 1 & 0 \\ 0 & q_{22} \end{bmatrix} - \begin{bmatrix} 0 & 0 \\ 1 & 0 \end{bmatrix}\begin{bmatrix} k_{11} & k_{12} \\ k_{21} & k_{22} \end{bmatrix} - \begin{bmatrix} k_{11} & k_{12} \\ k_{21} & k_{22} \end{bmatrix}\begin{bmatrix} 0 & 1 \\ 0 & 0 \end{bmatrix}$$

$$+ \begin{bmatrix} k_{11} & k_{12} \\ k_{21} & k_{22} \end{bmatrix}\begin{bmatrix} 0 \\ 1 \end{bmatrix}g^{-1}\begin{bmatrix} 0 & 1 \end{bmatrix}\begin{bmatrix} k_{11} & k_{12} \\ k_{21} & k_{22} \end{bmatrix} = \begin{bmatrix} 0 & 0 \\ 0 & 0 \end{bmatrix}$$

Three algebraic equations to be solved are

$$\frac{k_{12}^2}{g} - 1 = 0, \quad -k_{11} + \frac{k_{12}k_{22}}{g} = 0, \quad \text{and} \quad -2k_{12} + \frac{k_{22}^2}{g} - q_{22} = 0$$

The solutions are

$$k_{12} = k_{21} = \pm\sqrt{g}, \quad k_{22} = \pm\sqrt{g(q_{22} + 2k_{12})}, \quad \text{and} \quad k_{11} = \frac{k_{12}k_{22}}{g}$$

The performance functional $J(x(\cdot), u(\cdot)) = \frac{1}{2}\int_0^\infty (x_1^2 + q_{22}x_2^2 + gu^2)dt$, $q_{22} \geq 0$, $g > 0$ is positive definite. Therefore,
we have

$$k_{11} = \sqrt{q_{22} + 2\sqrt{g}}, \quad k_{12} = k_{21} = \sqrt{g}, \quad \text{and} \quad k_{22} = \sqrt{g(q_{22} + 2\sqrt{g})}$$

Thus, the control law is

$$u = -\frac{1}{g}\left(\sqrt{g}x_1 + \sqrt{g(q_{22} + 2\sqrt{g})}x_2\right) = -\frac{1}{\sqrt{g}}x_1 - \sqrt{\frac{q_{22} + 2\sqrt{g}}{g}}x_2$$

Having received the analytic solution, let us derive the feedback gains and eigenvalues using MATLAB. Applying the lqr function, one obtains the feedback gains and return function coefficients, as well as computes the eigenvalues. Letting $q_{22} = 10$ and $g = 100$, we have

```
>> A=[0 1;0 0],B=[0;1],
[K_feedback,K,Eigenvalues] = lqr(A,B,[1 0;0 1],[10])
A =

        0        1
        0        0
```

```
B =
     0
     1
K_feedback =
     0.3162       0.8558
K =
     2.7064       3.1623
     3.1623       8.5584
Eigenvalues =
   -0.4279 + 0.3648i
   -0.4279 - 0.3648i
```

Hence,

$$K = \begin{bmatrix} k_{11} & k_{12} \\ k_{21} & k_{22} \end{bmatrix} = \begin{bmatrix} 2.71 & 3.16 \\ 3.16 & 8.56 \end{bmatrix}, \quad k_{11} = 2.71, \quad k_{12} = k_{21} = 3.16, \quad k_{22} = 8.56$$

and the controller is given as

$$u = -0.32x_1 - 0.86x_2$$

The analytical and numerical results have been obtained, and the stability of the closed-loop microsystem, modeled as

$$\frac{dx_1}{dt} = x_2, \quad \frac{dx_2}{dt} = -0.32x_1 - 0.86x_2$$

is guaranteed. In fact, the eigenvalues have negative real parts. In particular, we found that the eigenvalues are –0.428 + 0.365i and –0.428 – 0.365i.

Example 9.4

Consider the microsystem described by the following equations of motion:

State-space equation:

$$\dot{x} = Ax + Bu = \begin{bmatrix} -0.1 & 6 & -1 & -10 \\ 0 & 4 & 1 & 0 \\ 0 & 3 & -5 & 0 \\ 0 & 0 & 1 & 0 \end{bmatrix} \begin{bmatrix} x_1 \\ x_2 \\ x_3 \\ x_4 \end{bmatrix} + \begin{bmatrix} 1 & 0 \\ -5 & 1 \\ -2 & 2 \\ 0 & 0 \end{bmatrix} \begin{bmatrix} u_1 \\ u_2 \end{bmatrix}$$

Output equation:

$$y = \begin{bmatrix} 0 & 0 & 0 & 1 \end{bmatrix} \begin{bmatrix} x_1 \\ x_2 \\ x_3 \\ x_4 \end{bmatrix} + \begin{bmatrix} 0 & 0 \end{bmatrix} \begin{bmatrix} u_1 \\ u_2 \end{bmatrix}$$

Find the control law minimizing the quadratic performance functional

$$J(x(\cdot),u(\cdot)) = \frac{1}{2}\int_0^\infty (x^T Q x + u^T G u)dt$$

$$= \frac{1}{2}\int_0^\infty \left(\begin{bmatrix} x_1 & x_2 & x_3 & x_4 \end{bmatrix} \begin{bmatrix} 10 & 0 & 0 & 0 \\ 0 & 10 & 0 & 0 \\ 0 & 0 & 10 & 0 \\ 0 & 0 & 0 & 10 \end{bmatrix} \begin{bmatrix} x_1 \\ x_2 \\ x_3 \\ x_4 \end{bmatrix} + \begin{bmatrix} u_1 & u_2 \end{bmatrix} \begin{bmatrix} 100 & 0 \\ 0 & 100 \end{bmatrix} \begin{bmatrix} u_1 \\ u_2 \end{bmatrix} \right) dt$$

$$= \frac{1}{2}\int_0^\infty \left(10x_1^2 + 10x_2^2 + 10x_3^2 + 10x_4^2 + 100u_1^2 + 100u_2^2 \right) dt$$

Simulate the resulting closed-loop microsystem using the lsim MATLAB function.

SOLUTION

The feedback control algorithm, as given in Equation 9.17, is $u = -G^{-1}B^T K x = -K_F x$. The following MATLAB m-file is used to find the numerical results needed (the comments are provided to assist the reader).

```
echo off; clear all; format short e;
% Constant-coefficient matrices A, B, C and D for microsystem
A=[-0.1    6    -1    -10;
       0    4     1      0;
       0    3    -5      0;
       0    0     1      0];
disp('eigenvalues_A'); disp(eig(A));
% Eigenvalues of the matrix A
B=[1     0;
  -5     1;
  -2     2;
   0     0];
C=[0 0 0 1]; D=[0 0 0 0];
% Weighting matrices Q and G
Q=[10 0    0    0;
    0 10    0    0;
    0  0   10    0;
    0  0    0   10];
G=[100    0;
     0  100];
% Feedback and return function coefficients, eigenvalues
[K_feedback,K,Eigenvalues]=lqr(A,B,Q,G);
disp('K_feedback'); disp(K_feedback);
disp('K'); disp(K);
disp('eigenvalues A-BK_feedback'); disp(Eigenvalues);
```

```
% Matrix of the closed-loop MEMS
A_closed_loop=A-B*K_feedback;
% Dynamics
t=0:0.02:10;
% Control inputs
uu=[10*ones(max(size(t)),4)];
[y,x]=lsim(A_closed_loop,B*K_feedback,C,D,uu,t);
plot(t,x);
title('Microsystem Dynamics, x1, x2, x3, x4');
xlabel('time [seconds]'); pause;
plot(t,y); pause;
plot(t,x(:,1),'-',t,x(:,2),'-',t,x(:,3),'-',
t,x(:,4),'-'); pause;
plot(t,x(:,1),'-'); pause;
plot(t,x(:,2),'-'); pause;
plot(t,x(:,3),'-'); pause;
plot(t,x(:,4),'-'); pause;
```

The feedback gain matrix K_F, the return function matrix K, and the eigenvalues of the closed-loop microsystems $(A - BG^{-1}B^T K) = (A - BK_F)$ are found as given above.

```
eigenvalues_A
 -1.0000e-001
            0
 -5.3218e+000
  4.3218e+000

K_feedback
 -2.8608e-001 -2.0838e+000   8.1629e-002   1.7167e+000
 -6.2334e-002  1.5481e-001   5.4208e-001   2.7384e+000
K
   7.1354e+000   1.0494e+001 -8.3639e+000 -5.1729e+001
   1.0494e+001   5.0848e+001 -1.7684e+001 -1.2431e+002
  -8.3639e+000 -1.7684e+001   3.5946e+001   1.9907e+002
  -5.1729e+001 -1.2431e+002   1.9907e+002   1.1468e+003
eigenvalues A-BK_feedback
 -5.4041e+000
 -3.8173e+000
 -2.5357e+000
 -5.5148e-001
```

Having found matrices K and K_F, we obtain the optimal control algorithm as

$$u = -K_F x = -\begin{bmatrix} 0.29 & -2.1 & 0.08 & 1.72 \\ 0.06 & 0.15 & 0.54 & 2.74 \end{bmatrix} \begin{bmatrix} x_1 \\ x_2 \\ x_3 \\ x_4 \end{bmatrix}$$

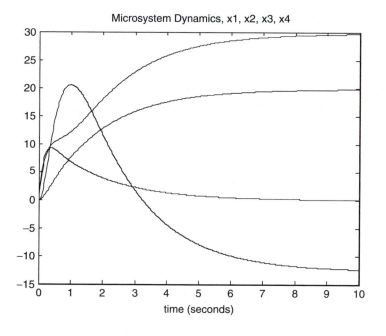

FIGURE 9.20
Evolution of the MEMS state variables.

The dynamics of the microsystem state variables if initial conditions are set to be

$$
\begin{bmatrix} x_{10} \\ x_{20} \\ x_{30} \\ x_{40} \end{bmatrix} = \begin{bmatrix} 0 \\ 0 \\ 0 \\ 0 \end{bmatrix} \quad \text{and} \quad \begin{bmatrix} u_1 \\ u_2 \end{bmatrix} = \begin{bmatrix} 5 \\ 5 \end{bmatrix}
$$

are plotted in Figure 9.20.

9.3.2 Tracking Problem for Linear MEMS and NEMS

We covered the stabilization problem and designed stabilizing linear control algorithms using the Hamilton-Jacobi theory. The tracking controllers must be synthesized using the tracking error $e(t)$.

The optimal tracking control problem can be formulated as:

For MEMS $\dot{x}(t) = Ax + Bu$ with the output equation $y(t) = Hx(t)$, synthesize the tracking optimal controller $u = \phi(e, x)$ by minimizing the performance functional.

The tracking error is $e(t) = r(t) - y(t)$. Using the output equation $y(t) = Hx(t)$, and for multivariable microsystems we have

$$
\begin{aligned}
e(t) &= Nr(t) - y(t) = Nr(t) - Hx(t) \\[2mm]
&= \begin{bmatrix}
n_{11} & 0 & \cdots & 0 & 0 \\
0 & n_{22} & \cdots & 0 & 0 \\
\vdots & \vdots & \ddots & \vdots & \vdots \\
0 & 0 & \cdots & n_{b-1b-1} & 0 \\
0 & 0 & \cdots & 0 & n_{bb}
\end{bmatrix}
\begin{bmatrix} r_1 \\ r_2 \\ \vdots \\ r_{b-1} \\ r_b \end{bmatrix}
- \begin{bmatrix}
h_{11} & h_{12} & \cdots & h_{1n-1} & h_{1n} \\
h_{21} & h_{22} & \cdots & h_{2n-1} & h_{2n} \\
\vdots & \vdots & \ddots & \vdots & \vdots \\
h_{b-11} & h_{b-12} & \cdots & h_{b-1n-1} & h_{b-1n} \\
h_{b1} & h_{b2} & \cdots & h_{bn-1} & h_{bn}
\end{bmatrix}
\begin{bmatrix} x_1 \\ x_2 \\ \vdots \\ x_{n-1} \\ x_n \end{bmatrix}
\end{aligned}
$$

where $N \in \mathbb{R}^{b \times b}$ and $H \in \mathbb{R}^{b \times n}$ are the constant-coefficient matrices.

Using $e(t) = Nr(t) - y(t)$, and denoting $e(t) = \dot{x}^{ref}(t)$, we consider the dynamics of the *exogenous* system

$$\dot{x}^{ref}(t) = Nr - y = Nr - Hx \tag{9.19}$$

Augmenting the system from Equation 9.8, $\dot{x}(t) = Ax + Bu$, $x(t_0) = x_0$, with Equation 9.19 yields

$$\dot{x}(t) = Ax + Bu, \quad y = Hx, \quad x_0(t_0) = x_0$$

$$\dot{x}^{ref}(t) = Nr - y = Nr - Hx$$

Hence, we have

$$\dot{x}_\Sigma(t) = A_\Sigma x_\Sigma + B_\Sigma u + N_\Sigma r, \quad y = Hx, \quad x_{\Sigma 0}(t_0) = x_{\Sigma 0} \tag{9.20}$$

where

$$x_\Sigma = \begin{bmatrix} x \\ x^{ref} \end{bmatrix} \in \mathbb{R}^c (c = n + b) \; A_\Sigma = \begin{bmatrix} A & 0 \\ -H & 0 \end{bmatrix} \in \mathbb{R}^{c \times c}, \; B_\Sigma = \begin{bmatrix} B \\ 0 \end{bmatrix} \in \mathbb{R}^{c \times m}, \; \text{and} \; N_\Sigma = \begin{bmatrix} 0 \\ N \end{bmatrix} \in \mathbb{R}^{c \times b}$$

The quadratic performance functional is given as

$$J\left(\begin{bmatrix} x(\cdot) \\ x^{ref}(\cdot) \end{bmatrix}, \; u(\cdot) \right) = \frac{1}{2} \int_{t_0}^{t_f} \left(\begin{bmatrix} x \\ x^{ref} \end{bmatrix}^T Q \begin{bmatrix} x \\ x^{ref} \end{bmatrix} + u^T Gu \right) dt \tag{9.21}$$

Using the quadratic functional (Equation 9.21), the Hamiltonian function is found to be

$$H\left(x_\Sigma, u, r, \frac{\partial V}{\partial x_\Sigma} \right) = \frac{1}{2} \left(x_\Sigma^T Q x_\Sigma + u^T Gu \right) + \left(\frac{\partial V}{\partial x_\Sigma} \right)^T (A_\Sigma x_\Sigma + B_\Sigma u + N_\Sigma r) \tag{9.22}$$

From Equation 9.22, using the first-order necessary condition for optimality, one finds

$$\frac{\partial H\left(x, u, r, \frac{\partial V}{\partial x_\Sigma} \right)}{\partial u} = u^T G + \left(\frac{\partial V}{\partial x_\Sigma} \right)^T B_\Sigma$$

Thus, the optimal control algorithm is

$$u = -G^{-1} B_\Sigma^T \frac{\partial V(x_\Sigma)}{\partial x_\Sigma} = -G^{-1} \begin{bmatrix} B \\ 0 \end{bmatrix}^T \frac{\partial V\left(\begin{bmatrix} x \\ x^{ref} \end{bmatrix} \right)}{\partial \begin{bmatrix} x \\ x^{ref} \end{bmatrix}} \tag{9.23}$$

The solution of the Hamilton-Jacobi-Bellman partial differential equation

$$-\frac{\partial V}{\partial t} = \frac{1}{2}x_\Sigma^T Q x_\Sigma + \left(\frac{\partial V}{\partial x_\Sigma}\right)^T A x_\Sigma - \frac{1}{2}\left(\frac{\partial V}{\partial x_\Sigma}\right)^T B_\Sigma G^{-1} B_\Sigma^T \frac{\partial V}{\partial x_\Sigma} \tag{9.24}$$

is satisfied by the quadratic return function. Thus,

$$V(x_\Sigma) = \frac{1}{2}x_\Sigma^T K(t)x_\Sigma \tag{9.25}$$

Making use of Equation 9.24 and Equation 9.25, one finds the Riccati equation to obtain the unknown symmetric matrix K. We have

$$-\dot{K} = Q + A_\Sigma^T K + K^T A_\Sigma - K^T B_\Sigma G^{-1} B_\Sigma^T K, \; K(t_f) = K_f \tag{9.26}$$

The controller is found from Equation 9.23 using Equation 9.25. In particular,

$$u = -G^{-1}B_\Sigma^T K x_\Sigma = -G^{-1}\begin{bmatrix} B \\ 0 \end{bmatrix}^T K \begin{bmatrix} x \\ x^{ref} \end{bmatrix} \tag{9.27}$$

From $\dot{x}^{ref}(t) = e(t)$, one has $x^{ref}(t) = \int e(t)dt$. Therefore, we obtain the integral control law

$$u(t) = -G^{-1}B_\Sigma^T K x_\Sigma(t) = -G^{-1}\begin{bmatrix} B \\ 0 \end{bmatrix}^T K \begin{bmatrix} x(t) \\ \int e(t)dt \end{bmatrix} \tag{9.28}$$

In the control algorithm (Equation 9.28), the integral of the error vector is used in addition to the state feedback.

9.3.3 Transformation Method and Tracking Control of Linear MEMS

The tracking control problem is solved by designing the proportional-integral control for MEMS and NEMS using the *transformation* method.[8] We define the tracking error vector as

$$e(t) = Nr(t) - y(t) = Nr(t) - Hx^{sys}(t)$$

Then, for linear microsystems, we have

$$\dot{e}(t) = N\dot{r}(t) - \dot{y}(t) = N\dot{r}(t) - H\dot{x}^{sys}(t) = N\dot{r}(t) - HA^{sys}x^{sys} - HB^{sys}u$$

Using the expanded state vector $x(t) = \begin{bmatrix} x^{sys}(t) \\ e(t) \end{bmatrix}$, one finds

$$\dot{x}(t) = \begin{bmatrix} \dot{x}^{sys}(t) \\ \dot{e}(t) \end{bmatrix} = \begin{bmatrix} A^{sys} & 0 \\ -HA^{sys} & 0 \end{bmatrix}\begin{bmatrix} x^{sys} \\ e \end{bmatrix} + \begin{bmatrix} B^{sys} \\ -HB^{sys} \end{bmatrix}u + \begin{bmatrix} 0 \\ N \end{bmatrix}\dot{r}$$

$$= Ax + Bu + \begin{bmatrix} 0 \\ N \end{bmatrix}\dot{r}, \; y = Hx^{sys}$$

Let us report the *space transformation* method. We introduce the z and v vectors as given by

$$z = \begin{bmatrix} x \\ u \end{bmatrix} \quad \text{and} \quad v = \dot{u}$$

Using z and v, one obtains the model as

$$\dot{z}(t) = \begin{bmatrix} A & B \\ 0 & 0 \end{bmatrix} z + \begin{bmatrix} 0 \\ I \end{bmatrix} v = A_z z + B_z v, \quad y = Hx^{sys}, \quad z(t_0) = z_0$$

Minimizing the quadratic functional

$$J = \int_{t_0}^{t_f} (z^T Q_z z + v^T G_z v) dt, \quad Q_z \in \mathbb{R}^{(n+m) \times (n+m)}, \quad Q_z \geq 0, \quad G \in \mathbb{R}^{m \times m}, \quad G > 0$$

the application of the first-order necessary condition for optimality (n1) gives

$$v = -G_z^{-1} B_z^T K z$$

The Riccati equation to find the unknown matrix $K \in \mathbb{R}^{(n+m) \times (n+m)}$ is

$$-\dot{K} = K A_z + A_z^T K - K B_z G_z^{-1} B_z^T K + Q_z, \quad K(t_f) = K_f$$

Hence, one has

$$\dot{u}(t) = -G_z^{-1} B_z^T K z = -G_z^{-1} \begin{bmatrix} 0 \\ I \end{bmatrix}^T \begin{bmatrix} K_{11} & K_{21}^T \\ K_{21} & K_{22} \end{bmatrix} \begin{bmatrix} x \\ u \end{bmatrix} = -G_z^{-1} K_{21} x - G_z^{-1} K_{22} u = K_{f1} x + K_{f2} u$$

From $\dot{x}(t) = Ax + Bu$, we have $u = B^{-1}(\dot{x}(t) - Ax)$. Thus,

$$u = B^{-1}(\dot{x}(t) - Ax) = (B^T B)^{-1} B^T (\dot{x}(t) - Ax)$$

One obtains

$$\dot{u}(t) = K_{f1} x + K_{f2} u = K_{f1} x + K_{f2}(B^T B)^{-1} B^T (\dot{x}(t) - Ax) = [K_{f1} - K_{f2}(B^T B)^{-1} B^T A] x(t)$$

$$+ K_{f2}(B^T B)^{-1} B^T \dot{x}(t) = (K_{f1} - K_{F1} A) x(t) + K_{F1} \dot{x}(t) = K_{F2} x(t) + K_{F1} \dot{x}(t)$$

The controller is derived from the following form:

$$u(t) = K_{F1} x(t) - K_{F1} x_0 + \int K_{F2} x(\tau) d\tau + u_0$$

The designed controller is the proportional-integral control law with state feedback because

$$x(t) = \begin{bmatrix} x^{sys}(t) \\ e(t) \end{bmatrix}$$

For nonlinear MEMS, the proposed procedure can be straightforwardly used. In particular, we have the following proportional-integral controller:

$$\dot{u}(t) = -G_z^{-1} B_z^T \frac{\partial V}{\partial z} = -G_z^{-1} \begin{bmatrix} 0 \\ I \end{bmatrix}^T \frac{\partial V(x,u)}{\partial [x\, u]^T}$$

where $V(x,u)$ is the nonquadratic return function.
  Nonquadratic functionals

$$J = \int_{t_0}^{t_f} \left(\sum_{i=0}^{\varsigma} \frac{2\eta+1}{2(ki+\eta+1)} \left[\frac{x^{sys \frac{ki+\eta+1}{2\eta+1}}}{e^{\frac{ki+\eta+1}{2\eta+1}}} \right]^T Q_{zi} \left[\frac{x^{sys \frac{ki+\eta+1}{2\eta+1}}}{e^{\frac{ki+\eta+1}{2\eta+1}}} \right] + v^T G_z v \right) dt$$

can be used to guarantee the desired tracking performance.

Example 9.5

Design the tracking controller for a microsystem with a PZT microactuator controlled by changing the applied voltage (V) using ICs. The equation of motion of a PZT microactuator is

$$m_e \frac{d^2 y}{dt^2} + b \frac{dy}{dt} + ky = kd_e V$$

where y is the microactuator displacement (output).

SOLUTION

The second-order differential equation of the microsystem with microactuator is

$$\frac{dy}{dt} = v,$$

$$\frac{dv}{dt} = -\frac{k}{m_e} y - \frac{b}{m_e} v + \frac{kd_e}{m_e} V$$

Consider the reference input $r(t)$. Using the desired microactuator displacement $r(t)$, the tracking error between the reference and the output is

$$e(t) = r(t) - y(t), \quad N = 1$$

Thus, we have the following system

$$\dot{x}^{sys} = A^{sys} x^{sys} + B^{sys} u$$

$$\dot{e} = N\dot{r} - HA^{sys} x^{sys} - HB^{sys} u$$

where

$$A^{sys} = \begin{bmatrix} 0 & 1 \\ -\dfrac{k}{m_e} & -\dfrac{b}{m_e} \end{bmatrix}, \quad B^{sys} = \begin{bmatrix} 0 \\ \dfrac{kd_e}{m_e} \end{bmatrix}, \quad \text{and} \quad H = [1 \quad 0]$$

One obtains

$$\dot{x} = \begin{bmatrix} \dot{x}^{sys} \\ \dot{e} \end{bmatrix} = \begin{bmatrix} A^{sys} & 0 \\ -HA^{sys} & 0 \end{bmatrix} \begin{bmatrix} x^{sys} \\ e \end{bmatrix} + \begin{bmatrix} B^{sys} \\ -HB^{sys} \end{bmatrix} u + \begin{bmatrix} 0 \\ N \end{bmatrix} \dot{r}, \quad y = \begin{bmatrix} H & 0 \end{bmatrix} \begin{bmatrix} x^{sys} \\ e \end{bmatrix}$$

Following the design procedure reported, using the *transformation* method, one has

$$z = \begin{bmatrix} x^{sys} \\ e \\ u \end{bmatrix}$$

Therefore, we obtain

$$\dot{u}(t) = -K_f z(t) = -K_f \begin{bmatrix} y(t) \\ v(t) \\ e(t) \\ V(t) \end{bmatrix}$$

The proportional-integral tracking controller is derived as

$$u(t) = K_{F1} \begin{bmatrix} y(t) \\ v(t) \\ e(t) \end{bmatrix} + \int K_{F2} \begin{bmatrix} y(\tau) \\ v(\tau) \\ e(\tau) \end{bmatrix} d\tau$$

Using the microactuator parameters

$$k = 3000, \quad b = 1, \quad d_e = 0.000001, \quad \text{and} \quad m_e = 0.02$$

the tracking controller is designed using the lqr MATLAB solver using the weighting matrices

$$Q_z = \begin{bmatrix} 1 & 0 & 0 & 0 \\ 0 & 1 & 0 & 0 \\ 0 & 0 & 1 \times 10^{10} & 0 \\ 0 & 0 & 0 & 1 \end{bmatrix}$$

and $G_z = 10$.

The closed-loop microactuator dynamics is documented in Figure 9.21.

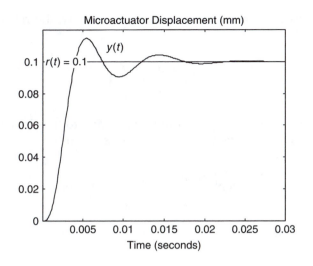

FIGURE 9.21
Closed-loop microsystem dynamics if $r(t) = 0.1$.

It should be emphasized that the differential equations that model the microsystem with the PZT microactuator

$$m_e \frac{d^2 y}{dt^2} + b \frac{dy}{dt} + ky = k(d_e V - z)$$

should be integrated with the hysteresis model $\dot{z} = \alpha d_e \dot{V} - \beta |\dot{V}| z - \gamma \dot{V} |z|$.

Thus, we have the state-space nonlinear model of PZT microactuators as

$$\frac{dy}{dt} = v$$
$$\frac{dv}{dt} = -\frac{k}{m_e} y - \frac{b}{m_e} v - \frac{k}{m_e} z + \frac{k d_e}{m_e} V$$
$$\frac{dz}{dt} = -\beta |\dot{V}| z + \alpha d_e \dot{V} - \lambda |z| \dot{V}$$

9.3.4 Time-Optimal Control of MEMS and NEMS

Time-optimal controllers for MEMS and NEMS can be designed using the functional

$$J = \frac{1}{2} \int_{t_0}^{t_f} \left(x_\Sigma^T Q x_\Sigma \right) dt$$

Taking note of the Hamilton-Jacobi equation

$$-\frac{\partial V}{\partial t} = \min_{-1 \leq u \leq 1} \left[\frac{1}{2} x_\Sigma^T Q x_\Sigma + \left(\frac{\partial V}{\partial x_\Sigma} \right)^T (A x_\Sigma + B_\Sigma u) \right]$$

and using the first-order necessary condition for optimality, the relay-type controller is found to be

$$u = -\text{sgn}\left(B_\Sigma^T \frac{\partial V}{\partial x_\Sigma} \right), \quad -1 \le u \le 1$$

This control algorithm cannot be applied to systems due to the chattering phenomenon. Therefore, relay-type control laws with dead zone, as given by

$$u = -\text{sgn}\left(B_\Sigma^T \frac{\partial V}{\partial x_\Sigma} \right)\Bigg|_{\text{dead zone}}, \quad -1 \le u \le 1$$

can be used in microsystems.

Example 9.6

Using Hamilton-Jacobi theory, synthesize the time-optimal control law for the microsystem described by the following differential equations

$$\dot{x}_1(t) = x_1^5 u_1 + x_2^7, \quad -1 \le u_1 \le 1$$

$$\dot{x}_2(t) = x_1^3 x_2^5 u_2, \quad -1 \le u_2 \le 1$$

using the performance functional

$$J = \int_{t_0}^{t_f} W_x(x)dt$$

SOLUTION

The Hamilton-Jacobi equation is

$$-\frac{\partial V}{\partial t} = \min_{u \in U}\left[W_x(x) + \left(\frac{\partial V}{\partial x}\right)^T (F(x) + B(x)u) \right]$$

$$= \min_{\substack{-1 \le u_1 \le 1 \\ -1 \le u_2 \le 1}}\left[W_x(x) + \frac{\partial V}{\partial x_1}\left(x_1^5 u_1 + x_2^7\right) + \frac{\partial V}{\partial x_2} x_1^3 x_2^5 u_2 \right]$$

From the first-order necessary condition for optimality, an optimal controller is found. In particular,

$$u_1 = -\text{sgn}\left(x_1^5 \frac{\partial V}{\partial x_1} \right) \quad \text{and} \quad u_2 = -\text{sgn}\left(x_1^3 x_2^5 \frac{\partial V}{\partial x_2} \right)$$

The Hamilton-Jacobi-Bellman partial differential equations is

$$-\frac{\partial V}{\partial t} = W_x(x) - \left| x_1^5 \frac{\partial V}{\partial x_1} \right| + \frac{\partial V}{\partial x_1} x_2^7 - \left| x_1^3 x_2^5 \frac{\partial V}{\partial x_2} \right|$$

Example 9.7

Design an optimal relay-type controller for the microsystem studied in Example 9.3. In particular, the equations of motion are

$$\dot{x}_1(t) = x_2, \quad \dot{x}_2(t) = u, \quad -1 \le u \le 1$$

SOLUTION

The control takes values $u = 1$ and $u = -1$.

If $u = 1$, from , $\dot{x}_1(t) = x_2$, $\dot{x}_2(t) = 1$ one has

$$\frac{dx_2}{dx_1} = \frac{1}{x_2}$$

The integration gives $x_2^2 = 2x_1 + c_1$.

If $u = -1$, from $\dot{x}_1(t) = x_2$, $\dot{x}_2(t) = -1$, we obtain

$$\frac{dx_2}{dx_1} = -\frac{1}{x_2}$$

The integration results in $x_2^2 = -2x_1 + c_2$.

Due to the switching action ($u = 1$ or $u = -1$), the switching curve is derived as a function of the state variables. The comparison of

$$x_2^2 = 2x_1 + c_1 \quad \text{and} \quad x_2^2 = -2x_1 + c_2$$

gives the switching curve, i.e., the switching curve is

$$-x_1 - \tfrac{1}{2}x_2|x_2| = 0$$

Since the control takes the values $u = 1$ and $u = -1$, and making use of the derived expression for the switching curve, one finds the time-optimal (relay) controller as

$$u = -\mathrm{sgn}\left(x_1 + \tfrac{1}{2}x_2|x_2|\right)$$

Having found the optimal control law using calculus by analyzing the solutions of the differential equations with the relay controller (closed-loop system switching), the Hamilton-Jacobi theory is applied. We minimize the functional

$$J = \int_{t_0}^{t_f} 1\,dt$$

From the Hamilton-Jacobi equation

$$-\frac{\partial V}{\partial t} = \min_{-1 \le u \le 1}\left[1 + \frac{\partial V}{\partial x_1}x_2 + \frac{\partial V}{\partial x_2}u\right]$$

an optimal controller derived using n1 is given as

$$u = -\text{sgn}\left(\frac{\partial V}{\partial x_2}\right)$$

The solution of the partial differential equation is

$$V(x_1, x_2) = k_{11}x_1^2 + k_{12}x_1x_2 + k_{22}x_2^3|x_2|$$

That is, the nonquadratic return function is used. The controller is found as

$$u = -\text{sgn}\left(x_1 + \tfrac{1}{2}x_2|x_2|\right)$$

The time-optimal controller designed using the Hamilton-Jacobi theory corresponds to the results obtained using the calculus of variations. The transient dynamics is analyzed. The switching curve, the phase-plane evolution of the variables, and the transient behavior for different initial conditions are documented in Figure 9.22.

Using different design methods, it was shown that an optimal controller is $u = -\text{sgn}(x_1 + \tfrac{1}{2}x_2|x_2|)$. From $\dot{x}_1(t) = x_2$, we can express the controller using proportional and derivative feedback attaining the analogy with PID controllers. In particular, we have

$$u = -\text{sgn}\left(x_1 + \tfrac{1}{2}x_2|x_2|\right) = -\text{sgn}\left(x_1 + \tfrac{1}{2}\dot{x}_1|\dot{x}_1|\right)$$

9.4   Sliding Mode Control of MEMS and NEMS

We have covered time-optimal control that results in relay-type controllers with switching surfaces. Sliding mode control has direct analogies to the time-optimal control. Soft- and hard-switching sliding mode control laws have been synthesized.[6,7] Sliding mode soft-switching algorithms provide superior performance, and the chattering phenomenon (typical for hard-switching relay-type and sliding mode control laws) is eliminated. To design controllers, we model the states and error dynamics of microsystems as

$$\dot{x}(t) = Ax + Bu, \quad -1 \leq u \leq 1$$

$$\dot{e}(t) = N\dot{r}(t) - HAx - HBu$$

The smooth-sliding manifold is

$$M = \left\{(t, x, e) \in R_{\geq 0} \times X \times E \mid v(t, x, e) = 0\right\}$$

$$= \bigcap_{j=1}^{m}\left\{(t, x, e) \in R_{\geq 0} \times X \times E \mid v_j(t, x, e) = 0\right\}$$

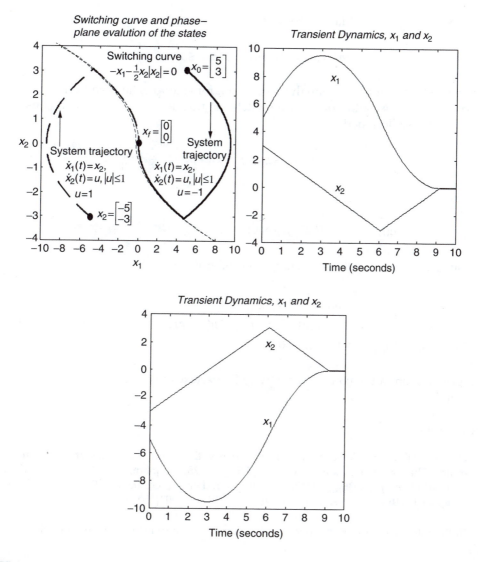

FIGURE 9.22
System evolution with time-optimal (switching) control.

The time-varying nonlinear switching surface is defined as

$$v(t,x,e) = K_{vxe}(t,x,e) = 0 \quad \text{or}$$

$$v(t,x,e) = [K_{vx}(t) \quad K_{ve}(t)]\begin{bmatrix} x(t) \\ e(t) \end{bmatrix} = K_{vx}(t)x(t) + K_{ve}(t)e(t) = 0$$

$$\begin{bmatrix} v_1(t,x,e) \\ \vdots \\ v_m(t,x,e) \end{bmatrix} = \begin{bmatrix} k_{vx11}(t) & \cdots & k_{vx1c}(t) & k_{ve11}(t) & \cdots & k_{ve1b}(t) \\ \vdots & \vdots & \vdots & \vdots & \vdots & \vdots \\ k_{vxm1}(t) & \cdots & k_{vxmc}(t) & k_{vem1}(t) & \cdots & k_{vemb}(t) \end{bmatrix} \begin{bmatrix} x_1(t) \\ \vdots \\ x_c(t) \\ e_1(t) \\ \vdots \\ e_b(t) \end{bmatrix} = 0$$

The soft-switching control law is given as

$$u(t,x,e) = -G\phi(v), \quad -1 \le u \le 1, \quad G > 0$$

where ϕ is the continuous real-analytic function, for example, tanh and erf.

In contrast, the discontinuous (hard-switching) tracking controllers with constant and varying gains are designed as

$$u(t,x,e) = -G\,\text{sgn}(v), \quad G > 0, \quad G \in \mathbb{R}^{m \times m}$$

or

$$u(t,x,e) = -G(t,x,e)\,\text{sgn}(v), \quad G(\cdot):\mathbb{R}_{\ge 0} \times \mathbb{R}^c \times \mathbb{R}^b \to \mathbb{R}^{m \times m}$$

The simplest hard-switching tracking controller is

$$u(t,x,e) = \begin{cases} u_{max}, & \forall v(t,x,e) > 0 \\ 0, & \forall v(t,x,e) = 0, \\ u_{min}, & \forall v(t,x,e) < 0 \end{cases} \quad u_{min} \le u(t,x,e) \le u_{max}, \quad u_{max} > 0, \quad u_{min} < 0$$

and a polyhedron in the control space with 2^m vertexes results.

Example 9.8

Design a sliding mode controller for a system with a permanent-magnet synchronous motor. The parameters are 40 V, $r_s(\cdot) \in [0.5_{T=20°C} \quad 0.75_{T=130°C}]$ohm, $L_{ss}(\cdot) \in [0.009 \quad 0.01]$ H, $L_{ls} = 0.001$ H, $\psi_m(\cdot) \in [0.036_{T=20°C} 0.055_{T=130°C}]$ V · sec/rad or $\psi_m(\cdot) \in [0.069_{T=20°C} 0.055_{T=130°C}]$ N · m/ A, $B_m = 0.000013$ N · m · sec/rad and $J(\cdot) \in [0.0001 \quad 0.0003]$ kg · m².[7]

In the rotor reference frame, permanent-magnet synchronous motors are modeled as

$$\frac{di_{qs}^r}{dt} = -\frac{r_s}{L_{ss}}i_{qs}^r - \frac{\psi_m}{L_{ss}}\omega_r - i_{ds}^r\omega_r + \frac{1}{L_{ss}}u_{qs}^r$$

$$\frac{di_{ds}^r}{dt} = -\frac{r_s}{L_{ss}}i_{ds}^r + i_{qs}^r\omega_r + \frac{1}{L_{ss}}u_{ds}^r$$

$$\frac{di_{os}^r}{dt} = -\frac{r_s}{L_{ls}}i_{os}^r + \frac{1}{L_{ls}}u_{os}^r$$

$$\frac{d\omega_r}{dt} = \frac{3P^2\psi_m}{8J}i_{qs}^r - \frac{B_m}{J}\omega_r - \frac{P}{2J}T_L$$

where $i_{qs}^r, i_{ds}^r, i_{os}^r$ and $u_{qs}^r, u_{ds}^r, u_{os}^r$ are the *quadrature-*, *direct-*, and *zero*-axis current and voltage components.

Examine whether it is possible to linearize these equations of motion in order to design the control law.

SOLUTION

The state and control variables are the *quadrature*, *direct*, and *zero* currents and voltages, as well as the angular velocity, i.e.,

$$x = \begin{bmatrix} i_{qs}^r \\ i_{ds}^r \\ i_{os}^r \\ \omega_r \end{bmatrix} \quad \text{and} \quad u = \begin{bmatrix} u_{qs}^r \\ u_{ds}^r \end{bmatrix}$$

The *quadrature-*, *direct-*, and *zero-*axis voltage and current components have a DC form.

Let us study the stability (the stability analysis is covered in Section 9.8 using the Lyapunov stability theory [see Example 9.15]). For the uncontrolled motor, we have $u_{qs}^r = u_{ds}^r = u_{os}^r = 0$. The total derivative of the positive-definite quadratic function

$$V\left(i_{qs}^r, i_{ds}^r, i_{os}^r, \omega_r\right) = \tfrac{1}{2} i_{qs}^{r\,2} + \tfrac{1}{2} i_{ds}^{r\,2} + \tfrac{1}{2} i_{os}^{r\,2} + \tfrac{1}{2} \omega_r^2$$

is found to be

$$\frac{dV(i_{qs}^r, i_{ds}^r, i_{os}^r, \omega_r)}{dt} = -\frac{r_s}{L_{ss}} i_{qs}^{r\,2} - \frac{r_s}{L_{ss}} i_{ds}^{r\,2} - \frac{r_s}{L_{ls}} i_{os}^{r\,2} - \frac{\psi_m(8J - 3P^2 L_{ss})}{8JL_{ss}} i_{qs}^r \omega_r - \frac{B_m}{J} \omega_r^2$$

The motor parameters are time-varying, and, in general, $r_s(\cdot) \in [r_{s\min} \; r_{s\max}]$, $L_{ss}(\cdot) \in [L_{ss\min} \; L_{ss\max}]$, $\psi_m(\cdot) \in [\psi_{m\min} \; \psi_{m\max}]$, and $J(\cdot) \in [J_{\min} \; J_{\max}]$. However, $r_{s\min} > 0$, $L_{ss\min} > 0$, $\psi_{m\min} > 0$, and $J_{\min} > 0$. Hence, the open-loop system is uniformly robustly asymptotically stable in the large because the total derivative of a positive-definite function $V(i_{qs}^r, i_{ds}^r, i_{os}^r, \omega_r)$ is negative (details are reported in Section 9.8). Thus, it is not necessary to apply the linearizing feedback to transform a nonlinear motor model into a linear one by canceling the beneficial internal nonlinearities $-i_{ds}^r \omega_r$ and $i_{qs}^r \omega_r$.

In general, due to the bounds imposed on the voltages, as well as variations of L_{ss}, one cannot attain the feedback linearization. More important is that from the electric machinery standpoint, to attain the balanced voltage set applied to the stator windings, one controls only u_{qs}^r while $u_{ds}^r = 0$. In fact, to guarantee the balanced operation, $u_{ds}^r = 0$.

Our goal is to design a soft-switching controller. The applied phase voltages are bounded by 40 V. The time-invariant linear and nonlinear *switching surfaces* of stabilizing controllers are obtained as functions of the state variables i_{qs}^r, i_{ds}^r, and ω_r. We have

linear switching surface

$$\upsilon\left(i_{qs}^r, i_{ds}^r, \omega_r\right) = -0.00049 i_{qs}^r - 0.00049 i_{ds}^r - 0.0014 \omega_r = 0$$

nonlinear switching surface

$$\upsilon\left(i_{qs}^r, i_{ds}^r, \omega_r\right) = -0.00049 i_{qs}^r - 0.00049 i_{ds}^r - 0.000017 \omega_r - 0.000025 \omega_r |\omega_r| = 0$$

It was emphasized that only *quadrature* voltage is regulated. Hence, one denotes

$$u = u_{qs}^r$$

A discontinuous hard-switching stabilizing controller is found to be

$$u = \mathrm{sgn}_{-40}^{+40} \upsilon\left(i_{qs}^r, i_{ds}^r, \omega_r\right) = \begin{cases} +40, & \upsilon\left(i_{qs}^r, i_{ds}^r, \omega_r\right) > 0 \\ 0, & \upsilon\left(i_{qs}^r, i_{ds}^r, \omega_r\right) = 0 \\ -40, & \upsilon\left(i_{qs}^r, i_{ds}^r, \omega_r\right) < 0 \end{cases}$$

To avoid the singularity, this discontinuous algorithm is *regularized* as

$$u = 40 \frac{\upsilon\left(i_{qs}^r, i_{ds}^r, \omega_r\right)}{\left|\upsilon\left(i_{qs}^r, i_{ds}^r, \omega_r\right)\right| + \varepsilon}, \quad \varepsilon = 0.0005$$

A soft-switching stabilizing controller is designed by making use of linear and nonlinear *switching surfaces* . In particular,

$$u = 40 \tanh^{1/9} \upsilon\left(i_{qs}^r, i_{ds}^r, \omega_r\right)$$

where $\upsilon(i_{qs}^r, i_{ds}^r, \omega_r) = -0.00049 i_{qs}^r - 0.00049 i_{ds}^r - 0.0014 \omega_r = 0$ is the linear switching surface, and $\upsilon(i_{qs}^r, i_{ds}^r, \omega_r) = -0.00049 i_{qs}^r - 0.00049 i_{ds}^r - 0.000017 \omega_r - 0.000025 \omega_r |\omega_r| = 0$ is the nonlinear switching surface.

   Numerical simulations are performed to study the state evolution due to initial conditions and control command. The three-dimensional plot for evolution of state variables i_{qs}^r, i_{ds}^r, and is shown in Figure 9.23. It is evident that the system evolves to the origin.

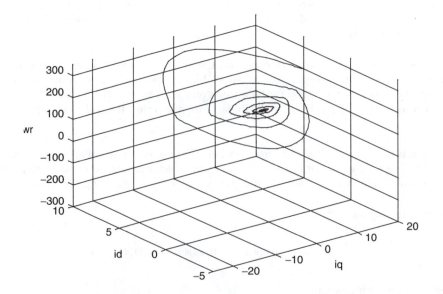

FIGURE 9.23
Three-dimensional state evolution due to initial conditions $\begin{bmatrix} i_{qs}^r(0) \\ i_{ds}^r(0) \\ \omega_r(0) \end{bmatrix} = \begin{bmatrix} 20 \\ 5 \\ 200 \end{bmatrix}$.

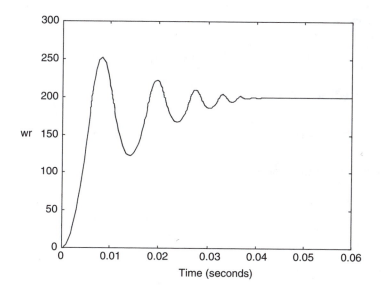

FIGURE 9.24
Transient dynamics of the motor angular velocity if.

The tracking controller is synthesized using the states i_{qs}^r, i_{ds}^r, and   and the tracking error $e = \omega_{reference} - \omega_r$. The nonlinear time-invariant switching surface is given by

$$v\left(i_{qs}^r, i_{ds}^r, \omega_r, e\right) = -0.0005 i_{qs}^r - 0.0005 i_{qs}^r - 0.00003\omega_r + 0.0015e + 0.0001e^3 = 0$$

A soft-switching tracking controller is found using the tanh function, and

$$u = 40 \tanh^{1/9} v\left(i_{qs}^r, i_{ds}^r, \omega_r, e\right)$$

Figure 9.24 illustrates the dynamics of a closed-loop system with the designed tracking controller. The reference angular velocity is assigned to be $\omega_{reference} = 200 \frac{rad}{sec}$.
Motor angular velocity ω_r, $[\frac{rad}{sec}]$
The simulation results illustrate that the settling time is 0.04 sec. Excellent dynamic performance of the synthesized system is evident, and the tracking error is zero. Due to soft switching, the singularity and sensitivity problems are avoided, robustness and stability are improved, and the chattering effect (high frequency switching) is eliminated.

9.4.1 Feedback Linearization and Control of Permanent-Magnet Synchronous Micromotors

It is observed that the *linearizability* condition is met. In fact, for a model

$$\frac{di_{qs}^r}{dt} = -\frac{r_s}{L_{ss}} i_{qs}^r - \frac{\psi_m}{L_{ss}} \omega_r - i_{ds}^r \omega_r + \frac{1}{L_{ss}} u_{qs}^r,$$

$$\frac{di_{ds}^r}{dt} = -\frac{r_s}{L_{ss}} i_{ds}^r + i_{qs}^r \omega_r + \frac{1}{L_{ss}} u_{ds}^r$$

$$\frac{d\omega_r}{dt} = \frac{3P^2 \psi_m}{8J} i_{qs}^r - \frac{B_m}{J} \omega_r$$

the controller can be represented as

$$u_{qs}^r = u_{qs}^{r\ lin} + u_{qs}^{r\ cont} \quad \text{and} \quad u_{ds}^r = u_{ds}^{r\ lin} + u_{ds}^{r\ cont}$$

The linearizing feedback is

$$u_{qs}^{r\ lin} = \left(L_{ls} + \tfrac{3}{2}\overline{L}_m \right) i_{ds}^r \omega_r \quad \text{and} \quad u_{ds}^{r\ lin} = -\left(L_{ls} + \tfrac{3}{2}\overline{L}_m \right) i_{qs}^r \omega_r$$

and the mathematical model is linearized to

$$\frac{di_{qs}^r}{dt} = -\frac{r_s}{L_{ls} + \tfrac{3}{2}\overline{L}_m} i_{qs}^r - \frac{\psi_m}{L_{ls} + \tfrac{3}{2}\overline{L}_m} \omega_r + \frac{1}{L_{ls} + \tfrac{3}{2}\overline{L}_m} u_{qs}^r$$

$$\frac{di_{ds}^r}{dt} = -\frac{r_s}{L_{ls} + \tfrac{3}{2}\overline{L}_m} i_{ds}^r + \frac{1}{L_{ls} + \tfrac{3}{2}\overline{L}_m} u_{ds}^r$$

$$\frac{d\omega_r}{dt} = \frac{3P^2\psi_m}{8J} i_{qs}^r - \frac{B_m}{J}\omega_r - \frac{P}{2J}T_L$$

The eigenvalues are found to be

$$\lambda_1 = -\frac{\left(2B_m L_{ss} + 2r_s J - \sqrt{4B_m^2 L_{ss}^2 - 8B_m L_{ss} r_s J + 4r_s^2 J^2 - 6L_{ss} J^3 \psi_m^2 P^2} \right)}{4L_{ss}J}$$

$$\lambda_2 = -\frac{\left(2B_m L_{ss} + 2r_s J + \sqrt{4B_m^2 L_{ss}^2 - 8B_m L_{ss} r_s J + 4r_s^2 J^2 - 6L_{ss} J^3 \psi_m^2 P^2} \right)}{4L_{ss}J}$$

and

$$\lambda_3 = -\frac{r_s}{L_{ss}}$$

The real parts of these eigenvalues are negative. Thus, the stability is guaranteed.

For the resulting linear system, one can design controllers. For example, let us apply the proportional stabilizing controller as given by

$$u_{qs}^{r\ cont} = -[k_{iq} \ \ k_{id} \ \ k_\omega] \begin{bmatrix} i_{qs}^r \\ i_{ds}^r \\ \omega_r \end{bmatrix} \quad \text{and} \quad u_{ds}^{r\ cont} = 0$$

The eigenvalues for the closed-loop system are found to be

$$\lambda_1 = -\frac{\left(2k_{iq}J + 2B_m L_{ss} + 2r_s J - \sqrt{4k_{iq}^2 J^2 - 8k_{iq}JB_m L_{ss} + 8k_{iq}J^2 r_s + 4B_m^2 L_{ss}^2 - 8B_m L_{ss} r_s J + 4r_s^2 J^2 - 6L_{ss}J^3\psi_m P^2(\psi_m + k_\omega)} \right)}{4L_{ss}J}$$

$$\lambda_2 = -\frac{\left(2k_{iq}J + 2B_mL_{ss} + 2r_sJ + \sqrt{4k_{iq}^2J^2 - 8k_{iq}JB_mL_{ss} + 8k_{iq}J^2r_s + 4B_m^2L_{ss}^2 - 8B_mL_{ss}r_sJ + 4r_s^2J^2 - 6L_{ss}^3\psi_mP^2(\psi_m + k_\omega)}\right)}{4L_{ss}J}$$

$$\lambda_3 = -\frac{r_s}{L_{ss}}$$

Hence, applying the linearizing nonlinear feedback and controller

$$u_{qs}^{r\ cont} = -[k_{iq}\ k_{id}\ k_\omega]\begin{bmatrix} i_{qs}^r \\ i_{ds}^r \\ \omega_r \end{bmatrix}, \quad u_{ds}^{r\ cont} = 0$$

one can tentatively conclude that stability, specified transient quantities, and dynamic performance are attained.

Mathematically, feedback linearization may reduce the complexity of the corresponding analysis and design. However, even from a mathematical standpoint, simplification and *optimum* performance would be achieved at the expense of large control efforts required due to the linearizing feedback. This leads to saturation. It must be emphasized that there does not exist the need to linearize because the open-loop system is uniformly asymptotically stable as was illustrated using the Lyapunov stability theory.

The most critical problem is that the linearizing feedback

$$u_{ds}^{r\ lin} = -\left(L_{ls} + \frac{3}{2}\overline{L}_m\right)i_{qs}^r\omega_r$$

cannot be applied due to the unbalance conditions. In fact, to guarantee the balanced operation, one supplies

$$u_{ds}^r = 0 \quad \text{and} \quad u_{0s}^r = 0$$

That is, nonlinear linearizing feedback $u_{ds}^{lin} = -(L_{ls} + \frac{3}{2}\overline{L}_m)i_{qs}^r\omega_r$ cannot be applied. Hence, the feedback linearizing controllers cannot be applied to control systems with synchronous micromotors. Other methods need to be developed to solve the motion control problem, methods that do not entail the applied voltages to the saturation limits to cancel beneficial nonlinearities $-i_{ds}^r\omega_r$ and $i_{qs}^r\omega_r$, methods that do not lead to unbalanced motor operation.

We derived the expressions for the eigenvalues. In pole placement design, the specification on the optimum (desired) transient responses in terms of system models and feedback coefficients is equivalent to the specification imposed on desired transfer functions of closed-loop systems. Clearly, the desired eigenvalues can be specified by the designer, and these eigenvalues are used to find the corresponding feedback gains. However, there is no guarantee that these eigenvalues can be achieved (most MEMS and NEMS cannot be linearized, and, even if one can mathematically perform the linearization, the saturation effect is a factor). Furthermore, the pole placement concept, while guaranteeing the desired location of the characteristic eigenvalues, can lead to positive feedback coefficients and unrobust closed-loop systems. Hence, the stability, robustness to parameter variations, and system performance must be examined.

9.5   Constrained Control of Nonlinear MEMS and NEMS

In general, MEMS and NEMS are modeled by nonlinear differential equations. Our goal is to minimize

$$J(x(\cdot), u(\cdot)) = \int_{t_0}^{t_f} (W_{xu}(x, u)) \, dt$$

subject to the system described by nonlinear differential equations

$$\dot{x}(t) = F(x) + B(x)u, \quad x(t_0) = x_0$$

The positive-definite and continuously differentiable integrand function $W_{xu}(\cdot)$: $\mathbb{R}^c \times \mathbb{R}^m \to \mathbb{R}_{\geq 0}$ is used. We assume that $F(\cdot)$: $\mathbb{R}_{\geq 0} \times \mathbb{R}^n \to \mathbb{R}^n$ and $B(\cdot)$: $\mathbb{R}_{\geq 0} \times \mathbb{R}^n \to \mathbb{R}^{n \times m}$ are continuous and Lipschitz.

To find an optimal control, the necessary conditions for optimality must be studied. The Hamiltonian is given by

$$H\left(x, u, \frac{\partial V}{\partial x}\right) = W_{xu}(x, u) + \left(\frac{\partial V}{\partial x}\right)^T (F(x) + B(x)u)$$

where $V(\cdot)$: $\mathbb{R}^n \to \mathbb{R}_{\geq 0}$ is the smooth and bounded return function, $V(0) = 0$.

The first- and second-order necessary conditions for optimality are

$$\frac{\partial H\left(x, u, \frac{\partial V}{\partial x}\right)}{\partial u} = 0 \quad \text{and} \quad \frac{\partial^2 H\left(x, u, \frac{\partial V}{\partial x}\right)}{\partial u \times \partial u^T} > 0$$

Using the first-order necessary condition for optimality, one derives the control function $u(\cdot)$:$(t_0, t_f) \to \mathbb{R}^m$, which minimizes the functional. Constrained optimization of microsystems is a topic of great practical interest. We consider the systems modeled by nonlinear differential equations

$$\dot{x}^{sys}(t) = F_s(x^{sys}) + B_s(x^{sys})u^{2w+1}, \quad y = H(x^{sys}), \quad u_{min} \leq u \leq u_{max}, \quad x^{sys}(t_0) = x_0^{sys}$$

where $x^{sys} \in X_s$ is the state vector, $u \in U$ is the vector of control inputs, $y \in Y$ is the measured output, and w is the nonnegative integer.

Using the Hamilton-Jacobi theory, the bounded controllers can be synthesized for continuous-time microsystems. To design the tracking controller, we augment the system and exogenous dynamics. Hence we have

$$\dot{x}^{sys}(t) = F_s(x^{sys}) + B_s(x^{sys})u^{2w+1}, \quad y = H(x^{sys}), \quad u_{min} \leq u \leq u_{max}, \quad x^{sys}(t_0) = x_0^{sys}$$

$$\dot{x}^{ref}(t) = Nr - y = Nr - H(x^{sys})$$

Using the augmented state vector $x = [\begin{smallmatrix} x^{sys} \\ x^{ref} \end{smallmatrix}] \in X$, one obtains

$$\dot{x}(t) = F(x,r) + B(x)u^{2w+1}, \quad u_{min} \le u \le u_{max}, \quad x(t_0) = x_0, \quad x = \begin{bmatrix} x^{sys} \\ x^{ref} \end{bmatrix},$$

$$F(x,r) = \begin{bmatrix} F_s(x^{sys}) \\ -H(x^{sys}) \end{bmatrix} + \begin{bmatrix} 0 \\ N \end{bmatrix} r, \quad B(x) = \begin{bmatrix} B_s(x^{sys}) \\ 0 \end{bmatrix}$$

We map the control bounds imposed by a bounded, integrable, one-to-one globally Lipschitz, vector-valued continuous function Φ. Our goal is to analytically design the bounded admissible state-feedback controller in the form $u = \Phi(x)$. The most common Φ are the algebraic and transcendental (exponential, hyperbolic, logarithmic, trigonometric) continuously differentiable, integrable, one-to-one functions. For example, the odd one-to-one integrable function tanh with domain $(-\infty, +\infty)$ maps the control bounds. This function has the corresponding inverse function $\tanh^{-1}$ with range $(-\infty, +\infty)$.

The performance cost to be minimized is given as

$$J = \int_{t_0}^{\infty} [W_x(x) + W_u(u)]dt = \int_{t_0}^{\infty} \left[W_x(x) + (2w+1) \int (\Phi^{-1}(u))^T G^{-1} \operatorname{diag}(u^{2w}) du \right] dt$$

where $G^{-1} \in \mathbb{R}^{m \times m}$ is the positive-definite diagonal matrix.

Performance integrands W_x and W_u are real-valued, positive-definite, and continuously differentiable integrand functions. Using the properties of Φ, one concludes that inverse function Φ^{-1} is integrable. Hence, the integral $\int (\Phi^{-1}(u))^T G^{-1} \operatorname{diag}(u^{2w}) du$ exists.

Example 9.9

Synthesize the performance functional in order to design a bounded controller for the following microsystem:

$$\frac{dx}{dt} = ax + bu^3, \quad u_{min} \le u \le u_{max}$$

SOLUTION

Making use of the performance integrand

$$W_u(u) = (2w+1) \int (\Phi^{-1}(u))^T G^{-1} \operatorname{diag}(u^{2w}) du$$

and applying the integrable tanh function, one has the following positive-definite integrand:

$$W_u(u) = 3 \int \tanh^{-1} u G^{-1} u^2 du = \frac{1}{3} u^3 \tanh^{-1} u + \frac{1}{6} u^2 + \frac{1}{6} \ln(1 - u^2), \quad G^{-1} = \frac{1}{3}$$

In general, the hyperbolic tangent can be widely used to map the saturation effect. For the single-input case, one has

$$W_u(u) = (2w+1)\int u^{2w} \tanh^{-1}\frac{u}{k}\,du = u^{2w+1}\tanh^{-1}\frac{u}{k} - k\int\frac{u^{2w+1}}{k^2-u^2}\,du \qquad \Box$$

First- and second-order necessary conditions for optimality that the control guarantees a minimum to the Hamiltonian

$$H = W_x(x) + (2w+1)\int (\Phi^{-1}(u))^T G^{-1}\text{diag}(u^{2w})\,du + \frac{\partial V(x)}{\partial x}^T [F(x,r)+B(x)u^{2w+1}]$$

are

$$\frac{\partial H}{\partial u} = 0 \quad \text{and} \quad \frac{\partial^2 H}{\partial u \times \partial u^T} > 0$$

The positive-definite return function is $V(x_0) = \inf_{u\in U} J(x_0,u) = \inf J(x_0,\Phi(\cdot)) \ge 0$, and this function is found using the Hamilton-Jacobi-Bellman equation

$$-\frac{\partial V}{\partial t} = \min_{u\in U}\left\{ W_x(x) + (2w+1)\int (\Phi^{-1}(u))^T G^{-1}\text{diag}(u^{2w})\,du \right.$$

$$\left. + \frac{\partial V(x)}{\partial x}^T [F(x,r)+B(x)u^{2w+1}] \right\}$$

The controller is derived by minimizing the nonquadratic functional. The first-order necessary condition leads us to a bounded control law as given by

$$u = -\Phi\left(GB(x)^T \frac{\partial V(x)}{\partial x} \right), \quad u \in U$$

The second-order necessary condition for optimality is met because the matrix G^{-1} is positive definite. Hence, a unique, bounded, real-analytic, and continuous control candidate is designed.

The solution of the functional equation should be found using nonquadratic return functions. To obtain V, the performance cost is evaluated at the allowed values of the states and control. Linear and nonlinear functionals admit the final values, and the minimum value of the nonquadratic cost is given by power-series forms

$$J_{\min} = \sum_{i=0}^{\eta} v(x_0)^{\frac{2(i+\gamma+1)}{2\gamma+1}}, \quad \eta = 0,1,2,\ldots, \quad \gamma = 0,1,2,\ldots$$

The solution of the partial differential equation is satisfied by a continuously differentiable positive-definite return function

$$V(x) = \sum_{i=0}^{\eta} \frac{2\gamma+1}{2(i+\gamma+1)}\left(x^{\frac{i+\gamma+1}{2\gamma+1}} \right)^T K_i x^{\frac{i+\gamma+1}{2\gamma+1}}$$

where matrices K_i are found by solving the Hamilton-Jacobi equation.

The nonlinear bounded controller is given as

$$u = -\Phi\left(GB(x)^T \sum_{i=0}^{\eta} \text{diag}\left[x(t)^{\frac{i-\gamma}{2\gamma+1}}\right]K_i(t)x(t)^{\frac{i+\gamma+1}{2\gamma+1}}\right),$$

$$\text{diag}\left[x(t)^{\frac{i-\gamma}{2\gamma+1}}\right] = \begin{bmatrix} x_1^{\frac{i-\gamma}{2\gamma+1}} & 0 & \cdots & 0 & 0 \\ 0 & x_2^{\frac{i-\gamma}{2\gamma+1}} & \cdots & 0 & 0 \\ \vdots & \vdots & \ddots & \vdots & \vdots \\ 0 & 0 & \cdots & x_{c-1}^{\frac{i-\gamma}{2\gamma+1}} & 0 \\ 0 & 0 & \vdots & 0 & x_c^{\frac{i-\gamma}{2\gamma+1}} \end{bmatrix}$$

9.6 Optimization of Microsystems Using Nonquadratic Performance Functionals

The Hamilton-Jacobi theory, maximum principle, dynamic programming, and Lyapunov concept provide the designer with a general setup to solve linear and nonlinear optimal control problems for MEMS and NEMS. We have illustrated that the general results can be derived using different performance functionals. In particular, quadratic and nonquadratic integrands have been applied. These functionals lead to solution of optimization problems. The importance of synthesis of performance functionals resides on the fact that the control laws are predefined by the functionals used. In fact, the closed-loop MEMS and NEMS performance (settling time, overshoot, evolution of the state, output, and control variables) is defined by the performance integrands used.

The closed-loop microsystem performance is optimal (with respect to the minimizing functional) and stability margins are assigned in the specific sense as implied by the performance functionals. The innovative performance integrands that allow one to measure MEMS and NEMS performance as well as to design bounded control laws were reported. The system optimality and performance depend to a large extent on the specifications imposed (desired steady-state and dynamic performance, e.g., settling and rise time, accuracy, static error, overshoot, bandwidth, etc.) and the inherent system capabilities, e.g., state and control bounds.

As was illustrated, to design the *admissible* (bounded) control laws, we minimize

$$J = \int_{t_0}^{t_f}\left[\frac{1}{2}x^T Q x + \int (\Phi^{-1}(u))^T G du\right]dt$$

using the bounded, integrable, one-to-one, real-analytic globally Lipschitz continuous function, $\Phi(\cdot):\mathbb{R}^n \to \mathbb{R}^m$, $\Phi \in U \subset \mathbb{R}^m$.

For Equation 9.7 and Equation 9.8, the minimization of this nonquadratic functional for linear and nonlinear MEMS and NEMS with $u_{\min} \le u \le u_{\max}$, $u \in U$, gives

$$-\frac{\partial V}{\partial t} = \min_{u \in U}\left\{\frac{1}{2}x^T Q x + \int (\Phi^{-1}(u))^T G du + \frac{\partial V}{\partial x}^T [F(x)+B(x)u]\right\} \quad \text{for} \quad \dot{x}(t) = Ax + Bu$$

and

$$-\frac{\partial V}{\partial t} = \min_{u \in U} \left\{ \tfrac{1}{2} x^T Q x + \int \left(\Phi^{-1}(u) \right)^T G du + \frac{\partial V}{\partial x}^T [F(x) + B(x)u] \right\} \quad \text{for}$$

$$\dot{x}(t) = F(x) + B(x)u$$

Using the first-order necessary condition for optimality, the *admissible* controllers are found as

$$u = -\Phi\left(G^{-1} B^T \frac{\partial V(x)}{\partial x} \right) \quad \text{and} \quad u = -\Phi\left(G^{-1} B(x)^T \frac{\partial V(x)}{\partial x} \right), \quad u \in U$$

These designed admissible control laws are bounded. Furthermore, the second-order necessary condition for optimality, as well as sufficient conditions, is satisfied. Having emphasized the results for linear and nonlinear microsystems in the design of bounded control laws, we focus our attention on the synthesis of performance functionals and design of control algorithms.

Consider a class of linear MEMS modeled by Equation 9.8. We propose to apply the following performance functional

$$J = \int_{t_0}^{t_f} \tfrac{1}{2} \left[\omega(x)^T Q \omega(x) + \dot{\omega}(x)^T P \dot{\omega}(x) \right] dt \tag{9.29}$$

where $\omega(\cdot) : \mathbb{R}^n \to \mathbb{R}_{\geq 0}$ is the differentiable real-analytic continuous function, and $Q \in \mathbb{R}^{n \times n}$ and $P \in \mathbb{R}^{n \times n}$ are the positive-definite diagonal weighting matrices.

Using Equation 9.29, the system transient performance and stability are specified by two integrands: $\omega(x)^T Q \omega(x)$ and $\dot{\omega}(x)^T P \dot{\omega}(x)$. These integrands are given as the nonlinear functions of the states and the rate of the variables changes. It is evident that the performance functional (Equation 9.29) depends on the microsystem dynamics (states and control variables), control efforts, energy, etc. In particular, taking note of Equation 9.8, we have

$$\dot{\omega}(x) = \frac{\partial \omega}{\partial x} \dot{x} = \frac{\partial \omega}{\partial x} (Ax + Bu)$$

Hence, making use of Equation 9.29, one has the following form of the functional that we will minimize to derive the optimal controller:

$$J = \int_{t_0}^{t_f} \tfrac{1}{2} \left[\omega(x)^T Q \omega(x) + \dot{x}^T \frac{\partial \omega}{\partial x}^T P \frac{\partial \omega}{\partial x} \dot{x} \right] dt$$

$$= \int_{t_0}^{t_f} \tfrac{1}{2} \left[\omega(x)^T Q \omega(x) + (Ax + Bu)^T \frac{\partial \omega}{\partial x}^T P \frac{\partial \omega}{\partial x} (Ax + Bu) \right] dt \tag{9.30}$$

The $\omega(x)$ is the differentiable and integrable real-valued continuous function. For example, we can apply
$\omega(x) = x$ (this leads to the quadratic integrand function that was used)

$$\omega(x) = x^3, \quad \omega(x) = x^5, \quad \text{or} \quad \omega(x) = e^{-x}$$

For linear microsystems (Equation 9.8) and the performance functional (Equation 9.29), the Hamiltonian function is

$$H\left(x, u, \frac{\partial V}{\partial x}\right) = \tfrac{1}{2}\omega(x)^T Q\omega(x) + \tfrac{1}{2}\dot\omega(x)^T P\dot\omega(x) + \frac{\partial V}{\partial x}^T (Ax + Bu) \qquad (9.31)$$

The application of the first-order condition for optimality $\partial H/\partial u = 0$ gives the following *optimal* controller:

$$u = -\left(B^T \frac{\partial \omega}{\partial x}^T P \frac{\partial \omega}{\partial x} B\right)^{-1} B^T \left(\frac{\partial \omega}{\partial x}^T P \frac{\partial \omega}{\partial x} Ax + \frac{\partial V}{\partial x}\right) \qquad (9.32)$$

The second-order condition for optimality $\partial^2 H/(\partial u \times \partial u^T) > 0$ is guaranteed. In particular, from Equation 9.31, one has

$$\frac{\partial^2 H}{\partial u \times \partial u^T} = B^T \frac{\partial \omega}{\partial x}^T P \frac{\partial \omega}{\partial x} B > 0$$

because $\omega(x)$ is chosen such that $(\partial \omega/\partial x)B$ has full rank, and $P > 0$.

Performance integrands $\omega(x)^T Q\omega(x)$ and $\dot\omega(x)^T P\dot\omega(x)$ are used in Equation 9.29. The synthesis of the performance integrands results in the integrable and differentiable function $\omega(x)$. For example, applying, $\omega(x) = x$, we have

$$J = \int_{t_0}^{t_f} \tfrac{1}{2}[x^T Qx + (Ax + Bu)^T P(Ax + Bu)]dt \qquad (9.33)$$

From Equation 9.33, one obtains the functional equation

$$-\frac{\partial V}{\partial t} = \min_u \left\{ \tfrac{1}{2}[x^T Qx + (Ax + Bu)^T P(Ax + Bu)] + \frac{\partial V}{\partial x}^T (Ax + Bu) \right\} \qquad (9.34)$$

The *optimal* controller is found using the first-order necessary condition for optimality. In particular,

$$u = -(B^T PB)^{-1} B^T \left(PAx + \frac{\partial V}{\partial x} \right) \qquad (9.35)$$

The solution of the functional equation (Equation 9.34) is given by the quadratic return function (Equation 9.13) $V = \tfrac{1}{2}x^T Kx$, where the unknown symmetric matrix $K \in \mathbb{R}^{n \times n}$ is found by solving the following nonlinear differential equation:

$$-\dot{K} = Q - KB(B^T PB)^{-1}B^T K, \quad K(t_f) = K_f \tag{9.36}$$

Making use of Equation 9.35 and $V = \frac{1}{2}x^T Kx$, we have the following linear controller:

$$u = -(B^T PB)^{-1}B^T (PA + K)x \tag{9.37}$$

This controller is different from the conventional linear quadric control law (Equation 9.17). Furthermore, the equations to compute matrix K are different (see Equation 9.16 and Equation 9.36). The classical and reported results are compared using Table 9.1.

We study nonlinear dynamic microsystems (Equation 9.7). The performance functional is synthesized in the following form:

$$J = \int_{t_0}^{t_f} \frac{1}{2}\left[\omega(x)^T Q\omega(x) + [F(x) + B(x)u]^T \frac{\partial \omega}{\partial x}^T P \frac{\partial \omega}{\partial x}[F(x) + B(x)u]\right]dt \tag{9.38}$$

For system (Equation 9.7) and performance functional (Equation 9.38), we have the following Hamiltonian function:

$$H\left(x, u, \frac{\partial V}{\partial x}\right) = \frac{1}{2}\omega(x)^T Q\omega(x) + \frac{1}{2}[F(x) + B(x)u]^T \frac{\partial \omega}{\partial x}^T P \frac{\partial \omega}{\partial x}[F(x) + B(x)u]$$

$$+ \frac{\partial V}{\partial x}^T [F(x) + B(x)u] \tag{9.39}$$

The positive-definite return function $V(\cong): \mathbb{R}^n \to \mathbb{R}_{\geq 0}$ satisfies the following differential equation:

$$-\frac{\partial V}{\partial t} = \min_{u}\left\{\frac{1}{2}\omega(x)^T Q\omega(x) + \frac{1}{2}[F(x) + B(x)u]^T \frac{\partial \omega}{\partial x}^T P \frac{\partial \omega}{\partial x}[F(x) + B(x)u]\right.$$

$$\left. + \frac{\partial V}{\partial x}^T [F(x) + B(x)u]\right\} \tag{9.40}$$

TABLE 9.1

Comparison for Linear Microsystems $\dot{x}(t) = Ax + Bu$

Performance Functionals	Classical quadratic: $J = \int_{t_0}^{t_f} \frac{1}{2}(x^T Qx + u^T Gu)dt$
	Proposed (generalized): $J = \int_{t_0}^{t_f} \frac{1}{2}[\omega(x)^T Q\dot\omega(x) + \dot\omega(x)^T P\dot\omega(x)]dt$
Control	Classical quadratic: $u = -G^{-1}B^T Kx$
	Proposed (generalized): $u = -\left(B^T \frac{\partial \omega}{\partial x}^T P \frac{\partial \omega}{\partial x} B\right)^{-1} B^T \left(\frac{\partial \omega}{\partial x}^T P \frac{\partial \omega}{\partial x} Ax + \frac{\partial V}{\partial x}\right)$
	and $u = -\left(B^T PB\right)^{-1}B^T (PA + K)x$ for $\omega(x) = x$
Riccati Equations	Classical quadratic: $-\dot{K} = Q + A^T K + K^T A - K^T BG^{-1}B^T K$
	Proposed (generalized): $-\dot{K} = Q - KB\left(B^T PB\right)^{-1}B^T K$ for $\omega(x) = x$

From Equation 9.40, taking note of the first-order condition for optimality, one finds the control law as given by

$$u = -\left(B(x)^T \frac{\partial \omega}{\partial x}^T P \frac{\partial \omega}{\partial x} B(x) \right)^{-1} B(x)^T \left(\frac{\partial \omega}{\partial x}^T P \frac{\partial \omega}{\partial x} F(x) + \frac{\partial V}{\partial x} \right) \tag{9.41}$$

The controller (Equation 9.41) is an *optimal* control because the second-order necessary condition for optimality is met because

$$\frac{\partial^2 H}{\partial u \times \partial u^T} = B(x)^T \frac{\partial \omega}{\partial x}^T P \frac{\partial \omega}{\partial x} B(x) > 0$$

The solution of the partial differential equation (Equation 9.40) is satisfied by the return function

$$V(x) = \sum_{i=0}^{\eta} \frac{2\gamma+1}{2(i+\gamma+1)} \left(x^{\frac{i+\gamma+1}{2\gamma+1}} \right)^T K_i x^{\frac{i+\gamma+1}{2\gamma+1}}, \quad \eta = 0, 1, 2, \dots$$

Example 9.10

For the first-order system considered in Example 9.1

$$\frac{dx}{dt} = ax + bu$$

obtain the generalized performance functional assigning $\omega(x) = x$, $Q = 1$ and $P = 1$. Derive the optimal control. Study the stability of the closed-loop system.

SOLUTION

Using Equation 9.39, the performance functional is found to be

$$J = \int_{t_0}^{\infty} \frac{1}{2} \left[Q\omega(x)^2 + P\left(\frac{\partial \omega}{\partial x}\right)^2 (a^2 x^2 + 2abxu + b^2 u^2) \right] dt$$

$$= \int_{t_0}^{\infty} \frac{1}{2} (x^2 + a^2 x^2 + 2abxu + b^2 u^2) dt$$

Making use of the quadratic return function $V = \frac{1}{2} k x^2$, the *optimal* linear controller (Equation 9.32) is

$$u = -\frac{1}{b}(a + k)x$$

Solving the differential equation (Equation 9.36) $-\dot{k} = 1 - k^2$, we obtain $k = 1$. Thus, the controller is $u = -(1/b)(a+1)x$. The closed-loop system is stable and evolves as $dx/dt = -x$.

Example 9.11

For the system $dx/dt = ax + bu$, minimize the performance functional (Equation 9.39). Derive the performance integrands to synthesize the time-optimal controller. Note that this microsystem was studied in Example 9.1 and Example 9.10.

SOLUTION

In the performance functional (Equation 9.39),

$$J = \int_{t_0}^{\infty} \frac{1}{2} \left[Q\omega(x)^2 + P\left(\frac{\partial \omega}{\partial x}\right)^2 (a^2x^2 + 2abxu + b^2u^2) \right] dt$$

we synthesize the following nonquadratic integrand

$$\omega(x) = \tanh x$$

The nonquadratic performance functional is found to be

$$J = \int_{t_0}^{\infty} \frac{1}{2} [\tanh^2 x + \text{sech}^4 x(a^2x^2 + 2abxu + b^2u^2)] dt$$

In this functional, we assign the weighting coefficients as $Q = 1$ and $P = 1$.

This performance functional with the synthesized integrands can be straightforwardly examined. For $x \ll 1$, (small state variations), $\tanh^2 x \approx x^2$ and $\text{sech}^4 x \approx 1$. That is, if $x \ll 1$, the performance functional used can be related to the

$$J = \int_{t_0}^{\infty} \frac{1}{2} \left[Q\omega(x)^2 + P\frac{\partial^2 \omega}{\partial x^2} (a^2x^2 + 2abxu + b^2u^2) \right] dt$$

i.e., equivalent to the performance functional used in Example 9.10 when we set $\omega(x) = x$.

If $x \gg 1$ (large state variations), we have $\tanh^2 x \approx 1$ and $\text{sech}^4 x \approx 0$. Hence, the performance functional if $x \gg 1$ is

$$J \approx \frac{1}{2} \int_{t_0}^{\infty} dt$$

This performance functional is used to solve the time-optimal (minimum time) problem.

Taking note of the first-order necessary condition for optimality, one finds an *optimal* controller as

$$u = -\frac{a}{b}x - \frac{1}{b\,\text{sech}^4 x}\frac{\partial V}{\partial x}$$

The functional equation to be solved is

$$-\frac{\partial V}{\partial t} = \frac{1}{2}\tanh^2 x - \frac{1}{2\operatorname{sech}^4 x}\frac{\partial^2 V}{\partial x^2}$$

Letting $V = \frac{1}{2}kx^2$, we have

$$u = -\frac{a}{b}x - \frac{1}{b\operatorname{sech}^4 x}kx$$

It should be emphasized that the quadratic return function $V = \frac{1}{2}kx^2$ approximates the solution of the nonlinear functional partial differential equation. In general, nonquadratic return functions must be used. However, for illustrative purposes we apply the quadratic return function.

The closed-loop system evolves as

$$\frac{dx}{dt} = -\frac{k}{\operatorname{sech}^4 x}x$$

If $x \ll 1$, $\operatorname{sech}^4 x \approx 1$, and thus,

$$u = -\frac{a+k}{b}x$$

Then, the system dynamics is $dx/dt = -kx$.

For $x \gg 1$ we have the high feedback gain to bring the closed-loop system to the origin.

9.7 Hamilton-Jacobi Theory and Quantum Mechanics

This section studies several key problems in analysis, control, and optimization of nanosystems. From the optimization perspective, it is illustrated that the Schrödinger equation can be derived using the Hamilton-Jacobi principle.[8,11] The importance of the results is that the Schrödinger equation was obtained from the closed-loop solution through optimization of the functional. This establishes the relationship between the Hamilton-Jacobi theory and quantum mechanics.

The Hamiltonian for a particle is given as

$$\underset{v,p}{\Delta}\int_{t_0}^{t_f}[pv - H(t,x,p)]dt = 0$$

where $\underset{v,p}{\Delta}$ is the variation of the succeeding expression with respect to v and p.

Using the optimal (stationary) value of the integral

$$\int_{t_0}^{t_f}[pv - H(t,x,p)]dt$$

as denoted by $V(t,x)$, one has

$$\underset{v,p}{S}\left[\frac{dV}{dt} + pv - H(t,x,p)\right] = \underset{v,p}{S}\left[\frac{\partial V}{\partial t} + \frac{\partial V}{\partial x}v + pv - H(t,x,p)\right] = 0$$

where $\underset{v,p}{S}$ is the optimal (stationary) value obtained by varying v and p.
Therefore, we have

$$p = -\frac{\partial V}{\partial x}, \quad \dot{p} = -\frac{\partial V}{\partial x}\frac{\partial V}{\partial t} - v\frac{\partial^2 V}{\partial x^2} = -\frac{\partial H}{\partial x} + \frac{\partial H}{\partial p}\frac{\partial p}{\partial x}$$

$$v = \frac{\partial H}{\partial p}, \quad H = \frac{\partial V}{\partial t}$$

The variables x and p are independent, and therefore, the Hamiltonian equations of motion

$$\dot{p} = -\frac{\partial H}{\partial x}, \quad \dot{x} = \frac{\partial H}{\partial p}$$

result.

The complex functions are used in quantum mechanics, and we replace the displacement x by the complex variable q, and instead of velocity v we have $dq = vdt + ndz$, where n is complex, $n = -i\hbar(1/m)$, and z is the white noise (normalized Wiener process).

Therefore, one obtains

$$\underset{v,p}{\Delta E}\int_{t_0}^{t_f}[pv - H(t,q,p)]dt = 0$$

where E denotes the expectation.

Making use of the variational principle, from $Edz = 0$, $Edz^2 = dt$, and $Edq^2 = n^2dt$, we have

$$\underset{v,p}{S}\left[\frac{dV}{dt} + pv + v\frac{\partial V}{\partial q} + \tfrac{1}{2}n^2\frac{\partial^2 V}{\partial q^2} - H(t,q,p)\right] = 0$$

The minimization gives the following Hamiltonian equations:

$$p = -\frac{\partial V}{\partial q}, \quad v = \frac{\partial H}{\partial p}, \quad H = \frac{\partial V}{\partial t} - \frac{i\hbar}{2m}\frac{\partial^2 V}{\partial q^2}$$

Letting $V = i\hbar \log \Psi$ and taking note of $p = -i\hbar(\partial/\partial q)$, one finds

$$p = -\frac{\partial V}{\partial q} = \Psi^{-1}p\Psi, \quad i\hbar\frac{\partial^2 V}{\partial q^2} = \Psi^{-1}p^2\Psi - p^2$$

Thus, we obtain the Schrödinger equation

$$E\Psi = i\hbar\frac{\partial}{\partial t}\Psi(t,q) = \left[-\frac{\hbar^2}{2m}\frac{\partial^2}{\partial q^2} + \Pi(q)\right]\Psi(t,q) = \left[\frac{p^2}{2m} + \Pi(q)\right]\Psi(t,q) = H\Psi$$

The wave function $\Psi = e^{\frac{V}{i\hbar}}$ is the solution of the Schrödinger equation. The Schrödinger equation was derived using the Hamilton-Jacobi principle.[11] Furthermore, it was illustrated that the Schrödinger equation was found by minimizing the functional. It is very important that the Schrödinger equation was derived from the closed-loop solution through optimization of the functional because it establishes the relationship between the Hamilton-Jacobi theory and quantum mechanics. The Schrödinger equation leads to the solution in the form of wave functions, while the Hamiltonian-Jacobi concept results in the optimal cost function. Furthermore, it was shown that the Schrödinger equation results from a closed-loop solution of

$$\underset{v,p}{\Delta}E\int_{t_0}^{t_f}[pv - H(t,x,p)]dt = 0$$

and the closed-loop solution represents the goal-seeking behavior (dynamics) of nature.

9.8 Lyapunov Stability Theory in Analysis and Control of MEMS and NEMS

The MEMS dynamics are described by nonlinear differential equations. In particular,

$$\dot{x}(t) = F(x,r,d) + B(x)u, \quad y = H(x), \quad u_{min} \leq u \leq u_{max}, \quad x(t_0) = x_0 \tag{9.42}$$

where x is the state vector (displacement, position, velocity, current, voltage, etc.), u is the bounded control vector (supplied voltage), r and y are the measured reference and output vectors, and d is the disturbance vector (load and friction torques, etc.).

In Equation 9.42, the state-space equation is $\dot{x}(t) = F(x,r,d) + B(x)u$, while the output equation is $y = H(x)$. The control bounds are represented as $u_{min} \leq u \leq u_{max}$.

Let us examine the stability of time-varying nonlinear dynamic systems described by

$$\dot{x}(t) = F(t,x), \quad x(t_0) = x_0, \quad t \geq 0$$

The following theorem is formulated.

Theorem 9.1

Consider the microsystem described by nonlinear differential equations

$$\dot{x}(t) = F(t,x), \quad x(t_0) = x_0, \quad t \geq 0$$

If there exists a positive-definite scalar function $V(t, x)$, called the Lyapunov function, with continuous first-order partial derivatives with respect to t and x

$$\frac{dV}{dt} = \frac{\partial V}{\partial t} + \left(\frac{\partial V}{\partial x}\right)^T \frac{dx}{dt} = \frac{\partial V}{\partial t} + \left(\frac{\partial V}{\partial x}\right)^T F(t, x)$$

then (1) The equilibrium state of $\dot{x}(t) = F(t, x)$ is stable if the total derivative of the positive-definite function $V(t, x) > 0$ is $dV/dt \leq 0$.

(2) The equilibrium state of $\dot{x}(t) = F(t, x)$ is uniformly stable if the total derivative of the positive-definite decreasing function $V(t, x) > 0$ is $dV/dt \leq 0$.

(3) The equilibrium state of $\dot{x}(t) = F(t, x)$ is uniformly asymptotically stable in the large if the total derivative $V(t, x) > 0$ of is negative definite, that is, $dV/dt < 0$.

(4) The equilibrium state of $\dot{x}(t) = F(t, x)$ is exponentially stable in the large if there exist the K_∞-functions $\rho_1(\cdot)$ and $\rho_2(\cdot)$ and K-function $\rho_3(\cdot)$ such that

$$\rho_1(\|x\|) \leq V(t, x) \leq \rho_2(\|x\|) \quad \text{and} \quad \frac{dV(x)}{dt} \leq -\rho_3(\|x\|)$$

Examples are studied to illustrate how this theorem can be straightforwardly applied.

Example 9.12

Study the stability of a microstructure that is modeled by $\dot{x}(t) = F(x)$. In particular, two nonlinear time-invariant differential equations are

$$\dot{x}_1(t) = -x_1^5 - x_1^3 x_2^4, \quad \dot{x}_2(t) = -x_2^9, \quad t \geq 0$$

These differential equations describe the microstructure end-to-end unforced dynamics.

SOLUTION

A scalar positive-definite function is expressed in the quadratic form as

$$V(x_1, x_2) = \tfrac{1}{2}\left(x_1^2 + x_2^2\right)$$

The total derivative is found to be

$$\frac{dV(x_1, x_2)}{dt} = \left(\frac{\partial V}{\partial x}\right)^T \frac{dx}{dt} = \left(\frac{\partial V}{\partial x}\right)^T F(x) = \frac{\partial V}{\partial x_1}\left(-x_1^5 - x_1^3 x_2^4\right)$$

$$+ \frac{\partial V}{\partial x_2}\left(-x_2^9\right) = -x_1^6 - x_1^4 x_2^4 - x_2^{10}$$

Thus, we have

$$\frac{dV(x_1, x_2)}{dt} < 0$$

The total derivative of $V(x_1, x_2) > 0$ is negative definite.

Therefore, the equilibrium state of the microstructure is uniformly asymptotically stable.

Example 9.13

Study the stability of a microstructure if the mathematical model of the end-to-end behavior is given by the time-varying nonlinear differential equation $\dot{x}(t) = F(t, x)$. In particular,

$$\dot{x}_1(t) = -x_1 + x_2^3, \quad \dot{x}_2(t) = -e^{-10t}x_1 x_2^2 - 5x_2 - x_2^3, \quad t \geq 0$$

SOLUTION

A scalar positive-definite function is chosen in the quadratic form as

$$V(t, x_1, x_2) = \tfrac{1}{2}\left(x_1^2 + e^{10t}x_2^2\right), V(t, x_1, x_2) > 0$$

The total derivative is given by

$$\frac{dV(t, x_1, x_2)}{dt} = \frac{\partial V}{\partial t} + \frac{\partial V}{\partial x_1}\left(-x_1 + x_2^3\right) + \frac{\partial V}{\partial x_2}\left(-e^{-10t}x_1 x_2^2 - 5x_2 - x_2^3\right) = -x_1^2 - e^{10t}x_2^4$$

Therefore, the total derivative is negative definite, i.e.,

$$\frac{dV(x_1, x_2)}{dt} < 0$$

Hence, making use of the theorem, one concludes that the equilibrium state is uniformly asymptotically stable.

Example 9.14

Study the stability of the microsystem that is described by the differential equations

$$\dot{x}_1(t) = -x_1 + x_2$$

$$\dot{x}_2(t) = -x_1 - x_2 - x_2|x_2|, \quad t \geq 0$$

SOLUTION

The positive-definite scalar Lyapunov candidate is chosen in the following form:

$$V(x_1, x_2) = \tfrac{1}{2}\left(x_1^2 + x_2^2\right)$$

Thus,

$$V(x_1, x_2) > 0.$$

The total derivative is

$$\frac{dV(x_1, x_2)}{dt} = x_1\dot{x}_1 + x_2\dot{x}_2 = -x_1^2 - x_2^2\left(1 + |x_2|\right)$$

Therefore,

$$\frac{dV(x_1, x_2)}{dt} < 0$$

Hence, the equilibrium state of the microsystem is uniformly asymptotically stable, and the quadratic function $V(x_1, x_2) = \frac{1}{2}(x_1^2 + x_2^2)$ is the Lyapunov function.

Example 9.15   STABILITY OF THE SYNCHRONOUS MICROMOTOR

Consider a microdrive actuated by the permanent-magnet synchronous micromotor. This micromotor was examined in Example 9.8. In the rotor reference frame, the mathematical model ($T_L = 0$) is given as

$$\frac{di_{qs}^r}{dt} = -\frac{r_s}{L_{ls} + \frac{3}{2}\overline{L}_m} i_{qs}^r - \frac{\psi_m}{L_{ls} + \frac{3}{2}\overline{L}_m}\omega_r - i_{ds}^r\omega_r + \frac{1}{L_{ls} + \frac{3}{2}\overline{L}_m} u_{qs}^r$$

$$\frac{di_{ds}^r}{dt} = -\frac{r_s}{L_{ls} + \frac{3}{2}\overline{L}_m} i_{ds}^r + i_{qs}^r\omega_r + \frac{1}{L_{ls} + \frac{3}{2}\overline{L}_m} u_{ds}^r$$

$$\frac{d\omega_r}{dt} = \frac{3P^2\psi_m}{8J} i_{qs}^r - \frac{B_m}{J}\omega_r$$

Study the microdrive stability, allowing for the following cases:

$u_{qs}^r = 0$ and $u_{ds}^r = 0$ (open-loop microsystem)

$u_{qs}^r \neq 0$, $u_{qs}^r = -k_\omega\omega_r$, and $u_{ds}^r = 0$ (closed-loop microsystem)

SOLUTION

For the open-loop microdrive, we have $u_{qs}^r = 0$ and $u_{ds}^r = 0$. Hence, the differential equations $\dot{x}(t) = F(x)$ are rewritten as

$$\frac{di_{qs}^r}{dt} = -\frac{r_s}{L_{ls} + \frac{3}{2}\overline{L}_m} i_{qs}^r - \frac{\psi_m}{L_{ls} + \frac{3}{2}\overline{L}_m}\omega_r - i_{ds}^r\omega_r \, , \qquad \frac{di_{ds}^r}{dt} = -\frac{r_s}{L_{ls} + \frac{3}{2}\overline{L}_m} i_{ds}^r + i_{qs}^r\omega_r ,$$

$$\frac{d\omega_r}{dt} = \frac{3P^2\psi_m}{8J} i_{qs}^r - \frac{B_m}{J}\omega_r$$

The state-space and output equations (Equation 9.42) model MEMS. We illustrate the application of the matrix notations for the permanent-magnet synchronous micromotor. Using the three differential equations given above, rewritten in the state-space form $\dot{x}(t) = F(x) = Ax + F_N(x)$, one obtains

$$\dot{x}(t) = \begin{bmatrix} -\dfrac{r_s}{L_{ls} + \frac{3}{2}\overline{L}_m} & 0 & -\dfrac{\psi_m}{L_{ls} + \frac{3}{2}\overline{L}_m} \\[3ex] 0 & -\dfrac{r_s}{L_{ls} + \frac{3}{2}\overline{L}_m} & 0 \\[3ex] \dfrac{3P^2\psi_m}{8J} & 0 & -\dfrac{B_m}{J} \end{bmatrix} \begin{bmatrix} i_{qs}^r \\[2ex] i_{ds}^r \\[2ex] \omega_r \end{bmatrix} + \begin{bmatrix} -i_{ds}^r\omega_r \\[2ex] i_{qs}^r\omega_r \\[2ex] 0 \end{bmatrix}$$

where

$$
A = \begin{bmatrix} -\dfrac{r_s}{L_{ls}+\frac{3}{2}\overline{L}_m} & 0 & -\dfrac{\psi_m}{L_{ls}+\frac{3}{2}\overline{L}_m} \\[2ex] 0 & -\dfrac{r_s}{L_{ls}+\frac{3}{2}\overline{L}_m} & 0 \\[2ex] \dfrac{3P^2\psi_m}{8J} & 0 & -\dfrac{B_m}{J} \end{bmatrix} \quad \text{and} \quad F_N(x) = \begin{bmatrix} -i_{ds}^r\omega_r \\[2ex] i_{qs}^r\omega_r \\[2ex] 0 \end{bmatrix}
$$

Using the quadratic positive-definite Lyapunov function

$$
V(i_{qs}^r, i_{ds}^r, \omega_r) = \tfrac{1}{2}\left(i_{qs}^{r\,2} + i_{ds}^{r\,2} + \omega_r^2 \right)
$$

the expression for the total derivative is found to be

$$
\frac{dV(i_{qs}^r, i_{ds}^r, \omega_r)}{dt} = -\frac{r_s}{L_{ss}}\left(i_{qs}^{r\,2} + i_{ds}^{r\,2} \right) - \frac{B_m}{J}\omega_r^2 - \frac{8J\psi_m - 3P^2 L_{ss}\psi_m}{8JL_{ss}}i_{qs}^r\omega_r
$$

Thus,

$$
\frac{dV\left(i_{qs}^r, i_{ds}^r, \omega_r \right)}{dt} < 0
$$

One concludes that the equilibrium state of an open-loop microdrive is uniformly asymptotically stable.

Consider the closed-loop microsystem. To guarantee balanced operation, we set the following *quadrature* and *direct* voltages:

$$
u_{qs}^r = -k_\omega \omega_r \quad \text{and} \quad u_{ds}^r = 0
$$

That is, the quadrature voltage is proportional to the angular velocity, and k_ω is the proportional feedback gain. Thus, the following differential equations result for the closed-loop microdrive:

$$
\frac{di_{qs}^r}{dt} = -\frac{r_s}{L_{ls}+\frac{3}{2}\overline{L}_m}i_{qs}^r - \frac{\psi_m}{L_{ls}+\frac{3}{2}\overline{L}_m}\omega_r - i_{ds}^r\omega_r - \frac{1}{L_{ls}+\frac{3}{2}\overline{L}_m}k_\omega\omega_r
$$

$$
\frac{di_{ds}^r}{dt} = -\frac{r_s}{L_{ls}+\frac{3}{2}\overline{L}_m}i_{ds}^r + i_{qs}^r\omega_r
$$

$$
\frac{d\omega_r}{dt} = \frac{3P^2\psi_m}{8J}i_{qs}^r - \frac{B_m}{J}\omega_r
$$

In the state-space form, one has

$$
\dot{x}(t) = \begin{bmatrix} -\dfrac{r_s}{L_{ls}+\frac{3}{2}\overline{L}_m} & 0 & -\dfrac{\psi_m+k_\omega}{L_{ls}+\frac{3}{2}\overline{L}_m} \\[2ex] 0 & -\dfrac{r_s}{L_{ls}+\frac{3}{2}\overline{L}_m} & 0 \\[2ex] \dfrac{3P^2\psi_m}{8J} & 0 & -\dfrac{B_m}{J} \end{bmatrix}\begin{bmatrix} i_{qs}^r \\[2ex] i_{ds}^r \\[2ex] \omega_r \end{bmatrix} + \begin{bmatrix} -i_{ds}^r\omega_r \\[2ex] i_{qs}^r\omega_r \\[2ex] 0 \end{bmatrix}
$$

Taking note of the quadratic positive-definite Lyapunov function

$$V\left(i_{qs}^r, i_{ds}^r, \omega_r\right) = \frac{1}{2}\left(i_{qs}^{r\,2} + i_{ds}^{r\,2} + \omega_r^2\right)$$

we obtain

$$\frac{dV\left(i_{qs}^r, i_{ds}^r, \omega_r\right)}{dt} = -\frac{r_s}{L_{ss}}\left(i_{qs}^{r\,2} + i_{ds}^{r\,2}\right) - \frac{B_m}{J}\omega_r^2 - \frac{8J(\psi_m + k_\omega) - 3P^2 L_{ss}\psi_m}{8JL_{ss}}i_{qs}^r\omega_r$$

Hence, $V(i_{qs}^r, i_{ds}^r, \omega_r) > 0$ and

$$\frac{dV\left(i_{qs}^r, i_{ds}^r, \omega_r\right)}{dt} < 0$$

Therefore, the conditions for asymptotic stability are guaranteed.

It is obvious that the rate of decreasing of

$$\frac{dV\left(i_{qs}^r, i_{ds}^r, \omega_r\right)}{dt}$$

affects the microdrive dynamics and feedback coefficient. The derived expression for

$$\frac{dV\left(i_{qs}^r, i_{ds}^r, \omega_r\right)}{dt}$$

clearly illustrates the role of the proportional feedback gain k_ω.

It has been shown that dynamic systems can be controlled to attain the desired transient dynamics, stability margins, etc. Let us study how to solve the motion control problem with the ultimate goal to synthesize tracking controllers applying Lyapunov's stability theory. Using the reference (command) vector $r(t)$ and the system output $y(t)$, the tracking error (which ideally must be zero) is $e(t) = Nr(t) - y(t)$. For example, if one assigns the reference angular velocity, the tracking error is found as the difference between the command and actual angular velocities, i.e., $e(t) = r(t) - y(t) = \omega_{ref}(t) - \omega_r(t)$.

The Lyapunov theory is applied to derive the *admissible* control laws. That is, the admissible bounded controller should be designed as a continuous function within the constrained control set

$$U = \{u \in \mathbb{R}^m : u_{min} \leq u \leq u_{max}, \quad u_{min} < 0, \quad u_{max} > 0\} \subset \mathbb{R}^m$$

In Section 9.2, the PID bounded controllers were given by Equation 9.2 and Equation 9.3. The bounded PID control laws can be expressed using the Lyapunov function. Most importantly, one can find the coefficients of the controllers by making use of the criteria imposed on the Lyapunov function, for example,

$$\frac{dV(t, x, e)}{dt} < 0$$

Specifically, the feedback gains are derived by solving the nonlinear matrix inequality.

Using the Lyapunov candidate $V(t, x, e)$, the bounded PID-type controller with the state feedback extension is expressed as

$$u = \text{sat}_{u_{\min}}^{u_{\max}} \left(G_x(t)B(x)^T \frac{\partial V(t,x,e)}{\partial x} + G_x(t)B(e)^T \frac{\partial V(t,x,e)}{\partial e} \right) \tag{9.43}$$

where $G_x(\cdot): \mathbb{R}_{\geq 0} \to \mathbb{R}^{m \times m}$ and $G_e(\cdot): \mathbb{R}_{\geq 0} \to \mathbb{R}^{m \times m}$ are the bounded symmetric weighting matrix-functions, $G_x > 0$ and $G_e > 0$; and $V(\cdot): \mathbb{R}_{\geq 0} \times \mathbb{R}^n \times \mathbb{R}^b \to \mathbb{R}_{\geq 0}$ is the continuously differentiable real-analytic function with respect to $x \in X$ and $e \in E$.

Theorem 9.2

Consider the closed-loop microelectromechanical systems (Equation 9.42) with controller (Equation 9.43).

Then, 1) Solutions of the closed-loop system are uniformly ultimately bounded.

2) The equilibrium point is exponentially stable in the convex and compact state evolution set $X(X_0, U, R, D) \subset \mathbb{R}^n$.

3) Tracking is ensured and disturbance attenuation is guaranteed in the state-error evolution set $XE(X_0, E_0, U, R, D) \subset \mathbb{R}^n \times \mathbb{R}^b$.

If there exists a continuous differentiable function $V(t,x,e)$ in XE such that for all $x \in X$, $e \in E$, $u \in U$, $r \in R$, and $d \in D$ on (t_0, ∞)

(i)
$$\rho_1 \|x\| + \rho_2 \|e\| \leq V(t,x,e) \leq \rho_3 \|x\| + \rho_4 \|e\|$$

and for the closed-loop system that is described by (Equation 9.42 and Equation 9.43) the following inequality holds:

(ii)
$$\frac{dV(t,x,e)}{dt} \leq -\rho_5 \|x\| - \rho_6 \|e\|$$

Here, $\rho_1(\cdot): \mathbb{R}_{\geq 0} \to \mathbb{R}_{\geq 0}$, $\rho_2(\cdot): \mathbb{R}_{\geq 0} \to \mathbb{R}_{\geq 0}$, $\rho_3(\cdot): \mathbb{R}_{\geq 0} \to \mathbb{R}_{\geq 0}$, and $\rho_4(\cdot): \mathbb{R}_{\geq 0} \to \mathbb{R}_{\geq 0}$ are the K_∞-functions, and $\rho_5(\cdot): \mathbb{R}_{\geq 0} \to \mathbb{R}_{\geq 0}$ and $\rho_6(\cdot): \mathbb{R}_{\geq 0} \to \mathbb{R}_{\geq 0}$ are the K-functions.

The major problem is to design the Lyapunov candidate functions. The quadratic Lyapunov function has already been applied. The nonquadratic Lyapunov candidates can be designed by applying the following equation:

$$V(t,x,e) = \sum_{i=0}^{\eta} \frac{2\gamma+1}{2(i+\gamma+1)} \left(x^{\frac{i+\gamma+1}{2\gamma+1}} \right)^T K_{xi}(t) x^{\frac{i+\gamma+1}{2\gamma+1}}$$

$$+ \sum_{i=0}^{\varsigma} \frac{2\beta+1}{2(i+\beta+1)} \left(e^{\frac{i+\beta+1}{2\beta+1}} \right)^T K_{ei}(t) e^{\frac{i+\beta+1}{2\beta+1}} \tag{9.44}$$

To design the Lyapunov functions, the nonnegative integers were used. In particular, $\eta = 0,1,2,...$, $\gamma = 0,1,2,...$, and $\varsigma = 0,1,2,...$, $\beta = 0,1,2,...$,

From Equation 9.43 and Equation 9.44, one obtains a bounded *admissible* proportional controller

$$u = \text{sat}_{u_{\min}}^{u_{\max}}\left(G_x(t)B(x)^T \sum_{i=0}^{\eta} \text{diag}\left[x^{\frac{i-\gamma}{2\gamma+1}} \right] K_{xi}(t)x^{\frac{i+\gamma+1}{2\gamma+1}} \right.$$

$$\left. + G_e(t)B(e)^T \sum_{i=0}^{\varsigma} \text{diag}\left[e^{\frac{i-\beta}{2\beta+1}} \right] K_{ei}(t)e^{\frac{i+\beta+1}{2\beta+1}} \right) \qquad (9.45)$$

Here, $K_{xi}(\cdot)\colon \mathbb{R}_{\geq 0} \to \mathbb{R}^{n \times n}$ and $K_{ei}(\cdot)\colon \mathbb{R}_{\geq 0} \to \mathbb{R}^{b \times b}$ are the matrix functions.

It is evident that by assigning the integers to be zero, the well-known quadratic Lyapunov candidate results, i.e., $V(t,x,e) = \frac{1}{2}x^T K_{x0}(t)x + \frac{1}{2}e^T K_{e0}(t)e$. Using this Lyapunov candidate, the bounded controller is found to be

$$u = \text{sat}_{u_{\min}}^{u_{\max}}(G_x(t)B(x)^T K_{x0}(t)x + G_e(t)B(e)^T K_{e0}(t)e)$$

By making use of the microsystem dynamics, the total derivative of the Lyapunov candidate $V(t,x,e,)$ is obtained. Solving the inequality

$$\frac{dV(t,x,e)}{dt} \leq -\rho_5\|x\| - \rho_6\|e\|$$

or

$$\frac{dV(t,x,e)}{dt} \leq 0$$

one finds the feedback coefficients.[7]

Example 9.16

Consider a microdrive actuated by a permanent-magnet DC micromotor with IC (*step-down* converter) as shown in Figure 9.25. Find the control algorithm.

SOLUTION

Using the Kirchhoff laws and the averaging concept, we have the follow-ing nonlinear state-space model:

$$\begin{bmatrix} \dfrac{du_a}{dt} \\ \dfrac{di_L}{dt} \\ \dfrac{di_a}{dt} \\ \dfrac{d\omega_r}{dt} \end{bmatrix} = \begin{bmatrix} 0 & \dfrac{1}{C_L} & -\dfrac{1}{C_L} & 0 \\ -\dfrac{1}{L_L} & 0 & 0 & 0 \\ \dfrac{1}{L_a} & 0 & -\dfrac{r_a}{L_a} & -\dfrac{k_a}{L_a} \\ 0 & 0 & \dfrac{k_a}{J} & -\dfrac{B_m}{J} \end{bmatrix} \begin{bmatrix} u_a \\ i_L \\ i_a \\ \omega_r \end{bmatrix} + \begin{bmatrix} 0 \\ \left(\dfrac{V_d}{L_L u_{t\max}} - \dfrac{r_d}{L_L u_{t\max}} i_L \right) \\ 0 \\ 0 \end{bmatrix} u_c - \begin{bmatrix} 0 \\ 0 \\ 0 \\ \dfrac{1}{J} \end{bmatrix} T_L, \quad u_c \in [0 \quad 10]\,V$$

A bounded control law should be synthesized.

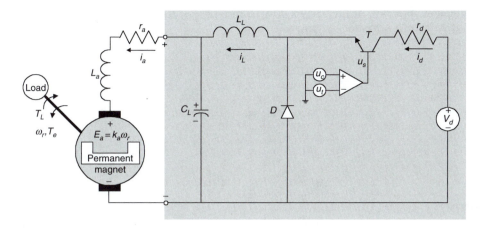

FIGURE 9.25
Permanent-magnet DC micromotor with *step-down* converter.

From Equation 9.44, one finds the nonquadratic function $V(e,x)$ as

$$V(e,x) = \tfrac{1}{2}k_{e0}e^2 + \tfrac{1}{4}k_{e1}e^4 + \tfrac{1}{2}k_{ei0}e^2 + \tfrac{1}{4}k_{ei1}e^4 + \tfrac{1}{2}[u_a \ \ i_L \ \ i_a \ \ \omega_r]K_{x0}\begin{bmatrix} u_a \\ i_L \\ i_a \\ \omega_r \end{bmatrix} + \tfrac{1}{2}[u_a \ \ i_L \ \ i_a \ \ \omega_r]K\begin{bmatrix} u_a \\ i_L \\ i_a \\ \omega_r \end{bmatrix}$$

where $K_{x0} \in \mathbb{R}^{4\times4}$.

Therefore, from Equation 9.43, one obtains

$$u = \mathrm{sat}_0^{+10}\Big(k_1 e + k_2 e^3 + k_3 \int e\, dt + k_4 \int e^3 dt - k_5 u_a - k_6 i_L - k_7 i_a - k_8 \omega_r$$

$$- (k_9 u_a + k_{10} i_L + k_{11} i_a + k_{12} \omega_r) i_L \Big)$$

If the criteria (*i*) and (*ii*) imposed on the Lyapunov pair are guaranteed, one concludes that the stability conditions are satisfied. The positive-definite nonquadratic function was used. The feedback gains must be found by solving the inequality $dV(e,x)/dt < 0$, which represents the second condition imposed on the Lyapunov function. The following inequality is solved:

$$\frac{dV(e,x)}{dt} \le -\tfrac{1}{2}\|e\|^2 - \tfrac{1}{4}\|e\|^4 - \tfrac{1}{2}\|x\|^2$$

to find the feedback coefficients in the designed controller.

Many examples of the design of tracking controllers for electromechanical systems are reported.[7]

Example 9.17

Study the flip-chip microsystem that consists of an eight-layered lead magnesium niobate actuator (3-mm diameter, 0.25-mm thickness), actuated by a monolithic high-voltage switching regulator, $-1 \leq u \leq 1$.[8] Find the control law. A set of differential equations to model the microactuator dynamics is

$$\frac{dF_y}{dt} = -9472F_y + 13740F_y u + 48593u$$

$$\frac{dv_y}{dt} = 947F_y - 94100v_y - 2609v_y^{1/3} - 2750x_y$$

$$\frac{dx_y}{dt} = v_y$$

SOLUTION

The control is bounded. In particular, we have $-1 \leq u \leq 1$. The error is the difference between the reference and microactuator position. That is,

$$e(t) = r(t) - y(t)$$

where $y(t) = x_y$ and $r(t) = r_y(t)$.

Using Equation 9.44, we have

$$V(e,x) = \tfrac{1}{2}k_{e0}e^2 + \tfrac{1}{4}k_{e1}e^4 + \tfrac{1}{2}k_{ei0}e^2 + \tfrac{1}{4}k_{ei1}e^4 + \tfrac{1}{2}[F_y \ v_y \ x_y]K_{xo}\begin{bmatrix} F_y \\ v_y \\ x_y \end{bmatrix}$$

Applying the design procedure, a bounded control law is synthesized as

$$u = \mathrm{sat}_{-1}^{+1}\left(94827e + 2614e^3 + 4458\int e\,dt + 817\int e^3\,dt\right)$$

The feedback gains were found by solving the inequality

$$\frac{dV(e,x)}{dt} \leq -\|e\|^2 - \|e\|^4 - \|x\|^2$$

The criteria imposed on the Lyapunov pair are satisfied. In fact,

$$V(e,x) > \quad \text{and} \quad \frac{dV(e,x)}{dt} \leq 0$$

Hence, the bounded control law guarantees stability and ensures tracking. The experimental validation of stability and tracking is important. The controller is tested and examined. Figure 9.26 illustrates the transient dynamics for the position if the reference signal (desired position) is assigned to be $r_y(t) = 4 \times 10^{-6}\sin 1000t$. Figure 9.26 also documents the dynamics if the reference is $r_y(t) = const = 4 \times 10^{-6}$. From these end-to-end transient dynamics, it is evident that the stability is guaranteed, desired performance is achieved, and the output precisely follows the reference position $r_y(t)$. In general, the sensor accuracy must be studied by implementing control laws. In fact, the tracking error $e(t)$ is used to derive the control efforts (duty cycle).

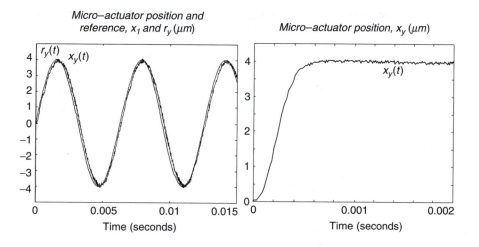

FIGURE 9.26
Transient output dynamics if $r_y(t) = 4 \times 10^{-6} \sin 1000t$ and $r_y(t) = const = 4 \times 10^{-6}$.

Example 9.18

Consider a flip-chip microservo with permanent-magnet stepper micromotor controlled by ICs.[8] Design the tracking control algorithm making use of the control bounds. The mathematical model in the *ab* variables is given in the form of nonlinear differential equations. In particular,

$$\frac{di_{as}}{dt} = -\frac{r_s}{L_{ss}}i_{as} + \frac{RT\psi_m}{L_{ss}}\omega_{rm}\sin(RT\theta_{rm}) + \frac{1}{L_{ss}}u_{as}$$

$$\frac{di_{bs}}{dt} = -\frac{r_s}{L_{ss}}i_{bs} - \frac{RT\psi_m}{L_{ss}}\omega_{rm}\cos(RT\theta_{rm}) + \frac{1}{L_{ss}}u_{bs}$$

$$\frac{d\omega_{rm}}{dt} = -\frac{RT\psi_m}{J}[i_{as}\sin(RT\theta_{rm}) + i_{bs}\cos(RT\theta_{rm})] - \frac{B_m}{J}\omega_{rm} - \frac{1}{J}T_L$$

$$\frac{d\theta_{rm}}{dt} = \omega_{rm}$$

Stepper micromotor parameters are $RT = 6$, $r_s = 60$ ohm, $\psi_m = 0.0064$ N · m/A, $L_{ss} = 0.05$ H, $B_m = 1.3 \times 10^{-7}$ N · m · sec/rad, and $J = 1.8 \times 10^{-8}$ kg · m². The phase voltages are bounded as $u_{min} \le u_{as} \le u_{max}$ and $u_{min} \le u_{bs} \le u_{max}$, where $u_{min} = -12$ V and $= 12$ V.

SOLUTION

The nonlinear controller is given as

$$u = \begin{bmatrix} u_{as} \\ u_{bs} \end{bmatrix} = \begin{bmatrix} -\sin(RT\theta_{rm}) & 0 \\ 0 & \cos(RT\theta_{rm}) \end{bmatrix} \text{sat}_{u_{min}}^{u_{max}}\left(G_x(t)B^T \frac{\partial V(t,x,e)}{\partial x} + G_e(t)B^T \frac{\partial V(t,x,e)}{\partial e} \right)$$

The rotor displacement is denoted as $\theta_{rm}(t)$, and the output is $y(t) = \theta_{rm}(t)$. Thus, the tracking error is $e(t) = r(t) - y(t)$. The Lyapunov candidate is found as

$$V(e,x) = \tfrac{3}{4} K_{e0} e^{4/3} + \tfrac{1}{2} K_{e1} e^2 + \tfrac{3}{4} K_{ei0} e^{4/3} + \tfrac{1}{2} K_{ei1} e^2 + \tfrac{1}{2} [i_{as} \quad i_{bs} \quad \omega_{rm} \quad \theta_{rm}] K_{x0} \begin{bmatrix} i_{as} \\ i_{bs} \\ \omega_{rm} \\ \theta_{rm} \end{bmatrix}$$

By solving

$$\frac{dV(e,x)}{dt} \leq -\|e\|^{4/3} - \|e\|^2 - \|x\|^2$$

a bounded controller is found as

$$u_{as} = -\sin(RT\theta_{rm})\operatorname{sat}_{-12}^{+12}\left(14e + 2.9e^{1/3} + \frac{1}{s}6.1e + \frac{1}{s}4.3e^{1/3} \right)$$

$$u_{bs} = \cos(RT\theta_{rm})\operatorname{sat}_{-12}^{+12}\left(14e + 2.9e^{1/3} + \frac{1}{s}6.1e + \frac{1}{s}4.3e^{1/3} \right)$$

The sufficient conditions for stability are satisfied because $V(e,x)$ and

$$\frac{dV(e,x)}{dt} < 0$$

Figure 9.27 and Figure 9.28 document the microservo dynamics if the reference (command) displacement was assigned to be 0.5 and 1 rad. From analytical and experimental results, one concludes that the robust stability and tracking are guaranteed.

9.9   Digital Control of MEMS and NEMS

9.9.1   Introduction to Digital Control and Transfer Function Concept

In microsystems, digital ICs can be utilized to implement control algorithms making use of the sensor data. Furthermore, diagnostics, filtering, data acquisition, and other tasks can be performed using discrete mathematics and digital computing. Therefore, digital control algorithms must be designed, and discrete-time models for microsystems are studied in this section.

To model MEMS and NEMS in the discrete-time domain, usually continuous-time mathematical models are derived. In fact, the majority of micro- and nanosystems and devices are analog. Correspondingly, the discretization of differential equations must be performed. The continuous-time signal $e(t)$ is sampled with the sampling period T_s, and the continuous- and discrete-time domains are related as $t = kT_s$, where k is the integer. The difference between continuous- and discrete-time domains should be emphasized. It was shown that using cornerstone physical laws, the evolution of MEMS and NEMS

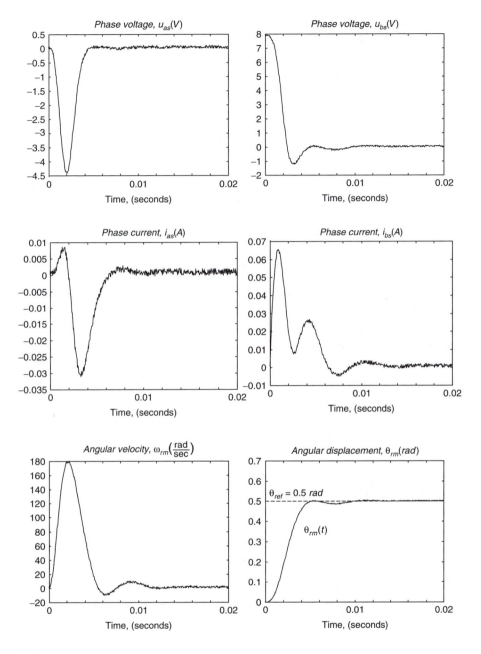

FIGURE 9.27
Transient output dynamics of microservo.

and their components are described by differential equations. Solving these differential equations, one finds the evolution of the state variables in the continuous-time domain. In contrast, discrete-time microsystem components can described by difference equations (digital controllers, digital sensors, etc.), and the evolution of these digital systems can be found in the discrete-time domain. Many MEMS and NEMS can be examined as hybrid systems, i.e., the systems integrate analog and digital components, devices, and subsystems.

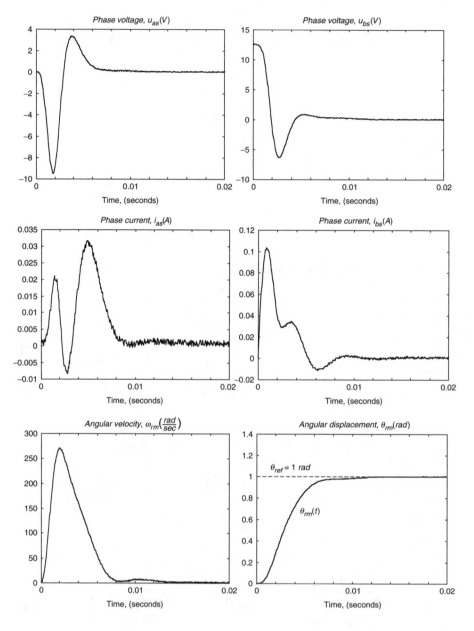

FIGURE 9.28
Transient output dynamics of microservo.

To perform the analysis and design of digital controllers, the differential equations are discretized because the use and design of digital controllers leads to the use of discrete-time and hybrid modeling and analysis paradigms.

Example 9.19

For the first-order linear constant-coefficient differential equation $dx/dt = -ax(t) + bu(t)$, derive the discrete-time model in the form of a difference equation. Find the z-domain transfer function. That is, discretize the differential equation in order to model the system in the discrete-time domain.

SOLUTION

From $dx/dt = -ax(t) + bu(t)$, performing discretization, we have

$$\left.\frac{dx}{dt}\right|_{t=kT_s} = -ax(kT_s) + bu(kT_s)$$

For a small sampling period T_s, the forward rectangular rule (Euler approximation) gives

$$\frac{dx}{dt} \approx \frac{x(t+T_s) - x(t)}{T_s}$$

Thus,

$$\left.\frac{dx}{dt}\right|_{t=kT_s} = \frac{x(kT_s + T_s) - x(kT_s)}{T_s}$$

Therefore, making use of the forward difference, one obtains

$$\frac{x(kT_s + T_s) - x(kT_s)}{T_s} = -ax(kT_s) + bu(kT_s)$$

We denote $x(t)$ and $u(t)$ at discrete instances t_k and t_{k+1} as

$$x_k = x(t)\big|_{t=kT_s}, \quad x_{k+1} = x(t)\big|_{t=(k+1)T_s} \quad \text{and} \quad u_k = u(t)\big|_{t=kT_s}$$

Therefore, one obtains

$$\frac{x_{k+1} - x_k}{T_s} = -ax_k + bu_k$$

where $x_{k+1} = x[(k+1)T_s]$, $x_k = x(kT_s)$, and $u_k = u(kT_s)$.

Thus, the following difference equation results:

$$x_{k+1} = (1 - aT_s)x_k + bT_s u_k$$

This difference equation can be written as

$$x_k = (1 - aT_s)x_{k-1} + bT_s u_{k-1}$$

From the obtained difference equation, the transfer function results. In particular,

$$G(z) = \frac{X(z)}{U(z)} = \frac{bT_s z^{-1}}{1 - (1 - aT_s)z^{-1}} = \frac{bT_s}{z - (1 - aT_s)}$$

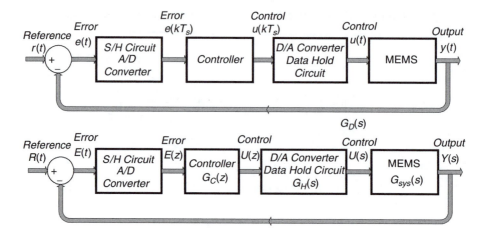

FIGURE 9.29
Block diagrams of hybrid microsystems with digital controllers.

That is, the continuous-time system is represented in the discrete-time domain by the difference equation found and in the z-domain by the transfer function obtained.

Hybrid microsystems integrate analog and digital components (see Figure 9.29). Mathematical modeling of systems and their components (including A/D and D/A converters, data hold circuits, etc.) should be studied. We assume that the microsystem dynamics is modeled by linear constant-coefficient (time-invariant) differential equations. The closed-loop system was documented in Figure 9.29 using the transfer function of system $G_{sys}(S)$, data hold circuit $G_H(S)$, and digital controller $G_C(S)$. To convert the discrete-time signals from microcontrollers and DSPs into piecewise continuous signals to be fed to ICs, distinct data hold circuits are used. Zero- and first-order data hold circuits are typically implemented to avoid the complexity and time delay associated with the application of high-order data hold circuits. The generic N-order data hold circuit with the zero-order data hold is documented in Figure 9.30.

For the zero-order data hold, the piecewise continuous data hold output is

$$h(t) = \sum_{k=0}^{\infty} e(kT_s)[1(t - kT_s) - 1(t - (k+1)T_s)]$$

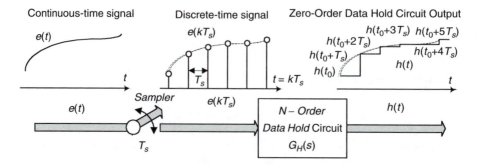

FIGURE 9.30
Sampler and N-order data hold; zero-order data hold circuit.

The output of the zero-order data hold is the piecewise continuous signal that is equal to the last sampled value (the value of the continuous-time signal at $t = kT_s$) until the next sampled value is available. This is illustrated in Figure 9.30. In general, we have the following mathematical representation:

$$h(kT_s + t) = h(kT_s) = e(t)\big|_{t=kT_s} \quad \text{for} \quad 0 \le t < T_s$$

Using two Laplace transforms $L[1(t)] = 1/s$ and $L[1(t - kT_s)] = e^{-kT_s s}/s$, one finds

$$L[h(t)] = \sum_{k=0}^{\infty} e(kT_s) \frac{e^{-kT_s s} - e^{-(k+1)T_s s}}{s} = \frac{1 - e^{-T_s s}}{s} \sum_{k=0}^{\infty} e(kT_s) e^{-kT_s s}$$

Taking note of the Laplace transform for the signal $e(kT_s)$, as given by

$$E_{sampled}(s) = \sum_{k=0}^{\infty} e(kT_s) e^{-kT_s s}$$

one obtains the transfer function of the zero-order data hold using the following relation:

$$L[h(t)] = \frac{1 - e^{-T_s s}}{s} \sum_{k=0}^{\infty} e(kT_s) e^{-kT_s s}$$

In particular, the transfer function of the zero-order data hold is $G_H(s) = 1 - (e^{-T_s s}/s)$.

The first-order data hold, which is also commonly used and performs the direct linear extrapolation, can be expressed in the time domain as

$$h(t) = 1(t) + \frac{t}{T_s} 1(t) - \frac{t - T_s}{T_s} 1(t - T_s) - 1(t - T_s)$$

This expression gives the output signal of the first-order hold in the time domain. The transfer function (s-domain representation) is

$$G_H(s) = \frac{1}{s} + \frac{1}{T_s s^2} - \frac{1}{T_s s^2} e^{-T_s s} - \frac{1}{s} e^{-T_s s} = (1 - e^{-T_s s}) \frac{T_s s + 1}{T_s s^2}$$

The system–hold augmented transfer function $G_D(s)$ for the dynamic systems $G_{sys}(s)$ and the data hold circuit $G_H(s)$ is

$$G_D(s) = G_H(s) G_{sys}(s)$$

The table of transforms for distinct continuous- and discrete-time signals, as well as their s- and z-domain representations, is reported in Table 9.2.

Example 9.20

Using the analog PID controller, derive the z-domain representations for digital P, PI, and PID controllers.

TABLE 9.2

Signals and *s*- and *z*-Transform Table

Laplace Transform $X(S)$	Time-Domain Signal $x(t)$	Time-Domain Signal $X(kT_s)$	z-Transform $X(z)$
$\dfrac{1}{s}$	Unit step $1(t)$	$1(kT_s)$	$\dfrac{1}{1-z^{-1}} = \dfrac{z}{z-1}$
$\dfrac{1}{s^2}$	$t1(t)$	$kT_s1(kT_s)$	$\dfrac{T_sz^{-1}}{(1-z^{-1})^2} = \dfrac{T_sz}{(z-1)^2}$
$\dfrac{2}{s^3}$	$t^21(t)$	$(kT_s)^21(kT_s)$	$\dfrac{T_s^2z^{-1}(1+z^{-1})}{(1-z^{-1})^3}$
$\dfrac{6}{s^4}$	$t^31(t)$	$(kT_s)^31(kT_s)$	$\dfrac{T_s^3z^{-1}(1+4z^{-1}+z^{-2})}{(1-z^{-1})^4}$
$\dfrac{24}{s^5}$	$t^41(t)$	$(kT_s)^41(kT_s)$	$\dfrac{T_s^4z^{-1}(1+11z^{-1}+11z^{-2}+z^{-3})}{(1-z^{-1})^5}$
$\dfrac{1}{s+a}$	$e^{-at}1(t)$	$e^{-akT_s}1(kT_s)$	$\dfrac{1}{1-e^{-aT_s}z^{-1}}$
$\dfrac{a}{s(s+a)}$	$(1-e^{-at})1(t)$	$(1-e^{-akT_s})1(kT_s)$	$\dfrac{(1-e^{-aT_s})z^{-1}}{(1-z^{-1})(1-e^{-aT_s}z^{-1})}$
$\dfrac{b-a}{(s+a)(s+b)}$	$(e^{-at}-e^{-bt})1(t)$	$(e^{-akT_s}-e^{-bkT_s})1(kT_s)$	$\dfrac{(e^{-aT_s}-e^{-bT_s})z^{-1}}{(1-e^{-aT_s}z^{-1})(1-e^{-bT_s}z^{-1})}$
$\dfrac{1}{(s+a)^2}$	$te^{-at}1(t)$	$kT_se^{-akT_s}1(kT_s)$	$\dfrac{T_se^{-aT_s}z^{-1}}{(1-e^{-aT_s}z^{-1})^2}$
$\dfrac{s}{(s+a)^2}$	$(1-at)e^{-at}1(t)$	$(1-akT_s)e^{-akT_s}1(kT_s)$	$\dfrac{1-(1+aT_s)e^{-aT_s}z^{-1}}{(1-e^{-aT_s}z^{-1})^2}$
$\dfrac{\omega_0}{s^2+\omega_0^2}$	$\sin(\omega_0t)1(t)$	$\sin(\omega_0kT_s)1(kT_s)$	$\dfrac{z^{-1}\sin(\omega_0T_s)}{1-2z^{-1}\cos(\omega_0T_s)+z^-}$
$\dfrac{s}{s^2+\omega_0^2}$	$\cos(\omega_0t)1(t)$	$\cos(\omega_0kT_s)1(kT_s)$	$\dfrac{1-z^{-1}\cos(\omega_0T_s)}{1-2z^{-1}\cos(\omega_0T_s)+z^{-2}}$
$\dfrac{\omega_0}{(s+a)^2+\omega_0^2}$	$e^{-at}\sin(\omega_0t)1(t)$	$e^{-akT_s}\sin(\omega_0kT_s)1(kT_s)$	$\dfrac{e^{-akT_s}z^{-1}\sin(\omega_0T_s)}{1-2e^{-aT_s}z^{-1}\cos(\omega_0T_s)+e^{-2aT_s}z^{-2}}$
$\dfrac{s+a}{(s+a)^2+\omega_0^2}$	$e^{-at}\cos(\omega_0t)1(t)$	$e^{-akT_s}\cos(\omega_0kT_s)1(kT_s)$	$\dfrac{1-e^{-aT_s}z^{-1}\cos(\omega_0T_s)}{1-2e^{-aT_s}z^{-1}\cos(\omega_0T_s)+e^{-2aT_s}z^{-2}}$

SOLUTION

The analog PID controller is $u(t) = k_pe(t) + k_i[e(t)/s] + k_dse(t)$, and the transfer function is

$$G_{PID}(s) = \frac{U(s)}{E(s)} = \frac{k_ds^2 + k_ps + k_i}{s}$$

For the proportional analog controller, one obtains

$$u_p(t) = k_pe(t) \quad \text{and} \quad G_P(s) = \frac{U_p(s)}{E(s)} = k_p$$

Thus, the proportional digital control law is

$$u_p(kT_s) = k_p e(kT_s) \quad \text{and} \quad G_P(z) = \frac{U_p(z)}{E(z)} = k_p$$

The integral $u_i(t) = k_i[e(t)/s]$ and derivative $u_d(t) = k_d s e(t)$ terms of the PID controller, with transfer functions

$$G_I(s) = \frac{U_I(s)}{E(s)} = \frac{k_i}{s} \quad \text{and} \quad G_D(s) = \frac{U_D(s)}{E(s)} = k_d s$$

can be discretized and represented in the z-domain. In particular, the z-transform table, given in Table 9.2, is used. For the integral part, using the Euler approximation, the transfer function is

$$G_i(z) = \frac{U_i(z)}{E(z)} = \frac{T_s}{1 - z^{-1}} = \frac{T_s z}{z - 1}$$

By using the trapezoidal approximation to find the derivative term, the first difference results, and

$$G_d(z) = \frac{U_d(z)}{E(z)} = \frac{1 - z^{-1}}{T_s} = \frac{z - 1}{T_s z}$$

By performing the summation of the derived transfer function terms, the P, PI, and PID controllers result.

There exist a great variety of analog PID-type controllers with the corresponding transfer functions $G_{PID}(s)$. For example, PID controller $u(t) = k_p e(t) + ki[e(t)/s] + k_d s e(t)$ has the following transfer functions:

$$G_{PID}(s) = \frac{U(s)}{E(s)} = \frac{k_d s^2 + k_p s + k_i}{s}$$

One finds the z-domain representation of the control signal as $U(z)$. The transfer functions of digital PID controllers can be straightforwardly derived. In the *error* form (the error signal is used to calculate the control), the following expressions result:

$$U(z) = \left(k_{dp} + \frac{k_{di}}{1 - z^{-1}} + k_{dd}(1 - z^{-1}) \right) E(z)$$

and

$$G_{PID}(z) = \frac{U(z)}{E(z)} = k_{dp} + \frac{k_{di}}{1 - z^{-1}} + k_{dd}(1 - z^{-1})$$

Hence, we have

$$G_{PID}(z) = \frac{(k_{dp} + k_{di} + k_{dd})z^2 - (k_{dp} + 2k_{dd})z + k_{dd}}{z^2 - z}$$

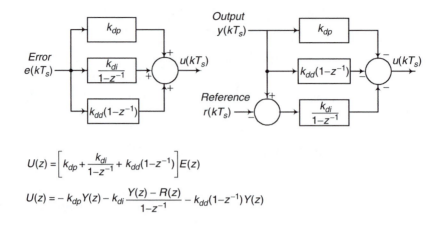

$$U(z) = \left[k_{dp} + \frac{k_{di}}{1-z^{-1}} + k_{dd}(1-z^{-1}) \right] E(z)$$

$$U(z) = -k_{dp} Y(z) - k_{di} \frac{Y(z) - R(z)}{1-z^{-1}} - k_{dd}(1-z^{-1}) Y(z)$$

FIGURE 9.31
Representations of *error* and *reference-output* forms of the digital controllers.

The *reference-output* form of the digital PID control law is

$$U(z) = -k_{dp} Y(z) - k_{di} \frac{Y(z) - R(z)}{1-z^{-1}} - k_{dd}(1-z^{-1}) Y(z)$$

These results can be visualized. For the *error* and *reference-output* forms, the block diagrams of the digital PID controllers are illustrated in Figure 9.31.

The feedback gains k_{dp}, k_{di}, and k_{dd} of the digital control laws are found by using the proportional, integral, and derivative coefficients of the analog PID controller (k_p, k_i, and k_d) and the sampling period T_s. In particular, one applies the following formulas:

$$k_{dp} = k_p - \frac{1}{2} k_{di}, \; k_{di} = T_s k_i, \quad \text{and} \quad k_{dd} = \frac{k_d}{T_s}$$

Using microcontrollers and DSPs, the PID controller can be implemented as

$$u(kT_s) = \underbrace{k_p e(kT_s)}_{Proportional} + \underbrace{\tfrac{1}{2} k_i T_s \sum_{i=1}^{k} [e((i-1)T_s) + e(iT_s)]}_{Integral} + \underbrace{\frac{k_d}{T_s} [e(kT_s) - e((k-1)T_s)]}_{Derivative}$$

To find the transfer function for systems and controllers in the z-domain, the Tustin approximation is applied to the transfer functions in the s-domain. In particular, from $z = e^{sT_s}$, we have $s = [1/T_s]\ln(z)$. The Tustin approximation is

$$\ln(z) \approx 2\frac{z-1}{z+1} = 2\frac{1-z^{-1}}{1+z^{-1}}$$

This approximation is obtained by truncating the series expansion of ln (z), and, in general,

$$\ln(z) = 2\left[\frac{z-1}{z+1} + \frac{1}{3}\left(\frac{z-1}{z+1}\right)^3 + \frac{1}{5}\left(\frac{z-1}{z+1}\right)^5 + \cdots \right], z > 0$$

Thus, we have

$$s = \frac{1}{T_s}\ln(z) \approx \frac{2}{T_s}\frac{z-1}{z+1} = \frac{2}{T_s}\frac{1-z^{-1}}{1+z^{-1}}$$

Example 9.21

Derive the expression for $G_{PID}(z)$ using the Tustin approximation if

$$G_{PID}(s) = \frac{U(s)}{E(s)} = \frac{k_d s^2 + k_p s + k_i}{s}$$

Find the expression to implement the digital controller, that is, derive $u(k)$.

SOLUTION

From

$$G_{PID}(s) = \frac{U(s)}{E(s)} = \frac{k_d s^2 + k_p s + k_i}{s}$$

making use of

$$s \approx \frac{2}{T_s}\frac{1-z^{-1}}{1+z^{-1}}$$

we have

$$G_{PID}(z) = \frac{U(z)}{E(z)} = \frac{k_d\left(\dfrac{2}{T_s}\dfrac{1-z^{-1}}{1+z^{-1}}\right)^2 + k_p \dfrac{2}{T_s}\dfrac{1-z^{-1}}{1+z^{-1}} + k_i}{\dfrac{2}{T_s}\dfrac{1-z^{-1}}{1+z^{-1}}}$$

$$= \frac{\left(2k_p T_s + k_i T_s^2 + 4k_d\right) + \left(2k_i T_s^2 - 8k_d\right)z^{-1} + \left(-2k_p T_s + k_i T_s^2 + 4k_d\right)z^{-2}}{2T_s\left(1-z^{-2}\right)}$$

Thus,

$$U(z) - U(z)z^{-2} = k_{e0}E(z) + k_{e1}E(z)z^{-1} + k_{e2}E(z)z^{-2}$$

where $k_{e0} = k_p + \frac{1}{2}k_i T_s + 2k_d/T_s$, $k_{e1} = k_i T_s - 4k_d/T_s$, and $k_{e2} = -k_p + \frac{1}{2}k_i T_s + 2k_d/T_s$
To implement the digital controller, the following control signal must be supplied:

$$u(k) = u(k-2) + k_{e0}e(k) + k_{e1}e(k-1) + k_{e2}e(k-2)$$

That is, to implement the analog PID controller as a digital control law, one uses the following components: $e(k)$, $e(k-1)$, $e(k-2)$, and $u\ (k-2)$.

The closed-loop system with digital controller $G_C(z)$ is illustrated in Figure 9.32.

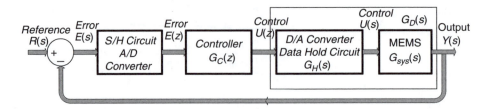

FIGURE 9.32
Block diagrams of the closed-loop MEMS with digital controllers.

The transfer function of the closed-loop microsystems with digital controllers $G_C(z)$ is

$$G(z) = \frac{Y(z)}{R(z)} = \frac{G_C(z)G_D(z)}{1 + G_C(z)G_D(z)}$$

If the digital PID controller is implemented, one has

$$G(z) = \frac{Y(z)}{R(z)} = \frac{G_{PID}(z)G_D(z)}{1 + G_{PID}(z)G_D(z)}$$

9.9.2   Control of Digital Microsystems with Permanent-Magnet DC Micromotors

Consider microsystems actuated by permanent-magnet DC micromotors (analog control was studied in Section 9.2.2). Our goal is to design the digital PID controllers, simulate and analyze microsystems, and demonstrate the results. To be specific, we will study the microstage reported in Figure 9.9. Our goal is to design digital control laws to attain the desired performance (fast displacement, zero tracking error, etc.).

Taking note of three differential equations that were derived to describe the evolution of the microsystem

$$\frac{di_a}{dt} = -\frac{r_a}{L_a}i_a - \frac{k_a}{L_a}\omega_r + \frac{1}{L_a}u_a \,,$$

$$\frac{d\omega_r}{dt} = \frac{1}{J}(T_e - T_{viscous} - T_L) = \frac{1}{J}(k_a i_a - B_m \omega_r - T_L)\,, \quad \frac{d\theta_r}{dt} = \omega_r$$

the block diagram of the closed-loop system with digital PID controller, converters, and data hold circuit is documented in Figure 9.33.

Augmenting the transfer function of the open-loop microsystem

$$G_{sys}(s) = \frac{Y(s)}{U(s)} = \frac{k_{gear}k_a}{s\left(L_a J s^2 + (r_a J + L_a B_m)s + r_a B_m + k_a^2\right)}$$

with the transfer function of the zero-order data hold

$$G_H(s) = \frac{1 - e^{-T_s s}}{s}$$

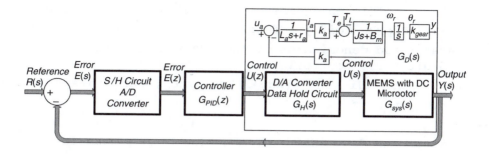

FIGURE 9.33
Block diagrams of the closed-loop microsystems with the digital PID controller.

one has the augmented transfer function. In particular,

$$G_D(s) = G_H(s)G_{sys}(s) = \frac{1 - e^{-T_s s}}{s} \frac{k_{gear}k_a}{s\left(L_a J s^2 + (r_a J + L_a B_m)s + r_a B_m + k_a^2\right)}$$

The rated armature voltage for the motor studied is ±30 V. The DC micromotor parameters are r_a = 200 ohm, L_a = 0.002 H, k_a = 0.2 V · sec/rad, k_a = 0.2 N · m/A, J = 0.00000002 kg · m², and B_m = 0.00000005 N · m · sec/rad.

The transfer function in the z-domain $G_D(z)$ is found from $G_D(s)$ by making use of the c2dm function. The filter function is used to model the dynamics. The following MATLAB m-file was developed:

```
% MEMS parameters
ra=200; La=0.002; ka=0.2; J=0.00000002; Bm=0.00000005;
kgear=0.01;
% Numerator and denominator of the open-loop MEMS
% transfer function
format short e
num_s=[ka*kgear];
den_s=[ra*J (ra*Bm+ka^2) 0];
% armature inductance is neglected
den_s=[La*J (ra*J+La*Bm) (ra*Bm+ka^2) 0];
% armature inductance is not neglected
num_s, den_s
pause;
% Numerator and denominator of the transfer function
% with zero-order data hold GD(z)
Ts=0.0005; % sampling time (sampling period) Ts
[num_dz,den_dz]=c2dm(num_s,den_s,Ts,'zoh');
num_dz, den_dz
pause;
% Feedback coefficient gains of the analog PID controller
kp=25000; ki=250; kd=25;
```

```
% Feedback coefficient gains of the digital PID controller
kdi=Ts*ki; kdp=kp-kdi/2; kdd=kd/Ts;
% Numerator and denominator of the transfer function of
% the digital PID controller
num_pidz=[(kdp+kdi+kdd) -(kdp+2*kdd) kdd]; den_pidz=[1 -1 0];
num_pidz, den_pidz
pause;
% Numerator and denominator of the closed-loop transfer
% function G(z)
num_z=conv(num_pidz,num_dz);
den_z=conv(den_pidz,den_dz)+conv(num_pidz,num_dz);
num_z, den_z
pause;
% Samples, t=k*Ts
k_final=100; k=0:1:k_final;
% Reference input r(t)=1.5 rad
ref=1.5; % reference (command) input is 1.5 rad
r=ref*ones(1,k_final+1);
% modeling of the servo-system output y(k)
y=filter(num_z,den_z,r);
% Plotting statement
plot(k,y,'o',k,y,'--');
title('MEMS Output: Angular Displacement y(k), [rad]');
xlabel('Discrete time k, t=kTs [seconds]');
ylabel('Displacement, y');
disp('end')
```

Having found the numerator and denominator of $G_{sys}(s)$ to be

```
num_s =
   2.0000e-003
den_s =
   4.0000e-011   4.0001e-006   4.0010e-002                    0
```

by neglecting the armature inductance, the following transfer function results for the open-loop system:

$$G_{sys}(s) = \frac{Y(s)}{U(s)} = \frac{2 \times 10^{-3}}{s(4 \times 10^{-11} s^2 + 4 \times 10^{-6} s + 4 \times 10^{-2})}$$

The sampling period is assigned to be 0.001 second, i.e., $T_s = 0.0001$ sec. The transfer function $G_D(z)$ in the z-domain is found using the reported MATLAB numerical results.

```
num_dz =
          0   1.6464e-006   1.7064e-006   2.6300e-008
den_dz =
   1.0000e+000  -1.3240e+000   3.2409e-001  -4.5389e-005
```

We have

$$G_D(z) = \frac{1.65 \times 10^{-6} z^2 + 1.71 \times 10^{-6} z + 2.63 \times 10^{-8}}{z^3 - 1.32 z^2 + 0.32 z - 4.53 \times 10^{-5}}$$

The transfer function of the digital PID controller is

$$G_{PID}(z) = \frac{(k_{dp} + k_{di} + k_{dd}) z^2 - (k_{dp} + 2 k_{dd}) z + k_{dp}}{z^2 - z}, \quad k_{dp} = k_p - \frac{1}{2} k_{di},$$

$$k_{di} = T_s k_i \quad \text{and} \quad k_{dd} = \frac{k_d}{T_s}$$

The feedback gains of the analog PID controller were assigned to be $k_p = 25000$, $k_i = 250$, and $k_d = 25$. The feedback coefficients of the digital controller are found using the equations $k_{dp} = k_p - \frac{1}{2} k_{di}$, $k_{di} = T_s k_i$, and $k_{dd} = k_d / T_s$. The numerator and denominator of the transfer function

$$G_{PID}(z) = \frac{(k_{dp} + k_{di} + k_{dd}) z^2 - (k_{dp} + 2 k_{dd}) z + k_{dp}}{z^2 - z}$$

are found. In particular, from

```
num_pidz =
    2.7500e+005  -5.2500e+005   2.5000e+005
den_pidz =
        1        -1         0
```

we have

$$G_{PID}(z) = \frac{2.75 \times 10^5 z^2 - 5.25 \times 10^5 z + 2.5 \times 10^5}{z^2 - z}$$

The transfer function of the closed-loop system is given by

$$G(z) = \frac{Y(z)}{R(z)} = \frac{G_{PID}(z) G_D(z)}{1 + G_{PID}(z) G_D(z)}$$

The following numerical results are found:

```
num_z =
         0   4.5277e-001  -3.9511e-001  -4.7703e-001   4.1280e-001
 6.5750e-003
den_z =
 1.0000e+000  -1.8713e+000   1.2530e+000  -8.0117e-001   4.1284e-001
 6.5750e-003
```

Thus, we have

$$G(z) = \frac{0.453 z^4 - 0.395 z^3 - 0.477 z^2 + 0.413 z + 6.58 \times 10^{-3}}{z^5 - 1.87 z^4 + 1.25 z^3 - 0.8 z^2 + 0.413 z + 6.58 \times 10^{-3}}$$

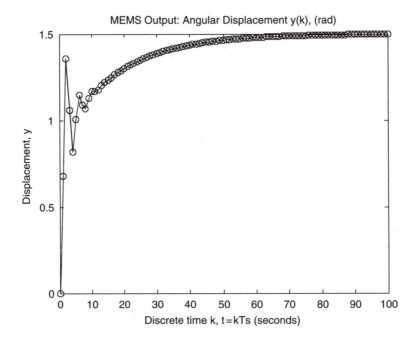

FIGURE 9.34
Output dynamics of the microsystem with digital PID controller.

The output dynamics for the reference input $r(kT_s) = 1.5$ rad, $k \geq 0$ is shown in Figure 9.34.
   The settling time is $k_{settling} T_s = 60 \times 0.0001 = 0.006$ sec, and there is no overshoot.
   It must be emphasized that the sampling time significantly influences the system dynamics. Let us increase the sampling time; for example, let $T_s = 0.0005$ sec. Using the MATLAB file reported for the sampling time 0.005 sec, we have

```
num_z =

          0   1.5011e+000  -2.1350e+000   3.8947e-001   2.4447e-001
1.4616e-005

den_z =
  1.0000e+000  -5.0250e-001  -1.1279e+000   3.8590e-001   2.4447e-001
1.4616e-005
```

That is, the closed-loop transfer function is found to be

$$G(z) = \frac{1.5z^4 - 2.14z^3 + 0.39z^2 + 0.244z + 1.46 \times 10^{-5}}{z^5 - 0.503z^4 - 1.13z^3 + 0.39z^2 + 0.244z + 1.46 \times 10^{-5}}$$

The output of the microservosystem $y(kt_s)$ is plotted in Figure 9.35.
   The settling time is $k_{settling} T_s = 40 \times 0.0005 = 0.002$ sec, and the overshoot is 50%.
   Thus, the sampling time significantly influences the closed-loop system performance. If $T_s = 0.001$ sec, the closed-loop system becomes unstable. However, as was illustrated in Section 9.2, the control bounds must be integrated in the analysis. The stability analysis (system is stable or unstable) of the closed-loop system without system nonlinearities does not provide accurate results. The stable simplified systems can be unstable as all nonlinearities are integrated, while unstable simplified systems can be stable as one examines nonlinearities and constraints.

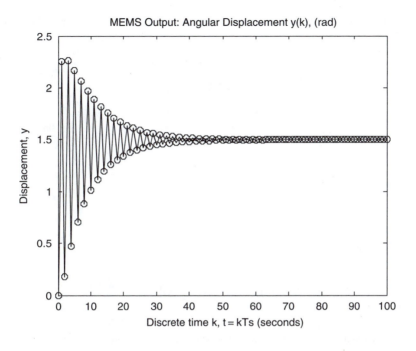

FIGURE 9.35
Output of the microservosystem with digital PID controller, $T_s = 0.0005$ sec.

The analytical results are important in addition to numerical solution and modeling, which were performed in the MATLAB environment. Making note of the transfer functions $G_{sys}(s)$ and $G_H(s)$, one finds

$$G_D(s) = G_H(s)G_{sys}(s) = \frac{1-e^{-T_s s}}{s} \frac{k_{gear}k_a}{s\left(L_a Js^2 + (r_a J + L_a B_m)s + r_a B_m + k_a^2\right)}$$

Assuming that the eigenvalues are real and distinct and the zero-order data hold circuit is used, for the hold circuit we have

$$G_H(s) = \frac{1-e^{-T_s s}}{s}$$

Thus, one obtains

$$G_D(s) = G_H(s)G_{sys}(s) = \frac{1-e^{-T_s s}}{s} \frac{k}{s(T_1 s + 1)(T_2 s + 1)}$$

Here, $k = k_{gear}k_a$, and the time constants are found from

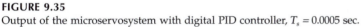

$$L_a Js^2 + (r_a J + L_a B_m)s + r_a B_m + k_a^2.$$

The partial fraction expansion is performed letting $T_1 \neq T_2$. This situation is the most common in practice. Assuming that the eigenvalues $-1/T$ and $-1/T_2$ are distinct (nonrepeated), from Heaviside's expansion formula, we have

$$\frac{k}{s^2(T_1 s + 1)(T_2 s + 1)} = \frac{c_1}{s} + \frac{c_2}{s^2} + \frac{c_3}{T_1 s + 1} + \frac{c_4}{T_2 s + 1}$$

where C_1, C_2, C_3, and C_4 are the unknown coefficients that are calculated using

$$c_1 = \frac{d}{ds}\left(\frac{ks^2}{s^2(T_1s+1)(T_2s+1)}\right)\Bigg|_{s=0} = \frac{-k(2T_1T_2s+T_1+T_2)}{(T_1T_2s^2+(T_1+T_2)s+1)}\Bigg|_{s=0} = -k(T_1+T_2)$$

$$c_2 = \frac{k}{(T_1s+1)(T_2s+1)}\Bigg|_{s=0} = k\,, c_3 = \frac{k}{s^2(T_2s+1)}\Bigg|_{s=-\frac{1}{T_1}} = \frac{kT_1^3}{T_1-T_2}\,, \quad \text{and}$$

$$c_4 = \frac{k}{s^2(T_1s+1)}\Bigg|_{s=-\frac{1}{T_2}} = \frac{kT_2^3}{T_2-T_1}$$

One can find $G_D(z)$ as

$$G_D(z) = Z[G_D(s)] = Z[G_H(s)G_{sys}(s)] = \frac{z-1}{z}Z\left[\frac{G_{sys}(s)}{s}\right]$$

Using the z-transform table, we have the explicit expression for $G_D(z)$:

$$G_D(z) = \frac{z-1}{z}Z\left[\frac{k}{s^2(T_1s+1)(T_2s+1)}\right] = k\frac{z-1}{z}Z\left[-\frac{T_1+T_2}{s}+\frac{1}{s^2}+\frac{T_1^3}{T_1-T_2}\left(\frac{1}{T_1s+1}\right)+\frac{T_2^3}{T_2-T_1}\frac{1}{(T_2s+1)}\right]$$

$$= k\frac{z-1}{z}\left(-(T_1+T_2)\frac{z}{z-1}+\frac{T_s z}{(z-1)^2}+\frac{T_1^2}{T_1-T_2}\frac{z}{z-e^{-T_s/T_1}}+\frac{T_2^2}{T_2-T_1}\frac{z}{z-e^{-T_s/T_2}}\right)$$

Taking note of

$$G_D(z) = k\left(-T_1-T_2+T_s\frac{1}{z-1}+\frac{T_1^2}{T_1-T_2}\frac{z-1}{z-e^{-T_s/T_1}}+\frac{T_2^2}{T_2-T_1}\frac{z-1}{z-e^{-T_s/T_2}}\right)$$

and

$$G_{PID}(z) = \frac{(k_{dp}+k_{di}+k_{dd})z^2-(k_{dp}+2k_{dd})z+k_{dd}}{z^2-z}$$

the transfer function of the closed-loop system results. In particular,

$$G(z) = \frac{Y(z)}{R(z)} = \frac{G_{PID}(z)G_D(z)}{1+G_{PID}(z)G_D(z)}$$

The analytic solution $y(k)$ can be straightforwardly derived for different system parameters, distinct feedback gains (that vary to attain the desired dynamics), changed sampling time, distinct waveforms of the reference inputs, etc. Thus, in addition to numerical results, more general analytic studies can be carried out. Unfortunately, due to the fact that MEMS and NEMS are nonlinear (bounded applied voltage, nonlinear magnetization, etc.), the results obtained, assuming that the system dynamics is described by linear differential equations, are not valid. As a result, the designer performs numerical analysis integrating system nonlinearities.

9.9.3 Control of Linear Discrete-Time MEMS and NEMS Using the Hamilton-Jacobi Theory

Consider discrete-time microsystems modeled by the state-space difference equations

$$x_{n+1} = A_n x_n + B_n u_n, \quad n \geq 0$$

Different performance indexes can be applied to optimize the closed-loop system dynamics. For example, the quadratic performance index to be minimized is

$$J = \sum_{n=0}^{N-1} \left[x_n^T Q_n x_n + u_n^T G_n u_n \right], \quad Q_n \geq 0, \quad G_n > 0$$

Making use of the Hamilton-Jacobi theory, the goal is to find the controller that guarantees that the value of the performance index is minimum or maximum. This minimization problem is solved using the Hamilton-Jacobi concept. For linear dynamic systems and quadratic performance indexes, the solution of the Hamilton-Jacobi-Bellman equation

$$V(x_n) = \min_{u_n} \left[x_n^T Q_n x_n + u_n^T G_n u_n + V(x_{n+1}) \right]$$

is satisfied by the quadratic return function $V(x_n) = x_n^T K_n x_n$.

Thus, by making use of this quadratic return function, we have

$$
\begin{aligned}
V(x_n) &= \min_{u_n} \left[x_n^T Q_n x_n + u_n^T G_n u_n + (A_n x_n + B_n u_n)^T K_{n+1} (A_n x_n + B_n u_n) \right] \\
&= \min_{u_n} \left[x_n^T Q_n x_n + u_n^T G_n u_n + x_n^T A_n^T K_{n+1} A_n x_n + x_n^T A_n^T K_{n+1} B_n u_n \right. \\
&\quad \left. + u_n^T B_n^T K_{n+1} A_n x_n + u_n^T B_n^T K_{n+1} B_n u_n \right]
\end{aligned}
$$

The first-order necessary condition for optimality gives

$$u_n^T G_n + x_n^T A_n^T K_{n+1} B_n + u_n^T B_n^T K_{n+1} B_n = 0$$

Hence, the optimal controller is

$$u_n = -\left(G_n + B_n^T K_{n+1} B_n \right)^{-1} B_n^T K_{n+1} A_n x_n$$

The second-order necessary condition for optimality is guaranteed. In fact,

$$\frac{\partial^2 H(x_n, u_n, V(x_{n+1}))}{\partial u_n \times \partial u_n^T} = \frac{\partial^2 \left(u_n^T G_n u_n + u_n^T B_n^T K_{n+1} B_n u_n \right)}{\partial u_n \times \partial u_n^T} = 2G_n + 2B_n^T K_{n+1} B_n > 0$$

Using the controller derived, one finds

$$x_n^T K_n x_n = x_n^T Q_n x_n + x_n^T A_n^T K_{n+1} A_n x_n - x_n^T A_n^T K_{n+1} B_n \left(G_n + B_n^T K_{n+1} B_n \right)^{-1} B_n K_{n+1} A_n x$$

Thus, the difference equation to find the unknown symmetric matrix of the quadratic return function is

$$K_n = Q_n + A_n^T K_{n+1} A_n - A_n^T K_{n+1} B_n \left(G_n + B_n^T K_{n+1} B_n \right)^{-1} B_n K_{n+1} A_n$$

If in the performance index $N = \infty$, we have

$$J = \sum_{n=0}^{\infty} \left[x_n^T Q_n x_n + u_n^T G_n u_n \right], \ Q_n \geq 0, \quad G_n > 0$$

The optimal control law is

$$u_n = -\left(G_n + B_n^T K_n B_n \right)^{-1} B_n^T K_n A_n x_n$$

where the unknown symmetric matrix K_n is found by solving the following nonlinear equation:

$$-K_n + Q_n + A_n^T K_n A_n - A_n^T K_n B_n \left(G_n + B_n^T K_n B_n \right)^{-1} B_n K_n A_n = 0, \quad K_n = K_n^T$$

Matrix K_n is positive definite, and the MATLAB `dlqr` function is used to find the feedback matrix $(G_n + B_n^T K_n B_n)^{-1} B_n^T K_n A_n$, return function matrix K_n, and eigenvalues. In particular,

```
>> help dlqr
DLQR  Linear-quadratic regulator design for discrete-time systems.
[K,S,E] = DLQR(A,B,Q,R,N)  calculates the optimal gain matrix K
such that the state-feedback law  u[n] = -Kx[n]  minimizes the
cost function
        J = Sum {x'Qx + u'Ru + 2*x'Nu}
subject to the state dynamics   x[n+1] = Ax[n] + Bu[n].
The matrix N is set to zero when omitted.  Also returned are the
Riccati equation solution S and the closed-loop eigenvalues E:
                          -1
A'SA - S - (A'SB+N)(R+B'SB) (B'SA+N') + Q = 0,  E = EIG(A-B*K).
```

The closed-loop systems is expressed as

$$x_{n+1} = A_n x_n + B_n u_n , \text{ where } u_n = -\left(G_n + B_n^T K_{n+1} B_n \right)^{-1} B_n^T K_{n+1} A_n x_n$$

Thus, we have the system dynamics as

$$x_{n+1} = \left[A_n - B_n \left(G_n + B_n^T K_{n+1} B_n \right)^{-1} B_n^T K_{n+1} A_n \right] x_n$$

Example 9.22

For the second-order discrete-time system

$$x_{n+1} = \begin{bmatrix} x_{1n+1} \\ x_{2n+1} \end{bmatrix} = A_n x_n + B_n u_n = \begin{bmatrix} 1 & 2 \\ 0 & 3 \end{bmatrix} \begin{bmatrix} x_{1n} \\ x_{2n} \end{bmatrix} + \begin{bmatrix} 4 & 5 \\ 6 & 7 \end{bmatrix} \begin{bmatrix} u_{1n} \\ u_{2n} \end{bmatrix}$$

find the controller minimizing the performance index

$$J = \sum_{n=0}^{\infty} \left[x_n^T Q_n x_n + u_n^T G_n u_n \right]$$

$$= \sum_{n=0}^{\infty} \left[\begin{bmatrix} x_{1n} & x_{2n} \end{bmatrix} \begin{bmatrix} 10 & 0 \\ 0 & 10 \end{bmatrix} \begin{bmatrix} x_{1n} \\ x_{2n} \end{bmatrix} + \begin{bmatrix} u_{1n} & u_{2n} \end{bmatrix} \begin{bmatrix} 5 & 0 \\ 0 & 5 \end{bmatrix} \begin{bmatrix} u_{1n} \\ u_{2n} \end{bmatrix} \right]$$

$$= \sum_{n=0}^{\infty} \left(10x_{1n}^2 + 10x_{2n}^2 + 5u_{1n}^2 + 5u_{2n}^2 \right)$$

using the dlqr solver.

SOLUTION

We have the following MATLAB script (statement) to solve the problem:

```
A=[1 2; 0 3];B=[4 5; 6 7];
Q=10*eye(size(A)); G=5*eye(size(B));
[Kfeedback, Kn, Eigenvalues] = dlqr(A,B,Q,G)
```

The following numerical solution results:

```
Kfeedback =
 -2.8456e-001   2.3091e-001
  3.3224e-001   2.2197e-001
Kn =
  1.9963e+001  -7.1140e-001
 -7.1140e-001   1.0577e+001
Eigenvalues =
  5.2192e-001
  1.5903e-002
```

Thus, the unknown matrix K_n is found to be

$$K_n = \begin{bmatrix} 20 & -0.711 \\ -0.711 & 10.6 \end{bmatrix}$$

Hence, the controller is found as

$$u_n = -\left(G_n + B_n^T K_{n+1} B_n \right)^{-1} B_n^T K_{n+1} A_n x = - \begin{bmatrix} -0.285 & 0.231 \\ 0.332 & 0.222 \end{bmatrix} \begin{bmatrix} x_{1n} \\ x_{2n} \end{bmatrix}$$

$$u_{1n} = 0.285x_{1n} - 0.231x_{2n}, \quad u_{2n} = -0.332x_{1n} - 0.222x_{2n}$$

The system is stable because the eigenvalues are 0.522 and 0.0159, i.e., the eigenvalues are within the unit circle.

Example 9.23

For the third-order microsystems modeled as

$$x_{n+1} = \begin{bmatrix} x_{1n+1} \\ x_{2n+1} \\ x_{3n+1} \end{bmatrix} = A_n x_n + B_n u_n = \begin{bmatrix} 1 & 1 & 2 \\ 3 & 3 & 4 \\ 5 & 5 & 6 \end{bmatrix} \begin{bmatrix} x_{1n} \\ x_{2n} \\ x_{2n} \end{bmatrix} + \begin{bmatrix} 10 \\ 20 \\ 30 \end{bmatrix} u_n$$

find an optimal controller minimizing the quadratic performance index

$$J = \sum_{n=0}^{\infty} \left[x_n^T Q_n x_n + u_n^T G_n u_n \right]$$

$$= \sum_{n=0}^{\infty} \left[\begin{bmatrix} x_{1n} & x_{2n} & x_{3n} \end{bmatrix} \begin{bmatrix} 1 & 0 & 0 \\ 0 & 10 & 0 \\ 0 & 0 & 100 \end{bmatrix} \begin{bmatrix} x_{1n} \\ x_{2n} \\ x_{3n} \end{bmatrix} + 1000 u_n^2 \right]$$

$$= \sum_{n=0}^{\infty} \left(x_{1n}^2 + 10 x_{2n}^2 + 100 x_{3n}^2 + 1000 u_n^2 \right)$$

and simulate the closed-loop system in the MATLAB environment. The matrices of the output equation are

$C = [100 \ 10 \ 1]$ and $H = [0]$

SOLUTION

We enter the system matrices by typing in the Command widow

```
>> A=[1 1 2; 3 3 4; 5 5 6]; B=[10; 20; 30];
```

The weighting matrices are entered as

```
>> Q=eye(size(A)); Q(2,2)=10; Q(3,3)=100; G=1000;
```

The controller, feedback coefficients, and eigenvalues are found using the dlqr solver. In particular,

```
>> [Kfeedback,Kn,Eigenvalies]=dlqr(A,B,Q,G)
```

The following results are displayed in the Command widow:

```
Kfeedback =
   1.5536e-001   1.5536e-001   1.9913e-001
Kn =
   4.8425e+001   4.7425e+001   3.1073e+001
   4.7425e+001   5.7425e+001   3.1073e+001
   3.1073e+001   3.1073e+001   1.3983e+002
```

```
Eigenvalies =
  -6.7355e-001
   3.8806e-002
   4.5752e-016
```

Thus,

$$K_n = \begin{bmatrix} 48.4 & 47.4 & 31.1 \\ 47.4 & 57.4 & 31.1 \\ 31.1 & 31.1 & 140 \end{bmatrix}$$

The controller is synthesized as

$$u_n = -0.155x_{1n} - 0.155x_{2n} - 0.2x_{3n}$$

The dynamic performance of the closed-loop system, which is stable (eigenvalues are within the unit circle), is simulated using the MATLAB filter function. In particular, having derived the closed-loop system dynamics in the form of the linear difference equation

$$x_{n+1} = \left[A_n - B_n \left(G_n + B_n^T K_{n+1} B_n \right)^{-1} B_n^T K_{n+1} A_n \right] x_n$$

one finds the numerator and denominator of the transfer function in the z-domain. Then, the dynamics is calculated. The MATLAB statement is

```
A_closed=A-B*Kfeedback;C=[100 10 1]; H=[0];
[num,den]=ss2tf(A_closed,B,C,H);
k=0:1:20;r=2*[ones(1,21)];x=filter(num,den,r);
plot(k,x,'-',k,x,'o',k,100*r,'+')
```

The simulation results (system output "o" and reference r "+") are documented in Figure 9.36. Please note that the scale for the reference was increased by a factor of 100 for plotting purposes.

It must be emphasized that the system output does not follow the reference input $r(k)$ because the stabilization problem was solved. It was illustrated that the tracking control problem must be approached and solved to guarantee that the system output follows the reference command.

9.9.4 Constrained Optimization of Discrete-Time MEMS and NEMS

Due to the constraints imposed on controls, the designer must synthesize bounded control laws. In this section, the constrained optimization problem for multivariable discrete-time MEMS and NEMS is covered. In particular, the constrained digital controllers are synthesized using the Hamilton-Jacobi theory. We study discrete-time microsystems with bounded control modeled as

$$x_{n+1} = A_n x_n + B_n u_n, \quad x_{n0} \in X_0, \quad u_n \in U, \quad n \geq 0$$

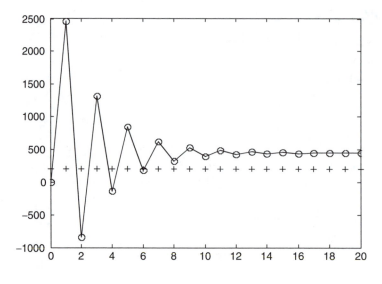

FIGURE 9.36
Output dynamics of the closed-loop microsystem.

The nonquadratic performance index to be minimized is

$$J = \sum_{n=0}^{N-1} \left[x_n^T Q_n x_n + 2 \int (\phi^{-1}(u_n))^T G_n du_n - u_n^T B_n^T K_{n+1} B_n u_n \right]$$

Performance indexes must be positive definite. Hence, to attain the positive definiteness,

$$\left[x_n^T Q_n x_n + 2 \int (\phi^{-1}(u_n))^T G_n du_n \right] > u_n^T B_n^T K_{n+1} B_n u_n \text{ for all } x_n \in X \text{ and } u_n \in U$$

The Hamilton-Jacobi-Bellman recursive equation is

$$V(x_n) = \min_{u_n \in U} \left[x_n^T Q_n x_n + 2 \int (\phi^{-1}(u_n))^T G_n du_n - u_n^T B_n^T K_{n+1} B_n u_n + V(x_{n+1}) \right]$$

Using the quadratic return function $V(x_n) = x_n^T K_n x_n$, one finds

$$x_n^T K_n x_n = \min_{u_n \in U} \left[x_n^T Q_n x_n + 2 \int (\phi^{-1}(u_n))^T G_n du_n - u_n^T B_n^T K_{n+1} B_n u_n \right.$$
$$\left. + (A_n x_n + B_n u_n)^T K_{n+1} (A_n x_n + B_n u_n) \right]$$

The first- and second-order necessary conditions for optimality were used, and the bounded controller results as

$$u_n = -\phi \left(G_n^{-1} B_n^T K_{n+1} A_n x_n \right), \quad u_n \in U$$

It is evident that

$$\frac{\partial^2\left(2\int\left(\phi^{-1}(u_n)\right)^T G_n du_n\right)}{\partial u_n \times \partial u_n^T}$$

is positive definite because one-to-one integrable functions ϕ and ϕ^{-1} lie in the first and third quadrants and weighting matrix G_n is positive definite. Thus, the second-order necessary condition for optimality is satisfied, and

$$\frac{\partial\left(\phi^{-1}(u_n)\right)^T}{\partial u_n} G_n > 0$$

We have

$$x_n^T K_n x_n = x_n^T Q_n x_n + 2\int x_n^T A_n^T K_{n+1} B_n d\left(\phi\left(G_n^{-1}B_n^T K_{n+1}A_n x_n\right)\right)$$

$$+ x_n^T A_n^T K_{n+1}A_n x_n - 2x_n^T A_n^T K_{n+1}B_n \phi\left(G_n^{-1}B_n^T K_{n+1}A_n x_n\right)$$

where

$$2\int x_n^T A_n^T K_{n+1}B_n d\left(\phi\left(G_n^{-1}B_n^T K_{n+1}A_n x_n\right)\right) = 2x_n^T A_n^T K_{n+1}B_n \phi\left(G_n^{-1}B_n^T K_{n+1}A_n x_n\right)$$

$$- 2\int\left(\phi\left(G_n^{-1}B_n^T K_{n+1}A_n x_n\right)\right)^T d\left(B_n^T K_{n+1}A_n x_n\right)$$

Hence, the unknown matrix $K_{n+1} \in \mathbb{R}^{c \times c}$ is found by solving

$$x_n^T K_n x_n = x_n^T Q_n x_n + x_n^T A_n^T K_{n+1}A_n x_n - 2\int\left(\phi\left(G_n^{-1}B_n^T K_{n+1}A_n x_n\right)\right)^T d\left(B_n^T K_{n+1}A_n x_n\right)$$

Describing the control bounds imposed by the continuous integrable one-to-one bounded functions $\phi \in U$, one finds the term

$$2\int\left(\phi\left(G_n^{-1}B_n^T K_{n+1}A_n x_n\right)\right)^T d\left(B_n^T K_{n+1}A_n x_n\right).$$

For example, using the tanh function to map the saturation-type constraints, one obtains

$$\int \tanh z\, dz = \log \cosh z \text{ and } \int \tanh^g z\, dz = -\frac{\tanh^{g-1} z}{g-1} + \int \tanh^{g-2} z\, dz, g \neq 1$$

Matrix K_{n+1} is found by solving the recursive equation, and feedback gains result.
Minimizing the nonquadratic performance index, we derived the bounded control law as

$$u_n = -\phi\left(G_n^{-1}B_n^T K_n A_n x_n\right), \quad u_n \in U$$

The *admissibility* concept, based on the Lyapunov stability theory, is applied to verify the stability of the resulting closed-loop system if the open-loop system is unstable. The resulting closed-loop system evolves in X. A subset of the admissible domain of stability $S \subset \mathbb{R}^c$ is found by using the Lyapunov stability theory $S = \{x_n \in \mathbb{R}^c : x_{n0} \in X_0, \ u_n \in U | V(0) = 0,$ $V(x_n) > 0, \ \Delta V(x_n) < 0, \ \forall x_n \in X(X_0, U)\}$.

The region of attraction can be studied, and S is an *invariant* domain. It should be emphasized that the quadratic Lyapunov function is applied, and $V(x_n) = x_n^T K_n x_n$ is positive definite if $K_n > 0$. Hence, the first difference, as given by

$$\Delta V(x_n) = V(x_{n+1}) - V(x_n) = x_n^T A_n^T K_{n+1} A_n x_n - 2x_n^T A_n^T K_{n+1} B_n \phi\left(G_n^{-1} B_n^T K_{n+1} A_n x_n\right)$$

$$+ \phi\left(G_n^{-1} B_n^T K_{n+1} A_n x_n\right)^T B_n^T K_{n+1} B_n \phi\left(G_n^{-1} B_n^T K_{n+1} A_n x_n\right) - x_n^T K_n x_n$$

must be negative definite for all $x_n \in X$ to ensure stability.

The evolution of the closed-loop microsystems depends on the initial conditions and constraints. In particular, it was illustrated that $X(X_0, U)$. The *sufficiency* analysis of stability is performed by studying sets $S \subset \mathbb{R}^c$ and $X(X_0, U) \subset \mathbb{R}^c$. Stability is guaranteed if $X \subseteq S$.

Linear discrete-time systems have been examined. The constrained optimization problem must be solved for nonlinear microsystems that are modeled by nonlinear difference equations. The Hamilton-Jacobi theory is applied to design bounded controllers using nonquadratic performance indexes that are minimized. In particular, we study nonlinear discrete-time MEMS and NEMS described as

$$x_{n+1} = F(x_n) + B(x_n)u_n, \quad u_{n\min} \le u_n \le u_{n\max}$$

To design a nonlinear admissible controller $u_n \in U$, we map the imposed control bounds by a continuous integrable one-to-one bounded function $\phi \in U$. The nonquadratic performance index is

$$J = \sum_{n=0}^{N-1} \left[x_n^T Q_n x_n - u_n^T B(x_n)^T K_{n+1} B(x_n) u_n + 2 \int (\phi^{-1}(u_n))^T G_n du_n \right]$$

The integrand $2\int(\phi^{-1}(u_n)^T G_n du_n)$ is positive definite because the integrable one-to-one function ϕ lies in the first and third quadrants, integrable function ϕ^{-1} exists, and weighting matrix G_n is assigned to be positive definite. The positive definiteness of the performance index is guaranteed if

$$\left(x_n^T Q_n x_n + 2 \int (\phi^{-1}(u_n))^T G_n du_n \right) > u_n^T B(x_n)^T K_{n+1} B(x_n) u_n$$

$$\text{for all} \quad x_n \in X \quad \text{and} \quad u_n \in U$$

The positive definiteness of the performance index can be studied as positive-definite symmetric matrix K_{n+1} is found. The inequality, which guarantees the positive definiteness of the performance index in X and U, can be easily guaranteed by assigning the weighting matrices Q_n and G_n.

Using the Hamilton-Jacobi theory, the first- and second-order necessary conditions for optimality are applied. For the quadratic return function $V(x_n) = x_n^T K_n x_n$, we have

$$V(x_{n+1}) = x_{n+1}^T K_{n+1} x_{n+1} = \left(F(x_n) + B(x_n)u_n\right)^T K_{n+1}\left(F(x_n) + B(x_n)u_n\right)$$

and, therefore,

$$x_n^T K_n x_n = \min_{u_n \in U} \left[x_n^T Q_n x_n - u_n^T B(x_n)^T K_{n+1} B(x_n) u_n + 2 \int (\phi^{-1}(u_n))^T G_n du_n \right.$$

$$\left. + (F(x_n) + B(x_n)u_n)^T K_{n+1}(F(x_n) + B(x_n)u_n) \right]$$

Using the first-order necessary condition for optimality, a bounded control law results, i.e.,

$$u_n = -\phi\left(G_n^{-1}B(x_n)^T K_{n+1}F(x_n)\right), \quad u_n \in U$$

The

$$\frac{\partial^2 \left(2\int (\phi^{-1}(u_n))^T G_n du_n \right)}{\partial u_n \times \partial u_n^T}$$

is positive definite because ϕ and ϕ^{-1} lie in the first and third quadrants, and G_n is positive definite. Thus, the second-order necessary condition for optimality is satisfied.

Using the derived bounded controller, $u_n = -\phi\left(G_n^{-1}B(x_n)^T K_{n+1}F(x_n)\right)$, we have the following recursive equation:

$$x_n^T K_n x_n = x_n^T Q_n x_n + 2\int F(x_n)^T K_{n+1}B(x_n)d\left(\phi\left(G_n^{-1}B(x_n)^T K_{n+1}F(x_n)\right)\right)$$

$$+ F(x_n)^T K_{n+1}F(x_n) - 2F(x_n)^T K_{n+1}B(x_n)\phi\left(G_n^{-1}B(x_n)^T K_{n+1}F(x_n)\right)$$

Then, integration by parts gives

$$2\int F(x_n)^T K_{n+1}B(x_n)d\left(\phi\left(G_n^{-1}B(x_n)^T K_{n+1}F(x_n)\right)\right)$$

$$= 2F(x_n)^T K_{n+1}B(x_n)\phi\left(G_n^{-1}B(x_n)^T K_{n+1}F(x_n)\right)$$

$$- 2\int \left(\phi\left(G_n^{-1}B(x_n)^T K_{n+1}F(x_n)\right)\right)^T d\left(B(x_n)^T K_{n+1}F(x_n)\right)$$

The equation to find the unknown symmetric matrix $K_{n+1} \in \mathbb{R}^{n \times n}$ is

$$x_n^T K_n x_n = x_n^T Q_n x_n + F(x_n)^T K_{n+1}F(x_n)$$

$$- 2\int \left(\phi\left(G_n^{-1}B(x_n)^T K_{n+1}F(x_n)\right)\right)^T d(B(x_n)^T K_{n+1}F(x_n))$$

By mapping the control bounds imposed by the continuous integrable one-to-one bounded functions $\phi \in U$, one finds the term $2\int(\phi(G_n^{-1}B(x_n)^T K_{n+1}F(x_n)))^T d(B(x_n)^T K_{n+1}F(x_n))$. For example, using the hyperbolic tangent that describes the saturation-type constraints, we have

$$\int \tanh z \, dz = \log \cosh z \quad \text{and} \quad \int \tanh^g z \, dz = -\frac{\tanh^{g-1} z}{g-1} + \int \tanh^{g-2} z \, dz, \quad g \neq 1$$

If the open-loop system is unstable, the admissibility concept must be applied to verify the stability of the resulting closed-loop system. The closed-loop system evolves in $X \subset \mathbb{R}^c$, and

$$\left\{ x_{n+1} = F(x_n) - B(x_n)\phi\left(G_n^{-1}B(x_n)^T K_{n+1}F(x_n)\right), \, x_{n0} \in X_0 \right\} \in X(X_0, U)$$

Using the Lyapunov stability theory, the domain of stability $S \subset \mathbb{R}^c$ is found by applying the sufficient conditions under which the discrete-time system is stable. The positive-definite quadratic function $V(x_n)$ is used. To guarantee the stability, the first difference

$$\Delta V(x_n) = V(x_{n+1}) - V(x_n) = F(x_n)^T K_{n+1}F(x_n) - 2F(x_n)^T K_{n+1}B(x_n)\phi\left(G_n^{-1}B(x_n)^T K_{n+1}F(x_n)\right)$$

$$+ \phi\left(G_n^{-1}B(x_n)^T K_{n+1}F(x_n)\right)^T B(x_n)^T K_{n+1}B(x_n)\phi\left(G_n^{-1}B(x_n)^T K_{n+1}F(x_n)\right) - x_n^T K_n x_n$$

must be negative definite for all $x_n \in X$. That is, we have

$$S = \left\{ x_n \in \mathbb{R}^c : x_{n0} \in X_0, \quad u \in U \big| V(0) = 0, \quad V(x_n) > 0, \, \Delta V(x_n) < 0, \quad \forall x \in X(X_0, U) \right\}$$

The *sufficiency* analysis is performed by studying $S \subset \mathbb{R}^c$ and $X(X_0, U) \subset \mathbb{R}^n$. The constrained optimization problem is solved via the bounded admissible control law designed, and for open-loop unstable system, stability is guaranteed if $X \subseteq S$.

9.9.5 Tracking Control of Discrete-Time Microsystems

We study microsystems modeled by the following difference equation in the state-space form:

$$x_{n+1}^{system} = A_n x_n^{system} + B_n u_n, \quad x_{n0}^{system} \in X_0, \quad u_n \in U, n \geq 0$$

The output equation is $y_n = H_n x_n^{system}$.

Taking note of the *exogenous* system, we have

$$x_n^{ref} = x_{n-1}^{ref} + r_n - y_n$$

Thus, one finds

$$x_{n+1}^{ref} = x_n^{ref} + r_{n+1} - y_{n+1} = x_n^{ref} + r_{n+1} - H_n\left(A_n x_n^{system} + B_n u_n\right)$$

Hence, an augmented model is given by

$$x_{n+1} = \begin{bmatrix} x_{n+1}^{system} \\ x_{n+1}^{ref} \end{bmatrix} = \begin{bmatrix} A_n & 0 \\ -H_n A_n & I_n \end{bmatrix} x_n + \begin{bmatrix} B_n \\ -H_n B_n \end{bmatrix} u_n + \begin{bmatrix} 0 \\ I_n \end{bmatrix} r_{n+1}, \quad x_n = \begin{bmatrix} x_n^{system} \\ x_n^{ref} \end{bmatrix}$$

To synthesize the bounded controller, the nonquadratic performance index to be minimized is

$$J = \sum_{n=0}^{N-1} \left[x_n^T Q_n x_n + 2\int (\phi^{-1}(u_n))^T G_n du_n - u_n^T \begin{bmatrix} B_n \\ -H_n B_n \end{bmatrix}^T K_{n+1} \begin{bmatrix} B_n \\ -H_n B_n \end{bmatrix} u_n \right]$$

Using the quadratic return function $V(x_n) = x_n^T K_n x_n$, from the Hamilton-Jacobi equation

$$x_n^T K_n x_n = \min_{u_n \in U}\left[x_n^T Q_n x_n + 2\int (\phi^{-1}(u_n))^T G_n du_n - u_n^T \begin{bmatrix} B_n \\ -H_n B_n \end{bmatrix}^T K_{n+1} \begin{bmatrix} B_n \\ -H_n B_n \end{bmatrix} u_n \right.$$

$$\left. + \left(\begin{bmatrix} A_n & 0 \\ -H_n A_n & I_n \end{bmatrix} x_n + \begin{bmatrix} B_n \\ -H_n B_n \end{bmatrix} u_n \right)^T K_{n+1} \left(\begin{bmatrix} A_n & 0 \\ -H_n A_n & I_n \end{bmatrix} x_n + \begin{bmatrix} B_n \\ -H_n B_n \end{bmatrix} u_n \right) \right]$$

using the first-order necessary condition for optimality, one obtains the following bounded tracking controller:

$$u_n = -\phi\left(G_n^{-1} \begin{bmatrix} B_n \\ -H_n B_n \end{bmatrix}^T K_{n+1} \begin{bmatrix} A_n & 0 \\ -H_n A_n & I_n \end{bmatrix} x_n \right), \quad u_n \in U.$$

The unknown matrix K_{n+1} is found by solving the following equation:

$$x_n^T K_n x_n = x_n^T Q_n x_n + x_n^T \begin{bmatrix} A_n & 0 \\ -H_n A_n & I_n \end{bmatrix}^T K_{n+1} \begin{bmatrix} A_n & 0 \\ -H_n A_n & I_n \end{bmatrix} x_n$$

$$- 2\int \left(\phi\left(G_n^{-1} \begin{bmatrix} B_n \\ -H_n B_n \end{bmatrix}^T K_{n+1} \begin{bmatrix} A_n & 0 \\ -H_n A_n & I_n \end{bmatrix} x_n \right) \right)^T d\left(\begin{bmatrix} B_n \\ -H_n B_n \end{bmatrix}^T K_{n+1} \begin{bmatrix} A_n & 0 \\ -H_n A_n & I_n \end{bmatrix} x_n \right)$$

The tracking control problem can be solved. Let us study nonlinear discrete-time micro-systems that are described by nonlinear difference equations

$$x_s(k+1) = F_s(x_s(k)) + B_s(x_s(k))u(k), \quad y(k) = Hx_s(k)$$

The reference vector $r(k)$ is considered to solve the motion control problem, and the tracking error vector is $e(k) = r(k) - y(k)$. The *exosystem* state equation is

$$x_i(k) = x_i(k-1) + r(k) - y(k)$$

and

$$x_i(k+1) = x_i(k) + r(k+1) - y(k+1) = -Hx_s(k+1) + x_i(k) + r(k+1)$$

The following augmented state-space difference equation results:

$$x(k+1) = \begin{bmatrix} x_s(k+1) \\ x_i(k+1) \end{bmatrix} = F(x(k)) + B(x(k))u(k) + B_r r(k+1),$$

$$F(x(k)) = \begin{bmatrix} F_s(x_s(k)) \\ -HF_s(x_s(k)) + Ix_i(k) \end{bmatrix}, \quad B(x(k)) = \begin{bmatrix} B_s(x_s(k)) \\ -HB_s(x_s(k)) \end{bmatrix}, \quad B_r = \begin{bmatrix} 0 \\ I \end{bmatrix}$$

A control law will be designed by minimizing the nonquadratic performance index

$$J = \sum_{k=0}^{N-1}\left[\sum_{i=0}^{\varsigma}\left(x(k)^{\frac{i+\beta+1}{2\beta+1}}\right)^{T} Q_i x(k)^{\frac{i+\beta+1}{2\beta+1}} + u(k)^T G u(k)\right]$$

where ς and β are the real integers assigned by the designer, $\varsigma = 0,1,2,\ldots$, $\beta = 0,1,2,\ldots$.
The Hamilton-Jacobi-Bellman recursive equation is given as

$$V(x(k)) = \min_{u(k)}\left[\sum_{i=0}^{\varsigma}\left(x(k)^{\frac{i+\beta+1}{2\beta+1}}\right)^{T} Q_i x(k)^{\frac{i+\beta+1}{2\beta+1}} + u(k)^T G u(k) + V(x(k+1))\right]$$

where the positive-definite return function, which represents the minimum value of the performance index, is assigned in the quadratic form $V(x(k)) = x(k)^T K(k)x(k)$, where $K(k)$ is the unknown matrix.
Using the first-order necessary condition for optimality, one obtains a nonlinear controller

$$u(k) = -(B(x(k))^T K(k+1)B(x(k)) + G)^{-1}B(x(k)x(k))^T K(k+1)F(x(k))$$

It is evident that the second-order necessary condition for optimality is guaranteed because matrix G is assigned to be positive definite, i.e.,

$$\frac{\partial^2(u(k)^T G u(k))}{\partial u(k) \times \partial u(k)^T} = G > 0$$

The unknown matrix K is found by solving the following equation:

$$x(k)^T K(k)x(k) = \sum_{i=0}^{\varsigma}\left(x(k)^{\frac{i+\beta+1}{2\beta+1}}\right)^{T} Q_i x(k)^{\frac{i+\beta+1}{2\beta+1}} + F(x(k))^T K(k+1)F(x(k))$$

$$- F(x(k))^T K(k+1)B(x(k))\left(B(x(k))^T K(k+1)B(x(k)) + G\right)^{-1}$$

$$\times B(x(k))^T K(k+1)F(x(k))$$

Homework Problems

1. Why should one control MEMS and NEMS?

2. What specifications are imposed on MEMS and NEMS in the behavioral domain?

3. Explain the differences between bounded and unbounded controllers. Provide examples. Explain how control bounds influence the system performance in the behavioral domain.

4. What are the challenges in the design of bounded controllers? Can one analytically solve constrained problems for high-order microsystems? How should the designer approach and solve control problems for high-order microsystems?

5. Let the controller be expressed as $u = k_{p1}e + k_{p2}|e^2| + k_{p3}e^3 + k_i\int edt$. Explain what feedback terms are used. Conclude whether this controller can be applied, justifying your results. Explain why one should use the controller as $u = \mathbf{sat}(k_{p1}e + k_{p2}|e^2| + k_{p3}e^3 + k_i\int edt)$ to examine the microsystem performance. Propose

the terms to improve the system performance. Also, provide the terms that degrade the system performance and lead to unstable closed-loop systems.

6. Let the performance functional be given as

$$J = \min_{t,x,e} \int_0^\infty \left(x^2 + e^4 + t|e| \right) dt$$

Explain what performance characteristics are specified and how. Justify the results. Let more strict specifications be imposed on the settling time and tracking error. Propose the additional integrand in the performance functional.

7. For the second-order microsystem $dx_1/dt = -x_1 + x_2$, $dx_2/dt = x_1 + u$, find a control law that minimizes the quadratic functional $j = \frac{1}{2}\int_0^\infty (x_1^2 + 2x_2^2 + 3u^2)dt$ applying MATLAB. Study the closed-loop system stability.

8. Using the Lyapunov stability theory, study the stability of the microsystem that is described by the differential equations

$$\dot{x}_1(t) = -x_1 + 10x_2$$

$$\dot{x}_2(t) = -10x_1 - x_2^7, \quad t \geq 0$$

9. Let the microsystem be described by the differential equations

$$\dot{x}_1(t) = -x_1 + 10x_2$$

$$\dot{x}_2(t) = x_1 + u$$

Derive (synthesize) the control law that will stabilize this system. Using the Lyapunov stability theory, prove the stability of the closed-loop system.

References

1. Athans, M. and Falb, P.L., *Optimal Control: An Introduction to the Theory and its Applications*, McGraw-Hill Book Company, New York, 1966.
2. Dorf, R.C. and Bishop, R.H., *Modern Control Systems*, Addison-Wesley Publishing Company, Reading, MA, 1995.
3. Franklin, J.F., Powell, J.D., and Emami-Naeini, A., *Feedback Control of Dynamic Systems*, Addison- Wesley Publishing Company, Reading, MA, 1994.
4. Khalil, H.K., *Nonlinear Systems*, Prentice-Hall, Inc., NJ, 1996.
5. Kuo, B.C., *Automatic Control Systems*, Prentice Hall, Englewood Cliffs, NJ, 1995.
6. Levine, W.S., Ed., *Control Handbook*, CRC Press, FL, 1996.
7. Lyshevski, S.E., *Control Systems Theory with Engineering Applications*, Birkhauser, Boston, MA, 2000.
8. Lyshevski, S.E., *MEMS and NEMS: Systems, Devices, and Structures*, CRC Press, Boca Raton, FL, 2002.
9. Ogata, K., *Discrete-Time Control Systems*, Prentice-Hall, Upper Saddle River, NJ, 1995.
10. Ogata, K., *Modern Control Engineering*, Prentice-Hall, Upper Saddle River, NJ, 1997.
11. Rosenbrock, H.H., A stochastic variational principle for quantum mechanics, *Phys. Lett.*, 100A, 343–346, 1986.

10

Examples in Synthesis, Analysis, Design, and Fabrication of MEMS

10.1 Introduction

In many applications (from medicine and biotechnology to aerospace and security), the use of nano- and microscale structures, devices, and systems is very important.[1-4] This chapter discusses the analysis, modeling, design, and fabrication of electromagnetic-based microscale structures, devices, and MEMS. It is obvious that to attain our objectives and goals, the synergy of multidisciplinary engineering, science, and technology must be utilized. In particular, electromagnetic theory and mechanics compose the fundamentals for analysis, modeling, simulation, design, and optimization, while fabrication is based on the micromachining and high-aspect-ratio techniques and processes. As was illustrated in Chapter 3 and Chapter 4, micromachining is the extension of the CMOS technologies developed to fabricate ICs. For many years developments in MEMS have been concentrated on the fabrication of microstructures, adopting, modifying, and enhancing silicon-based technologies commonly used in ICs. The reason for the refining of conventional microelectronic technologies, development of novel techniques, design of new processes, and application of new materials and chemicals is simple: In general microstructures are three-dimensional with frequently required high aspect ratios and large structural heights. In contrast two-dimensional (planar) microelectronic devices were fabricated utilizing CMOS.

It was documented in Chapter 4 that various silicon structures can be made through bulk silicon micromachining (using wet or dry processes) or through surface micromachining. Metallic micromolding techniques, based on photolithographic processes, are also widely used to fabricate microstructures. Molds are created in polymer films (usually photoresist) on planar surfaces, and then filled by electrodepositing metal. (Electrodeposition plays a key role in the fabrication of the microstructures and microdevices that are the components of MEMS.) High-aspect-ratio technologies use optical, e-beam, and x-ray lithography to create the hundreds of micron high trenches in polymethylmethacrylate resist on the electroplating base (seed layer). Electrodeposition of magnetic materials and conductors, plating, etching, and lift-off are commonly used to fabricate microscale structures and devices.

Although it is recognized that the ability to use and refine existing microelectronic fabrication technologies, techniques, and materials is very important, and the development of affordable high-yield processes to fabricate MEMS is a key factor in the rapid growth of microsystems, other emerging issues arise. In particular, devising (synthesis), design, modeling, analysis, and optimization of novel MEMS are extremely important. Therefore, MEMS theory and microengineering fundamentals have recently been expanded to thoroughly study other critical problems such as system-level synthesis, CAD, integration, classification, analysis, modeling, prototyping, optimization, and simulation. This chapter covers some important examples in fabrication, analysis, and design of electromagnetic microstructures, micro-devices, and MEMS. The descriptions of the fabrication processes are given, modeling and analysis issues are emphasized, and the design is performed. Finally computer-aided design issues are addressed.

10.2 Analysis of Energy Conversion and MEMS Performance from Materials and Fabrication Viewpoints

Energy conversion takes place in micro- and nanoelectromechanical motion devices. This chapter examines electromagnetic motion microdevices that convert electrical energy to mechanical energy and vice versa. Hence, energy conversion is our particular interest. It is evident that fabrication technologies, processes, and materials affect the energy conversion in microsystems and microdevices. One optimizes the processes based on the specification and requirements performing the system design with the ultimate objective to achieve optimal overall performance. Analysis and design of high-performance NEMS and MEMS require a coherent analysis of electromagnetic and electro-mechanical motion device characteristics, energy conversion, materials, and fabrication technologies. Analysis and optimization of steady-state and dynamic characteristics depend significantly on an intimate knowledge of energy conversion. Analytical and numerical studies, that is, quantitative and qualitative analysis, must be performed using energy conversion principles. These speed up the design and allow one to optimize systems, approaching performance limits that can be achieved.

Fundamental principles of energy conversion applicable to electromagnetic motion microdevices must be studied first to provide basic fundamentals. The physics of energy conversion must be researched and used to study the statics and dynamics of the electromechanical motion devices. Using the principle of conservation of energy we can formulate that for lossless micro- and nanoelectromechanical motion devices (in the conservative system, no energy is lost through friction, heat, or other irreversible energy conversion), the sum of the instantaneous kinetic and potential energies of the system remains constant. The energy conversion is represented in Figure 10.1.

To find the total energy stored in the magnetic field, the following formula should be applied:

$$W_m = \frac{1}{2}\int_v \vec{B} \cdot \vec{H} dv$$

where $\vec{B}$ and $\vec{H}$ are related using the permeability μ, $\vec{B} = \mu\vec{H}$.

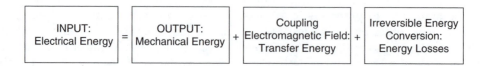

FIGURE 10.1
Energy transfer in electromechanical micro- and nanosystems.

The material becomes magnetized in response to the external field $\vec{H}$, and the dimensionless magnetic susceptibility χ_m or relative permeability μ_r are used. We have

$$\vec{B} = \mu\vec{H} = \mu_0(1 + \chi_m)\vec{H} = \mu_0\mu_r\vec{H} = \mu\vec{H}$$

Based on the value of the magnetic susceptibility χ_m, the materials are classified as follows:

- Nonmagnetic, $\chi_m = 0$, and thus $\mu_r = 1$
- Diamagnetic, $\chi_m \approx -1 \times 10^{-5}$ (for example, $\chi_m = -9.5 \times 10^{-6}$ for copper, $\chi_m = -3.2 \times 10^{-5}$ for gold, and $\chi_m = -2.6 \times 10^{-5}$ for silver)
- Paramagnetic, $\chi_m \approx 1 \times 10^{-4}$ (for example, $\chi_m = 1.4 \times 10^{-3}$ for Fe_2O_3, and $\chi_m = 1.7 \times 10^{-3}$ for Cr_2O_3)
- Ferromagnetic, $|\chi_m| \gg 1$ (iron, nickel, and cobalt, Neodymium Iron Boron and Samarium Cobalt permanent magnets). Ferromagnetic materials exhibit high magnetizability and are divided in hard (alnico, rare-earth elements, copper-nickel alloy, and other alloys) and soft (iron, nickel, cobalt, and their alloys) materials.

The magnetization behavior of the ferromagnetic materials is mapped by the magnetization curve, where H is the externally applied magnetic field, and B is total magnetic flux density in the medium. Typical B-H curves for hard and soft ferromagnetic materials are given in Figure 10.2, respectively.

The B versus H curve allows one to establish the energy analysis. Assume that initially $B_0 = 0$ and $H_0 = 0$. Let H increase from $H_0 = 0$ to H_{max}. Then B increases from $B_0 = 0$ until the maximum value of B, denoted as B_{max}, is reached. If H then decreases to H_{min}, B decreases to B_{min} through the remanent value B_r (the so-called the residual magnetic flux density) along the different curve. (See Figure 10.2.) For variations of H, $H \in [H_{min}\ H_{max}]$, B changes within the hysteresis loop, and $B \in [B_{min}\ B_{max}]$. Figure 10.2 illustrates typical curves representing the dependence of magnetic induction B on magnetic field H for a ferromagnetic material. When H is first applied, B follows curve a as the favorably oriented magnetic domains grow. This curve reaches the saturation. When H is then reduced, B follows curve b but retains a finite value (the remanence Br) at $H = 0$. In

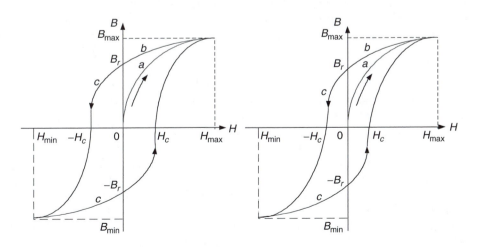

FIGURE 10.2
B-H curves for hard and soft ferromagnetic materials.

order to demagnetize the material, a negative field $-H_c$ (H_c is called the coercive field or coercivity) must be applied. As H is further decreased and then increased to complete the cycle (curve c), a hysteresis loop is formed. The area within this loop is a measure of the energy loss per cycle for a unit volume of the material.

In the per-unit volume, the applied field energy is

$$W_F = \oint_B H dB$$

while the stored energy is expressed as

$$W_c = \oint_H B dH$$

In the volume v, we have the following expressions for the field and stored energy

$$W_F = v\oint_B H dB \quad \text{and} \quad W_c = v\oint_H B dH$$

A complete B versus H loop should be considered, and the equations for field and stored energy represent the areas enclosed by the corresponding curve. It should be emphasized that each point of the B versus H curve represents the total energy. In ferromagnetic materials that are used in MEMS, time-varying magnetic flux produces core losses that consist of hysteresis losses (due to the hysteresis loop of the B-H curve) and the eddy-current losses, which are proportional to the current frequency and lamination thickness. The area of the hysteresis loop is related to the hysteresis losses. Soft ferromagnetic materials have narrow hysteresis loop and they are easily magnetized and demagnetized. Therefore, the lower hysteresis losses, compared with hard ferromagnetic materials, result.

For microelectromechanical motion devices, the flux linkages are plotted versus the current because the current and flux linkages are used rather than the flux intensity and flux density. In electric micromachines almost all energy is stored in the air gap. Using the fact that the air is a conservative medium, one concludes that the coupling field is lossless. Figure 10.2 illustrates the nonlinear magnetizing characteristic. Using the current and flux linkages, the energy stored in the magnetic field is $W_F = \oint_\psi di\,\psi$ (area above the magnetization curve), while the coenergy is expressed as $W_c = \oint_i \psi di$ (area under the magnetization curve). The total energy is $W_F + W_c = \oint_\psi di\,\psi + \oint_i \psi di = \psi i$. The properties of magnetic and ferromagnetic materials and composites widely used in MEMS must be reported. Table 10.1 reports the initial permeability μ_i, maximum relative permeability $\mu_{r\,\text{max}}$, coersivity (coercive force) H_c, saturation polarization J_s, hysteresis loss per cycle W_h, and Curie temperature T_C for high-permeability bulk metals and alloys. The designer must be aware that deposited materials will have different characteristics, as will be documented in this chapter.

Table 10.2 reports the remanence B_r, flux coersivity (coercive force) H_{Fc}, intrinsic coercivity (coercive force) H_{Ic}, maximum energy product BH_{max}, Curie temperature T_C, and the maximum operating temperature (T_{max}) for hard permanent magnets. The saturation magnetization M_0 is reported at $T = 293$ K, Curie temperature T_C, and the crystal system for some ferromagnetic composites are given in Table 10.3. In general these characteristics are affected by the size of media dimensions, magnets, fabrication processes, etc. Therefore, the designer should not blindly use the data reported in the literature and find the specific information of interest for the examined cases.

TABLE 10.1

Magnetic Properties of High-Permeability Soft Metals and Alloys

Material	Composition (%)	μ_i/μ_0	$\mu_{r\,max}/\mu_0$	H_c [A/m]	J_s [T]	W_h [J/m³]	T_C [K]
Iron	$Fe_{99\%}$	200	6000	70	2.16	500	1043
Iron	$Fe_{99.9\%}$	25000	350,000	0.8	2.16	60	1043
Silicon-iron	$Fe_{96\%}Si_{4\%}$	500	7000	40	1.95	50–150	1008
Silicon-iron (110) [001]	$Fe_{97\%}Si_{3\%}$	9000	40,000	12	2.01	35–140	1015
Silicon-iron (100) <100>	$Fe_{97\%}Si_{3\%}$		100,000	6	2.01		1015
Steel	$Fe_{99.4\%}C_{0.1\%}Si_{0.1\%}Mn_{0.4\%}$	800	1100	200			
Hypernik	$Fe_{50\%}Ni_{50\%}$	4000	70,000	4	1.60	22	753
Deltamax [100] <100>	$Fe_{50\%}Ni_{50\%}$	500	200,000	16	1.55		773
Isoperm [100] <100>	$Fe_{50\%}Ni_{50\%}$	90	100	480	1.60		
78 Permalloy	$Ni_{78\%}Fe_{22\%}$	4000	100,000	4	1.05	50	651
Supermalloy	$Ni_{79\%}Fe_{16\%}Mo_{5\%}$	100,000	1,000,000	0.15	0.79	2	673
Mumetal	$Ni_{77\%}Fe_{16\%}Cu_{5\%}Cr_{2\%}$	20,000	100,000	4	0.75	20	673
Hyperco	$Fe_{64\%}Co_{35\%}Cr_{0.5\%}$	650	10,000	80	2.42	300	1243
Permendur	$Fe_{50\%}Co_{50\%}$	500	6000	160	2.46	1200	1253
2V Permendur	$Fe_{49\%}Co_{49\%}V_{2\%}$	800	4000	160	2.45	600	1253
Supermendur	$Fe_{49\%}Co_{49\%}V_{2\%}$		60,000	16	2.40	1150	1253
25 Perminvar	$Ni_{45\%}Fe_{30\%}Co_{25\%}$	400	2000	100	1.55		
7 Perminvar	$Ni_{70\%}Fe_{23\%}Co_{7\%}$	850	4000	50	1.25		
Perminvar (magnetically annealed)	$Ni_{43\%}Fe_{34\%}Co_{23\%}$		400,000	2.4	1.50		
Alfenol(Alperm)	$Fe_{84\%}Al_{16\%}$	3000	55,000	3.2	0.8		723
Alfer	$Fe_{87\%}Al_{13\%}$	700	3700	53	1.20		673
Aluminum-Iron	$Fe_{96.5\%}Al_{3.5\%}$	500	19,000	24	1.90		
Sendust	$Fe_{85\%}Si_{10\%}Al_{5\%}$	36,000	120,000	1.6	0.89		753

Source: Lide, D.R., Ed., *Handbook of Chemistry and Physics,* 83rd Edition, CRC Press, Boca Raton, FL, 2002. Dorf, R.C., Ed., *Handbook of Engineering Tables,* CRC Press, Boca Raton, FL, 2003. McCurrie, R.A., *Structure and Properties of Ferromagnetic Materials,* Academic Press, London, 1994. Gray, D.E., Ed., *American Institute of Physics Handbook,* 3rd Edition, McGraw Hill, New York, 1972.

TABLE 10.2

Magnetic Properties of High-Permeability Hard Metals and Alloys

Composite and Composition	B_r [T]	H_{Fc} [A/m]	H_{Ic} [A/m]	$(BH)_{max}$ [kJ/m³]	T_C [°C]	T_{max} [°C]
Alnico1: 20Ni,12Al,5Co	0.72		35	25		
Alnico2: 17Ni,10Al,12.5Co,6Cu	0.72		40–50	13–14		
Alnico3: 24–30Ni,12–14Al,0–3Cu	0.5–0.6		40–54	10		
Alnico4: 21–28Ni,11–13Al,3–5Co,2–4Cu	0.55–0.75		36–56	11–12		
Alnico5: 14Ni,8Al,24Co,3Cu	1.25	53	54	40	850	520
Alnico6: 16Ni,8Al,24Co,3Cu,2Ti	1.05		75	52		
Alnico8: 15Ni,7Al,35Co,4Cu,5Ti	0.83	1.6	160	45		
Alnico9: 15Ni,7Al,35Co,4Cu,5Ti	1.10	1.45	1.45	75	850	520
Alnico12: 13.5Ni,8Al,24.5Co,2Nb	1.20		64	76.8		
Ferroxdur: $BaFe_{12}O_{19}$	0.4	1.6	192	29	450	400
$SrFe_{12}O_{19}$	0.4	2.95	3.3	30	450	400
$LaCo_5$	0.91			164	567	
$CeCo_5$	0.77			117	380	
$PrCo_5$	1.20			286	620	
$NdCo_5$	1.22			295	637	
$SmCo_5$	1.00	7.9	696	196	700	250
$Sm(Co_{0.76}Fe_{0.10}Cu_{0.14})_{6.8}$	1.04	4.8	5	212	800	300
$Sm(Co_{0.65}Fe_{0.28}Cu_{0.05}Zr_{0.02})_{7.7}$	1.2	10	16	264	800	300
$Nd_2Fe_{14}B$ (sintered)	1.22	8.4	1120	280	300	100
Vicalloy II: Fe,52Co,14V	1.0	42		28	700	500
Fe,24Cr,15Co,3Mo (anisotropic)	1.54	67		76	630	500
Chromindur II: Fe,28Cr,10.5Co	0.98	32		16	630	500
Fe,23Cr,15Co,3V,2Ti	1.35	4		44	630	500
Fe, 36Co	1.04		18	8		
Co (rare-earth)	0.87		638	144		
Cunife: Cu,20Ni,20Fe	0.55	4		12	410	350
Cunico: Cu,21Ni,29Fe	0.34	0.5		8		
Pt,23Co	0.64	4		76	480	350
Mn,29.5Al,0.5C (anisotropic)	0.61	2.16	2.4	56	300	120

Source: Lide, D.R., Ed., *Handbook of Chemistry and Physics*, 83rd Edition, CRC Press, Boca Raton, FL, 2002. Dorf, R.C., Ed., *Handbook of Engineering Tables*, CRC Press, Boca Raton, FL, 2003. McCurrie, R.A., *Structure and Properties of Ferromagnetic Materials*, Academic Press, London, 1994. Gray, D.E., Ed., *American Institute of Physics Handbook*, 3rd Edition, McGraw Hill, New York, 1972. Jiles, D., *Magnetism and Magnetic Materials*, Chapman & Hall, London, 1991.

For ferrites commonly used in microdevices and the saturation magnetic polarization J_s, Curie temperature and line width ΔH are very important. Table 10.4 reports ferrite properties.

In MEMS the fabrication of thin-film magnetic microstructures requires deposition of conductors, insulators, magnets, and other materials with the specified properties needed to attain the goals, that is, design efficient MEMS capable to perform the desired functions and tasks. To fabricate MEMS and NEMS, different materials are used. Some bulk material constants (conductivity σ, resistivity ρ at 20°C, relative permeability μ_r, thermal expansion t_e, and dielectric constant–relative permittivity ε_r) are given in Table 10.5 in SI units.

Although MEMS and microdevice topologies and configurations vary (see the MEMS classification concept documented in Chapter 5), in general electromagnetic microtransducers have closed-ended, open-ended, or integrated electromagnetic systems. As an example, Figure 10.3 illustrates the microtoroid and the linear micromotor with the closed-ended and open-ended electromagnetic systems, respectively. The copper microwindings and ferromagnetic cores (microstructures made using different magnetic materials) can be fabricated through electroplating, patterning, planarization, and other fabrication processes. Figure 10.3 depicts the electroplated circular copper conductors that form the windings (10 μm wide and thick with 10 μm spacing) deposited on the insulated layer of the ferromagnetic core.

TABLE 10.3

Properties of Ferromagnetic Composites

Composite	Crystal System	M_0 [Gauss]	T_C [K]
MnB	orthorh(FeB)	152	578
MnAs	hex(FeB)	670	318
MnBi	hex(FeB)	620	630
MnSb	hex(FeB)	710	587
M_4N		183	743
MnSi	cub(FeSi)		34
CrTe	hex(NiAs)	247	339
$CrBr_3$	hex(BiI_3)	270	37
CrI_3	hex(BiI_3)		68
CrO_2	tetr(TiO_2)	515	386
EuO	Cub	1910 (at $T = 0$ K)	77
EuS	Cub	1184 (at $T = 0$ K)	16.5
$GdCl_3$	Orthorh	550 (at $T = 0$ K)	2.2
FeB	Orthorh		598
Fe_2B	Tetr ($CuAl_2$)		1043
$FeBe_5$	cub($MgCu_2$)		75
Fe_3C	Orthorh		483
FeP	orthorh (MnP)		215

Source: Lide, D.R., Ed., *Handbook of Chemistry and Physics*, 83rd Edition, CRC Press, Boca Raton, FL, 2002. Dorf, R.C., Ed., *Handbook of Engineering Tables*, CRC Press, Boca Raton, FL, 2003. Kittel, C., *Introduction to Solid State Physics*, John Wiley and Sons, New York, 1986. Ashcroft, N.W. and Mermin, N.D., *Solid State Physics*, Holt, Rinehart, and Winston, New York, 1976.

The comprehensive electromagnetic analysis must be performed for microscale structures, devices, and systems. For example, the torque (force) developed and the voltage induced by microtransducers depend on the inductance. The microdevice efficiency is a function of the winding resistance (resistivity of the coils deposited vary), eddy currents, hysteresis, etc. Studying the microtoroid, consider a circular path of radius R in a plane normal

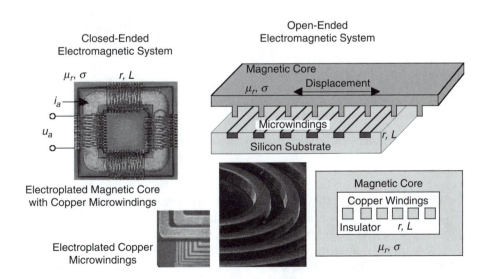

FIGURE 10.3
Closed-ended and open-ended electromagnetic systems in microtransducers (toroidal microstructure with the insulated copper circular conductors wound around the magnetic material and linear micromotor) with ferromagnetic cores (electroplated thin films).

TABLE 10.4

Ferrite Properties

Material	J_s [T]	T_C [°C]	ΔH [kA/m]	Applications
γ-Fe_2O_3 (spinel)	0.52	575		
Fe_3O_4 (spinel)	0.60	585		
$NiFe_2O_4$ (spinel)	0.34	575	350	MEMS microwave devices
$MgFe_2O_4$ (spinel)	0.14	440	70	
$NiZnFe_2O_4$ (spinel)	0.50	375	120	MEMS transformer cores
$MnFe_2O_4$ (spinel)	0.50	300	50	MEMS microwave devices
$NiCoFe_2O_4$ (spinel)	0.31	590	140	MEMS microwave devices
$NiCoAlFe_2O_4$ (spinel)	0.15	450	330	MEMS microwave devices
$NiAl_{0.35}Fe_{1.65}O_4$ (spinel)	0.12	430	67	MEMS microwave devices
$NiAlFe_2O_4$ (spinel)	0.05	1860	32	MEMS microwave devices
$Mg_{0.9}Mn_{0.1}Fe_2O_4$ (spinel)	0.25	290	56	MEMS microwave devices
$Ni_{0.5}Zn_{0.5}Al_{0.8}Fe_{1.2}O_4$ (spinel)	0.14		17	MEMS microwave devices
$CuFe_2O_4$ (spinel)	0.17	455		MEMS electromechanical transducers
$CoFe_2O_4$ (spinel)	0.53	520		MEMS electromechanical transducers
$LiFe_5O_8$ (spinel)	0.39	670		MEMS microwave devices
$Y_3Fe_5O_{12}$ (garnet)	0.178	280	55	MEMS microwave devices
$Y_3Fe_5O_{12}$ (single crystal, garnet)	0.178	292	0.5	MEMS microwave devices
$(Y,Al)_3Fe_5O_{12}$ (garnet)	0.12	250	80	MEMS microwave devices
$(Y,Gd)_3Fe_5O_{12}$ (garnet)	0.06	250	150	MEMS microwave devices
$Sm_3Fe_5O_{12}$ (garnet)	0.170	305		MEMS microwave devices
$Eu_3Fe_5O_{12}$ (garnet)	0.116	293		MEMS microwave devices
$GdFe_5O_{12}$ (garnet)	0.017	291		MEMS microwave devices
$BaFe_{12}O_{19}$ (hexagonal crystal)	0.45	430	1.5	Permanent magnets
$Ba_3Co_2Fe_{24}O_{41}$ (hexagonal crystal)	0.34	470	12	MEMS microwave devices
$Ba_2Zn_2Fe_{12}O_{22}$ (hexagonal crystal)	0.28	130	25	MEMS microwave devices
$Ba_3Co_{1.35}Zn_{0.65}Fe_{24}O_{41}$ (hexagonal crystal)	390	16		MEMS microwave devices
$Ba_2Ni_2Fe_{12}O_{22}$ (hexagonal crystal)	0.16	500	8	MEMS microwave devices
$SrFe_{12}O_{19}$ (hexagonal crystal)	0.4	450		Permanent magnets

Source: Lide, D.R., Ed., *Handbook of Chemistry and Physics*, 83rd Edition, CRC Press, Boca Raton, FL, 2002. Dorf, R.C., Ed., *Handbook of Engineering Tables*, CRC Press, Boca Raton, FL, 2003. R.A., McCurrie, *Structure and Properties of Ferromagnetic Materials*, Academic Press, London, 1994.

to the axis. The magnetic flux intensity is calculated using the following formula: $\oint_s H \cdot ds = 2\pi R H = Ni$, where N is the number of turns. Thus, one has $H = Ni/2\pi R$. The value of H is a function of the R, and therefore the field is not uniform.

Microwindings must guarantee adequate inductance in the limited footprint area with minimal resistance. For example, in the microscale transducers and power converters, 0.5 μH (or higher) inductance is required at high frequency (from 1 to 10 MHz). Compared with the conventional minidevices, thin-film electromagnetic microtransducers have lower efficiency due to higher resistivity of microcoils (thin films), eddy currents, hysteresis, fringing effect, and other undesirable phenomena. These phenomena usually have the

TABLE 10.5

Material Constants

Material	Silver	Copper	Gold	Al	Tungsten	Zinc	Nickel	Iron
$\sigma \times 10^7$	6.2	5.8	4.1	3.8	1.82	1.67	1.45	1
ρ	1.6×10^{-8}	1.7×10^{-8}	2.4×10^{-8}	2.6×10^{-8}	5.5×10^{-8}	6×10^{-8}	6.9×10^{-8}	1×10^{-7}
μ_r	0.99998	0.999999	0.99999	1.00002	1.00008	500 nonlinear	600 nonlinear	900–280,000 nonlinear
$t_e \times 10^{-6}$	19	16.7–17	14	23–25	4.6	N/a	N/a	N/a

Material	Si	SiO$_2$	Si$_3$N$_4$	SiC	GaAs	Ge
e_r	11.8	3.8–3.9	7.6	6.5	11–13	16
$t_e \times 10^{-6}$	2.5–2.65	0.5–0.51	2.7	3–3.7	5.7–6.9	2.2–5
$\sigma \times 10^7$	1.5	0.01	N/a	2.2	0.5	0.7

Note: The data reported strongly depends on the size, fabrication processes and technologies, operating envelope, and environmental conditions, and may significantly vary from the reported.

secondary effects in the miniscale and conventional electromechanical devices, and therefore these phenomena in conventional systems are usually neglected. The inductance can be increased by ensuring a large number of turns, using core magnetic materials with high relative permeability, increasing the cross-sectional core area, and decreasing the path length. In fact, at low frequency, the formula for inductance is

$$L = \frac{\mu_0 \mu_r N^2 A}{l}$$

where μ_r is the relative permeability of the core material; A is the cross-sectional area of the ferromagnetic core; and l is the magnetic path length.

Using the reluctance $\Re = l / \mu_0 \mu_r A$, one has $L = N^2 / \Re$.

For the electromagnetic microtransducers, the flux is a very important variable of interest. Using the net current, one has

$$\Phi = \frac{Ni}{\Re}$$

It is important to recall that the inductance is related to the energy stored in the magnetic field, and

$$L = \frac{2W_m}{i^2} = \frac{1}{i^2} \int_v \mathbf{B} \cdot \mathbf{H} dv$$

Thus, one has

$$L = \frac{1}{i^2} \int_v \mathbf{B} \cdot \mathbf{H} dv = \frac{1}{i^2} \int_v \mathbf{H} \cdot (\nabla \times \mathbf{A}) dv = \frac{1}{i^2} \int_v \mathbf{A} \cdot \mathbf{J} dv = \frac{1}{i} \oint_l \mathbf{A} \cdot d\mathbf{l} = \frac{1}{i} \oint_s \mathbf{B} \cdot d\mathbf{s} = \frac{\Phi}{i}$$

or

$$L = \frac{N\Phi}{i}$$

We found that the inductance is the function of the number of turns, flux, and current.

Making use of $L = \mu_0 \mu_r N^2 A / l$, one concludes that the inductance increases as a function of the squared number of turns. However, a large number of turns require the high turn density fabrication (small track width and spacing in order to place many turns in a given footprint area). However, reducing the track width leads to increase of the conductor resistance decreasing the efficiency. Hence, the trade-off between inductance and winding resistance must be studied. The DC resistance is found as $R = \rho_c (l_c / A_c)$, where ρ_c is the conductor resistivity, which is a nonlinear function of thickness; l_c is the conductor length; and A_c is the conductor cross-sectional area.

To achieve low resistance, one must deposit thick conductors with the thickness in the order of tens of micrometers. The most feasible process for deposition of conductors is electroplating. High-aspect-ratio processes ensure thick conductors and small track widths and spaces. (High-aspect-ratio conductors have a high thickness-to-width ratio.) However, the footprint area is limited not allowing one to achieve a large conductor cross-sectional area. High inductance value can also be achieved increasing the ferromagnetic core cross-sectional area using thick magnetic cores with large A. However, most magnetic materials used are thin film metal alloys, which generally have characteristics not as good as the bulk ferromagnetic materials commonly used in electromechanical and electromagnetic devices. These result in the eddy current and undesirable hysteresis effects that increase the core losses and decrease the inductance. It should be emphasized that eddy currents must be minimized. In addition, the size of MEMS is usually specified, and therefore the limits on the maximum cross-sectional area are imposed.

As illustrated ferromagnetic cores and microwindings are key components of micro-structures, and different magnetic and conductor materials and processes to fabricate microtransducers are employed. Commonly the permalloy (nickel $Ni_{80\%}$–iron $Fe_{20\%}$ alloy) thin films are used. Permalloy as well as other alloys (e.g., $Ni_{x\%}Fe_{100-x\%}$, $Ni_{x\%}Co_{100-x\%}$, $Ni_{x\%}Fe_{y\%}Mo_{100-x-y\%}$, amorphous cobalt-phosphorus, and others) are soft magnetic materials that can be made through electrodeposition and other deposition processes. In general the deposits have nonuniform composition and nonuniform thickness due to the electric current nonuniformity over the electrodeposition area and other phenomena. For example, hydrodynamic effects in the electrolyte increase nonuniformity. These nonuniformities are reduced by applying specific chemicals. The inductance and losses remain constant up to a certain frequency (which is a function of the layer thickness, materials, fabrication processes, etc.). In the high-frequency operating regimes, the inductance rapidly decreases and the losses increase due to the eddy current and hysteresis effects. For example, for the microinductor fabricated using permalloy ($Ni_{80\%}Fe_{20\%}$) thin-film ferromagnetic core with copper microwindings, the inductance decreases rapidly above 1 MHz, 3 MHz, and 6 MHz for the 10 μm, 8 μm, and 5 μm thick layers, respectively. The skin depth of the ferromagnetic core thin films is a function of the magnetic properties and the frequency f. In particular, the skin depth is found as $\delta = \sqrt{1/\pi f \mu \sigma}$, where μ and σ are the permeability and conductivity of the ferromagnetic core material.

The total power losses are found using the Poynting vector, $\Xi = E \times H$, and the total power loss can be approximately derived using the expression for the power crossing the conductor surface within the area, that is,

$$P_{average} = \int_s \Xi_{average} ds = \frac{1}{4} \int_s \sigma \delta E_0^2 e^{-2/\delta \sqrt{\pi f \mu \sigma}} ds.$$

The skin depth (depth of penetration) is available. For bulk copper as well as for copper microcoils with thickness greater than 10 μm deposited using optimized processes, we have $\delta_{Cu} = 0.066/\sqrt{f}$.

In general the inductance begins to decrease when the ratio of the lamination thickness to skin depth is greater than one. Thus, the lamination thickness must be less than skin depth at the operating frequency f to attain the high inductance value. In order to illustrate the need to comprehensively study microinductors, we analyze the toroidal microinductor (1 mm by 1 mm, 3 μm core thickness, 2000 permeability). The inductance and winding resistance are analyzed as the functions of the operating frequency. Modeling results indicate that the inductance remains constant up to 100 kHz and decreases for higher frequency. The resistance increases significantly at frequencies higher than 150 kHz. The copper microconductor thickness is 2 μm, and the DC winding resistance is 10 ohm.) The decreased inductance and increased resistance at high frequency are due to hysteresis and eddy current effects. Therefore, CAD packages and MEMS computational environments must integrate nonlinear phenomena and secondary effects to attain high-fidelity analysis and simulation.

The skin depth of the magnetic core materials depend on permeability and conductivity. The $Ni_{x\%}Fe_{100-x\%}$ thin films have a relative permeability in the range of 600 to 2000, and the resistivity is from 18 to 23 μohm-cm. It should be emphasized that the materials with high resistivity have low eddy current losses and allow one to deposit thicker layers as the skin depth is high. Therefore, high-resistivity magnetic materials are under consideration, and the electroplated FeCo thin films have 100–130 μohm-cm resistivity. Other high-resistivity materials (deposited by sputtering) are FeZrO and CoHfTaPd. In general sputtering has advantages for the deposition of laminated layers of magnetic and insulating materials, because magnetic and insulating composites can be deposited in the same process step. Electroplating, as a technique for deposition of laminated multilayered structures, requires different processes to deposit magnetic and insulating materials (layers).

Three commonly used materials that are ferromagnetic at room temperature are iron, nickel, and cobalt. Ferromagnetic materials lose their ferromagnetic characteristics above the Curie temperature, which for iron is 770°C (1043°K). Thus, alloys of these metals (and other compounds) are also ferromagnetic, e.g., alnico (aluminum–nickel–cobalt alloy with small amount of copper). Some alloys of nonferromagnetic metals are ferromagnetic, for example, bismuth–manganese and copper–manganese–tin alloys. At low temperature, some rare-earth elements (gadolinium, dysprosium, and others) are ferromagnetic.

Ferrimagnetic materials (which have smaller response on an external magnetic field than ferromagnetic materials) are also examined. Ferrites (group of ferromagnetic materials) have low conductivity. To fabricate microstructures and devices with low eddy currents, iron oxide magnetite (Fe_3O_4) and nickel ferrite ($NiFe_2O_4$) are used.

The fabrication of ICs, which are one of the major functional components of MEMS, are reported in the literature.[1] The goal of this chapter is to examine microscale structures and devices (microtransducers) controlled by ICs. Therefore, the fabrication of microstructures and microtransducers should be covered. Microtransducers (which integrate stationary and movable ferromagnetic cores, windings, bearing, etc.) are much more complex than microstructures. Therefore, the fabrication of microinductors is covered first. The major processes involved in the electromagnetic microtransducer and microinductor fabrication are etching and electroplating magnetic vias and through-holes, and then fabricating the inductor-type microstructures on top of the through-hole wafer using multilayer thick photoresist processes.[5-7] For example, let us use the silicon substrate (100-oriented n-type double-sided polished silicon wafers) with a thin layer of thermally grown silicon dioxide (SiO_2). Through-holes are patterned on the topside of the Si-SiO_2 wafer (photolithography process), and then etched in the KOH system (different etch rate can be achieved based on the concentration and temperature). The wafer then is removed from the KOH system with 20–30 μm of silicon remaining to be etched. A Ti-Cu or Cr-Cu seed layer (20–40 nm and 400–500 nm thickness, respectively) is deposited on the backside of the wafer using electron beam evaporation. The copper acts as the electroplating seed layer, while a titanium (or chromium) layer is used to improve adhesion of the copper layer to the silicon wafer. On the copper seed layer, a protective NiFe thin film layer is electroplated directly above the through-holes to attain protection and stability. The through-holes are fully etched again (in the KOH system). Then the remaining SiO_2 is stripped (using the BHF solution) to reveal the backside metal layers. Then the titanium adhesion layer is etched in the HF solution. (If the Cr-Cu seed layer is used, chromium can be removed using the $K_3Fe(CN)_6$–NaOH solution.) This allows the electroplating of through-holes from the exposed copper seed layer. The empty through-holes are electroplated with the NiFe thin film. This forms the magnetic vias. Because the KOH-based etching process is crystallographically dependent, the sidewalls of the electroplating mold are the 111-oriented crystal planes (54.7° angular orientation to the surface). As a result of these 54.7° angularly oriented sidewalls, the electroplating can be nonuniform. To overcome this problem, the through-holes can be over-plated and polished to the surface level.[5-7] After the through-hole overplating and polishing, the seed layer is removed, and 10–20 μm coat (e.g., polyimide PI2611) is spun on the backside and cured at 300°C to cover the protective NiFe layer. At this time, the microinductor can be fabricated on the topside of the wafer. In particular, the microcoils are fabricated on top of the through-hole wafer with the specified ferromagnetic core geometry (e.g., plate, toroidal, horseshoe, or other shapes) parallel to the surface of the wafer. The microcoils must be wound around the ferromagnetic core to form the electromagnetic system. Therefore, the additional structural layers are needed. (For example, the first level is the conductors that are the bottom segments of each micro-coil turn, the second level includes the ferromagnetic core and vertical conductors that connect the top and bottom of each microcoil turn segment, and the third level consists the top conductors that are connected to the electrical vias, and thus form microcoil turns wound around the ferromagnetic core.) It is obvious that the insulating (dielectric) layers are required to insulate the core and microcoils. The fabrication can be performed applying the electron beam evaporation of the Ti-Cu seed layer, and then 25–35 μm electroplating molds are formed. AZ-4000 photoresist can be used. The copper microcoils are electroplated on the top of the mold through electroplating. After electroplating is completed, the photoresist is removed with acetone. Then the seed layer is removed. Copper is etched in the H_2SO_4 solution, while the titanium adhesion layer is etched by the HF solution. A new layer of the AZ-4000 photoresist is spun on the wafer to insulate the bottom conductors from the core. The vias openings are patterned at the ends of the conductors, and the photoresist is cured forming the insulating layer. In addition to insulation, the hard curing leads to reflow of the photoresist serving the planarization purpose needed to pattern additional layers. Another seed layer is deposited from which electrical vias and ferromagnetic core are patterned and electroplated. This leads to two lithography sequential steps, and the electrical vias (electroplated Cu) and ferromagnetic core (NiFe thin film) are electroplated using the same seed layer. After the vias and ferromagnetic core are completed, the photoresist and seed layers are removed. Then the hard curing is performed. The top microconductors are patterned and deposited from another seed layer using the same process as explained above for the bottom microconductors. The detailed description of the processes described and the fabricated microtransducers are available in References 5–7. We have outlined the fabrication of microinductors because this technique can be adopted and used to fabricate microtransducers. It also must be emphasized that the analysis and design must be performed using the equations of motion.

10.3 Analysis of Translational Microtransducers

In this section, we study translational microtransducers. Figure 10.4 illustrates a microelectromechanical device (translational microtransducer) with a stationary member (ferromagnetic core with windings) and movable member (microplunger) that can be fabricated using the micromachining technologies. Our goal is to perform the analysis and modeling of the microtransducer developing the lumped-parameter mathematical model. That is, the goal is to derive the differential equations that model the microtransducer steady-state and dynamic behavior.

Applying Newton's second law for translational motion, we have

$$F(t) = B_v \frac{dx}{dt} + (k_{s1}x + k_{s2}x^2) + F_e(t) = m\frac{d^2x}{dt^2}$$

where x denotes the microplunger displacement; m is the mass of a movable member (microplunger); B_v is the viscous friction coefficient; k_{s1} and k_{s2} are the spring constants;

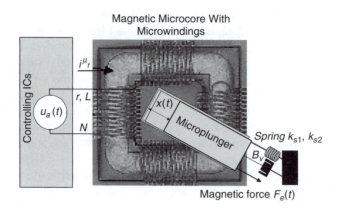

FIGURE 10.4
Schematic of the microtransducer with controlling ICs.

and $F_e(t)$ is the magnetic force,

$$F_e(i,x) = \frac{\partial W_c(i,x)}{\partial x}$$

The restoring/stretching force exerted by the spring is given by $(k_{s1}x + k_{s2}x^2)$.
Assuming that the magnetic system is linear, the coenergy is $W_c(i,x) = \frac{1}{2}L(x)i^2$.
Thus, the electromagnetic force developed is found to be $F_e(i,x) = \frac{1}{2}i^2\frac{dL(x)}{dx}$.
In this formula, the analytic expression for the term $dL(x)/dx$ must be derived. The inductance, as a nonlinear function of the displacement, is

$$L(x) = \frac{N^2}{\Re_f + \Re_g} = \frac{N^2\mu_f\mu_0 A_f A_g}{A_g l_f + A_f \mu_f(x+2d)}$$

where $\Re_f$ and $\Re_g$ are the reluctances of the magnetic material and airgap; A_f and A_g are the cross-sectional areas; and l_f and $(x + 2d)$ are the lengths of the magnetic material and the air gap.
Therefore,

$$\frac{dL}{dx} = -\frac{N^2\mu_f^2\mu_0 A_f^2 A_g}{[A_g l_f + A_f \mu_f(x+2d)]^2}$$

Hence, one finds

$$F_e(i,x) = -\frac{N^2\mu_f^2\mu_0 A_f^2 A_g}{2[A_g l_f + A_f \mu_f(x+2d)]^2}i^2$$

Using Kirchhoff's law, the voltage equation for the electric circuit is

$$u_a = ri + \frac{d\psi}{dt}$$

The flux linkage ψ is given as $\psi = L(x)i$. Thus, one obtains

$$u_a = ri + L(x)\frac{di}{dt} + i\frac{dL(x)}{dx}\frac{dx}{dt}$$

In this equation, $dx/dt = v$.

The following nonlinear differential equation for the circuitry results:

$$\frac{di}{dt} = -\frac{r}{L(x)}i + \frac{1}{L(x)}\frac{N^2\mu_f^2\mu_0 A_f^2 A_g}{[A_g l_f + A_f\mu_f(x+2d)]^2}iv + \frac{1}{L(x)}u_a$$

In this equation, the second term is the emf.

Augmenting this equation with the differential equation with the torsional-mechanical dynamics,

$$F(t) = B_v\frac{dx}{dt} + (k_{s1}x + k_{s2}x^2) + F_e(t) = m\frac{d^2x}{dt^2}$$

three nonlinear differential equations for the considered translational microtransducer are found to be

$$\frac{di}{dt} = -\frac{r[A_g l_f + A_f\mu_f(x+2d)]}{N^2\mu_f\mu_0 A_f A_g}i + \frac{\mu_f A_f}{A_g l_f + A_f\mu_f(x+2d)}iv + \frac{A_g l_f + A_f\mu_f(x+2d)}{N^2\mu_f\mu_0 A_f A_g}u_a$$

$$\frac{dv}{dt} = \frac{N^2\mu_f^2\mu_0 A_f^2 A_g}{2m[A_g l_f + A_f\mu_f(x+2d)]^2}i^2 - \frac{1}{m}(k_{s1}x + k_{s2}x^2) - \frac{B_v}{m}v$$

$$\frac{dx}{dt} = v$$

The derived differential equations represent the lumped-parameter mathematical model of the microtransducer. In general the high-fidelity mathematical modeling and analysis must be performed integrating nonlinearities (for example, nonlinear magnetic characteristics and hysteresis) and secondary effects. However, the lumped-parameter mathematical models as given in the form of nonlinear differential equations have been validated for microtransducers. It is found that the major phenomena and effects are modeled for the current, velocity, and displacement. Secondary effects such as Coulomb friction, hysteresis and eddy currents, fringing effect, and other phenomena have not been modeled and analyzed. At low frequency (up to 1 MHz), the lumped-parameter modeling provides one with the capabilities to attain reliable preliminary steady-state and dynamic analysis using primary circuitry and mechanical variables. It is also important to emphasize that the voltage, applied to the microwinding, is regulated by ICs. The majority of ICs to control micro-transducers are designed using pulse-width-modulation topologies. The switching frequency of ICs is usually 1 MHz or higher. Therefore, as was shown, it is very important to study the microtransducer performance at a high operating frequency. This can be performed using Maxwell's equations, which will lead to the high-fidelity mathematical models.[3] The parameters in the derived mathematical model (resistance, permeability, number of turns, mass, areas, length, etc.) can be found as the micro-transducer sizing and materials are known. It also must be emphasized that fabrication processes significantly affect the choice of materials and dimensions.

10.4 Single-Phase Reluctance Micromotors: Modeling, Analysis, and Control

Consider the single-phase reluctance micromachined micromotors illustrated in Figure 10.5. The emphasis is concentrated on the analysis, modeling, and control of reluctance micromotors in the rotational microactuator applications. Therefore, mathematical models must be found. The lumped-parameter modeling paradigm is based on the use of the circuitry (voltage and current) and mechanical (velocity and displacement) variables to derive the differential equations using Newton's and Kirchhoff's laws. In these differential equations, the micromotor parameters are used. In particular, for the studied micromotor, the parameters are the stator resistance r_s, the magnetizing inductances in the quadrature and direct axes L_{mq} and L_{md} (it is evident that the air gap depends on the rotor position, and thus the magnetizing inductance varies as the function of the rotor displacement, see Figure 10.5), the average magnetizing inductance $\bar{L}_m$, the leakage inductance L_{ls}, the moment of inertia J, and the viscous friction coefficient B_m.

The expression for the electromagnetic torque was derived in Chapter 6 as $T_e = L_{\Delta m} i_{as}^2 \sin 2\theta_r$, where $L_{\Delta m}$ is the half-magnitude of the sinusoidal magnetizing inductance L_m variations, $L_m(\theta_r) = \bar{L}_m - L_{\Delta m} \cos 2\theta_r$.

To develop the electromagnetic torque, the current i_{as} must be fed as a function of the rotor angular displacement θ_r. If $i_{as} = i_M \mathrm{Re}(\sqrt{\sin 2\theta_r})$, then

$$T_{e\,average} = \frac{1}{\pi} \int_0^\pi L_{\Delta m} i_{as}^2 \sin 2\theta_r d\theta_r = \tfrac{1}{4} L_{\Delta m} i_M^2$$

The micromotor under consideration is the synchronous micromachine. The obtained expression for the phase current is very important to control the synchronous microtransducers. In particular, the Hall-effect position sensor should be used to measure the rotor displacement, and the ICs must feed the phase current as a nonlinear function of θ_r. Furthermore, the electromagnetic torque is controlled by changing the current magnitude i_M. The mathematical model of the single-phase reluctance micromotor is found using

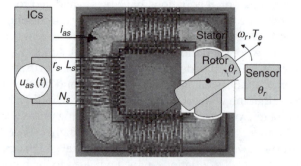

FIGURE 10.5
Single-phase reluctance micromotor with ICs and Hall-sensor to measure the rotor displacement.

Kirchhoff's and Newton's second laws. In particular, we have the following differential equations:

$$u_{as} = r_s i_{as} + \frac{d\psi_{as}}{dt} \text{ (circuitry equation—Kirchhoff's law)}$$

$$T_e - B_m \omega_r - T_L = J \frac{d^2\theta_r}{dt^2} \text{ (torsional-mechanical equation—Newton's law)}$$

Here the electrical angular velocity ω_r and displacement θ_r are used as the mechanical system variables.

From $u_{as} = r_s i_{as} + (d\psi_{as}/dt)$, using the flux linkage equation $\psi_{as} = (L_{is} + \bar{L}_m - L_{\Delta m} \cos 2\theta_r)i_{as}$, one obtains the circuitry dynamics equation with the mechanical variable θ_r. In addition, the torsional-mechanical equation integrates the electromagnetic torque, which is the function of the electrical variable i_{as}. Making use of the circuitry and torsional-mechanical dynamics, one finds a set of three first-order nonlinear differential equations that models single-phase reluctance micromotors:

$$\frac{di_{as}}{dt} = -\frac{r_s}{L_{ls} + \bar{L}_m - L_{\Delta m} \cos 2\theta_r} i_{as} - \frac{2L_{\Delta m}}{L_{ls} + \bar{L}_m - L_{\Delta m} \cos 2\theta_r} i_{as} \omega_r \sin 2\theta_r$$

$$+ \frac{1}{L_{ls} + \bar{L}_m - L_{\Delta m} \cos 2\theta_r} u_{as}$$

$$\frac{d\omega_r}{dt} = \frac{1}{J}\left(L_{\Delta m} i_{as}^2 \sin 2\theta_r - B_m \omega_r - T_L\right)$$

$$\frac{d\theta_r}{dt} = \omega_r$$

As the mathematical model is found and the micromotor parameters are measured, nonlinear simulation and analysis can be straightforwardly performed to study the dynamic responses and analyze the micromotor performance characteristics. In particular, the resistance, inductances, moment of inertia, viscous friction coefficient, and other parameters can be directly measured or identified based on micromotor testing. The steady-state and dynamic analysis based on the lumped-parameter mathematical model is straightforward. However, the lumped-parameter mathematical models simplify the analysis, and thus these models must be compared with the experimental data to validate the results. Microfabrication processes to make reluctance micromotors are described later in this chapter. The disadvantage of single-phase reluctance micromotors are high torque ripple, vibration, noise, low reliability, etc.

10.5 Microfabrication Topics

Electromechanical microstructures and microtransducers can be fabricated through deposition of the conductors (coils and windings), ferromagnetic core, insulation layers, and other microstructures (movable and stationary members and their components including bearing). The order of the processes, materials, and sequential steps are different depending on the MEMS that must be devised, designed, analyzed, and optimized first. This section is aimed to provide the reader with the basic fabrication features and processes involved in the MEMS fabrication.

10.5.1 Microcoils/Microwindings Fabrication through the Copper, Nickel, and Aluminum Electrodeposition

Among many important processes to fabricate MEMS, the deposition is the most critical one. Therefore, let us focus our attention on the deposition through electroplating. The reader should be familiar with the basics of electrochemistry as far as electrolytic processes are concerned. In particular, purification (copper, zinc, cobalt, nickel, and other metals) and electroplating processes are discussed and documented in undergraduate introductory Chemistry textbooks. Let us recall the major principles. When aqueous solutions are electrolyzed using metal electrodes, an electrode will be oxidized if its oxidation potential is greater than for water. As examples, nickel and copper are oxidized more readily than water, and the reactions are

Nickel
$$Ni~(s) \rightarrow Ni^{2+}~(aq) + 2e^-, E_{ox} = 0.28~V$$
$$2H_2O~(l) \rightarrow 4H^+~(aq) + O_2~(g) + 4e^-, E_{ox} = -1.23~V$$
Copper
$$Cu~(s) \rightarrow Cu^{2+}~(aq) + 2e^-, E_{ox} = -0.34~V$$
$$2H_2O~(l) \rightarrow 4H^+~(aq) + O_2~(g) + 4e^-, E_{ox} = -1.23~V$$

where s, aq, l, and g denote the solid, aqueous, liquid, and gas states, respectively.

If the anode is made from nickel in an electrolytic cell, nickel metal is oxidized as the anode reaction. If Ni^{2+} (aq) is the solution, it is reduced at the cathode in the preference to reduction of water. As current flows, nickel dissolves from the anode and deposits on the cathode. The reactions are

$$Ni~(s) \rightarrow Ni^{2+}~(aq) + 2e^-~(anode)$$

and

$$Ni^{2+}~(aq) + 2e^- \rightarrow Ni~(s)~(cathode)$$

Most metals (chromium, iron, cobalt, nickel, copper, zinc, silver, gold, and others) used to fabricate MEMS are the transition metals that occupy the "d" block of the periodic table. Important physical properties of these metals were listed in Table 10.1 to Table 10.5. The chemistry of transition metals (reactions involved and chemicals used) is very important due to the significance of electroplating, etching, and other fabrication processes. In this book, most practical and effective processes, techniques, and materials (chemicals) are covered.

The conductors (microcoils to make windings) in microstructures and microtransducers can be fabricated by electrodepositing the copper and other low-resistivity metals. Electrodeposition of metals is made by immersing a conductive surface in a solution containing ions of the metal to be deposited. The surface is electrically connected to an external power supply, and current is fed through the surface into the solution. In general, the reaction of the metal ions ($Metal^{x+}$) with x electrons (xe^-) to form metal (Metal) is

$$Metal^{x+} + xe^- = Metal$$

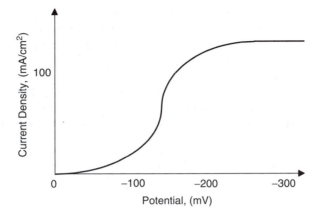

FIGURE 10.6
Current-voltage curve for a copper electroplating process.

To electrodeposit copper on the silicon wafer, the wafer is typically coated with a thin conductive layer of copper (seed layer) and immersed in a solution containing cupric ions. Electrical contact is made to the seed layer, and current is flowed (passed) such that the reaction $Cu^{2+} + 2e^- \rightarrow Cu$ occurs at the wafer surface. The wafer, which is electrically interacted such that the metal ions are changed to metal atoms, is the cathode. Another electrically active surface (anode) is the conductive solution to make the electrical path. At the anode, the oxidation reaction occurs that balances the current flow at the cathode, thus maintaining the electric neutrality. In the case of copper electroplating, all cupric ions removed from solution at the wafer cathode are replaced by dissolution from the copper anode. According to the Faraday law of electrolysis, in the absence of secondary reactions, the current delivered to a conductive surface during electroplating is proportional to the quantity of the metal deposited. Thus, the metal deposited can be controlled varying the electroplating current (current density) and the electrodeposition time. If the applied voltage potential is zero, there is an equilibrium. Once potential is changed by the external power source, the current flows and electroplating results. For electroplating the current is approximated by an exponential Tafel equation. Figure 10.6 illustrates a current-voltage curve for a copper electrodeposition process. The electrodeposition rate is controlled by changing the reaction rate kinetics. The nonlinear current-voltage dependence results in the need for a special electroplating cell design to attain the uniform voltage potentials across the wafer surface.

As the voltage potential increases, the mass transfer effects become predominant, and the current saturation (current density limit) is reached. The elements reacting the cathode Cu^{2+} do not reach the interface at a rate sufficient to sustain the high rate of reaction. Usually, electroplating processes are performed at the 30–50% current density limit to avoid undesirable electrodeposition effects. To ensure that the rate of mass transfer of electroactive elements to the interface is large compared to the reaction rate and the desired uniformity is achieved, the rates of migration diffusion and convection must be controlled. Convection is the most important phase of the mass transfer and can vary from stagnant to laminar or turbulent flow. It includes impinging flow caused by the solution pumping, undesirable flows due to substrate movement, and flows resulting from density variations. Electroplating can be carried out using a constant current, a constant voltage, or variable waveforms of current or voltage at different temperature. Using a constant current, accurate control of the mass of the electrodeposited metal can be obtained. Electroplating using variable waveforms requires more complex equipment to control, but is meaningful in maintaining the specific thickness distributions and desired film properties. The wave-forms are periodic (rectangular, sinusoidal, trapezoidal, or triangular) forward and reverse pulses with variable (peak, high, average, low, or controlled) magnitude of the forward and reverse current or voltage. In addition, the duty cycle (the ratio of the forward and reverse time periods) can be controlled.

From the electrochemistry viewpoint, the commonly used solutions for copper electroplating are acidic (copper sulfate in fluoborate pyrophosphate bath) and alkaline (cyanide in not-cyanide bath). Different additives can be added to achieved the desired characteristics.

The hydrated Cu ions reaction is

$$Cu^{2+} \rightarrow Cu(H_2O)_6{}^{2+}$$

and the cathode reactions are

$$Cu^{2+} + 2e^- \rightarrow Cu, \; Cu^{2+} + e^- \rightarrow Cu^+, \; Cu^+ + e^- \rightarrow Cu,$$

$$2Cu^+ \rightarrow Cu^{2+} + Cu, \; H^+ + e^- \rightarrow \tfrac{1}{2}H_2$$

The copper electroplating solution commonly used is $CuSO_4$–$5H_2O$ (250 g/l) and H_2SO_4 (25 ml/l). The basic processes are shown in Figure 10.7. The description of the sequential steps is given. For electrodeposition of copper, Silicon, Kapton, and other substrates can be

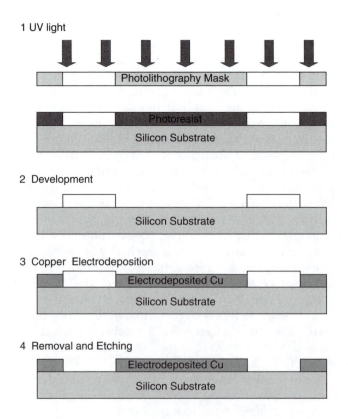

FIGURE 10.7
Electrodeposition of copper.

used. After clearing, the silicon substrate is covered with a 5–10-nm chromium or titanium and a 100–200-nm copper seed layer by sputtering. The copper microcoils (microstructures) are patterned using UV photolithography. The AZ-4562 photoresist can be spincoated and prebaked on a ramped hot plate at 90–100°C (ramp 30–40% with initial temperature 20–25°C) for one hour. Then the photoresist is exposed in the Karl Suss Contact Masker with the energy 1200–1800 mJ-cm². The development is released in 1:4 diluted alkaline solution (AZ-400) for 4–6 min. This gives the photoresist thickness 15–25 μm. Copper is electroplated with a three-electrode system with a copper anode and a saturated calomel reference electrode. The current power supply is the Perkin Elmer Current Source e.g., &G 263. The Shipley sulfate bath with the 5–10 ml/l brightener to smooth the deposit can be used. The elec-trodeposition is per-formed at 20–25°C with the magnetic stirring and the DC current density from 40 to 60 mA/cm². This current density leads to smooth copper thin films with 5–10-nm rms roughness for the 10-μm thickness of the deposited copper thin film. The resistivity of the electrode-posited copper thin film (microcoils) is 1.6–1.8 μohm-cm (close to the bulk copper resistivity). After the deposition, the photoresist is removed.

As was emphasized, to reduce the losses in the copper microwindings, it is desirable to deposit the thicker conductors. Therefore, a thick photoresist (tens of μm) is required instead of the standard photoresist. For the positive photoresists (AZ-4000), 20–30 μm thickness with good structural resolution can be obtained. When thicker layers are desired, the negative photoresists (e.g., EPON Resin SU-8, which is the negative epoxy-based photoresist, MicroChem Co.) have advantages because higher aspect ratio can be achieved than with the positive photoresists. In the one-step process, the aspect ratio 20:1 with straight side walls can be achieved.

In particular, SU-8 (formulated in GBL) and SU-8 2000 (formulated in cyclopentanone) are chemically amplified, epoxy-based negative resists that can be used for thin films with thicknesses from 1 μm to 200 μm. These SU-8 and SU-8 2000 resists have been widely used because they have high functionality, high optical transparency, and are sensitive to near UV radiation. High aspect ratio and straight sidewalls can be formed in thick films by contact-proximity or projection printing. It should be emphasized that SU-8 resist is highly resistant to solvents, acids, and bases and has excellent thermal stability, making it well suited for applications when microstructures are a permanent part of the microdevice.

Different industrial electrodeposition systems are available to perform electroplating. Usu-ally the systems are available for the wafer size up to 6 inches. The following metals and alloys can be electroplated using the same station: copper, nickel, palladium, gold, tin, lead, iron, silver, nickel-iron (permalloy), palladium-nickel, and others. Different rated current is used (for example, from –1 to 1 A), and the maximum peak current can be 3–5 times greater than the rated current. The current is stabilized usually within 0.001 A, and the current ripple is less than 0.5–1%. Filtration (cartridge with 1 μm or less features), temperature regulation, current waveform specification, duty cycle control, and ventilation feature are guaranteed. Furthermore, the industrial electrodeposition systems are computerized; see Figure 10.8. The electroplating is performed using anode, cathode, and solution. Let us describe the electro-plating process by depositing the nickel. The electroplating bath and electrodeposition process are illustrated in Figure 10.8.

The anode is made from nickel, and the cathode is made from the conductive material. The solution consists of the nickel Ni^{2+}, hydrogen H^+, and sulfate SO_4^{2-} ions. As the voltage is applied, the positive ions in the solution are attracted to the negative cathode. The nickel ions Ni^{2+} that reach the cathode gain electrons and are deposited (electroplated) on the surface of the cathode forming the nickel electrodeposit. At the same time, nickel is electrochemically etched from the nickel anode to produce ions for the aqueous solution and electrons for the power source. The hydrogen ions H^+ gain electrons from the cathode

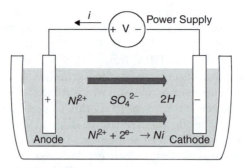

FIGURE 10.8
Electroplating bath and nickel electrodeposition.

and form hydrogen gas leading to bubbles. The resulting hydrogen gas formation is undesirable because it slows the electroplating process because current is not fully used for electrodeposition. Furthermore, bubbles degrade the electrodeposits' uniformity. To fabricate Ni-based microstructures by electrodeposition, a conductive plating base (seed layer is the sputtered nickel) and electrodeposit patterning are involved. Usually the electrodeposit is patterned by an additive process (selective electrodeposition) instead of a subtractive process (etching). The recipes (chemicals) for chemical baths to electrodeposit materials are different. For example, for nickel deposition, the solution can be made using nickel sulfate (250 g/l), nickel chloride (10 g/l), boric acid (30 g/l), ferrous sulfate (10 g/l) and saccharin (5 g/l).

The solutions for deposition of magnetic thin-film alloys, layers of which can be made through electrodeposition, will be given later.

It must be emphasized that commonly used magnetic materials and conductors do not adhere well to silicon. Therefore, as was described, the adhesion layers (e.g., titanium Ti or chromium Cr) are deposited on the silicon surface prior to the magnetic material electroplating.

The electrodeposition rate is proportional to the current density, and therefore the uniform current density at the substrate seed layer is needed to attain the uniform thickness of the electrodeposit. To achieve the selective electrodeposition, portions of the seed layer are covered with the resist (the current density at the mask edges is nonuniform degrading electroplating). In addition to the current density, the deposition rate is also a nonlinear function of temperature, solution (chemicals), pH, direct/reverse current or voltage waveforms magnitude, waveform pulse shapes (sinusoidal, rectangular, trapezoidal, etc.), duty ratio, plating area, etc. The simplified equations to calculate the thickness and electrodeposition time for the specified materials are

$$\text{Thickness}_{material} = \frac{\text{Time}_{electroplating} \times \text{Current}_{density} \times \text{Weight}_{molecular}}{\text{Faradey}_{constant} \times \text{Density}_{material} \times \text{Electron}_{number}}$$

$$\text{Time}_{electroplating} = \frac{\text{Thickness}_{material} \times \text{Faradey}_{constant} \times \text{Density}_{material} \times \text{Electron}_{number}}{\text{Current}_{density} \times \text{Weight}_{molecular}}$$

It was emphasized that electroplating is used to deposit thin-film conductors and magnetic materials. However, microtransducers need the insulation layers; otherwise the core and coils as well as multilayer microcoils themselves will be short-circuited. Furthermore, the seed layers are embedded in microfabrication processes. As the ferromagnetic core is fabricated on top of the microcoils (or microcoils are made on the core), the seed layer is difficult to remove because it was placed at the bottom or at the center of the microstructure. The mesh seed layer can serve as the electroplating seed layer for the lower conductors, and as the microstructure is made, the edges of the mesh seed layer can be exposed and removed through plasma etching.[6] Thus, the microcoils are insulated. It should be emphasized that relatively high-aspect-ratio techniques must be used to fabricate the ferro-mag-netic core and microcoils, and patterning as well as surface planarization issues must be addressed.

Ferromagnetic cores in microstructures and microtransducers must be made. For example, the electroplated $Ni_{x\%}Fe_{100-x\%}$ thin films, such as permalloy $Ni_{80\%}Fe_{20\%}$, can be deposited to form the ferromagnetic core of microtransducers (actuators and sensors), inductors, transformers, switches, etc. The basic processes and sequential steps used are similar to the processes for copper electrodeposition, and the electroplating is done in the electroplating bath. The windings (microcoils) must be insulated from cores, and therefore the insulation layers must be deposited. The insulating materials used to insulate the windings from the core are benzocyclobutene, polyimide (PI-2611), etc. For example, cyclotene 7200-35 is photosensitive and can be patterned through photolithography. The benzocyclobutene, used as the photoresist, offers good planarization, pattern properties, and stability at low temperatures. It exhibits negligible hydrophilia.

The sketched fabrication process with sequential steps to make the electromagnetic microtransducer with movable and stationary members is illustrated in Figure 10.9. On the silicon substrate, the chromium-copper-chromium (Cr-Cu-Cr) mesh seed layer is deposited (through electron-beam evaporation) forming a seed layer for electroplating. The insulation layer (polyimide Dupont PI-2611) is spun on the top of the mesh seed layer to form the electroplating molds. Several coats can be done to obtain the desired thickness of the polyimide molds. One coat usually results in 8–12 μm insulation layer thickness. After coating, the polyimide is cured (at 280–310°C) in nitrogen for 1 hour. A thin aluminum layer is deposited on top of the cured polyimide to form a hard mask for dry etching.

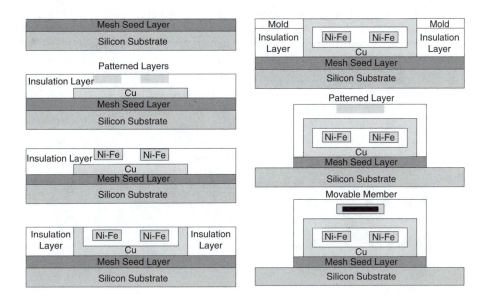

FIGURE 10.9
Basic fabrication sequential steps for the microtransducer fabrication.

Molds for the lower conductors are patterned and plasma etched until the seed layer is exposed. After etching the aluminum (hard mask) and chromium (top Cr-Cu-Cr seed layer), the molds are filled with the electroplated copper through the described copper electroplating process. One coat of polyimide insulates the lower conductors and the core. Thus, insulation is achieved. The seed layer is deposited, mesh-patterned, coated with polyimide, and hard-cured. The aluminum thin layers (hard mask for dry etching) are deposited, and the mold for the ferromagnetic cores is patterned and etched until the seed layer is exposed. After etching the aluminum (hard mask) and the chromium (top Cr-Cu-Cr seed layer), the mold is filled with the electroplated $Ni_{x\%}Fe_{100-x\%}$ thin films (electroplating process). The desired composition and thickness of the $Ni_{x\%}Fe_{100-x\%}$ thin films can be achieved, as will be described later. One coat of the insulation layer (polyimide) is spincast and cured to insulate the core and upper conductors. The via holes are patterned in the sputtered aluminum layer (hard mask) and etched through the polyimide layer using oxygen plasma. The vias are filled with the electroplated copper (electroplating process). The copper-chromium seed layer is deposited, and the molds for the upper conductors are formed using thick photoresist. The molds are filled with the electroplated copper and removed. Then the gap for the movable member is made using the conventional processes. After removing the seed layer, the passivation layer (polyimide) is coated and cured to protect the top conductors. The polyimide is masked and etched to the silicon substrate. The bottom mesh seed layer is wet etched, and the microtransducer (with the ICs to control it) is diced and sealed.

The electrodeposition process must be optimized to attain uniformity with the ultimate goal to fabricate high-performance MEMS. As an example, Figure 10.10 illustrates electrodeposited copper on the silicon substrate with the seed layer. In most cases one cannot guarantee the ideal geometry and dimensions of a microstructure. To deposit windings, one can form deep grooves (25–75 μm) in silicon by anisotropic etching in KOH, etch the trenches (10–20 μm deep), electroplate conductors (Au, Cu, or other metals) to form the windings, and finally elec-troplate Cu and PbSn to form the bumps (if needed). (See Figure 10.10.)

Electroplated aluminum is the needed material to fabricate microstructures and microdevices. In particular, aluminum can be used as the conductor to fabricate microcoils as well as mechanical microstructures (gears, bearing, pins, reflecting surfaces, etc.). Advanced techniques and processes for the electrodeposition of aluminum are documented in Reference 9. Using the micromolding processes, high aspect ratio can be achieved. In CMOS and surface micromachining, single-crystal and polycrystalline silicon, silicon-based compounds, polymers, as well as various electroplated and physical vapor deposited metals (including aluminum) are used. The need for integrity and compliance of materials leads to application of different materials based on their electrochemomechanical properties and thermal characteristics. Aluminum (which has high thermal conductivity and corrosion resistance, low resistivity and neutron absorption, stable mechanical properties, etc.) has been widely used to fabricate ICs. It is used as conductors and sacrificial layers. Thus, it is important to

FIGURE 10.10
Electrodeposited copper resist pattern used as an etching mask for etching the trenches, etched trenches (10 μm width), and windings with PbSn bumps.

develop the aluminum-compliant surface micromachining processes. The electrolytic solutions used to electroplate aluminum are based on organic solvents. One of the major problems is to find compatible molding materials to fabricate high-aspect-ratio microstructures.

Aluminum can be electrodeposited from inorganic and organic fused salt mixtures, as well as from solutions of aluminum compounds in certain organic solvents. Aluminum is more chemically active than hydrogen, and therefore cannot be electrodeposited from solutions that contain water or other compounds with the acidic hydrogen. Fused salt baths result in inherent thermal distortion due to residual stresses in thin films as well as other drawbacks.[9] In contrast, the aluminum chloride-lithium aluminum–hydride ethereal baths provide low-stress thin films. The National Bureau of Standards' hydride process for aluminum electroplating was commercialized. The aluminum electroplating solution composition is: $AlCl_3$ (aluminum chloride, 400 g/l) and $LiAlH_4$ (lithium aluminum hydride, 15 g/l)— electroplating is done at temperatures ranging from 10°C to 60°C.[9]

Due to the nature of the bath, safety issues during electroplating must be strictly obeyed.

The electroplating must be carried out in an inert atmosphere (for example, in a sealed glove box with dry nitrogen as the ambient gas). Care must be taken mixing the electrolytic solution (protecting it from ignition and spark sources) and using the electrolytic solution. To mix the electrolytic solution in the nitrogen-filled sealed glove box, the $AlCl_3$ is slowly added to diethyl ether. This is an exothermic reaction, which produces heat. Thus, cooling is recommended during mixing to minimize evaporation and to permit rapid addition of the aluminum chloride. Then the $LiAlH_4$ is slowly added while mixing the solution.[9] For aluminum electroplating, the typical current density, 10–15 mA/cm[2,] results in the electroplating rate of 0.8–1.2 μm/min. The resistivity of the electrolytic solution is in the range of 95–110 ohm-cm. To fabricate micromolded electroplated aluminum microstructures, the material (including aluminum) must have desirable properties (high-aspect-ratio molds and removal) associated with conventional electroforming materials, as well as properties desirable for nonconventional electroplating processes (e.g., the ability to withstand solvent-based electrolytic solutions and chemical resistance to perform preprocessing required for electroplating). Thus, the standard photoresist electroplating mold processes cannot be utilized to make aluminum-based microstructures. The polyimide materials have the properties necessary to withstand aluminum electroplating conditions. The micromolding processes used to make the aluminum microstructures use photosensitive (photocrosslinked on exposure) and nonphotosensitive polyimides.[9] The nonphotosensitive polyimide process involves plasma or reactive ion etching to produce molds with the desired side-wall profiles and geometries. The major process for fabrication of electroplated microstructures using photosensitive polyimide is similar to the LIGA processes.[2,4,10] However, the photosensitive polyimide is used as the electroplating mask instead of polymethylmethacrylate, and the UV exposure source is used instead of the x-ray synchrotron radiation. The electroplating system consisting of the adhesion–seed–protective layers is deposited on the silicon substrate as shown in Figure 10.11.

After the metal is deposited on the substrate, the antireflective coating is spun in the vacuum spinning station. The photosensitive polyimide is spun on top of the antireflective coating. The photosensitive polyimide is soft-baked in the oven and patterned. Conventional microelectronic alignment and exposure stations can be used. The polyimide is developed and rinsed in solvent-based solutions to create patterned molds in the thick polyimide films. After the polyimide molds are made, the antireactive coating is removed in oxygen plasma. The protective metal layer overlying the electroplating seed layer is removed. Chromium is etched using the $HCl:H_2O$ 1:1 solution. Two additional steps are required to prepare the seed layer for aluminum electroplating: (1) the sample is rinsed in isopropyl alcohol or methanol to remove residual water vapor from the surface; (2) the sample is dipped in a

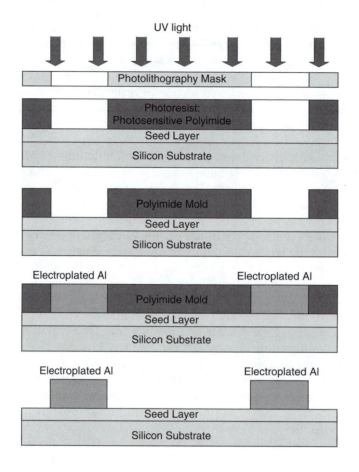

FIGURE 10.11
Fabrication of aluminum microstructures.

solution of salicylic acid (100 g/l solution) in ether. (This solution prepares the surface of the metal seed layer for electrodeposition removing residual metal oxides.) Then the sample is transferred to the electrolytic cell after removal from the salicylic acid solution. Electroplated aluminum is deposited to the surface of the mold. After completion of the electroplating process, the substrate is rinsed in alcohol to remove residues. The polyimide mold can be removed through wet etching of the polyimide using strippers or dry etching in oxygen-based plasmas.[9] Gold, copper, aluminum, and nickel can be used as seed layers for aluminum-based electroplated microstructures. Electroplating using aluminum and nickel seed layers is sensitive to the oxidation effects.

Achieving high-aspect-ratio for aluminum-based microstructures is of great importance, and electroplating of aluminum using polymer micromolds can be viewed as the primary technique to make these microstructures. Paper 9 reports that the developed refined process allows one to make aluminum microstructures with the same dimensions as the polyimide used as the mold. High-aspect-ratio is achieved utilizing the inversion characteristic of the process. In general, high-aspect-ratio polyimide microstructures are simpler to fabricate than high-aspect-ratio trenches. The fabrication starts with the basic process using photosensitive polyimide as the molding material through which aluminum is electroplated. For example, the electroplating metal system consisted of a 30-nm titanium adhesion layer between the silicon (or ceramic) substrate and the electroplating seed layer, 100 nm of either gold or copper as the electroplating seed layer, and an overlying 100-nm chromium layer to protect the electroplating seed layer during processes leading to the final electroplating. The polyimide mold is spun on at the specified thickness and photolithographically defined in the pattern required for the desired microstructure geometry and shape. After aluminum is electroplated on the top of the polyimide molds, the polyimide is removed, leaving the free-standing (released) aluminum microstructure. The seed layer is exposed (after the polyimide is removed), and the overlying chromium protective layer is removed using reactive ion etching.

10.5.2 $Ni_{x\%}Fe_{100-x\%}$ Thin-Film Electrodeposition

As was reported, the ferromagnetic core of microstructures and microtransducers must be fabricated. Two major challenges in the fabrication of high-performance microstructures are to make electroplated magnetic thin films with good magnetic properties as well as planarize microstructures (with primary emphasis on stationary and movable members). Electroplating and micromolding techniques and processes are used to deposit NiFe alloys ($Ni_{x\%}Fe_{100-x\%}$ thin films). In conventional electromechanical motion devices, the $Ni_{80\%}Fe_{20\%}$ alloy is called permalloy, while $Ni_{50\%}Fe_{50\%}$ is called orthonol.

Let us document the deposition process. To deposit $Ni_{x\%}Fe_{100-x\%}$ thin films, the silicon wafer is covered with a seed layer (for example, for Cr-Cu-Cr seed layer, one may use 15–25 nm chromium, 100–200 nm copper, and 25–50 nm chromium) deposited using electron beam evaporation. The photoresist layer (e.g., 10–20 μm Shipley STR-1110) is deposited on the seed layer and patterned. Then the electrodeposition of the $Ni_{x\%}Fe_{100-x\%}$ is performed at a particular temperature (usually 20–30°C) using a two-electrode system, and in general the current density is in the range of 1 to 30 mA/cm². The temperature and pH should be maintained within the recommended values. High pH causes highly stressed NiFe thin films, and low pH reduces leveling and causes chemical dissolving of the iron anodes, resulting in disruption of the bath equilibrium and nonuniformity. High temperature leads to hazy deposits, and low temperature causes high current density burning. Thus, many trade-offs must be taken into account.

For deposition, the pulse-width modulation (with varied waveforms, different forward and reverse magnitudes, and controlled duty cycle) can be used applying commercial or in-house power supplies. Denoting the duty cycle length as T, the forward and reverse pulse lengths are denoted as T_f and T_r. The pulse length T is usually in the range of 5–20 μsec, and the duty cycle (ratio T_f/T_r) can be varied from 1 to 0.1. The ratio T_f/T_r influences the percentage of Ni in the $Ni_{x\%}Fe_{100-x\%}$ thin films. For example, the composition of $Ni_{x\%}Fe_{100-x\%}$ can be regulated based on the desired properties that will be discussed later. However, varying the ratio T_f/T_r, the changes of the Ni are relatively modest (from 79 to 85%), and therefore other parameters vary to attain the desired composition.

The nickel (and iron) composition is a function of the current density, and Figure 10.12 illustrates the nickel (iron) composition in the $Ni_{x\%}Fe_{100-x\%}$ thin films.

The $Ni_{80\%}Fe_{20\%}$ thin films of different thickness (which is a function of the electrodeposition time) are usually made at the current density of 14–16 mA/cm². This range of the current density can be used to fabricate various thicknesses of permalloy thin films (from 500 nm to 50 μm). The rms value of the thin film roughness is 4–7 nm for the 30-μm thickness. To guarantee good surface quality, the current density should be kept at the specified range, and usually to change the composition of the $Ni_{x\%}Fe_{100-x\%}$ thin films, the reverse current is controlled.

To attain a good deposit of the permalloy, the electroplating bath may contain $NiSO_4$ (0.7 mol/l), $FeSO_4$ (0.03 mol/l), $NiCl_2$ (0.02 mol/l), saccharine (0.016 mol/l) as leveler (to reduce the residual stress allowing the fabrication of thicker films), and boric acid (0.4 mol/l). Other chemicals can be added and concentrations can be different.

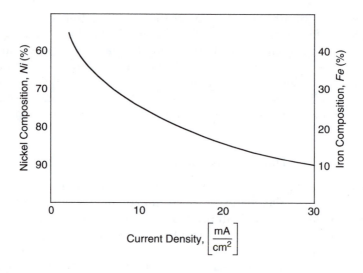

FIGURE 10.12
Nickel and iron compositions in $Ni_{x\%}Fe_{100-x\%}$ thin films as the functions of the current density.

The air agitation and saccharin are added to reduce internal stress and to keep the Fe composition stable. The deposition rate varies linearly as a function of the current density (obeying the Faraday law), and the electrodeposition slope is 100–150 nm-cm²/min-mA. The permalloy thin film density is 9 g/cm³ (as for the bulk permalloy).

The magnetic properties of the $Ni_{80\%}Fe_{20\%}$ (permalloy) thin films are studied, and the field coercivity H_c is a function of the thickness. For example, $H_c = 650$ A/m for 150 nm thickness and $H_c = 30$ A/m for 600-nm films.

Other possible solutions for electroplating the $Ni_{80\%}Fe_{20\%}$ (deposited at 25°C) and $Ni_{50\%}Fe_{50\%}$ (deposited at 55°C) thin films are

- $Ni_{80\%}Fe_{20\%}$: $NiSO_4$–$6H_2O$ (200 g/l), $FeSO_4$–$7H_2O$ (9 g/l), $NiCl_2$–$6H_2O$ (5 g/l), H_3BO_3 (27 g/l), saccharine (3 g/l), and pH (2.5–3.5)
- $Ni_{50\%}Fe_{50\%}$: $NiSO_4$–$6H_2O$ (170 g/l), $FeSO_4$–$7H_2O$ (80 g/l), $NiCl_2$–$6H_2O$ (138 g/l), H_3BO_3 (50 g/l), saccharine (3 g/l), and pH (3.5–4.5)

To electroplate $Ni_{x\%}Fe_{100-x\%}$ thin films, various additives (chemicals) and components (available from M&T Chemicals and other suppliers) can be used to optimize (control) the thin-film properties and characteristics. The objective is to control the internal stress and ductility of the deposit, keep the iron content solublized, obtain bright film, leveling of the process, attain the desired surface roughness, and most importantly guarantee the desired magnetic properties.

In general permalloy thin films have optimal magnetic properties at the following composition of nickel and iron: 80.5% Ni and 19.5% Fe. Thin films with minimal magnetostriction usually have optimal coercivity and permeability properties. For $Ni_{80.5\%}Fe_{19.5\%}$ thin films, the magnetostriction has zero crossing, the coercivity is 20 A/m or higher (coercivity is a non-linear function of the film thickness), and the permeability is from 600 to 2000. Varying the composition of Fe and Ni, the characteristics of the $Ni_{x\%}Fe_{100-x\%}$ thin films can be changed. The composition of the $Ni_{x\%}Fe_{100-x\%}$ thin films is controlled by changing the current density, T_f/T_r ratio (duty cycle), bath temperature (varying the temperature, the composition of Ni can be varied from 75 to 92%), reverse current (varying the reverse current in the range 0–1 A, the composition of Ni can be changed from 72 to 90%), air agitation of the solution,

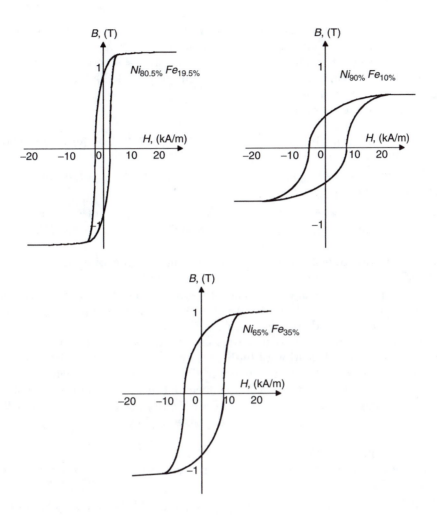

FIGURE 10.13
B-H curves for different $Ni_{x\%}Fe_{100-x\%}$ permalloy thin films.

paddle frequency (0.1–1 Hz), forward and reverse pulses waveforms, etc. The *B-H* curves for three different $Ni_{x\%}Fe_{100-x\%}$ thin films are illustrated in Figure 10.13. The $Ni_{80.5\%}Fe_{19.5\%}$ thin films have the saturation flux density 1.2 T, remanence $B_r = 0.26$ T-A/m, and the relative permeability 600–2000.

Other electroplated permanent magnets (NiFeMo, NiCo, CoNiMnP, and others) and micromachined polymer magnets exhibit good magnetic properties and can be used as an alternative solution to the $Ni_{x\%}Fe_{100-x\%}$ thin films widely used.

10.5.3 NiFeMo and NiCo Thin-Film Electrodeposition

To attain the desired magnetic properties (flux density, coercivity, permeability, etc.) and thickness, different thin film alloys can be used. Magnetic materials are primarily used to fabricate stationary and movable members of microtransducers. Based on the microstructures and microtransducers design, applications, and operating envelopes (temperature, shocks, radiation, humidity, etc.), different thin films can be applied. The $Ni_{x\%}Fe_{100-x\%}$ thin films can be effectively used, and the desired magnetic properties can be readily achieved varying the composition of Ni. For sensors the designer usually maximizes the flux density and permeability and minimizes the coercivity. The $Ni_{x\%}Fe_{100-x\%}$ thin films have a flux density of up to 1.2 T, coercivity of 20 (for permalloy) to 500 A/m, and permeability of 600 to 2000. It was emphasized that the magnetic properties depend on the thickness and other factors. Having emphasized the magnetic properties of the $Ni_{x\%}Fe_{100-x\%}$ thin films, let us perform the comparison.

It was reported in the literature that:[11]

- $Ni_{79\%}Fe_{17\%}Mo_{4\%}$ thin films have a flux density of 0.7 T, coercivity of 5 A/m, and permeability of 3400.

- $Ni_{85\%}Fe_{14\%}Mo_{1\%}$ thin films have a flux density of 1–1.1 T, coercivity of 8–300 A/m, and permeability of 3000–20000.

- $Ni_{50\%}Co_{50\%}$ thin films have a flux density of 0.95–1.1 T, coercivity of 1200–1500 A/m, and permeability of 100–150. ($Ni_{79\%}Co_{21\%}$ thin films have a permeability of 20.)

In general, high flux density, low coercivity, and high permeability lead to high-performance MEMS. However, other issues (affordability, compliance, integrity, operating envelope, fabrication processes, etc.) must also be add-ressed before making a final choice. The magnetic characteristics, in addition to the film thickness, are significantly influenced by the fabrication processes and chemicals used.

The ferromagnetic core in microstructures and microtransducers must be made. Two major challenges in the fabrication of high-performance microstructures and microtransducers are to make electroplated magnetic thin films with good magnetic properties and to planarize the stationary and movable members. Electroplating and micromolding techniques and processes are used to deposit NiFe, NiFeMo, NiCo, and other thin films. For example, the $Ni_{80\%}Fe_{20\%}$, NiFeMo, and NiCo (deposited at 25°C) electroplating solutions are

- $Ni_{80\%}Fe_{20\%}$: $NiSO_4$–$6H_2O$ (200 g/l), $FeSO_4$–$7H_2O$ (9 g/l), $NiCl_2$–$6H_2O$ (5 g/l), H_3BO_3 (27 g/l), and saccharine (3 g/l). The current density is 10–25 mA/cm^2. Nickel foil is used as the anode.

- NiFeMo: $NiSO_4$–$6H_2O$ (60 g/l), $FeSO_4$–$7H_2O$ (4 g/l), Na_2MoO_4–$2H_2O$ (2 g/l), NaCl (10 g/l), citrid acid (66 g/l), and saccharine (3 g/l). The current density is 10–30 mA/cm^2. Nickel foil is used as the anode.

- $Ni_{50\%}Co_{50\%}$: $NiSO_4$–$6H_2O$ (300 g/l), $NiCl_2$–$6H_2O$ (50 g/l), $CoSO_4$–$7H_2O$ (30 g/l), H_3BO_3 (30 g/l), sodium lauryl sulfate (0.1 g/l), and saccharine (1.5 g/l). The current density is 10–25 mA/cm^2. Nickel or cobalt can be used as the anode.

The most important feature is that the $Ni_{x\%}Fe_{100-x\%}$–NiFeMo–NiCo thin films (multilayer nanocomposites) can be fabricated shaping the magnetic properties of the resulting materials to attain the desired performance characteristics through design and fabrication processes.

10.5.4 Micromachined Polymer Magnets

Electromagnetic microtransducers can be devised, modeled, analyzed, designed, optimized, and then fabricated. Micromachined permanent-magnet thin films, including polymer magnets (magnetically hard ceramic ferrite powder embedded in epoxy resin), can be used. This section is focused on the application of the micromachined polymer magnets. Different forms and geometry of polymer magnets are available. Thin-film disks and plates are uniquely suitable for microtransducer applications, for example, to actuate the switches in microscale logic devices, to displace mirrors in optical devices and optical MEMS, etc. In fact, permanent magnets are used in rotational and translational (linear) microtransducers, microsensors, microswitches, etc. The polymer magnets have thicknesses ranging from hundreds of micrometers to several millimeters. Excellent magnetic properties can be achieved. For example, the micromachined polymer permanent-magnet disk with 80% strontium ferrite concentration (4 mm diameter and 90 μm thickness), magnetized normal to the thin-film plane (in the thickness direction), has the intrinsic coercivity $H_{ci} = 320,000$ A/m and a residual induction $B_r = 0.06$ T.[12] Polymer magnets with thickness up to several millimeters can be fabricated by the low-temperature processes. To make permanent magnets, Hoosier Magnetics Co. strontium ferrite powder (1.1–1.5 μm grain size) and Shell epoxy resin (cured at 80°C for 2 hours) can be used.[12] The polymer matrix contains a bisphenol-A–based epoxy resin diluted with cresylglycidyl ether, and the aliphatic amidoamine is used for curing. To prepare the polymer magnet composites, the strontium ferrite powder is mixed with the epoxy resin in the ball-mill rotating system. After the aliphatic amidoamine is added, the epoxy is deposited and patterned using screen-printing. Then the magnet is cured at 80°C for 2 hours and magnetized in the desired direction. In addition to fabrication processes, one should study other issues, for example, the magnetization dynamics and permanent-magnet magnetic properties. The magnetic field in thin films are modeled, analyzed, and simulated solving differential equations, and the analytic and numerical results will be covered in this chapter.

10.5.5 Planarization

The planarization of the movable and stationary members is a very important issue. For example, in fabrication of microtransducers, the gaps between the magnetic poles (permanent magnets) and teeth must be filled in order to eliminate the forces without disintegration and degradation of thin films. Copper can be applied to fill the gaps. Negative epoxy-based photoresistive SU-8, which has superior intrinsic adhesion characteristics (chemically resistant and thermally stable up to 250°C), is widely used. The processing of the photoresistive epoxy resin starts with spin-coating in the dehumidified water at 150–250 rad/sec, resulting in a thickness of 15–25 μm. The softbrake can be made at 100°C, following by cooling. After the relaxation exposure, the final hardbrake process is carried out. To deposit the copper in the desired area, photomasks are made that cover all regions except the regions where the deposition is needed (for example, between the teeth). The copper is deposited using the electroplating process described. The teeth gap can be filled with other deposited materials, e.g., aluminum. However, aluminum deposition is more challenging than copper electroplating.

10.6 Magnetization Dynamics of Thin Films

The magnetic field, including the magnetization distribution, in thin films is modeled, analyzed, and simulated solving differential equations. The dynamic variables are the magnetic field density and intensity, magnetization, magnetization direction, wall position domain, etc. The thin films must be magnetized. Therefore, let us study the magnetization dynamics in thin films. To attain high-fidelity modeling, the magnetization dynamics in the angular coordinates are described by the Landay-Lifschitz-Gilbert equations:[13]

$$\frac{d\psi}{dt} = -\frac{\gamma}{M_s(1+\alpha^2)}\left(\sin^{-1}\psi\frac{\partial E(\theta,\psi)}{\partial\theta} + \alpha\frac{\partial E(\theta,\psi)}{\partial\psi}\right)$$

$$\frac{d\theta}{dt} = -\frac{\gamma\sin^{-1}\psi}{M_s(1+\alpha^2)}\left(\alpha\sin^{-1}\psi\frac{\partial E(\theta,\psi)}{\partial\theta} - \frac{\partial E(\theta,\psi)}{\partial\psi}\right)$$

FIGURE 10.14
Magnetic field in the permalloy thin-film micromagnet.

where M_s is the saturation magnetization; $E(\theta, \psi)$ is the total Gibb's thin film free energy density; and γ and α are the gyromagnetic and phenomenological constants.

The total energy consists of the magnetocrystalline anisotropy energy, the exchange energy, and the magnetostatic self-energy (stray field energy).[14] In particular, the Zeeman energy equation is

$$E = \int_v \left(\frac{k_{exh}}{J_s^2} \sum_{j=1}^{3} (\nabla J_j)^2 - \frac{k_J}{J_s^2}(\mathbf{a}_j \mathbf{J})^2 - \tfrac{1}{2}\mathbf{J}\cdot\mathbf{H}_D - \mathbf{J}\cdot\mathbf{H}_{ex} \right) dv$$

while the Gilbert equation is

$$\frac{\partial \mathbf{J}}{\partial t} = -\left| \gamma \right| \mathbf{J} \times \mathbf{H}_{eff} + \frac{\alpha}{J_s}\mathbf{J} \times \frac{\partial \mathbf{J}}{\partial t}$$

Here $\mathbf{J}$ is the magnetic polarization vector; $\mathbf{H}_D$ and $\mathbf{H}_{ex}$ are the demagnetizing and external magnetic fields; $\mathbf{H}_{eff}$ is the effective magnetic field (sum of the applied, demagnetization, and anisotropy fields); k_{exh} and k_J are the exchange and magnetocrystalline anisotropy constants; and $\mathbf{a}_j$ is the unit vector parallel to the uniaxial easy axis.

Using the vector notations, we have

$$\frac{d\mathbf{M}}{dt} = -\frac{\gamma}{1+\alpha^2}\left(\mathbf{M}\times\mathbf{H}_{eff} + \frac{\alpha}{M_s}\mathbf{M}\times(\mathbf{M}\times\mathbf{H}_{eff}) \right)$$

Thus, using the nonlinear differential equations given, the high-fidelity modeling and analysis of nanostructured nanocomposite permanent magnets can be performed using field and material quantities, parameters, constants, etc. For example, using the available software (developed by the National Institute of Standards and Technologies), the results of the three-dimensional simulation of the magnetic field in the permalloy thin-film micromagnet (50–100–5 nm) are illustrated in Figure 10.14.

The characteristics of nano- and microscale structures (multilayer nanocomposed thin-film layers) can be measured. The cryogenic microwave probe test-station for testing is illustrated in Figure 10.15.

10.7 Microstructures and Microtransducers with Permanent-Magnet: Micromirror Actuators

The electromagnetic microactuator (permanent magnet on the cantilever flexible beam and spiral planar windings controlled by ICs fabricated using CMOS-MEMS technology)

FIGURE 10.15
Cryogenic probe test-station.

is illustrated in Figure 10.16. In addition, the microstructure with eight cantilever beams and eight-by-eight fiber switching matrix is illustrated.

The electromagnetic microactuators can be made using conventional surface micromachining and CMOS fabrication technologies through electroplating, screen-printing, lamination

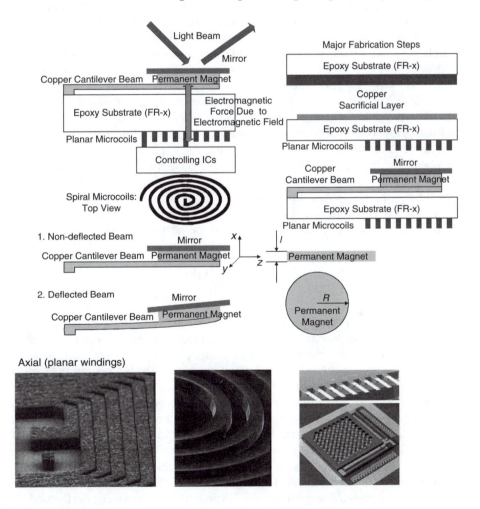

FIGURE 10.16
Electromagnetic microactuator with controlling ICs, microstructure with eight cantilever beams, and eight incoming–eight outgoing array of optical fiber (eight-by-eight fiber switching matrix).

processes, sacrificial layer techniques, photolithography, etching, etc. In particular, the electromagnetic microactuator studied can be made on the commercially available epoxy substrates (e.g., FR series) that have the one-sided laminated copper layer. The copper layer thickness (which can be 10 μm and higher) is defined by the admissible current density and the current value needed to establish the desired magnetic field to attain the specified mirror deflection, deflection rate, settling time, and other steady-state and dynamic characteristics. The spiral planar microcoils can be made on the one-sided laminated copper layer using photolithography and wet etching in the ferric chloride solution. The resulting x-μm thick N-turn microwinding will establish the magnetic field. The number of turns is a function of the footprint area available, thickness, spacing, outer-inner radii, geometry, fabrication techniques and processes used, etc. After fabrication of the planar microcoils, the cantilever beam with the permanent magnet and mirror is fabricated on the other side of the substrate. First a photoresist sacrificial layer is spin-coated and patterned on the substrate. Then, the Cr-Cu-Cr or Ti-Cu-Cr seed layer is deposited to perform the copper electroplating (if copper is used to fabricate the flexible cantilever structure). The second photoresist layer is spun and patterned to serve as a mold for the electroplating of the copper-based cantilever beam. The copper cantilever beam is electroplated in the copper-sulfate–based plating bath. After the electroplating, the photoresist plating mold and the seed layer are removed releasing the cantilever beam structure. Depending on the permanent magnet used, the corresponding fabricated processes must be done before or after releasing the beam. The permanent-magnet disk is positioned on the cantilever beam's free end. For example, the polymer magnet can be screen-printed, and after curing the epoxy magnet, the magnet is then magnetized by the external magnetic field. Then the cantilever beam with the fabricated mirror is released by removing the sacrificial photoresist layer using acetone. The studied electromagnetic microactuator is fabricated using low-cost (affordable), high-yield micromachining–CMOS technology, processes, and materials. The most attractive feature is the application of the planar microcoils, which can be easily made. The use of the polymer permanent magnets (which have good magnetic properties) allows one to design high-performance electromagnetic microactuators. It must be emphasized that polysilicon can be used to fabricate the cantilever beam, and other permanent magnets can be applied.

The vertical electromagnetic force F_{ze}, acting on the permanent magnet, is expressed as[12]

$$F_{ze} = M_z \int_v \frac{dH_z}{dz} dv$$

where M_z is the magenetization, and H_z is the vertical component of the magnetic field intensity produced by the planar microwindings. H_z is a nonlinear function of the current fed or voltage applied to the microwindings, number of turns, microcoils geometry, etc. Therefore, the thickness of the microcoils must be derived based on the maximum value of the current needed and the admissible current density.

The magnetically actuated cantilever microstructures were also studied in References 15 and 16. The expressions for the electromagnetic torque are found as the functions of the magnetic field using assumptions and simplifications that in general limit the applicability of the results. The differential equations that model the electromagnetic and torsional-mechanical dynamics can be derived. In particular, the equations for the electromagnetic field are found using electromagnetic theory. The electromagnetic field intensity H_z is controlled changing the current fed to the planar microwindings. The steady-state analysis, performed using the small-deflection theory,[17] is also valuable. The static deflection of the cantilever beam x can be found using the force and beam quantities. In particular, $x = (l^3/3EJ)F_n$, where l is the

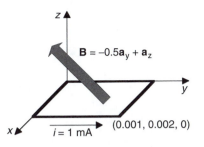

FIGURE 10.17
Rectangular planar loop in a uniform magnetic field
with flux density $\mathbf{B} = -0.5\mathbf{a}_y + \mathbf{a}_z$.

effective length of the beam; E is Young's (elasticity) modulus; J is the equivalent moment of inertia of the beam with permanent magnet and mirror, and for the stand-alone cantilever beam with the rectangular cross section $J = \frac{1}{12}wh^3$; w and h are the width and thickness of the beam; and F_n is the net force that is normal to the cantilever beam.

Assuming that the magnetic flux is constant through the magnetic plane (loop), the torque on a planar current loop of any size and shape in the uniform magnetic field is

$$\mathbf{T} = i\mathbf{s} \times \mathbf{B} = \mathbf{m} \times \mathbf{B}$$

where i is the current, and $\mathbf{m}$ is the magnetic dipole moment [A-m^2].

Thus, the torque on the current loop always tends to turn the loop to align the magnetic field produced by the loop with permanent-magnet magnetic field causing the resulting electromagnetic torque. For example, for the current loop shown in Figure 10.17, the torque is found to be

$$\mathbf{T} = i\mathbf{s} \times \mathbf{B} = \mathbf{m} \times \mathbf{B} = 1 \times 10^{-3}\left[(1 \times 10^{-3})(2 \times 10^{-3})\mathbf{a}_z\right] \times (-0.5\mathbf{a}_y + \mathbf{a}_z)$$

$$= 1 \times 10^{-9}\mathbf{a}_x \text{ N-m}$$

The electromagnetic force is found as $\mathbf{F} = \oint_i i d\mathbf{l} \times \mathbf{B}$.

In general, the magnetic field quantities are derived using

$$\mathbf{B} = \frac{\mu_0}{4\pi} i \oint_l \frac{d\mathbf{l} \times \mathbf{r}_0}{r^2} \quad \text{or} \quad \mathbf{H} = \frac{1}{4\pi} i \oint_l \frac{d\mathbf{l} \times \mathbf{r}_0}{r^2}$$

and the Ampere circuital law gives $\oint_i \mathbf{H} \cdot d\mathbf{l} = i_{total}$ or $\oint_i \mathbf{H} \cdot d\mathbf{l} = Ni$.

Making use these expressions and taking note of the variables defined in Figure 10.18, we have

$$\mathbf{H} = \frac{1}{4\pi} i \oint_l \frac{d\mathbf{l} \times \mathbf{r}_1}{r_1^3} \quad \text{and} \quad \mathbf{B} = \frac{\mu_0}{4\pi} i \oint_l \frac{d\mathbf{l} \times \mathbf{r}_1}{r_1^3}$$

where $d\mathbf{l} = \mathbf{a}_\phi a d\phi = (-\mathbf{a}_x \sin\phi + \mathbf{a}_y \cos\phi)a d\phi$ and $\mathbf{r}_1 = \mathbf{a}_x(x - a\cos\phi) + \mathbf{a}_y(y - a\sin\phi) + \mathbf{a}_z z$.
Hence, $d\mathbf{l} \times \mathbf{r}_1 = \left[\mathbf{a}_x z\cos\phi + \mathbf{a}_y z\sin\phi - \mathbf{a}_z(y\sin\phi + x\cos\phi - a)\right]a d\phi$.
Then, neglecting the small quantities ($a^2 \ll r^2$), we have

$$r_1^3 = (x^2 + y^2 + z^2 + a^2 - 2ax\cos\phi - 2ay\sin\phi)^{3/2} \approx r^3\left(1 - \frac{2ax}{r^2}\cos\phi - \frac{2ay}{r^2}\sin\phi\right)^{3/2}$$

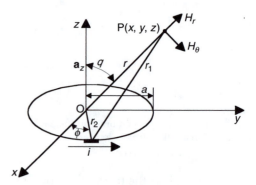

FIGURE 10.18
Planar current loop.

Therefore, one obtains

$$\frac{1}{r_1^3} = \frac{1}{r^3}\left(1 + \frac{3ax}{r^2}\cos\phi + \frac{3ay}{r^2}\sin\phi\right)$$

Thus, we have

$$\mathbf{B} = \frac{\mu_0 a}{4\pi} i \int_0^{2\pi} [\mathbf{a}_x z\cos\phi + \mathbf{a}_y z\sin\phi - \mathbf{a}_z(y\sin\phi + x\cos\phi - a)]\, a\frac{1}{r^3}$$

$$\times\left(1 + \frac{3ax}{r^2}\cos\phi + \frac{3ay}{r^2}\sin\phi\right)d\phi$$

$$= \frac{\mu_0 a^2}{4\pi r^3} i\left[\mathbf{a}_x\frac{3xz}{r^2} + \mathbf{a}_y\frac{3yz}{r^2} - \mathbf{a}_z\left(\frac{3x^2}{r^2} + \frac{3y^2}{r^2} - 2\right)\right]|$$

Furthermore, using the coordinate transformation equations, in the spherical coordinate system one has

$$\mathbf{B} = \frac{\mu_0 a^2}{4\pi r^3} i(2\mathbf{a}_r\cos\theta + \mathbf{a}_\theta\sin\theta)$$

We have the expressions for the far-field components:

$$B_r = \frac{\mu_0 a^2\cos\theta}{2\pi r^3}i, \quad B_\theta = \frac{\mu_0 a^2\sin\theta}{4\pi r^3}i \quad \text{and} \quad B_\phi = 0$$

It is evident that due to the symmetry about the z-axis, the magnetic flux density does not have the B_ϕ component.

Using the documented technique, one can easily find the magnetic vector potential. In general

$$\mathbf{A} = \frac{\mu_0}{4\pi} i\oint_l \frac{d\mathbf{l}}{r_1}$$

Assuming that $a^2 \ll r^2$, we have get the following expression:

$$\frac{1}{r_1} = \frac{1}{r}\left(1 + \frac{ax}{r^2}\cos\phi + \frac{ay}{r^2}\sin\phi\right)$$

Therefore,

$$\mathbf{A} = \frac{\mu_0 a}{4\pi}i\int_0^{2\pi}(-\mathbf{a}_x\sin\phi + \mathbf{a}_y\cos\phi)\frac{1}{r}\left(1 + \frac{ax}{r^2}\cos\phi + \frac{ay}{r^2}\sin\phi\right)d\phi$$

$$= \frac{\mu_0 a}{4\mu r^3}i(-\mathbf{a}_x y + \mathbf{a}_y x)$$

Hence, in the spherical coordinate system, we obtain

$$\mathbf{A} = (\mathbf{A}\cdot\mathbf{a}_r)\mathbf{a}_r + (\mathbf{A}\cdot\mathbf{a}_\phi)\mathbf{a}_\phi + (\mathbf{A}\cdot\mathbf{a}_\theta)\mathbf{a}_\theta = \frac{\mu_0 a}{4\pi r^2}i\mathbf{a}_\phi\sin\theta = A_\phi\mathbf{a}_\phi$$

It should be emphasized that the derived equations can be expressed using the magnetic dipole.

However, in the microtransducer studied, high-fidelity analysis should be performed. Hence, let us perform the comprehensive analysis.

The vector potential is found to be

$$A_\phi(r,\theta) = \frac{\mu_0 ai}{4\pi}\int_0^{2\pi}\frac{\cos\phi\, d\phi}{\sqrt{a^2 + r^2 - 2ar\sin\theta\cos\phi}}$$

and

$$B_r = \frac{1}{r\sin\theta}\frac{\partial(\sin\theta A_\phi)}{\partial\theta}, \quad B_\theta = -\frac{1}{r}\frac{\partial(rA_\phi)}{\partial r}, \quad B_\phi = 0$$

Making use of the following approximation,

$$A_\phi(r,\theta) = \frac{\mu_0 ai}{4\pi}\int_0^{2\pi}\frac{\cos\phi\, d\phi}{\sqrt{a^2 + r^2 - 2ar\sin\theta\cos\phi}} \approx \frac{\mu_0 a^2 r\sin\theta i}{4(a^2 + r^2)^{3/2}}\left(1 + \frac{15a^2r^2\sin^2\theta}{8(a^2 + r^2)^2} + \cdots\right)$$

one finds

$$B_r(r,\theta) = \frac{\mu_0 a^2\cos\theta i}{2(a^2 + r^2)^{3/2}}\left(1 + \frac{15a^2r^2\sin^2\theta}{4(a^2 + r^2)^2} + \cdots\right)$$

$$B_\theta(r,\theta) = -\frac{\mu_0 a^2\sin\theta i}{4(a^2 + r^2)^{5/2}}\left(2a^2 - r^2 + \frac{15a^2r^2\sin^2\theta(4a^2 - 3r^2)}{8(a^2 + r^2)^2} + \cdots\right)$$

We specify three regions:

Near the axis, $\theta \ll 1$
At the center, $r \ll a$
In far-field, $r \gg a$

The electromagnetic torque and field depend on the current in the microwindings and are nonlinear functions of the displacement.

Finally it is important to emphasize that the magnetic dipole moment is

$$\mathbf{m} = \pi a^2 i \mathbf{a}_z$$

where πa^2 is the planar area of the loop.

For sinusoidal field variation, using the equation for the vector magnetic potential $\nabla^2 \mathbf{A} + \omega_f^2 \mu\varepsilon \mathbf{A} = -\mu \mathbf{J}$ and the Lorentz condition $\nabla \cdot \mathbf{A} = -j\omega_f \mu\varepsilon V$, we have

$$\mathbf{A} = \frac{\mu}{4\pi}\int_v \frac{1}{r}\mathbf{J}e^{-j\beta r}dv, \quad \beta = \omega_f\sqrt{\mu\varepsilon}$$

Thus, for the current loop, the vector magnetic potential is given as

$$\mathbf{A} = \frac{\mu}{4\pi}\int_v \frac{1}{r_1}\mathbf{J}e^{-j\beta r_1}dv = \frac{\mu}{4\pi}\int_l \frac{1}{r_1}\mathbf{i}e^{-j\beta r_1}dl$$

Taking note of $e^{-j\beta r_1} = e^{-j\beta r}e^{-j\beta(r_1-r)} \approx e^{-j\beta r}[1 - j\beta(r_1 - r)]$, one has

$$\mathbf{A} = \frac{\mu a^2}{4r^2}i(1 + j\beta r)e^{-j\beta r}\sin\theta\,\mathbf{a}_\phi$$

The expression for the electromagnetic forces and torques must be derived to model and analyze the torsional-mechanical dynamics. Newton's laws of motion can be applied to study the mechanical dynamics in the Cartesian or other coordinate systems. For the translational motion in the x-axis, we previously used $dv/dt = (1/m)(F_e - F_L)$ and $dx/dt = v$ to model the translational torsional-mechanical dynamics of the electromagnetic microactuators using the electromagnetic force F_e and the load force F_L.

For the studied microactuator, the rotational motion can be studied, and the electromagnetic torque can be approximated as

$$T_e = 4R^2 t_{tf} M H_p \cos\theta$$

where R and t_{tf} are the radius and thickness of the permanent-magnet thin-film disk; M is the permanent-magnet thin film magnetization; H_p is the field produced by the planar windings; and θ is the displacement angle.

Then the microactuator rotational dynamics are given by

$$\frac{d\omega}{dt} = \frac{1}{J}(T_e - T_L) \quad \text{and} \quad \frac{d\theta}{dt} = \omega$$

where T_L is the load torque that integrates the friction and disturbances torques.

It should be emphasized that more complex and comprehensive mathematical models can be developed and used integrating the nonlinear electromagnetic and six-degree-of-freedom rotational-translational motions (torsional-mechanical dynamics) of the cantilever beam. As an illustration, we consider the high-fidelity modeling of the electromagnetic system.

10.7.1 Electromagnetic System Modeling in Microactuators with Permanent Magnets: High-Fidelity Modeling and Analysis

In this section, we focus our efforts to derive the expanded equations for the electromagnetic torque and force on cylindrical permanent-magnet thin films. See Figure 10.16. The permanent-magnet thin film is assumed to be uniformly magnetized, and the equations are developed for two orientations of the magnetization vector. In particular, the orientation is parallel to the axis of symmetry, and the orientation is perpendicular to this axis. Electromagnetic fields and gradients produced by the planar windings should be found at a point in inertial space that coincides with the origin of the permanent-magnet axis system in its initial alignment. Our ultimate goal is to control microactuators, and thus high-fidelity mathematical models (which will result in viable analysis, control, and optimization) must be derived.[18] To attain our objective, the complete equations for the electromagnetic torque and force on a cylindrical permanent-magnet thin films are found.

The following notations are used: A, R, and l are the area, radius, and length of the cylindrical permanent magnet; $\mathbf{B}$ is the magnetic flux density vector; $\mathbf{B}_e$ is the expanded magnetic flux density vector; $[\partial \mathbf{B}]$ is the matrix of field gradients [T/m]; $[\partial \mathbf{B}_e]$ is the matrix of expanded field gradients [T/m]; $\mathbf{F}$ and $\mathbf{T}$ are the total force and torque vectors on the permanent-magnet thin film; i is the current in the planar microwinding; $\mathbf{m}$ is the magnetic moment vector [A-m²]; $\mathbf{M}$ is the magnetization vector [A/m]; $\mathbf{r}$ is the position vector (x, y, z are the coordinates in the Cartesian system),

$$\mathbf{r} = \begin{bmatrix} x \\ y \\ z \end{bmatrix}$$

T_r is the inertial coordinate vector-transformation matrix; W and Π are the work and potential energy; θ is the Euler orientation for the 3-2-1 rotation sequence; ∇ is the gradient operator; $_{ij}$ is the partial derivative of i component in j-direction; $_{(ij)k}$ is the partial derivative of ij partial derivative in k-direction; and ‾ (bar over a variable) indicates that it is referenced to the microactuator coordinates.

10.7.2 Electromagnetic Torques and Forces: Preliminaries

The equations for the electromagnetic torque and force on a cylindrical permanent-magnet thin film are found by integrating the equations for torques and forces on an incremental volume of the permanent-magnet thin film with magnetic moment $M dv$ over the volume. Figure 10.16 illustrates the microactuator with the cylindrical permanent-magnet thin film in the coordinate system, which consists of a set of orthogonal body-fixed axes that are initially aligned with a set of orthogonal x-, y-, and z-axes fixed in the inertial space.

The equations for the electromagnetic torque and force on an infinitesimal current can be derived using the fundamental relationship for the force on a current-carrying-conductor element in a uniform magnetic field. In particular, for a planar current loop (planar microwinding) with constant current i in the uniform magnetic field $\mathbf{B}$ (vector $\mathbf{B}$ gives the magnitude and direction of the flux density of the external field), the force on an element $d\mathbf{l}$ of the conductor is found using the Lorentz force law:

$$\mathbf{F} = \oint_{l} i\, d\mathbf{l} \times \mathbf{B}$$

Assuming that the magnetic flux is constant through the magnetic loop, the torque on a planar current loop of any size and shape in the uniform magnetic field is

$$\mathbf{T} = i\oint_{l} \mathbf{r} \times (d\mathbf{l} \times \mathbf{B}) = i\oint_{l} \left((\mathbf{r}\cdot\mathbf{B}) d\mathbf{l} - \mathbf{B}\oint_{l}\mathbf{r}\cdot d\mathbf{l} \right)$$

Using Stokes's theorem, one has

$$\mathbf{T} = i\left(\int_{s} d\mathbf{A} \times \nabla(\mathbf{r}\cdot\mathbf{B}) - \mathbf{B}\int_{s}(\nabla\times\mathbf{r})\cdot d\mathbf{A} \right) = i\int_{s} d\mathbf{A}\times\mathbf{B}$$

or

$$\mathbf{T} = i\mathbf{A}\times\mathbf{B} = \mathbf{m}\times\mathbf{B}$$

The electromagnetic torque $\mathbf{T}$ acts on the infinitesimal current loop in a direction to align the magnetic moment $\mathbf{m}$ with the external field $\mathbf{B}$, and if $\mathbf{m}$ and $\mathbf{B}$ are misaligned by the angle θ, we have

$$\mathbf{T} = \mathbf{m}\mathbf{B}\sin\theta$$

The incremental potential energy and work are found as

$$dW = d\Pi = \mathbf{T}d\theta = m\mathbf{B}\sin\theta d\theta \quad \text{and} \quad W = \Pi = -m\mathbf{B}\cos\theta = -\mathbf{m}\cdot\mathbf{B}$$

Using the electromagnetic force, we have

$$dW = -d\Pi = \mathbf{F}\cdot d\mathbf{r} = -\nabla\Pi\cdot d\mathbf{r} \quad \text{and} \quad \mathbf{F} = -\nabla\Pi = \nabla(\mathbf{m}\cdot\mathbf{B}) = (\mathbf{m}\cdot\nabla)\mathbf{B}$$

10.7.3   Coordinate Systems and Electromagnetic Fields

The transformation from the inertial coordinates to the permanent-magnet coordinates is

$$\bar{\mathbf{r}} = T_r \mathbf{r}$$

$$= \begin{bmatrix} \cos\theta_y\cos\theta_z & \cos\theta_y\sin\theta_z & -\sin\theta_y \\ \sin\theta_x\sin\theta_y\cos\theta_z - \cos\theta_x\sin\theta_z & \sin\theta_x\sin\theta_y\sin\theta_z + \cos\theta_x\sin\theta_z & \sin\theta_x\cos\theta_y \\ \cos\theta_x\sin\theta_y\cos\theta_z + \sin\theta_x\sin\theta_z & \cos\theta_x\sin\theta_y\sin\theta_z - \sin\theta_x\cos\theta_z & \cos\theta_x\cos\theta_y \end{bmatrix} \begin{bmatrix} x \\ y \\ z \end{bmatrix}$$

$$\mathbf{r} = \begin{bmatrix} x \\ y \\ z \end{bmatrix}, \quad \bar{\mathbf{r}} = \begin{bmatrix} \bar{x} \\ \bar{y} \\ \bar{z} \end{bmatrix}$$

Using the transformation matrix

$$T_r = \begin{bmatrix} \cos\theta_y \cos\theta_z & \cos\theta_y \sin\theta_z & -\sin\theta_y \\ \sin\theta_x \sin\theta_y \cos\theta_z - \cos\theta_x \sin\theta_z & \sin\theta_x \sin\theta_y \sin\theta_z + \cos\theta_x \sin\theta_z & \sin\theta_x \cos\theta_y \\ \cos\theta_x \sin\theta_y \cos\theta_z + \sin\theta_x \sin\theta_z & \cos\theta_x \sin\theta_y \sin\theta_z - \sin\theta_x \cos\theta_z & \cos\theta_x \cos\theta_y \end{bmatrix}$$

if the deflections are small, we have

$$T_{rs} = \begin{bmatrix} 1 & \theta_z & -\theta_y \\ -\theta_z & 1 & \theta_x \\ \theta_y & -\theta_x & 1 \end{bmatrix}$$

We use the 3-2-1 orthogonal transformation matrix for the *z-y-x* Euler rotation sequence, and $\theta_x, \theta_y, \theta_z$ are the rotation Euler angles about the *x-*, *y-*, and *z*-axes.

The field **B** and gradients of **B** are produced by the microcoils fixed in the inertial frame and expressed assuming that the electromagnetic fields can be described by the second-order Taylor series. Expanding **B** about the origin of the *x-y-z* system as a Taylor series, we have[18]

$$\mathbf{B}_e = \mathbf{B} + (\mathbf{r} \cdot \nabla)\mathbf{B} + \tfrac{1}{2}(\mathbf{r} \cdot \nabla)^2 \mathbf{B} \quad \text{or}$$

$$B_{ei} = B_i + \frac{\partial B_i}{\partial \mathbf{r}}\mathbf{r} + \tfrac{1}{2}\mathbf{r}^T \frac{\partial^2 B_i}{\partial \mathbf{r}^2}\mathbf{r}$$

where

$$\frac{\partial B_i}{\partial \mathbf{r}} = \begin{bmatrix} \dfrac{\partial B_i}{\partial x} & \dfrac{\partial B_i}{\partial y} & \dfrac{\partial B_i}{\partial z} \end{bmatrix} \quad \text{and} \quad \frac{\partial^2 B_i}{\partial \mathbf{r}^2} = \begin{bmatrix} \partial\dfrac{\partial B_i}{\partial x}{\partial x} & \partial\dfrac{\partial B_i}{\partial x}{\partial y} & \partial\dfrac{\partial B_i}{\partial x}{\partial z} \\[2em] \partial\dfrac{\partial B_i}{\partial y}{\partial x} & \partial\dfrac{\partial B_i}{\partial y}{\partial y} & \partial\dfrac{\partial B_i}{\partial y}{\partial z} \\[2em] \partial\dfrac{\partial B_i}{\partial z}{\partial x} & \partial\dfrac{\partial B_i}{\partial z}{\partial y} & \partial\dfrac{\partial B_i}{\partial z}{\partial z} \end{bmatrix}$$

We denote $B_{ij} = \partial B_i / \partial j$ and

$$B_{(ij)k} = \frac{\partial \dfrac{\partial B_i}{\partial j}}{\partial k}$$

Then $\partial B_i / \partial \mathbf{r} = \begin{bmatrix} B_{ix} & B_{iy} & B_{iz} \end{bmatrix}$.

$$\frac{\partial^2 B_i}{\partial \mathbf{r}^2} = \begin{bmatrix} B_{(ix)x} & B_{(ix)y} & B_{(ix)z} \\ B_{(iy)x} & B_{(iy)y} & B_{(iy)z} \\ B_{(iz)x} & B_{(iz)y} & B_{(iz)z} \end{bmatrix}$$

Hence, the first-order gradients are given as

$$B_{eij} = B_{ij} + \frac{\partial \dfrac{\partial B_i}{\partial j}}{\partial \mathbf{r}} \mathbf{r} = B_{ij} + \begin{bmatrix} B_{(ij)x} & B_{(ij)y} & B_{(ij)z} \end{bmatrix} \mathbf{r}$$

The expanded fields are expressed in the permanent-magnet coordinates as

$$\overline{\mathbf{B}}_e = \overline{\mathbf{B}} + (\overline{\mathbf{r}} \cdot \overline{\nabla})\overline{\mathbf{B}} + \tfrac{1}{2}(\overline{\mathbf{r}} \cdot \overline{\nabla})^2 \overline{\mathbf{B}}$$

where $\overline{\mathbf{B}} = T_r \mathbf{B}$ and $\overline{\nabla} = T_r \nabla$.
   Using $\mathbf{r} = T_r^T \overline{\mathbf{r}}$, one has

$$B_{ei} = B_i + \frac{\partial B_i}{\partial \mathbf{r}} T_r^T \overline{\mathbf{r}} + \tfrac{1}{2}\overline{\mathbf{r}}^T T_r \frac{\partial^2 B_i}{\partial \mathbf{r}^2} T_r^T \overline{\mathbf{r}}, \quad \text{and} \quad \overline{\mathbf{B}}_e = T_r \begin{bmatrix} B_x + \dfrac{\partial B_x}{\partial \mathbf{r}} T_r^T \overline{\mathbf{r}} + \tfrac{1}{2}\overline{\mathbf{r}}^T T_r \dfrac{\partial^2 B_x}{\partial \mathbf{r}^2} T_r^T \overline{\mathbf{r}} \\ B_y + \dfrac{\partial B_y}{\partial \mathbf{r}} T_r^T \overline{\mathbf{r}} + \tfrac{1}{2}\overline{\mathbf{r}}^T T_r \dfrac{\partial^2 B_y}{\partial \mathbf{r}^2} T_r^T \overline{\mathbf{r}} \\ B_z + \dfrac{\partial B_z}{\partial \mathbf{r}} T_r^T \overline{\mathbf{r}} + \tfrac{1}{2}\overline{\mathbf{r}}^T T_r \dfrac{\partial^2 B_z}{\partial \mathbf{r}^2} T_r^T \overline{\mathbf{r}} \end{bmatrix}$$

10.7.4 Electromagnetic Torques and Forces

Now let us derive the fields and gradients at any point in the permanent magnet using the second-order Taylor series approximation. To eliminate the transformations between the inertial and permanent magnet coordinate systems and simplify the second-order negligibly small components, we assume that the relative motion between the magnet and the reference inertial coordinate is zero and the T_{rs} transformation matrix is used. Otherwise, the second-order gradient terms will be very cumbersome.

The magnetization (the magnetic moment per unit volume) is constant over the volume of the permanent-magnet thin films, and $\mathbf{m} = \mathbf{M}v$.

Assuming that the magnetic flux is constant, the total electromagnetic torque and force on a planar current loop (microwinding) in the uniform magnetic field are

$$\overline{\mathbf{T}} = \int_v (\overline{\mathbf{M}} \times \mathbf{B}_e + \overline{\mathbf{r}} \times (\overline{\mathbf{M}} \cdot \nabla)\mathbf{B}_e)\,dv \quad \text{and} \quad \overline{\mathbf{F}} = \int_v (\overline{\mathbf{M}} \cdot \nabla)\mathbf{B}_e\,dv$$

where

$$(\overline{\mathbf{M}} \cdot \nabla)\mathbf{B}_e = [\partial \mathbf{B}_e]\overline{\mathbf{M}} = \begin{bmatrix} B_{exx} & B_{exy} & B_{exz} \\ B_{eyx} & B_{eyy} & B_{eyz} \\ B_{ezx} & B_{ezy} & B_{ezz} \end{bmatrix} \begin{bmatrix} M_{\bar{x}} \\ M_{\bar{y}} \\ M_{\bar{z}} \end{bmatrix}$$

Case 1: Magnetization Along the Axis of Symmetry

For orientation of the magnetization vector along the axis of symmetry (*x*-axis) of the permanent-magnet thin films, we have

$$(\overline{\mathbf{M}} \cdot \nabla)\mathbf{B}_e = [\partial \mathbf{B}_e]\overline{\mathbf{M}} = M_{\bar{x}} \begin{bmatrix} B_{exx} \\ B_{exy} \\ B_{exz} \end{bmatrix}$$

Thus, in the expression $\overline{\mathbf{T}} = \int_v (\overline{\mathbf{M}} \times \mathbf{B}_e + \overline{\mathbf{r}} \times (\overline{\mathbf{M}} \cdot \nabla)\mathbf{B}_e) dv$, the terms are

$$\overline{\mathbf{r}} \times (\overline{\mathbf{M}} \cdot \nabla)\mathbf{B}_e = M_{\bar{x}} \begin{bmatrix} -B_{exy}\bar{z} + B_{exz}\bar{y} \\ B_{exx}\bar{z} - B_{exz}\bar{x} \\ -B_{exx}\bar{y} + B_{exy}\bar{x} \end{bmatrix} \quad \text{and} \quad \overline{\mathbf{M}} \times \mathbf{B}_e = M_{\bar{x}} \begin{bmatrix} 0 \\ -B_{ez} \\ B_{ey} \end{bmatrix}$$

Therefore,

$$T_{\bar{x}} = M_{\bar{x}} \int_v (B_{exz}\bar{y} - B_{exy}\bar{z}) dv$$

$$T_{\bar{y}} = -M_{\bar{x}} \int_v B_{ez} dv + M_{\bar{x}} \int_v (B_{exx}\bar{z} - B_{exz}\bar{x}) dv$$

and

$$T_{\bar{z}} = M_{\bar{x}} \int_v B_{ey} dv + M_{\bar{x}} \int_v (B_{exy}\bar{x} - B_{exx}\bar{y}) dv$$

The terms in the derived equations must be evaluated.

Let us find the analytic expression for the electromagnetic torque $T_{\bar{x}}$. In particular, we have

$$\int_v B_{exz}\bar{y} dv = B_{xz} \int_v \bar{y} dv + B_{(xx)z} \int_v \bar{x}\bar{y} dv + B_{(xy)z} \int_v \bar{y}^2 dv + B_{(xz)z} \int_v \bar{z}\bar{y} dv$$

where

$$\int_v \bar{y} dv = 0, \quad \int_v \bar{x}\bar{y} dv = 0, \quad \int_v \bar{z}\bar{y} dv = 0 \quad \text{and}$$

$$\int_v \bar{y}^2 dv = \int_{-\frac{1}{2}l}^{\frac{1}{2}l} \int_{-R}^{R} \int_{-\sqrt{R^2-\bar{z}^2}}^{\sqrt{R^2-\bar{z}^2}} \bar{y}^2 d\bar{y} d\bar{z} d\bar{x} = \tfrac{1}{4}\pi l R^4 = \tfrac{1}{4} v R^4$$

Therefore, $M_{\bar{x}} \int_v B_{exz}\bar{y} dv = M_{\bar{x}} \tfrac{1}{4} B_{(xy)z} v R^4$.

Furthermore, $M_{\bar{x}} \int_v B_{exy} \bar{z} dv = M_{\bar{x}} \frac{1}{4} B_{(xy)_z} v R^4$. Thus, for $T_{\bar{x}}$, one has

$$T_{\bar{x}} = M_{\bar{x}} \int_v (B_{exz} \bar{y} - B_{exy} \bar{z}) dv = M_{\bar{x}} \left(\tfrac{1}{4} B_{(xy)_z} v R^4 - \tfrac{1}{4} B_{(xy)_z} v R^4 \right) = 0$$

Then, for $T_{\bar{y}}$, we obtain

$$T_{\bar{y}} = -M_{\bar{x}} \int_v B_{ez} dv + M_{\bar{x}} \int_v (B_{exx} \bar{z} - B_{exz} \bar{x}) dv$$

$$= M_{\bar{x}} \left[-\left(B_z + B_{(zx)x} \tfrac{1}{24} l^2 + B_{(zy)y} \tfrac{1}{8} R^2 + B_{(zz)z} \tfrac{1}{8} R^2 \right) v + B_{(xx)z} \left(\tfrac{1}{4} R^2 - \tfrac{1}{12} l^2 \right) v \right]$$

$$= -v M_{\bar{x}} \left(B_z + B_{(xx)z} \left(\tfrac{1}{4} R^2 - \tfrac{1}{8} l^2 \right) + B_{(yy)z} \tfrac{1}{8} R^2 + B_{(zz)z} \tfrac{1}{4} R^2 \right)$$

Finally, we obtain the expression for T_z as

$$T_{\bar{z}} = M_{\bar{x}} \int_v B_{ey} dv + M_{\bar{x}} \int_v (B_{exy} \bar{x} - B_{exx} \bar{y}) dv$$

$$= v M_{\bar{x}} \left(B_y + B_{(xx)y} \left(\tfrac{1}{8} l^2 - \tfrac{1}{4} R^2 \right) - B_{(yy)y} \tfrac{1}{8} R^2 - B_{(yz)z} \tfrac{1}{8} R^2 \right)$$

Thus, the following electromagnetic torque equations result

$$T_{\bar{x}} = 0$$

$$T_{\bar{y}} = -v M_{\bar{x}} \left(B_z + B_{(xx)z} \left(\tfrac{1}{4} R^2 - \tfrac{1}{8} l^2 \right) + B_{(yy)z} \tfrac{1}{8} R^2 + B_{(zz)z} \tfrac{1}{4} R^2 \right)$$

$$T_{\bar{z}} = v M_{\bar{x}} \left(B_y + B_{(xx)y} \left(\tfrac{1}{8} l^2 - \tfrac{1}{4} R^2 \right) - B_{(yy)y} \tfrac{1}{8} R^2 - B_{(yz)z} \tfrac{1}{8} R^2 \right)$$

The electromagnetic forces are found as well. In particular, from

$$F_{\bar{x}} = M_{\bar{x}} \int_v B_{exx} dv, \quad F_{\bar{y}} = M_{\bar{x}} \int_v B_{exy} dv \quad \text{and} \quad F_{\bar{z}} = M_{\bar{x}} \int_v B_{exz} dv$$

using the expressions for the expanded magnetic fluxes, $\int_v B_{exx} dv = \int_v (B_{xx} + B_{(xx)x} \bar{x} + B_{(xx)y} \bar{y} + B_{(xx)z} \bar{z}) dv$, and performing the integration, one has the following expressions for the electromagnetic forces as the function of the magnetic field:

$$F_{\bar{x}} = v M_{\bar{x}} B_{xx}, \quad F_{\bar{y}} = v M_{\bar{x}} B_{xy}, \quad F_{\bar{z}} = v M_{\bar{x}} B_{xz}$$

The documented expressions for torques and forces, obtained as the nonlinear functions of the permanent-magnet parameters (magnetization, area, radius, and length) and magnetic field variables, allow one to model six-degree-of-freedom torsional-mechanical dynamics.

Case 2: Magnetization Perpendicular to the Axis of Symmetry

For orientation of the magnetization vector perpendicular to the axis of symmetry, the following equation is used to find the electromagnetic torque

$$\overline{\mathbf{T}} = \int_v (\overline{\mathbf{M}} \times \mathbf{B}_e + \overline{\mathbf{r}} \times (\overline{\mathbf{M}} \cdot \nabla) \mathbf{B}_e) \, dv$$

where

$$(\overline{\mathbf{M}} \cdot \nabla) \mathbf{B}_e = [\partial \mathbf{B}_e] \overline{\mathbf{M}} = M_{\overline{z}} \begin{bmatrix} B_{exz} \\ B_{eyz} \\ B_{ezz} \end{bmatrix}$$

$$\overline{\mathbf{r}} \times (\overline{\mathbf{M}} \cdot \nabla) \mathbf{B}_e = M_{\overline{z}} \begin{bmatrix} -B_{eyz}\overline{z} + B_{ezz}\overline{y} \\ B_{exz}\overline{z} - B_{ezz}\overline{x} \\ -B_{exz}\overline{y} + B_{eyz}\overline{x} \end{bmatrix}$$

$$\overline{\mathbf{M}} \times \mathbf{B}_e = M_{\overline{z}} \begin{bmatrix} -B_{ey} \\ B_{ex} \\ 0 \end{bmatrix}$$

Thus,

$$T_{\overline{x}} = -M_{\overline{z}} \int_v B_{ey} \, dv + M_{\overline{z}} \int_v (B_{exz}\overline{y} - B_{eyz}\overline{z}) \, dv$$

$$T_{\overline{y}} = M_{\overline{z}} \int_v B_{ex} \, dv + M_{\overline{z}} \int_v (B_{exz}\overline{z} - B_{ezz}\overline{x}) \, dv$$

and

$$T_{\overline{z}} = M_{\overline{z}} \int_v (B_{eyz}\overline{x} - B_{exz}\overline{y}) \, dv$$

Expressing the fluxes and performing the integration, we have the following expressions for the torque components as the function of the magnetic field

$$T_{\overline{x}} = -v M_{\overline{z}} \left(B_y + B_{(xx)y} \frac{1}{24} l^2 + B_{(yy)y} \frac{1}{8} R^2 + B_{(yz)z} \frac{1}{8} R^2 \right)$$

$$T_{\overline{y}} = v M_{\overline{z}} \left(B_x + B_{(xz)z} \left(\frac{3}{8} R^2 - \frac{1}{12} l^2 \right) + B_{(xx)x} \frac{1}{24} l^2 + B_{(xy)y} \frac{1}{8} R^2 \right)$$

$$T_{\overline{z}} = v M_{\overline{z}} B_{(xy)z} \left(\frac{1}{12} l^2 - \frac{1}{4} R^2 \right)$$

The electromagnetic forces are found to be

$$F_{\bar{x}} = M_{\bar{z}} \int_v B_{exz}dv = vM_{\bar{z}}B_{xz}, \quad F_{\bar{y}} = M_{\bar{z}} \int_v B_{eyz}dv = vM_{\bar{z}}B_{yz}$$

$$F_{\bar{z}} = M_{\bar{z}} \int_v B_{ezz}dv = vM_{\bar{z}}B_{zz}$$

Thus, the expressions for the electromagnetic force and torque components are derived. These equations provide one with the clear perspective how to model, analyze, and control the electromagnetic forces and torques changing the applied magnetic field because the terms $B_{ij} = \partial B_i/\partial j$ and

$$B_{(ij)k} = \frac{\partial \dfrac{\partial B_i}{\partial j}}{\partial k}$$

can be viewed as the control variables. The electromagnetic field (B_{ij} and $B_{(ij)k}$) is controlled by regulating the current in the planar microwindings and designing the microwindings (or other radiating energy microdevices). As was discussed, the derived forces and torques must be used in the torsional-mechanical equations of motion for the microactuator, and in general the six-degree-of-freedom microactuator mechanical dynamics result. These mechanical equations of motion are easily integrated with the derived electromagnetic equations, and closed-loop systems can be designed to attain the desired microactuator performance. These equations guide us to the importance of electromagnetic features in the modeling, analysis, and design of microactuators.

10.7.5   Some Other Aspects of Microactuator Design and Optimization

In addition to the electromagnetic-mechanical (electromechanical) analysis and design, other design and optimization problems are involved. As an example, let us focus our attention on the planar windings. The ideal planar microwindings must produce the maximum electromagnetic field, minimizing the footprint area and taking into the consideration the material characteristics, operating conditions, applications, power requirements, and many other factors. Many planar winding parameters and characteristics can be optimized. For example, the DC resistance must be minimized to improve the efficiency, increase the flux, decrease the losses, etc. To attain good performance, in general, microwindings have the concentric circular current path and no interconnect resistances. For N-turn winding, the total DC resistance r_t is found to be

$$r_t = \frac{2\pi\rho}{t_w} \sum_{k=1}^{N} \frac{1}{\ln\left(r_{Ok}\big/r_{Ik}\right)}$$

where ρ is the winding material resistivity; t_w is the winding thickness; and r_{Ok} and r_{Ik} are the outer and inner radii of the k turn winding.

To achieve the lowest resistance, the planar winding radii can be optimized by minimizing the resistance. The minimum resistance is denoted as $r_{t\min}$. In particular, making use of first- and second-order necessary conditions for minimization, one has

$$\frac{dr_t}{dr_w} = 0 \quad \text{and} \quad \frac{d^2 r_t}{dr_w^2} > 0$$

where r_w is the inner or outer radius of an arbitrary turn of the optimized planar windings from the standpoint of minimizing the resistance.

Then the minimum value of the microcoil resistance is given by

$$r_{t\min} = \frac{2\pi\rho}{t_w} \frac{N}{\left(r_{OR}\middle/r_{IR}\right)}$$

where r_{OR} and r_{IR} are the outer and inner radii of the windings (that is, $r_{O\,N\text{-th microcoil}}$ and $r_{I\,1\text{st microcoil}}$).

Thus, using the number of turns and turn-to-turn spacing, the outer and inner radii of the k turn winding are found as

$$\frac{r_{Ok}}{r_{Ik}} = \left(\frac{r_{OR}}{r_{IR}}\right)^{1/N}$$

For spiral windings, the *averaging (equivalency)* concept should be used because the outer and inner radii are the functions of the planar angle as shown in Figure 10.19. The width of the N-th microcoil is specified by the rated voltage, current density versus maximum current density, fabrication technologies, material characteristics, etc. These planar coils can be fabricated. Figure 10.19 documents the coil as its width varies from 2 to 8 μm.

10.8 Reluctance Electromagnetic Micromotors

Surface micromachining and CMOS technologies (photolithography, deposition, electroplating, molding, planarization, and other processes) are used to fabricate rotational and translational microtransducers. Microtransducers integrate magnetic materials (core), permanent magnets, windings, bearing, etc. It was emphasized that the NiFe thin films (as well other magnetic and nonmagnetic materials) and copper coils can be electroplated, and the fabrication processes were discussed. Manageable results are reported in References 2, 4, 19, and 20. In general the NiFe and copper electroplating baths, electroplating seed layers, and through-mask plating techniques are similar to those used to fabricate inductive thin-film heads. High-aspect-ratio optical lithography and x-ray lithography can be used to form the various resist layers needed. The stators and rotors can be fabricated separately or sequentially depending on the microtransducers devised and fabrication technologies used. For example, if stator and rotor are made separately, the microrotor is released from the substrate and installed on the shaft made as a part of the stator fabrication. High-aspect-ratio microstructure and electroplating-through-mask processes are the major steps to fabricating synchronous (including permanent-magnet and reluctance), induction, and other electromagnetic

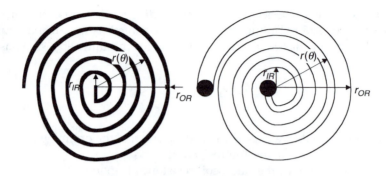

FIGURE 10.19
Planar spiral microwinding.

microtransducers. These microtransducers can be used as high-precision microscale drives and actuators in automotive, aerospace, power, manufacturing, biomedical, biotechnology, and other applications.

Shown in Figure 10.5, a synchronous reluctance micromotor has stationary and movable members. The stator and rotor are batch-fabricated microstructures made using similar processes and materials because the same magnetic electroplated material can be used. However, the copper microwindings are electroplated on the stator. Microtransducers must be made through the developed batch-fabrication processes sequence (series of lithographic, electroplating, and planarization steps). As studied before, the process starts by the sputter-deposition of the Ti-Cu-Ti seed layer on an oxidized silicon wafer (Si-SiO$_2$ substrate), lithographically forming the plate-through (patterned) mask for the bottom part of the copper microwindings, removing the upper Ti layer, and electroplating the copper in the exposed seed layer regions. The photoresist and seed layer are removed from the regions between the electroplated structure, and the insulation layer (polymer dielectric) is deposited. Then the surface is planarized. A thin patterned polymer dielectric is formed on top of the copper microcoils to insulate the windings from the ferromagnetic core (stator), leaving apertures at the end of each copper curl to form the microwinding in the subsequent steps. The second Ti-Cu-Ti seed layer is deposited, and the photoresist is patterned. The upper Ti is etched, and the NiFe thin film is electroplated. Removing the photoresist and seed layer, the NiFe thin film ferromagnetic core (stator) is covered by the insulation layer (dielectric). Microcoils are connected and insulated to form the desired microwinding structure. The rotor is fabricated by electroplating the NiFe thin film on the lithographically patterned Si-SiO$_2$ substrate precoated with a copper layer. The adhesion layer under the copper is Ti or Ta. The rotor can be detached through the chemical dissolution of the copper layer and installed on the shaft located in the center of the stator cavity.

The microshaft and support pins should be made during the photolithography and electroplating processes.[20]

It was emphasized that the seed layers for the electrodeposited copper and NiFe thin films consist of the sputtered Cu (100–300 nm) on the adhesion layer (Ti or Ta). A layer of Ti is deposited on the Cu to enhance the resist adhesion. The fabrication processes are based on high-aspect-ratio optical lithography using positive (novolac-type) and negative optical photoresists. Although the positive photoresists are attractive due to easy removal following electrodeposition, it is difficult to form large thickness and uniformity with respect to sidewall angle. The negative photoresist consists of solutions of different concentrations of Shell EPON Negative epoxy-based photoresist resin SU-8 and is photosensitized with a commercially available triaryl sulfonium salt (Cyravure UVI, Union Carbide Co). Using negative photoresist, straight sidewall profiles (vertical sidewall with 85–90° angle) can be obtained for a photoresist thickness up to 130 μm for the single layers of resist on the copper and NiFe layers. In contrast, the positive resist (high-viscosity novolac, Shipley SJR 3740 or AZ P4620, Hoechst Celanese Co.), can be applied using a multilayer spin method to achieve a thickness up to 70 μm. This photoresist can be exposed using UV light (patterning can be performed using contact printing on the UV Mask Aligner), and photoresist is removed without affecting the underlying structure. In particular, the hexamethyl disiloxane can be used to obtain good adhesion of the photoresist to the underlying electroplating copper seed layer. The baking process is necessary to stabilize the photoresist prior to UV light exposure.

The rotors and the stator are made using the NiFe thin films. The electrode-position is carried out in the horizontal paddle cell connected to a plating solution reservoir. The solution should be continuously filtered (0.2 μm Millipore Co. Filters can be used). To stabilize the temperature, the plating reservoir should have an automatic temperature-control system because the temperature must be 25°C ± 0.1°C. The solution pH (3 ± 0.1) is controlled by adding the HCl. The ferrous ion concentration should be kept constant adding a ferrous sulfate solution (pH of 2) at the rate required to match the iron consumption rate. The nickel plate serves as the anode, and the stainless steel plate serves as an auxiliary electrode, which is coplanar with the cathode. The auxiliary electrode reduces current nonuniformity at the edge of the cathode improving the deposit thickness uniformity across the cathode (particularly at the edges). The current fed to the auxiliary electrode depends on the current density at the cathode. The current density commonly used is 10–15 mA/cm^2 to guarantee the uniform electrodeposition. The NiFe thin film layers can be planarized by polishing.

Electroplated copper is used to make microcoils, rotor support pins, and other microstructures (if needed). Electroplating processes, which guarantee the uniformity of the copper thickness, are commercially available and were described in this chapter.

It was emphasized that the insulation is very important in the microtransducer fabrication because the core must be insulated from the copper microcoils. Insulator layers must exhibit low stress and should be planarizable. Various dielectric materials (polyimide, hard-baked photoresist, and epoxy resin) have been studied from insulation and planarization viewpoints. The use of polyimide requires the barrier layer (electrolessly deposited material, e.g., NiP) to prevent a chemical reaction with copper during polyimide curing. In addition, voids formed due to the difficulty of removing solvent from the deep narrow regions may result in short-circuits between the coils and core. The solventless epoxy resin eliminates the problem of voids, reduces stress, and does not require curing at high temperatures. The temperature used for polyimides is 360–400°C. The cured epoxy resin (dielectric material) is polished to planarize the surface and to expose copper and NiFe thin films. Through spin-coating, the planarized substrates with 5–7-μm epoxy resin layer can be made. The epoxy is patterned, covering microcoils to attain the insulation. After curing of the insulation layer, a seed layer is deposited, followed by photoresist. The insulation layer creates a nonplanar surface. However, this not a serious problem for the subsequent photoresist process. The difference in height between the core and the studs after NiFe thin film electrodeposition is removed through polishing following epoxy backfill.

The micromotor sizing depends on the specifications and requirements (for example, the equation for the electromagnetic torque leads to the specified current, inductances, airgap, cross-sectional area, active magnetic length, etc.) as well as the processes and materials that significantly influence the magnetic and thermal characteristics, resistivity, etc. Processes and material integration, as well as the integrity of the consequential steps, are key to fabricate micromotors. Microstators can be fabricated with 10–30-μm thick copper microconductors and 50-μm NiFe thin film using optical lithography. Magnetic characteristics, resistivity, and inductance are the functions of the Cu and NiFe thickness. To fabricate microtransducers, different numbers of masks are needed. In particular, the studied microtransducer requires five mask levels, as well as additional masks for reactive ion etching of the rotor cavity at the end of the fabrication, to make the bearing. Electrodeposition processes are also involved. The rotor cavity in the stator is opened using reactive ion etching with masking to ensure selectivity (epoxy resin dielectric removal rate in the range of 0.4–0.6 μm/min). Attention must be concentrated on the minimization of heat during sputter deposition of the electroplated seed layer. In fact, one must prevent the void formation at the Cu–insulator region because heat can cause cracks and voids in the thin insulation layer between the bottom Cu and NiFe. The effect of the thermal expansion mismatch must be minimized. During the reactive ion etching (to open the rotor cavity in the stator), the heating has to be minimized as well.

Thorough design can be performed optimizing the microtransducer characteristics by optimizing different parameters (sizing rotor and stator, shaping stator and rotor geometry, varying the airgap, deriving ferromagnetic core and permanent-magnet thickness, designing nanocomposite thin films, optimizing the magnetic properties of the magnetic materials such as permeability and hysteresis to minimize losses, minimizing the resistivity of the microwindings, maximizing the flux, reducing flux leakage and fringing effects, minimizing the friction, decreasing the torque ripple, attenuating viboacoustic phenomena, eliminating resonance effects, etc.). Many nonlinear phenomena cannot be modeled, analyzed, and even reasonably evaluated using the lumped-parameter mathematical models. Therefore, Maxwell's equations must be used integrating the basic nonlinear electromagnetic and torsional-mechanical phenomena, effects, and features.

Microtransducers can be fabricated using optical and x-ray lithography. Though the electromagnetic microtransducer design and operating principles are similar, different final design leads to many distinct features. As a result, fabrication processes and materials are different. For example, the size of microtransducers and microstructures is different, and high-aspect-ratio processes are usually involved. For optical lithography fabrication, the photoresist for electroplating the ferromagnetic core and conductor vias can be applied by spin coating. Using the LIGA process, the semirigid polymethylmethacrylate (PMMA) sheet can be glued to the substrate with PMMA, and the insulation of the core from the top and bottom copper conductors is needed. In fact, the majority of microtransducers have the bottom copper conductors level, the NiFe level, the top copper conductors level, and insulation layers. These three levels can be made using the Ti-Cu-Ti seed layer, spin-coating, and thermally curing the PMMA adhesion layer. Gluing a PMMA sheet (millimeter-thick Perspex CQ), fly cutting the PMMA to the desired thickness (100–250 μm), and x-ray exposure are carried out. These processes are followed by resist-pattern development, pretreatment of the surface (remove Ti), electroplating of Cu or NiFe thin films, and polishing of the PMMA resist. Then the AZ optical photoresist is applied, UV-exposing and developing of the photoresist are made, and copper is electroplated. Following removal of the AZ photoresist, PMMA, and seed layers (through chemical etching or ion milling processes), the SiO_2-filled epoxy dielectric is applied and thermally cured. The excess dielectric is removed by polishing to expose the copper in preparation for the seed layer deposition and x-ray lithography as the needed steps to form the next level.

PMMA positive resists are based on special grades of polymethyl methacrylate designed to provide high contrast, high resolution for e-beam, deep UV (220–250 nm), and x-ray lithographic processes. Standard PMMA resist has 495,000 and 950,000 molecular weights (MW) in a wide range of film thicknesses formulated in chlorobenzene, or the safer solvent anisole. However, 50,000, 100,000, 200,000 and 2.2 million MW are available. Copolymer resists are based on a mixture of PMMA and methacrylic acid (usually from 8% to 20%). Copolymer MMA can be used in combination with PMMA in bilayer lift-off resist processes where independent control of size and shape of each resist layer is needed. Standard copolymer resists are formulated in the safer solvent ethyl lactate and are available in a wide range of film thicknesses.

The same sequential steps are performed to fabricate the NiFe thin film (second level). The connection of the top and bottom copper microconductors are made. After polishing the PMMA resist with respect to NiFe, the spacer pads (20–25 μm high) are electroplated using optically patterned photoresist. Following encapsulation of the NiFe with SiO_2-filled dielectric and polishing it to the NiFe thin film, the top copper conductor level is fabricated using the x-ray–patterned PMMA resist. The x-ray lithography leads to the vertical sidewall profile leading to the high aspect ratio microstructures. The encapsulated epoxy resin SiO_2-filled dielectric can be removed from the central rotor region through the reactive ion etching. The use of the epoxy resin SiO_2-filled dielectric complicates the etching process (compared with the optical lithography) because the etch rate of the silicon dioxide is slower than that of the epoxy resin. The rotor is electroplated on a separate substrate through the x-ray–exposed PMMA, detached, and integrated with bearing following reactive ion etching. The rotor cavity on the stator is made.

In general the x-ray lithography fabrication, which allows one to make thicker microstructures and larger microtransducers, leads to the following challenges:[20] (1) Thicker layers lead to thermal mismatch problems between copper microcoils, NiFe thin films, core, and insulator-dielectric. The thermal mismatch in the vertical direction is a serious issue—a perfectly planarized magnetic layer embedded in a polymer with high coefficient of thermal expansion can suffer delamination at the vertical metal-polymer interface when heated to 180°C during PMMA curing. This can cause cracking and delamination of the seed layer. This problem can be relaxed using a dielectric containing low-thermoexpansion composites, e.g., silicon dioxide. However, using the dielectric containing inorganic materials to reduce the thermal expansion mismatch results in changes of the reactive ion etching processes used to open the cavities to make the bearing for rotational microtransducers. (2) Thick (100–400 μm) PMMA resist creates alignment problems due to the difficulty of accurately registering the mask to alignment marks under the thick resist. The processes for alignment through PMMA up to 400 μm thick with ±5 μm accuracy are developed by IBM.[20] (3) Plating on the thick PMMA requires cautious design of the step sequences (surface pretreatment, PMMA development, and wetting initiation). (4) Solvents and chemicals used in the PMMA processing must be compatible with underlying microstructure materials. (5) Lamination of the PMMA sheet requires the minimization of the irregularities in the planarized wafer surface. Planarization leads to the development of appropriate polishing techniques to deal with the various composite surfaces.

It should be emphasized that SiO_2 has been used as an insulating material in microelectronics. For example, SiO_2 is used as the gate dielectric of MOSFETs. Fabrication process design, integration, optimization, and materials selection are the key in fabrication of electromagnetic microtransducers. The major fabrication problems that the designer faces are deposition of microconductors and magnetic thin films, planarization, core-insulator/dielectric-conductor thermal expansion mismatch, voids (including those caused by coefficient of thermal expansion mismatch), residue-free development in thick optically exposed photoresist and x-ray–exposed PMMA resist, stress-free films, etc.

10.9 Micromachined Polycrystalline Silicon Carbide Micromotors

Multilayer fabrication processes at low temperature and micromolding techniques were developed to fabricate SiC microstructures and salient-pole micromotors that can be used at very high temperature (400°C and higher).[21,22] This was done through the SiC surface micromachining.

Throughout the book it has been emphasized that the MATLAB environment can be effectively used for CAD. The application of MATLAB is illustrated for the simple problem of data fitting. The bulk modulus of SiC versus temperature (in °C) is given as

$$B = 203_{T=20}, \ 200_{T=250}, \ 197_{T=500}, \ 194_{T=750}, \ 191_{T=1000}, \ 188_{T=1200}, \ 186_{T=1400}, \ 184_{T=1500}.$$

The interpolation is performed using the `spline` MATLAB solver (spline fit function). The MATLAB file used is

```
T=[20 250 500 750 1000 1200 1400 1500];% Temperature
B=[203 200 197 194 191 188 186 184];  % Bulk Modulus Data
Tinterpol=20:10:1500;
Binterpol=spline(T,B,Tinterpol);
plot(T,B,'o',Tinterpol,Binterpol,'-');
xlabel('Temperature, deg C');
ylabel('Bulk Modulus, GPa');
title('Temperature-Bulk Modulus Data and Spline Interpolation');
```

The resulting temperature–bulk modulus plot of the interpolated spline data (solid line) and the data values used are given in Figure 10.20.

Advantages of SiC micromachining and SiC technologies (high temperature and ruggedness) should be weighed against fabrication drawbacks, because new processes must be designed and optimized. Reactive ion etching is used to pattern SiC thin films. But many problems such as masking, low etch rates, and poor etch selectivity must be addressed and resolved. Single-layer reactive ion etching–based polycrystalline SiC surface micromachining processes using polysilicon or SiO as the sacrificial layer are known[21,22]. In addition, the micromolding processes (used to fabricated polysilicon molds in conjunction with polycrystalline SiC film deposition and mechanical polishing to pattern polycrystalline SiC films) are introduced. The micromolding process can be used for single- and multilayer SiC surface micromachining.

The micromotor fabrication processes are illustrated in Figure 10.21. A 5–10 μm thick sacrificial molding polysilicon is deposited through the LPCVD on a 3–5 μm sacrificial thermal oxide. The rotor-stator mold formation can be made on the polished (chemical-mechanical polishing) polysilicon surface, enabling 2-μm fabrication features using standard lithography and reactive ion etching. After mold formation and delineation, the SiC is deposited on the wafer using an atmospheric pressure chemical vapor deposition reactor. In particular, the phosphorus-doped (*n*-type) polycrystalline SiC films are deposited on the SiO sacrificial layers at 1050°C at a 0.5–1 μm/hour rate. Deposition is not selective, and SiC will be deposited on the surfaces of the polysilicon molds as well. Mechanical polishing of SiC is needed to

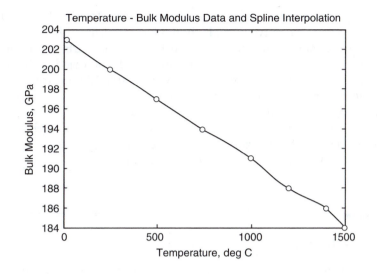

FIGURE 10.20
Temperature–bulk modulus data and its spline interpolation.

expose the polysilicon and planarize the wafer surface. The polishing can be done with 3 μm diameter diamond suspension, 360 N normal force, and 15 rad/sec pad rotation.[21,22] The removal rate of SiC is reported to be 100 nm/min. The wafers are polished until the top surface of the polysilicon mold is exposed. Polishing must be stopped at once due to the fast polishing rate. The flange mold is fabricated through the polysilicon and the sacrificial oxide etching (using KOH and BHF, respectively). The 0.5-μm bearing clearance low-temperature oxide is deposited and annealed at 1000°C. Then the 1-μm polycrystalline SiC film is deposited and patterned by reactive ion etching to make the bearing. The release begins with the etching

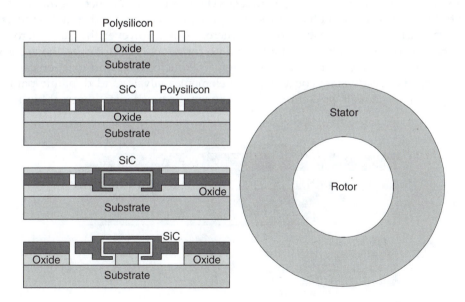

FIGURE 10.21
Fabrication of the SiC micromotors: cross-sectional schematics.

(in BHF solution) to strip the left-over bearing clearance oxide. The sacrificial mold is removed by etching (KOH system) the polysilicon. It should be emphasized that the SiC and SiO are not etched during the mold removal step. Then the moving parts of the micromotor are released. The micromotor is rinsed in water and methanol, and dried with an air jet.

Using this fabrication process, a micromotor with a 100–150-μm rotor diameter, 2-μm airgap, and 21-μm bearing radius was fabricated and tested.[21,22] The rated voltage was 100 V and the maximum angular velocity was 30 rad/sec. For silicon and polysilicon micromotors, two of the most critical problems are the bearing and ruggedness. The application of SiC reduces the friction and improves the ruggedness. These contribute to the reliability of the SiC-based fabricated micromachines.

10.10 Axial Electromagnetic Micromotors

The major problem is to devise novel microtransducers in order to lessen fabrication difficulties, and guarantee affordability, efficiency, reliability, and controllability of MEMS. In fact, the electrostatic and planar micromotors fabricated and tested to date are found to be inadequate for a wide range of applications due to difficulties associated and cost. Therefore, this section is devoted to devising novel affordable rotational micromotors.

Figure 10.22 illustrates a devised axial topology micromotor that has a closed-ended electromagnetic system. The stator is made on the substrate with deposited microwindings. Printed copper coils can be made using the fabrication processes described as well as using double-sided substrate with the one-sided deposited copper thin film through conventional photolithography processes. The bearing post is fabricated on the stator substrate and the bearing hold is a part of the rotor microstructure. The rotor with permanent-magnet thin films rotates due to the electromagnetic torque developed. It is important to emphasize that stator and rotor are made using the conventional well-developed processes and materials.

It is evident that conventional silicon and SiC technologies can be used. The documented micromotor has a great number of advantages. The most critical benefit is the fabrication simplicity. In fact, axial micromotors can be straightforwardly fabricated, and this will enable their wide applications as microactuators and microsensors. However, the axial micromotors must be designed and optimized to attain good performance. The optimization is based on electromagnetic, mechanical, and thermal design. The micromotor optimization can be carried out using the steady-state concept (finite element analysis) and dynamic paradigms (lumped-parameter models or complete electromagnetic-mechanical-thermal high-fidelity

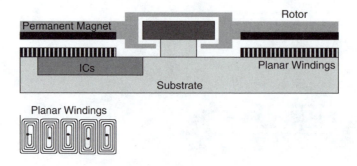

FIGURE 10.22
Slotless axial electromagnetic micromotor (cross-sectional schematics) with controlling ICs.

mathematical models derived as a set of partial differential equations using Maxwell's, torsional-mechanical, and heat equations). In general the nonlinear optimization problems need to be addressed, formulated, and solved to guarantee the superior microtransducer performance. In addition to the microtransducer design, one must concentrate on the ICs and controller design. In particular, the circuitry is designed based on the converter and inverter topologies (e.g., hard- and soft-switching, one-, two-, or four-quadrant, etc.), filters and sensors used, rated voltage and current, etc. From the control prospective, the electromagnetic features must be thoroughly examined. For example, the electromagnetic micromotor studied is the synchronous micromachine. Therefore, to develop the electromagnetic torque, the voltages applied to the stator windings must be supplied as the functions of the rotor angular displacement. Therefore, the Hall-effect sensors must be used, or the so-called sensorless controllers (the rotor position is observed or estimated using the directly measured variables) must be designed and implemented using ICs. This brief discussion illustrates a wide spectrum of fundamental problems involved in the design of integrated microtransducers with controlling and signal processing ICs.

10.11 Cognitive Computer-Aided Design of MEMS

The critical focus themes in microsystems development and implementation are rapid synthesis, design, and prototyping through synergetic multidisciplinary system-level research.[3] Let us discuss the taxonomy of MEMS devising and optimization, which is relevant to cognitive study, classification, and synthesis of electromechanical systems in general, including MEMS and many NEMS. This problem has been partially covered in Chapter 5. In contrast, this section mainly emphasizes MEMS. The simplest example is devising MEMS-based microtransducers that comprise a number of microstructures. Thus, the microtransducer can be defined in terms of microstructures needed to build it. Hence, one can use the Boolean theory to describe microtransducers in terms of microstructures. Similar reasoning can be used in describing the fabrication processes that can be represented as the sequential steps with specific materials, techniques and processes applied. Different levels of description can be researched. For example, electrodeposition processes are influenced by the chemicals used and the process parameters (current density, temperature, pH, etc.). Rather than emphasize the well-developed fabrication technologies and process (materials and processes characterization are very important problems which were studied and documented in References 1–7, 9–12, 15, 16, and 19–22), our attention will be primarily concentrated on the device- and system-level synthesis and analysis. Devising MEMS is the closed evolutionary process to study possible system-level evolutions based on synergetic integration of microscale structures, devices, and other components in the unified functional core. The ability to devise and optimize MEMS to a large extent depends on the computational efficiency, adaptability, functionality, integrity, compliance, robustness, flexibility, prototypeability, visualability, interactability, decision making, and intelligence of the computer-based design tools and environments. It is likely that the fundamental theory and applied experimental results in conjunction with high-performance interacting software (e.g., MATLAB and MATEMATICA) will allow the designer to devise (synthesize), prototype, design, model, simulate, analyze, and optimize MEMS. Furthermore, synergetic quantitative synthesis and symbolic descriptions can be efficiently used when searching and evaluating possible organizations, architectures, configurations, topologies, geometries, and other descriptive features providing the evolutionary potential needed.

Although general, systematic, and straightforward approaches can be developed to optimize microtransducers, radiating energy and optical microdevices, and ICs, it is unlikely that general devising tools based on abstract concepts can be effectively applied for MEMS. This is due to a great variety of possible MEMS solutions, architectures, subsystem-device-structure organizations, and electromagnetic-optical-mechanical and electromechanical features. However, restricting the number of the possible solutions based on the specifications and requirements, the MEMS synthesis can be performed (using informatics theory, artificial intelligence, knowledge-based libraries, and expert techniques) emphasizing the electromagnetic-optical-mechanical-thermal phenomena, and the numerical optimization and design of complex MEMS can be accomplished through CAD. It was emphasized that the conceptual design of MEMS integrates devising (synthesis), prototyping, and analysis with consecutive design, optimization, and verification tasks.

Currently there does not exist a CAD environment that can perform functional synthesis, modeling, analysis, design, or optimization for even relatively simple MEMS and their components. CAD environments and tools, such as SPICE and VHDL, have been developed only to design and model application-specific ICs that are the important components in microsystems. The corresponding CAD packages allow one to design analog, digital, and hybrid ICs, integrating fabrication technologies using materials and processes databases, e.g., semiconductor and other materials, etchants with etching rates, electroplating and lithography processes, etc. Promising developments in the development of MEMS CAD tools have been reported[3,23–25]. However, further joint synergetic efforts are needed to perform integrated mechanoelectromagnetic analysis and design. The ultimate goal is to progress beyond the three-dimensional representation (drawing) of MEMS (which has some degree of merit and is needed as the synthesis, design, and optimizations tasks are completed) and steady-state analysis, to devising and optimizing MEMS using data-intensive analysis and heterogeneous synthesis. The hierarchical synthesis, design, optimization, and fabrication flow of the MEMS-CAD diagram is illustrated in Figure 10.23.

As illustrated in Figure 10.23, top-level specifications are used to synthesize (discover) novel MEMS, and then to devise electromechanical and electrooptomechanical (MOEMS) subsystems, components, and microdevices. High-level integrated synthesis must be carried out using databases and intelligent libraries of microscale subsystems, devices, and structures. The current MEMS design tools allow one to perform steady-state analysis and design of a limited number of microstructures and microdevices. Some CAD tools were extended to simulate and analyze electromechanical microdevices using linear and nonlinear differential equations applying the lumped-parameter mathematical modeling and simulation paradigms. For example, three-dimensional modeling and simulations of the cantilever beam were performed[23,24]. However, MEMS synthesis and classification, as well as analysis and optimization, depend on nonlinear phenomena. To the best of the author's knowledge, these complex phenomena and effects have not been integrated into the design. The reusability and leverage of the extensive CAD developed for minitransducers are limited due to the critical need to devise novel microsystems and devices (for example, axial-flux or disk topologies), six-degree-of-freedom microactuators, sensorless control, etc. The secondary phenomena and effects, usually neglected in conventional miniscale electromechanical motion devices (modeled using lumped-parameter models and analyzed using finite element analysis techniques), cannot be ignored. In general, microsystem dynamics must be thoroughly studied applying high-fidelity mathematical models. Therefore finite element analysis cannot be viewed as an enabling approach. It is the author's hope that analytic and numerical results, documented in this book, will allow us to progress toward the MEMS CAD developments.

In general, for microscale structures, devices, and systems, evolutionary synthesis developments can be represented as the X-design flow map documented in Figure 10.24. This map

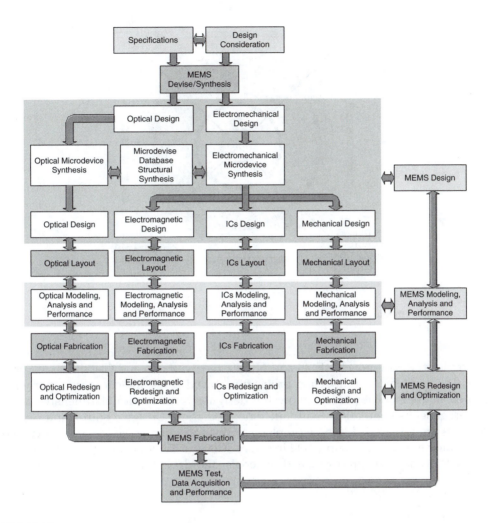

FIGURE 10.23
CAD for MEMS.

illustrates the synthesis flow from devising (synthesis) to modeling–analysis–simulation–design, from modeling–analysis–simulation–design to optimization–refining, and finally from optimization–refining to fabrication–testing sequential evolutionary processes. The proposed X-map consists of four domains of MEMS representation: (1) devising (synthesis), (2) modeling–analysis–simulation–design, (3) optimization–refining, and (4) fabrication–testing.

The desired degree of abstraction in the synthesis of new MEMS and NEMS requires one to apply this X-design flow map to devise novel MEMS and NEMS that integrate novel high-performance micro- and nanoscale structures and devices as components. The failure to verify the design of any MEMS component in the early phases at least causes the failure to design high-performance MEMS and leads to redesign. The interaction between four domains (devising, modeling–analysis–simulation–design, optimization–refining, and fabrication–testing) allows one to guarantee bidirectional top-down and bottom-up design features applying low-level component data to high-level design, and using high-level requirements to devise and design low-level components. The X-design flow map ensures hierarchy, modularity, locality, integrity, and other important features that allow one to design complex MEMS and NEMS. Using the results reported in this book, heterogeneous CAD can be developed.

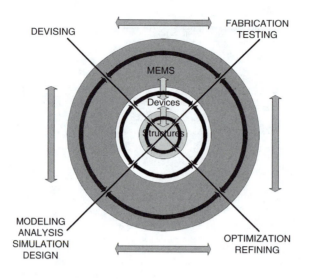

FIGURE 10.24
X-design flow map with four domains.

For example, to devise novel MEMS and NEMS, the Synthesis and Classification Solver is applied. The mathematical modeling problem is solved using the concepts reported. The developed mathematical models allow the designer to perform data-intensive analysis and heterogeneous simulations with outcome prediction. Design, control, and optimization aspects are researched. (These problems with application to MEMS were documented and illustrated in previous chapters.) Finally, as was illustrated, using different fabrication technologies, techniques, processes, and materials, affordable high-performance MEMS can be made.

Homework Problems

1. Develop surface micromachining processes to fabricate an axial topology permanent-magnet stepper micromachine with a rotor diameter of 100 μm. You may design and develop the sequential fabrication processes to fabricate the permanent-magnet or reluctance stepper motors.

2. Examine the basic operating features for permanent-magnet stepper micromotors and propose the corresponding ICs to control this motion microdevice. Explain the ICs' architecture and basic features.

3. Devise a deformable array-based micromirror. (A cantilever-like deformable membrane can be used.) The micromirror should be controlled in three dimensions to deflect light. Devise and design the radiating energy devices (antennae) to deflect the mirror surface. Study the micromirror performance. Develop the surface micromachining processes and propose the materials to fabricate the high-performance array-based micromirror.

4. Develop a surface micromachining process to fabricate a reluctance micromachine (rotor diameter of 100 μm) with the sensors to measure the rotor displacement.

Examine how to integrate micromachine– sensor–ICs as a functional MEMS. Examine how to control the reluctance microactuator and estimate the performance. Make conclusions regarding the efficiency of reluctance motion microdevices.

References

1. Campbell, S.A., *The Science and Engineering of Microelectronic Fabrication*, Oxford University Press, New York, 2001.
2. Kovacs, G.T.A., *Micromachined Transducers Sourcebook*, WCB McGraw-Hill, Boston, 1998.
3. Lyshevski, S.E., *Nano- and Micro-Electromechanical Systems: Fundamental of Micro- and Nano-Engineering*, CRC Press, Boca Raton, FL, 1999.
4. Madou, M., *Fundamentals of Microfabrication*, CRC Press, Boca Raton, FL, 1997.
5. Kim, Y.-J. and Allen, M.G., Surface micromachined solenoid inductors for high frequency applications, *EE Trans. Components, Packaging, and Manufacturing Technology*, part C, 21, 1, 26–33, 1998.
6. Park, J.Y. and Allen, M.G., Integrated electroplated micromachined magnetic devices using low temperature fabrication processes, *EE Trans. Electronics Packaging Manufacturing*, 23, 1, 48–55, 2000.
7. Sadler, D.J., Liakopoulos, T.M., and Ahn, C.H., A universal electromagnetic microactuator using magnetic interconnection concepts, *Jour. Microelectromechanical Systems*, 9, 4, 460–468, 2000.
8. Lyshevski, S.E., *Electromechanical Systems, Electric Machines, and Applied Mechatronics*, CRC Press, Boca Raton, FL, 1999.
9. Frazier, A.B. and Allen, M.G., Uses of electroplated aluminum for the development of microstructures and micromachining processes, *Jour. Microelectromechanical Systems*, 6, 2, 91–98, 1997.
10. Guckel, H., Christenson, T.R., Skrobis, K.J., Klein, J., and Karnowsky, M., Design and testing of planar magnetic micromotors fabricated by deep x-ray lithography and electroplating, in *Technical Digest of International Conference on Solid-State Sensors and Actuators, Transducers 93*, Yokohama, Japan, pp. 60–64, 1993.
11. Taylor, W.P., Schneider, M., Baltes, H., and Allen, M.G., Electroplated soft magnetic materials for microsensors and microactuators, in *Proceedings of the Conference on Solid-State Sensors and Actuators, Transducers 97*, Chicago, IL, pp. 1445–1448, 1997.
12. Lagorce, L.K., Brand, O., and Allen, M.G., Magnetic microactuators based on polymer magnets, *Jour. Microelectromechanical Systems*, 8, 1, 2–9, 1999.
13. Smith, D.O., Static and dynamic behavior in thin permalloy films, *Jour. Appl. Phys., 29*, 2, 264–273, 1958.
14. Suss, D., Schreft, T., and Fidler, J., Micromagnetics simulation of high energy density permanent magnets, *EE Trans. Magnetics*, 36, 5, 3282–3284, 2000.
15. Judy, J.W. and Muller, R.S., Magnetically actuated, addressable microstructures, *Jour. Microelectromechanical Systems*, 6, 3, 249–256, 1997.
16. Yi, Y.W. and Liu, C., Magnetic actuation of hinged microstructures, *Jour. Microelectromechanical Systems*, 8, 1, 10–17, 1999.
17. Gere, J.M. and Timoshenko, S.P., *Mechanics of Materials*, PWS Press, 1997.
18. Groom, N.J. and Britcher, C.P., A description of a laboratory model magnetic suspension test fixture with large angular capability, in *Proceedings of the Conference on Control Applications, NASA Technical Paper* – 1997, vol. 1, pp. 454–459, 1992.
19. Ahn, C.H., Kim, Y.J., and Allen, M.G., A planar variable reluctance magnetic micromotor with fully integrated stator and coils, *Jour. Microelectromechanical Systems,* 2, 4, 165–173, 1993.
20. O'Sullivan, E.J., Cooper, E.I., Romankiw, L.T., Kwietniak, K.T., Trouilloud, P.L., Horkans, J., Jahnes, C.V., Babich, I.V., Krongelb, S., Hegde, S.G., Tornello, J.A., LaBianca, N.C., Cotte, J.M., and Chainer, T.J., Integrated, variable-reluctance magnetic minimotor, *IBM Jour. Research and Development,* 42, 5, 1998.

21. Yasseen, A.A., Wu, C.H., Zorman, C.A., and Mehregany, M., Fabrication and testing of surface micromachined polycrystalline SiC micromotors, *EE Trans. Electron Device Letters, 21*, 4, 164–166, 2000.
22. Yasseen, A.A., Zorman, C.A., and Mehregany, M., Surface micromachining of polycrystalline silicon carbide films microfabricated molds of SiO and polysilicon, *Jour. Microelectromechanical Systems, 8*, 1, 237–242, 1999.
23. Bai, Z., Bindel, D., Clark, J.V., Demmel, J., Pister, K.S.J., and Zhou, N., New numerical techniques and tools in Sugar for 3D MEMS simulation, in *Proceedings of the Conference on Modeling and Simulation of Microsystems*, Hilton Head Island, SC, pp. 31–34, 2001.
24. Clark, J.V., Zhou, N., Bindel, D., Schenato, L., Wu, W., Demmel, J., and Pister, K.S.J., 3D MEMS simulation modeling using modified nodal analysis, in *Proceedings Of Microscale Systems: Mechanics and Measurements Symposium*, Orlando, FL, pp. 68–75, 2000.
25. Mukherjee, T., Fedder, G.K., Ramaswamy, D., and White, J., Emerging simulation approaches for micromachined devices, *EE Trans. Computer-Aided Design of Integrated Circuits and Systems, 19*, 12, 1572–1589, 2000.

Index

3D+ modeling and simulation, 68, 454

A

AAA protein superfamily, 275
Accelerometers, 88
  uses for, 155
Acceptors, 478
Accuracy, 542, 595
Accurate pointing devices, piezotransducers in, 379
ACh molecules, 60
Acoustic control, accelerometers in, 155
Active optimization-control problem, mathematical
     formulation of, 328
Actuation, 36
  induced-strain, 377
Actuators, 70
  control of, 78
  electrostatic, energy density of, 66
  microscale, 81
Adaptability, 35
Adaptive decision making, 56
Adaptive self-organization, 36
Additional potential, 479
Adenine, 82
Admissibility concept, 638
Admissible control laws, 595, 608
Admissible current density, 678
Admissible time-invariant control law, 563
Advanced shifting, 375
ADXL202 accelerometer, 151
Aerospace, impact of micro- and nanotechnology
     on, 14
Aerospace technology, actuators, 70
Algorithmic codes, 159
Alignment, 91
  local, 178
Alkoxide hydrolysis, 392
Allowed energies, 473
Aluminum, 87
  electrodeposition of, 662, 668
  mechanically robust alloys, 92
Amino acids, 61, 171, 176, 516, 534
  linking of, 168
Ampere-Maxwell equation, 195
Analysis, 35, 417, 645
Anisotropic wet etching, 103, 115
Annealing, 122
Anodic bonding, 103, 114
Antiferromagnetic coupling, multilayered thin
     films, 73

Ants, 1
Application-specific toolboxes, 51
Arbitrary reference frame, dynamics of induction
     micromotors in, 295
Architectronics, 57
Aristotle, 12
Armature winding, 262
Armchair carbon nanotubes, 501
Arsenic, 111
  covalent bonds of, 478
Artificial intelligence, 37, 141
Aspratate, 61
Assembly, 417
  MEMS, 113
Assessment analysis tools, 37
Atomic mass unit, 3
Atomic radius, 19
Atomic structures, 2
  quantum mechanics and, 415
Atomic weight, 20
Atomic-level dynamics, study of using wave function,
     454
Atomic-scale motion, 426
Atomic-scale positional assembly, 417
Atomic-scale systems, 418
Atomistic modeling, 462
Atoms, 2
  impurity, 478
  mathematical modeling of with many
     electrons, 452
  size of, 27, 488
ATP, 7, 40, 136
  binding consensus, 275
Atto, 1
Autocorrelation analysis, 180
  sequence measures, 183
Automated synthesis, 49
Automotive systems, impact of micro- and
     nanotechnology on, 14
Averaging concept for spiral windings, 691
Avionics, impact of micro- and nanotechnology on, 14
Axial electromagnetic micromotors, 698
Axial micromachines, DNA-based, 167
Axial nanomotors, 7
Axial topology motion devices, 53
Axial topology permanent-magnet synchronous
     micromachines, 314
  mathematical models of, 318
Axial topology transducers, 149
Axonal-dendritic chemical synaptic
     transmission, 533

705

Ion milling, 92
Ionic current, 60
Ionic gradient, 60
Iron, 656
Isotopes, 415
Isotropic wet etching, 103, 115

J

Jellyfish, 62

K

Karl Suss mask aligners, 101
Kinetic energy, 238
Kirchhoff's laws, 229, 539, 660
   modeling of induction micromotors using, 281
   voltage, 197, 264, 334, 340, 372
Knowledge domain, 141
Knowledge-based libraries, 37
KOH-based etching process, 116, 656
Kohn-Sham one-electron equations, 454
Kroneker delta, 382
Kronig-Penney model, 470
Kroto, Sir Harry, 85, 506

L

L-Edit, 105
$La_2@C_{xx}$ metallofullerenes, 513
Lagrange equations of motion, 33, 229, 243, 259, 418
   dynamics of permanent-magnet synchronous
     micromotors, 352
Lagrangian mechanics, 539
Laminar flow, 663
Lamination processes, 677
Landay-Lifschitz-Gilbert equations, 675
Langevin, Paul, 378
Language, 175
Lattices, sets and, 146
Legendre equation, 447
Lenz's law, 197
Leucippus, 4
Lift-off stenciling techniques, 102
LIGA processes, 86, 114, 127, 261, 669, 694
   fabrication of MEMS with, 69
Light, wave behavior of, 415
Light-harvesting complex, 40
Linear analog PID control law, 546
Linear discrete-time MEMS and NEMS, control of
   using Hamilton-Jacobi theory, 631
Linear micromotors, 650
Linear optimization, 156
Linear piezoelectricity theory, 383
Linear programming, 149, 158
Linear quadratic regulator, 546
Linearizability condition, 589
Linearly constrained optimization problem, 160

Linkage groups, 84
Lippmann, Gabriel, 377
Liquid-phase epitaxial thin film fabrication, 392
Lithography, 91, 127
   deep-UV processes, 99
   tools, 77
Local alignments, 178
Localized additional potential, 479
London's dispersion forces, 534
Lorenz equation, 60
Lorenz force law, 215, 326
Low-entropy structure design, 36
Low-pressure chemical vapor deposition. *See* LPCVD
Lowenstam, Heinz A., 30
LPCVD, 77, 89, 92, 121
   low-stress, 92
Lumped-parameter mathematical model, 46, 217, 316,
     657, 659, 660
   augmented, 227
   axial topology permanent-magnet synchronous
     micromachines, 319
   micromotor optimization using, 698
   permanent-magnet synchronous micromotors, 374
   piezotransducers, 390
   state-space form of, 349
   synchronous reluctance micromotors, 334, 338
Lyapunov concept, 595
Lyapunov exponent, energy dependence of, 521
Lyapunov stability theory, 591, 608, 638
   in analysis and control of MEMS and NEMS, 603

M

M-theory, 27
M. magnetotacticum, 31
Machine performance, 44
Machine variables
   dynamics of induction micromotors in, 290
   mathematical modeling with, 369
Magnetic fields, 493
Magnetic films, 92
Magnetic materials, properties of, 648
Magnetic susceptibility, 647
Magnetic tunnel junctions, 88
Magnetic vias, 656
Magnetization dynamics, thin films, 675
Magnetoactive materials, 377
Magnetomotive force, 198
Magnetosomes, 30
Magnetostriction, 316
Magnetostrictive compounds, 377
Magnetotactic bacteria, 30
Magnets, micromachined polymer, 674
Main group elements, 20
Manufacturing, impact of micro- and nanotechnology
   on, 14
MAPLE, 160
Mass transfer effects, 663
Materials, effect of on energy conversion, 646
MATHEMATICA, 160, 390, 699